International Conference on Frustration in Condensed Matter 2011

(ICFCM)

Journal of Physics: Conference Series Volume 320

Sendai, Japan
11-14 January 2011

Editors:

Hiroko Aruga Katori
Taka-hisa Arima
Hikaru Kawamura
Shigeki Fujiyama

ISBN: 978-1-61839-290-9
ISSN: 1742-6588

Printed from e-media with permission by:

Curran Associates, Inc.
57 Morehouse Lane
Red Hook, NY 12571

Some format issues inherent in the e-media version may also appear in this print version.

Printed by Curran Associates, Inc. (2011)

For permission requests, please contact the Institute of Physics
at the address below.

Institute of Physics
Dirac House, Temple Back
Bristol BS1 6BE UK

Phone: 44 1 17 929 7481
Fax: 44 1 17 920 0979

techtracking@iop.org

Additional copies of this publication are available from:

Curran Associates, Inc.
57 Morehouse Lane
Red Hook, NY 12571 USA
Phone: 845-758-0400
Fax: 845-758-2634
Email: curran@proceedings.com
Web: www.proceedings.com

TABLE OF CONTENTS

LOCALIZED SPIN SYSTEMS

ELECTRONIC SYSTEMS

MULTIFERROICS AND FERROELECTRICS

PREFACE

The International Conference on Frustration in Condensed Matter (ICFCM) was held in Sendai, Japan, from 11–14 January 2011, at Sendai International Center. The conference was held under the auspices of the MEXT project 'Novel State of Matter Induced by Frustration'. Frustration represents the situation where several optimization conditions compete with each other so that a system cannot satisfy them simultaneously. Indeed, frustration is ubiquitous in nature, not just limited to the traditional field of frustrated magnetism. It has now become a hot topic in the entire field of condensed matter physics, and is recognized as an important and fundamental concept which plays important roles in a variety of systems, including magnets, metals, superconductors, multiferroics, and dielectrics. In such frustrated systems, novel effects due to enhanced fluctuations arise, often leading to an exotic order and a new thermodynamic phase, gigantic responses, and novel dynamics.

The ICFCM conference covered a variety of frustration-related phenomena in condensed matter physics, including:

- Novel order and spin liquids in geometrically frustrated magnets;
- Coupling between spin, lattice, charge and orbital degrees of freedom;
- Frustration and chirality;
- Itinerant frustrated systems;
- Multiferroics; and
- Ferroelectrics and relaxors, etc.

The conference attracted 188 participants from all over the world, including 19 invited speakers, both domestic and from abroad. During the conference, many stimulating discussions were had both inside and outside the conference room. From among the presentations given in the conference, 91 contributions are included in this proceedings. I am sure that the readers will enjoy these contributions and learn a lot from them.

Hikaru Kawamura
Conference Chairman
Osaka University, Japan

Editors

Hiroko Aruga Katori (Chief) *Tokyo University of Agriculture and Technology, Japan*
Hikaru Kawamura *Osaka University, Japan*
Taka-hisa Arima *The University of Tokyo, Japan*
Shigeki Fujiyama *RIKEN, Japan*

Dates

Issue 1 (2011)

International Conference on Frustration in Condensed Matter (ICFCM) IOP Publishing
Journal of Physics: Conference Series **320** (2011) 012001 doi:10.1088/1742-6596/320/1/012001

Collective Phenomena in the $LiHo_xY_{1-x}F_4$ Quantum Ising Magnet: Recent Progress and Open Questions

Michel J P Gingras[1,2,†] **and Patrik Henelius**[3,*]

[1]Department of Physics & Astronomy, University of Waterloo, Waterloo, Ontario, N2L 3G1, Canada

[2]Canadian Institute for Advanced Research, 180 Dundas Street West, Suite 1400, Toronto, Ontario, M5G 1Z8, Canada

[3]Department of Theoretical Physics, Royal Institute of Technology, SE-106 91 Stockholm, Sweden

E-mail: †gingras@gandalf.uwaterloo.ca, *henelius@kth.se

Abstract. In $LiHo_xY_{1-x}F_4$, the magnetic Holmium Ho^{3+} ions behave as effective Ising spins that can point parallel or antiparallel to the crystalline c-axis. The predominant inter-Ho^{3+} interaction is dipolar, while the Y^{3+} ions are non-magnetic. The application of a magnetic field B_x transverse to the c-axis Ising direction leads to quantum spin-flip fluctuations, making this material a rare physical realization of the celebrated *transverse field Ising model.* The problems of classical and transverse-field-induced quantum phase transitions in $LiHo_xY_{1-x}F_4$ in the dipolar ferromagnetic ($x = 1$), diluted ferromagnetic ($0.25 \lesssim x < 1$) and highly diluted $x \lesssim 0.25$ dipolar spin glass regimes have attracted much experimental and theoretical interest over the past twenty-five years. Two questions have received particular attention: (i) is there an *antiglass* (quantum disordered) phase at low Ho^{3+} concentration and (ii) what is the mechanism responsible for the fast B_x-induced destruction of the ferromagnetic ($0.25 \lesssim x < 1$) and spin glass ($x \lesssim 0.25$) phases? This paper reviews some of the recent theoretical and experimental progress in our understanding of the collective phenomena at play in $LiHo_xY_{1-x}F_4$, in both zero and nonzero B_x.

1. The pure $LiHoF_4$ material

1.1. *Dipolar ferromagnetism*

The insulating rare-earth compound $LiHoF_4$ has the scheelite structure depicted in Fig. 1 [1]. In this material, only the Holmium ions, Ho^{3+}, are magnetic. $LiHoF_4$ forms a tetragonal structure (space group $C_{4h}^6 - I4_1/a$) with lattice constants $a = 5.175$ Å and $c = 10.75$ Å. There are 4 magnetic Ho^{3+} ions per unit cell with fractional coordinates $(0, 0, \frac{1}{2})$, $(0, \frac{1}{2}, \frac{3}{4})$, $(\frac{1}{2}, \frac{1}{2}, 0)$ and $(\frac{1}{2}, 0, \frac{1}{4})$. In Fig. 1, the lines connecting the central Ho^{3+} ion with neighbouring F^- ions and then to the nearest Ho^{3+} ions illustrate the superexchange pathways between nearest-neighbour Ho^{3+} cations [2]. As in most magnetic rare-earth compounds, the unfilled $4f$ electronic orbitals of the Ho^{3+} ions are not very spatially extended. This makes the exchange interaction weak and causes magnetostatic dipolar couplings to be the strongest inter-ionic interactions. Indeed, $LiHoF_4$ is a dipolar Ising ferromagnet with a magnetic moment of $\sim 7\mu_B$ per Ho^{3+} ion [2] and with a critical temperature $T_c \approx 1.53$ K [2]. In the mid-seventies, crystal-field calculations [3], susceptibility [4] and spectroscopic measurements [5] showed that $LiHoF_4$ is an Ising-like

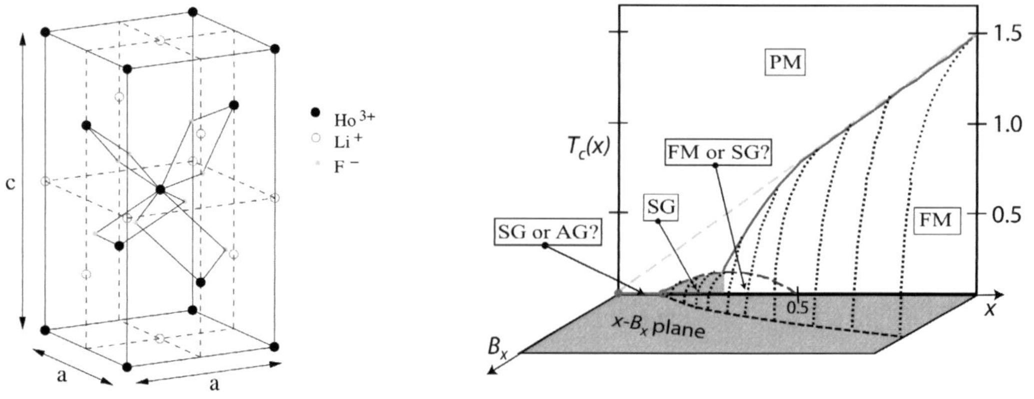

Figure 1. The left panel shows the crystalline tetragonal structure of $LiHoF_4$ with lattice spacing $a = 5.175$ Å and $c = 10.75$ Å. The thin solid lines illustrate the superexchange pathways between magnetic Ho^{3+} ions (black circles) mediated via fluorine F^- ions (green circles). In $LiHo_xY_{1-x}F_4$, the magnetic Ho^{3+} ions are substituted randomly by non-magnetic Y^{3+} ions, with the lattice structure remaining the same for all x. The global Ising direction is along the c axis and a magnetic field B_x is applied perpendicular to that axis in transverse field experiments. The right panel shows a schematic temperature (T) – dilution (x) – transverse field (B_x) phase diagram of $LiHo_xY_{1-x}F_4$. At high temperature T or large B_x, the system is in a paramagnetic (PM) phase. For $x = 1$, $LiHoF_4$ is an Ising dipolar ferromagnet (FM) below $T_c = 1.53$ K. Upon the substitution of Ho^{3+} by Y^{3+}, the FM phase persists down to $x \sim 0.25$. Upon cooling the diluted FM for $0.25 \lesssim x \lesssim 0.5$, magnetic susceptibility measurements find the development of a "ferroglass" regime (FM or SG?). As x decreases, random frustration builds in and, for $x \lesssim 0.25$, a dipolar Ising spin glass (SG) phase develops (shaded blue region). Whether the dipolar SG exists down to $x = 0^+$ or an exotic quantum disordered "antiglass" (AG) phase occurs at $x > 0$ (red line segment) is not yet resolved and is the subject of much controversy. The application of a transverse field B_x introduces quantum fluctuations causing a reduction of $T_c(x)$ (surface delineated by dotted lines) which, ultimately, drives a zero temperature quantum phase transition at zero temperature (dash-line curve in the $x - B_x$ plane).

dipolar ferromagnet with a highly anisotropic g-tensor ($g_\perp \approx 0$ and $g_\parallel \approx 14$). The dipolar interaction has a long and interesting history and this physical realization rekindled interest in dipolar ferromagnetism. Luttinger and Tisza had found that the ground state of a dipolar ferromagnet depends on both the lattice structure and sample shape [6]. However, taking the demagnetization field into account, Griffiths showed that, in the absence of an applied magnetic field, the free energy is independent of the sample shape [7]. Using renormalization group arguments, Larkin and Khmelnitskii [8] and Aharony [9] showed that $d^* = 3$ is the upper critical dimension for mean-field critical behaviour in a dipolar ferromagnet. This observation led to significant experimental efforts to find the expected logarithmic corrections to mean-field theory in $LiHoF_4$ [10, 11, 12, 13, 14]. The demagnetization field also renders the ferromagnetic transition quite peculiar in the presence of dipolar interactions [4] as these oppose a uniform magnetization and leads to the formation of domains below the transition [5, 15]. The size and shape of the domains depend on an energy balance between surface and bulk contributions [16]. For $LiHoF_4$, the ground state domain configuration is expected to consist of parallel antiferromagnetically aligned sheets [17, 18, 19]. The size of the domains increases with decreasing demagnetization factor and, in the limit of infinitely long thin samples, a state of uniform magnetization is possible. Also, for thick samples, a branched pattern is predicted in the domain structure for a dipolar Ising ferromagnet [17, 18]. $LiHoF_4$ is an excellent test case for theories of domain formation due to the availability of an accurate microscopic model of the material (see below). It remains an open experimental challenge to verify the nature of the predicted rich domain structure.

1.2. Microscopic model and transverse field Ising model for $LiHoF_4$

The magnetic properties of $LiHoF_4$ are determined by the $4f^{10}$ electrons of the Ho^{3+} ions. Applying Hund's rules results in a 17-fold degenerate $^5I_8$ electronic ground state. The crystal field from the surrounding ions, described by a crystal field potential $V_{\rm CF}$, lifts this degeneracy and gives rise to a two-fold non-Kramers degenerate ground state, $|\psi_0^{\pm}\rangle$, and an excited singlet, $|\psi_{\rm e}\rangle$, at an energy of approximately 10 K above $|\psi_0^{\pm}\rangle$ [3, 5]. The minimal microscopic Hamiltonian in the presence of a transverse magnetic field $\vec{B} = B_x\hat{x}$ can be written as [20]:

$$\begin{aligned} H &= \sum_i V_{\rm CF}(\vec{J}_i) - \frac{\mu_0}{4\pi} g_L\mu_B \sum_i B_x J_i^x + \frac{1}{2}\frac{\mu_0}{4\pi}(g_L\mu_B)^2 \sum_{i\neq j} \mathcal{L}_{ij}^{\mu\nu} J_i^{\mu} J_j^{\nu} \\ &+ \mathcal{J}_{\rm ex}\sum_{\langle i,j\rangle} \vec{J}_i\cdot\vec{J}_j + A\sum_i \vec{I}_i\cdot\vec{J}_i, \end{aligned} \tag{1}$$

where $\mu,\nu = x,y,z$. $\mathcal{L}_{ij}^{\mu\nu}$ is the magnetic dipole interaction, $\mathcal{L}_{ij}^{\mu\nu} = [\delta^{\mu\nu}|\vec{r}_{ij}|^2 - 3(\vec{r}_{ij})^{\mu}(\vec{r}_{ij})^{\nu}]/|\vec{r}_{ij}|^5$, $\mathcal{J}_{\rm ex}$ is the nearest-neighbour exchange interaction. A is the strength of the hyperfine interaction and $\vec{I}_i$ ($I = 7/2$) is the total angular angular momentum vector of the Ho-nucleus at the i-th site. Disregarding the hyperfine interactions, an effective spin 1/2 model for $LiHoF_4$ from the above microscopic model of Eq. (1) was derived in Ref. [20] and revisited in Ref. [21]. Without hyperfine interactions, the single site Hamiltonian (first two terms in Eq. (1)) can be diagonalized numerically for arbitrary value of the transverse field B_x. The ground state doublet is split by B_x, with an energy $\Delta(B_x)$ between the two states. Meanwhile, the excited $|\psi_{\rm e}\rangle$ singlet state remains more than 10 K above the split doublet [20, 21]. Consequently, the $\vec{J}_i$ operators can be projected onto the two-dimensional subspace of the two lowest energy eigenstates via the relationship $J_i^{\mu} = C_{\mu 0} + \sum_{\nu=x,y,z} C_{\mu\nu}(B_x)\sigma_i^{\nu}$, hence providing a description in terms of effective pseudospin-1/2 operators [20, 21]. The strongest effective interaction is $\propto (C_{zz})^2\mathcal{L}_{ij}^{zz}\sigma_i^z\sigma_j^z$ and, to an accuracy of a few percent [20, 21], the effective model can be written (up to a field dependent constant) as

$$\begin{aligned} H_{\rm Ising} &= -\frac{1}{2}\Delta\sum_i \sigma_i^x + \frac{1}{2}\frac{\mu_0}{4\pi}(g_L\mu_B C_{zz})^2 \sum_{i\neq j}\mathcal{L}_{ij}^{zz}\sigma_i^z\sigma_j^z \\ &+ \mathcal{J}_{\rm ex}(C_{zz})^2 \sum_{\langle i,j\rangle}\sigma_i^z\sigma_j^z. \end{aligned} \tag{2}$$

Note that the $C_{\mu\nu}$ parameters and the *effective* transverse field Γ, with $\Gamma \equiv \Delta/2$, depend on the applied physical transverse field B_x. In the limit of small B_x, $\Delta(B_x) \propto B_x^2$ [20, 21]. In Eq. (2), $\Delta/2$ plays the role of an effective transverse field Γ acting on the x-component of the pseudospin $\vec{\sigma}_i$ – hence the realization of a transverse field Ising model (TFIM) in $LiHoF_4$ [22].

For $B_x = 0$ ($\Delta = 0$), the effective dipolar model (2) has been analyzed using mean-field theory [2, 4] and classical Monte Carlo simulations. Numerical Monte Carlo simulations are complicated by the difficult nature of the conditionally-convergent dipolar lattice sums. An early Monte Carlo simulation using free boundary conditions for rather small system sizes found a critical temperature of 1.89 K [23]. To obtain better accuracy with long range dipolar interactions, there are two standard approaches that can be implemented: Ewald summation [19, 21, 24, 25, 26, 27] and cavity methods [20, 28]. An early Monte Carlo simulation using the Ewald summation method to a dipolar Ising ferromagnet on a body-centered lattice was carried out by Xu *et al.* [24, 25]. Monte Carlo simulations using the cavity method [20] and the Ewald method [19, 21] have determined the critical temperature for the dipolar model in Eq. (2) with $\Delta = 0$ on the $LiHoF_4$ lattice and estimated the nearest-neighbour (antiferromagnetic) exchange interaction $\mathcal{J}_{ex}$ needed to lower the Monte Carlo T_c value in order to match it to the experimental value of 1.53 K [2]. A more recent study, using the Ewald summation method and system sizes up to 32 000 dipoles, found that the magnetization, specific heat and susceptibility do match experimental results at a quantitative level [19]. The pure $LiHoF_4$ material can therefore be viewed as one of the best realizations of a dipolar Ising model. A review of materials that behave like Ising systems can be found in Ref. [29].

1.3. Hyperfine interaction

As in other Ho-based materials [30, 31], there is a strong hyperfine coupling ($A = 0.039$ K) between the electronic ($J = 8$) and nuclear ($I = 7/2$) moments of Ho^{3+} in $LiHoF_4$. The hyperfine contribution dominates the specific heat below 0.5 K [2], and leads to a substantial increase in the critical transverse field at low temperatures [22]. As we shall discuss in Section 2, the hyperfine interaction also strongly influences the phase diagram of the dilute compound $LiHo_xY_{1-x}F_4$ in a transverse field [32, 33]. The hyperfine interaction is therefore important and may not be omitted in a proper quantitative description of the low-temperature properties of $LiHo_xY_{1-x}F_4$.

1.4. Transverse field effects

At a continuous phase transition, the order parameter fluctuates coherently over increasing length scales as the critical point is approached. At finite temperatures, the fluctuations are classical in nature sufficiently close to the critical temperature. In some systems, at temperatures near absolute zero, a phase transition can be induced by tuning an external parameter such as a magnetic field, pressure, doping or disorder. In such a case, the fluctuations are quantum mechanical in nature and the transition is called a *quantum phase transition* (QPT) [34, 35]. The archetypical model for a quantum phase transition is the *transverse field Ising model* (TFIM) [36, 37]. There are, however, not many real Ising magnets with experimentally accessible QPTs [38] since the required transverse field usually far exceeds laboratory fields (currently less than 50 T for DC fields). This would be the case for dipolar Ising spin ice materials [39], but not for the $Dy(OH)_3$ and $Ho(OH)_3$ dipolar Ising ferromagnets [38]. Due (i) to the weak dipolar interaction and (ii) the low-lying singlet $|\psi_e\rangle$ above the ground Ising doublet $|\psi_0^{\pm}\rangle$, $LiHoF_4$ is one of very few magnetic systems where one can observe the vanishing of the magnetization as a transverse field is applied [22]. Interestingly, recent neutron scattering measurements of $LiHoF_4$ have revealed that the relatively strong hyperfine interactions in Ho mask the true QPT of the hybridized electronuclear critical mode, and the energy gap in the solely electronic degrees of freedom remains finite as the system passes through the QPT [40].

Given the accuracy of the effective Ising model for $LiHoF_4$ in the absence of a transverse field, the material would seem to constitute an ideal testing ground for the inclusion of non-commuting quantum-mechanical terms in realistic spin models of magnetic materials. As outlined above, there is a well defined procedure to describe the effect of the transverse field B_x in $LiHoF_4$ within a TFIM [20, 21]. However, a quantum Monte Carlo calculation [20] of the critical temperature, T_c, vs B_x phase diagram reveals only a qualitative agreement with the experiment of Ref. [22]. In the low temperature region (below 0.5 K), the effect of the hyperfine interactions on the $B_x - T$ phase diagram becomes important, and this is only approximately taken into account in the simulation of Ref. [20]. Recent theoretical work proposes an alternative way to incorporate the effect of the hyperfine interactions in an effective model of $LiHoF_4$ in nonzero B_x [33]. Notwithstanding the importance of hyperfine effects at low temperature, one notes that *even*

Journal of Physics: Conference Series **320** (2011) 012001 doi:10.1088/1742-6596/320/1/012001

close to the classical critical temperature of $T_c = 1.53$ K, the experimental [22] phase boundary rises much more steeply than in the model of Eq. (2) [20, 21]. To compound the puzzle, a very recent experiment using dilatometry [41] confirms the previous $T_c(B_x)$ experimental phase diagram obtained on the basis of susceptibility measurements [22]. The reason for this stability of the ordered state with respect to weak B_x is an open question. Possible explanations include magnetostriction effects, small terms that are neglected in the effective model of Eq. (2) (see Refs. [20, 21] for a discussion of those terms) or anisotropic exchange terms, such as $\mathcal{J}_{\rm ex}^{uv}(r_{ij})J_i^u J_j^v$, altogether ignored in the microscopic model of Eq. (1). While the latter interactions may largely be inconsequential at the classical transition, when projected within the manifold of $|\psi_0^{\pm}\rangle$ Ising doublets when $B_x = 0$ [42], these could perhaps end up affecting the $T_c(B_x \neq 0)$ phase diagram when $B_x \neq 0$.

2. $LiHo_xY_{1-x}F_4$ – transverse field and random field physics

2.1. Disordered magnetic systems

All real materials contain a certain amount of frozen-in random disorder such as impurities, interstitials and crystalline defects. One may then ask how the thermodynamic properties and the stability of the ground state of an otherwise idealized pure system are affected by various types of random disorder [43, 44]. Since the mid 1970s, magnetic systems have proven to be almost ideal paradigms to study the role of random disorder in condensed matter systems. Examples of cornerstone questions in the field of random disordered systems are: Does a transition that is first order in a pure system remains first order in the presence of random disorder [45]? For a second order transition, does an arbitrary small amount of random disorder change the critical exponents and hence modify the universality class [46]? Does the long range order exhibited by a pure system survive arbitrary small *random fields* conjugate to the order parameter [47]? When the disorder is large enough and the interactions are in strong competition and lead to *random frustration*, is there a thermodynamic transition at nonzero temperature to a *spin glass* state [48, 49]? Under what conditions does a random magnet exhibit nonanalytical behavior above its critical temperature, in the so-called Griffiths phase, which is a remnant of the long range order phase displayed by the parent disorder-free system [50]? The above questions were originally by and large investigated for classical phase transitions at nonzero temperature. In the late 1980s, those and similar questions started to attract considerable interest in the context of quantum phase transitions [34, 35]. Here again, the $LiHoF_4$ material, or more precisely, its Y^{3+} diamagnetically diluted $LiHo_xY_{1-x}F_4$ variant, has proven to be a most interesting material to investigate quantum phase transitions in a disordered system that is tuned by an external parameter. As in the pure $LiHoF_4$ material, the parameter tuning the level of quantum (spin flip) fluctuations in $LiHo_xY_{1-x}F_4$ is a magnetic field B_x applied perpendicularly to the Ho^{3+} Ising spin direction.

2.2. Transverse field and random field effects in $LiHo_xY_{1-x}F_4$

The $LiHo_xY_{1-x}F_4$ compound forms a solid solution from the pure dipolar ferromagnetic $LiHoF_4$ to the non-magnetic $LiYF_4$ without a change of the crystalline structure and, thanks to the close ionic radius of Ho^{3+} and Y^{3+}, $LiHo_xY_{1-x}F_4$ displays a minimal dependence of the a and c lattice parameters on x. The critical ferromagnetic transition temperature, $T_c(x)$, decreases as the magnetic Ho^{3+} ions are substituted by non-magnetic Y^{3+}. For $0.25 \lesssim x < 1$, $LiHo_xY_{1-x}F_4$ can be described as a *diluted ferromagnet* (see Fig. 1). As discussed in Section 1.1, the main interactions between the Ho^{3+} ions are magnetostatic dipole-dipole interactions. The dipolar interaction is inherently frustrated since its sign, that of $\mathcal{L}_{ij}^{\mu\nu}$ in Eq. (1), depends on the orientation of the magnetic moments with respect to $\vec{r}_{ij} \equiv \vec{r}_j - \vec{r}_i$, the position vector of ion j relative to ion i. Hence, as a result of the random substitution $Ho^{3+} \to Y^{3+}$, *random frustration* builds up as x decreases until, for $x \approx 0.25$, the diluted ferromagnetic state gives

Journal of Physics: Conference Series **320** (2011) 012001 doi:10.1088/1742-6596/320/1/012001

way to a *dipolar spin glass* phase (see. Fig. 1). We note that there has been much debate as to whether such a spin glass phase occurs for $x \lesssim 0.25$ in $LiHo_xY_{1-x}F_4$ [51, 52, 53]. We proceed in this section with the assumption that a thermodynamic spin glass phase does exist, at least for $0.15 \lesssim x \lesssim 0.25$. We return in Section 3 to the question of the existence of a spin glass phase in $LiHo_xY_{1-x}F_4$ for $x \lesssim 0.25$, below the concentration where the diluted ferromagnetic phase has disappeared (see Fig. 1).

The essence of the research on transverse field induced quantum fluctuations and quantum phase transitions in $LiHo_xY_{1-x}F_4$ originally lay in the expectation that this material may be a physical realization of a transverse field Ising model (TFIM) with Hamiltonian

$$H_{\rm TFIM} = -\sum_{i>j} \Omega_{ij}^{zz} \sigma_i^z \sigma_j^z - \Gamma \sum_i \epsilon_i \sigma_i^x , \tag{3}$$

with effective transverse field Γ. In Eq. (3), the randomly frustrated Ω_{ij}^{zz} couplings model, or rather mimic, the site-diluted $\mathcal{L}_{ij}^{zz} \to \epsilon_i \epsilon_j \mathcal{L}_{ij}^{zz}$ in Eq. (2), where $\epsilon_i = 1$ when the site i is occupied by a magnetic Ho^{3+} ion and $\epsilon_i = 0$ when it is occupied by non-magnetic Y^{3+}. In theoretical studies, important simplifications have been made by taking a Gaussian distribution for Ω_{ij}^{zz} of mean Ω_0 and standard deviation Ω and setting $\epsilon_i = 1\ \forall i$ [54, 55, 56, 57]. Of particular interest in these studies was the prediction of Griffiths phase physics near the QPT both in two and three dimensional spin glass versions ($\Omega_0 = 0$) of Eq. (3) [56].

Two puzzles were identified when the behaviour of $LiHo_xY_{1-x}F_4$ subject to $B_x \neq 0$ was first compared with expectations from theoretical and numerical studies. Firstly, while $dT_c/dx \propto x$ in the diluted ferromagnetic (FM) regime ($0.25 \lesssim x < 1.0$, see Fig. 1), as in mean-field theory [58], the rate at which $T_c(B_x)$ is reduced by B_x gets progressively faster than mean-field theory predicts as the Ho^{3+} concentration x decreases [59]. This means that, in comparison with the energy scale for FM order that determines the classical transition at $T_c(B_x = 0, x)$, B_x becomes more efficient at destroying FM order the lower the concentration x is [59]. Secondly, for $B_x = 0$, $LiHo_{0.167}Y_{0.833}F_4$ has generally been believed to exhibit a conventional spin glass (SG) transition [51] (see however Ref. [52, 53]), with the transition signaled by a nonlinear magnetic susceptibility, χ_3, diverging at T_g as $\chi_3(T) \propto (T - T_g)^{-\gamma}$ [48]. However, $\chi_3(T)$ in $LiHo_{0.167}Y_{0.833}F_4$ becomes less singular as B_x is increased from $B_x = 0$ [60, 61]. This suggests that, in contrast with numerical work [55, 56], no B_x-driven quantum phase transition between a polarized quantum paramagnetic state and a SG state occurs as $T \to 0$.

The resolution of these two puzzles lies in the observation that, in the properly derived effective low-energy theory akin to Eq. (2) when the occupation factor $\epsilon_i \neq 1$ [62, 63], the transverse field B_x induces not only quantum fluctuations, but also *random fields* [43, 44, 47], h_i^z, that couple linearly to the $\hat{z}$ component of $\vec{\sigma}_i$ [32, 33, 62, 63, 64, 65]. There are two key physical ingredients responsible for these random fields. Firstly, the magnetic field B_x, perpendicular to the c-axis Ising direction, induces a finite expectation value of the Ho^{3+} magnetic moments along $\hat{x}$, $\langle \mu_i^x \rangle = g\mu_B \langle J_i^x \rangle$, at *all* temperatures. Quantum mechanically, this moment arises from the B_x-induced admixing of the (initially strictly Ising) $|\psi_0^{\pm}\rangle$ ground doublet with the excited crystal field states. Secondly, the bare microscopic Hamiltonian in Eq. (1), as opposed to the effective TFIM of Eq. (2) and its simplified version in Eq. (3), contains off-diagonal (anisotropic) $\mathcal{L}_{ij}^{\mu\nu}$ ($\mu \neq \nu$) terms of dipolar origin that couple the various components J_i^{μ} of $\vec{J}_i$. In pure $LiHoF_4$, perfect mirror symmetries make the lattice sum $h_i^z \equiv \sum_j \mathcal{L}_{ij}^{zx} \langle J_j^x \rangle$ vanish identically. For diluted $LiHo_xY_{1-x}F_4$, h_i^z no longer vanishes; the transverse magnetic moments $\langle \mu_i^x \rangle$ induce, via the random $\epsilon_i \mathcal{L}_{ij}^{zx} \langle J_i^x \rangle$ terms, internal random fields that act on the z component of $\vec{J}_i$, J_i^z. In other words, diluted $LiHo_xY_{1-x}F_4$ subject to a transverse field B_x maps onto a TFIM with the addition of *correlated random fields* [62, 63]. For $0.25 \lesssim x < 1$, this makes $LiHo_xY_{1-x}F_4$ a unique example of a *random field Ising ferromagnet model* [43, 44, 47], albeit with the random

Journal of Physics: Conference Series **320** (2011) 012001 doi:10.1088/1742-6596/320/1/012001

fields being correlated in $LiHo_xY_{1-x}F_4$. Indeed, the experimental procedure for creating such a random field Ising system is convoluted and typically considers a diluted Ising antiferromagnet in a uniform longitudinal field B_z [66, 67]. Analytical [33, 62, 63, 64, 65] and numerical [62, 63, 64] studies have shown, rather convincingly, that the presence of B_x-induced random fields h_i^z can explain the two aforementioned puzzles observed in $LiHo_xY_{1-x}F_4$ in nonzero B_x, in both the diluted FM regime and in the SG regime. In particular, the B_x-induced random fields h_i^z eliminate the divergence of the nonlinear susceptibility χ_3 in the spin glass regime $x \lesssim 0.25$ [32, 62, 64].

2.3. Open questions regarding the random field physics in $LiHo_xY_{1-x}F_4$
Because of space constraint, we have focused above on the role of the induced random fields on the phase transitions that occur in the diluted ferromagnetic and spin glass regimes. As the critical (ferromagnetic or spin glass) temperature decreases, and drops below approximately 0.5 K, the role of the hyperfine interaction in inhibiting quantum fluctuations and strongly renormalizing the critical transverse field cannot be ignored [32, 33]. While progress has been achieved as to how to proceed [33], work remains to be done to investigate via quantitatively accurate calculations (e.g. quantum Monte Carlo simulations) the role of hyperfine interactions on the global $x - T - B_x$ phase diagram of $LiHo_xY_{1-x}F_4$.

In the context of the physics of induced random fields, it is worthwhile to note that theoretical work predicts that in the simplest TFIM of Eq. (3), the addition of uncorrelated random fields should destroy the zero temperature quantum criticality and drive the renormalization group flow towards the fluctuationless (classical) zero temperature random field fixed point [68]. It would be interesting to investigate the implication of that result for the *correlated* h_i^z random fields discussed above and that exist in $LiHo_xY_{1-x}F_4$ in nonzero B_x, as suggested by some experiments [69]. Finally, we note that the presence of induced random fields has been invoked to interpret the transverse field and temperature dependence (e.g. scaling behavior) of the magnetic susceptibility in $LiHo_xY_{1-x}F_4$ ($x = 0.44$). In particular, these random fields have been postulated to give rise to Griffiths' singularity effects in the paramagnetic regime at temperatures just above T_c [69]. It would be useful to explore this claim further. In fact, a systematic experimental investigation of the fascinating physics of random field systems in the diluted ferromagnetic regime of $LiHo_xY_{1-x}F_4$ and subject to a nonzero transverse field B_x is probably warranted.

3. $LiHo_xY_{1-x}F_4$ – spin glass vs antiglass

3.1. Spin glass vs antiglass physics in $LiHo_xY_{1-x}F_4$
The previous section discussed the properties of $LiHo_xY_{1-x}F_4$ in nonzero transverse field B_x. We now turn to the question of the existence of a spin glass phase in $LiHo_xY_{1-x}F_4$ for $x \lesssim 0.25$ and in zero transverse field ($B_x = 0$).

The $x - T$ phase diagram of $LiHo_xY_{1-x}F_4$ in the limit of large dilution ($x \lesssim 0.25$) has been and remains a subject of much controversy. Theoretical studies of the dilute Ising dipolar model predict that a spin-glass state may be favored over ferromagnetic order at high dilution [24, 70]. Since there is no percolation threshold for long range dipolar couplings, a thermodynamic spin glass state is expected to persist all the way to $x = 0^+$ [70]. Furthermore, a study of the quantum corrections to the Ising model arising from interaction-induced virtual crystal field fluctuations finds that these quantum effects should be small in $LiHo_xY_{1-x}F_4$ [42]. This is in contrast to the $Tb_2Ti_2O_7$ and $Tb_2Sn_2O_7$ pyrochlore Ising magnets [71, 72] and, therefore, are unlikely to destroy the spin glass phase in $LiHo_xY_{1-x}F_4$ at finite Ho^{3+} concentration. Consequently, in light of the accumulated evidence for a thermodynamic spin glass phase transition in the three-dimensional Edwards-Anderson (EA) Ising spin glass model [49], one would also expect such a transition in $LiHo_xY_{1-x}F_4$ for $0 < x \ll 1$ and $B_x = 0$ since the EA model and dipolar systems should be in the same universality class for a spin glass transition [73].

Measurements of the imaginary part of the magnetic susceptibility, χ'', in $LiHo_{0.045}Y_{0.955}F_4$ by Reich *et al.* [74, 75] was the first to challenge the above theoretical picture of a spin glass phase down to $x = 0^+$. In a typical glassy system, the frequency distribution of χ'' broadens upon lowering the temperature [48]. In contrast, the authors of Refs. [74, 75] found a distribution that *narrows* as the temperature is reduced, indicative of an overall speed-up of the dynamics and an "antiglass" (quantum disordered) ground state. The measurement was later repeated [76] to lower temperatures (50 mK), still showing an evident narrowing of the spectrum. The manifestation of dissipationless dynamics of a broad inhomogeneous distribution of quantum oscillators within the inferred antiglass regime was further demonstrated through (i) hole burning in the AC susceptibility spectrum [76] and, (ii) perhaps most interestingly, via the observation of persistent oscillations for up to 30 seconds after an external AC field is turned off [76]. It has been argued that this antiglass behaviour may originate from a quantum mechanical entanglement of the magnetic dipole moments and the formation of a sort of random-singlet state [77]. However, as mentioned above, the quantum effect corrections to the otherwise perfect Ising nature of the single-ion crystal field ground doublet are expected to be very small in $LiHo_xY_{1-x}F_4$ [42]. Also, while hyperfine interactions do strongly renormalize the critical transverse field in transverse field experiments (see Section 2) [32, 33], at the same time, preliminary calculations suggest that these interactions may not be efficient at quenching the spin glass phase at a Ho^{3+} concentration of order of 5% [42]. It is therefore not clear how the entanglement mechanism proposed in Ref.[77] on the basis of the ad-hoc model considered therein can really proceed efficiently when starting from the microscopic model of Eq. (1).

Apart from the question of the existence of an antiglass phase, there have also been contradictory experimental reports on the spin glass transition in the regime $0.1 \lesssim x \lesssim 0.2$, where it was generally believed to be well established. Specifically, Jönsson *et al.* found that the non-linear susceptibility χ_3 does not diverge for samples with $x = 4.5\%$ and, surprisingly, for $x = 16.5\%$ as well [52]. These results prompted Jönsson *et al.* to conclude that there is no thermodynamic spin glass phase at all in the compound. In response to this work, Ancona-Torres *et al.* re-examined the non-linear susceptibility for $x = 16.7\%$ and $x = 19.8\%$ and concluded that χ_3 does diverge [51]. A possible explanation put forward in Ref. [51] for the discrepancy between the works of Ref. [52] and Ref. [51] is that the sweep rates of the magnetic field in the work of Jönsson *et al.* [52] were too fast and the applied magnetic fields too strong to expose the true spin glass critical behaviour of the system. To compound the debate, Jönsson and co-workers [53] continue to maintain that evidence for a spin glass transition in $LiHo_xY_{1-x}F_4$ for $x = 16.5\%$ is inconclusive, unlike the claims of Ancona-Torres *et al.* [51].

Several numerical studies have aimed at finding a spin-glass transition in effective dipolar Ising models mimicking the dilute compound $LiHo_xY_{1-x}F_4$. The original work of Xu *et al.* only identified the disappearance of the long range ferromagnetic order for a diluted dipolar body-centered cubic lattice and did not investigate the spin glass freezing [24]. The traditional way to identify the glass transition has been to calculate the Edwards-Anderson overlap between two replicas, #1 and #2, $q = 1/N \sum_i \sigma_{i,1}^z \sigma_{i,2}^z$, and to determine the temperature at which the corresponding Binder ratios, $g = 1 - [\langle q^4 \rangle]/3[\langle q^2 \rangle]^2$, for different system sizes intersect. However, several Monte Carlo studies [26, 27] have been unable to detect a spin glass transition on the basis of such intersection, or other finite-size scaling properties of q (see Ref. [78]). These "first generation" Monte Carlo simulations of diluted dipole-coupled Ising spins did therefore suggest that the absence of a thermodynamic spin glass transition in $LiHo_xY_{1-x}F_4$ at small x could possibly have a *classical* origin. However, as we discuss in the following section, more recent experimental and numerical studies reveal a different picture.

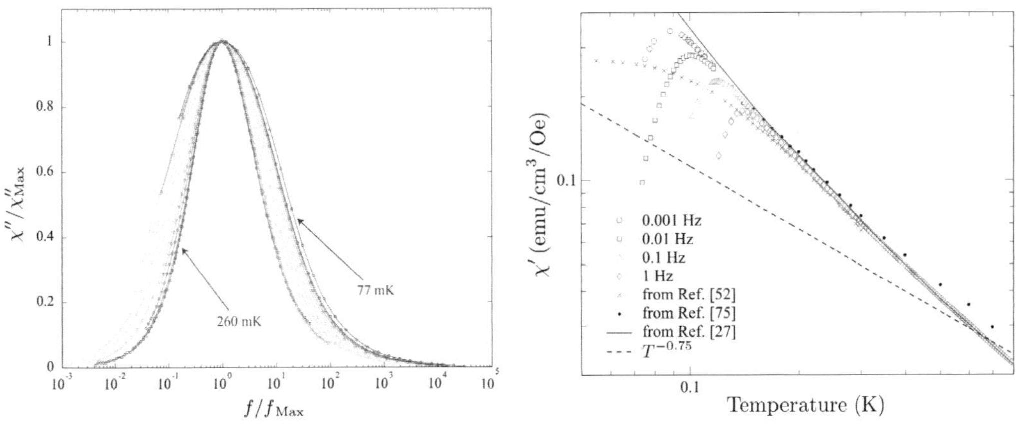

Figure 2. The left panel shows a plot of the imaginary part of the AC susceptibility, χ", as a function of frequency, f, normalized to its maximum value $\chi"_{\rm max}$ at frequency $f_{\rm max}$, and for temperatures between 77 mK and 260 mK. The figure shows a broadening of the spectrum below and above the peak frequency $f_{\rm max}$ at which $\chi"(f)$ peaks (adapted from Ref. [79]). The right panel shows the real part of the AC susceptibility, $\chi'(f)$, as a function of temperature. One notes the general good agreement with the $\lim_{f\to 0}\chi"(f)$ data set from Ref. [75] down to a temperature $\sim$ 180 mk as well as with the DC susceptibility data determined via classical Monte Carlo [27]. The figure, adapted from Ref. [79], illustrates the disagreement with the results of Jönsson *et al.* in Ref. [52] as well as with the $T^{-0.75}$ behaviour reported in Ref. [77] and argued to support an entangled collective singlet ground state.

3.2. Recent developments in spin glass physics in $LiHo_xY_{1-x}F_4$

Since much of the evidence for the antiglass phase is based on the scaling of the imaginary part of the AC susceptibility, the reproducibility of these measurements is very important. Recent measurements by Quilliam *et al.* [79] contradict the earlier results [74, 75, 76] and find that the frequency distribution of χ" *broadens* as the temperature is reduced also in the high dilution limit ($x = 4.5\%$) as shown in the left panel of Fig. 2. Furthermore, the quasi-static susceptibility, χ_0, reported in Ref. [79] does not show the unusual $T^{-0.75}$ divergence found earlier [77]. We note, perplexingly, as noted in Refs. [27, 79], that the quasi-static susceptibility χ_0 reported by Ghosh *et al.* in Ref. [77] does not agree with the previous results [74, 75] by the same group on, presumably, the very same sample. Meanwhile, the experimental χ_0 data of Quilliam *et al.* [79] agree reasonably well with the early results of Refs. [74, 75] down to a temperature of approximately 180 mK, as well as with the DC susceptibility found in a *classical* Monte Carlo simulation of Eq. (2) with $B_x = 0$ [27], as shown in the right panel of Fig. 2. The work of Ref. [79] would appear to give a strong indication that there is conventional spin glass freezing in the high dilution limit of $LiHo_xY_{1-x}F_4$. Furthermore, the specific heat data reported in Ref. [80] do not display any of the unusual features that were previously interpreted as a telltale signature of entangled magnetic moments within a quantum disordered antiglass state [77]. Note that the specific heat determined via classical Monte Carlo simulations [27] does not agree particularly well with either set of experimental data. This discrepancy may be in part due to the difficulty in taking into account accurately of the hyperfine contribution to the experimental specific heat data.

The reason for the difference between the AC susceptibility and specific heat experimental results of Refs. [74, 75, 76, 77] and Refs. [79, 80] remains an open question. Due to the very slow dynamics, all these measurements are technically very demanding. However, we believe that it would be important to settle the question of the nature of the high dilution phase in $LiHo_xY_{1-x}F_4$. Perhaps the study of other highly diluted Ising magnetic materials could help shed some light on the physics at play in $LiHo_xY_{1-x}F_4$. In this context, a study of the highly diluted Ising spin ice materials $(Dy_xY_{1-x})_2Ti_2O_7$ and $(Ho_xY_{1-x})_2Ti_2O_7$ [81] as well as the proposed $Ho_xY_{1-x}(OH)_3$ and $Dy_xY_{1-x}(OH)_3$ TFIM materials [38] to compare with $LiHo_xY_{1-x}F_4$ could prove interesting and instructive.

The failure of previous Monte Carlo simulations [26, 27, 78] to find a thermodynamic spin glass transition in the dipolar model would seem to have been, at least partially, explained recently. As in three-dimensional Edward-Anderson Ising spin glass model [49], the spin glass correlation length ξ_{sg} turns out to be a much better indicator of a spin glass transition than the aforementioned Binder ratio g. A recent Monte Carlo study [82] that focuses on ξ_{sg} confirms that this is also the case also for the dipolar Ising model. In Ref. [82], a finite-size crossing of ξ_{sg} divided by the linear system size L, ξ_{sg}/L, is observed, providing fair evidence for a finite temperature spin-glass transition, in agreement with the most recent AC susceptibility experiments [79]. However, the critical temperature determined by the crossing of ξ_{sg}/L is slightly different for ξ_{sg} along the a and c axes. This indicates that the finite size corrections to scaling are not under control for the system sizes considered and simulations on larger system sizes are therefore needed. A more recent numerical study argues for a quasi-long-range ordered spin glass phase in a system of diluted Ising dipoles [83].

3.3. *Open questions regarding the spin glass physics in $LiHo_xY_{1-x}F_4$*

The single most important open question is the nature of the magnetic state in the limit of high dilution ($x < 0.1$). Is it a traditional spin glass, as suggested by the recent measurements of Quilliam *et al.* [79] or is it an unusual antiglass spin liquid [74, 75, 76, 77]? One may even ask whether there is a spin glass phase at all for $x \lesssim 0.25$ [51, 52, 53]. It would appear that the only way to resolve this issue is through more experiments since the computational effort to perform Monte Carlo simulations of large systems of magnetic moments interacting with long range $1/r^3$ dipolar interactions is rather prohibitive, and finite-size scaling corrections are significant [82]. The "sample quality" (e.g. impurity, homogeneous Y^{3+} disorder, etc) may be of importance in this highly dilute limit and this issue should be examined carefully. Sample shape and demagnetization effects should not affect the existence of the spin glass phase, but this should nevertheless be investigated. There is a significant qualitative difference between the static and dynamic susceptibility measurements reported by three research groups (Refs. [51, 74, 75, 76, 77], Refs. [52, 53] and Refs. [79, 80]), and a new and independent series of measurements could help shed light on this outstanding question. It appears that the nonlinear susceptibility results of Ref. [52] are being dismissed [51, 79] on the basis of measurements that do not really probe the spin glass critical regime. However, disagreement on this interpretation remains [53]. The hole-burning and persistent oscillations observed by Ghosh *et al.* [76] are also indicative of a very unusual physical state, yet they have not been reported or reproduced by another group as far as we know. To reiterate, we believe that at this stage, in order for the field to move forward, more experiments that carefully explore the critical regime, notwithstanding the highly technical burden associated with such experiments, are very much needed.

On the numerical side, it would be of interest to go beyond the classical Monte Carlo simulations [26, 27, 82] that have so far been applied to the dilute system. The effects of the hyperfine interactions and off-diagonal interactions between the $\vec{\sigma}_i$ pseudospins are challenging to incorporate in a quantum Monte Carlo simulation since they lead to a sign problem. Nevertheless, it would be important to invest some effort in this problem in order to achieve more quantitative comparisons with experiments.

Journal of Physics: Conference Series **320** (2011) 012001 doi:10.1088/1742-6596/320/1/012001

4. Conclusion

Due to the existence of a well-established two-state Ising model for the rare-earth compound $LiHo_xY_{1-x}F_4$, this material constitutes a unique model magnet where theory, experiments and simulations can be compared and contrasted at a quantitative level. Despite the apparent simplicity of the underlying model for the parent compound, the introduction of disorder through dilution and quantum effects through a transverse magnetic field leads to a complex system which continues to challenge experimentalists and theorists alike. The two foremost outstanding questions are whether a thermodynamic spin glass phase survives in the limit of high dilution of the magnetic Ho^{3+} ions and the reason for the unexpected stability of the ferromagnetic ground state of pure $LiHoF_4$ with respect to an applied transverse field. Continued research into this remarkable realization of the Ising model will hopefully continue to contribute pushing forward the forefront of research on such diverse topics as random fields, coherent oscillations and tunable quantum fluctuations.

References

[1] Wyckoff R W G 1965 *Crystal structures* vol III (NewYork:Interscience)
[2] Mennenga G, deJongh L J and Huiskamp W J 1984 *J. Magn. Magn. Mater.* **44** 59
[3] Hansen P E, Johansson T and Nevald R 1975 *Phys. Rev. B* **12** 5315
[4] Cooke A H, Jones D A, Silva J F A and Wells M R 1975 *J. Phys. C: Sol. St. Phys.* **8** 4083
[5] Battison J E, Kasten A, Leask M J M, Lowry J B and Wanklyn B M 1975 *J. Phys. C: Sol. St. Phys.* **8** 4089
[6] Luttinger J M and Tisza L 1946 *Phys. Rev.* **70** 954
[7] Griffiths R 1968 *Phys. Rev* **176** 655
[8] Larkin A I and Khmelnitskii D E 1969 *Sov. Phys. JETP* **29** 1123
[9] Aharony A 1973 *Phys. Rev. B* **8** 3363
[10] Griffin J A, Huster M and Folweiler R J 1980 *Phys. Rev. B* **22** 4370–4378
[11] Beauvillain P, Renard J P, Laursen I and Walker P J 1978 *Phys. Rev. B* **18** 3360–3368
[12] Ahlers G, Kornblit A and Guggenheim H 1975 *Phy. Rev. Lett.* **34** 1227
[13] Frowein R, Kötzler J and Assmus W 1979 *Phys. Rev. Lett.* **42** 739–742
[14] Nikkel J and Ellman B 2001 *Phys. Rev. B* **64** 214420
[15] Meyer P, Pommier J and Ferre J 1989 *SPIE Electro-optic and magneto-optic materials and applications* vol 1126 p 93
[16] Kittel C 1949 *Rev. Mod. Phys.* **21** 541
[17] Gabay M and Garel T 1984 *J. Physique Lett.* **45** 989
[18] Gabay M and Garel T 1985 *J. Physique* **46** 5
[19] Biltmo A and Henelius P 2009 *Europhys. Lett.* **97** 27007
[20] Chakraborty P B, Henelius P, Kjonsberg H, Sandvik A W and Girvin S M 2004 *Phys. Rev. B* **70** 144411
[21] Tabei S M A, Gingras M J P, Kao Y J and Yavors'kii T 2008 *Phys. Rev. B* **78** 184408
[22] Bitko D, Rosenbaum T F and Aeppli G 1996 *Phys. Rev. Lett.* **77** 940–943
[23] Jensen S J K and Kjaer K 1989 *J. Phys.: Condens. Matter* **1** 2361
[24] Xu H, Bergersen B, Niedermayer F and Racz Z 1991 *J. Phys: Condens. Matter* **3** 4999
[25] Xu H, Bergersen B and Racz Z 1992 *J. Phys: Condens. Matter* **4** 2035
[26] Biltmo A and Henelius P 2007 *Phys. Rev. B* **76** 054423
[27] Biltmo A and Henelius P 2008 *Phys. Rev. B.* **78** 054437
[28] Kretschmer R and Binder K 1979 *Z. Physik B* **34** 375
[29] Wolf W P 2000 *Braz. Jour. Phys.* **30** 794
[30] Krusius M, Anderson A C and Holmström B 1969 *Phys. Rev.* **177** 910–916
[31] Bramwell S T, Harris M J, den Hertog B C, Gingras M J P, Gardner J S, McMorrow D F, Wildes A R, Cornelius A L, Champion J D M, Melko R G and Fennell T 2001 *Phys. Rev. Lett.* **87** 047205
[32] Schechter M and Stamp P C E 2005 *Phys. Rev. Lett.* **95** 267208
[33] Schechter M and Stamp P C E 2008 *Phys. Rev. B* **78** 054438
[34] Sondhi S L, Girvin S M, Carini J P and Shahar D 1997 *Rev. Mod Phys.* **69** 315
[35] Sachdev S 1999 *Quantum Phase Transitions* (Cambridge University Press)
[36] de Gennes P G 1963 *Solid State Commun.* **1** 132
[37] Young A P 1975 *J. Phys. C* **8** L309
[38] Stasiak P and Gingras M J P 2008 *Phys. Rev. B* **78** 224412
[39] Ruff J P C, Melko R G and Gingras M J P 2005 *Phys. Rev. Lett.* **95** 097202

[40] Rønnow H M, Parthasarathy R, Jensen J, Aeppli G, Rosenbaum T F and McMorrow D F 2005 *Science* **308** 389–392
[41] Dunn J L, Stahl C, Reshitnyk Y, Sim W and Hill R W 2010 *arXiv:1005.1935*
[42] Chin A and Eastham P R 2008 *Quantum Magnetism* NATO Science for Peace and Security Series B - Physics and Biophysics ed B Barbara and Y Imry and G Sawatzky and P C E Stamp (Springer) p 57
[43] Grinstein G M 1984 *Fundemental Problems in Statistical Mechanics VI: Proceedings of the 1984 Trondheim Summer School* ed E G D Cohen (North Holland) p 57
[44] Fisher D S, Grinstein G M and Khurana A 1988 *Physics Today* **41** 56
[45] Imry Y and Wortis M 1979 *Phys. Rev. B* **19** 3580
[46] Harris A B 1974 *J. Phys. C* **7** 1671
[47] Imry Y and k Ma S 1975 *Phys. Rev. Lett.* **35** 1399
[48] Mydosh J A 1993 *Spin Glasses: An Experimental Introduction* (Taylor & Francis, London)
[49] Kawashima N and Rieger H 2004 *Frustrated Spin Systems* (World Scientific) chap 6
[50] Griffiths R B 1969 *Phys. Rev. Lett.* **23** 17
[51] Ancona-Torres C, Silevitch D M, Aeppli G and Rosenbaum T F 2008 *Phys. Rev. Lett.* **101** 057201
[52] Jönsson P E, Mathieu R, Wernsdorfer W, Tkachuk A and Barbara B 2007 *Phys. Rev. Lett.* **56** 256403
[53] Jönsson P E, Mathieu R, Wernsdorfer W, Tkachuk A and Barbara B 2008 *arXiv:0803.1357*
[54] Miller J and Huse D A 1993 *Phys. Rev. Lett.* **70** 3147
[55] Guo M, Bhatt R N and Huse D A 1994 *Phys. Rev. Lett.* **72** 4137
[56] Guo M, Bhatt R N and Huse D A 1996 *Phys. Rev. B.* **54** 3336
[57] Kim D H and Kim J J 2002 *Phys. Rev. B.* **66** 054432
[58] Brooke J, Bitko D, Rosenbaum T F and Aeppli G 1999 *Science* **284** 779
[59] Brooke J 2000 *Ph.D. thesis* (U. of Chicago)
[60] Wu W, Bitko D, Rosenbaum T F and Aeppli G 1993 *Phys. Rev. Lett.* **71** 1919
[61] Wu W 1992 *Ph.D. thesis* (U. of Chicago)
[62] Tabei S M A, Gingras M J P, Kao Y J, Stasiak P and Fortin J Y 2006 *Phys. Rev. Lett* **97** 237203
[63] Tabei S M A, Vernay F and Gingras M J P 2008 *Phys. Rev. B* **77** 014432
[64] Schechter M and Laflorencie N 2006 *Phys. Rev. Lett.* **97** 137204
[65] Schechter M 2008 *Phys. Rev. B* **77** 020401(R)
[66] Fishman S and Aharony A 1979 *J. Phys. C* **12** L729
[67] Cardy J L 1984 *Phys. Rev. B.* **29** 505
[68] Senthil T 1998 *Phys. Rev. B* **57** 8375
[69] Silevitch D M, Bitko D, Brooke J, Ghosh S, Rosenbaum T F and Aeppli G 2007 *Nature* **448** 567
[70] Stephen M J and Aharony A 1981 *J. Phys. C* **14** 1665
[71] Molavian H R, Gingras M J P and Canals B 2007 *Phys. Rev. Lett.* **98** 157204
[72] McClarty P A, Stasiak P and Gingras M J P 2010 *arXiv:1011.5689*
[73] Bray A J, Moore M A and Young A P 2007 *Phys. Rev. Lett.* **98** 256403
[74] Reich D H, Rosenbaum T and Aeppli G 1987 *Phys. Rev. Lett.* **59** 1969
[75] Reich D H, Ellman B, Yang J, Rosenbaum T, Aeppli G and Belanger D P 1990 *Phys. Rev. B* **42** 4631
[76] Ghosh S, Parthasarathy R, Rosenbaum T F and Aeppli G 2002 *Science* **296** 2195
[77] Ghosh S, Rosenbaum T F, Aeppli G and Coppersmith S N 2003 *Nature* **425** 48
[78] Snider J and Yu C C 2005 *Phys. Rev. B.* **72** 214203
[79] Quilliam J A, Meng S, Mugford C G A and Kycia J B 2008 *Phys. Rev. Lett.* **101** 187204
[80] Quilliam J A, Mugford C G A, Gomez A, Kycia S W and Kycia J B 2007 *Phys. Rev. Lett.* **98** 037203
[81] Ke X, Freitas R S, Ueland B G, Lau G C, Dahlberg M L, Cava R J, Moessner R and Schiffer P 2007 *Phys. Rev. Lett.* **991** 137203
[82] Tam K M and Gingras M J P 2009 *Phys. Rev. Lett.* **103** 087202
[83] Alonso J J and Fernández J F 2010 *Phys. Rev. B* **81** 064408

Z_2-vortex order of frustrated Heisenberg antiferromagnets in two dimensions

Hikaru Kawamura

Department of Earth and Space Science, Faculty of Science, Osaka University, Toyonaka 560-0043, Japan

E-mail: kawamura@ess.sci.osaka-u.ac.jp

Abstract. We discuss the recent experimental data on various frustrated quasi-two-dimensional Heisenberg antiferromagnets from the viewpoint of the Z_2-vortex order, which include S=3/2 triangular-lattice antiferromagnet $NaCrO_2$, S=1 triangular-lattice antiferromagnet $NiGa_2S_4$, S=1/2 organic triangular-lattice antiferromagnets κ-(BEDT-TTF)$_2$Cu$_2$(CN)$_3$ and EtMe$_3$Sb[Pd(dmit)$_2$]$_2$, and S=1/2 kagome-lattice antiferromagnet volborthite $Cu_3V_2O_7(OH)_2{\cdot}2H_2O$, *etc.*

1. Introduction

Recently, geometrically frustrated magnets have attracted much interest because of their unconventional ordering behaviors, including the possible quantum spin-liquid state [1, 2, 3] and the novel ordered states like chiral ordered state [4, 5, 6, 7]. Such novel ordered states as well as the associated phase transitions are often borne by novel excitations inherent to frustrated systems. A Z_2-vortex, stabilized in a class of two-dimensional (2D) Heisenberg magnets with the locally noncollinear spin order [8], is a typical example of such novel excitations inherent to geometrically frustrated Heisenberg magnets. It was first identified in 1984 in the antiferromagnetic (AF) classical Heisenberg model on the 2D triangular lattice as a topologically stable point defect characterized by a two-valued topological quantum number [8]. In this Z_2-vortex, Heisenberg spins, or more precisely chirality vectors, circulate around a vortex core making a topologically stable vortex (Fig.1), whereas whether they circulate in clockwise or counter-clockwise fashion makes no distinction topologically. In other words, the Z_2-vortex has no characteristic winding number in sharp contrast to the standard vortex.

More recently, it was argued [9, 10, 11] that such a Z_2-vortex excitation might be responsible for certain anomalous behaviors observed experimentally in several triangular-lattice Heisenberg AFs such as S=3/2 $NaCrO_2$ [12, 13, 14, 15], S=1 $NiGa_2S_4$ [16, 17, 18, 19, 20, 21, 22, 23, 24, 25, 26] and S=1/2 organic compounds, κ-(BEDT-TTF)$_2$Cu$_2$(CN)$_3$ [27, 28, 29, 30, 31, 32, 33, 34, 35] or EtMe$_3$Sb[Pd(dmit)$_2$]$_2$ [36, 37, 38, 39]. In these materials, a spin-liquid-like behavior without the standard magnetic long-range order (LRO) was observed, while all of these compounds exhibit a weak but clear transition-like anomaly at a finite temperature. As a possible explanation of such an experimental anomaly, a Z_2-vortex driven topological transition was invoked [9, 10, 11]. Although relatively little attention has been paid in the literature to Z_2-vortices compared with, *e.g.*, the standard Z-vortex realized in the two-dimensional XY model [40, 41], it might have important relevance to the ordering of a variety of frustrated quasi-2D magnets than hitherto

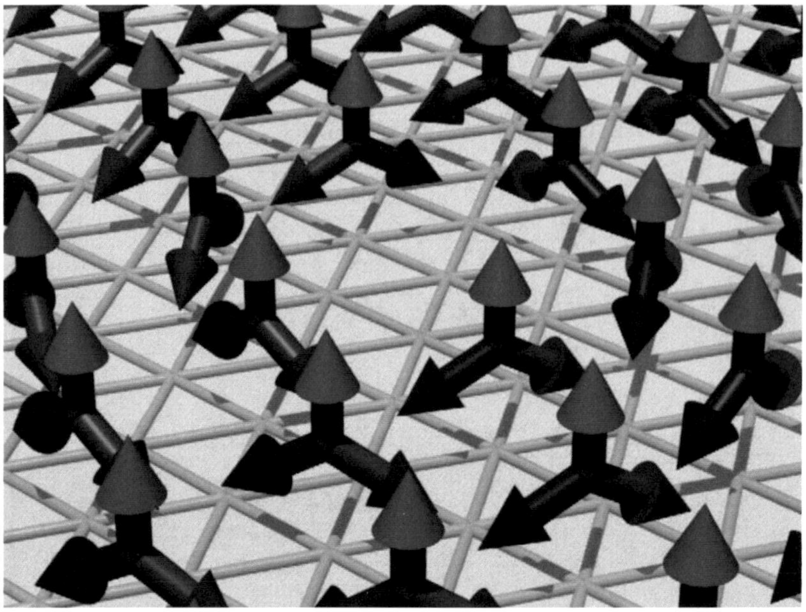

Figure 1. An illustration of a Z_2-vortex. Taken from [T. Okubo and H. Kawamura, J. Phys. Soc. Jpn. **79** (2010) 084706].

thought generally. The aim of the present article is to point out such possible relevance of Z_2-vortex excitations to the ordering process of various geometrically frustrated magnets under recent extensive study.

It was suggested in Ref.[8] that the frustrated 2D Heisenberg AF sustaining a Z_2-vortex might exhibit a thermodynamic phase transition at a finite temperature driven by the binding-unbinding of the Z_2-vortices. The transition is of topological character where the ergodicity is broken topologically in the sense that the phase space is narrowed from the doubly connected one to the singly connected one. An interesting observation is that the standard AF spin correlation length ξ_s is kept finite even at and below the Z_2-vortex transition point $T = T_v$. Nontopological excitations such as spinwaves which survive below T_v are enough to destroy the spin correlation to be paramagnetic with finite ξ_s. The low-temperature state characterized by the topologically broken ergodicity and by the finite (but often long) spin correlation length and correlation time is called a 'spin-gel' state [9]: See Fig.2.

Remarkable feature of the proposed Z_2-vortex transition might be the decoupling of the two length scales: At the Z_2-vortex transition temperature $T = T_v$, the vortex correlation length ξ_v corresponding to the mean separation of free Z_2-vortices diverges, while the spinwave correlation length ξ_{sw} is kept finite (Fig.2(left)) [9]. Reflecting the finiteness of ξ_{sw}, the full spin correlation length ξ_s is kept finite even at $T < T_v$. Possible decoupling behaviour discussed above might be in sharp contrast to the behavior of another vortex-driven transition, *i.e.*, the standard Kosterlitz-Thouless (KT) transition of the 2D XY model, where there exists only one diverging length scale and the spin correlation length stays infinite throughout the low-temperature phase [40, 41]. Finiteness of the spin correlation length even below T_v is a pronounced feature of the spin-gel state to be contrasted to the standard KT ordered state.

Finiteness of the spin correlation length ξ_s at $T = T_v$ has important consequences on the response of the system against weak perturbative interactions coupled to the spin, *e.g.*, magnetic field, magnetic anisotropy and interplane coupling [9, 10]. In the situation where ξ_s stays finite at the topological transition $T = T_v$, if the perturbation A is sufficiently small satisfying $A\xi_s(T = T_v)^2 < k_B T_v$, the Z_2-vortex transition and the low-temperature 'spin gel' state remains essentially the same as in the unperturbed ones: See Fig.2 for comparison of the two types of vortex transitions. The situation might be contrasted to that in the KT transition of the 2D XY model, where, reflecting the divergence of the spin correlation length, even an infinitesimal perturbation leads to the 3D magnetic long-range order in real magnets.

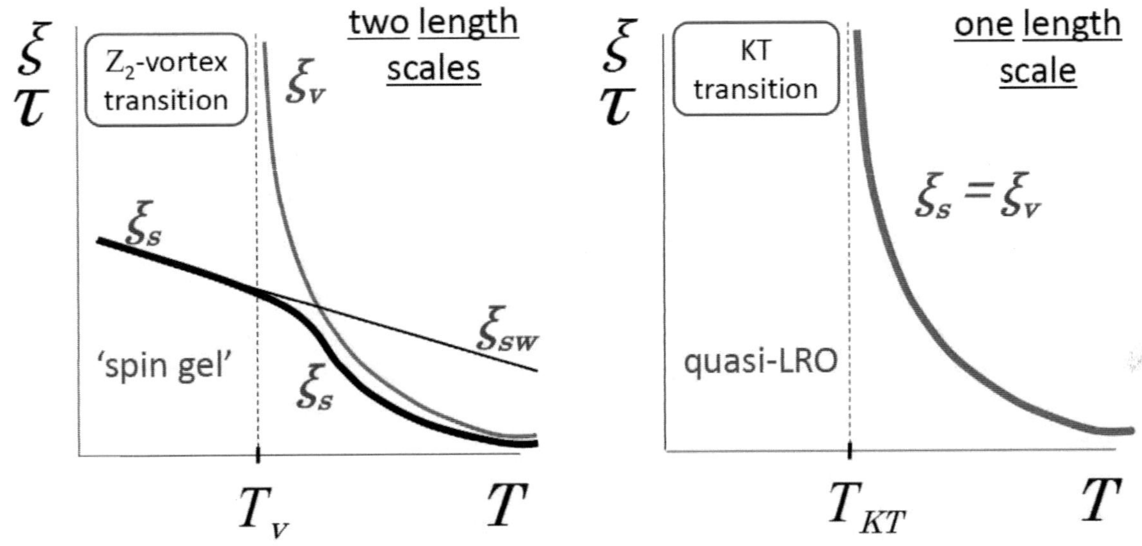

Figure 2. Schematic figure of the growth of the vortex correlation length ξ_v, the spinwave correlation length ξ_{sw} and the full spin correlation length ξ_s for each case of the Z_2-vortex transition of the frustrated Heisenberg model (left), and of the standard Kosterlitz-Thouless transition of the two-dimensional XY model (right).

We note that the question of whether the decoupling of the two length scales, each associated with the topological and nontopological degrees of freedom, really occurs or not is still at issue. If there is no such decoupling, the vortex binding-unbinding phenomenon might eventually become a sharp crossover rather than a genuine thermodynamic transition [42]. Recent numerical simulations of the 2D triangular-lattice classical Heisenberg AF has indicated that the Z_2-vortex binding-unbinding phenomenon looks like a thermodynamic phase transition, at least at the length scale of 10^3 lattice spacings [9]. Hence, it could be a very sharp crossover with enough experimental relevance, if not to be a genuine phase transition. In the present paper, we wish to discuss experimental consequences of such a Z_2-vortex binding-unbinding phenomenon of the frustrated 2D Heisenberg magnets, not entering into the question of whether it is a genuine phase transition or a sharp crossover.

2. The properties of the Z_2-vortex ordering

In this section, according to the recent theoretical analysis [9], we wish to summarize the expected properties of the Z_2-vortex and its binding-unbinding topological transition.

1) When one approaches $T = T_v$ from above, the vortex correlation length ξ_v diverges as

$$\xi_v \sim \exp\left[\left(\frac{A}{T - T_v}\right)^{\alpha}\right], \quad T > T_v, \tag{1}$$

with $\alpha < 1$. The standard spin correlation length grows rapidly toward T_v, but remains finite even at and below $T = T_v$, exhibiting only a weak essential singularity at $T = T_v$: See Fig.2(left).

2) Singular part of the free energy around T_v is given by $f_v \approx \xi_v^{-2}$.

3) The specific heat exhibits a weak essential singularity at $T = T_v$, while a non-singular peak or a hump appears slightly above T_v. The latter corresponds to the temperature where the maximum number of Z_2-vortex pairs dissociate giving rise to the maximum entropy change, while the transition temperature T_v corresponds to the temperature where the first Z_2-vortex pair dissociates.

4) The magnetic susceptibility also shows a week essential singularity at T_v.

5) The Z_2-vortex transition accompanies an anomaly in the spin dynamics. When ones approaches T_v from above, the spin correlation time grows sharply toward T_v, but not truly diverges at $T = T_v$, remaining finite even at and below T_v. Even below T_v, large and slow fluctuations should remain ('spin-gel' state).

6) The Z_2-vortex transition and the low-temperature spin-gel state are robust against sufficiently weak magnetic perturbations such as magnetic anisotropy, interplane coupling and applied magnetic fields. When these perturbations get larger beyond a critical value, which is determined by the spin correlation length at T_v, the system would change itself into the more standard ordered states such as the AF long-range ordered state or the frozen spin-glass state, instead of the spin-gel state.

7) The dynamical spin structure $S(q, \omega)$ exhibits a characteristic central peak originated from free Z_2-vortices above $T = T_v$ [11]. $S(q, \omega)$ also exhibits a characteristic intensity at finite ω along the zone boundary originated from Z_2-vortex pairs.

3. Experiments on frustrated Heisenberg antiferromagnets in two dimensions

We now wish to discuss possible implications of the Z_2-vortex ordering scenario summarized in the previous section to recent experiments on several quasi-2D frustrated Heisenberg AFs. Those include (a) S=3/2 triangular-lattice AF $NaCrO_2$, (b) S=1 triangular-lattice AF $NiGa_2S_4$, (c) S=1/2 organic triangular-lattice AFs κ-(BEDT-TTF)$_2$Cu$_2$(CN)$_3$ and EtMe$_3$Sb[Pd(dmit)$_2$]$_2$, (d) S=1/2 kagome-lattice AF $Cu_3V_2O_7(OH)_2 \cdot 2H_2O$, and (e) S=3/2 kagome-bilayer (or pyrochlore-slab) AF $SrCr_{8-x}Ga_{4+x}O_{19}$, *etc.*

(a) $NaCrO_2$

$NaCrO_2$ is a S=3/2 triangular-lattice Heisenberg AF with its Curie-Weiss temperature $T_{CW} \simeq 290$K. This magnet does not exhibit the conventional AF LRO down to low temperature, while it exhibits a magnetic short-range order (SRO) characterized by an incommensurate wavevector close to the 120-degrees structure [12, 13, 14, 15, 44]. The origin of the incommensurate spin structure is ascribed either to an easy-axis-type magnetic anisotropy or to a weak interplane coupling. A clear but rounded specific-heat peak is observed at $T_{peak} = 41$K, whereas a transition-like dynamical anomaly is observed at $T_f \simeq 30$K, a temperature slightly below T_{peak}. At $T = T_f$, the spin dynamics is rapidly slowed down giving rise to a quasi-static internal field. The spin dynamics probed by NMR, ESR and μSR, however, is not completely frozen below T_f. Rather, the spins remain slowly fluctuating even below T_f unlike the conventional AF or the spin glass. Such a dynamically fluctuating ordered state extends over a wide temperature range down to 10K. The spin correlation length determined from neutron scattering is kept finite $\xi \simeq 20$ lattice spacings even below T_f, which is not resolution-limited.

These experimental features are fully consistent, at least qualitatively, with the Z_2-vortex scenario discussed in the previous section, if the experimental freezing temperature T_f is identified as T_v and the low-temperature state as the 'spin-gel' state. In particular, energetics seems appropriate. Recent Monte Carlo simulation on the triangular-lattice classical Heisenberg model with the nearest-neighbor AF coupling yielded $T_{peak}/T_{CW} \simeq 0.137$ and $T_v/T_{CW} \simeq 0.123$ [9], which are close to the corresponding experimental values for $NaCrO_2$, *i.e.*, 0.14 and 0.10, respectively. It remains to be seen whether the dynamical spin structure factor exhibits a characteristic behavior predicted as a fingerprint of the Z_2-vortex, the point 7 above [11].

Magnetic ordering of similar chromium triangular-lattice AFs like $HCrO_2$ [15, 43], $LiCrO_2$ [15, 43, 44, 45, 46, 47, 48], $CuCrO_2$ [47, 49, 50, 51], $AgCrO_2$ [47, 52] and $PdCrO_2$ [53, 54] were also investigated experimentally. In contrast to $NaCrO_2$, these materials exhibit a 3D AF LRO at a Neel temperature $T = T_N$ due to the weak interlayer coupling. Yet, since all these materials are magnetically quasi-2D, the Z_2-vortex might exist in the 2D regime realized in the temperature region slightly above T_N. A particularly interesting compound in this series might be $KCrO_2$, where no evidence of the 3D AF LRO was reported by neutron scattering as for $NaCrO_2$ [44]. Further experimental study on this compound is encouraged.

It was also reported that the two-dimensionality could strongly be enhanced in $Cu_{1-x}Ag_xCrO_2$, a compound obtained from $CuCrO_2$ by chemical substitution [55, 56]. Then, one might expect that the Z_2-vortex has a higher chance to be realized in $Cu_{1-x}Ag_xCrO_2$. Of course, in this compound, care has to be taken on the possible effect of quenched randomness.

(b) $NiGa_2S_4$

$NiGa_2S_4$ is a S=1 triangular-lattice Heisenberg AF with its Curie-Weiss temperature $T_{CW} \simeq$ 80K [16, 17, 18, 19, 20, 21, 22, 23, 24, 25, 26]. This material attracted considerable attention of researchers, because the low-temperature specific heat exhibits a T^2 behavior suggesting the existence of Goldstone modes associated with a broken continuous symmetry, whereas no magnetic LRO was observed down to low temperature [16]. The observed magnetic SRO is an incommensurate helical structure in the basal plane close to the 60-degrees structure. Such a spin structure is stabilized by the AF third-neighbor interaction J_3 which frustrates the ferromagnetic first-neighbor coupling J_1 with $J_1/J_3 \simeq -0.2$ [23].

Meanwhile, $NiGa_2S_4$, with an integer spin quantum number S=1 in contrast to a half-integer spin quantum number S=3/2 of $NaCrO_2$, exhibits a strikingly similar ordering behavior to that of $NaCrO_2$. Indeed, $NiGa_2S_4$ exhibits a clear but rounded specific-heat peak at T_{peak} = 12K [23] (the peak of C/T is located at 10K), whereas a transition-like dynamical anomaly is observed at $T^* \simeq$ 8.5K, a temperature slightly below T_{peak}, where the spin dynamics is rapidly slowed down. Note that quasi-static internal fields of *dipolar* nature is observed below T^*, at least at the time scale of NQR and muon measurements. The spin dynamics probed by NMR, NQR, ESR and μSR, however, is not completely frozen even below T^*. Rather, the spins remain slowly fluctuating at the time scale of MHz, unlike the conventional AF or the spin glass. Such a dynamically fluctuating ordered state extends over a wide temperature range down to 2K. The spin correlation length determined from neutron scattering is kept finite even at and below T^*, *i.e.*, $\xi \simeq 7$ lattice spacings [16]. Dynamical freezing at $T = T^*$ also accompanies a weak anomaly in the susceptibility. Anomaly in the susceptibility looks clearer than the one observed in $NaCrO_2$, presumably due to the fact that $NiGa_2S_4$ possesses a ferromagnetic nearest-neighbor coupling. Application of magnetic fields greater than $100 \sim 1000$ G gradually changes the slowly fluctuating state into the more conventional frozen state [20].

As for $NaCrO_2$, those experimental features of $NiGa_2S_4$ are fully consistent, at least qualitatively, with the Z_2-vortex ordering scenario, if the experimental freezing temperature T^* is identified as T_v and the low-temperature state as the spin-gel state. Concerning its energetics, $NiGa_2S_4$ is characterized by $T_{peak}/T_{CW} \simeq 0.15$ and $T_v/T_{CW} \simeq 0.11$, which are close to the corresponding theoretical values 0.133 and 0.123 [9]. Furthermore, a theoretical estimate of the crossover-field intensity $\sim$ 1000G obtained with the information of the spin correlation length $\xi \simeq 6$ lattice spacings at $T = T^*$, seems consistent with the μSR measurement [20].

In this Z_2-vortex picture, slow dynamics below T^* is primarily borne by spinwaves. Then, spinwaves would be responsible for the T^2 specific heat. Indeed, recent inelastic neutron-scattering measurements identified a damped spinwave excitation at a low temperature $T = 1.5$K [25]. The T^2 behavior of the specific heat sets in around T^*, which is consistent with such a picture. Ref.[57] accounted for the T^2 specific heat based on spinwave excitations of the noncollinear AF order of S=1 quantum magnets, neglecting the vortex degrees of freedom [57]. Vortex-free assumption of Ref.[57] is well justified below the Z_2-vortex transition.

Impurity effects on the low-temperature specific heat and on the transition-like anomaly T^* was studied in Ref.[24]. Upon substitution of various kinds of impurities, *i.e.*, $S = 0$ Zn^{2+}, $S = 2$ Fe^{2+}, $S = 3/2$ Co^{2+} and $S = 5/2$ Mn^{2+}, the transition-like anomaly of the pure compound at T^* gradually changed its character to that of the standard spin-glass transition in all cases studied, accompanied by a pronounced cusp and the deviation between the FC and ZFC data. By contrast, the low-temperature specific heat persistently exhibited a T^2 behavior scaled by the Curie-Weiss temperature for Zn^{2+} and Fe^{2+} with integer spins, but exhibited a different T-linear behavior characteristic of the standard spin glass for Co^{2+} and Mn^{2+} with half-integer spins. This observation seems to suggest that the nature of low-energy excitations depends significantly on the size of spins, *i.e.*, whether S being an integer or a half-integer. It remains to be seen whether the standard spinwave picture could account for such exotic properties or not. In this context, one may also notice that, in case of kagome-bilayer AF $SrCr_{8-x}Ga_{4+x}O_{19}$ (SCGO) to be discussed below, the low-temperature specific heat is observed to persistently exhibit a T^2 behavior below the spin-glass transition temperature, although the spin quantum number there is a half-integer ($S = 3/2$ for Cr^{3+}): See section 3(e) below.

Concerning the magnitude of the spin correlation length, quantitative discrepancy exists between the experimental value and the numerical value obtained for the simplest classical Heisenberg model with the nearest-neighbor AF coupling. Monte Carlo simulation yielded ξ of order 10^3 near $T = T_v$ [9], while it is only 8 or 20 experimentally. So, some mechanism, not taken into account in the simplest classical Heisenberg model, is required to explain the shortness of experimental ξ. It should be emphasized here that the magnitude of ξ is a nonuniversal property governed by the non-vortex physics, which could significantly be reduced, say, by quantum fluctuations, charge fluctuations, or further frustration effects associated with the interaction other than the main exchange couplings. Interestingly, recent ARPES study performed at higher temperatures (100$\sim$200K) has revealed that the low-energy charge fluctuation (hole dynamics) across the Mott gap is characterized by the wavevectors which are completely different from the wavevectors characterizing the mangnetic SRO at lower temperatures, the occurrence of spin-charge separation [26].

In fact, the specific heat of $NiGa_2S_4$ exhibits another pronounced peak at a temperature around $T \sim 100$K, an order of magnitude higher temperature than that at $T_{peak} = 10$K. This higher specific-heat peak might be related to charge fluctuations as investigated in Ref.[26]. Ref.[58] (see also Ref.[26]) suggested that the strong nematic correlations might set in around this higher specific-heat-peak temperature. However, the T-linear behavior in the specific heat mentioned above sets in at the lower specific-heat-peak temperature of 10K, not at the higher specific-heat-peak temperatureof of 100K, becoming vanishingly small (after the lattice contribution subtracted) in the temperature range between the two specific-heat-peak temperatures. Hence, strong nematic correlations, as are expected to accompany nematic-wave excitations giving rise to the T^2 specific heat, are unlikely to occur in the high-temperature regime of $\sim$100K.

Finally, the noncollinear AF order might explain another noticeable feature of experiments that the T^2 specific heat is quite robust against applied magnetic fields [16]. This experimental observation is rather surprising since applied fields reduce the Hamiltonian symmetry from O(3) to O(2), leading to smaller number of Goldstone modes, *i.e.*, from three to one. As discussed in Ref.[59], an interesting observation here is that the noncollinear AF ground state in magnetic fields is capable of keeping an "accidental" degeneracy not related to the Hamiltonian symmetry O(2), essentially of the same amount as in the zero-field case, when the ordered state is a commensurate one, *e.g.*, the 120-degrees or the 60-degrees structure. In fact, even in applied fields, the ground-state manifold still retains *three* continuous parameters which can be set freely, just as in the zero-field case. One of these three is of symmetry origin, *i.e.* a true Goldstone mode, while other two are not of symmetry origin (accidental), *i.e.* *pseudo*-Goldstone modes. Thus, at the classical level, such pseudo-Goldstone modes may account for the robustness of the low-temperature specific heat in applied fields, while this degeneracy would become approximate in quantum systems. The degeneracy would also become approximate when the ordering wavevector deviates from the commensurate position.

(c) κ-(BEDT-TTF)$_2$Cu$_2$(CN)$_3$ and EtMe$_3$Sb[Pd(dmit)$_2$]$_2$

S=1/2 organic triangular-lattice AFs κ-(BEDT-TTF)$_2$Cu$_2$(CN)$_3$ [27, 28, 29, 30, 31, 32, 33, 34, 35] and EtMe$_3$Sb[Pd(dmit)$_2$]$_2$ [36, 37, 38, 39] exhibit a spin-liquid behavior without the conventional magnetic LRO down to low temperature. The lattice here is not a regular triangular lattice, but is a slightly distorted one. The charge and the spin densities are spread over a dimer molecule so that the system intrinsically possesses the polarization degree of freedom within a molecule. Furthermore, these materials are Mott insulators lying close to the metal-insulator boundary so that intermolecular charge fluctuations might play an important role.

Recent measurements have revealed that the specific heat [32] and the thermal conductivity [39] of these organic compounds exhibit a T-linear behavior at low temperatures, often ascribed to the fermionic 'spinon'-like excitations [60, 61]. More precisely, for the case of κ-(BEDT-TTF)$_2$Cu$_2$(CN)$_3$, such a T-linear behavior was reported for the specific heat [32], while a small nonzero energy gap was reported for the thermal conductivity [33]. For EtMe$_3$Sb[Pd(dmit)$_2$]$_2$, even the thermal conductivity exhibits a T-linear behavior [39].

Recent measurements have also revealed a weak transition-like anomaly occurring at a finite temperature just above the temperature range where the T-linear behavior is observed in the specific heat or in the thermal conductivity. In case of κ-(BEDT-TTF)$_2$Cu$_2$(CN)$_3$, two anomalies (or crossovers) were observed, one at ~6K and the other at ~3K. The higher one at 6K is characterized by the hump (or the peak) of the specific heat [32] and of the thermal expansivity coefficient [34], with a minimum of the NMR T_1^{-1} [27] and of the in-plane thermal conductivity [33]. The lower one at 3K is characterized by a weak additional structure in the specific heat [32] and in the thermal expansivity coefficient [34], with a maximum of the in-plane thermal conductivity [33]. Similar anomaly or crossover is also observed for EtMe$_3$Sb[Pd(dmit)$_2$]$_2$ [37, 38, 39].

We note that the Z_2-vortex order is one possible candidate of the experimentally observed anomaly. In this correspondence, one identifies the lower anomaly as the Z_2-vortex transition temperature T_v and the higher anomaly as the specific-heat-peak temperature T_{peak}. Experimentally, the ratio T_{peak}/T_v comes around two in these organic compounds, which deviates somewhat from the values for NaCrO$_2$ and NiGa$_2$S$_4$ discussed above.

Of course, for the vortex to be a meaningful excitation, minimal amount of noncollinear spin SRO, at least of a few lattice spacings, is required, which, however, is a plausible situation in these compounds in view of the observed spectral broadening of NMR signal and the gapless behavior of the susceptibility.

κ-(BEDT-TTF)$_2$Cu$_2$(CN)$_3$ exhibits a dielectric anomaly in the temperature range 20K~40K [35], somewhat above the temperature where the transition-like anomaly is observed in the specific heat or in NMR T_1^{-1}. The AC dielectric constant exhibits a strongly frequency-dependent peak at the time scale of kHz in this temperature range. The observed eminent frequency dependence then suggests that the polarization degree of freedom becomes slowed down significantly in this temperature regime and tends to be frozen. It remains to be seen how this slowing down or freezing of the polarization degree of freedom is related or unrelated to the spin-liquid and the transition-like behaviors observed at lower temperatures. An extrapolation of the ω-dependent peak temperature to $\omega \to 0$ yields an estimate of the static dielectric transition temperature $T_c \sim 6$K, which happens to come close to the specific-heat hump temperature. On the basis of such an observation, Ref.[35] argued that the anomaly of the dielectric constant might be linked to the anomaly of the specific heat [35]. However, such coincidence should be taken with care, since the dielectric measurements are performed already at the macroscopic time scale (~kHz order) not much different from the time scale of thermal measurements and the observed peak temperature of the AC dielectric constant (20K~40K) is still considerably higher than the specific-heat hump temperature (6K). Thus, at the time scale of real measurements, the specific-heat hump is likely to be a phenomenon not directly related to the dielectric anomaly.

The T-linear specific heat and of the thermal conductivity observed at low temperatures cannot simply be explained by spinwaves. Some other excitations different from spinwaves or any type of Goldstone modes are definitely required here. Two possibilities may be given here: One is a fermionic 'spinon' excitation possibly realized in the RVB-like quantum spin-liquid state, as extensively discussed in the literature [60, 61]. The other possibility I wish to propose here is a more conservative one. As discussed above, since the polarization within a molecule has been dynamically frozen at temperatures higher than the transition temperature, and since the spin is likely to be coupled to the polarization in some way or other, randomly frozen polarization might act as a quenched randomness to the spin dynamics, which might give rise to low-energy spin excitations with a constant density of states at low energies, as often encountered in random spin systems. In this context, it might be intereseting to point out that Ref.[30] observed that applied magnetic fields induced inhomogenous magnetization in κ-(BEDT-TTF)$_2$Cu$_2$(CN)$_3$, which was ascribed to impurities or defects. Of course, one needs to explain the reason why the spin itself escapes the freezing in such a situation, in contrast to the standard spin-glass case. Further study is required to clarify the nature of low-energy excitations relevant to the low-temperature properties of these compounds.

Anyway, one should recall that the issue of the low-energy excitations responsible for the low-temperature behavior could more or less be independent of the issue of a finite-temperature transition. Within the Z_2-vortex ordering picture, in particular, any type of excitation possibly realized in the vortex-free sector in the phase space, can be a candidate of relevant low-energy excitation.

(d) $Cu_3V_2O_7(OH)_2 \cdot 2H_2O$

Volbothite $Cu_3V_2O_7(OH)_2\cdot$ $2H_2O$ is a S=1/2 Heisenberg AF on the kagome lattice [62, 63, 64, 65, 66, 67, 68, 69]. In this compound, the lattice is slightly distorted in a uniaxial manner, *i.e.*, a regular triangle being distorted to an isosceles triangle. The extent of the distortion, however, seems not so large. Vesignieite $BaCu_3V_2O_8(OH)_2$ is also a S=1/2 kagome-lattice Heisenberg AF with a small lattice distortion [70, 71]. In vesignieite, the lattice distortion is even less than in volborthite.

Volbothite also exhibits a spin-liquid-like behavior in the sense that no AF LRO is observed down to very low temperature [62, 63, 64]. Recent neutron scattering experiment has revealed the onset of the magnetic short-range order at low temperatures [65]. Interestingly, NMR and specific heat measurements have revealed that volborthite exhibits a transition-like anomaly at a finite temperature [68, 62, 63, 64, 72]. Indeed, the NMR relaxation rate T_1^{-1} exhibits a sharp peak suggestive of a thermodynamic phase transition at around $T = T^* = 0.9$K [62, 63], while the specific heat shows a hump or a kink at $T_K = 1.05$K [64, 72]. Interestingly, a T-linear behavior of the specific heat is observed below T^* together with the T^2-term, in spite of the fact that volborthite is a very good insulator. Similar behavior, *i.e.*, a combination of the T-linear term and the T^2 term, is observed also above T^*, but with different coefficients from those at $T < T^*$. A closer look at the data reveals that the peak temperature of NMR T_1^{-1} and the kink (hump) temperature of the specific heat are not the same, the former $T = 0.90$K being about

15% lower than the latter $T = 1.05$K. Such $\sim$ 15% difference reminds us of the T_v vs. T_{peak} relation associated with the Z_2-vortex ordering.

Below T^*, spins remain fluctuating but the dynamics becomes very slow, as evidenced by NMR T_2^{-1} measurements [63]. (Volbothite also exhibits a weak spin-glass-like freezing phenomenon at a temperature different from T^*, which, however, is of extrinsic origin [62, 63].) This slowly fluctuating low-temperature phase remains stable against weak magnetic fields. When an applied field intensity exceeds a critical value of $\sim$ 4.3T, the system exhibits a phase transition into the more conventional AF long-range ordered state [62, 63].

Vesignieite also exhibits a spin-liquid-like behavior at low temperatures [70, 71]. Although fuller experimental analysis is yet to be done, smaller lattice distortion of vesignieite makes this material a quite attractive model compound.

Many of the above features of volborthite, *i.e.*, [A] the onset of a sharp dynamical anomaly at a finite temperature T^*, [B] the specific-heat hump (kink) temperature located slightly above ($\sim$ 15%) the dynamical transition temperature T^*, [C] appearance of a slowly fluctuating low-temperature phase without the conventional AF nor spin-glass LRO at $T < T^*$, and [D] the fluctuating low-temperature state stabilized against weak applied fields, which is transformed into the conventional LR ordered state upon application of stronger fields, are all quite similar to the features observed in triangular-lattice Heisenberg AFs $NaCrO_2$ and $NiGa_2S_4$ discussed above, and are also fully consistent with the Z_2-vortex ordering picture.

The existence of the T-linear term in the low-temperature specific heat is a feature absent in $NaCrO_2$ and $NiGa_2S_4$ (though present in organic triangular AFs). The behavior of the low-temperature specific heat reflects the properties of the low-energy excitation of the system. If the relevant low-energy exciations are Goldstone modes like spinwaves, a T^2 term should arise as is observed in $NaCrO_2$ and $NiGa_2S_4$. Hence, the existence of the T-linear term indicates that the relevant low-energy excitation in volborthite is exotic. Indeed, Kagome AF has been known to sustain exotic localized low-energy excitations like the weathervane excitation, which might lead to the observed T-linear specific heat.

It should be noticed again, however, that the properties of the low-temperature phase, *e.g.*, the existence or nonexistence of the T-linear term in the low-temperate specific heat, is a property governed by the non-vortex physics, an issue independent of the Z_2-vortex transition itself. In other words, the Z_2-vortex ordering scenario is well compatible with various different scenarios of the low-temperature phase, as long as the relevant low-energy excitations are realizable in a restricted vortex-free (topologically simply-connected) phase space.

The other aspect to be considered is the role played by the Dzaloshinskii-Moriya (DM) interaction, which inevitably exists to certain extent in any kagome-lattice magnet. The DM interaction tends to lower the symmetry, lift the degeneracy of states, and thereby favor the ordering (usually favors the $q = 0$ state). Thus, its effect on the nature of low-energy excitations needs to be further examined.

(e) $SrCr_{8-x}Ga_{4+x}O_{19}$

$SrCr_{8-x}Ga_{4+x}O_{19}$ (SCGO) is a S=3/2 kagome-bilayer (or pyrochlore slab) system, extensively studied in 1990's as a typical model system of kagome-lattice Heisenberg AF [73, 74, 75, 76, 77, 78, 79, 80, 81, 82, 83]. Crystal structure of magnetic ions (Cr^{3+}) in this material, however, is not really a planar kagome, but rather a kagome-bilayer where two kagome layers are connected via a sparse triangular layer in between, *i.e.*, a pyrochlore slab along (111) [84, 85]. Thus, a constituent unit of the lattice is a tetrahedron rather than a triangle. Another point to be noticed is that a certain fraction of Cr ions are randomly diluted by nonmagnetic Ga so that the lattice contains sizable amount of disorder, which might affect the magnetic ordering of this material to some extent.

SCGO exhibits a rather sharp spin-glass-like transition at a finite temperature T_f characterized by a cusp in the susceptibility accompanied by the deviation between the FC and

ZFC data. Even the negative divergence of the nonlinear susceptibility, which is a characteristic of the standard spin-glass order, was observed [73]. This spin-glass-like transition, however, is unusual in the sense that (i) only a fraction of spin is frozen in the low-temperature phase and large spin fluctuations remain even below T_f, (ii) the low-temperature specific heat is proportional to T^2 suggestive of 2D spinwaves (or certain Goldstone modes) in sharp contrast to the standard spin-glass case where the low-temperature specific heat exhibits a T-linear behavior, and (iii) the T_f-value of SCGO tends to increase with decreasing the defect concentration, apparently suggesting that the defect might not be essential in the spin-glass-like transition of SCGO [82]. The specific heat exhibits a rounded peak slightly above T_f without an appreciable anomaly at $T = T_f$, as commonly observed in standard spin glasses. Very much similar behavior was also reported for other S=3/2 kagome-bilayer compound $Ba_2Sn_2ZnCr_{7p}Ga_{10-7p}O_{22}$ (BSZCGO) [86, 87, 88, 89].

Some of the above features of SCGO, *e.g.*, considerable fluctuations remaining in the ordered phase and the specific-heat peak slightly above T_f, seem common with the features observed in other frustrated magnets discussed above, $NaCrO_2$, $NiGa_2S_4$ and volborthite. The occurrence of a clear spin-glass transition in SCGO, accompanied by the divergence of the nonlinear susceptibility, is at odds with other magnets, however. The implication from the observation (iii) above might be the spin-glass transition of SCGO is not intrinsically defect-mediated.

One possibility in the context of the Z_2-vortex ordering scenario might be that the spin-glass transition of SCGO is essentially a Z_2-vortex transition, which takes a character of the spin-glass transition due to the presence of significant amount of defects (randomness). In other words, the Z_2-vortex transition serves as a "seed" of the experimentally observed spin-glass transition. In view of the fact that the observed spin-glass transition is quite pronounced accompanying the associated critical phenomena, it remains to be seen what role the Z_2-vortex is playing here. The true nature of the spin-glass transition of SCGO has remained elusive yet.

4. Summary and discussion

We discussed the recent experimental data on various frustrated quasi-2D Heisenberg AFs from the viewpoint of the Z_2-vortex order, which include (a) S=3/2 triangular-lattice AF $NaCrO_2$, (b) S=1 triangular-lattice AF $NaGa_2S_4$, (c) S=1/2 organic triangular-lattice AFs κ-(BEDT-TTF)$_2$Cu$_2$(CN)$_3$ and EtMe$_3$Sb[Pd(dmit)$_2$]$_2$, (d) S=1/2 kagome-lattice AF volborthite $Cu_3V_2O_7(OH)_2 \cdot 2H_2O$, and (e) S=3/2 kagome-bilayer AF $SrCr_{8-x}Ga_{4+x}O_{19}$.

These magnets exhibit the spin-liquid-like behavior without the conventional AF LRO at low temperatures. Furthermore, most of them exhibit the following features in common: [A] onset of a sharp dynamical anomaly at a finite temperature T^* without a strong anomaly in the specific heat, [B] the specific-heat exhibiting a hump (or a kink) slightly above the dynamical transition temperature T^*, [C] appearance of a slowly fluctuating low-temperature state without the conventional AF nor spin-glass order at $T < T^*$, and [D] the fluctuating low-temperature state stabilized against weak applied fields, which is transformed into the conventional LR ordered state upon application of stronger fields. These features are observed in common at least for [a] $NaCrO_2$, [b] $NiGa_2S_4$ and [d] volborthite. Note that these materials are already of quite a variety, either S=3/2, 1 or 1/2, and either triangular or (distorted) kagome. Hence, the above features [A]-[D] should possess universal origin independent of the spin quantum number and the detailed lattice structure. The Z_2-vortex ordering scenario meets such a universal criterion since the conditions required there are, (i) the system should be 2D, *i.e.*, weak enough interplane coupling, (ii) the system should be Heisenberg-like, *i.e.*, weak enough magnetic anisotropy, (iii) the existence of the noncollinear spin SRO arising from frustration. All compounds discussed above satisfy these criteria to certain extent, and could be candidates of the Z_2-vortex bearing system. Such universal features of the Z_2-vortex ordering scenario is a favorable factor in explaining why the exotic properties [A]-[D] are observed in common in a variety of 2D frustrated

Heisenberg AFs with different lattice geometries and spin quantum numbers.

S=1/2 organic compounds κ-(BEDT-TTF)$_2$Cu$_2$(CN)$_3$ and EtMe$_3$Sb[Pd(dmit)$_2$]$_2$ exhibit somewhat different properties from NaCrO$_2$, NiGa$_2$S$_4$ and volborthite discussed above. For example, T_{peak}/T^*-value takes a larger value of $\sim$ 2. Dynamical slowing down in the low-temperature phase below T^* seems less pronounced as compared with the one in the other compounds. As discussed in §3(c), the latter property as well as the T-linear low-temperature specific heat might reflect a novel type of low-energy excitations inherent to these compounds, but may well be compatible with the Z_2-vortex order. Even for S=3/2 bilayer-kagome magnet SCGO, Z_2-vortex might be playing some role. The open issue here is to clarify the nature of the observed spin-glass-like transition.

Overall, the Z_2-vortex ordering scenario seems to give consistent account of a number of exotic ordering properties observed in a variety of frustrated quasi-2D Heisenberg magnets in common. Hopefully, further experimental and theoretical studies such as inelastic neutron-scattering will reveal further novel features of the Z_2-vortex and other exotic excitations inherent to frustrated magnets.

Special thanks are due to T. Okubo for useful discussion and collaboration. Enlightening discussion with Z. Hiroi, S. Nakatsuji, Y. Nambu, N. Hagiwara, H. Yamaguchi, H. Ohta, S. Maegawa, T. Itou, H. Kikuchi, Y. Okamoto, C. Broholm, D. MacLaughlin, P. Mendels, R. Kajimoto, K. Tomiyasu, M. Matsuura, Y. Motome, R. Kato, K. Kanoda, Y. Nakazawa, M. Yamashita, C. Hotta and T. Arima is gratefully acknowledeged. This work is supported by Grand-in-Aid for scientific Research on Priority Areas "Novel States of Matter Induced by Frustration" (19052006).

[1] Anderson P W 1973 Mater. Res. Bull. **8** 153
[2] Lee P A 2008 Science **321** 1306
[3] Balents L 2010 Nature **464** 199
[4] Miyashita S and Shiba H 1984 J. Phys. Soc. Jpn. **53** 1145
[5] Onoda S and Nagaosa N 2007 Phys. Rev. Lett. **99** 027206.
[6] Okubo T and Kawamura H 2010 Phys. Rev. B **99** 014404
[7] Kawamura H 2010 J. Phys. Soc. Jpn. **79** 011007
[8] Kawamura H and Miyashita S 1984 J. Phys. Soc. Jpn. **53** 4138.
[9] Kawamura H, Yamamoto A and Okubo T, 2010 J. Phys. Soc. Jpn. **79** 023701
[10] Kawamura H and Yamamoto A 2007 J. Phys. Soc. Jpn. **76** 073704
[11] Okubo T and Kawamura H 2010 J. Phys. Soc. Jpn. **79** 084706
[12] Olariu A, Mendels P, Bert F, Ueland B G, Schiffer P, Berger R F and Cava R J 2006 Phys. Rev. Lett. **97** 167203
[13] Hsieh D, Qian D, Berger R F, Cava R J, Lynn J W, Huang Q, and Hasan M Z 2008 Physica. B **403** 1341
[14] Hsieh D, Qian D, Berger R F, Cava R J, Lynn J W, Huang Q and Hasan M Z 2008 J. Phys. Chem. Solids **69** 3174
[15] Hemmida M, Krug von Nidda H A, Büttgen N, Loidl A, Alexander L K, Nath R, Mahajan A V, Berger R F, Cava R J, Singh Y, and Johnston D C 2009 Phys. Rev. B **80** 054406
[16] Nakatsuji S, Nambu Y, Tonomura H, Sakai O, Jonas S, Broholm C, Tsunetsugu H, Qiu Y, and Maeno Y 2005 Science **309** 1697
[17] Nambu Y, Nakatsuji S and Maeno Y 2006 J. Phys. Soc. Jpn. **75** 043711
[18] Takeya H, Ishida K, Kitagawa K, Ihara Y, Onuma K, Maeno Y, Nambu Y, Nakatsuji S, MacLaughlin D E, Koda A and Kadono R 2008 Phys. Rev. B **77** 054429
[19] Yaouanc A, de Réotier P D, Chapuis Y, Marin C, Lapertot G, Cervellino A, and Amato A 2008 Phys. Rev. B **77** 092403
[20] MacLaughlin D E, Nambu Y, Nakatsuji S, Heffner R H, Shu L, Bernal O O and Ishida K 2008 Phys. Rev. B **78** 220403(R)
[21] Yamaguchi H, Kimura S, Hagiwara M, Nambu Y, Nakatsuji S, Maeno Y, and Kindo K 2008 Phys. Rev. B **78** 180404
[22] Yamaguchi H, Kimura S, Hagiwara M, Nambu Y, Nakatsuji S, Maeno Y, Matsuo A and Kindo K 2010 J. Phys. Soc. Jpn. **79** 054710

[23] Nakatsuji S, Nambu Y and Onoda S 2010 J. Phys. Soc. Jpn. **79** 011003
[24] Nambu Y, Nakatsuji S, Maeno Y, Okudzeto E K and Chan J Y 2008 Phys. Rev. Lett. **101** 207204
[25] Stock C, Jonas S, Broholm C, Nakatsuji S, Nambu Y, Onuma K, Maeno Y and Chung J H 2010 Phys. Rev. Lett. **105** 037402
[26] Takubo K, Nambu Y, Nakatsuji S, Wakisaka Y, Sudayama T, Fournier D, Levy G, Damascelli A, Arita M, Namatame H, Taniguchi M and Mizokawa T 2010 Phys. Rev. Lett. **104** 226404
[27] Shimizu Y, Miyagawa K, Kanoda K, Maesato M and Saito G 2003 Phys. Rev. Lett. **91** 107001
[28] Kawamoto A, Honma Y and Kumagai K 2004 Phys. Rev. B **70** R060510
[29] Kurosaki Y, Shimizu Y, Miyagawa k, Kanoda K and Saito G 2005 Phys. Rev. Lett. **95** 177001
[30] Shimizu Y, Miyagawa K, Kanoda K, Maesato M and Saito G 2006 Phys. Rev. B **73** 140407
[31] Ohira S, Shimizu Y, Kanoda K and Saito G 2006 J. Low Temp. Phys. **142** 153
[32] Yamashita S, Nakazawa Y, Oguni M, Oshima Y, Nojiri H, Shimizu Y, Miyagawa K and Kanoda K 2008 Nature Physics **4** 459
[33] Yamashita M, Nakata N, Kasahara Y, Sasaki T, Yoneyama M, Kobayashi N, Fujimoto S, Shibauchi T and Matsuda Y 2009 Nature Physics **5** 44
[34] Manna R S, de Souza M, Bruhl A, Schlueter J A and Lang M 2010 Phys. Rev. Lett. **104** 016403
[35] Abdel-Jawad M, Terasaki I, Sasaki T, Yoneyama N, Kobayashi N, Uesu Y and Hotta C 2010 Phys. Rev. B **82** 125119
[36] Tamura M, Nakao A and Kato R 2006 J. Phys. Soc. Jpn. **75** 093701
[37] Itou T, Oyamada A, Maegawa S, Tamura M and Kato R2008 Phys. Rev. B **77** 104413
[38] Itou T, Oyamada A, Maegawa S and Kato R 2010 Nature Physics **6** (2010) 673
[39] Yamashita M, Nakata N, Senshu Y, Nagata M, Yamamoto H M, Kato R, Shibauchi T and Matsuda Y 2010 Science **328** 1246
[40] Kosterliz J M and Thouless D J 1973, J. Phys. C: Solid State Phys. **6** 1181
[41] Kosterliz J M and Thouless D J 1974 J. Phys. C: Solid State Phys. **7** 1046
[42] Southern B W and Young A P 1993 Phys. Rev. B **48** 13170
[43] Ajiro Y, Kikuchi H, Sugiyama S, Nakashima T, Shamoto S, Nakayama N, Kiyama M, Yamamoto N and Oka Y 1988 J. Phys. Soc. Jpn. **57** 2268
[44] Soubeyroux J L, Fruchart D, Delmas C and Le Flem G 1979 J. Mag. Mag. Mater. **14** 159
[45] Kadowaki H, Takei H and Motoya K 1995 J. Phys. Condens. Matter **7** 6869
[46] Alexander L K, Buttgen N, Nath R, Mahajan A V and Loidl A 2007 Phys. Rev. B **76** 064429
[47] Seki S, Onose Y and Tokura Y 2008 Phys. Rev. Lett. **101** 067204
[48] Olariu A, Mendels P, Bert F, Alexander L K, Mahajan A V, Hiller A D and Amato A 2009 Phys. Rev. B **79** 224401
[49] Kadowaki H, Kikuchi H and Ajiro Y 1990 J. Phys. Condens. Matter **2** 4485
[50] Poienar M, Damay F,Martin C, Hardy V, Maignan A and Andre G 2009 Phys. Rev. B **79** 014412
[51] Kimura K, Nakamura H, Kimura S, Hagiwara M and Kimura T 2009 Phys. Rev. Lett. **103** 107201
[52] Oohara Y, Mitsuda S, Yoshizawa H, Yaguchi N, Kuriyama H, Asano T and Mekata M 1994 J. Phys. Soc. Jpn. **63** 847
[53] Mekata M, Sugino T, Oohara A, Oohara Y and Yoshizawa H 1995 Physica B **213& 214** 221
[54] Takatsu H, Yonezawa S, Fujimoto S and Maeno Y 2010 Phys. Rev. Lett. **105** 137201
[55] Okuda T, Kishimoto T, Uto K, Hokazono T, Onose Y, Tokura Y, Kajimoto R and Matsuda M 2009 J. Phys. Soc. Jpn. **78** 013604
[56] Kajimoto R, Nakajima K, Ohira-Kawamura S, InamuraY, Kakurai K, Arai M, Hokazono T, Oozono A and Okuda T 2010 J. Phys. Soc. Jpn. **79** 123705
[57] Fujimoto S 2006 Phys. Rev. B **73** 184401
[58] Stoudenmire E M, Trebst S and Balents L 2009 Phys. Rev. B **79** 214436
[59] Kawamura H and Miyashita S 1985 J. Phys. Soc. Jpn. **54** 4530
[60] Lee S S, Lee P A and Senthil T 2007 Phys. Rev. Lett. **98** 067006
[61] Nave C P and Lee P A 2007 Phys. Rev. B **76** 235124
[62] Yoshida H, Okamoto Y, Tayama T, Sakakibara T, Tokunaga M, Matsuo A, Narumi Y, Kindo K, Yoshida M, Takigawa M and Hiroi Z 2009 J. Phys. Soc. Jpn. **78** 043704
[63] Yoshida M, Takigawa M, Yoshida H, Okamoto Y and Hiroi Z 2009 Phys. Rev. Lett. **103** 077207
[64] Yamashita S, Moriura T, Nakazawa Y, Yoshida H, Okamoto Y and Hiroi Z 2010 J. Phys. Soc. Jpn. **79** 083710
[65] Nilsen G J, Coomer F C, de Vries M A, Stewart J R, Deen P P, Harrison A and Ronnow H M 2010 arXiv:1001.2462
[66] Hiroi Z, Hanawa M, Kobayashi N, Nohara M, Takagi H, Kato Y and Takigawa M 2001 J. Phys. Soc. Jpn. **70** 3377
[67] Fukaya A, Fukamoto Y, Ito T, Larkin M I, Savici A T, Uemura Y J, Kyriakou P P, Luke G M, Rovers M T,

Kojima K M, Keren A, Hanawa M and Hiroi Z 2003 Phys. Rev. Lett. **91** 207603
[68] Bert F, Bono D, Mendels P, Ladieu F, Duc F, Trombe J-C and Millet P 2005 Phys. Rev. Lett. **95** 087203
[69] Bert F, Bono D, Mendels P, Trombe J-C, Millet, Amato A, Baines C and Hiller A 2004 J. Phys. Condens. Matter **16** S829
[70] Okamoto Y, Yoshida H and Hiroi Z 2009 J. Phys. Soc. Jpn. **78** 033701
[71] Zhang W, Ohta H, Okubo S, Fujisawa M, Sakurai T, Okamoto Y, Yoshida H and Hiroi Z 2010 J. Phys. Soc. Jpn. **79** 023708
[72] Hiroi Z, private communication
[73] Ramirez A P, Espinosa G P and Cooper A S 1990 Phys. Rev. Lett. **64** 2070 ; 1992 Phys. Rev. B **45** 2505
[74] Broholm C, Aeppli G, Espinosa G P and Cooper A S 1990 Phys. Rev. Lett. **65** 3173
[75] Martinez B, 1992 Phys. Rev. B **46** 10786
[76] Martinez B, Labarta A, Rodriguez-Solá R and Obradors X 1994 Phys. Rev. B **50** 15779
[77] Uemura Y J *et al.* 1994 Phys. Rev. Lett. **73** 3306
[78] Lee S H, Broholm C, Aeppli G, Perring T G, Hessen B and Taylor A 1996 Phys. Rev. Lett. **76** 4424
[79] Keren A, Mendels P, Horvatic M, Ferrer F, Uemura Y J, Mekata M and Asano T 1998 Phys. Rev. B**57** 10745
[80] Ramirez A P, Hessen B and Winklemann M 2000 Phys. Rev. Lett. **84** 2957
[81] Mendels P, Keren A, Limot L, Mekata M, Collin G and Horvatic M 2000 Phys. Rev. Lett. **85** 3496
[82] Mekata M and Yamada Y 2001 Can J. Phys. **79** 1421
[83] Limot L, Mendels P, Collin G, Mondelli C, Ouladdiaf B, Mutka H, Blanchard N and Mekata M 2002 Phys. Rev. B **65** 144447
[84] Kawamura H and Arimori Y 2002 Phys. Rev. Lett. **88** 077202
[85] Arimori Y and Kawamura H 2001 J. Phys. Soc. Jpn. **70** 3695
[86] Hagemann I S, Huang Q, Gao X P A, Ramirez A P and Cava R J 2001 Phys. Rev. Lett. **86** 894
[87] Bono D, Mendels P, Collin G and Blanchard N 2004 Phys. Rev. Lett. **92** 217202
[88] Bono D, Mendels P, Collin G, Blanchard N, Bert F, Amoto A, Bines C and Hiller A D 2004 Phys. Rev. Lett. **93** 187201
[89] Bonnet P, Payen C, Mutka H, Danot M, Fabritchnyi P, Stewart J R, Mellergard A and Ritter C 2004 J. Phys. Condens. Matter **16** S835

Journal of Physics: Conference Series **320** (2011) 012003 doi:10.1088/1742-6596/320/1/012003

Novel magnetic order in the kagome lattice of volborthite

Zenji Hiroi, Hiroyuki Yoshida, Yoshihiko Okamoto, Makoto Yoshida

Institute for Solid State Physics, University of Tokyo, 5-1-5 Kashiwanoha, Kashiwa, Chiba 277-8581, Japan

E-mail: hiroi@issp.u-tokyo.ac.jp

Abstract. The ground-state magnetic properties of the spin-1/2 quasi-kagome compound volborthite are discussed. Volborthite exhibits a peculiar phase transition to an unconventional ordered phase at 1 K. This phase is characterized by dense low-energy excitations and extremely slow spin fluctuations together with spin-wave like excitations, as evidenced by means of heat capacity and ^{51}V NMR. To understand these features, we consider an exotic long-range order based on the 120° $\sqrt{3} \times \sqrt{3}$ spin structure, which allows local modes in the hexagons of the kagome lattice.

1. Introduction

The physics of the Heisenberg spins on the kagome lattice has been extensively studied for the last few decades [1]. Due to strong frustration, it is believed that there is an opportunity to discover an unprecedented ground state (GS) which lies between conventional long-range order (LRO) and a singlet-based state with a spin gap (Fig. 1). Possibly, the state is to be called a spin liquid [2]. A classical antiferromagnet with large spins may prefer the former GS with ordering temperature (T_N) decreasing with increasing the empirical frustration factor f [3], $-\Theta_{CW}/T_N$, where Θ_{CW} is the Curie-Weiss temperature. By contrast, a quantum spin system tends to stabilize by forming spin singlet pairs and thus assumes a gapped state such as the resonating-valence-bond (RVB) state proposed by Fazekas and Anderson [4]. Recent exact diagonalization calculations on finite clusters found either a small ($\sim J/20$) or almost vanishing gap for the spin-1/2 Heisenberg kagome antiferromagnet (KAFM) [5, 6], which suggests that a long-range RVB (LR-RVB) state consisting of pairs of far-separated spins is possibly realized. Alternatively, a valence bond crystal (VBC) with a tiny gap may be stable [7]. In any regard, the GS of the KAFM certainly exists very close to the quantum critical point (QCP) between the two GSs, making it intriguing but difficult to study both theoretically and experimentally.

We have been focusing our attention on the material volborthite $Cu_3V_2O_7(OH)_2 \cdot 2H_2O$ which comprises Cu^{2+} ions carrying spin 1/2 on a distorted kagome lattice made of isosceles triangles (Fig. 2) [8]. The compound is considered to be a candidate for the spin-1/2 Heisenberg KAFM, though there is a distortion in magnetic couplings in the kagome lattice [9]. Moreover, there is a controversial issue on the nature of the magnetic couplings in volborthite: recent density-functional-theory calculations claim that the kagome lattice consists of frustrated J_1-J_2 chains together with third spins in between and thus can be far from the anisotropic kagome model [10]. The great advantage of volborthite over other compounds like herbertsmithite [11] and vesignieite [12] is that one can prepare high quality samples containing fewer impurity spins, i.e., only 0.07% of the total spin [13]. This allows one to investigate the intrinsic properties of the compound down to very low temperatures. High lattice coverage is crucial to study compounds as expected to exist near a quantum critical point, because even a small perturbation by defects may influence the GS seriously; a glassy state is often induced, masking intrinsic properties [14].

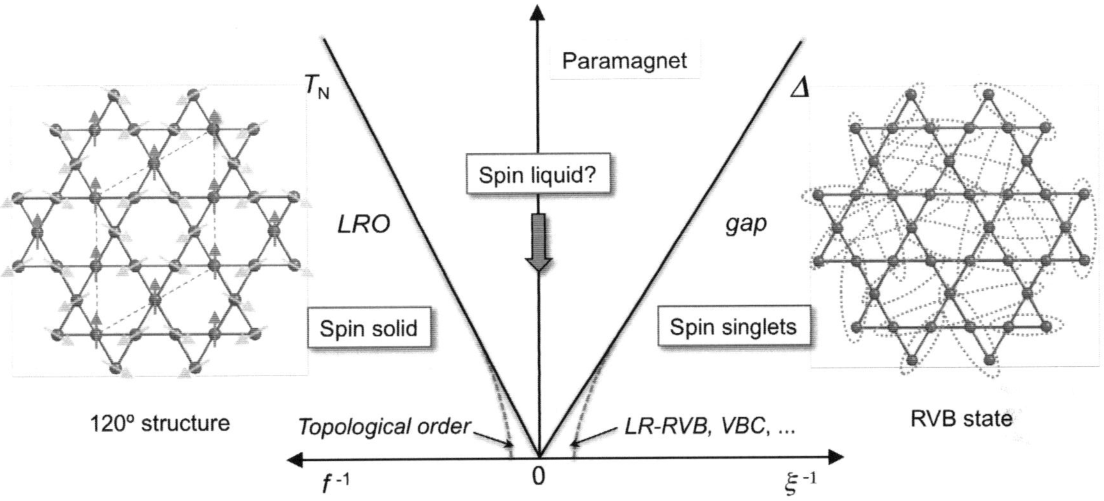

Figure 1. Conceptual phase diagram for frustrated spin systems in two dimension such as the kagome antifferomagnet. Spin liquids may appear around a quantum critical point where long-range order (LRO) disappears as the empirical frustration factor f (= $-\Theta_{CW}/T_N$) becomes large with decreasing interlayer couplings or increasing quantum fluctuations, and the spin gap Δ tends to vanish as the magnetic correlation length ξ becomes large. Otherwise, a certain exotic LRO containing topological excitations, a long-range resonating-valence-bond (LR-RVB) state or a valence bond crystal (VBC) could be stabilized as the ground state in this region.

The magnetic susceptibility, χ, of volborthite shows a broad peak at 20 K, with no anomaly down to 60 mK, indicating the absence of conventional LRO. Furthermore, it approaches a large, finite value at $T = 0$, providing evidence for a gapless GS. In contrast, ^{51}V-NMR measurements reveal a magnetic transition at 1 K to a peculiar phase which is characterized by the presence of dense low energy excitations and unusually slow spin dynamics [13, 15]. These results strongly suggest that the GS of volborthite is neither a simple LRO nor a gapped state. Moreover, three magnetization steps are observed at magnetic fields of 4.3 T, 25.5 T and 46 T in measurements preformed up to 55 T [13]. The first step at 4.3 T is accompanied by a field-induced phase transition to a more magnetically ordered phase, as evidenced by ^{51}V-NMR measurements [16]. Very recent magnetization measurements up to 68 T found a magnetization plateau at 0.39 of the saturation magnetization, which is significantly larger than the 1/3 as expected for a conventional up-up-down configuration on every triangular plaquette [17]. The magnetic field-temperature phase diagram determined by magnetization and ^{51}V NMR is depicted in Fig. 2, which includes phases I, II, III, IV and V at low temperatures on increasing magnetic field.

In the present paper, we focus on the GS of volborthite as well as its response to weak magnetic fields below $\mu_0 H_{s1}$ = 4.3 T. Particularly, the origin of the 1 K transition is investigated, trying to understand the nature of magnetic order in phase I. We find another transition or crossover at 2 T, above which the magnetic properties of phase I is significantly modified. We will propose a model based on the 120° spin structure with a $\sqrt{3} \times \sqrt{3}$ unit to interpret these results.

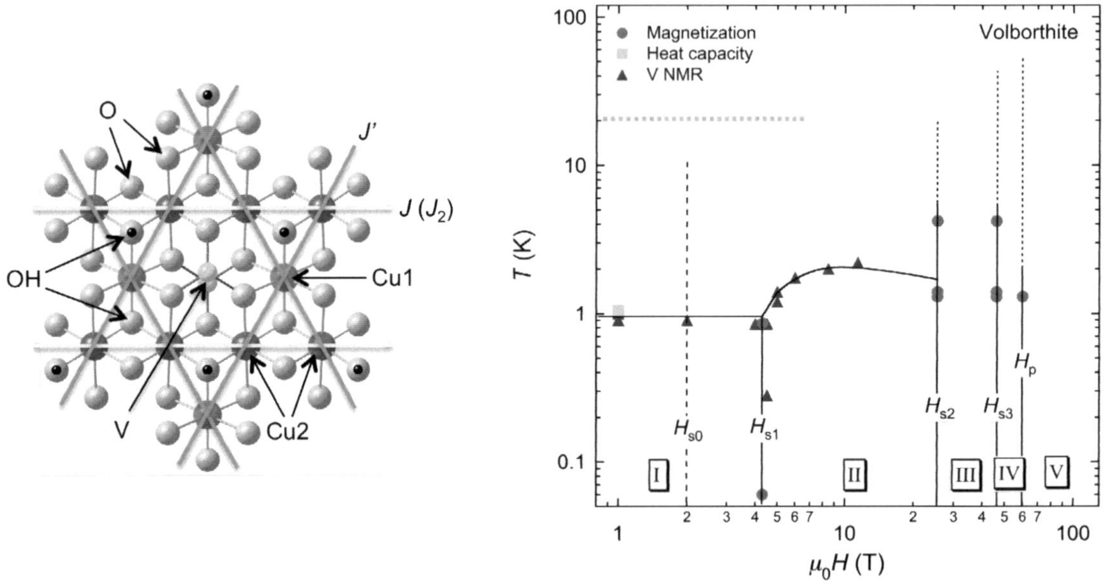

Figure 2. Crystal structure and H-T phase diagram of volborthite. A V atom located above or below each hexagon made of six Cu atoms, which is slightly shifted from the centre of the hexagon, probes the hyperfine field from the neighboring Cu spins. The phase diagram includes at least four phases I-IV separated by the three magnetization steps at μ_0H_{s1} = 4.3 T, μ_0H_{s2} = 25.5 T and μ_0H_{s3} = 46 T, followed by phase V showing a magnetization plateau above $\mu_0H_p \sim$ 60 T. There is another characteristic field at $\mu_0H_{s0} \sim$ 2 T.

2. Results

2.1. 1 K transition

The presence of a transition at low temperature in volborthite was first found by Bert *et al.* in their ^{51}V NMR experiments [15]. They observed a broadening of the NMR spectrum below 2 K and a peak in the V nuclear spin relaxation rate ($1/T_1$). Unfortunately, however, the 1 K transition was not clearly detected, because the sample contained an amount of disorder that caused an additional broadening of the spectrum. Later, in an improved sample with less disorder, M. Yoshida *et al.* [16] found a sudden broadening below 1 K in the ^{51}V NMR spectrum, as well as a sharp peak in $1/T_1$ that was nearly an order of magnitude larger than that reported by Bert *et al.* Since the transition becomes more pronounced as disorder is reduced, it must reflect an intrinsic property of volborthite. The transition temperature determined by the NMR experiments is insensitive to magnetic field below H_{s1}, as shown in the phase diagram of Fig. 2. In addition, the Lorentzian shaped NMR spectrum observed below 1 K indicates a spatially modulated spin structure such as a spin-density-wave (SDW) state or substantially short-ranged magnetic order [16]. On the other hand, μSR experiments by Fukata *et al.* found a slowing down of Cu spin fluctuations on cooling towards 1 K, followed by slow and nearly T-independent spin fluctuations persisting down to 50 mK [18].

The $1/T_1$ data at a magnetic field of 1 T obtained by M. Yoshida *et al.* is reproduced in Fig. 3 [16], showing a sharp peak at T^* = 0.90 K but no divergence, such as that observed in $NiGa_2S_4$ at 8.5 K [19]. Moreover, the $1/T_1$ is proportional to T below 0.5 K, providing strong evidence for the presence of dense low energy excitations, as in a fermionic system like an ordinary metal. To our knowledge, no frustrated spin compounds have exhibited such T-linear behavior in $1/T_1$; the $1/T_1$ of $NiGa_2S_4$ is proportional to T^3 [19] and that of κ-$(ET)_2Cu_2(CN)_3$ follows a power-law decay of $T^{1.5}$ at low temperatures [20].

Heat capacity measured by Nakazawa *et al.* is plotted in Fig. 3 [21]. Magnetic heat capacity, C_m, data obtained at magnetic fields of 0 and 1 T overlap to each other completely and show a kink at 1.05 K. Thus, there is a weak but well-defined thermodynamic phase transition that is not of second order but of higher order: we call this temperature T_k. Note that T_k is higher by 0.15 K than T^* for the $1/T_1$ peak, suggesting that slowing down of spins is not induced at the phase transition, but develops at lower temperatures after a narrow fluctuating crossover regime extending below T_k. This reminds us of a similar observation for the triangular Heisenberg antiferromagnet $NaCrO_2$, where the peak in $1/T_1$ is located at ~25 K well below T_N = 40-50 K, as determined by means of magnetization and heat capacity [22]. The importance of these observations has been pointed out and interpreted based on Z_2-vortex order by H. Kawamura [23]

The temperature dependence of the heat capacity provides important information about the GS and the nature of the transition [21]. C_m/T is proportional to T with a finite intersect toward T = 0 both above and below T_k. This implies that C_m unusually has two contributions: $C_m = \gamma T + \beta T^2$. Note that phonon contributions have already been subtracted from raw data and only magnetic contributions are included in this insulating compound. Below T_k the slope in C_m/T increases suddenly and the intersect is much reduced. Since the data is limited above 0.8 K, it is difficult to obtain a reliable value for the low-temperature intersect. Assuming the same T dependence as above T_k yields γ = 15 mJ K^{-2} mol-Cu^{-1}, which is less than half of the extrapolated value of 40.1 mJ K^{-2} mol-Cu^{-1} for the high-temperature state [21]. Further heat capacity measurements down to lower temperatures are in progress to attain a reliable value. In any regard, it is clear that finite γ remains at low temperatures, which is consistent with the finite $1/T_1T$ obtained from V NMR experiments. Thus, the most pronounced difference between the high- and low-temperature states concerns the form of the 'density-of-states'.

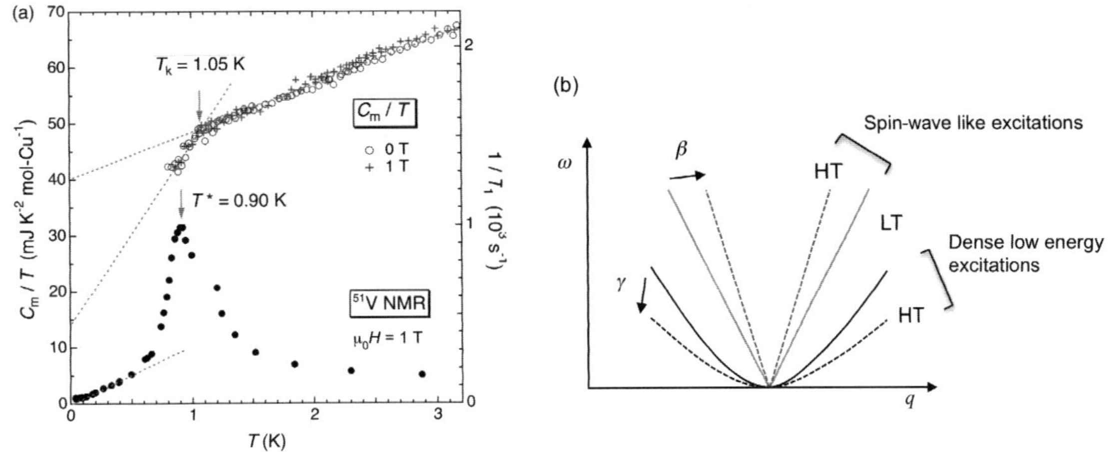

Figure 3. (a) Magnetic heat capacity [19] and the relaxation rate, $1/T_1$, from ^{51}V NMR experiments [16] and (b) schematic representations for the dispersion relations expected from their temperature dependences.

The second term in the heat capacity, proportional to T^2, is typically observed for spin-wave excitations in two-dimensional antiferromagnets. It is rather surprising to find such a conventional term in the unconventional GS of volborthite. It is to be noted, however, that many frustrated spin compounds have exhibited similar T^2 behavior even in the absence of conventional LRO [1, 3]. For example, it is well known that the spin-3/2 pyrochlore-slab (or kagome) compound $SrCr_{9p}Ga_{12-9p}O_{19}$ (SCGO) shows a clear T^2 heat capacity below 4 K in the spin-glass state [24]; this is also the case for deuteronium jarosite [25] and $NiGa_2S_4$, both without LRO [26]. Therefore, common physics must exist behind in these frustrated spin systems. In addition, the T-linear term has never been observed in other compounds except for a triangular lattice molecular compound κ-$(ET)_2Cu_2(CN)_3$ [27].

In order to explain these two terms in heat capacity, it is necessary to assume two independent branches in the dispersion relation, as schematically depicted in Fig. 3(b). Generally, a term with the n-th power of T in heat capacity implies the existence of a branch having energy proportional to the d/n-th power of momentum transfer q, where d is the dimension of the system. Thus, the T-linear and T^2 terms correspond to branches with $\omega \approx q^2$ and q, respectively. The former gives dense low energy excitations and the latter spin-wave excitations. We must now construct a reasonable model that can explain this unusual coexistence of the two branches. Furthermore, this model must give an explanation as to why γ decreases and β increases on cooling below T_k.

2.2. Anomaly in magnetization at 2 T

There is a weak anomaly in magnetization at a magnetic field of 2 T before the first magnetization step sets in [13]. Figure 4 shows an M-H curve measured at the very low temperature of 60 mK, which first increases linearly, deviates upward slightly at around 2 T, and further rises up at 4.3 T. A tiny initial rise observed near H = 0 comes from about 0.1 % of almost free impurity spins. It is therefore not intrinsic. The deviation from the initial linear relation at 2 T is clearly detected in the derivative curve shown in the top of Fig. 4. This deviation is not due to the 'tail' of the first magnetization step, as the step begins at 3.7 T, where a kink exists in the derivative curve. The 2 T anomaly exists even at a relatively high temperature of 1.62 K, when the magnetization step has already disappeared.

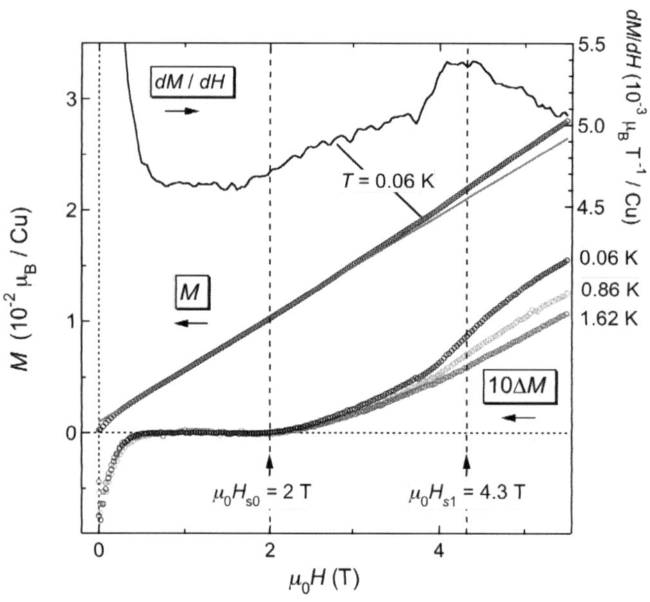

Figure 4. Magnetization versus magnetic field curve measured at 60 mK [13]. A derivative curve is shown in the top. Magnetizations measured at 0.06, 0.86 and 1.62 K after subtraction of an initial linear component and multiplication by a factor of 10 are also shown.

We also measured magnetization at higher temperatures (Fig. 5); the 2 T anomaly is clearly observed at 2 and 5 K, and even at 10 K, and then tends to vanish on further heating. On the other hand, M-T curves measured at H = 1 and 2 T overlap almost completely in the whole T range below 30 K, as shown in Fig. 5(b), while deviations are observed above 3 T, particularly at T < 10 K, which is consistent with the behavior observed in the M-H curves. Possibly, this deviation starts to grow below T_p = 20 K at a broad peak in magnetic susceptibility, indicating that it is not related to the 1 K transition but the development of magnetic correlations below T_p. Hence, the 2 T anomaly is essentially different from the magnetization steps at higher fields. The critical field H_{s0} is plotted in the H-T phase diagram of Fig. 2.

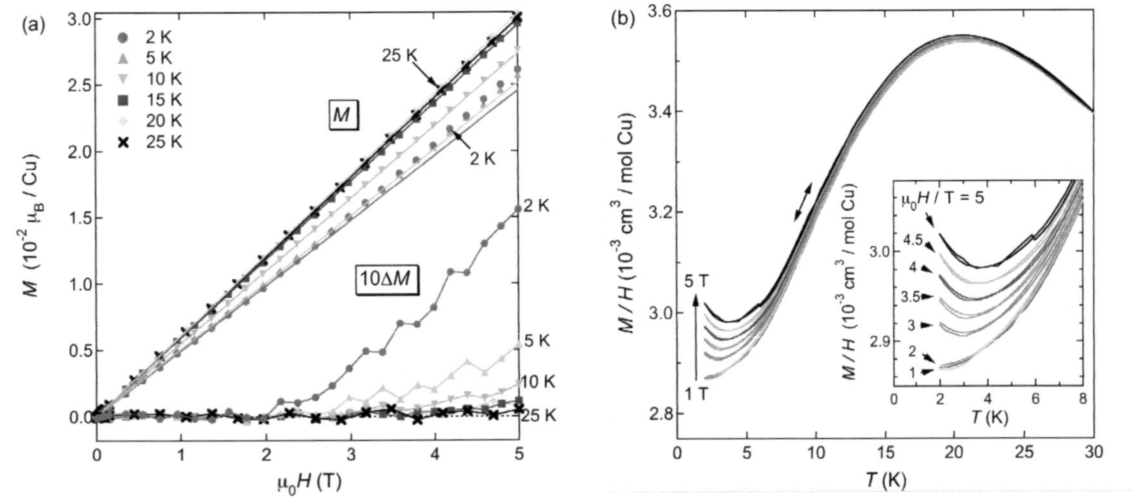

Figure 5. (a) *M-H* curves measured at 2, 5, 10, 15, 20 and 25 K in a SQUID magnetometer upon increasing *H*. Corresponding data after subtraction of initial linear components are also plotted. (b) Temperature dependence of magnetic susceptibility measured at 1, 2, 3, 3.5, 4, 4.5 and 5 T on heating and cooling.

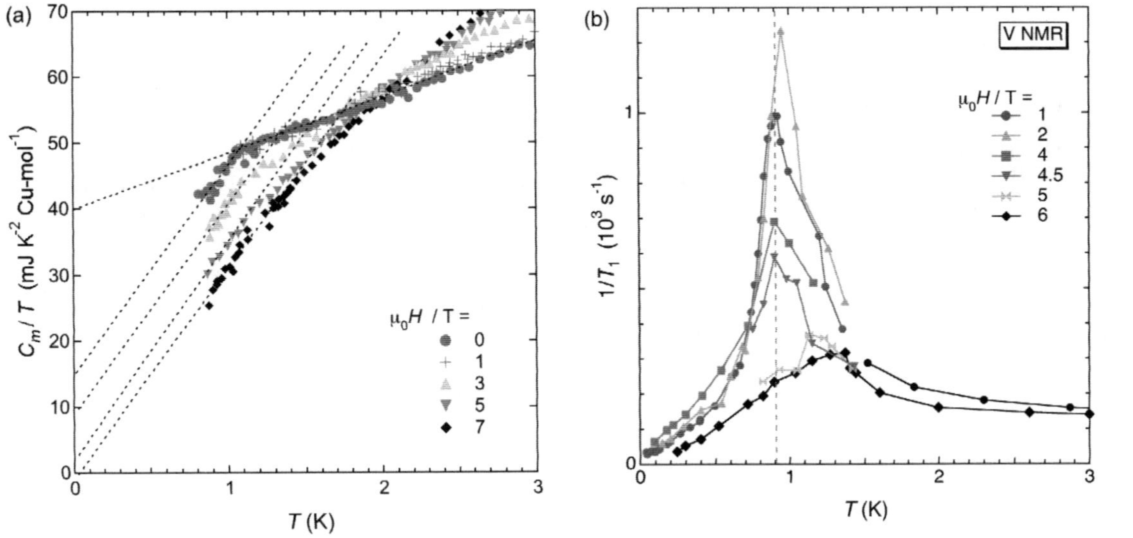

Figure 6. (a) Magnetic heat capacity measured by Yamashita *et al.* in magnetic fields of 0, 1, 3, 5 and 7 T [19] and (b) ^{51}V NMR relaxation rate, $1/T_1$, measured by M. Yoshida *et al.* in magnetic fields of 1, 2, 4, 4.5, 5 and 6 T [16].

One may notice that this 2 T anomaly resembles what is observed at a spin flop transition that often occurs in LRO Heisenberg spin systems with weak anisotropy. However, this should not be the case, because the anomaly exists at high temperatures even in the paramagnetic state. Alternatively, the *M-H* and *M-T* curves of Figs. 4 and 5 are reminiscent of similar behavior observed in spin-gapped compounds such as $TlCuCl_3$, where a field-induced magnetic transition occurs at $H_g = \Delta/(g\mu_B)$ when the gap collapses. This has been interpreted as a Bose-Einstein condensation of magnons [28]. We will discuss on this issue later.

It is important to investigate how this 2 T anomaly affects heat capacity and NMR $1/T_1$. Figure 6 shows their H dependences [16, 21]. C_m/T remains the same at 0 and 1 T, but shifts downward at low temperatures on further increasing H, with the 1 K kink being obscured. Note that γ for the low-temperature phase decreases gradually, if one assumes the T-linear dependence, and seems to vanish at 5 or 7 T, close to the boundary between phases I and II, while the slope β remains almost constant. The $1/T_1$ curve measured at 2 T shows a large (even larger than that at 1 T) peak at around 0.9 K and exactly follows the 1 T curve at low temperatures with the same magnitude of $1/T_1T$. The 0.9 K peak survives at 4 and 4.5 T, though it becomes small, and the T-linear terms at low temperature remain. At 5 or 6 T, in contrast, the peak shits to a higher temperature (~1.3 K), which corresponds to the transition to phase II on cooling, and the T-linear term disappears; the 6 T data below 1 K is proportional to $T^{1.5}$ [16]. Therefore, the magnetic state at zero field appears robust below a critical field of approximately 2 T, while the density of the low energy excitations are gradually reduced with increasing H above 2 T and may completely disappear at H_{s1}.

3. Discussion

Here we discuss the magnetic state of volborthite in phase I, at low magnetic field. Summarizing the related observations obtained so far: (1) magnetic susceptibility decreases rather steeply below $T_p \sim 20$ K, (2) a high-order magnetic phase transition sets in at $T_k = 1.05$ K, (3) NMR $1/T_1$ exhibits a peak at a slightly lower temperature of $T^* = 0.90$ K, (4) the Lorentzian shape of V NMR spectrum observed below T^* suggests a SDW-like order, (5) two kinds of excitations exist both above and below T_k; dense low energy excitations characterized by T-linear contributions in heat capacity as well as in $1/T_1$, and 2D spin-wave like excitations, (6) γ decreases and β increases below T_k, and (7) a change in the magnetic state occurs at $\mu_0H_{s0} \sim 2$ T, above which the density of the low energy excitations starts to decrease before vanishing at H_{s1}. In order to explain these features, we will consider a simple model based on the $\sqrt{3} \times \sqrt{3}$-type LRO of spins.

3.1. Gap or LRO

One theoretical prediction for the spin-1/2 KAFM is a LR-RVB state with a tiny singlet-triplet gap filled by a singlet continuum [5]. Thus, one would not expect gapped behavior in the heat capacity but only in the magnetic susceptibility. Since the magnitude of a spin gap is inversely proportional to the magnetic correlation length, ξ, or the separation of paired spins, the LR-RVB state contains both far-separated singlet pairs as well as near-neighbor ones, as depicted in Fig. 1. On the other hand, once LRO sets in due to some perturbations (defects, anisotropies, etc.), the 120° spin structures of the $\sqrt{3} \times \sqrt{3}$ type would be selected by the order-by-disorder mechanism [1]. In the $\sqrt{3} \times \sqrt{3}$-type coplanar LRO, three sublattice spins on every triangular plaquette form a 120° angle and are arranged on the kagome lattice as depicted in Fig. 1 and Fig. 7(a). It is, in other words, a staggered order of the scalar chirality associated with each triangular plaquette. However, this state always remains imperfect even at $T = 0$, because a local zero-energy mode on every hexagon, called the weathervane (W) mode, can destroy the order by introducing chirality domains [1, 29, 30]. The W mode is composed of a coherent out-of-plane rotation of six spins belonging to one hexagon (Fig. 7(b)) and costs no energy, if only nearest-neighbor antiferromagnetic couplings are considered. The $\sqrt{3} \times \sqrt{3}$ structure may be less stable for spin 1/2 than in the classical limit owing to quantum fluctuations [31].

Apparently, there is no spin gap in volborthite. However, the magnetic susceptibility decreases below T_p towards a 20%-reduced value at $T = 0$. This reduction is small for a gapped system but larger than expected for SRO in 2D. Thus, it is plausible to assume that certain singlet correlations develop below T_p. In fact, diffuse neutron scattering experiments find nearest-neighbor antiferromagnetic or singlet correlations at 15 and 5 K [32].

On this issue, it is illustrative to consult previous results on spin ladder compounds such as $SrCu_2O_3$ [33] and $LaCuO_{2.5}$ [34]. The former has a well-defined singlet GS with a large gap of $\Delta \sim 420$

K, while the latter falls into LRO at 125 K [35, 36]. Nevertheless, the χ of $LaCuO_{2.5}$ exhibits very similar gapped behavior with $\Delta \sim 470$ K as $SrCu_2O_3$. This is because $LaCuO_{2.5}$, having a larger interladder coupling, exists close to the QCP in the LRO side. Quantum Monte Carlo simulations by M. Troyer *et al.* found a decoupled ladder regime at intermediate temperatures that shows pseudogap-type behavior similar to uncoupled ladders [37]. In such a case, singlet correlations develop below $T \sim \Delta/k_B$, but LRO of reduced spins is finally reached on further cooling.

By analogy, the χ of volborthite must represent a development of singlet correlations corresponding to a pseudogap of the order of $J/2$. If so, the 2 T anomaly might correspond to the collapse of the lowest bounds of the pseudogap, that is 2 T ~ 1.5 K ~ $J/50$. Recent ESR experiments by H. Ohta *et al.* seems to be in line with this scenario: they found a variation of the g factor below T_p, suggesting the development of a large spin gap, and also a small gap excitation of 40 GHz ~ 1.9 K at a low temperature of 1.8 K [38]. Resemblance to spin-gapped compounds such as $TlCuCl_3$ in the M-H and M-T curves, as mentioned before, supports this speculation. Hence, the GS of volborthite could have been a LR-RVB state or other spin liquid states that have a density-of-states in an approximate range between $J/50$ and $J/2$, if LRO was not reached down to $T = 0$.

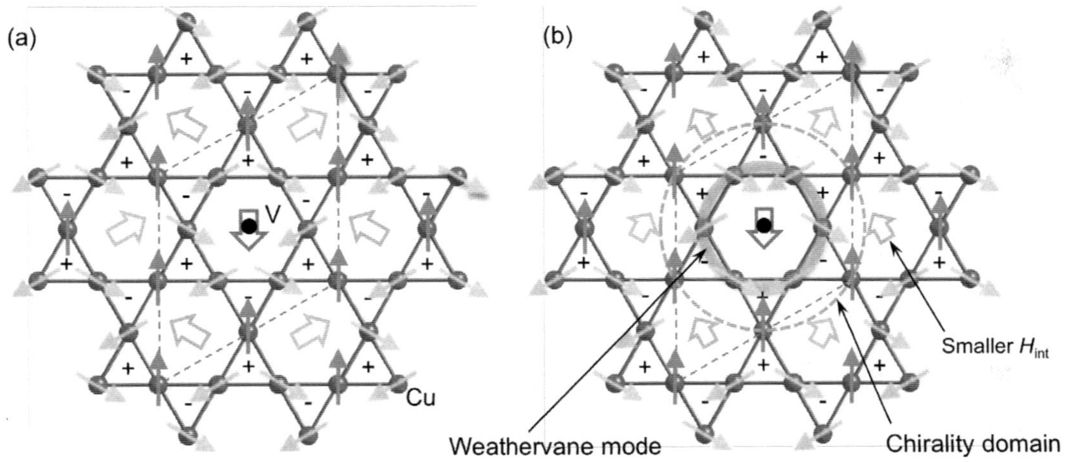

Figure 7. (a) Ideal √3 × √3 spin structure. The three sublattices, indicated by blue, green and yellow arrows on the lattice points, form 120° angles with respect to each other. '+' or '-' on each triangle represents the sign of the scalar chirality. A large arrow inside a hexagon shows the magnitude and direction of the internal magnetic field H_{int} from the surrounding six Cu spins at a V nucleus located nearly above or below the center of the hexagon (Fig. 2). (b) Snapshot of a spin arrangement after a weathervane mode within the central hexagon. When six spins on the centre hexagon are collectively rotated by 120°, as a result of the 180° out-of-plane precession around the axis of the outer spins on the kagome star, a chirality domain is generated, as shown by the broken circle, and the H_{int} at the V sites in the surrounding hexagons is reduced by 1/√3.

3.2. Magnetic ground state of volborthite

Very recent neutron scattering experiments on volborthite by Nilsen *et al.* found semi-classical SRO developing below 5 K and accompanying spin-wave like excitations [32]. Although the spin configuration at the lowest temperature was not determined, the observed SRO seems partly compatible with the √3 × √3 structure with a short magnetic correlation length. On the other hand, Bert *et al.* observed a Gaussian like frozen component in their ^{51}V NMR experiments, accounting for 80%

of ^{51}V nuclei [15], which is distinguished from an ordinary powder pattern for LRO. This was attributed to the formation of a $\sqrt{3} \times \sqrt{3}$ structure with disordered chirality.

The ^{51}V nucleus is located above or below the center of each hexagon (not exactly the center as shown in Fig. 2) and experiences an internal magnetic field H_{int} from neighboring six Cu spins. As illustrated in Fig. 7(b), once each spin on the hexagon at the center of the figure is rotated by 120° due to the W mode (180° precession about the axis of the outer spins in star, but only 120° rotation), thus generating a chirality domain, the V nuclei in the neighboring six hexagons should feel a reduced $(1/\sqrt{3})$ H_{int} from surrounding Cu spins [15]. This is because all three sublattice spins now coexist on each of these hexagons. Note that the internal field at the V nucleus in the center hexagon remains the same as before: the hexagon only contains two spin orientations. Assuming a random distribution of chirality frozen through this process may result in such a Gaussian like spectrum. Thus, the NMR results may be consistent with classical SRO based on the $\sqrt{3} \times \sqrt{3}$ structure [15]. On one hand, they seem to be inconsistent with implications from the previous μSR experiments, which found slow and nearly T-independent spin fluctuations at low temperatures from 1 K to 50 mK [18]. This contradiction may be due to the fact that the muon is possibly not in a symmetric position in the lattice and feels different spin fluctuations from neighboring Cu spins.

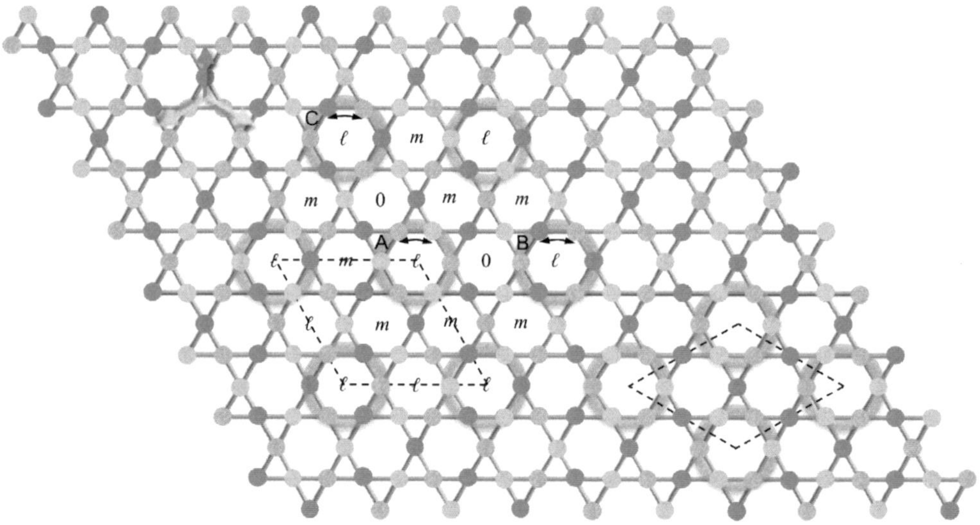

Figure 8. Schematic representation of possible arrangement of weathervane (W) modes or hexagon doublets (HDs). The three flavours of spin in the original $\sqrt{3} \times \sqrt{3}$ structure are represented by dots with different colors. A thick circle on a hexagon represents a W mode or a HD. A $2a \times 2a$ arrangement of HDs is depicted near the centre of the figure, and a $\sqrt{3}a \times \sqrt{3}a$ one in the lower right. Spins belonging to the HDs can fluctuate, while other spins surrounding the hexagons to complete the kagome stars are completely fixed in their directions and so perfectly long-range ordered. It is assumed in the snapshot of the $2a \times 2a$ arrangement that the dots in the hexagons marked by A, B and C have rotated by 60° from their original positions, which reduces the H_{int} of the surrounding hexagons to 'm' or '0'. 'l', 'm' and '0' inside some hexagons mean large, medium and zero internal magnetic fields at the V site, respectively. Depending on the mutual spin arrangements, the H_{int} can vary over time and be distributed.

The Lorentzian-like NMR spectrum observed below 1 K by M. Yoshida *et al.* must then have the same origin [16]. It is likely that the W modes are moving around quickly in the developing SRO at high temperatures, so that the NMR signal remains sharp without static H_{int} at the V site within the NMR timescale. Below T_k = 1.05 K, as LRO sets in, their movement slows down and may freeze to generate static but distributed internal magnetic fields, as in an SDW state. However, it is more likely from the viewpoint of entropy that the W modes are regularly arrayed in the kagome net at $T = 0$.

Let's consider one possible arrangement of W modes, as illustrated in Fig. 8, in which they form a triangular lattice with a $2a \times 2a$ unit. This is the pattern which results in maximum density, assuming no overlap between kagome stars (hexagon plus surrounding six lattice points). Another possible arrangement with higher density by one-third has a $\sqrt{3}a \times \sqrt{3}a$ unit, which is attained by relaxing the previous condition, but allowing no overlap between hexagons. Considering unusual characteristics for the former pattern, H_{int} on the V nucleus in every W mode is always the same as in the perfect LRO, irrespective of the rotation or the exchange of spins on the hexagon, while that in the intervening hexagons can vary among 1, $1/\sqrt{3}$, and 0, depending on the configurations in the neighboring two W modes. Taking this into account, one would expect a continuous distribution in H_{int} giving a Gaussian or Lorentzian like ^{51}V NMR spectrum even. Note that, still in the ordered arrangement of W modes, there remains a degree of freedom inside each W mode. We believe that this sort of coexistence of order and disorder or fluctuations such as shown in Fig. 8 represents one interesting aspect of the kagome physics.

3.3. Magnetic excitations

The origin of the dense low-energy excitations observed in C_m and ^{51}V NMR can be the W mode, though such a classical mode may not survive down to low temperatures in the presence of anisotropies, instead, freezing with a gap. It would be interesting if there is a quantum analogue for the W mode. We will consider two distinct spin arrangements for one hexagon or one kagome star, which appear in the course of the out-of-plane rotation when the spin plane of the kagome star coincides with that of the rest, to be totally coplanar. This is the case depicted in Fig. 7. This means that there are two degrees of freedom per hexagon carrying entropy of $k_B\ln 2$, which results in huge entropy of 1/3 of $R\ln 2$ (1.92 J K^{-1}) in total. Let's call this degree of freedom a hexagon doublet (HD).

We think that the observed T-linear behavior in heat capacity and $1/T_1$ comes from flipping between the two configurations of this hexagon doublet. Assuming that the total entropy of $(1/3)R\ln 2$ is spread over a wide temperature range such as J/k_B or T_p, one would expect $\gamma = 25$ or 96 mJ K^{-2} mol-Cu^{-1}, reasonably close to 40 mJ K^{-2} mol-Cu^{-1} as observed above T_k. The reduction of γ below T_k suggests that the number of HDs has decreased through the LRO of spins or by the order of HDs.

It is important, moreover, to notice that spins outside HDs are fixed in their directions and perfectly ordered in the long-range way, which should give spin-wave excitations, as in fact detected in heat capacity as well as in the neutron scattering experiments [32]. The T^2 term also exists above T_k with a smaller coefficient, indicating that the spin lattice is harder at high temperatures. The presence of spin-wave like excitations above T_k may be related to "disguised" spin waves theoretically predicted in the coplanar or nematic state for the classical Heisenberg KAFM [39, 40]; possibly, spin-wave excitations may exist both above and below T_k with the stiffness changing across T_k, as a result of the LRO. Since the T^2 heat capacity has also been observed in other kagome and triangular compounds, similar spin-wave like excitations must be commonly involved in 2D frustrated spin systems existing near a QCP.

Concerning the presence of a pseudogap as a trace or a fingerprint of a LR-RVB state, states filling up the gap are obviously magnetic, not nonmagnetic as expected from theory, and can be associated with true or "disguised" spin waves and also with the degree of freedom from the HD. The observed reduction in γ above 2 T may indicate that the density of the low energy excitations is reduced by the contamination of triplet states pulled down by the magnetic field.

3.4. Implications for states in magnetic fields

What is happening under higher magnetic fields, particularly at the three magnetization steps, remains mysterious and may be investigated in line with the present model. The W mode should play a role as far as the three-sublattice spins in the $\sqrt{3} \times \sqrt{3}$ spin structure remain equivalent. On the other hand, above the second magnetization step where magnetization is largely increased, a ferrimagnetic state must be responsible with two sublattice spins (e.g. green and yellow spins in Fig. 7) form a smaller angle than 120°, as predicted for a distorted kagome system [41]. There is another mystery concerning volborthite as well as vesignieite, that is, a large enhancement in magnetization from 1/3 at the magnetization plateau [17] Finally, we would like to point out that a similar hexagon-based phenomenon has been theoretically predicted for the quantum KAFM at a high magnetic field corresponding to $3J$: a jump in magnetization takes place from 7/9 to 1 (fully polarized state), when the density of local magnons on the hexagon reaches a maximum at a $\sqrt{3} \times \sqrt{3}$ unit [42].

4. Concluding remarks

In summary, the magnetic ground state of the spin-1/2 quasi-kagome antiferromagnet volborthite has been investigated, and an exotic long-range order based on the $\sqrt{3} \times \sqrt{3}$ spin structure has been proposed. Irrespective of the details of the ordering, it is apparent that volborthite lies in the left side of the phase diagram in Fig. 1, which seems to contradict all theoretical models thus far considered for spin-1/2 Heisenberg KAFM. The reason for this is not clear at the moment, but may be related to the distortion of the kagome lattice or deviations from the ideal nearest-neighbor kagome model. However, we would like to point out that our discussion here has assumed only the presence of a local $\sqrt{3} \times \sqrt{3}$ spin structure, independent of microscopic magnetic couplings, and thus should not be specific to volborthite.

One interesting issue of frustration physics is to find an optimal route to lift macroscopic degeneracy arising from geometrical frustration. Even though quantum fluctuations are strong enough to render a 'liquid' state viable for the spin-1/2 KAFM, it seems difficult to attain such a liquid state in real compounds. It is well known now that local excitations appear in some spinel compounds having the pyrochlore lattice [43, 44], which may be introduced to avoid frustration and stabilize LRO. Provided that the incorporation of local modes such as the weathervane mode to the kagome lattice is realistic, we think that such an exotic LRO as mentioned here could be an alternative to spin liquids in the presence of some perturbations. The introduction of certain local collective modes into LRO of spins might be a general scheme for highly frustrated spin systems to reach non-degenerate GSs. Nevertheless, it is to be noted that there must be a fingerprint for the unrealized, true GS at elevated temperatures above T_N or at excited states, allowing us to study its nature experimentally. Since there still remains uncertainty and considerable mystery about the GS of the spin-1/2 KAFM, further systematic study on various kagome compounds as well as theoretical models is required.

We thank M. Takigawa, Y. Nakazawa, G. J. Nilsen, Y. Tomiyasu and P. Mendels for helpful discussion. This work was supported by Grant-in-Aids for Scientific Research on Priority Areas "Novel States of Matter Induced by Frustration" (19052003) given by the MEXT, Japan.

References

[1] Mendels P and Wills A S 2011 *Introduction to Frustrated Magnetism* ed by Lacroix C Mendels P and Mila F (Berlin: Springer-Verlag) pp 207-239
[2] Balents L 2010 *Nature* **464** 199
[3] Ramirez A P 1994 *Annu. Rev. Mater. Sci.* **24** 453
[4] Fazekas P and Anderson P W 1974 *Philos. Mag.* **30** 423
[5] Waldtmann C Everts H-U Bernu B Lhuillier C Sindzingre P Lecheminant P and Pierre L 1998 *Eur. Phys. J. B* **2** 501
[6] Sindzingre P and Lhuillier C 2009 *EPL* **88** 27009
[7] Singh R R P and Huse D A 2007 *Phys. Rev. B* **76** 180407 and references therein
[8] Lafontaine M A Bail A L and Férey G 1990 *J. Solid State Chem.* **85** 220
[9] Hiroi Z Kobayashi N Hanawa M Nohara M Takagi H Kato Y and Takigawa M 2001 *J. Phys. Soc. Jpn.* **70** 3377
[10] Janson O Richter J Sindzingre P and Rosner H 2010 *Phys. Rev. B* **82** 104434
[11] Shores M P Nytko E A Bartlett B M and Nocera D G 2005 *J. Am. Chem. Soc.* **127** 13462
[12] Okamoto Y Yoshida H and Hiroi Z 2009 *J. Phys. Soc. Jpn.* **78** 033701

[13] Yoshida H Okamoto Y Tayama T Sakakibara T Tokunaga M Matsuo A Narumi Y Kindo K Yoshida M Takigawa M and Hiroi Z 2009 *J. Phys. Soc. Jpn.* **78** 043704
[14] Hiroi Z Yoshida H Okamoto Y and Takigawa M 2009 *J. Phys.: Conf. Series* **145** 012002
[15] Bert F Bono D Mendels P Ladieu F Duc F Trombe J-C and Millet P 2005 *Phys. Rev. Lett.* **95** 087203
[16] Yoshida M Takigawa M Yoshida H Okamoto Y and Hiroi Z 2009 *Phys. Rev. Lett.* **103** 077207
[17] Okamoto Y Tokunaga M Yoshida H and Hiroi Z 2010 *submitted to Phys. Rev. Lett.*
[18] Fukaya A Fudamoto Y Gat I M Ito T Larkin M I Savici A T Uemura Y J Kyriakou P P Luke G M Rovers M T Kojima K M Keren A Hanawa M and Hiroi Z 2003 *Phys. Rev. Lett.* **91** 207603
[19] Takeya H Ishida K Kitagawa K Ihara Y Onuma K Maeno Y Nambu Y Nakatsuji S MacLaughlin D E Koda A and Kadono R 2008 *Phys. Rev. B* **77** 054429
[20] Shimizu Y Miyagawa K Kanoda K Maesato M and Saito G 2006 *Phys. Rev. B* **73** 140407
[21] Yamashita S Moriura T Nakazawa Y Yoshida H Okamoto Y and Hiroi Z 2010 *J. Phys. Soc. Jpn.* **79** 083710
[22] Olariu A Mendels P Bert F Ueland B G Schiffer P Berger R F and Cava R J 2006 *Phys. Rev. Lett.* **97** 167203
[23] Kawamura H *to be published in J. Phys.: Conf. Series*
[24] Ramirez A P Espinosa G P and Cooper A S 1992 *Phys. Rev. B* **45** 2505
[25] Fak B Coomer F C Harrison A Visser D and Zhitomirsky M E 2008 *Epl* **81** 17006
[26] Nakatsuji S Nambu Y Tonomura H Sakai O Jonas S Broholm C Tsunetsugu H Qiu Y and Maeno Y 2005 *Science* **309** 1697
[27] Yamashita S Nakazawa Y Oguni M Oshima Y Nojiri H Shimizu Y Miyagawa K and Kanoda K 2008 *Nat. Phys.* **4** 459
[28] Nikuni T Oshikawa M Oosawa A and Tanaka H 2000 *Phys. Rev. Lett.* **84** 5868
[29] Reimers J N and Berlinsky A J 1993 *Phys. Rev. B* **48** 9539
[30] Zhitomirsky M E 2008 *Phys. Rev. B* **78** 094423
[31] Lecheminant P Bernu B Lhuillier C Pierre L and Sindzingre P 1997 *Phys. Rev. B* **56** 2521
[32] Nilsen G J Coomer F C Vries M A d Stewart J R Deen P P Harrison A and Ronnow H M 2010 *arXiv:1001.2462*
[33] Azuma M Hiroi Z Takano M Ishida K and Kitaoka Y 1994 *Phys. Rev. Lett.* **73** 3463
[34] Hiroi Z 1996 *J. Solid State Chem.* **123** 223
[35] Kadono R Okajima H Yamashita A Ishii K Yokoo T Akimitsu J Kobayashi N Hiroi Z Takano M and Nagamine K 1996 *Phys. Rev. B* **54** R9628
[36] Matsumoto S Kitaoka Y Ishida K Asayama K Hiroi Z Kobayashi N and Takano M 1996 *Phys. Rev. B* **53** R11942
[37] Troyer M Zhitomirski M E and Ueda K 1997 *Phys. Rev. B* **55** R6117
[38] Ohta H Zhang W-m Okubo S Fujisawa M Sakurai T Okamoto Y Yoshida H and Hiroi Z 2010 *Phys. Status Solidi B* **247** 679
[39] Chalker J T Holdsworth P C W and Shender E F 1992 *Phys. Rev. Lett.* **68** 855
[40] Robert J Canals B Simonet V and Ballou R 2008 *Phys. Rev. Lett.* **101** 117207
[41] Stoudenmire E M and Balents L 2008 *Phys. Rev. B* **77** 174414
[42] Schulenburg J Honecker A Schnack J Richter J and Schmidt H J 2002 *Phys. Rev. Lett.* **88** 167207
[43] Lee S H Broholm C Ratcliff W Gasparovic G Huang Q Kim T H and Cheong S W 2002 *Nature* **418** 856
[44] Tomiyasu K Suzuki H Toki M Itoh S Matsuura M Aso N and Yamada K 2008 *Phys. Rev. Lett.* **101** 177401

Quantum kagome antiferromagnet : $ZnCu_3(OH)_6Cl_2$

Philippe Mendels, Fabrice Bert

Laboratoire de Physique des solides, Université Paris-Sud, UMR CNRS 8502, 91405 Orsay Cedex, France

E-mail: philippe.mendels@u-psud.fr

Abstract. Herbertsmithite, $ZnCu_3(OH)_6Cl_2$, is the prototype candidate for a spin liquid behavior on a geometrically perfect kagome lattice. Its discovery and the absence of any evidence for spin-freezing down to the lowest probed temperature to-date enable one to explore the properties of kagome-related physics in an unprecedented temperature range. We review its properties and discuss some open issues. A tentative comparison to models is also performed.

1. Introduction

Herbertsmithite $ZnCu_3(OH)_6Cl_2$, a rare mineral recently extracted and named after G.F. Herbert Smith (1872-1953) who discovered the mother atacamite family in 1906 [1], has been synthesized for the first time in 2005, in M.I.T. [2]. It has triggered a major renewal in the search of a quantum spin liquid on the kagome Heisenberg antiferromagnetic (KHA) lattice and opened the possibility to directly expose theories to experiments for the KHA ground state which has eluded any definitive conclusion for the last twenty years. Altogether with triangular compounds $\kappa(ET)_2Cu_2(CN)_3$ [3] and more recently, $(ET)Me_3Sb[Pd(dmit)_2]_2$ [4], it combines low dimensionality, a quantum character associated with spins 1/2 and the absence of magnetic freezing which was coined as the "end to the drought of spin liquids" [5, 6]. Herbertsmithite is the first experimental example of a kagome lattice where no order has been found at any temperature well below J through all experimental techniques [7, 8]. This illustrates the power of the combination of an enhancement of quantum fluctuations for $S = 1/2$ spins with the frustration of antiferromagnetic interactions on the loosely connected kagome lattice to stabilize novel ground states of magnetic matter.

Most of the models point to an exotic ground state and two broad classes of spin liquids have been explored theoretically, in particular for the kagome antiferromagnet [10], (i) topological spin liquids which feature a spin gap, a topological order and fractional excitations [11]; (ii) algebraic spin liquids with gapless excitations in all spin sectors [12].

Here, we partially review some of the most relevant experimental properties of Herbertsmithite, and complement some aspects of our recent review on this compound [6]. We tentatively discuss which model could be consistent with these findings within the context of well identified deviations to the ideal Heisenberg kagome antiferromagnet.

2. Structure and magnetic couplings

Herbertsmithite $ZnCu_3(OH)_6Cl_2$ is the $x = 1$ end compound of the Zn-paratacamite family $Zn_xCu_{4-x}(OH)_6Cl_2$ [1, 2]. It can be viewed as a double variant of the parent clinoatacamite

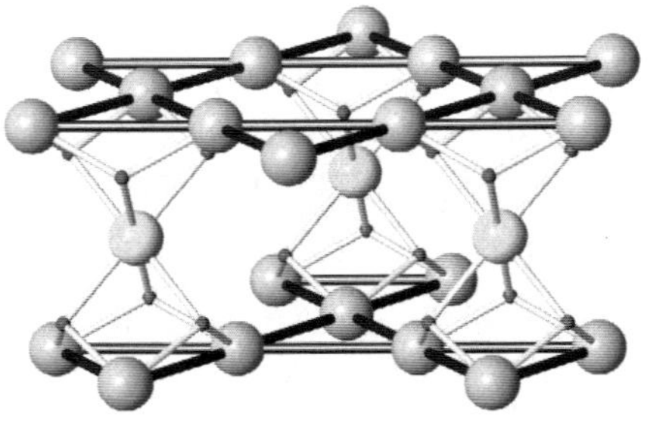

Figure 1. Simplified structure of ideal Herbertsmithite where only Cu (kagome top and bottom layers), Zn (intermediate triangular layer) and O which bridge between two Cu in the kagome planes, have been represented.

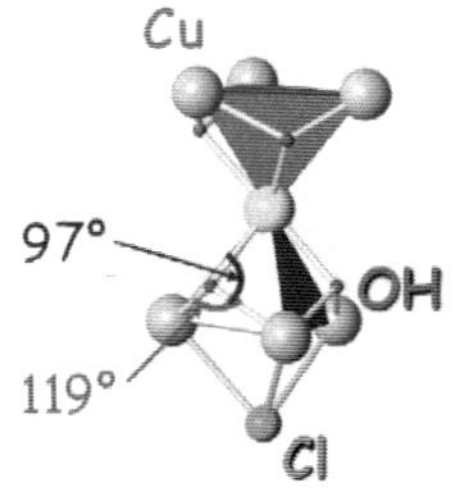

Figure 2. Local structure emphasizing the coupling path within a kagome layer and between one Cu kagome site and one Cu substituted on the Zn site, named interplane site in the text.

compound ($x = 0$), situated at the other end of the family. Starting from the latter, a Jahn-Teller distorted $S = 1/2$ pyrochlore, the symmetry first relaxes from monoclinic ($P2_1/n$) to rhombohedral ($R\overline{3}m$) around $x = 0.33$, leading to a perfect kagome lattice in the a-b plane with *isotropic* planar interactions; then, in the c- elongated $x > 0.33$ pyrochlore structure, the magnetic bridge along the c-axis between a-b kagome planes is progressively suppressed by replacing the apical Cu^{2+} by a diamagnetic Zn^{2+}.

In Herbertsmithite, the 119° Cu-OH-Cu bond-angle yields a moderate planar super-exchange interaction $J \simeq 180(10)$ K in comparison with other cuprates such as High T_c's where the Cu-O-Cu bonding is linear ($J \sim 1000 - 2000$ K). Note that the former value is consistent with that observed in cubanes where a similar OH bridge links adjacent coppers [13]. Both neutron low-T data on the $x = 0$ compound and a simple Curie-Weiss analysis of the variation of the high-T susceptibility with x, suggest that the interaction between apical and planar Cu is weakly ferromagnetic and estimated to be $J' \sim 0.1J$. This is also in line with Goodenough-Kanamori rules which predict a weak ferromagnetic exchange for a Cu-O-Cu angle of 96.7° and with the presence of a weak ferromagnetic component in the frozen phase of Zn depleted ($x < 0.66$) compounds.

3. Categorizing non-ideality and the related impact on physical properties

The dominant term in the Hamiltonian is certainly the Heisenberg exchange interaction between neighboring spins within the kagome layer (see above). Yet, frustrated antiferromagnets are known to yield non-spin liquid ground states under small perturbations. In the case of Herbertsmithite where both disorder and Dzyaloshinkii-Moriya (DM) anisotropy are at play, the absence of any kind of order or freezing seems to be quite surprisingly robust.

3.1. Dzyaloshinkii-Moriya interactions

In the absence of any macroscopic high-T susceptibility results published on single crystals, ESR proves to be the most appropriate technique to investigate the local spin anisotropy. The inclusion in the Hamiltonian of a Dzyaloshinkii-Moriya interaction $\overrightarrow{D}_{ij}.\overrightarrow{S}_i \times \overrightarrow{S}_j$ is necessary since there is no inversion symmetry between two adjacent Cu. $\overrightarrow{D}$ has two components, one perpendicular to the planes, D_z, and one planar D_p which act differently on the susceptibility and the specific heat. The effects are quite non linear and the large value of D_p invoked quite early after the discovery of Herbertsmithite to explain the rapid increase of the susceptibility at low T [14, 15] is far beyond the limit set by the ESR results. The small measured value of

$D_p \sim 0.01J$ and sizeable $D_z \sim 0.04 - 0.08J$ [16, 17] yield a fairly small correction, even negative, to the susceptibility as compared to the Curie-like upturn. Given this, the simple difference with series expansion at temperatures as high as $J/2$ argues for an extra term dominated by -if not only due to- out of plane defects [18]. Definitely, the latter is responsible for the drastic low-T upturn of the susceptibility.

Studies of the phase diagram of a quantum Heisenberg kagome lattice with DM interactions show that an original spin liquid phase can survive, at variance with the classical case where any minute amount of DM anisotropy will lead to long-range order [19]. For spins $S = 1/2$ a quantum critical point indeed appears for $D/J \sim 0.1$ which separates the liquid and the Néel states [20, 21, 22]. A peculiar variation of the susceptibility $\chi \sim T^{-0.53}$ at the QCP is expected [22].

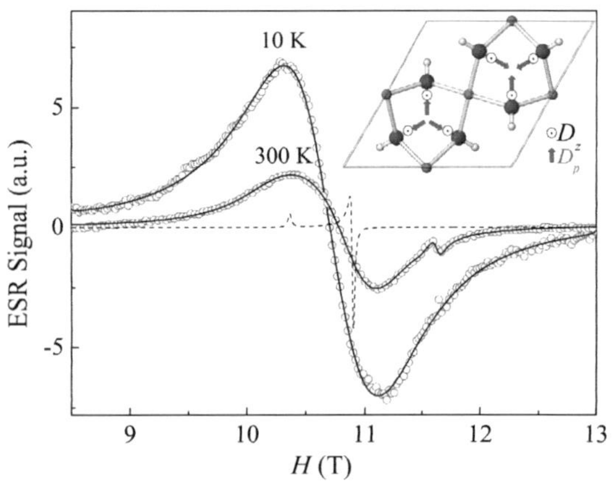

Figure 3. ESR spectra taken at 300 and 10 K. The broad linewidth is due to a large anisotropy. The dashed line shows the anisotropy of the g-shift tensor, hence the minute deviation from a pure Heisenberg character of the spins, quite common for Cu^{2+} ions. In the inset the Dzyaloshinskii vectors are represented. The Cu and O are respectively represented by small and large full circles (red and blue online). Adapted from ref. [16].

3.2. Out-of-plane defects

The existence of out of plane magnetic defects due to the substitution of Cu^{2+} $S = 1/2$ onto the Zn^{2+} non magnetic triangular layer has been the subject of controversies for a few years but this issue seems now settled. Based on electrostatic Jahn-Teller effect arguments, Cu^{2+} was indeed expected to preferentially sit in the distorted octahedral environment of the kagome planes. For a perfect stoichiometry of three Cu to one Zn, which seems to be achieved according to Inductively Coupled Plasma Atomic Emission Spectroscopy (ICP-AES) characterization, this would lead to a a perfect Cu^{2+} kagome plane and a non-magnetic Zn^{2+} site in between the planes [2]. There is now a consensus that this argument cannot be used as an all and nothing argument.

- Low-T-macroscopic measurements can indeed be described consistently within a scenario where the Zn site is occupied by 15-20% Cu^{2+}. The upturn of the susceptibility [8], the partial saturation of the high field magnetization measured up to 14 T at 1.7 K [23], the linear field dependence of a Schottky anomaly in the specific heat [24] are strong indications of a Zeeman splitting of quasi-free $S = 1/2$ moments.
- The low energy magnetic spectral weight in inelastic neutron scattering shifts linearly with field [25].
- There were long-standing evidences that neutron diffraction spectra from various groups were best refined using a 15-20% occupation of the Zn site by Cu, which was also acknowledged recently by the MIT group. Yet the shallow minimum of the R- factor was casting some doubt on this determination. An additional very convincing argument came

from the X-ray anomalous scattering data on a single crystal which allow to better quantify the level of Zn and Cu present at each site, namely 15(2)% of Cu was consistently found on the inter-kagome site [26].

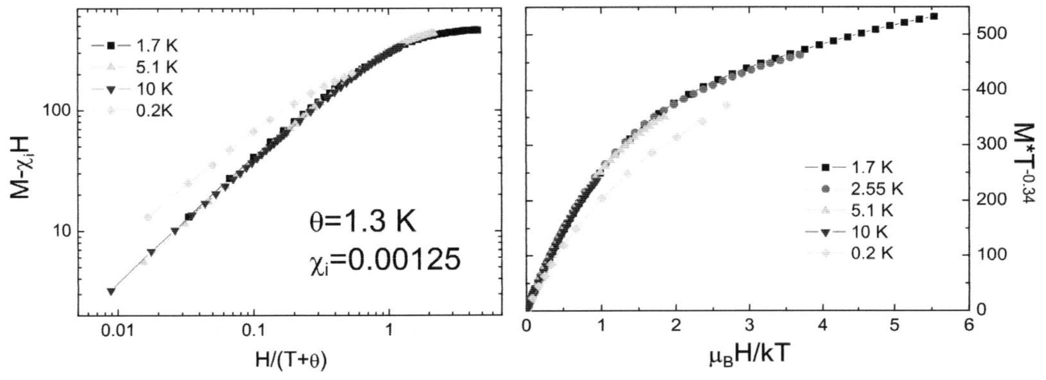

Figure 4. (Color on line) Two scalings for the low-T magnetization, M, measured up to 14 Teslas. The 0.2 K data deviate from the scaling. Left: the contribution from the kagome planes which remains linear at high field has been subtracted. The remaining part is associated with quasi-free spins and the Curie-Weiss like scaling which works well down to $\sim$ 1 K, suggests some magnetic interaction between the spins of the order of 1 K. Right:following the idea of a quantum critical behavior developed in [27],an equally good scaling is obtained. Yet, the fitted part does correspond to the now well established contribution from Cu defects on the Zn site which rules out the adaptability of a DM induced quantum critical scenario $M\ T^{-0.53}$ [22] for these data.

Given this, one has to wonder how these quasi-free spins might couple with the kagome planes and how they might impact on the physics. As stated in the previous section, a small ferromagnetic interaction at least one order of magnitude smaller than the in-plane interaction should be anticipated. Surprisingly, susceptibility data and magnetization experiments at low-T do not show any sign of such a ferromagnetic interaction but rather point at a weak antiferromagnetic effective coupling of the order of 1 K between these "lone" spins [23]. Since a simple dipolar coupling would yield too weak an interaction of 10 mK for 15% $S = 1/2$ moments, one is led to assume that this effective AF interaction might reveal some indirect coupling through the Cu's of the kagome planes. Using a reciprocity argument, one might infer that the lone spins have little influence on the kagome planes physics down to $\sim$ 1 K.

One should as well note in passing that it has been proposed at odds to interprete this low-T behavior of the susceptibility as an evidence for a quantum critical behavior under the assumption that it could be safely attributed to the kagome plane physics [27]. Given the arguments of the previous section - the low-T macroscopic susceptibility is dominated by the inter-plane channel defect-, the latter condition certainly does not apply here. Fig. 4 shows the comparison between a Curie-Weiss analysis of the susceptibility and the scaling proposed in ref. [27]. They equally work well for $T > 1$ K but fail to interprete the 0.2 K data which signals some different physics of the defects in the low-T range. The muon relaxation rate also displays some peculiar change around 1K. Such an analysis in terms of quantum criticality therefore can be safely replaced by the more sound scenario where a weak effective interaction couples these moments.

Journal of Physics: Conference Series **320** (2011) 012004 doi:10.1088/1742-6596/320/1/012004

3.3. In-plane vacancies : an open issue?

While 15-20% of Cu occupation of the inter-plane site was found, depending on the technique used and likely on the samples, ICP-AES analyses have always reliably supported the idea of a perfect global stoichiometry of 1.00:3.00 Zn/Cu ratio. This leads to the straightforward conclusion that intersite mixing occurs and a 5-7% amount of Zn^{2+} occupies the kagome planes, then creating spin vacancies.

This seems to be quite disputed by the recent X-ray anomalous scattering data [26], where an upper limit of 1(3)%Zn on the kagome layer was found. On the other hand, pushing to the extreme limit of 5% error on the 1.00:3.00 Zn/Cu ratio from ICP-AES reported on the same sample, would yield around 4% Zn on the kagome sites if 15% of the Zn sites are occupied with copper. We see that we are at the limit of the consistency between the two sets of data taken on the same sample, which likely require further inspections.

Two other arguments based on ^{17}O NMR spectra give evidence for the existence of in-plane defects [29].

- Typical spectra are displayed in Fig. 5. From a detailed analysis of the marked singularities of the powder lineshape which positions are governed by the magnetic shift and quadrupolar effects, *two different sites* can be identified. The most intense line at high T, named main(M), is associated with O sites linked to two Cu sites in the kagome planes. The other line (D) which is very sharp at low-T can be consistently interpreted as coming from O sites linked with one Cu and one non-magnetic Zn in the kagome planes. It is indeed half shifted at high-T as compared to the (M) line and corresponds to 20% spectral weight, *ie* a 5% Zn occupation of the kagome planes. Its sharpness at low T is an additional indication that it corresponds to very specific sites in the kagome planes, the susceptibility of which is not at all related to the Curie-like contribution of the out of plane defects.
- A broadening of the spectrum of the main line (M) is observed at low T. Would the kagome plane be free of any defect, one could still argue that:
  (i) The DM anisotropy could be responsible of this broadening on a powder spectrum. In the absence of any calculated field orientation dependence of the susceptibility due to DM anisotropic interaction, we crudely used the calculated susceptibility for an isolated dimer in the presence of DM interaction performed for $SrCu_2(BO_3)_2$ [28]. Since O probes two adjacent sites, the non staggered susceptibility should give birth to a powder shape which maximum relative width would be of the order of $\sim g\mu_B H D^2/J^3$. This represents $\sim 10^{-4}$ contribution to the shift, much too small.
  (ii) The g-factor anisotropy [16] could be responsible for some broadening. Its value $< 17\%$ is not enough to explain such a huge low-T broadening and that should scale with the shift which is not the case.
  (iii) Out-of plane defects could be responsible for such a width. An asymmetric broadening $\sim 1/T$ would be expected through a direct coupling of these defects to the planes, which is not the case. Such a coupling is also ruled out by the features of the (D) line reviewed above.

  One can therefore safely conclude that the susceptibility in the kagome planes is not homogeneous and the ^{17}O linewidth is a clear signature of the existence of a *staggered response around spinless defects.*

Based on this conclusion the relative 4:1 intensities of the (M) to (D) line argues in favor of 5% defects in the kagome planes for the investigated sample. These are likely due to 5% spin vacancies but one cannot of course discard that the (D) line might result from local structural in-plane distortions due to Cu substitutions on the Zn site. Another honest still open possibility is that the samples investigated in NMR and those studied in X-ray are somewhat different.

Journal of Physics: Conference Series **320** (2011) 012004 doi:10.1088/1742-6596/320/1/012004

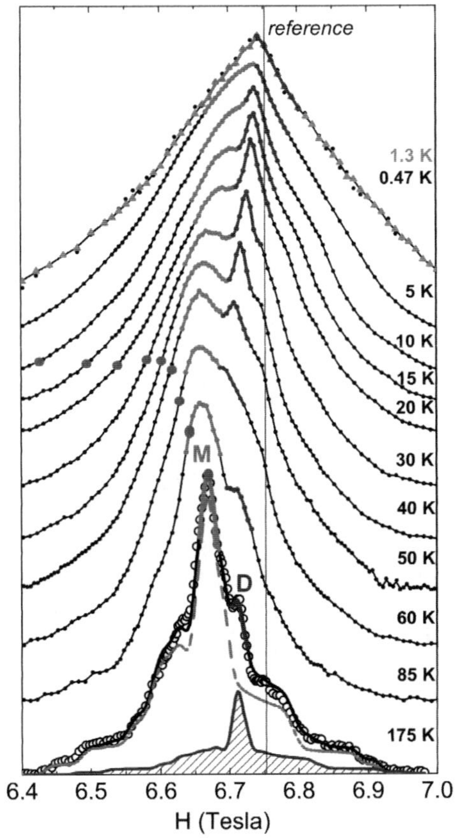

Figure 5. (Color online) ^{17}O NMR spectra from 175 K to 0.47 K from ref. [29]. The shift reference is indicated by the vertical line. (M) and (D) refer to the two oxygen sites in the kagome planes as described in the text. Large full green circles indicate the position of the center of the line that would be expected from χ_{macro}. At 175 K, a detailed analysis was performed and two contributions are evident from the displayed fits (see text and section 4). The hatched area represents the contribution from (D) oxygens sitting nearby one Zn defect in the kagome plane.

4. Evidences for a gapless ground state

Given the dominant contribution of out of plane defects in macroscopic measurements, only site-selective probes can give some insight into the physics of the kagome planes. There are certainly two ideal probes of the kagome physics, Cu and O nuclei. Unfortunately Cu has never been detected in the whole T-range and the detection of a signal which was reported in [30] is certainly encouraging but which Cu, either in-plane or out-of-plane, is detected is still an open question. O therefore appears to be the most secure probe. From a simple comparison to other probes at high-T, where the macroscopic susceptibility is dominated by the kagome planes, O is found to be one order of magnitude at least better coupled than Cl [30] and two orders of magnitude better than implanted muons in μSR experiments [32].

4.1. Kagome susceptibility from ^{17}O NMR [29, 31]

Tracking the position of the main line gives the shift, hence one can extract the hereafter called "intrinsic" susceptibility from 300 K$\sim$ 1.7J down to 0.45 K$\sim$ $J/400$. A broad maximum is observed between the high-T Curie-Weiss regime and the low-T regime where the susceptibility monotonously decreases between $J/3$ and $J/20$ to finally level-off at a finite value. The behavior in the intermediate T-range can certainly be associated with a strengthening of the short-range antiferromagnetic correlations. Fig. 6 illustrates the huge difference between the intrinsic

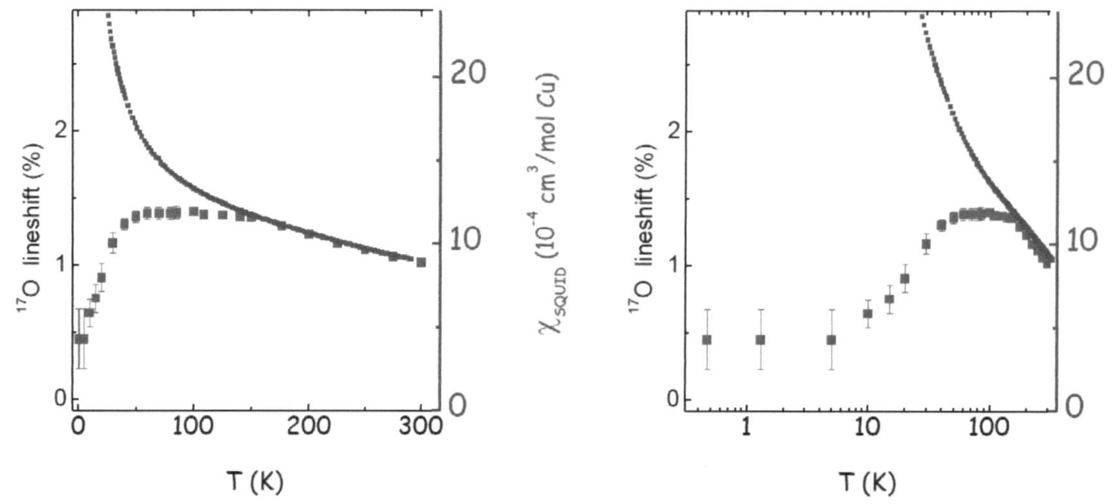

Figure 6. (Color online) Thermal variation of the susceptibility of the kagome planes measured from the shift of the main ^{17}O NMR line (red open symbols) compared to the macroscopic susceptibility (green solid line). The red dashed line is a guide to the eye for the low-T shift.

susceptibility of the kagome planes and the macroscopic susceptibility which is dominated at low T by the inter-plane defects.

Using the DM calculations in a dimer model cited above gives a strong indication that the DM interaction is not in an obvious manner the origin of the non zero susceptibility at $T = 0$ and points to a gapless ground state in Herbertsmithite ... unless for some not obvious reason J should be rescaled to a much weaker effective value, e.g. 0.1 J. This calls for further realistic inclusions of the DM interaction into the kagome Heisenberg hamiltonian.

4.2. Dynamical susceptibility

Both NMR T_1 relaxation measurements [29, 30] and inelastic neutron scattering (INS) [8, 24] give also a strong support to a gapless scenario, at least in the field and temperature ranges which have been probed. NMR T_1 measurements taken using various probes agree well from 30 K down to 1.2 K. ^{17}O data extend down to 0.47 K under an applied field of 6.6 Tesla where a gapless picture is found to hold with a power law behavior $T_1^{-1} \sim T^{0.71(5)}$. Although less sensitive, INS data taken down to 35 mK without any applied field consistently reveal that $\chi''(\omega)$ increases at low T with an uncommon divergent $\omega^{-0.7(3)}$ law.
One should as well note that a very different T-behavior is observed in μSR experiments. A tiny increase of the relaxation rate is found below 1 K, to reach a plateau under zero applied field [7]. Since the dipolar coupling of the muon to nearby moments is an important channel for relaxation, these dynamical effects might be associated with interplane Cu.

5. A tentative comparison to models

Finding the ground state of the kagome Heisenberg antiferromagnet has represented a "Holy Grail" in the field of frustrated quantum magnetism. Two scenarios have been in balance since the early days, that of a spin liquid, gapped or ungapped, where quantum fluctuations melt away any form of magnetic order and that of a honeycomb valence bond crystal combining

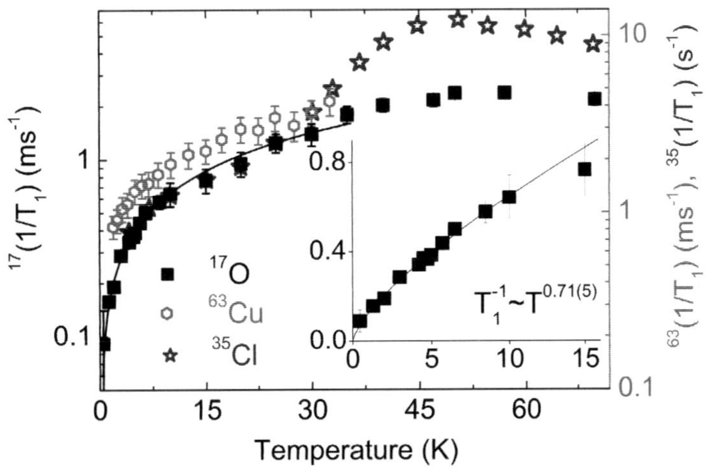

Figure 7. (Color online) Semi-log plot of the relaxation rates obtained for all probe nuclei, ^{17}O, ^{63}Cu and ^{35}Cl. Inset: Linear plot of T_1^{-1} with a power law fit for the low T ^{17}O data. Adapted from ref. [29, 30].

dimer bonds within a 36-sites unit cell. The winner is of course that of lowest energy. As an example, the Valence Bond Crystal (VBC) model was found to be the best in 2008 through dimer series expansions [33] whereas in late 2010 a Density Matrix Renormalization Group (DMRG) approach strongly supports a spin liquid ground state with a gap, with a possible value of $J/20$ [34].

On the experimental side, Herbertsmithite does not favor any of these scenarios. Neither a gap is observed much below $J/100$, nor singlets are detected. The central difficulty is to understand better how much the ground state is perturbed by the presence of in-plane defects and DM anisotropy. We have selected a few proposed scenarios and discuss below whether the experimental data might cope with them or not in the realistic experimental situation.

- *Dirac gapless spin liquid:* Using a fermionic approach in the context of fractional excitations allows to derive a specific heat $C \sim T^2$, a susceptibility $\chi \sim T$, and the NMR relaxation rate $1/T_1 \sim T^\eta$ [35, 36]. Although the low-T shift roughly scales linearly with T below $J/6$, the unstability of such a spin liquid against DM anisotropy rather argues against such a scenario.
- *Valence bond crystal/glass:* Nothing in the data collected on Herbertsmithite supports the idea of a simple VBC state [33], which would combine dimer bonds as proposed for the honeycomb 36 sites unit-cell. Including DM interactions certainly relieves the singlet ground state constrain and could lead to some agreement with the experimental data but some effective interaction $\sim 0.1J$ should be at play to explain the susceptibility, which remains unclear. Interestingly, in presence of disorder, a valence bond glass (VBG) could be stabilized where disorder might restore some susceptibility [37]. In this VBG framework, Raman scattering should give a very characteristic response. The defects are also predicted to yield free spins not simply localized singlets. Yet the recent DMRG calculations seem to disqualify such a valence bond based state, see below.

- *Gapped spin liquid:* This is the present winner in energy as found in a recent DMRG approach but a gap is predicted of the order of $J/20$ far too high to be consistent with the present experimental findings.
- *What about the message from exact diagonalizations (ED)?* Besides the increase of the size of samples which can be studied, the computing progress enables now to calculate a large number of levels in any S sector of clusters up to $N = 36$ spins. This opens the possibility of a detailed analysis of the magnetism of the cluster. Numerical data of various sizes were recently scrutinized and lead to the conclusion that the singlet-triplet gap could vanish at the thermodynamic limit although as early as 1998, it was thought to be of the order of $J/20$ [38, 39]. The analysis of the minimum energy in each spin sector leads to the conclusion that the KHA could be gapless with a susceptibility consistent with the NMR results, although 4 times larger [40].
  DM interactions and disorder can also be included in ED calculations [41, 42]. As in most models, a singlet is found to localize in the vicinity of a spin vacancy - a relief of frustration on the corresponding triangle - and DM anisotropy again restores a small susceptibility. The longer distance response is staggered. Both features reproduce qualitatively the NMR spectra which respectively signal a specific low-T response in the vicinity of the spin vacancy and a broadening of the line which cannot be ascribed only to DM interactions without defects.

6. Conclusion

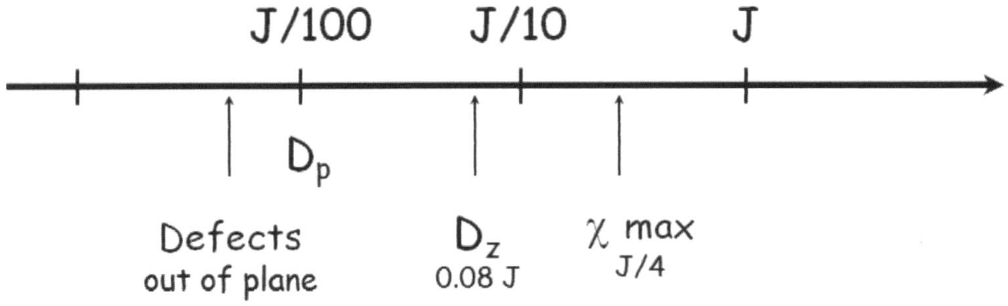

Figure 8. Plot of the various energy scales at play in Herbertsmithite.

In this paper, we aimed at clarifying some features associated with the defects. Out of plane defects seem not to affect much the physics of the kagome plane but they certainly obscure the macroscopic response. Including DM interaction in models, controlling the in-plane defects in order to extrapolate to a perfect kagome lattice are certainly the way to go for joining theory and experiment. At the present stage, the major result is that Herbertsmithite is a gapless spin liquid under the field and temperature conditions used up to now in experiments. Once the prehistorical age of Herbertsmithite will be over, spinless vacancies might play the role of a smoking gun for the physics of these compounds. Certainly much work is needed to reach that stage and the first synthesis of millimetric crystals are encouraging news for future research in the field ... and to the most pessimistic reader who would be tempted to argue that Herbertsmithite is too disordered to shed light on the KHA ground state, EDs give a rather contrasting view that many experimental results can already be reproduced at least at a qualitative level! We certainly are at the stage where the complexity of the compound is well appreciated and one can now expect some considerable step forward in the materials improvement and the physical discussions on this compound.

[1] Braithwaite R S W , Mereiter K, Paar W and Clark A 2004 *Mineral. Mag.* **68** 527
[2] Shores M P, Nytko E A, Bartlett B M and Nocera D G 2005 *J. A. Chem. Soc.* **127** 13462
[3] Shimizu Y, Miyagawa K, Kanoda K, Maesato M, Saito G 2003 *Phys. Rev. Lett.* **91** 107001
[4] Itou T, Oyamada A, Maegawa S and Kato R, 2010 *Nat. Phys* **6** 673
[5] Lee P A 2008 *Science, Perspectives* **321** 1306
[6] Mendels P and Bert F 2010 *J. Phys. Soc. Jpn.* **79** 011001
[7] Mendels P, Bert F, de Vries M A, Olariu A, Harrison A, Duc F, Trombe J C, Lord J S, Amato A and Baines C 2007 *Phys. Rev. Lett.* **98** 077204
[8] Helton J S, Matan K, Shores M P, Nytko E A, Bartlett B M, Yoshida Y, Takano Y, Suslov A, Qiu Y, Chung J H, Nocera D G and Lee Y S 2007 *Phys. Rev. Lett.* **98** 107204
[9] Levy B 2007 *Physics Today* **16** (Feb. 2007)
[10] Lacroix C, Mendels M and Mila F 2011 *Introduction to Frustrated Magnetism*, Springer Series in Solid-State Sciences **164** (Berlin: Springer)
[11] Kivelson S A, Rokhsar D S and Sethna J P 1987 *Phys. Rev. B* **35** 88865; Moessner R and Sondhi S L 2001 *Phys. Rev. Lett.* **86** 1881
[12] Hastings M B 2001 *Phys. Rev. B* **63** 014413; Hermele M, Senthil T and Fisher M P A 2005 *Phys. Rev. B* **72** 104404
[13] Wills A S and Henry J Y 2008 *J. Phys. Cond. Mat.*, **20** 472206
[14] Rigol M and Singh R R P 2007 *Phys. Rev. Lett.* **98** 207204
[15] Rigol M and Singh R R P 2007 *Phys. Rev. B* **76** 184403
[16] Zorko A, Nellutla S, van Tol J, Brunel L C, Bert F, Duc F, Trombe J C, de Vries M C, Harrison A and Mendels P 2008 *Phys. Rev. Lett.* **101** 026405; Zorko A , Nellutla S, van Tol J , Brunel L C, Bert F, Duc F, Trombe J-C and Mendels P 2009 *J. Phys.: Conf. Ser.* **145** 012014
[17] El Shawish S, Cépas O, and Miyashita S 2010 *Phys. Rev. B* **81** 224421
[18] Misguich G and Sindzingre P 2007 *Eur. Phys. J. B* **59** 305
[19] Elhajal M, Canals B and Lacroix C 2002 *Phys. Rev. B* **66** 014422
[20] Cépas O, Fong C M, Leung P W and Lhuillier C 2008 *Phys. Rev. B* **78** 140405
[21] Messio L, Cépas O and Lhuillier C 2010 *Phys. Rev. B* **81** 064428
[22] Huh Y, Fritz L, and Sachdev S 2010 *Phys. Rev. B* **81** 144432
[23] Bert F, Nakamae S, Ladieu F, L'Hote D, Bonville P, Duc F, Trombe J C and Mendels P 2007 *Phys. Rev. B* **76** 132411
[24] de Vries M A, Kamenev K V, Kockelmann W A, Sanchez-Benitez J and Harrison A 2008 *Phys. Rev. Lett.* **100** 157205
[25] Kim J H, Ji S, Lee S H, Lake B, Yildirim T, Nojiri H, Kikuchi H, Habicht K, Qiu Y and Kiefer K 2008 *Phys. Rev. Lett.* **101** 107201
[26] Freedman D E, Han T H, Prodi A, Mller P, Huang Q-Z, Chen Y-S, Webb S M, Lee Y S, McQueen T M and Nocera D G 2010 *J. A. Chem. Soc.* **132** 16185
[27] Helton J S, Matan K, Shores M P, Nytko E A, Bartlett B M, Qiu Y, Nocera D G, and Lee Y S 2010 *Phys. Rev. Lett.* **104** 147201
[28] Miyahara S, Fouet J -B, Manmana S R, Noack R M, Mayaffre H, Sheikin I, Berthier C, and Mila F 2007 *Phys. Rev. B* **75** 184402
[29] Olariu A, Mendels P, Bert F, Duc F F, Trombe J C, de Vries M and Harrison A 2008 *Phys. Rev. Lett.* 2008 **100** 087202
[30] Imai T, Nytko E A, Bartlett B M, Shores M P and Nocera D G 2008 *Phys. Rev. Lett.* **100** 077203
[31] Bert F, Olariu A, Zorko A, Mendels P, Trombe J C, Duc F, de Vries M A, Harrisson A, Hillier A D, Lord J, Amato A and Baines C 2009 *J. of Physics* **145** 012004
[32] Ofer O, Keren A, Nytko E A, Shores M P, Bartlett B M, Nocera D G, Baines C and Amato A 2009 *Phys. Rev.B.* **79** 134424
[33] Singh R R P and Huse D A 2007 *Phys. Rev. B* **76** 180407; ibid 2008 **77** 144415 2008
[34] Yan S, Huse D A and White S R 2010 arXiv:1011.611
[35] Ran Y, Hermele M Lee P A and Wen X G 2007 *Phys. Rev. Lett.* **98** 117205; ibid, *Phys. Rev. B* **77** 224413
[36] Ran Y, Ko W-H, Lee P A and Wen X G 2009 *Phys. Rev. Lett.* **102** 047205
[37] Singh R R P 2010 *Phys. Rev. Lett.* **104** 177203
[38] Lecheminant P, Bernu B, Lhuillier C, Pierre L and Sindzingre P 1997 *Phys. Rev. B* **56** 2521
[39] Waldtmann C, Everts H U, Bernu B, Lhuillier C, Sindzingre P, Lecheminant P and Pierre L, *Eur. Phys. J.* **2** 501
[40] Sindzingre P and Lhuillier C, *Eur. Phys. Lett.* 2009 **88** 27009

[41] Dommange S, Mambrini M, Normand B and Mila F 2003 *Phys. Rev.B* **68** 224416
[42] Rousochatzakis I, Manmana S R, Laeuchli A M, Normand B and Mila F 2009 *Phys. Rev. B* **79** 214415

Frustrated S=3/2 honeycomb antiferromagnet $Bi_3Mn_4O_{12}(NO_3)$

M Azuma,*[1] M Matsuda,[2] N Onishi,[1] S Olga,[1] Y Kusano,[3] M Tokunaga,[4] Y Shimakawa[1] and N Kumada[5]
[1]Institute for Chemical Research, Kyoto University, Uji 611-0011, Japan
[2]Oak Ridge National Laboratory, Oak Ridge, Tennessee 37831, USA
[3]Department of Applied Arts and Design, Kurashiki University of Science and the Arts, 2640 Nishinoura, Tsurajima-cho, Kurashiki-shi, Okayama 712-8505, Japan
[4]Institute for Solid State Physics, University of Tokyo, Kashiwa, Chiba 277-8581, Japan
[5]Department of Research Interdisciplinary Graduate School of Medicine and Engineering, University of Yamanashi, Miyamae-Cho 7, Kofu, 400-8511 Japan

E-mail: mazuma@msl.titech.ac.jp

Abstract. Magnetic properties of a S=3/2 honeycomb antiferromagnet $Bi_3Mn_4O_{12}(NO_3)$ are discussed. Susceptibility and specific heat measurements showed that this system is free from a long range ordering at least down to 0.4 K in a low magnetic field because of the geometrical frustration caused by the competition between the nearest and the next nearest interactions. On the other hand, magnetization and neutron diffraction studies revealed a field-induced long range ordering at around 10 K and 6 T.

1. Introduction

The magnetic properties of a solid reflect the arrangement of the magnetic ions in its crystal structure. Low-dimensional antiferromagnets exhibit a variety of ground states depending on the spin number and the spin configuration [1, 2]. The discovery of new S = 1/2 kagome antiferromagnet without lattice distortion [3] and the confirmation of the absence of long-range ordering (LRO) in the S = 1/2 triangle antiferromagnet $NiGa_2S_4$ [4] have drawn interest to the frustrated two-dimensional magnets. Honeycomb lattice can be regarded as a triangular derivative as well as Kagome lattice and is obtained by removing 1/3 of spins from a triangular lattice. Since an ideal antiferromagnetic (AF) honeycomb system with only the AF nearest-neighbor interaction (J_1) is free from frustration, the ground state is long-range ordering. However the system is frustrated if the next-nearest interaction (J_2) is also AF as shown in Fig. 3(b). According to theoretical calculations, honeycomb antiferromagnet does not show long range ordering when $J_2/J_1 > 1/6$ (for classical spin) or > 0.15 (for S = 3/2) [5,6].
Hydrothermal synthesis can produce crystal structures with cation environments and oxidation states unusual for high-temperature solid-state reaction conditions, and three new allotropic forms of bismuth oxide were prepared under hydrothermal conditions [7–9] because this synthesis method stabilizes the distorted crystal structures due to the $6s^2$ lone pairs of Bi^{3+} and the covalent Bi-O bonds. A further advantage of hydrothermal reaction is its ability to produce compounds otherwise unstable at high temperatures, such as those containing water or other gas species. Bismuth oxynitrates, such as $BiMO_2(NO_3)$ (M = Sr, Ba), have flat NO_3^- groups parallel to the c axis but with bond angles different from 120° [11] . Since NO^{3-} groups separate the M-O networks in these compounds, bismuth oxynitrates with magnetic ions are expected to show two-dimensional magnetism. Planar and regular

* To whom any correspondence should be addressed. Present address: Materials and Structures Laboratory, Tokyo Institute of Technology, 4259 Nagatsuta, Midori-ku, Yokohama 226-8053, Japan

Published under licence by IOP Publishing Ltd

NO_3^- groups with 120° O-N-O angles are also expected to act as a template for forming magnetic ions into triangular and triangle-derived lattices where magnetic frustration takes place [2].
In this paper we report the first example of bismuth-manganese oxynitrate $Bi_3Mn_4O_{12}(NO_3)$ prepared by the hydrothermal method [10]. The edge-sharing MnO_6 octahedra form honeycomb lattices that are well separated by Bi and NO^{3-} layers. Despite the relatively large Weiss constant of −252 K, no long-range antiferromagnetic (AF) ordering was found at temperatures down to 0.4 K because of the magnetic frustration due to the presence of next-nearest AF interaction. More interestingly, a peculiar magnetic field induced long-range ordering was found by a neutron diffraction study.

2. Experimental

The sample powder was prepared from a mixture of $Mn(NO_3)_2 \cdot 6H_2O$, $NaBiO_3$ by hydrothermal reaction at 543 K for 7 days in 70 mL Teflon-lined autoclaves. SEM pictures were taken using a JEOL JSM-6500F, and TEM and ED pictures were taken with a Topcon EM-002B. Neutron powder diffraction (NPD) measurement for structural determination was made, using a high-resolution powder diffractometer (HRPD) at the JRR-3 reactor of the JAEA site in Tokai, at a constant wavelength (λ = 1.82307 Å) in a vanadium container at room temperature. NPD in magnetic field was measured with triple-axis spectrometers TAS-2 with the fixed incident neutron energy of 14.7 meV and the energy resolutions (ΔE) of 1.7 Magnetic susceptibility was measured with a SQUID magnetometer (Quantum Design MPMS XL) on heating after zero field cooling and on cooling in a magnetic field of 1000 Oe between 2 and 400 K. The specific heat of a solidified pellet was measured at temperatures between 2 and 300 K by a relaxation method with a Quantum Design PPMS. The magnetizaion data were collected by extraction method with PPMS.

3. Results and Discussion

3.1. Structure refinement

The hydrothermal reaction yielded a fine black powder consisting of particles no more than about one micron across. A hexagonal platelet shape suggesting hexagonal or rhombohedral symmetry was observed by SEM as shown in Fig. 1 (a). Indexing of the electronic diffraction patterns provided the approximate dimensions of the hexagonal cell: $a \approx 5$ Å and $c \approx 13$ Å (Fig. 1 (b) and (c)). As no systematic absences were observed, the space group had to be chosen from 16 possible groups.

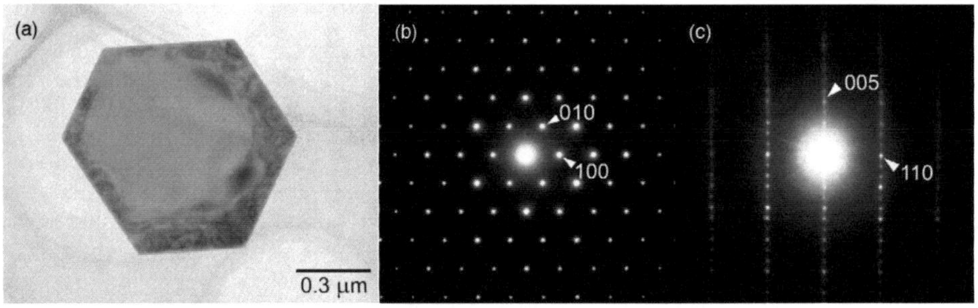

Figure 1. SEM image of the powder obtained by hydrothermal reaction (a). Electron diffraction patterns of hexagonal $Bi_3Mn_4O_{12}(NO_3)$ (b, c).

When analyzing the x-ray diffraction data (not shown) with the lowest hexagonal symmetry group (*P*6), a combination of the direct method of structure solution from powder pattern implemented in EXPO2004 and the direct-space method implemented in FOX allowed us to locate heavy atoms in a structural model similar to $PbSb_2O_6$ (with bismuth in lead sites and with manganese in antimony sites) but having a *c* parameter approximately three times longer. The hexagonal cell includes two $PbSb_2O_6$-like layers separated by a light-atoms layer unsolved from x-ray data. Since the presence of NO_3^- groups in the structure was suggested by FT-IR measurements, similar to the signal seen in FT-IR spectroscopy results for $YBi_2O_4NO_3$ [12],the neutron diffraction data were analyzed on a model with NO_3^- groups between the two $PbSb_2O_6$-like layers. Figure 2 shows the NPD pattern with the result of the Rietveld refinement. Further reduction of the space group to *P*3 was necessary to place the flat NO_3^- group whose presence was indicated by the direct-space analysis of the neutron diffraction data.

The absence of inversion symmetry was verified by means of optical second harmonic generation measurement.

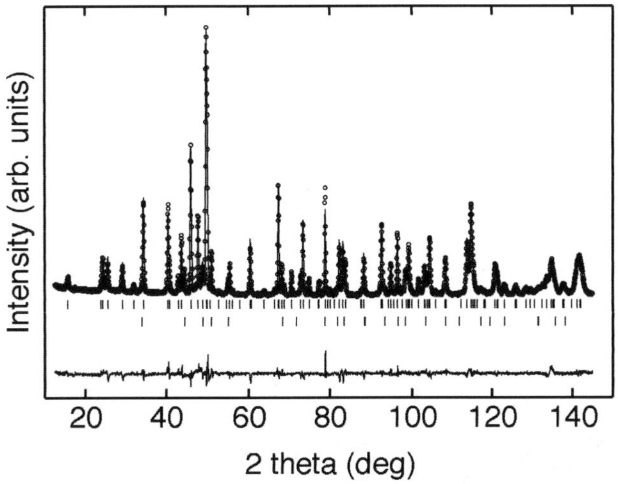

Figure 2. Rietveld plot of neutron data for $Bi_3Mn_4O_{12}(NO_3)$. Observed (circles), calculated (line) and difference (lower line) profiles are shown together with Bragg markers for the $Bi_3Mn_4O_{12}(NO_3)$ phase and MnO_2.

The refined crystal structure is illustrated in Fig. 3. The structural parameters refined from neutron data can be found elsewhere [11]. Since no sign of off-stoichiometry was observed, the occupancy factors were fixed to unity. The refinement showed the presence of 6.7 wt% MnO_2 as a secondary phase. To our knowledge, the novel bismuth manganese oxynitrate represents a new structural type. The bond valence sums calculated from the refined bond lengths listed in Table 1 indicate the following valence state: $Bi^{3+}{}_3Mn^{4+}{}_4O^{2-}{}_{12}(NO_3)^-$. The Mn^{4+} oxidation state is consistent with the formation of MnO_2 as the secondary phase. The manganese atoms have octahedral environment, slightly compressed along the *c* axis of the unit cell. The edge-sharing MnO_6 octahedra form honeycomb lattices as schematically illustrated in the inset of Fig. 3(b). The bismuth atoms, on the other hand, are in distorted environments. Bi1 is also in an octahedron but is shifted from the center. Bi2 and Bi3 have umbrella-like coordinations. They have three short Bi2(3)-O2(3) bonds on one side and six long Bi2(3)-O5(5) bonds on the other side. These environmental distortions might be explained by the space needed for the lone pairs of Bi^{3+}. The distance between $PbSb_2O_6$-like layers and NO_3 layers is quite large: the Bi2-O5 distance is 3.1 Å and the Bi3-O5 distance is 3.2 Å. Since each of these distances is significantly greater than the sum of the ionic radii, the bonding should be rather weak.

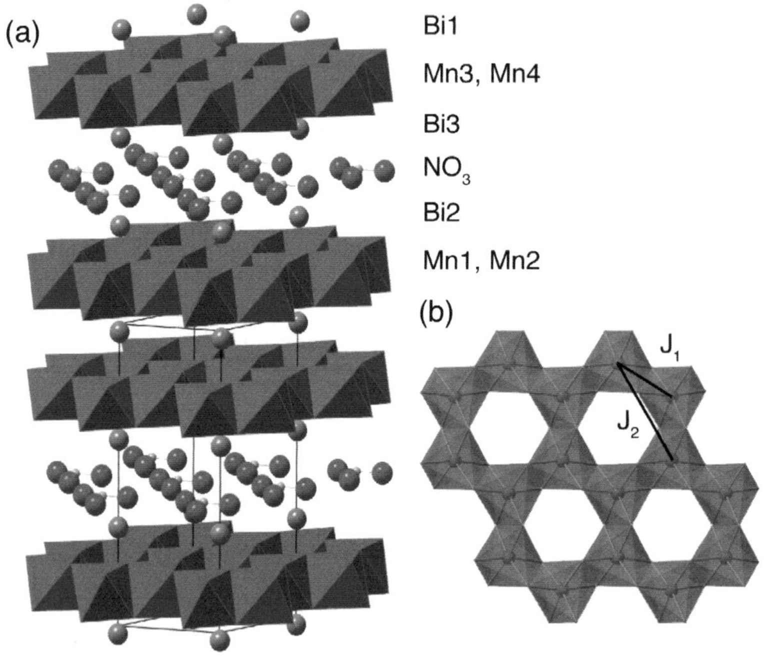

Figure 3. Refined crystal structure of $Bi_3Mn_4O_{12}(NO_3)$ (a) and antiferromagnetic pathways in a honeycomb layer (b).

Table 1. Selected bond distances

Bond	r (Å)	Bond	r (Å)
Bi(1) – O(1) (× 3)	2.29(5)	Mn(1) – O(1) (× 3)	1.82(5)
– O(4) (× 3)	2.41(6)	O(2) (× 3)	2.00(6)
Bi(2) – O(2) (× 3)	2.18(5)	Mn(2) – O(1) (× 3)	1.84(6)
– O(5) (× 3)	3.06(5)	O(2) (× 3)	1.97(6)
– O(5) (× 3)	3.08(5)	Mn(3) – O(3) (× 3)	1.89(5)
Bi(3) – O(3) (× 3)	2.12(5)	O(4) (× 3)	1.86(5)
– O(5) (× 3)	3.22(6)	Mn(4) – O(3) (× 3)	1.86(5)
– O(5) (× 3)	3.23(6)	O(4) (× 3)	1.89(6)
N – O(5) (× 3)	1.23(4)	Mn(1) – O(1) (× 3)	1.82(5)

3.2. Magnetic measurements in a low field

Because the $Bi_3Mn_4O_{12}(NO_3)$ pellet solidified at 6 GPa was highly resistive, the system can be regarded as a localized spin system. The temperature dependence of the magnetic susceptibility on a sample free from MnO_2 shown in Fig. 4(a) shows a broad maximum centered at about 70 K. This broad maximum is a characteristic feature of a low-dimensional antiferromagnet and is consistent with the two-dimensional crystal structure. Fitting the data between 300 and 400 K to the Curie-Weiss law $\chi = C/(T-\theta) + \chi_0$ where C is the Curie constant, θ is the Weiss constant, and χ_0 is the temperature-independent term gave C = 2.07, θ = −252 K, and χ_0 = -8.73×10^{-4} emu/mol. This relatively large negative χ_0 should be attributed to the presence of heavy element, Bi. The Curie constant is as expected for a Mn^{4+} (S = 3/2) system and is thus consistent with the 4+ oxidation state determined for Mn in the structure analysis. Despite the large AF Weiss constant, no sign of LRO was observed. The deviation of ZFC and FC data are not an indication of the LRO because no corresponding anomaly is evident in the specific heat data (Fig. 4(b)). It should be attributed to the spin glass freezing as confirmed by μSR measurement [13]. Figure 4(b) shows the temperature dependence of the specific heat divided by the temperature (C_P/T). Since there is no nonmagnetic reference, the lattice contribution is not subtracted. The C_P/T data show a broad maximum at around 40 K and decrease to zero with decreasing temperature. No other peak is found between 0.4 and 300 K, confirming the absence of the LRO.

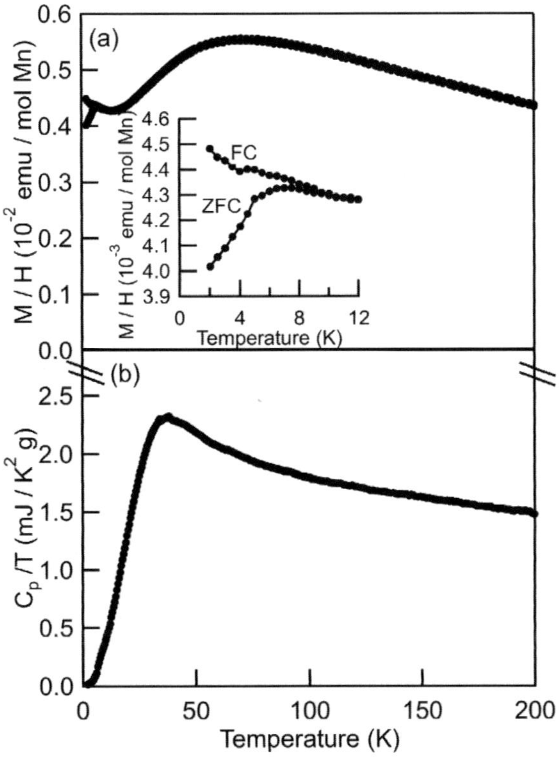

Figure 4. (a) Temperature dependence of the magnetic susceptibility of $Bi_3Mn_4O_{12}(NO_3)$ in zero field cooling and field cooling conditions. Inset shows the magnified view between 2K and 12 K. (b) Total specific heat divided by temperature of $Bi_3Mn_4O_{12}(NO_3)$ solidified pellet.

$Bi_3Mn_4O_{12}(NO_3)$ showed a two-dimensional magnetism consistent with the determined crystal structure. Because of its threefold symmetry, the honeycomb lattice of Mn^{4+} (S = 3/2) is uniform, without any distortion. It should also be pointed out that the orbital ordering as proposed for a S = 1/2 Kagome compound $Cu_3V_2O_7(OH)_2 \cdot 2H_2O$ [14] doesn't take place in this compound because of the d^3 nature of Mn^{4+} ion. Despite the AF Weiss temperature of −252 K, no LRO was observed down to 0.4 K. The ground state of an ideal AF honeycomb lattice system with only the AF nearest-neighbor interaction is LRO because the system is free from frustration. If next-nearest interaction is also AF, however, the system is frustrated [5,6]. In the present compound the nearest AF interaction mediated by the orthogonal Mn-O-Mn bond is rather weak. Considering the Weiss temperature of −252 K and the number of the nearest-neighbor spins, 3, one would expect the exchange interaction to be smaller than 100 K. The next-nearest interaction mediated by the Mn-O-O-Mn bond is therefore not negligible. The current compound is the first example of frustrated honeycomb antiferromagnet without long-range ordering. The ground states of such systems with classical [5] and quantum [6] spins had been investigated theoretically, but no actual compound had been known. It was predicted that the S = 3/2 honeycomb system with $J_2 / J_1 > 0.16$ would not exhibit a long-range ordering [6]. Further quantitative analyses of the susceptibility and specific heat data are desirable.

3.3. High-magnetic field behaviour

Although $Bi_3Mn_4O_{12}(NO_3)$ has no long-range ordering, magnetization shows a small jump of ~0.1μ_B/f.u. around 6-13 T below 20 K, as shown in Fig. 5(a), indicating a transition to a metamagnetic phase. In order to clarify the nature of the magnetic field-induced phase, we performed neutron scattering experiments under applied magnetic fields [15].

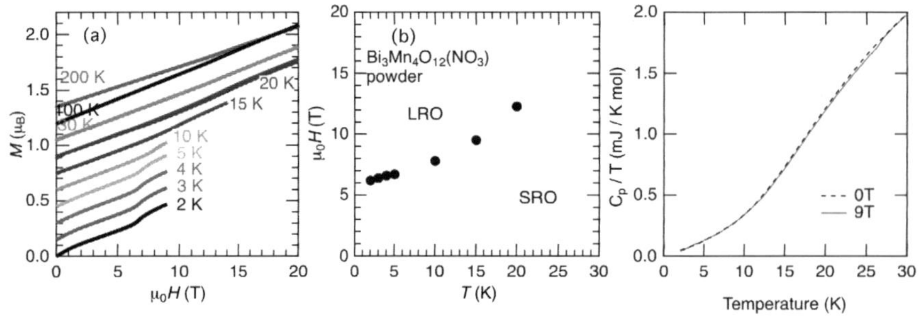

Figure 5. (a) Magnetization versus magnetic field in dc (T <10 K) and pulsed fields (T >15 K) for $Bi_3Mn_4O_{12}(NO_3)$ in a temperature range between 2 and 200 K. (b) Temperature-magnetic field phase diagram. LRO and SRO stand for the long-range and short-range ordered phases, respectively. (c) Total specific heat divided by temperature measured in 0 and 9 T.

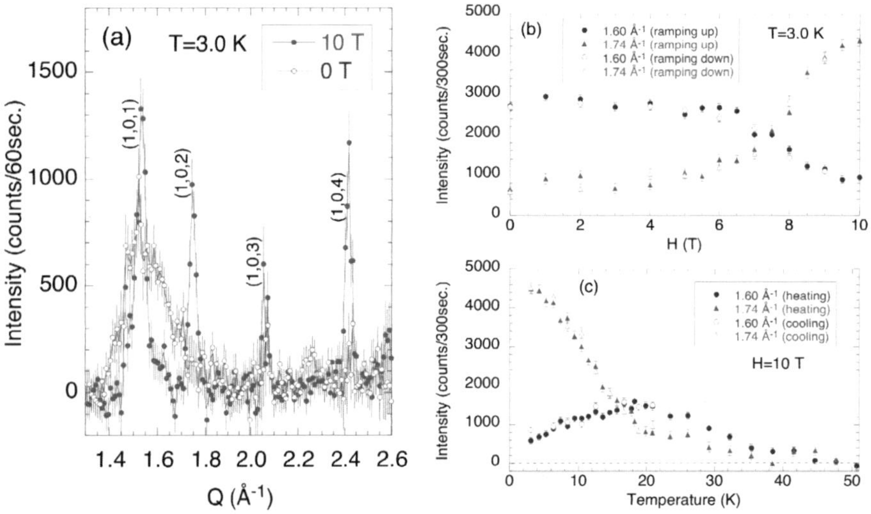

Figure 6. (a) Neutron powder diffraction patterns in $Bi_3Mn_4O_{12}(NO_3)$ with ΔE ~1.7 meV at H=0 and 10 T and at T=3 K. At 10 T the broad magnetic peak is reduced and sharp magnetic peaks appear. The magnetic field (b) and temperature dependences (c) of the elastic magnetic signal at Q=1.60 and 1.74 $Å^{-1}$. The signals at 1.60 and 1.74 $Å^{-1}$ mostly originate from the short-range order and the 3D long-range order, respectively. In all the data shown here, the background signal was subtracted.

Figure 6(a) shows the neutron powder diffraction patterns in $Bi_3Mn_4O_{12}(NO_3)$ at H=0 and 10 T. The spectrum at 0 T shows a broad peak originating from the short-range antiferromagnetic order as suggested by susceptibility and specific heat measurements. With application of magnetic field at T=3 K, the broad magnetic intensity is reduced and resolution-limited sharp magnetic Bragg peaks appear above H ~6T, as shown in Fig. 6(b). This result indicates that long-range magnetic order is induced in magnetic field. Figure 6(c) shows the temperature dependence of the elastic magnetic signals at Q=1.60 and 1.74 $Å^{-1}$ at H=10 T. The signal originating from the long-range order vanishes and that from the short-range order recovers around 18 K. Since the most intense peak is the (1, 0, 1) reflection, the nearest-neighbor spins align antiparallel both in the *ab* plane and along the *c* axis. Therefore, there is no frustration in the magnetic structure. Since the width of the magnetic field-induced sharp peak is almost the same as the instrumental resolution, it is evident that the magnetic structure is long-ranged and commensurate. From the results of the magnetic structural analysis, we determined that the spins lie in the ab plane and that the ordered moment is ~1.8 μ_B at 3 K. The moment size is about 2/3 of the full moment 3 μ_B for the Mn^{4+} ion. As shown in Fig. 6(a), there still exists a broad signal around 1.6 $Å^{-1}$ at 10 T. This is the remnant of the broad peak observed in the zero magnetic field. This suggests that the field-induced transition depend on magnetic field direction and that only the magnetic moments perpendicular to the magnetic field contribute to the ordering and the others remain disordered. It is puzzling why the transition depends on the field direction, although the compound is an isotropic Heisenberg system. The magnetic field-temperature phase diagram is summarized in Fig. 5(b). We observed the magnetic field-induced phase transition, in which the short-range order expands into the long-range Néel order by applying magnetic field. This indicates that this compound is located near the boundary of the ordered and disordered phases with J_2/J_1 close to 0.15. Accordingly the specific heat measured in a magnetic field of 9T shown in Fig. 5 (c) doesn't show any peak. The entropy change at the ordering is undetectably small. The transition temperature increases with increasing magnetic field in this compound as shown in Fig. 5(b). This indicates that the collinear Néel state becomes more stabilized at higher temperatures. This may suggest that thermal fluctuations stabilize the Néel order, where "order by disorder" mechanism is at work [16].

References

[1] Johnston D C 1997 *Handbook of Magnetic Materials*, vol. 10; Elsevier: Amsterdam, 1
[2] Ramirez A P 1994 *Annu. Rev. Mater. Sci.* **24** 453
[3] Shores M P, Nytko E A, Bartlett B M and Nocera D G 2005 *J. Am. Chem. Soc.* **127** 13462
[4] Nakatsuji S, Nambu Y, Tonomura H, Sakai O, Jonas S, Broholm C, Tsunetsugu H, Qiu and Y, Maeno Y, 2005 *Science* **309** 1697
[5] Katsura S, Ide T and Morita Y 1986 *J. Stat. Phys.* **42**, 381
[6] Takano K 2006 *Phys. Rev. B* **74**, 140402(R)
[7] Kumada N, Kinomura N, Woodward P M and Sleight A W 1995 *J. Solid State Chem.* **116** 281
[8] Kinomura N and Kumada N 1995 *Mater. Res. Bull.* **30** 129
[9] Kumada N and Kinomura N 1999 *Mat. Res. Soc. Symp. Proc.* **547** 227
[10] Ziegler P. I, Grigoraviciute I, Gibson K, Glaser J, Kareiva A and Meyer H J 2004 *J. Solid State Chem.* **177** 3610
[11] Olga S, Azuma M, Kumada N, Kusano Y, Matsuda M, Shimakawa Y, Takei T, Yonesaki Y and Kinomura N 2009 J. Am. Chem. Soc. **131**, 8313
[12] Kumada N, Takahashi N, Kinomura N and Sleight A W 1998 *J. Solid State Chem.* **139** 321
[13] Azuma M, Onshi N, Oka K, Shimakawa Y, Hiraishi M, Miyazaki M, Masuda T, Koda A, Kojima K M and R. Kadono, 2010 *KEK-MSL Progress Report 2009*, 26
[14] Okamoto Y, Yoshida H and Hiroi Z 2009 *J. Phys. Soc. Jpn.* **78** 033701.
[15] Matsuda M, Azuma M, Tokunaga M, Shimakawa Y and Kumada N, 2010 *Phys. Rev. Lett.* **105** 187201
[16] Okumura S, Kawamura H, Okubo T and Motome Y 2010 *J. Phys. Soc. Jpn.* **79** 114705

Low temperature phase diagram for the pyrochlore compound $Tb_2Ti_2O_7$

P Bonville[1], I Mirebeau[2], A Gukasov[2], S Petit[2], J Robert[2]

[1] CEA, Centre de Saclay, DSM/IRAMIS/Service de Physique de l'Etat Condensé, 91191 Gif-sur-Yvette, France

[2] CEA, Centre de Saclay, DSM/IRAMIS/Laboratoire Léon Brillouin, 91191 Gif-sur-Yvette, France

E-mail: pierre.bonville@cea.fr

Abstract. The pyrochlore material $Tb_2Ti_2O_7$, although possessing a negative paramagnetic Curie temperature witnessing the presence of antiferromagnetic exchange interactions, shows no long range magnetic order of Tb^{3+} moments down to 0.05 K. The reason for this behaviour is not thoroughly understood up to now. We present here a model based on the presence of a distortion from trigonal symmetry at low temperature, making this material a two-singlet system. Application of a 4-site mean field theory allows us to construct the phase diagram for this system and to understand how, in the presence of a distortion, $Tb_2Ti_2O_7$ can remain in a paramagnetic spin liquid phase down to T=0. Although the existence of a static distortion at low temperature in these materials is not yet convincingly assessed, we believe that our model can give useful hints towards an understanding of this interesting material.

1. Introduction

Frustration in materials crystallizing in the pyrochlore lattice has been a subject of interest since the pioneering work of J.Villain [1], and an intense burst of activity in the field has taken place in the last ten years (for a review, see Ref.[2]). Among the various compounds, the titanates $R_2Ti_2O_7$ (R=rare earth) occupy a special position since many different ground states are observed for the different R^{3+} ions and an understanding of their basic properties is now achieved [2]. The Tb titanate $Tb_2Ti_2O_7$ stands as an exception, since its most salient property, i.e. the absence of long range magnetic ordering of the Tb^{3+} moments down to at least 0.05 K [3], remains unexplained. Numerous tentatives were carried out to explain why this material remains in a spin-liquid state [4]. Its peculiar crystal electric field (CEF) level scheme, consisting of two close non-Kramers doublets separated by about 18 K and well isolated from the rest of the levels [5, 6], has been invoked to play an important role: by taking it properly into account, the spin correlations in the paramagnetic phase could be understood [7] and the formation of a "quantum spin-ice" phase was also put forward [8]. Indeed, the ground doublet is rather anisotropic, which is one of the requirements for spin-ice behaviour [9], but the effective interionic interaction is antiferromagnetic [3], which precludes a description in terms of canonical spin-ice. All these models predict a **k**=0 Néel order at low temperature, indicating that some essential ingredient is lacking to obtain a correct description of the physics of $Tb_2Ti_2O_7$.

2. The singlet ground state systems

The Tb^{3+} ion is a non-Kramers (or Van Vleck) ion, with an even number (8) of electrons in its $4f$ shell. Its ground spin-orbit multiplet has a total momentum J=6 and a Landé factor g_J=3/2. A specific property of non-Kramers ions is that the (2J+1)-fold degeneracy of this multiplet can be entirely lifted by a crystal electric field (CEF) of sufficiently low symmetry. By contrast, for Kramers ions, with an odd number of $4f$ electrons, an electric field cannot lift completely the spin-orbit degeneracy, even in low symmetry environments, and Kramers doublets are the eigenstates of the CEF interaction. Therefore, the low temperature properties of non-Kramers ions like Tb^{3+}, with a doublet ground state, are very sensitive to small perturbations of the CEF interaction (impurities, dislocations, vacancies) which lift the degeneracy of the doublet into two "non-magnetic" singlets, usually in a random way. The term "non-magnetic" refers to the fact that the mean value of the total momentum **J** in these singlets is zero. This picture actually holds for dilute rare earth ions, where no other interaction is at play at low temperature. In concentrated rare earth systems, the interionic exchange (and/or dipole-dipole) interaction tends to establish long range magnetic order, with a "magnetic" ground state, below a critical temperature. In the case of a singlet ground state system, there occurs thus a competition between crystal field and exchange, which has long been recognised and studied [10, 11, 12]. The main result of these early works is that magnetic long range ordering (LRO) in such systems can only settle in if the exchange is larger than a critical value:

$$\mathcal{J}_c = \frac{\Delta}{4\alpha^2}, \tag{1}$$

where Δ is the energy separation between the singlets and α the matrix element of J_z linking the two singlets. A non-Kramers rare earth material where the CEF interaction yields a singlet ground state can thus remain in a paramagnetic state down to very low temperature (no magnetic LRO) if the exchange integral is not large enough. Nevertheless, the exchange interaction induces short range dynamic spin correlations in this phase, hence the name Spin Liquid (SL) which is often given to this state.

3. The low temperature distortion in $Tb_2Ti_2O_7$ and the inelastic neutron spectrum

Recently, high resolution x-ray scattering measurements in $Tb_2Ti_2O_7$ [13] have evidenced anisotropic broadenings of some Bragg peaks below 15 K which were interpreted as due to cubic to tetragonal structural fluctuations, precursors to a T=0 cooperative Jahn-Teller (JT) transition. A cooperative Jahn-Teller transition can occur for non-Kramers ions [14] with degenerate ground state and consists of a symmetry lowering which lifts the degeneracy and lowers the energy of the system. In $Tb_2Ti_2O_7$, although a true JT transition is not observed down to 0.3 K, the distortion from trigonal symmetry has a sizeable coherence length of about 20 lattice cells at the lowest temperature. Another intriguing experimental result is the presence of a low energy excitation ($\simeq$ 0.2 meV) in the low temperature inelastic neutron spectra in zero field [6] and in finite fields [15, 16]. This line cannot arise from the sole trigonal symmetry CEF interaction appropriate for the rare earth site in the pyrochlore lattice (D_{3d}), since the first excitation has an energy of 1.45 meV in $Tb_2Ti_2O_7$. The presence of a distortion from trigonal symmetry at the rare earth site, that could be coherent over several lattice cells and during a finite time, can then be envisaged as a possible cause for the two above mentioned experimental anomalies at low temperature. In particular, the two lowest energy states would be two singlets separated by a small energy gap. The presence of a structural distortion in $Tb_2Ti_2O_7$ has already been invoked for explaining the "apparent isolation" of Tb tetrahedra from the point of vue of magnetic correlations [17], but another work [18] claims for the absence of a static distortion in $Tb_2Ti_2O_7$ from μSR measurements. We will discuss the relevance of the μSR findings to our work in a later section.

In this work, we shall rely on the x-ray and neutron scattering works mentioned above to explore the consequences of the presence of a small distortion from trigonal symmetry at the Tb site in $Tb_2Ti_2O_7$. We shall assume a distortion with tetragonal symmetry along a cubic axis, which can be considered as static during a finite time, following the suggestion in Ref.[13], stating that "cubic-to-tetragonal structural fluctuations" are present below 15 K in $Tb_2Ti_2O_7$. This may not correspond to the actual symmetry breaking, but we argue that the true symmetry lowering does not play an important role in the problem. Thus, at low temperature, the degeneracy of the ground doublet is lifted and we are now dealing with a two singlet system. In earlier studies [5, 6], the Tb^{3+} ground doublet in $Tb_2Ti_2O_7$ was shown to be close to the state $|J = 6; J_z = \pm 4\rangle$. Assuming here for simplicity that these are the true trigonal CEF wavefunctions, then the eigenstates in the presence of the distortion are the symmetric and antisymmetric wavefunctions:

$$\psi_a = \frac{1}{\sqrt{2}}[|+4\rangle - |-4\rangle] \text{ and } \psi_s = \frac{1}{\sqrt{2}}[|+4\rangle + |-4\rangle]. \quad (2)$$

The energy separation δ between the singlets depends on the particular form of the symmetry breaking distortion. We write the tetragonal CEF perturbation as:

$$\mathcal{H}_{\text{dist}} = D_Q J_Z^2 = \frac{D_Q}{3}[2J_x^2 + J_z^2 + \sqrt{2}(J_x J_z + J_z J_x)], \quad (3)$$

where OZ is the axis of the tetragonal distortion (chosen as the cubic <001> axis). For the local frame, the local anisotropy axis Oz is the cubic <111> axis associated with each Tb site, and Ox is the cubic $< \bar{1}\bar{1}2 >$ axis.

In order to determine the value of D_Q, we proceed to the computation of the powder inelastic neutron scattering spectrum at 1.6 K. This spectrum reveals a sizeable low energy line near 0.05 THz (0.2 meV), on the side of the elastic line, for large Q values (left panel of Fig.1). We attribute this line to the transition between the two singlets arising from the symmetry breaking. The spectrum, calculated using the CEF parameters as in Ref.[6] for the trigonal term, shows

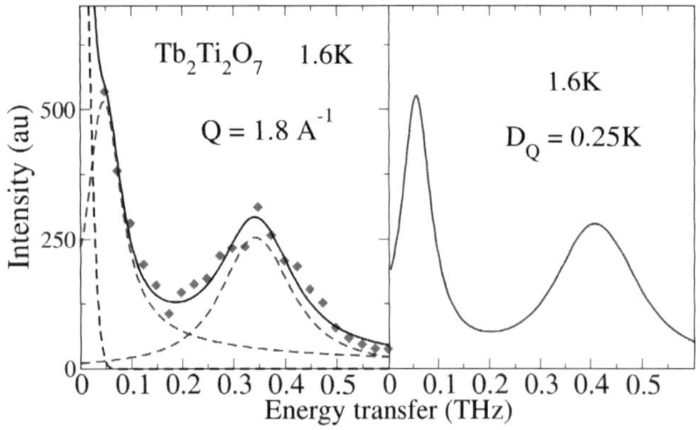

Figure 1. Neutron inelastic spectrum in $Tb_2Ti_2O_7$ at 1.6 K: **Left:** experimental data for Q=1.8 Å^{-1} from Ref.[6]; the dashed lines are empirical fits to Lorentzian shaped peaks; **Right:** spectrum calculated using the trigonal symmetry CEF with a tetragonal distortion D_Q=0.25 K (see text) and Lorentzian shaped lines. The linewidth (HWHM) is 0.025 THz (resp. 0.075 THz) for the smaller (resp. higher) energy line.

that a value D_Q=0.25 K reproduces correctly the position of this low energy transition (right panel of Fig.1). The eigenstates of type (2) yield a sizeable matrix element of J_z between the two singlets, resulting in a non-extinct line since the transition line intensity is proportional to the square of this matrix element. Our single ion mean field calculation cannot integrate the Q-dependence due to exchange, and thus it cannot reproduce the dispersion of the CEF transition energies. The trigonal CEF parameters were chosen so as to yield an agreement with the mean energy value, which explains why the calculation slightly overestimates the energy of the line near 0.35 THz (1.45 meV). Since the characteristic time for these inelastic neutron data is about 10^{-11} s, the lifetime of the distortion must be longer than 10^{-11} s.

4. The phase diagrams in case of isotropic exchange

We proceed by computing the phase diagram for antiferromagnetic exchange, with the CEF scheme of $Tb_2Ti_2O_7$, obtained when varying the exchange integral, or rather the molecular field constant λ (negative), which is related to the usual exchange constant $\mathcal{J}$ between true spins by $\mathcal{J} = (\frac{g_J}{g_J-1})^2\ \lambda\ \mu_B{}^2$. For this purpose, we obtain the magnetic configuration on one tetrahedron of the pyrochlore lattice by performing a self-consistent mean field calculation involving the 4 Tb ions of the tetrahedron. We take into account the short range exchange interaction with the 6 nearest neighbours of a given Tb ion. This exchange coupling can be chosen isotropic (one λ value) or anisotropic (a three-component $\tilde{\lambda}$ tensor). We have also included the infinite range dipolar interaction through Ewald summations [19]. Since we consider only one tetrahedron, our calculation can simulate only long range ordered magnetic structures *with a propagation vector* $\mathbf{k} = 0$. In the search for the ground configuration, we also impose the constraint that all 4 sites must be magnetically equivalent. We will compare the phase diagram obtained first in the absence of tetragonal distortion, then including the symmetry breaking. In the presence of the tetragonal distortion, one should observe a two singlet-like behaviour akin to that described in Refs.[10, 11, 12], i.e. there should be a range of λ values where the system remains a spin liquid state down to very low temperature.

In the absence of distortion, the obtained phase diagram is shown in Fig.2. As $|\lambda|$ decreases, the ground state is first an antiferromagnetic (AF) phase ("all spins in or all spins out"), then, above a critical value $\lambda_c = -0.10\,\mathrm{T}/\mu_B$, an Ordered Spin Ice (OSI) phase ("two spins in and two spins out"), which differs from the standard Spin Ice by its long range order character. In the region with an AF ground state, there is a limited range: $-0.175 \leq \lambda \leq -0.10\,\mathrm{T}/\mu_B$ where the system undergoes another transition to an OSI phase. For positive λ values (ferromagnetic exchange), an OSI phase is also found. In the OSI phase, the Tb^{3+} moment lies in the $(1\bar{1}0)$ plane, i.e. between the cubic [001] and [110] axes. Thus, for all λ values, the system shows magnetic Long Range Ordering, as expected since the ground Tb^{3+} CEF state is a doublet. This phase diagram can be compared with that of Ref.[20], where the moments are described by

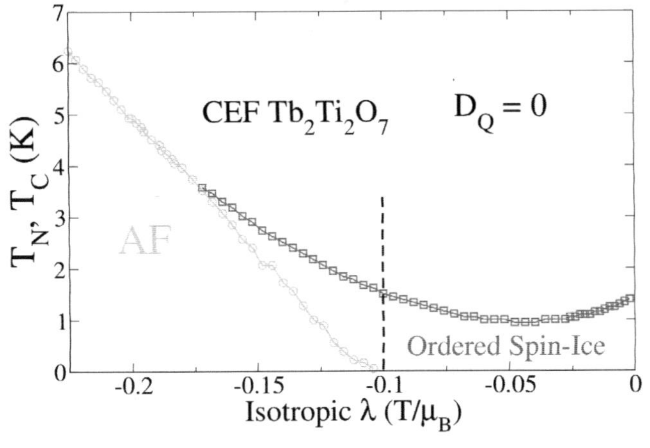

Figure 2. (T,λ) phase diagram for the trigonal CEF level scheme (no distortion) of $Tb_2Ti_2O_7$ as a function of the isotropic molecular field constant λ. The green (resp. red) line and points are the transition line to an AF (resp. OSI) phase. In the range $-0.175\,\mathrm{T}/\mu_B < \lambda < \lambda_c = -0.10\,\mathrm{T}/\mu_B$, two transitions occur as temperature decreases.

Ising spins in the infinite anisotropy limit. In this latter work, in terms of reduced exchange J_{nn} and dipolar D_{nn} constants, the critical value for crossing from AF to Spin Ice (non-ordered) is found to be $J_{nn}/D_{nn} \simeq -0.9$. In the present work, we have a fixed value of D_{nn}= 0.85 K using a moment value of 6 μ_B, and the equivalent critical value of J_{nn}/D_{nn} for crossing from AF to OSI

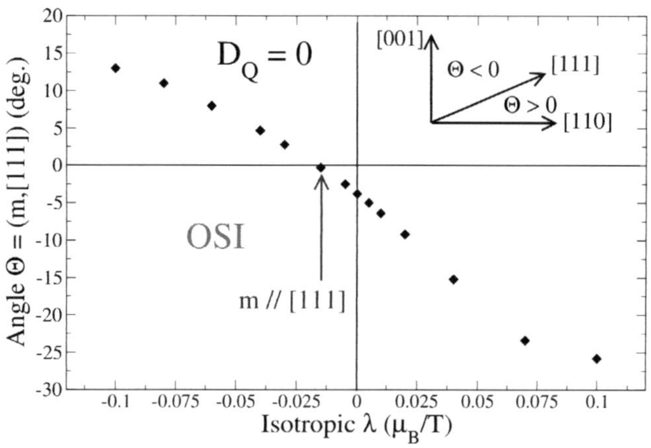

Figure 3. Variation of the angle Θ between the Tb^{3+} moment and the [111] axis in the Ordered Spin Ice phase, using the CEF level scheme of $Tb_2Ti_2O_7$ with no distortion from trigonal symmetry (D_Q=0), for both AF and ferromagnetic exchange. The crossover from $\Theta > 0$ to $\Theta < 0$ occurs for a non-zero λ value $\lambda_{cr} = -0.015\,\mathrm{T}/\mu_\mathrm{B}$.

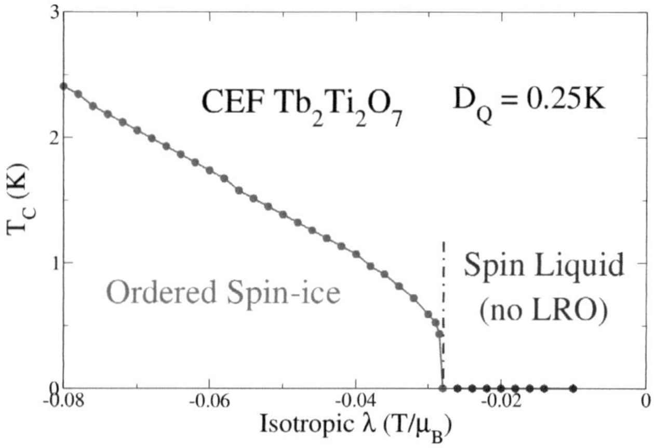

Figure 4. (T,λ) phase diagram for the trigonal CEF level scheme of $Tb_2Ti_2O_7$ with a tetragonal distortion of type (3), as a function of the isotropic molecular field constant λ. The red (resp. blue) line and points are the transition line to an OSI (resp. Spin Liquid or "no LRO") phase. The critical value is $\lambda_{SL} = -0.028\,\mathrm{T}/\mu_\mathrm{B}$.

is -2.5. We believe the differences between the phase diagrams obtained with the two models stem from the neglection of the finite CEF anisotropy in Ref.[20].

An interesting feature of the OSI phase is that the Tb moment direction is "tilted" with respect to the easy <111> axis. On Figure 3 is represented the variation of the angle Θ=(**m**, [111]) as a function of the molecular field constant, in the OSI phase. Our convention is that Θ is taken negative if the Tb^{3+} moment lies between [001] and [111], and positive for **m** lying between [111] and [110]. For the standard (non ordered) Spin Ice, $\Theta = 0$. One observes a continuous decrease of Θ from a position between the [110] and [111] axes for AF exchange, towards the [001] axis for positive λ values. The configuration with **m** // [111] ($\Theta = 0$) is obtained for a small non-zero value $\lambda_{cr} = -0.015\,\mathrm{T}/\mu_\mathrm{B}$. These findings, for $\lambda > 0$, are in agreement with those of Ref.[21], where the competition between exchange and CEF anisotropy was investigated in a pyrochlore lattice, with ferromagnetic interionic exchange. For finite anisotropy, an OSI phase was obtained with $\Theta < 0$ ranging from -54.7° (**m** // [001]) for infinitely large λ, to 0 for infinitely large anisotropy (Spin Ice). In the present work, the anisotropy has a fixed value determined by the CEF interaction and the $\Theta = 0$ limit cannot be reached for ferromagnetic exchange. It is reached for small AF exchange due to the continuous reorientation of **m** towards [110] as λ decreases.

We proceed now to the phase diagram including the tetragonal distortion (3), with D_Q=0.25 K, shown in Fig.4. The main difference with the zero distortion phase diagram is the appearance of a new critical value $\lambda_{SL} = -0.028\,\mathrm{T}/\mu_\mathrm{B}$ above which the spontaneous Tb^{3+} moment is zero, at least down to 0.05 K which is the lowest temperature of our calculation. This

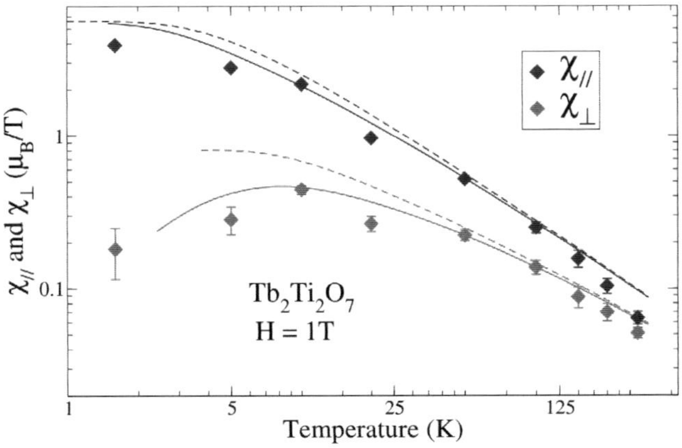

Figure 5. Thermal variation of the local susceptibilities $\chi_{//}(T)$ and $\chi_{\perp}(T)$ from Ref.[24] measured with a field of 1 T and calculated curves. The dashed lines correspond to an isotropic exchange value $\lambda = -0.025\,\mathrm{T}/\mu_\mathrm{B}$, and the solid lines to an anisotropic exchange tensor $\lambda_a = -0.04$, $\lambda_b = -0.09$ and $\lambda_c = -0.07\,\mathrm{T}/\mu_\mathrm{B}$.

phase has therefore a Spin Liquid (SL), or "no LRO", ground state and corresponds to the case studied in Ref.[10] where exchange is not strong enough so as to induce a long range ordered magnetic state. Due to the complexity of the pyrochlore lattice, with 4 magnetic sublattices, and to that inherent to the dipolar interaction, an approximate analytical expression for the critical exchange value could not be obtained, similar to expression (1) derived in Ref.[10], but work is in progress in that direction. The presence of a symmetry lowering at the Tb site in $Tb_2Ti_2O_7$, making this material a two singlet system at low temperature, could therefore stand at the origin for the lack of magnetic LRO down to very low temperature. This holds provided the experimental value of the molecular field constant corresponds to the SL phase; an estimation for λ in $Tb_2Ti_2O_7$, from the Curie-Weiss behaviour of the powder magnetic susceptibility, has been given in Ref.[6]: $\lambda \simeq -0.05\,\mathrm{T}/\mu_\mathrm{B}$[23]. This value lies not far from, but outside the range giving rise to the SL phase, as seen in Fig.4. This phase diagram could be relevant for $Tb_2Sn_2O_7$ where an OSI phase is observed below 0.9 K and where $\Theta \simeq 13°$ [22]. Work is currently in progress in this direction.

5. Anisotropic exchange

Up to now, we have considered that exchange is isotropic in $Tb_2Ti_2O_7$, but recent investigations [24, 25] have put forward the presence of anisotropic exchange in the pyrochlore titanates. Anisotropic exchange has been known to be present in rare earth insulating compounds since the first developments of rare earth studies [26, 27, 28]. We will use here a simple model where anisotropy is introduced through the principal values λ_a, λ_b and λ_c of the molecular field tensor $\tilde{\lambda}$. We assume that the $\tilde{\lambda}$ tensor is diagonal in the {**a**,**b**,**c**} frame where the **c**-axis is along the bond joining two rare earth ions. The local susceptibility data in single crystals of the rare earth titanates, obtained from neutron diffraction experiments, were analysed in Ref.[24] in terms of a single site self-consistent calculation, and the dipolar interaction was not included. We give here a new analysis in terms of the present model: the experimental data and the calculated curves are shown in Fig.5. The dashed curves in this figure were calculated with an isotropic molecular field constant $\lambda = -0.025\,\mathrm{T}/\mu_\mathrm{B}$ corresponding to the largest $|\lambda|$ value in the SL phase of Fig.4. The agreement with experiment is not very good at low temperature, especially for $\chi_{\perp}(T)$. By introducing exchange anisotropy, the agreement with experiment is improved; the susceptibility data however is not very sensitive to the λ_i values. A good match to the data is obtained with λ_a in the range -0.06 - $-0.04\,\mathrm{T}/\mu_\mathrm{B}$, λ_b in the range -0.12 - $-0.08\,\mathrm{T}/\mu_\mathrm{B}$ and $\lambda_c \simeq -0.07\,\mathrm{T}/\mu_\mathrm{B}$. We note that the small distortion from trigonal symmetry is not relevant for these susceptibility curves since its effect is sizeable only below 1-2 K, as will be shown in the next section.

Next, the phase diagram at T=0 as a function of the components of the $\tilde{\lambda}$ tensor has been

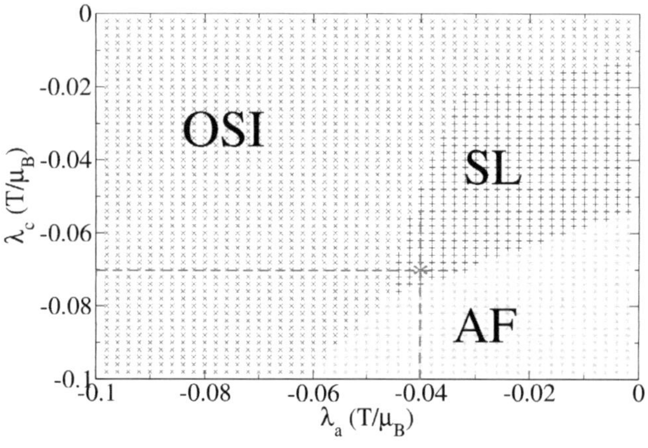

Figure 6. (λ_c, λ_a) phase diagram at T=0 for $\lambda_b = -0.09\,\mathrm{T}/\mu_\mathrm{B}$, for the trigonal CEF level scheme of $Tb_2Ti_2O_7$ with a tetragonal distortion of type (3) and anisotropic exchange. The pink star marks the point in the λ-space ($\lambda_a = -0.04$, $\lambda_b = -0.09$ and $\lambda_c = -0.07\,\mathrm{T}/\mu_\mathrm{B}$) which could represent the position of $Tb_2Ti_2O_7$.

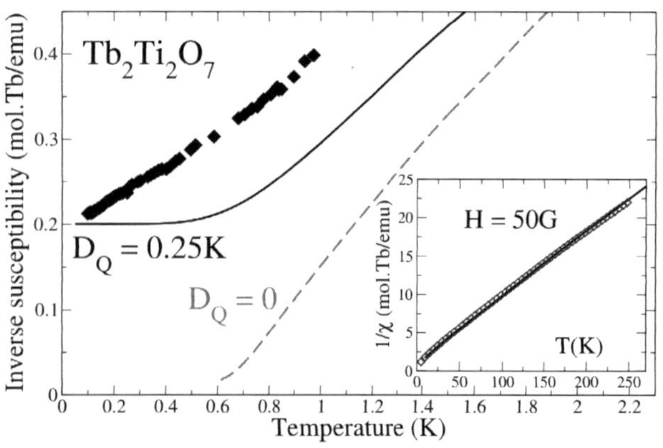

Figure 7. Low temperature part of the thermal variation of the powder inverse susceptibility in $Tb_2Ti_2O_7$. The data are from Ref.[30], the solid line is the calculation using the present model with a tetragonal distortion and an anisotropic molecular field tensor, and the dashed red line is the calculation without tetragonal distortion. Insert: high temperature part of the inverse powder susceptibility. The solid line is a calculation with the anisotropic $\tilde{\lambda}$ tensor derived above (see text).

computed. We find that, for not too small values of the λ_i, it does not depend strongly on λ_b. We therefore show in Fig.6 a cut of the 3-dimensional phase diagram at the value $\lambda_b = -0.09\,\mathrm{T}/\mu_\mathrm{B}$. For large $|\lambda_a|$ values, an OSI phase (in red) is obtained, and for smaller $|\lambda_a|$ values, a SL phase (in blue) is obtained for small λ_c and an AF phase (in green) for large λ_c values. The SL region of the phase diagram is relatively large, and one can mark a point in this region (pink star in Fig.6) whose coordinates are $\lambda_a = -0.04$, $\lambda_b = -0.09$ and $\lambda_c = -0.07\,\mathrm{T}/\mu_\mathrm{B}$ and which could represent the molecular field tensor in $Tb_2Ti_2O_7$. Indeed, calculations of the low temperature properties using this particular tensor yield good agreement: i) with the local susceptibility data as discussed above, ii) with the field evolution of the magnetic structure as discussed in Ref.[29], especially with the observation of a spin-melting of some Tb moments at a critical field of about 1.5 T, and they yield a qualitative agreement with the very low temperature susceptibility data as discussed below.

6. The low temperature susceptibility

For magnetic systems with a singlet ground state, separated from the excited states by an energy Δ, the magnetic susceptibility is expected to show a so-called Van-Vleck plateau for temperatures $T < \Delta/k_\mathrm{B}$ (in case of a Spin Liquid ground state). The static magnetic susceptibility $\chi(T)$ has been measured in a powder $Tb_2Ti_2O_7$ sample, down to very low temperature (0.01 K) in Ref.[30]. These data are reported (as $1/\chi$) in Fig.7, together with the calculations using the model developed in the present work: the black solid line represents the curve obtained with a tetragonal distortion ($D_Q = 0.25\,\mathrm{K}$) from the trigonal local symmetry, the red dashed curve without distortion ($D_Q = 0$). The expected Van-Vleck plateau for $D_Q = 0.25\,\mathrm{K}$ is clearly not observed in the data, which follow a kind of Curie-Weiss law down to 0.1 K, but the experimental value of $1/\chi(0.1\,\mathrm{K})$ matches well with the calculated plateau value. The curve obtained without distortion (red dashed line) is rather far from the data points. The high temperature Curie-Weiss law shown in the insert of Fig.7 is well reproduced with the anisotropic molecular field tensor determined above. We have so far no explanation for the discrepancy between the low temperature data and our calculation. It is worth noticing that, below 0.1 K, irreversibilities are observed in Ref.[30] with respect to FC and ZFC measurements procedures, showing that the ground state could be more intricate than described here. The spins I=3/2 of the ^{159}Tb nucleus could also play a role, since the hyperfine interaction is evidenced in specific heat measurements below 0.4 K [31]. The problem of the very low temperature state of the Tb^{3+} ion should then be determined by considering a hamiltonian with two terms of comparable magnitude: the distortion from trigonal symmetry and the hyperfine interaction, which is present on each site since ^{159}Tb is monoisotopic.

The μSR measurements presented in Ref.[18] conclude to the "lack of static lattice distortion in $Tb_2Ti_2O_7$", which seems to contradict our statements, although we believe the distortion is probably dynamic. Actually, the μSR data show that the width of the dipolar field distribution experienced by the μ^+ is approximately independent of temperature. This means that no extra width is observable that would be caused by a distribution of distortions developing at low temperature. But this is fully compatible with our picture of a single distortion arising below 15 K, probably during a finite time and over a finite distance. Another conclusion of this μSR study is that the μ^+ Knight-shift, which is proportional to the Tb^{3+} magnetic susceptibility, saturates below about 1 K. This is in agreement with our prediction of a Van-Vleck like behaviour for $\chi(T)$ below about 1 K (see Fig.7). The time window of the μSR technique being about 1 μs, this observation means that the lifetime of the distortion is longer than 1 μs.

7. Conclusion and perspectives

This work introduces a spontaneous crystal field distortion, due to an incipient Jahn-Teller transition, as a possible mechanism to account for the spin liquid ground state of $Tb_2Ti_2O_7$, unexplained so far. Our current studies suggest that this mechanism is also at play in the parent $Tb_2Sn_2O_7$ Ordered Spin Ice, where it could account for specific aspects of the magnetic structure and spin excitations. The spin lattice coupling, which plays a key role to relieve magnetic frustration in spinels and multiferroics with transition metals [32, 33], is here at play in terbium pyrochlores due to the peculiar Tb crystal field. Several points remain to be clarified, both from experimental and theoretical points of view. Experimentally, the precise nature of the distortion, its length and time scale remain to be determined, noticing that there is no direct observation of a symmetry lowering, and that only its consequences on microscopic magnetism are accessible to experiment. Theoretically, the influence of a distortion on the cooperative spin fluctuations (Q-dependence of the crystal field levels and/or spin waves) could be checked, among other consequences.

In contrast with spinels and multiferroics, in $Tb_2Ti_2O_7$, the ground state selected by the spontaneous distortion is not magnetically ordered. It is stabilized in a region of the phase diagram which also involves anisotropic exchange and dipolar contributions to the energy. In $Tb_2Ti_2O_7$, the spin liquid state induced by the distortion in zero magnetic field at ambient pressure has a counterpart in the long range antiferromagnetic orders induced by a magnetic field and/or an applied stress. Both phenomena are a manifestation of the spin-lattice coupling, whose effect is enhanced by the proximity of the first excited crystal field level of the Tb^{3+} ion. The magnetic field couples to the Tb^{3+} electric environment by mixing the CF levels (or not) depending on its orientation with respect to the < 111 > local axis. This induces a giant and anisotropic magnetostriction arising from a field induced distortion. In such a case, one expects the symmetry of the distortion to be intimately correlated with the symmetry of the magnetic structure. It could also provide some hints to elucidate the influence of the spin lattice coupling in this system.

References

[1] Villain J 1979 *Z.Phys.* B **33** 31
[2] Gardner J S, Gingras M J P and Greedan J E 2010 *Rev. Mod. Phys.* **82** 53
[3] Gardner J S *et al* 1999 *Phys. Rev. Lett.* **82** 1012
[4] Balents L 2010 *Nature* **464** 199
[5] Gingras M J P *et al* 2000 *Phys.Rev.* B **62** 6496
[6] Mirebeau I, Bonville P and Hennion M 2007 *Phys.Rev.* B **76** 184436
[7] Kao Y J *et al* 2003 *Phys. Rev.* B **68** 172407
[8] Molavian H R, Gingras M J P and Canals B 2007 *Phys. Rev. Lett.* **98** 157204
[9] Harris M J *et al* 1997 *Phys. Rev. Lett.* **79** 2554
[10] Bleaney B 1963 *Proc. Roy. Soc.* (London) **276A** 19
[11] Grover B 1965 *Phys. Rev.* **140** 1944
[12] Cooper B R and Vogt O 1971 *Jour. Physique* (Paris) **32** C1-958
[13] Ruff J P C *et al* 2007 *Phys. Rev. Lett.* **99** 237202
[14] Gehring G A and Gehring K A 1975 *Rep. Prog. Phys.* **38** 1
[15] Rule K C *et al* 2006 *Phys. Rev. Lett.* **96** 177201
[16] Rule K C and Bonville P 2009 *J. Phys.: Conference Series* **145** 012027
[17] Curnoe S H 2008 *Phys. Rev.* B **78** 094418
[18] Ofer O, Keren A and Baines Ch 2007 *J. Phys.: Condens. Matter* **19** 145270
[19] Ewald P P 1921 *Ann.Phys. (Lpz.)* **64** 253
[20] den Hertog B C and Gingras M J P 2000 *Phys. Rev. Lett.* **84** 3430
[21] Champion J D M, Bramwell S T, Holdsworth P C W and Harris M J 2002 *Europhys. Lett.* **57** 93
[22] Mirebeau I *et al* 2005 *Phys. Rev. Lett.* **94** 246402
[23] The value for λ given in Ref.[6] has to be divided by 6 to match the definition in the present work.
[24] Cao H B *et al* 2009 *Phys. Rev. Lett.* **103** 056402
[25] Malkin B Z *et al* 2010 *J. Phys.: Condens. Matter* **22** 276003
[26] Baker J M 1971 *Rev. Mod. Phys.* **34** 109
[27] Wolf W P 1971 *J. Physique (France)* **32** C1-26
[28] Rotter M *et al* 2000 *Eur. Phys. J. B* **14** 29
[29] Sazonov A P *et al* 2010 *Phys. Rev. B* **82** 174406
[30] Luo G, Hess S T and Corruccini L R 2001 *Physics Letters* A **291** 306
[31] Siddhartan R *et al* 1999 *Phys. Rev. Lett.* **83** 1854
[32] Yamashita Y and Ueda K 2000 *Phys. Rev. Lett.* **85** 4960
[33] Tchernyshyov O, Moessner R and Sondhi S L 2002 *Phys. Rev. Lett.* **88** 067203

Journal of Physics: Conference Series **320** (2011) 012007 doi:10.1088/1742-6596/320/1/012007

Ferrimagnetism of the Heisenberg model on the Kagome strip lattice

Tokuro Shimokawa and Hiroki Nakano

Graduate School of Material Science, University of Hyogo, Kouto, Kamigori, Ako 678-1297, Japan

E-mail: rk09s002@stkt.u-hyogo.ac.jp

Abstract. We study ground-state properties of the $S = 1/2$ Heisenberg model on the Kagome strip lattice by using exact-diagonalization and density matrix renormalization group methods. When the strength of magnetic frustration is tuned, we find from our numerical calculations that there exists an intermediate phase between the non-magnetic phase and the ferrimagnetic phase of the well-known Lieb-Mattis type and that the local magnetization shows an inocommensurate modulation with long-distance periodicity in the intermediate phase. Since the present model shares the same lattice structure in its inner part with the spatially anisotropic two-dimensional Kagome lattice in which an intermediate phase with a similar behavior is certainly observed, the intermediate phases of these models are the ferrimagnetic one of the unconventional non-Lieb-Mattis type.

1. Introduction

Ferrimagnetism is one of the most important phenomena in the field of magnetism. From studies on ferrimagnetism in recent years, a new type of ferrimagnetism was reported in several one-dimensional models with frustration[1, 2, 3, 4, 5, 6, 7, 8], which is different from the conventional ferrimagnetic state of the Lieb-Mattis (LM) type; the nontrivial ferrimagnetism of the new type is called the non-Lieb-Mattis (NLM) type.

The occurrence of the LM ferrimagnetism is understood from a viewpoint of the Marshall-Lieb-Mattis theorem[9, 10]. This theorem indicates that the spontaneous magnetization in the LM ferrimagnetic state is fixed to a simple fraction of the saturated magnetization. In the ferrimagnetism of the NLM type, on the other hand, the spontaneous magnetization changes continuously when the strength of the frustration is changed. The incommensurate modulation with long-distance periodicity of the local magnetization in the ground state is a characteristic behavior of the NLM ferrimagnetism. However, the mechanism of the occurrence of the NLM ferrimagnetism is still unclear.

Quite recently, it was found that the Heisenberg model on the spatially anisotropic Kagome lattice depicted in Fig.1(a) shows an intermediate-magnetization state between the LM ferrimagnetism and the non-magnetic state of the isotropic Kagome lattice as the first candidate of the NLM among the two-dimensional systems[11]. In this intermediate-magnetization state, it was confirmed that the spontaneous magnetization changes continuously when the strength of the frustration is changed and also that the local magnetization shows large dependence on the position of a site. However, it is difficult to judge certainly whether this behavior of the local

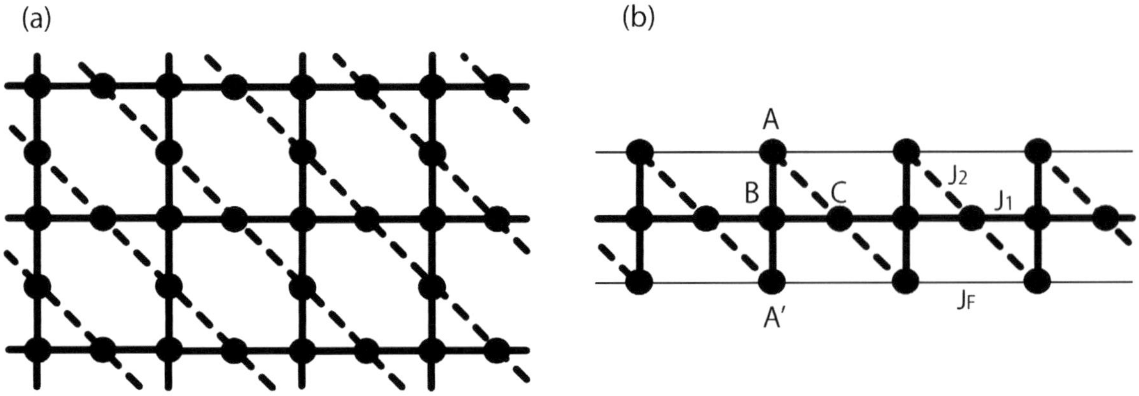

Figure 1. (a) Structure of spatially anisotropic Kagome lattice. (b) Structure of Kagome strip lattice with antiferromagnetic bonds J_1 (bold line) and J_2 (dotted line), and ferromagnetic bond $J_{\rm F}$ (thin line). An $S = 1/2$ spin is located at each site denoted by a circle. Sublattices are represented by A, A′, B and C in each unit cell.

magnetization corresponds to the incommensurate modulation of the NLM ferrimagnetism or not because available system sizes are not sufficiently large in such a two-dimensional system.

In this study, we examine the ground-state properties of Heisenberg model on the Kagome strip lattice depicted in Fig.1(b) instead of the two-dimensioanl lattice of Fig.1(a), where the inner part of the strip lattice of Fig.1(b) is common to a part of the two-dimensional lattice of Fig.1(a). We show that the model on the strip lattice of Fig.1(b) reveals the NLM ferrimagnetism in the ground state. We also present our result of the local magnetization in the NLM ferimagnetic ground state, which clearly shows the incommensurate modulation with long-distance periodicity.

This paper is organized as follows. In the next section, we present the model Hamiltonian on the strip lattice depicted in Fig.1(b) and the methods of our numerical calculations. After we examine the ground-state properties in the classical model of the same Hamiltonian in §3, we present results of those of the quantum Hamiltonian in §4. The final section is devoted to summary and some remarks.

2. Model and Method

The Hamiltonian of the $S = 1/2$ Heisenberg model on the Kagome strip lattice depicted in Fig.1(b) is given by

$$\begin{aligned}\mathcal{H} &= J_1 \sum_i [\mathbf{S}_{i,\rm A} \cdot \mathbf{S}_{i,\mathbf{B}} + \mathbf{S}_{i,\mathbf{B}} \cdot \mathbf{S}_{i,\rm A'} + \mathbf{S}_{i,\mathbf{B}} \cdot \mathbf{S}_{i,\mathbf{C}} + \mathbf{S}_{i,\mathbf{B}} \cdot \mathbf{S}_{i-1,\mathbf{C}}] \\ &+ J_2 \sum_i [\mathbf{S}_{i,\rm A} \cdot \mathbf{S}_{i,\mathbf{C}} + \mathbf{S}_{i,\mathbf{C}} \cdot \mathbf{S}_{i+1,\rm A'}] + J_{\rm F} \sum_i [\mathbf{S}_{i,\rm A} \cdot \mathbf{S}_{i+1,\rm A} + \mathbf{S}_{i,\rm A'} \cdot \mathbf{S}_{i+1,\rm A'}]. \qquad (1)\end{aligned}$$

Here $\mathbf{S}_{i,\xi}$ is an $S = 1/2$ spin operator at ξ sublattice in the unit cell i, where sublattices A, A′, B and C are depicted in Fig.1(b). The number of spin sites is denoted by N; we consider $N/4$ is an integer. Energies are measured in units of J_1; we take $J_1 = 1$ hereafter. Here we examine the region of $0 < J_2/J_1 < \infty$ and limit $J_{\rm F}/J_1 = -1$.

In this study, we use mainly two numerical methods; the numerical exact diagonalization (ED) method based on the Lanczos algorithm and the density matrix renormalization group (DMRG) method[12, 13]. We choose these two methods because it is well known that they are reliable when one investigates low-energy properties of one-dimensional quantum systems with frustration. Note that we use the "finite-system" DMRG method in the present research. We carefully choose a maximum retained states number (MS) and a sweep number (SW) in the DMRG calculations; MS and SW are given in the caption of Fig.3.

3. Classical system

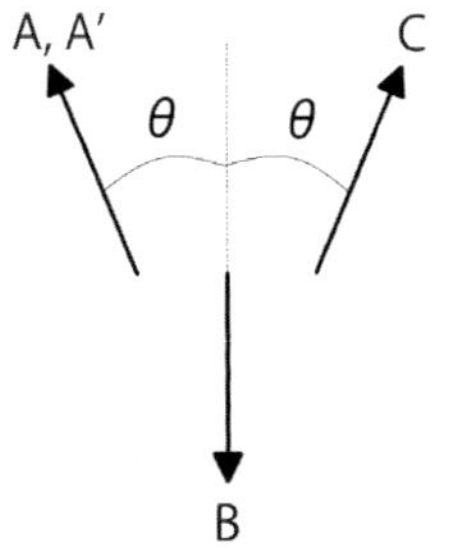

Figure 2. Classical spin configuration in ferrimagnetic state.

Before we present our results of the quantum system, we examine ground-state properties of the Hamiltonian (1) within a classical picture with a characteristic angle θ depicted in Fig.2, in which arrows denote classical spin vectors in a unit cell. Each classical vector has its length S. Note that the state of $\theta = 0$ shows the LM ferrimagnetism with $M/M_s = 1/2$, where M_s represents the saturated magnetization. For $J_F/J_1 = -1$, one can obtain the classical energy under the periodic boundary condition to be

$$E(J_2, \theta) = J_1 N S^2[-\cos(\theta) + (J_2/2J_1)\cos(2\theta) - 1/2]. \quad (2)$$

Minimizing $E(J_2, \theta)$ enable us to obtain the magnetization of the selected state. One finds that the LM ferrimagnetism of $\theta = 0$ with $M/M_s = 1/2$ is realized for $J_2/J_1 \le 1/2$. When J_2/J_1 is larger than $1/2$, the lowest energy is given by $J_1/J_2 = 2\cos(\theta)$. Therefore, the normalized magnetization is given by

$$M/M_s = (\frac{3}{2}\frac{J_1}{J_2} - 1)/4. \quad (3)$$

Finally, the ground state becomes non-magnetic at $J_1/J_2 = 2/3$. This result of the magnetization is depicted in Fig.3(b) and will be compared the results of the quantum system from our numerical calculations in the next section.

4. Quantum system

First, let us explain the way to obtain the spontaneous magnetization M in the ground state of the quantum system with isotropic Heisenberg interactions. Calculations by ED or DMRG methods provide us with the lowest energy $E(N, S^z_{tot}, J_2/J_1)$ in each subspace of Hilbert space divided by S^z_{tot}, where S^z_{tot} is the z-component of the total spin. For example, we present our results of the $N = 96$ system from our DMRG calculations in Fig.3(a), in which energies of each value of S^z_{tot} are depicted for $J_2/J_1 = 0.5$ and $J_2/J_1 = 0.58$. The spontaneous magnetization M is determined by the following procedure. One finds the lowest energy among $E(N, S^z_{tot}, J_2/J_1)$ for all the case of S^z_{tot}. Among states having the lowest-energy level, the highest value of S^z_{tot} is the spontaneous magnetization. (See arrows displayed in Fig.3(a).)

Let us now present results of the J_1/J_2-dependence of M/M_s in Fig.3(b). The magnetization of the classical arguement discussed in the above section is accompanied. We successfully find the existence of the intermediate region between the region of $M/M_s = 1/2$ and the region

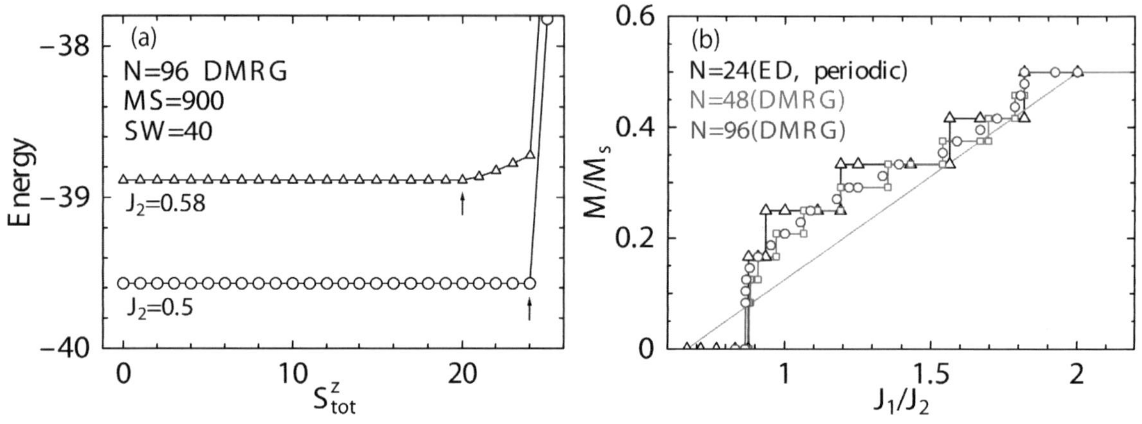

Figure 3. (a) Dependences of the lowest energy on $S^z_{\rm tot}$. Results of $J_2/J_1 = 0.5$ and 0.58 for the system size of $N = 96$ are presented. Arrows indicate the values of the spontaneous magnetization M in each J_2/J_1. The position of an arrow is given from the highest value of $S^z_{\rm tot}$ among those which give the common lowest energy. (b) J_1/J_2-dependence of $M/M_{\rm s}$ obtained from ED calculations for $N = 24$ (triangle) under the periodic boundary condition and DMRG calculation for $N = 48$ (square) and 96 (circle) under the open boundary condition. Note that we take J_1/J_2 instead of J_2/J_1 as the abscissa because the classical magnetization given by eq. (3) exhibits a linear dependence on J_1/J_2. The magnetization within the classical picture is drawn by the green line. Note that for $N = 96$ in (a) and (b), we use $MS = 900$ and $SW = 40$ and that for $N = 48$ in (b), we use $MS = 400$ and $SW = 20$.

of $M/M_{\rm s} = 0$. Since the present model for $J_2/J_1 = 0$ meets conditions of the Marshall-Lieb-Mattis theorem, the LM ferrimangetism is realized; thus the consequence of the theorem gives us $M/M_{\rm s} = 1/2$. The region of $J_1/J_2 \gtrsim 2$ with $M/M_{\rm s} = 1/2$ is directly connected to the LM ferrimagnetic state. Therefore, the region of $M/M_{\rm s} = 1/2$ is regarded as the one of the LM ferrimagnetism. In the limit of $J_2/J_1 \to \infty$, on the other hand, three spins of $\mathbf{S}_{i,{\rm A}}, \mathbf{S}_{i,{\rm C}}$ and $\mathbf{S}_{i+1,{\rm A'}}$ form a single $S = 1/2$ spin in the low-energy state; the system totally forms the $S = 1/2$ Δ chain in one dimension and has the non-magnetic ground state. The properties of the system for finite but large J_2/J_1 are nontrivial at least now. The present results in Fig.3(b) indicate that the system undergoes a phase transition between the magnetically ordered state and the magnetically disordered state at around $J_1/J_2 \sim 0.8$.

Next, we study N dependences of the following two phase boundaries: the boundary between the phases of $M/M_{\rm s} = 1/2$ and $0 < M/M_{\rm s} < 1/2$ denoted by the ratio $r_1 = J_2/J_1$ and the boundary between the phases of $0 < M/M_{\rm s} < 1/2$ and $M/M_{\rm s} = 0$ denoted by the ratio of J_2/J_1 as r_2. We present results of $N = 16, 20$ and 24 from ED calculations under the periodic and the open boundary conditions and those of $N = 48$ and 96 from DMRG calculations under the open boundary condition. Results of r_1 from ED and DMRG calculations are consistent with each other irrespective of the difference of boundary conditions. One finds that the ratio r_1 converges to $r_1 \sim 0.55$ in the limit of $N \to \infty$. Although the size dependence of r_2 is relatively larger than r_1, r_2 seems to go to the common value of $r_2 \sim 1.18$ in the thermodynamic limit. These results suggest that the presence of the region of intermediate-magetization state with $0 < M/M_{\rm s} < 1/2$ is evident in the thermodynamic limit.

Let us study local magnetization $\langle S^z_{i,\xi}\rangle$ within the subspace of the highest $S^z_{\rm tot}$ corresponding to the spontaneous magnetization M, where $\langle Q\rangle$ denotes the expectation value of a physical quantity Q. We investigate $\langle S^z_{i,\xi}\rangle$ in the two phases of $0 < M/M_{\rm s} < 1/2$ and $M/M_{\rm s} = 1/2$.

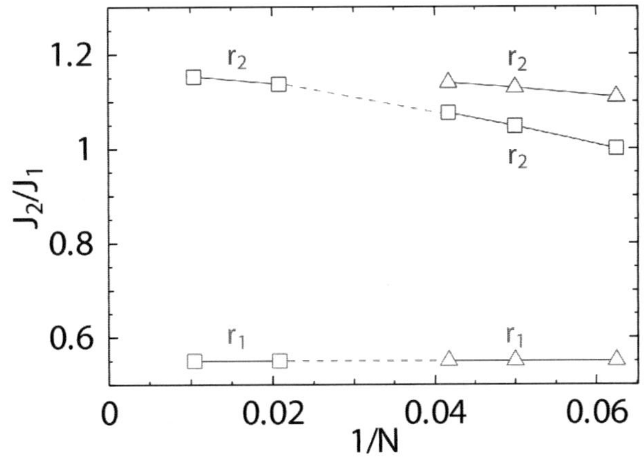

Figure 4. Size dependence of the phase boundary. The boundary between the phase of $M/M_s = 1/2$ and $0 < M/M_s < 1/2$ is denoted by r_1. The boundary between the phase of $M/M_s = 0$ and $0 < M/M_s < 1/2$ is represented by r_2. Triangles (Squares) denote the results under the periodic (open) boundary condition. Results for $N = 16, 20$ and 24 are obtained from our ED calculations and those for $N = 48$ and 96 are obtained from our DMRG calculations. The ratios r_1 and r_2 seem to be converged to ~ 0.55 and ~ 1.18 in the thermodynamic limit, respectively. The dotted lines are drawn as guides for eyes between the ED and DMRG data.

Results of $\langle S^z_{i,\xi} \rangle$ of the system of $N = 96$, namely $i = 1, 2, \cdots, 24$, at each sublattice A, A′, B and C are depicted in Fig.5(a) and Fig.5(b) for $J_2/J_1 = 0.5$ and 0.58 respectively. In Fig.5(a), one can observe the uniform behavior of upward directions at sublattice A, A′, and C and a downward direction at sublattice B although small boundary effects appear near the edges of the system. In Fig.5(b), on the other hand, we find an incommensurate modulation with long-distance periodicity clearly. This characteristic behavior of the local magnetization is a common phenomenon of the NLM ferrimagnetism in some one-dimensional quantum frustrated spin systems[4, 5]. Therefore we conclude that the phase of $0 < M/M_s < 1/2$ is the NLM

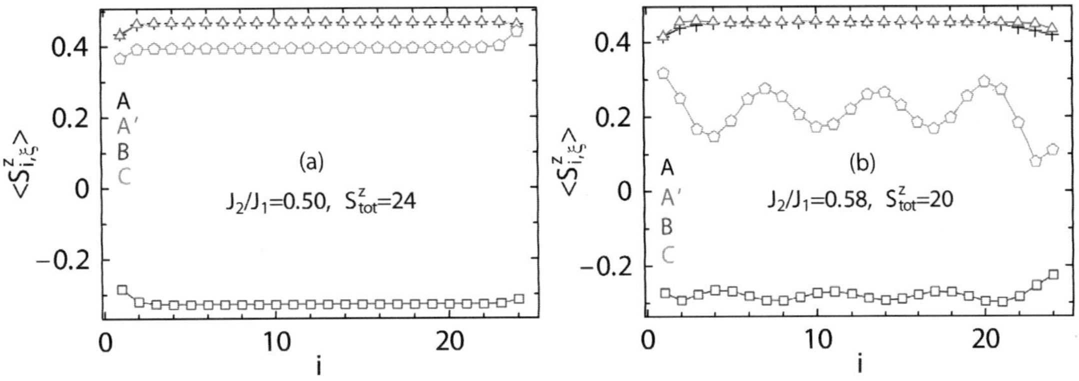

Figure 5. Local magnetization $\langle S^z_{i,\xi} \rangle$ at each sublattice, A (cross), A′ (triangle), B (square) and C (pentagon). Panels (a) and (b) are results for $J_2/J_1 = 0.5$ and 0.58 respectively. These results are obtained from our DMRG calculations for $N = 96$ ($i = 1, 2, \cdots, 24$).

ferrimagnetism.

We briefly comment the case of $-1 < J_F \leq 0$. We confirm that the width of the intermediate-mangetization region is getting narrower when $|J_F|$ is decreased and that the region disappears before J_F reaches $J_F = 0$. (not shown in this paper). This suggests that the interaction of J_F stabilizes the intermediate phase as a substitute of a part which we cut off from the two-dimensional Kagome lattice depicted in Fig.1(a).

5. Summary and remarks

We have studied the ground state properties of $S = 1/2$ Heisenberg model on the Kagome strip lattice by using ED and DMRG methods. We have found the intermediate-magnetization phase between the non-magnetic phase and LM ferrimagnetic phase. In the intermediate-magnetization state, local magnetization is found to show an incommensurate modulation with long-distance periodicity. This characteristic behavior indicates that the intermediate-magnetization state is the ferrimagnetic state of the NLM type. If we remind that a symptom of the NLM ferrimagnetism is observed in the two-dimensional system on the spatially anisotropic Kagome lattice[11] although the reported results are not direct evidences, the phenomenon observed in the two-dimensional case is also the NLM ferrimagnetism presumably.

Finally, we mention the case of three-leg ladder with diagonal interactions which is a part of the spatially anisotropic triangular lattice[15]. In this case, intermediate phase is found between the ferrimagnetic and the non-magnetic Luttinger-liquid phases; but the intermediate phase is not NLM ferrimagnetic. This is different from the NLM ferrimagnetic intermediate phase in the present study.

Acknowledgments

We wish to thank Prof. T. Sakai for fruitful discussions. This work was partly supported by Grant-in-Aid (No.20340096) from the Ministry of Education, Culture, Sports, Science and Technology of Japan. This work was partly supported by Grant-in-Aid (No. 22014012) for Scientific Research and Priority Areas "Novel States of Matter Induced by Frustration" from the Ministry of Education, Culture, Sports, Science and Technology of Japan. Diagonalization calculations in the present work were carried out based on TITPACK Version 2 coded by H. Nishimori. DMRG calculations were carried out by using ALPS DMRG application[14]. A part of the calculations was done at the Supercomputer Center, Institute for Solid State Physics, University of Tokyo.

References

[1] Sachdev S and Senthil T 1996 *Ann. Phys.* **251** 76
[2] Bartosch L, Kollar M and Kopietz P 2003 *Phys. Rev. B* **67** 092403
[3] Ivanov N B and Richter J 2004 *Phys. Rev. B* **69** 214420
[4] Yoshikawa S and Miyashita S 2005 *J. Phys. Soc. Jpn. Suppl.* **74** 71
[5] Hida K 2007 *J. Phys.: Condens. Matter* **19** 145225
[6] Hida K and Takano K 2008 *Phys. Rev. B* **78** 064407
[7] Montenegro-Filho R R and Coutinho-Filho M D 2008 *Phys. Rev. B* **78** 014418
[8] Shimokawa T and Nakano H *J. Phys. Soc. Jpn.* **80** 043703
[9] Lieb E and Mattis D 1962 *J. Math. Phys.* **3** 749
[10] Marshall W 1955 *Proc. Roy. Soc. A* **232** 48
[11] Nakano H, Shimokawa T and Sakai T *J. Phys. Soc. Jpn.* **80** 033709
[12] White S R 1992 *Phys. Rev. Lett.* **69** 2863
[13] White S R 1993 *Phys. Rev. B* **48** 10345
[14] Albuquerque *et al.* 2007 *J. Magn. Magn. Mater.* **310** 1187 (see also http://alps.comp-phys.org).
[15] Abe S, Sakai T, Okamoto K and Tsutsui K submitted to *J. Phys.: Conf. Series*, proceedings of ICFCM 2011

Singlet ground states of the bilinear-biquadratic exchange Hamiltonian with reflection symmetry

Kengo Tanaka
Department of Comprehensive Studies for Life Planning, Koran Women's Junior College, Fukuoka 811-1311, Japan
E-mail: tanaka@koran.ac.jp

Abstract. We apply the Lieb-Schupp method to the isotropic spin-S Hamiltonian with bilinear $(-J/S^2)$ and biquadratic $(-J'/S^4)$ exchange interactions on finite lattices with reflection-symmetry to prove that finite volume ground states of the models are spin singlet in the parameter region $J' \geq 2S^2J$, $J' \geq 0$.

1. Introduction

Frustration in magnetic systems may arise due to the lattice geometries and competing interactions. Typical models of the geometrically frustrated magnets are antiferromagnetic Heisenberg model on triangular, kagome, or pyrochlore lattice, and one with competing interactions are the axial-next-nearest-neighbor Heisenberg model or the multiple-spin exchange model. These quantum magnets are known to possess interesting phenomena which are quite different from non-frustrated magnetic systems, but the accurate method of theoretical investigations for the properties of frustrated quantum magnets are less limited than non-frustrated and/or classical cases.

The Marshall-Lieb-Mattis theorem is one of the most famous exact results of the quantum spin systems, and it proves ordering energy levels for a large class of the Heisenberg model on finite size bipartite lattices. In the case of antiferromagnetic Heisenberg model this theorem states that the model on a bipartite lattice which contains the same number of sublattice points has a unique spin singlet ground state. But this theorem is not applicable to the model on non-bipartite lattices (frustrated lattices). In 1999 Lieb-Schupp succeeded in a proof of the spin singlet ground states of the model on two dimensional frustrated pyrochlore checkerboard lattice [1, 2]. Their method use reflection symmetry of Hamiltonian, on the other hand, the Marshall-Lieb-Mattis theorem is based on the Perron-Frobenius theorem. So their method can be applied to a class of frustrated spin systems satisfying reflection symmetry, but can not conclude the uniqueness of the ground state.

In this study we treat the spin-S isotropic bilinear-biquadratic exchange Hamiltonian:

$$\mathcal{H} = -\frac{J}{S^2}\sum_{\langle x,y\rangle}\mathbf{S}(x)\cdot\mathbf{S}(y) - \frac{J'}{S^4}\sum_{\langle x,y\rangle}(\mathbf{S}(x)\cdot\mathbf{S}(y))^2 \qquad (1)$$

as one of the models with competing interactions. In the case of Hamiltonian (1) with $S = 1$ on the linear chain, infinite volume ground state of Hamiltonian (1) in the region $J' > \pm J > 0$ is considered to be dimer phase. Competition between bilinear and biquadratic interactions may induce broken translational invariance in spite that Hamiltonian (1) has translational invariance. On the other hand, finite volume ground state in $J' > J$, $J' \geq 0$ $(J \neq 0)$ including that dimer phase region is proved to be unique spin singlet with the Marshall-Lieb-Mattis type argument [3, 4, 5]. For $S > 1$ the same result is established in the region $0 \leq J' \leq -SJ/2(S-1)$ [3].

Our purpose is to show that, with the Lieb-Schupp method, finite volume ground states of Hamiltonian (1) are spin singlet in the region $J' \geq 2S^2 J$, $J' \geq 0$.

2. Model and its reflection symmetry

Let us consider finite lattices $\Lambda = \Lambda_{\mathrm{L}} \cup \Lambda_{\mathrm{C}} \cup \Lambda_{\mathrm{R}}$ with reflection symmetry which have an even number of independent sites and the property that it can be split in two equal parts Λ_{L} and Λ_{R} that are mirror images of one another about a symmetry plane without sites that cuts bonds $\langle x, y\rangle$ with $x \in \Lambda_{\mathrm{L}}$ and $y \in \Lambda_{\mathrm{R}}$ which is denoted by Λ_{C} [2]. With each sites $x \in \Lambda$, we associate the spin-S operator $\mathbf{S}(x) = (S_1(x), S_2(x), S_3(x))$ with $S(S+1)$ and satisfies the commutation relation $[S_i(x), S_j(y)] = \mathrm{i}\epsilon_{ijk} S_k(x)\delta_{xy}$. Here we use the usual basis in which $S_3(x)$ is diagonalized, $S_1(x)$ and $S_3(x)$ have real matrix elements, and $S_2(x)$ pure imaginary. In this study we consider that the spin model described by Hamiltonian (1) on Λ. The summation in Hamiltonian (1) is all over nearest-neighbor pairs in Λ.

For the following discussion let us introduce quadrupole operators:

$$O_{2,0}(x) = \frac{1}{2}\left[3S_3^2(x) - S(S+1)\right], \tag{2}$$

$$O_{2,\pm 1}(x) = \mp\sqrt{\frac{3}{2}}\frac{1}{2}\left(S_3(x)S^{\pm}(x) + S^{\pm}(x)S_3(x)\right), \tag{3}$$

$$O_{2,\pm 2}(x) = \sqrt{\frac{3}{8}}\left(S^{\pm}(x)\right)^2, \tag{4}$$

where $S^+(x)$ and $S^-(x)$ are the usual spin raising and lowering operators defined by

$$S^{\pm}(x) = S_1(x) \pm \mathrm{i}S_2(x). \tag{5}$$

Quadrupole operators satisfy the relation:

$$\sum_{q=-2}^{2} O_{2,q}(x)O_{2,q}^{\dagger}(y) = \frac{3}{2}\left(\mathbf{S}(x)\cdot\mathbf{S}(y)\right)^2 + \frac{3}{4}\mathbf{S}(x)\cdot\mathbf{S}(y) - \frac{1}{2}S^2(S+1)^2, \tag{6}$$

and then Hamiltonian (1) is written as

$$\mathcal{H} = J_{\mathrm{D}}\sum_{\langle x,y\rangle}\mathbf{S}(x)\cdot\mathbf{S}(y) - J_{\mathrm{Q}}\sum_{\langle x,y\rangle}\sum_{q=-2}^{2} O_{2,q}(x)O_{2,q}^{\dagger}(y) \tag{7}$$

with

$$J_{\mathrm{D}} = -\frac{J}{S^2} + \frac{1}{2}\frac{J'}{S^4}, \quad J_{\mathrm{Q}} = \frac{2}{3}\frac{J'}{S^4}. \tag{8}$$

Now we decompose Hamiltonian (7) as follows.

$$\mathcal{H} = \mathcal{H}_{\mathrm{L}} + \mathcal{H}_{\mathrm{R}} + \mathcal{H}_{\mathrm{C}}. \tag{9}$$

Here $\mathcal{H}_{\mathrm{L}}$, $\mathcal{H}_{\mathrm{R}}$, and $\mathcal{H}_{\mathrm{C}}$ is a collection of interactions on $\langle x, y\rangle \in \Lambda_{\mathrm{L}}$, Λ_{R}, and Λ_{C}, respectively. The right-hand side of equation (9) is unitarily transformed to

$$U^{\dagger}\mathcal{H}U = \mathcal{H}_{\mathrm{L}} + \mathcal{H}_{\mathrm{R}} - J_{\mathrm{D}} \sum_{\langle x,y\rangle\in\Lambda_{\mathrm{C}}} \sum_{i=1}^{3} T_i(x)T_i(y) - J_{\mathrm{Q}} \sum_{\langle x,y\rangle\in\Lambda_{\mathrm{C}}} \sum_{q=-2}^{2} O_{2,q}(x)O_{2,q}(y) \tag{10}$$

with

$$U = \exp\left(\mathrm{i}\pi \sum_{x\in\Lambda_{\mathrm{R}}} S_2(x)\right) \tag{11}$$

and $T_1(x) = S_1(x)$, $\mathrm{i}T_2(x) = S_2(x)$, $T_3(x) = S_3(x)$, where we have used

$$U^{\dagger}S_1(x)U = -S_1(x), \tag{12}$$
$$U^{\dagger}S_2(x)U = S_2(x), \tag{13}$$
$$U^{\dagger}S_3(x)U = -S_3(x), \tag{14}$$
$$U^{\dagger}O_{2,q}^{\dagger}(x)U = O_{2,q}(x), \tag{15}$$

for $x \in \Lambda_{\mathrm{R}}$.

Let us write a ground state

$$\psi = \sum_{\alpha,\beta} c_{\alpha\beta}\psi_{\alpha}^{\mathrm{L}} \otimes \psi_{\beta}^{\mathrm{R}} \tag{16}$$

of this Hamiltonian with a coefficient matrix $c_{\alpha\beta}$, where $\psi_{\alpha}^{\mathrm{L}}$ form a real orthonormal basis of S_3 eigenstate for $\mathcal{H}_{\mathrm{L}}$ and $\psi_{\beta}^{\mathrm{R}}$ is the corresponding state for $\mathcal{H}_{\mathrm{R}}$.

Let us consider the ground state energy of Hamiltonian (10), we have

$$\begin{aligned}\langle\psi|\mathcal{H}|\psi\rangle &= \mathrm{Tr}cc^{\dagger}\mathcal{H}_{\mathrm{L}} + \mathrm{Tr}c^{\dagger}c\mathcal{H}_{\mathrm{R}} \\ &- J_{\mathrm{D}} \sum_{\langle x,y\rangle\in\Lambda_{\mathrm{C}}} \sum_{i=1}^{3} \mathrm{Tr}c^{\dagger}T_i(x)cT_i^{\dagger}(y) - J_{\mathrm{Q}} \sum_{\langle x,y\rangle\in\Lambda_{\mathrm{C}}} \sum_{q=-2}^{2} \mathrm{Tr}c^{\dagger}O_{2,q}(x)cO_{2,q}^{\dagger}(y).\end{aligned} \tag{17}$$

We note that Hamiltonian (10) is left-right symmetric and operators appearing in it have real matrix elements. Then following the arguments in the papers [1, 2, 6, 7], we can see that an eigenstate with Hermite coefficient matrix $c = c^{\dagger}$ is also a ground state, and the third and forth term in the right-hand side of equation (17) are written as

$$-J_{\mathrm{D}} \sum_{\langle x,y\rangle\in\Lambda_{\mathrm{C}}} \sum_{i=1}^{3} \sum_{k,l} c_k c_l \left|(T_i(x))_{kl}\right|^2 - J_{\mathrm{Q}} \sum_{\langle x,y\rangle\in\Lambda_{\mathrm{C}}} \sum_{q=-2}^{2} \sum_{k,l} c_k c_l \left|(O_{2,q}(x))_{kl}\right|^2 \tag{18}$$

in the diagonal basis of c. This expression is bounded below by

$$-J_{\mathrm{D}} \sum_{\langle x,y\rangle\in\Lambda_{\mathrm{C}}} \sum_{i=1}^{3} \sum_{k,l} |c_k|\,|c_l| \left|(T_i(x))_{kl}\right|^2 - J_{\mathrm{Q}} \sum_{\langle x,y\rangle\in\Lambda_{\mathrm{C}}} \sum_{q=-2}^{2} \sum_{k,l} |c_k|\,|c_l| \left|(O_{2,q}(x))_{kl}\right|^2, \tag{19}$$

if $J_{\mathrm{D}} \geq 0$, $J_{\mathrm{Q}} \geq 0$. So we confirm that an eigenstate with positive semidefinite coefficient matrix $|c| = \sqrt{c^2}$ is a ground state.

3. Spin singlet ground state

In this section, at first we will prove that a ground state with positive semidefinite coefficient matrix $|c|$ is spin singlet.

Now we consider a tensor product of spin singlet state

$$\psi_0 = \bigotimes_{x\in\Lambda_{\rm L}} \sum_{M=-S}^{S} (-1)^{S-M} |S, M\rangle_x \otimes |S, -M\rangle_{x'} \tag{20}$$

where $x' \in \Lambda_{\rm R}$ is the reflection of $x \in \Lambda_{\rm L}$ in the symmetry plane on $\Lambda_{\rm C}$ and M is eigenvalue of $S_3(x)$. We can easily check the overlap of ψ_0 with a ground state with positive semidefinite coefficient matrix $|c|$ takes non-zero value. Furthermore, since ψ_0 has total spin zero and total spin is a good quantum number of this Hamiltonian, this facts imply that there is at least one ground state with total spin zero in the region $J_{\rm D} \geq 0$, $J_{\rm Q} \geq 0$.

Next, we will show that all ground states of this model are spin singlet. Let $b(x)$ be a real valued function on each site x. Then we consider the Hamiltonian in magnetic field of arbitrary amplitude for each site given by

$$\begin{aligned}\widetilde{\mathcal{H}}(b(x)) &= -J_{\rm D} \sum_{\langle x,y\rangle} \Big\{ T_1(x)T_1(y) + T_2(x)T_2(y) + (T_3(x) - b(x))\,(T_3(y) - b(y)) \\ &+ \frac{1}{2}\left[T_3^2(x) - (T_3(x) - b(x))^2 + T_3^2(y) - (T_3(y) - b(y))^2\right]\Big\} \\ &- J_{\rm Q} \sum_{\langle x,y\rangle} O_{2,q}(x) O_{2,q}(y).\end{aligned} \tag{21}$$

Following the argument of Kennedy-Lieb-Shastry [7, 8], we get inequality

$$E(b(x)) \geq E(0) \tag{22}$$

concerning for the ground state energy of $\widetilde{\mathcal{H}}(b)$ in $J_{\rm D} \geq 0$, $J_{\rm Q} \geq 0$.

When we choose

$$b(x) = \begin{cases} b & x \in \Lambda_{\rm L} \\ -b & x \in \Lambda_{\rm R}, \end{cases} \tag{23}$$

we see

$$U\widetilde{\mathcal{H}}(b)U^\dagger = \mathcal{H} - 2bJ_{\rm D} \sum_{x\in\Lambda_{\rm C}} S_3(x). \tag{24}$$

Let $\psi(b)$ be a ground state of $\widetilde{\mathcal{H}}(b)$ .By the variational principle and inequality (22), we have

$$\left\langle \psi(0) \left| \widetilde{\mathcal{H}}(b) \right| \psi(0) \right\rangle \geq \left\langle \psi(b) \left| \widetilde{\mathcal{H}}(b) \right| \psi(b) \right\rangle = E(b) \geq E(0), \tag{25}$$

which leads to

$$-2bJ_{\rm D} \sum_{x\in\Lambda_{\rm C}} \langle \psi(0)\, | S_3(x) |\, \psi(0) \rangle = 0. \tag{26}$$

This result is independent of value of b. Thus we can conclude $\sum_{x\in\Lambda_{\rm C}} \langle \psi(0)\, | S_3(x) |\, \psi(0) \rangle = 0$. We impose a periodic boundary condition in the direction perpendicular to the symmetry plane on $\Lambda_{\rm C}$, and then we have $S_3^{\rm tot} = \sum_{x\in\Lambda} \langle \psi(0)\, | S_3(x) |\, \psi(0) \rangle = 0$. This result also holds for $S_1^{\rm tot}$ and $S_2^{\rm tot}$. So it implies that all ground states have total spin zero.

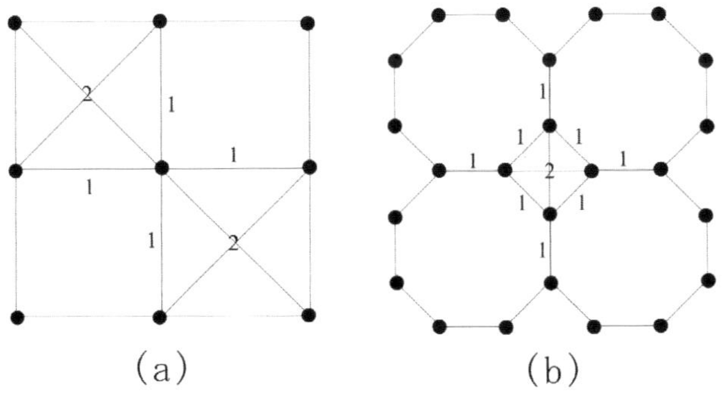

Figure 1. Example of Λ containing the structure of plaquettes with crossing bond. (a) pyrochlore checkerboard lattice. (b) 1/5-depleted lattice (CaV_4O_9). '1' and '2' in figures is denoted by first and second neighbor interaction, respectively.

4. Summary and Discussions

We have proved that ground states of Hamiltonian (1) on reflection symmetric lattices Λ with a periodic boundary condition in the one direction have total spin zero in the region $J_D \geq 0$, $J_Q \geq 0$ or $J' \geq 2S^2J$, $J' \geq 0$.

For $S = 1$ the results of Marshall-Lieb-Mattis type argument based on the Perron-Frobenius theorem state that the ground state is unique spin singlet in the region $J' > J, J' \geq 0 (J \neq 0)$ [3, 4, 5]. Our result can not conclude the uniqueness of the ground state, but these results of previous studies assure it in that parameter region. At $J' > 0, J = 0$, there may exist multiple degeneracy on the ground state [5]. Our results also conclude that these degenerate ground states are spin singlet.

For $S > 1$ the result in reference [3] gives that the ground state is unique spin singlet in the region $0 \leq J' \leq -SJ/2(S-1)$. Our results extend the region which one can conclude singlet ground states without uniqueness. In the region $J > 0, J' > 0$ the Marshall-Lieb-Mattis type argument can not give any conclusions as far as we know. This argument works well when we can find suitable unitary transformation which leads to signs of off-diagonal matrix elements of irreducible unitarily transformed Hamiltonian satisfying the Perron-Frobenius theorem. But there is no systematic method available to find it so far. The results for $S = 1$ in reference [4, 5] are established by using special commutation relations between quadrupole operators which hold only for $S = 1$. So straightforward extension of the Marshall-Lieb-Mattis type argument to the case of $S > 1$ seems to be difficult.

By the variational principle, we can easily check that the ground state becomes saturated ferromagnetic state for $0 < J' < SJ$ [9]. For $S > 1$ in the region $SJ < J' < 2S^2J$ what types of ground states realize are not clear. Furthermore the Lieb-Schupp method can not conclude the uniqueness of the ground state. We hope that these problems will be resolved by other technique.

Finally we shall comment on the extension of our results to the case of geometrically frustrated lattice. In section 2 we explained the properties of Λ, whose typical examples are d-dimensional hypercubic and honeycomb lattices. Figure 1-(b) without crossing bonds (1/5-depleted lattice) also clearly has reflection symmetry. In this case we can consider a symmetry plane which cuts plaquettes without sites. Our results for Hamiltonian (1) on the reflection symmetric lattices with bipartiteness can be extended to the case of the lattices containing the structure of plaquettes with crossing bond. Spin models on the square lattice with second neighbor interaction which are the so called '$J_1 - J_2$ model' and figure 1-(a) is its checkerboard version. Figure 1-(b) is 1/5-depleted lattice with the structure of plaquettes with crossing bond. Here we set $J_{\rm D}^1(J_{\rm Q}^1)$ and $J_{\rm D}^2(J_{\rm Q}^2)$ is first and second neighbor dipole (quadrupole) interaction, respectively. Through a similar procedure in section 2 and 3, we can conclude that our results also hold in the region $|J_{\rm D}^2| \leq J_{\rm D}^1/2$ and $|J_{\rm Q}^2| \leq J_{\rm Q}^1/2$. These conditions originate from the expression of interaction terms on plaquettes with crossing bond satisfying reflection symmetry so as to get a similar lower bound to (19) [1, 2, 10] and a similar inequality to (22) [7, 8, 10].

References

[1] E. H. Lieb and P. Schupp 1999 *Phys. Rev. Lett.* **83** 5362.
[2] E. H. Lieb and P. Schupp 2000 *Physica. A* **279** 978.
[3] R. G. Munro 1976 *Phys. Rev. B* **13** 4875.
[4] J. B. Parkinson 1979 *J. Phys. C: Solid State Phys.* **10** 1735.
[5] A. Tanaka and T. Idogaki 1997 *Phys. Rev. B* **56** 10774.
[6] E. H. Lieb 1989 *Phys. Rev. Lett.* **62** 1201.
[7] T. Kennedy, E. H. Lieb, and B. S. Shastry, 1988 *J. Stat. Phys.* **53** 1019.
[8] K. Tanaka, A. Tanaka, and T. Idogaki, 2001 *J. Phys. A: Math. Gene.* **34** 8767
[9] J. Aksamit 1980 *J. Phys. C: Solid State Phys.* **13** L871
[10] T. Kishi and K. Kubo 1989 *J. Phys. Soc. Jpn.* **58** 2547

Journal of Physics: Conference Series **320** (2011) 012009 doi:10.1088/1742-6596/320/1/012009

Loop algorithm for classical antiferromagnetic Heisenberg models with biquadratic interactions

Hiroshi Shinaoka[1,2], Yusuke Tomita[1] and Yukitoshi Motome[3]

[1]Institute for Solid State Physics, University of Tokyo, Kashiwanoha, Kashiwa, Chiba, 277-8581, Japan

[2]Present address: Nanosystem Research Institute, AIST, Tsukuba 305-8568, Japan

[3]Department of Applied Physics, University of Tokyo, 7-3-1 Hongo, Bunkyo-ku, Tokyo 113-8656, Japan

E-mail: h.shinaoka@aist.go.jp

Abstract. Monte Carlo simulation using the standard single-spin flip algorithm often fails to sample over the entire configuration space at low temperatures for frustrated spin systems. A typical example is a class of spin-ice type Ising models. In this case, the difficulty can be avoided by introducing a global-flip algorithm, the loop algorithm. Similar difficulty is encountered in $O(3)$ Heisenberg models in the presence of biquadratic interaction. The loop algorithm, however, is not straightforwardly applied to this case, since the system does not have a priori spin-anisotropy axis for constructing the loops. We propose an extension of the loop algorithm to the bilinear-biquadratic models. The efficiency is tested for three different ways to flip spins on a loop in Monte Carlo simulation. We show that the most efficient method depends on the strength of the biquadratic interaction.

1. Introduction

Monte Carlo (MC) simulation is a powerful tool for investigating thermodynamic properties of classical spin models [1]. The standard single-spin flip algorithm is widely used because it is simple and applicable to systems with any type of interactions. However, it is well known that it suffers from slow relaxation in many cases. For example, the single-spin-flip dynamics exhibits critical slowing down near a continuous transition temperature where the correlation length diverges. Such difficulty is avoided by using a nonlocal, global update, such as the Swendsen-Wang cluster algorithm for Ising systems [2] or its Wolff's extension to Heisenberg systems [3]. These cluster algorithms accelerate the relaxation by flipping many spins at once, i.e., updating the spin configuration drastically.

Besides the critical slowing down, the single-spin flip algorithm often suffers from dynamical freezing at low temperature (T) when the ground state has macroscopic degeneracy under the influence of geometrical frustration. A typical example is an antiferromagnetic Ising model on a pyrochlore lattice. The pyrochlore lattice is a three-dimensional frustrated structure given by a corner-sharing network of tetrahedra, as shown in Fig. 1(a). When the antiferromagnetic exchange interaction is limited to nearest-neighbor sites, no long-range ordering occurs down to zero T and the ground state has macroscopic degeneracy [4]. The degenerate manifold is identified by a collection of local constraints enforcing two spins pointing up and two spins pointing down in every tetrahedron, as exemplified in Fig. 1(a). This 'two-up two-down' constraint is called the ice rule because of an analogy to the constraint on positions of protons in hexagonal ice [5, 6]. Similar situation was recently discovered in the so-called spin-ice systems, in which the Ising-like spins point along the local $\langle 111 \rangle$ axes and interact with each other by ferromagnetic exchange interaction and dipole interaction [7, 8]. Because the degenerate 'ice-rule' configurations are separated by large energy barriers of the order of the dominant interaction scale J, the single-spin flip does not work at low $T \ll J$. The difficulty is avoided by introducing a global flip called the loop flip, in which one reverses all Ising spins on a specific closed loop passing through tetrahedra [9, 10, 11, 12, 13]; the loop is chosen so that the spins are up and down (inward and outward in the spin-ice problem) alternatively along the loop [see the hexagon in Fig. 1(a) as an example]. This loop flip enables to transform an ice-rule state to another ice-rule state bypassing the energy barriers.

The difficulty remains even when the Ising discreteness is relaxed and spins can fluctuate, as long as the ground-state manifold retains a multivalley structure with large energy barriers. Such situation is seen in variants of pyrochlore antiferromagnets, such as classical Heisenberg models in the presence of the single-ion easy-axis anisotropy [14, 15] and the biquadratic interaction [16]. In contrast to the Ising case, however, it is nontrivial how to define the loop with alternating spins. Moreover, the loop flip procedure is not unique in the Heisenberg spin case because of the continuous degrees of freedom. Recently, the authors extended the loop algorithm to the Heisenberg spin systems with single-ion anisotropy by defining 'colors' of spins as black and white, which is a natural generalization of up and down in the antiferromagnetic Ising case, in terms of the projection of spins on the anisotropy axis [15]. We tested different ways to flip spins on a formed loop and showed that the efficiency strongly depends on the method. In the models with the biquadratic interaction, however, because of the $O(3)$ spin rotational symmetry, we have no explicit anisotropy axis to project spins on. Therefore, an extension of the loop algorithm to this class of models is not straightforward.

In this paper, we develop an extension of the loop algorithm to classical antiferromagnetic Heisenberg spin systems with biquadratic interaction in which the spin-ice type manifold emerges at low T. We propose a way to define the projection axis for constructing loops and test three different ways to flip a formed loop. We apply the algorithm to a nearest-neighbor bilinear-biquadratic model, and compare the efficiency of the three methods. Interested readers are referred to Ref. [17] for further application of the present algorithm to a bond-disordered bilinear-biquadratic model.

2. Extension of the loop algorithm to bilinear-biquadratic spin systems

2.1. Model

In this section, we extend the loop algorithm to classical antiferromagnetic Heisenberg models with biquadratic interactions. We start with a Hamiltonian of a simple form:

$$\mathcal{H} = \sum_{\langle i,j \rangle} \left\{ J \left(\vec{S}_i \cdot \vec{S}_j \right) - b \left(\vec{S}_i \cdot \vec{S}_j \right)^2 \right\}, \tag{1}$$

where $\vec{S}_i$ denotes a classical Heisenberg spin at site i on the pyrochlore lattice [Fig. 1(a)] (we take $|\vec{S}_i| = 1$) and b is the coupling constant of the biquadratic interaction. Here we consider $b > 0$ which favors collinear spin configurations. We note that such 'ferro'-type biquadratic interaction originates in the spin-lattice coupling as well as quantum and thermal fluctuations. We consider the antiferromagnetic exchange interaction $J > 0$, and take the energy unit as $J = 1$. The sum runs over the nearest-neighbor bonds of the pyrochlore lattice. The following algorithm is applicable to more general spin-ice type models on other frustrated lattices with farther-neighbor or bond-dependent interactions, and site-dependent anisotropy.

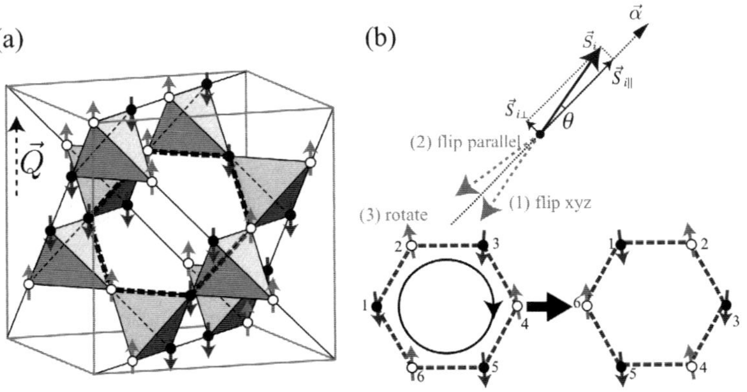

Figure 1. (a) The pyrochlore lattice composed of a three-dimensional network of corner-sharing tetrahedra. A 16-site cubic unit cell is shown. Spins are denoted by solid arrows. The broken arrow $\vec{Q}$ shows a common axis selected by b in the nematic phase. Black (filled) circles represent spins in the direction of $\vec{Q}$, while white (open) circles represent spins in the opposite direction. The spin configuration is an example of the spin-ice type states in which the 'two-up two-down' local constraint is satisfied in every tetrahedron. The hexagon with a bold dashed line denotes one of the shortest loops on which a flip of all spins (black and white) transforms the ice-rule state to another ice-rule state. (b) Three different ways to flip black and white: (1) *flip xyz*, (2) *flip parallel*, and (3) *rotate*. See the text for details.

When $b = 0$, the model given by Eq. (1) exhibits no long-range ordering down to $T = 0$, and the ground state is given by a collection of local constraints that enforces the sum of $\vec{S}_i$ to be zero in every tetrahedron. Consequently, the ground-state manifold has continuous macroscopic degeneracy [18, 19, 20]. For $b > 0$, the present model exhibits a weak first-order transition at $T_c \sim b$ to a nematic state in which spins spontaneously select a common axis $\vec{Q}$ without selecting their directions on it [16]. Hence, the ground state for $b > 0$ is identified by a collection of spin-ice type local constraints: in every tetrahedron, two out of four spins are aligned parallel to each other and the other two are antiparallel to them — 'two-up two-down' configuration [see Fig. 1(a)]. The ground-state manifold develops a multivalley structure whose minima correspond to different spin-ice-type configurations separated by large energy barriers of the order of b and J.

2.2. *Algorithm*

In the previous paper, the authors developed an extension of the loop algorithm to be applicable to a family of classical Heisenberg antiferromagnets with single-ion anisotropy, in which the single-spin flip algorithm suffers from slow relaxation due to the formation of the spin-ice type manifold [15]. In the extended loop algorithm, (i) we first project all the spins onto the anisotropy axis to assign black and white colors, (ii) next, construct a loop of alternating black and white, and (iii) flip all the spins on the loop. For the present bilinear-biquadratic model, a similar difficulty from slow relaxation is anticipated because the low-T state develops the spin-ice type degeneracy as mentioned above. However, it is not straightforward to apply the extended loop algorithm since the present model retains $O(3)$ spin rotational symmetry and does not have any explicit anisotropy axis to project the spins on. It is necessary to deduce the common axis $\vec{Q}$ selected by b for each MC sample.

Here, we propose the following procedure to define the projection axis. We first pick up a set of N_{T} tetrahedra $\{\mathcal{T}_m\}$ ($m = 1, \cdots, N_{\mathrm{T}}$) randomly from the whole system. Starting from an initial guess $\vec{\alpha}_0$ [we take $\vec{\alpha}_0 = (0, 0, 1)$], the normalized projection axis $\vec{\alpha}$ is obtained iteratively by

$$\vec{\alpha}_{n+1} \propto \sum_{i \in \{\mathcal{T}_m\}} \text{sign}(\vec{S}_i \cdot \vec{\alpha}_n)\vec{S}_i. \tag{2}$$

Here the sum is taken over all spins belonging to the selected tetrahedra $\{\mathcal{T}_m\}$, and n ($= 0, 1, \cdots, n_{\text{max}} - 1$) is the index of the iteration. For larger N_{T} and n_{max}, the resultant $\vec{\alpha} = \vec{\alpha}_{n_{\text{max}}}$ gives a better approximation of $\vec{Q}$. In practice, we take $N_{\text{T}} = 24$ and $n_{\text{max}} = 6$ in the following calculations for the system size with $N_{\text{s}} = 16 \times 8^3$ spins. We confirm that the loop flip is efficiently performed for these conditions, as demonstrated below. Once $\vec{\alpha}$ is defined in this way, the step (ii) and (iii) are done in the same way as in Ref. [15]. It should be noted that, to ensure the detailed balance, loops must be constructed avoiding the tetrahedra included in $\{\mathcal{T}_m\}$ as well as defect tetrahedra in which the ice rule is violated. Otherwise, the loop flip becomes irreversible because the flip changes $\vec{\alpha}$.

In the extended loop algorithm, the loop flip procedure to reverse all colors on a loop is not unique because spins can thermally fluctuate [15]. To choose an efficient method, careful consideration on the energy change is necessary. In Ref. [15], the authors tested two different ways: (1) *flip xyz* and (2) *flip parallel*. In *flip xyz*, all three Cartesian components of $\vec{S}_i$ are reversed as $\vec{S}_i \to -\vec{S}_i$, while in *flip parallel*, only parallel components $\vec{S}_{i\parallel}$ are reversed as $\vec{S}_i \to \vec{S}_i - 2(\vec{S}_i \cdot \vec{\alpha})\vec{\alpha}$ [see Fig. 1(b)]. The previous study revealed that, in the case of the single-ion anisotropy, only the acceptance rate for *flip parallel* can become one (rejection free) in the limit of $T \to 0$; the acceptance rate for *flip xyz* converges to a finite value less than unity as $T \to 0$ because of the effect of thermal fluctuations on the transverse component of spins [15]. In addition to these two updates, in this paper, we introduce another way to reverse colors, i.e., (3) *rotate*. In *rotate*, one translates every spin to the neighboring site on the loop simultaneously in the direction in which the loop was formed [see Fig. 1(b)]. In the next section, we try the three different methods and demonstrate that the most efficient method depends on the model parameter b.

3. Benchmark

In this section, we apply the extended loop algorithm to the model given by Eq. (1). We demonstrate the efficiency of loop flips in MC simulations and compare the efficiency of the three methods. In the following, we show the MC results for the systems size with $N_{\text{s}} = 16 \times 8^3$ spins under periodic boundary conditions. To retain the ergodicity, we use the loop flip together with the single-spin flip. One MC step consists of single-spin flips, followed by loop flips with either *flip xyz*, *flip parallel*, or *rotate*. In the single-spin flips, we randomly choose a new spin state on the unit sphere for each spin following a procedure proposed by Marsaglia [21].

At low T compared to b and J, spin configurations are enforced to satisfy the 'two-up two-down' ice rule, and the acceptance rate of the single-spin flip, P_{single}, is suppressed. This is demonstrated in Fig. 2. On the other hand, the probability that a closed loop is successfully formed, P_{loop}, steeply increases below the nematic transition temperature $T_{\text{c}} \sim b$, indicating that almost all tetrahedra start to follow the ice rule below T_{c}. [For $b > J$, the ice rule is weakly violated in the range of $T \gtrsim J$ even below T_{c}, but gradually satisfied for $T \lesssim J$, as exemplified in Fig. 2(d).] The acceptance rate of flips of a formed loop also increases below T_{c} and remains finite as $T \to 0$; here, P_{xyz}, P_{parallel}, and P_{rotate} are the rate for *flip xyz*, *flip parallel*, and *rotate*, respectively. The total acceptance rate of the loop flip is given by the product as $P_{\text{loop}} \times P_{xyz}$, $P_{\text{loop}} \times P_{\text{parallel}}$, and $P_{\text{rotate}} \times P_{\text{parallel}}$ for each method. Hence the acceptance rate of the loop flip sharply increases at $T < T_{\text{c}}$, compensating the decrease of P_{single}. These T dependences are qualitatively the same for the wide range of b, as shown in Fig. 2.

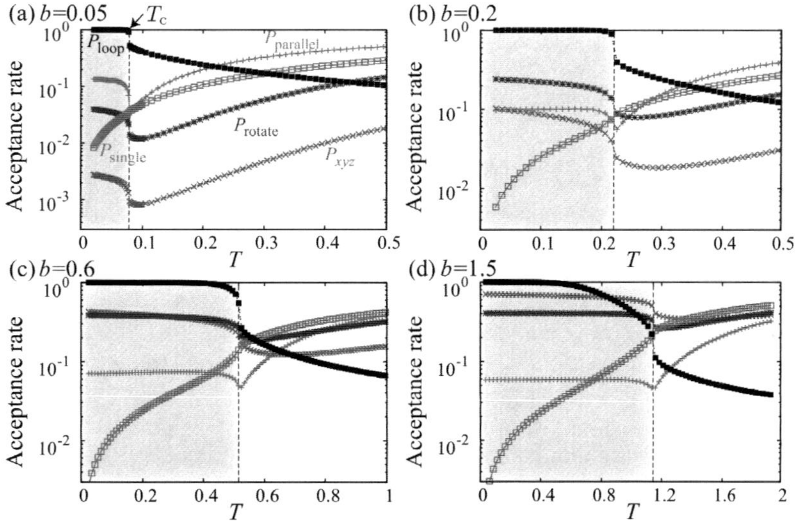

Figure 2. Temperature dependences of the acceptance rates of the single-spin flip ($P_{\rm single}$), the probability of formation of closed loops ($P_{\rm loop}$), the acceptance rates of flip of a formed loop by *flip xyz* (P_{xyz}), *flip parallel* ($P_{\rm parallel}$), and *rotate* ($P_{\rm rotate}$). The data are calculated at (a) $b = 0.05$, (b) 0.2, (c) 0.6, and (d) 1.5. The vertical broken lines denote the nematic transition temperature $T_{\rm c}$ estimated by the peak position in the specific heat (not shown).

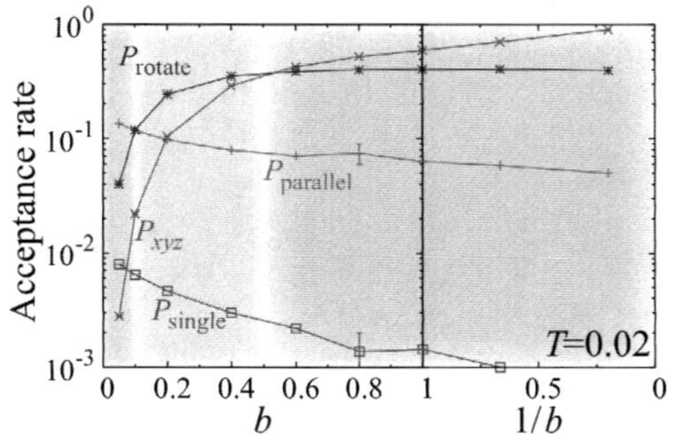

Figure 3. b dependence of $P_{\rm parallel}$, P_{xyz}, $P_{\rm rotate}$, and $P_{\rm single}$ at $T = 0.02$.

As demonstrated in the previous study for the models with single-ion anisotropy [15], the efficiency of the loop flip at low T depends on the method. Furthermore, in the present case with the biquadratic interaction b, the efficiency at low T strongly depends on b. The result is presented in Fig. 3. For small $b < 0.1$, *flip parallel* is most efficient, while it is taken over by *rotate* in the intermediate range of $0.1 < b < 0.5$, and finally by *flip xyz* for $b > 0.5$.

The difference of the efficiency is understood by the following consideration on the energy change by the flips. Considering a given state at a finite T well below $T_{\rm c}$, its energy measured from the ground-state energy is given by $E = E_J + E_b$, where E_J and E_b are the energies corresponding to the first and second terms in Eq. (1), respectively. Both E_J and E_b are of the order of T at low T. The three loop flips change the two contributions in different ways. The *flip xyz* changes E_J by a certain fraction ΔE_J but conserves E_b. Meanwhile, *flip parallel* and *rotate* change both of E_J and E_b by ΔE_J and ΔE_b, respectively. First we consider the large b limit, where $E \simeq E_b \propto T$ and $E_J/E_b \propto b^{-1}$. For *flip xyz* in which $\Delta E_b = 0$, we obtain $\Delta E/T = \Delta E_J/T \propto b^{-1} \to 0$ as $b \to \infty$. Since the acceptance rate is given by $\exp(-\Delta E/T)$, this consideration gives $\lim_{b\to+\infty} P_{xyz} \to 1$, that is, *flip xyz* becomes rejection free as $b \to +\infty$.

This is consistent with the behavior in Fig. 3. On the contrary, *flip parallel* and *rotate* cannot become rejection free for $b \to \infty$ because they change E_b by $O(T)$; P_{parallel} and P_{rotate} converge to finite values less than unity at $b \to \infty$, respectively. Note that *rotate* does not change a half of the nearest-neighbor bond energies. This may account for why $P_{\text{rotate}} > P_{\text{parallel}}$ at $b \to +\infty$.

For smaller b, spins deviate from the common axis $\vec{Q}$ with larger angle. Considering that $\vec{Q}$ is set by $E_b \sim T$ and E_b is proportional to $b\theta^2$ (θ is a typical deviation angle), we obtain $\theta = O(\sqrt{T/b})$ for $b \ll J$ [22]. Since *flip parallel* changes E_J and E_b by $O(J\theta^4)$ and $O(b\theta^2)$, respectively [15], we obtain $\Delta E/T = \Delta E_b/T = O(1)$ at low $T \ll T_c$ for *flip parallel*. This indicates that P_{parallel} does not vanish even at $b \to 0$ in the nematic phase. Meanwhile, the other two methods change the energy as $\Delta E/T = \Delta E_J/T = O(J\theta^2/T) = O(J/b)$ at low T and $b \ll J$ [15]. This suggests that their acceptance rates vanish as $b \to 0$ in contrast to *flip parallel*. Therefore, *flip parallel* becomes most efficient at $b \to 0$. These considerations are consistent with the numerical results shown in Fig. 3. In the intermediate regime, i.e, $0.1 < b < 0.5$, *rotate* is superior to the other two, presumably because of a remnant of the advantage of *rotate* over *flip parallel* at $b \to \infty$.

4. Summary

In this paper, we have extended the loop algorithm to the classical antiferromagnetic Heisenberg spin models with biquadratic interaction which have spin-ice type ground-state degeneracy. The efficiency of the extended loop algorithm has been demonstrated in Monte Carlo simulations. We have examined three different ways of loop flips, *flip xyz*, *flip parallel*, and *rotate*, and compared their efficiency. We have shown that the most efficient method depends on the strength of the biquadratic interaction b. This b dependence has been explained by considering effects of thermal fluctuations on the energy changes by the loop flips.

This work was supported by Grant-in-Aids (No. 19052008), Global COE Program "the Physical Sciences Frontier", and HPCI Strategic Program, from MEXT, Japan.

References

[1] Landau D P and Binder K 2000 *A guide to Monte Carlo simulation in statistical physics* (Cambridge: Cambridge Univ. Press)
[2] Swendsen R H and Wang J S 1987 *Phys. Rev. Lett.* **58** 86–88
[3] Wolff U 1989 *Phys. Rev. Lett.* **62** 361–364
[4] Anderson P W 1956 *Phys. Rev.* **102** 1008–1013
[5] Bernal J D and Fowlers R H 1933 *J. Chem. Phys.* **1** 515–548
[6] Pauling L 1935 *J. Am. Chem. Soc.* **57** 2680
[7] Harris M J, Bramwell S T, McMorrow D F, Zeiske T and Godfrey K W 1997 *Phys. Rev. Lett.* **79** 2554–2557
[8] Ramirez A P, Hayashi A, Cava R J, Siddharthan R and Shastry B S 1999 *Nature* **399** 333–335
[9] Rahman A and Stillinger F H 1972 *J. Chem. Phys.* **57** 4009
[10] Yanagawa A and Nagle J F 1979 *Chem. Phys.* **43** 329
[11] Barkema G T and Newman M E J 1998 *Phys. Rev.* E **57** 1155–1166
[12] Melko R G, den Hertog B C and Gingras M J P 2001 *Phys. Rev. Lett.* **87** 067203
[13] Melko R G and Gingras M J P 2004 *J. Phys.: Condens. Matter* **16** R1277
[14] Champion J D M, Bramwell S T, Holdsworth P C W and Harris M J 2002 *Europhys. Lett.* **57** 93
[15] Shinaoka H and Motome Y 2010 *Phys. Rev.* B **82** 134420
[16] Shannon N, Penc K and Motome Y 2010 *Phys. Rev.* B **81** 184409
[17] Shinaoka H, Tomita Y and Motome Y 2011 *Phys. Rev. Lett.* **107** 047204
[18] Reimers J N 1992 *Phys. Rev.* B **45** 7287–7294
[19] Moessner R and Chalker J T 1998 *Phys. Rev. Lett* **80** 2929–2932
[20] Moessner R and Chalker J T 1998 *Phys. Rev.* B **58** 12049–12062
[21] Marsaglia G 1972 *The Annals of Mathematical Statistics* **43** 645–646
[22] In the case of the single-ion anisotropy considered in Ref. [15], we assumed that $\theta \propto T$. It should be corrected as $\theta = O(\sqrt{T/D_{\text{I}}})$ (D_{I} is the anisotropy). This, however, does not alter the arguments in the previous study.

International Conference on Frustration in Condensed Matter (ICFCM) IOP Publishing
Journal of Physics: Conference Series **320** (2011) 012010 doi:10.1088/1742-6596/320/1/012010

Numerical exact diagonalization study of triangulated kagome Heisenberg spin system

M Isoda[1], **H Nakano**[2] **and T Sakai**[2,3]

[1] Department of Physics, Faculty of Education, Kagawa University, Saiwai-cho 1-1, Takamatsu 760-8522, Japan
[2] Graduate School of Material Science, University of Hyogo, 3-2-1 Kouto, Kamigori-cho, Akou-gun, Hyogo 678-1297, Japan
[3] Japan Atomic Energy Agency, SPring-8, Sayo 679-5148, Japan

E-mail: `misoda@ed.kagawa-u.ac.jp`

Abstract. The thermodynamic properties and the magnetization under magnetic eld at zero temperature of the spin-1/2 Heisenberg model on triangulated kagome lattice, which composed of two sublattices, are investigated by numerical exact diagonalization method for 18 and up to 27 spin clusters, respectively. The temperature dependence of magnetic susceptibility may be reproduced qualitatively with the weak ferromagnetic intersublattice nearest neighbor exchange coupling constant J_{AB} as proposed by previous studies, although it is quantitatively insu cient in comparison with the experimental result. The magnetization under magnetic eld shows the 5/9 plateau for every examined values of J_{AB}, but 1/3 plateau, visible in the experiment, may be reproduced only for not so weak antiferromagnetic case of J_{AB}. This discrepancy for J_{AB} in those two qualities has also been recognized in previous studies on classical spin models and is not still resolved even for the present quantum spin system.

1. Introduction

A new type of two-dimensional spin frustration system has been found, which is referred to "triangle-in-triangle" or "triangulated" kagome lattice system. Therein the small triangles are formed in every triangle forming kagome lattice. The lattice structure is realized by Cu^{2+} ions with spin-1/2 in compounds $Cu_9X_2(cpa)_6$ nH_2O (cpa=carboxypentonic acid; X=F,Cl,Br)[1-3]. In the lattice structure, lattice sites occupied by Cu^{2+} ions are classi ed into two sublattices, a sublattice A forming smaller triangles and another sublattice B forming large triangle constructing a kagome lattice as shown in Fig.1. The unit cell of this lattice structure includes nine sites, as shown in Fig.1(a), and includes six A-sublattice sites and three B-sublattice ones.

All these compounds have a magnetic ground state, but any phase transition to long-range order has not been observed down to 1.7K[1]. The measurements of magnetic susceptibility and magnetization under magnetic eld have shown the peculiar properties. On the magnetic susceptibility its inverse reveals the clear bend deviating from high temperature Curie-Weiss like behavior[1-3]. On its properties, the Weiss temperature changes from -230K at high temperature to 6K at below 50K and equivalently the slope of inverse magnetic susceptibility becomes to more than ve times. On the latter, the magnetization measured under magnetic eld up to 38T shows the tendency of saturation to one third of full polarization value[2,3].

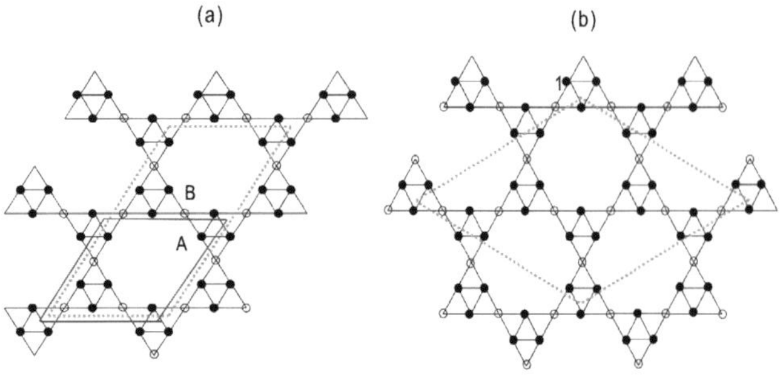

Figure 1. Lattice structure of triangulated kagome lattice. The 18 site cluster used for calculations is denoted by green-broken lines (a), where the unit cell of 9 sites is also shown by enclosing by brown-solid lines. Symbols A(solid circle) and B(open circle) denote A- and B-sublattice, respectively. The 27 site cluster used for calculation is shown in (b) enclosing by green-broken lines.

On the other side, theoretically, Ising model[4,5] and Ising-Heisenberg mixed model[6,7], where Heisenberg spin is resided on A-sublattice and Ising spin on B-sublattice, are treated exactly through a mapping to the ferromagnetic Ising model on the kagome lattice. In these studies, two types of magnetic interaction are taken into account, one for A-A interactions among small triangle J_{AA} and the other for nearest neighbor(nn) A-B interactions J_{AB}. The interaction J_{AA} is believed to be strongly antiferromagnetic[1,4,5,6], but J_{AB} still remains uncertainly even in its sign[5], in concern with the experiment of $Cu_9X_2(cpa)_6$ $n H_2O$. As an example, in comparison of thermodynamics of experimental results with those of Ising model, the temperature dependence of magnetic susceptibility multiplied by temperature, χT, supports a ferromagnetic coupling between A and B intersublattice nn spins J_{AB}[1,5], but oppositely the reproduction of 1/3 plateau in antiferromagnetic case in Ising model[2,5] might indicate the antiferromagnetic interaction of J_{AB}. Preceding these works, spin-1/2 quantum Heisenberg model was examined by the variational procedure based on the Bogoliubov inequality by Strečka[8]. In these theoretical investigations[6,8], the exchange coupling J_{AB} has been chosen as ferromagnetic and $J_{AB}/J_{AA} \simeq 0.025$. However the determination both of sign and magnitude of J_{AB} still remains without giving the consistent explanation for both experiments of magnetic susceptibility and magnetization.

According to the ESR investigation[9], it is reported that the g-value of Cu^{2+} in $Cu_9X_2(cpa)_6$ looks unusually isotropic. Then it is highly interested in how the magnetic susceptibility as a function of temperature and the 1/3 plateau in magnetization are varied in isotropic quantum spin model against those in the Ising and Ising-Heisenberg model.

In this study, the quantum Heisenberg model will be studied for the thermodynamics and the magnetization under magnetic eld with the variation of exchange coupling constants. We perform the numerical exact diagonalization study for the clusters of the lattice of $N = 18$ for nite temperature properties and up to $N = 27$ spins for the zero temperature property with periodic boundary condition. The chosen system sizes, which are twice and three times of lattice sites of unit cell, might be too small to derive reliable results for thermodynamic limit. However, the numerical diagonalization method may be a most trustworthy or only one numerical method for two-dimensional frustrated system, to our knowledge.

2. Numerical calculations

2.1. Thermodynamics

The present quantum spin Heisenberg model upon the underlying triangulated kagome lattice is given as

$$\mathcal{H} = J_{\mathrm{AA}} \sum_{<i,j>} \mathbf{S}_i^{\mathrm{A}} \quad \mathbf{S}_j^{\mathrm{A}} + J_{\mathrm{AB}} \sum_{<i,j>} \mathbf{S}_i^{\mathrm{A}} \quad \mathbf{S}_j^{\mathrm{B}} \quad H \sum_i (S_i^{z,\mathrm{A}} + S_i^{z,\mathrm{B}}), \tag{1}$$

where J_{AA} is the nn coupling constant among A-sublattice spins and J_{AB} the nn one between A-sublattice spin and B-sublattice one. The summation in the rst term should be performed for nn pairs of sites i and j, both belonging to A-sublattice, but it in the second term for nn pairs between two spins belonging to di erent sublattice. The quantum spin-1/2 operator on site i of sublattice-A is depicted as $\mathbf{S}_i^{\mathrm{A}}$ and the superscript z on the spin operator denotes the z-component of its operator. H is a magnetic eld applied along z direction. Through the paper, we use the conventional energy scale, $J' \equiv J_{\mathrm{AB}}/J_{\mathrm{AA}}$ and $h \equiv H/J_{\mathrm{AA}}$, and set as $J_{\mathrm{AA}} = 1$, assuming an antiferromagnetic interaction. The temperature T is also measured in this units of energy.

The eigen values for the Hamiltonian are obtained by using the TITPACK. For getting the thermodynamic quantities of magnetic susceptibility χ and speci c heat C, the full spectra of the small nite number of spins of $N = 18$ have been derived by the Householder algorithm. Based on the obtained spectra, the speci c heat and the magnetic susceptibility are calculated through the usual canonical thermal average procedure for various values of J' and the results are shown in Fig.2(a) and (b), respectively.

In Fig.2(a), the speci c heat shows two or three peaks (including shoulders in the cases of weakly antiferromagnetic $J' = 0.1, 0.3$). Such multiple peak structure is also found in kagome and triangular lattice quantum spin systems[10]. The highest-temperature peaks scarcely shift on the temperature axis regardless of J'. This fact has also been suggested in kagome and triangular lattice as due to a short range correlation[10]. The experimental speci c heat measurement for $\mathrm{Cu_9X_2(cpa)_6} \quad n\mathrm{H_2O}$ has not yet done as far as we know. The comparison our calculated result with experimental one has to be postponed.

In Fig.2(b), the inverse magnetic susceptibility and the magnetic susceptibility multiplied by temperature are shown. For the ferromagnetic interaction $J_{\mathrm{AB}}(< 0)$ even for small value as $J' = \quad 0.025$, the inverse magnetic susceptibility seems to vanish at $T = 0$ suggesting the magnetic ground state, but in antiferromagnetic cases $\chi\ ^{1}$ divergently increase toward zero temperature implying the singlet ground state. A prominent character is a downturn deviation below about $T \leq 0.5$ from high temperature Curie-Weiss like behavior which is distinctly observed in weak intersublattice coupling cases of $|J'| \quad 1$. Such character has been found in Monte Carlo (MC) simulation study for Ising spin[5] and has also been indicated by us previously as an inherent character of triangle-based frustrated lattice system for both models of Heisenberg[10] and Ising[11]. In those references, below the temperature, where $\chi\ ^{1}$ bends downward with decreasing temperature, a formation of a quantum doublet state on a triangle trimer of A-sublattice in triangulated kagome lattice might be developed as the short range correlation for weak intersublattice coupling $|J'| \quad 1$. This quantum trimer formation below the characteristic temperature may be represented as a classical-quantum crossover at the temperature. The higher temperature peak of speci c heat at around $T \quad 0.5$ emerged for every cases in Fig.2(a) may be related the characteristic temperature found in magnetic susceptibility[10,11].

The magnetic susceptibility multiplied by temperature, χT, as a function of temperature is also shown in Fig.2(b). In the ferromagnetic intersublattice coupling, the quantity χT has a

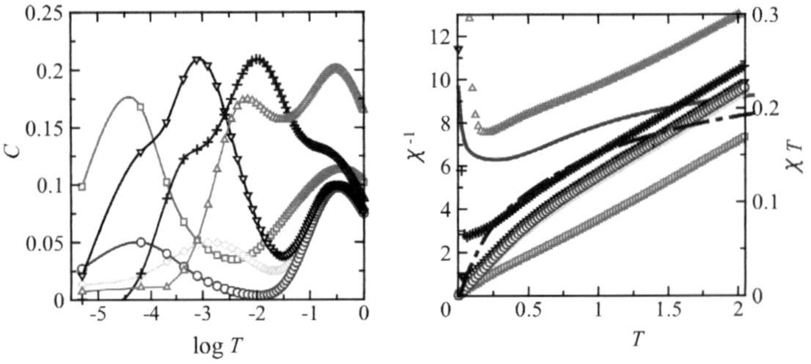

Figure 2. The speci c heat C as a function of logarithmic temperature (a) and the inverse magnetic susceptibility χ ^{1} (referred the left side axis) and the magnetic susceptibility multiplied by temperature χT (referred the right side axis) as a function of T (b) calculated by numerical exact diagonalization method for various values of J' on 18-spin cluster with periodic boundary condition. In both gures, the curves C and χ ^{1} for $J' =$ 1, 0.1, 0.025, 0.1, 0.3, 1 are depicted by red-square, green-diamond, blue-circle, black-reverse triangle, brown-cross and dark green-triangle, which are in order upward in (b). In (b), χT for $J' =$ 0.025 and 0.3 are depicted by a blue solid line and a brown dash-dot-dash line, respectively.

minimum and then steeply increase with decreasing temperature, but it monotonically decreases down to zero for antiferromagnetic $J_{\mathrm{AB}}(< 0)$. This di erence in the temperature dependence of χT between two cases di erent in sign of J' has been found for the Ising spin on the triangulated kagome lattice[5]. The experimental observation[1] shows a clear minimum well corresponding with the solid line in Fig.2(b) depicted as an example for weak ferromagnetic intersublattice coupling. The value of $J' =$ 0.025 is an evaluated one from experimentally observed magnetic susceptibility[6,8]. This fact seems to support the weak ferromagnetic coupling between A and B sublattices.

2.2. Magnetization under magnetic field

In next place, we investigate the magnetization under magnetic eld at zero temperature. For the purpose, we use the Lanczos algorithm for $N = 18$ and 27 spin clusters with the periodic boundary condition.

The obtained results for magnetization normalized by the full polarized value in each system size, m, as a function of magnetic eld h are devoted in Fig.3(a) and (b). An example for the case, where the 1/3 plateau is seen over a recognizable range in ratio to the eld strength for full polarization h 2 in the case of Fig.3(a), is shown in Fig.3(a) for $J' = 0.5$. The black solid line denotes the result for system size $N = 18$ and the red dash-dot-dash line for $N = 27$. In the case of $J' = 0.2$ (gure is not shown), the range of 1/3 plateau shrinks to about 0.2 su ciently smaller than the range of 5/9 plateau of about 1.2 and the full polarization eld of 1.7. As another example, the magnetization for an intermediate ferromagnetic coupling $J' =$ 0.5 is shown in Fig.3(b) for the same system sizes. Other some values for weak ferro- and antiferromagnetic interactions such as $J' = \pm 0.025$ were also examined for N=18 and 27 clusters. These cases of $J' = \pm 0.025$ are carefully examined, because the ferromagnetic case $J' =$ 0.025 corresponds to the value evaluated from the experiment of magnetic susceptibility and may well explain the temperature dependence of χT qualitatively as discussed in previous subsection. For the value of J', the 5/9 plateau is attained by very weak magnetic eld and survives rmly up to full polarization eld of about 1.5, but unfortunately 1/3 plateau does not realize over recognizable range.

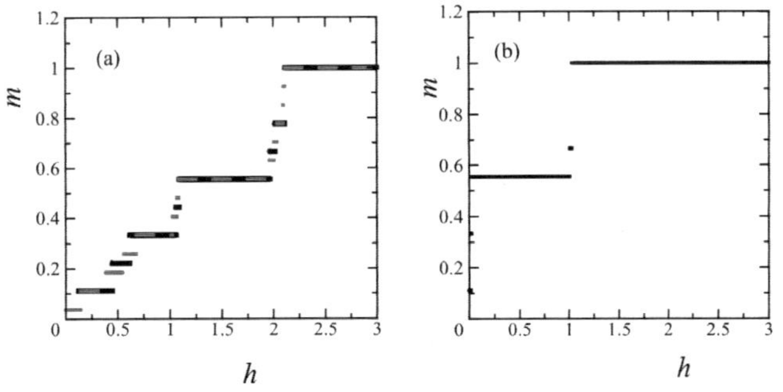

Figure 3. The magnetization m in units of fully polarized value under magnetic eld h. (a) is for $J' = 0.5$ as an example of intermediate antiferromagnetic intersublattice coupling. The black solid lines depict for $N = 18$ and the red dashed lines for $N = 27$. (b) is given for $J' =$ 0.5 as an example of other J' values in comparison with (a).

The emergence of 5/9 plateau has been recognized for corresponding value of $J' =$ 0.025 in Ref.8 for variational procedure for the same model as given in eq.(1). On the other side, the Ising spin MC simulation [5] has shown the 1/3 plateau for $J' = 0.5$. The present result well corresponds with these previous studies.

The spin state in 5/9 plateau found in this study may be readily understood for the present case of J' 1. Under the antiferromagnetic coupling J_{AA}, every A-site triangle is regarded as an e ective 1/2-spin. Thus six spins in A-site per unit cell contribute a magnetization $m_{\mathrm{A}} = 2/9$ under nite magnetic eld. Three spins in B-site, each bonded to four A-site spins weakly by J_{AB}, are aligned parallel to the magnetic eld, contributing $m_{\mathrm{B}} = 1/3$ per unit cell. Both of these gives $m = m_{\mathrm{A}} + m_{\mathrm{B}} = 5/9$. While, on the 1/3 plateau, the quantum spin state can not yet be determined through the present numerical study. As giving a brief insight, an classical Ising model has been examined. It is stabilized at the magnetic eld of $h \leq 2J'$ against the 5/9 plateau state only for the antiferromagnetic $J_{\mathrm{AB}}(J' > 0)$. There the spontaneous symmetry-breaking state, where a single B-site spin in a unit cell directs antiparallel to the magnetic eld and four A site spins coupled to this B-site spin are parallel to the magnetic eld, maintaining $S_z = 1/2$ for every A site trimer. This estimation of the transition eld between 1/3 and 5/9 plateau states seems to give a good estimation of the numerical results on quantum spins in Fig.3 (a), together with the sign of exchange coupling J_{AB}.

3. Conclusions and discussions

We have examined the spin-1/2 quantum Heisenberg spin model on the triangulated kagome lattice, by taking account of the intrasublattice nn interaction in A-sublattice and the intersublattice nn one, by a numerical exact diagonalization method. The speci c heat and the magnetic susceptibility have been calculated as a function of temperature for cluster-size of $N = 18$ and the magnetization at zero temperature as a function of magnetic eld for $N = 18$ and 27.

On the speci c heat, two or more peaks have been detected and the temperature at the highest temperature peak is insensitive on the intersublattice nn coupling. The peak might give the sign of the development of short range correlation up to nearly nn distance. Such multiple peaks are also detected in kagome and triangular lattice Heisenberg systems[10]. The experimental measurement of speci c heat has not been performed up to now, as far as we know, and then is highly desired.

The magnetic susceptibility shows the clear di erence in the low temperature behavior depending on the sign of intersublattice coupling constant $J_{\rm AB}$. For ferromagnetic case ($J_{\rm AB} < 0$), the inverse magnetic susceptibility vanishes to zero, but for antiferromagnetic case ($J_{\rm AB} > 0$) it increases divergently as approaching zero temperature. A characteristic feature is found in the case of weak intersublattice coupling $|J'|$ 1, the downward deviation from high temperature Curie-Weiss like temperature dependence below about $T \simeq 0.5$. In our result, the slope of inverse magnetic susceptibility increases up to about twice for the typical case of $J' =$ 0.025, which is the value proposed in some preceding issues[6,8]. However the experimental result shows larger change in slope more than four or ve times[2,3], suggesting the remarkable quantitative di erence. On another aspect, the magnetic susceptibility multiplied by temperature χT for $J' =$ 0.025 shows minimum and then increases as decreasing temperature, giving the qualitative consistency with the experiment[1].

The magnetization under magnetic eld at zero temperature has revealed the 5/9 plateau for every examined cases of $|J'| < 1$, but the 1/3 plateau visible in the experiment[2,3] has been reproduced not so weak J' under the assumption of $|J'| < 1$. The plateau ranges in both plateaus are not modi ed remarkably by changing the system size between $N = 18$ and 27 or by adding the small value of further neighbor interactions, although gures are not given. The elucidation of the quantum spin states of 1/3 and 5/9 plateaus should be perfomed in future.

The investigated quantum Heisenberg model has not suggested the qualitative di erence in the temperature dependence of susceptibility and the magnetic eld dependence of magnetization from the classical Ising[4,5] and the Ising-Heisenberg mixed models[6,7]. The contradiction both in its sign and magnitude of $J_{\rm AB}$ for the interpretation of two physical quantities discussed above is not still resolved and remains as a future problem. The careful estimation of further neighbor exchange interaction by the inspection of exchange path might be required or the additional mechanism like Dzyaloshinskii-Moriya interaction might be needed[12].

According to the recent survey on the magnetic plateau in quantum spin system, the plateau in spin-1/2 Heisenberg kagome system is realized like a ramp with nite slope[13]. In the triangulated kagome system, whether the observed plateaus are *real* plateaus or should be called as a ramp is a future problem.

Acknowledgments
Authors are grateful to Professors K. Okamoto and K. Ueda for valuable discussions. A part of this study was supported by Grant-in-Aid for Scienti c Research on Priority Areas "Novel States of Matter Induced by Frustration" (22014012). The computation in this work has been done using the facilities of the supercomputer Center, Institute for Solid State Physics, University of Tokyo and the diagonalization was based on TITPACK ver.2 coded by Professor H. Nishimori.

References

[1] Maruti S and ter Haar L W 1994 *J. Appl. Phys.* **75** 5949
[2] Mekata M, Abdulla M, Asano T, Kikuchi H, Goto T, Morishita T and Hori H 1998 *J. Magn. Magn. Mater.* **177-181** 731
[3] Mekata M, Abdulla M, Kubota M and Oohara Y 2001 *Can. J. Phys.* **79** 1409
[4] Zheng J and Sun G 2005 *Phys. Rev.* B**71** 052408
[5] Loh Y L, Yao D X and Carlson E W 2008 *Phys. Rev.* B**77** 134402
[6] Strečka J, Čanová L, Jaščur M and Hagiwara M 2008 *Phys. Rev.* B **78** 024427
[7] Yao D X, Loh Y L, Carlson E W and Ma M 2008 *Phys. Rev.* B **78** 024428
[8] Strečka J 2007 *J. Magn. Magn. Mater.* **316** e346
[9] Okubo S, Hayashi M, Kimura S, Ohta H, Motokawa M, Kikuchi H and Nagasawa H 1998 *Physica* B **246-247** 553
[10] Isoda M, Nakano H and Sakai T 2011 to be published in Mod. Phys. Lett. B
[11] Isoda M 2008 *J. Phys.: Condens. Matter.* **20** 315202
[12] Strečka J and Čanová L 2008 **145** *J. Phys.: Conf. Ser.* 012012
[13] Nakano H and Sakai T 2010 **79** *J. Phys. Soc. Jpn.* 053707

Self-consistent spin wave analysis of the magnetization plateau in triangular antiferromagnet

J Takano[1], H Tsunetsugu[1] and M E Zhitomirsky[2,3]

[1]Institute for Solid State Physics, University of Tokyo, Kashiwanoha 5-1-5, Kashiwa, Chiba 277-0882, Japan
[2]SPSMS, UMR-E9001, CEA-INAC/UJF, 17 rue des Martyrs, 38054 Grenoble Cedex 9, France
[3]Max-Planck-Institut für Physik Komplexer Systeme, Nöthnitzer str. 38, D-01187 Dresden, Germany

E-mail: takano@issp.u-tokyo.ac.jp

Abstract. We investigate theoretically the 1/3 magnetization plateau of the triangular-lattice Heisenberg antiferromagnet at zero temperature. Using the self-consistent spin-wave analysis, we show that the 1/3 plateau is stabilized in a finite range of magnetic fields: $0.844 < H/3SJ < 1.368$ for $S = 1/2$. By analyzing the Bose condensation of magnons, we demonstrate that the magnetization curve has logarithmic singularities at the both ends of the plateau and determine the prefactor of the singularity up to the leading order in $1/S$.

1. Introduction

The Heisenberg antiferromagnet on a triangular lattice has been intensively studied in the past as the basic example of geometric frustration in magnetism. One of the characteristic features of the triangular antiferromagnet is the magnetization process. Existence of the 1/3 magnetization plateau with spins forming the up-up-down structure has been theoretically established by spin wave [1], exact diagonalization [2], and coupled cluster expansion [3] studies. Recently, the 1/3-magnetization plateau has been observed in several magnetic materials, Cs_2CuBr_2 [4] with $S = 1/2$, $RbFe(MoO_4)_2$ [5] and $Rb_4Mn(MoO_4)_3$ [6] with $S = 5/2$. However, there is a quantitative discrepancy between analytical and numerical results about the plateau transition points. Chubukov *et al.* evaluated the plateau region in the leading order of $1/S$ as $1 \quad 0.084/S < H/3SJ < 1 + 0.215/S$ [1]. The exact diagonalization study [3] yields $0.92 < H/3SJ < 1.44$ for $S = 1/2$. While the upper plateau boundary is given rather accurately by the $1/S$ expansion, there is substantial discrepancy in the prediction of the lower transition point. Moreover, nature of the transitions at the endpoints of the plateau and corresponding critical behavior are not well understood by either analytical or numerical studies.

To improve quantitative underderstanding of the 1/3 plateau, we perform the self-consistent spin wave analysis in the plateau phase. We derive the plateau transition points, excitation spectrum inside the plateau phase and a singularity of magnetization curve at both ends of the plateau. Interaction between bosons has an essential role to stabilize the plateau phase, and we take it into account in the following two steps. First, we perform Hartree-Fock approximation on bosonic Hamiltonian defined in the vacuum of up-up-down state. We calculate the width of magnetization plateau phase with gapped excitations, which is qualitatively the same as the

Published under licence by IOP Publishing Ltd

previous study based on $1/S$ expansion [1]. Second, we analyze the singularity of the plateau transition. By taking successive magnon-magnon scatterings into account, we show a logarithmic singularity of magnetization curve at the endpoints of the plateau. Finally we compare our results with the other studies.

2. Self-consistent calculation in up-up-down vacuum

First we derive a bosonic Hamiltonian by Holstein-Primakoff transformation with the vacuum of up-up-down configuration, and analyze it using Hartree-Fock approximation. We obtain excitation spectrum with a finite energy gap, which is a linear function of magnetic field. We also determine the position of plateau transition by calculating the magnetic field where the gap closes.

We study the antiferromagnetic Heisenberg model on a triangular lattice in magnetic field

$$\mathcal{H} = J\sum_{\langle ij\rangle} \boldsymbol{S}_i\ \ \boldsymbol{S}_j\ \ \ 3SJh\sum_i S_i^z. \quad (J=1) \tag{1}$$

Note that classically the magnetization curve is straight and the up-up-down configuration is stable only at $h = 1$. To bosonize this Hamiltonian, we use the Holstein-Primakoff transformation introducing three types of boson operators

$$S_i^z = \ \ S + b_i^A,\ S_i^+ = \ \ b_i^{A\dagger}\sqrt{2S\ \ b_i^{A\dagger}b_i^A},\ S_i\ \ = \ \ \sqrt{2S\ \ b_i^{A\dagger}b_i^A}b_i^A,\ i\in A \text{ sublattice}$$

$$S_i^z = S\ \ b_i^a,\ S_i^+ = \sqrt{2S\ \ b_i^{a\dagger}b_i^a}b_i^a,\ S_i\ \ = b_i^{a\dagger}\sqrt{2S\ \ b_i^{a\dagger}b_i^a}.\ i\in\ a \text{ sublattice } (a=B,C) \tag{2}$$

Second we perform Hartree-Fock decoupling on quartic terms to derive an effective quadratic Hamiltonian. To do this, we introduce various kinds of boson-pair expectation values, which are self-consistently determined:

$$n_a \equiv \langle b_i^{a\dagger}b_i^a\rangle,\ m_{aa'} \equiv \langle b_i^{a\dagger}b_j^{a'}\rangle,\ \Delta_{aa'} \equiv \langle b_i^a b_j^{a'}\rangle, \tag{3}$$

where

$$\nu_{\boldsymbol{k}} = \frac{1}{3}\ \ e^{i\boldsymbol{k}\cdot\boldsymbol{\delta}_1} + e^{i\boldsymbol{k}\cdot\boldsymbol{\delta}_2} + e^{i\boldsymbol{k}\cdot\boldsymbol{\delta}_3}\ \ , \tag{4}$$

with lattice vectors $\boldsymbol{\delta}_1 = (1,0), \boldsymbol{\delta}_2 = (\ \ 1/2, \sqrt{3}/2), \boldsymbol{\delta}_3 = (\ \ 1/2,\ \ \sqrt{3}/2)$ and $(a,a') = (A,B),(B,C),(C,A)$. Some of these are related to each other and we use these relations here without explicit explanation. It is important to notice that the mean-field Hamiltonian commutes with total S^z operator and therefore with the Zeeman term. The mean-Field Hamiltonian can be thus decoupled as

$$\mathcal{H}_{\rm HF} = 3S\sum_{\boldsymbol{k}}\ \ {}^{\dagger}_{\boldsymbol{k}}\left[\hat{\mathcal{H}}_0(\boldsymbol{k})\ \ \ \delta h\hat{\Lambda} + \frac{1}{S}\hat{\mathcal{H}}_1(\boldsymbol{k})\right]\ \ {}_{\boldsymbol{k}} + \text{const.},\quad {}_{\boldsymbol{k}} = \ \ b_{\boldsymbol{k}}^A,\ b^{B\dagger}_{\ \ \boldsymbol{k}}, b^{C\dagger}_{\ \ \boldsymbol{k}}\ \ ^T. \tag{5}$$

Here $\hat{\mathcal{H}}_0(\boldsymbol{k})$ is quadratic term at magnetic field $h = 1$

$$\hat{\mathcal{H}}_0(\boldsymbol{k}) = \begin{pmatrix} 1 & \nu_{\boldsymbol{k}} & \nu_{\boldsymbol{k}}^* \\ \nu_{\boldsymbol{k}}^* & 1 & \nu_{\boldsymbol{k}} \\ \nu_{\boldsymbol{k}} & \nu_{\boldsymbol{k}}^* & 1 \end{pmatrix}. \tag{6}$$

$\delta h\hat{\Lambda}$ is the remaining Zeeman energy with $\delta h = h\ \ \ 1$, $\hat{\Lambda} = \text{diag}(1,\ \ 1,\ \ 1)$. $\hat{\mathcal{H}}_1$ denotes the mean field term

$$\hat{\mathcal{H}}_1(\boldsymbol{k}) = \begin{pmatrix} 2(n_B\ \ \ \Delta_{CA}) & \frac{1}{2}(n_A+n_B\ \ \ 2\Delta_{CA})\nu_{\boldsymbol{k}} & \frac{1}{2}(n_A+n_B\ \ \ 2\Delta_{CA})\nu_{\boldsymbol{k}}^* \\ \frac{1}{2}(n_A+n_B\ \ \ 2\Delta_{CA})\nu_{\boldsymbol{k}}^* & n_B\ \ \ n_A\ \ \ m_{BC}+\Delta_{CA} & (m_{BC}\ \ \ n_B)\nu_{\boldsymbol{k}} \\ \frac{1}{2}(n_A+n_B\ \ \ 2\Delta_{CA})\nu_{\boldsymbol{k}} & (m_{BC}\ \ \ n_B)\nu_{\boldsymbol{k}}^* & n_B\ \ \ n_A\ \ \ m_{BC}+\Delta_{CA} \end{pmatrix}. \tag{7}$$

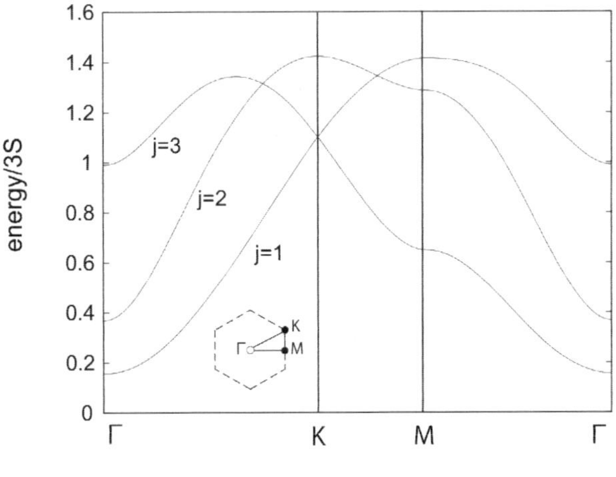

Figure 1. Magnon energy dispersion $\omega_j(\boldsymbol{k})$ at $h = 1$. j=2(1,3) corresponds to $S^+(S\ \ )$ mode.

We use the Bogoliubov transformation and diagonalize the quadratic Hamiltonian(5) as

$$\mathcal{H}_{\mathrm{HF}} = 3S \sum_{\boldsymbol{k}} \sum_{j=1}^{3} \omega_j(\boldsymbol{k}) \beta^{\dagger}_{j\boldsymbol{k}} \beta_{j\boldsymbol{k}}. \tag{8}$$

It is important that the Bogoliubov transformation is independent of magnetic field h and three modes consist of one S^+ mode ($j = 2$) and two $S\ \ $ modes ($j = 1, 3$), as the original boson operators defined by (2). Therefore, the only effect of applied field in the plateau phase is uniform shift of three magnon branches:

$$\omega_{1,3}(\boldsymbol{k}) = \omega_{1,3}^{(h=1)}(\boldsymbol{k}) + \delta h, \quad \omega_2(\boldsymbol{k}) = \omega_2^{(h=1)}(\boldsymbol{k}) \quad \delta h. \tag{9}$$

Plateau transition points h_{c1} and h_{c2} ($h_{c1} < h_{c2}$) are defined by the magnetic field where the lowest mode touches energy zero

$$h_{c1} = 1 \quad \omega_1^{(h=1)}(\boldsymbol{k} = 0), \quad h_{c2} = 1 + \omega_2^{(h=1)}(\boldsymbol{k} = 0). \tag{10}$$

Note that the two transition points are not symmetric about $h = 1$ due to the difference of energy gaps of the two modes.

Finally we calculate the spectrum and estimate the plateau transition points. We solve the self-consistent equations numerically and derive the magnon spectra $\{\omega_j(\boldsymbol{k})\}$. We show the spectra for the $S = 1/2$ case in fig.1. They are given by the absolute values of the three roots of the cubic equation

$$x^3 + 0.778x^2 + (\quad 1.921 + 1.654|\nu_{\boldsymbol{k}}|^2)x \quad 1.723 + 4.072|\nu_{\boldsymbol{k}}|^2 \quad 1.203(\nu_{\boldsymbol{k}}^3 + \nu_{\boldsymbol{k}}^{*3}) = 0, \tag{11}$$

where one positive root corresponds to S^+ mode ($j = 2$) and two negative roots correspond to $S\ \ $ modes ($j = 1, 3$).

Plateau transition points are calculated as explained above and the results are

$$(h_{c1}, h_{c2}) = \begin{cases} (0.844,\ 1.368) & S = 1/2 \\ (0.920,\ 1.195) & S = 1 \end{cases} . \tag{12}$$

The plateau width is thus 0.524 for $S = 1/2$. This is quite improved towards tha numerical value compared with the $1/S$ expansion result, although the values of transition points themselves are not so precise.

3. Analysis of plateau transition

We now analyze singularity of the plateau transitions. These transitions can be viewd as Bose condensation of one of the bosonic excitations derived in the previous section. We take into account successive magnon scatterings, which has dominant contribution to the self energy of the boson near the transition point, and show the presence of logarithmic singularity of magnetization curve.

When the energy gap closes, the Hartree-Fock solution derived in the previous section becomes unstable and Bose condensation occurs. This physically means the onset of canting of spin configuration. Our goal is to calculate the density of condensed bosons asymptotically correctly. To describe this singularity, we consider a state slightly outside the plateau phase and set $h = h_{cj} \pm \mu,\ j = 1, 2\ \mu \quad 1$. Since only the lowest mode ω_j is relevant to the critical behavior, we can neglect the other two modes and regard the system as consisting of only one type of bosons which have kinetic energy $\omega_j^{(h=h_{cj})}(\boldsymbol{k}) \quad \mu$. For convenience, we consider an equivalent alternative model: the ground canonical ensemble with chemical potential μ and gapless dispersion relation $\omega_j(\boldsymbol{k}) \quad \omega_j(\mathbf{0}) \equiv \epsilon(\boldsymbol{k})$. With this set-up, the density of condensed bosons is determined by Hugenholtz-Pines relation [7]

$$\Sigma_{11}(\boldsymbol{k}=0, \omega=0) \quad \Sigma_{12}(\boldsymbol{k}=0, \omega=0) = \mu, \tag{13}$$

where Σ_{11} and Σ_{12} denote normal and anomalous self energy, respectively. Note that this chemical potential μ is defined including the factor $1/3S$. As is clear from fig.1, the dispersion relation is quadratic near $\boldsymbol{k} = 0$, and can be written as $\epsilon(\boldsymbol{k}) \quad a\boldsymbol{k}^2$.

Let us consider which terms in the self energy are dominant. Ladder type contribution shown in fig.2(a) diverges as $\boldsymbol{k} \to 0,\ \omega \to 0$, and we have to correctly take their contribution into account to determine the critical behavior. Moreover, the Green's function in fig.2(a) should be calculated self-consistently to avoid a divergence, and it is necessary to take the terms illustrated in fig.2(b) into account. This approximation is also shown as fig.2(c), using vertex part determined by the Bethe-Salpeter equation fig.2(d). Hereafter we analyze the asymptotic behavior of magnetization as $\mu \to 0$, which is determined by the self-consistent equation for self energy fig.2(c)(d) and Hugenholtz-Pines relation (13) in the large S limit.

We further simplify the Green's functions in fig.2(d). Since the Green's function with small $\boldsymbol{k}$ and ω has dominant contribution in the vertex part, the self energy may be replaced by the limiting value:

$$\Sigma_{11}(\boldsymbol{k}, \omega) \to \Sigma_{11}(\mathbf{0}, 0),\ \Sigma_{12}(\boldsymbol{k}, \omega) \to \Sigma_{12}(\mathbf{0}, 0). \tag{14}$$

Moreover, as $\mu \to 0$, the normal vertex part shown in fig.2(e) predominates over the anomalous vertex part fig.2(f), and we neglect the latter. As a result, the normal and anomalous self energies determined by fig.2(c) satisfy $\Sigma_{11}(\mathbf{0}, 0) = 2\Sigma_{12}(\mathbf{0}, 0)$. Combining this with (13), we obtain

$$\Sigma_{11}(\mathbf{0}, 0) = 2\mu,\ \Sigma_{12}(\mathbf{0}, 0) = \mu. \tag{15}$$

These self energies are to be used to evaluate the Green's function in the vertex part fig.2(d).

Then we evaluate the normal vertex part with zero momentum, Γ. The Bethe-Salpeter equation shown in fig.2(d) is solved as

$$\Gamma = \frac{\Gamma^{(0)}}{1 \quad i\Gamma^{(0)} \sum_{\boldsymbol{k}} \int \frac{d\omega}{2\pi} G_{11}(\boldsymbol{k}, \omega) G_{11}(\quad \boldsymbol{k}, \quad \omega)}, \tag{16}$$

where the zeroth-order vertex is approximated by a constant:

$$\Gamma^{(0)}(\boldsymbol{k}_1, \boldsymbol{k}_2, \boldsymbol{k}_3, \boldsymbol{k}_4) = \Gamma^{(0)} \quad \Gamma^{(0)}(\mathbf{0}, \mathbf{0}, \mathbf{0}, \mathbf{0}). \tag{17}$$

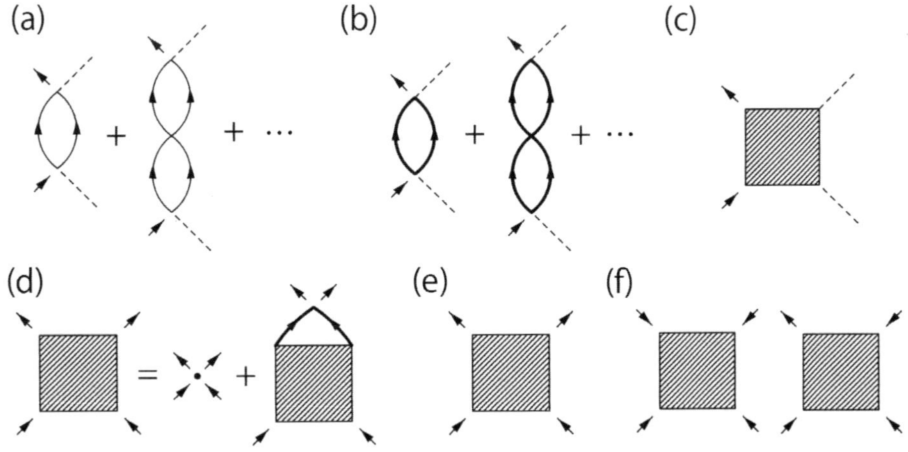

Figure 2. (a)Ladder type contribution to boson self energy. Dotted line denotes condensed boson at $\boldsymbol{k} = 0$. (b)Ladder type contribution with self-consistently determined Green's function. (c) Alternative expression of (b) using vertex part. (d)Bethe-Salpeter equation which determines the vertex part in (c). (e) Normal component of vertex part. (f) Anomalous component of vertex part.

The zeroth-order vertex $\Gamma^{(0)}$ can be calculated from the original bosonic Hamiltonian represented in terms of Bogoliubov quasiparticles derived in the previous section, and the result is $\Gamma^{(0)} = 3/SN$. Although the integral in (15) has an infrared divergence, we can cancel out this divergence by including the contribution of the anomalous Green's function G_{12}:

$$\int \frac{d\omega}{2\pi} \left[G_{11}(\boldsymbol{k},\omega) + G_{12}(\boldsymbol{k},\omega)\right]\left[G_{11}(\ \boldsymbol{k},\ \ \omega) + G_{12}(\ \boldsymbol{k},\ \ \omega)\right]. \tag{18}$$

The resulting vertex part Γ contains both the normal and anomalous parts. But we regard Γ as the normal part, since the contribution from the latter is negligible as $\mu \to 0$. Explicit calculation yields

$$\Gamma = \frac{3}{SN}\left(1 \quad \frac{3\sqrt{3}\ln\mu}{16\pi Sa}\right)^{1}. \tag{19}$$

Finally we determine the number of condensed bosons N_0 and then determine the singularity of magnetization. Self energy is determined by the vertex part as shown in fig.2(c) and we obtain

$$\Sigma_{11}(0) = 2N_0\Gamma = \frac{6N_0}{SN}\left(1 \quad \frac{3\sqrt{3}\ln\mu}{16\pi Sa}\right)^{1}. \tag{20}$$

Combining this with the first relation in (15), we obtain the number of condensed bosons:

$$N_0 = \frac{N}{3}\left(S\mu \quad \frac{3\sqrt{3}\mu\ln\mu}{16\pi a}\right) \tag{21}$$

The change of the ground state energy due to condensation is given by

$$\Delta E = 3S\left[\quad \mu N_0 + \frac{1}{2}\Gamma N_0^2\right]. \tag{22}$$

The change of magnetization per site is calculated by differentiating this as

$$\Delta M = \pm\frac{1}{3}\left[\ S\mu + \frac{3\sqrt{3}\mu\ln\mu}{16\pi a}\right]. \tag{23}$$

Where the upper(lower) sign corresponds to $h_{c1}(h_{c2})$. Inverse mass of excitation spectrum a can be calculated with the quadratic Hamiltonian (6) and the result is $a = 1/4 + O(1/S)$ for $h = h_{c1}$ and $a = 3/4 + O(1/S)$ for $h = h_{c2}$. Therefore the change of magnetization is

$$\Delta M/S = \begin{cases} \frac{1}{3}\Delta h + \frac{\sqrt{3}}{4S\pi}\Delta h\ln\Delta h & \text{at } h = h_{c1} \\ \frac{1}{3}\Delta h \quad \frac{\sqrt{3}}{12S\pi}\Delta h\ln\Delta h & \text{at } h = h_{c2} \end{cases}, \tag{24}$$

where $\Delta h = h_{c1} \quad h$ or $h \quad h_{c2}$.

This logarithmic singularity of magnetization curve generally appears near the saturation field of two-dimensional antiferromagnet [8], and has been studied through introducing infrared cutoff [9, 10], or using renormalization group [11]. Our result (24) agrees with those results up to the prefactor, if the boson mass is properly set.

4. Conclusion

In this paper we have performed spin wave analysis on the 1/3 magnetization plateau of the triangular-lattice Heisenberg antiferromagnet. We have proved that collinear configuration guarantees that the energy gap vanishes completely linearly in magnetic field. Our Hartree-Fock calculation predicted the transition points $H/(3JS) = 0.844$ and 1.368. Although there exist some discrepancies between those values and numerical results, the plateau width is substantially improved compared with the $1/S$ expansion. Regarding the singularity at the plateau transitions, we have demonstrated the presence of logarithmic term by taking acount of magnon multiple scatterings. This agrees with common behavior of bose condensation in two dimensions, but this has not been predicted for this problem yet, within the knowledge of the authors. We have succeeded in reproducing the correct logarithmic term up to prefactor by means of a method which is much simpler than other approaches. This may be also useful for further improvement by considering higher order terms.

Acknowledgements

This study was supported by Grant-in-Aid for Scientific Research on Priority Areas "Novel States of Matter Induced by Frustration" (19052003). Part of this work was performed within the Advanced Study Group Program on" Unconventional Magnetism in High Fields" at the Max-Planck Institute for the Physics of Complex Systems.

References

[1] Chubukov A V and Golosov G 1991 *J. Phys. Condens. Matter* **3** 69
[2] Honecker A 1999 *J. Phys. Condens. Matter* **11** 4967
[3] Farnell D J *et al.* 2009 *J. Phys. Condens. Matter* **21** 406002
[4] Ono T *et al.* 2003 *Phys. Rev.* B **67** 104431
[5] Svistov L E *et al.* 2003 *Phys. Rev.* B **67** 094434
[6] Ishii R *et al.* arXiv:0912.4796.
[7] Hugenholtz N and Pines D 1959 *Phys. Rev.* **116** 489
[8] Zhitomirsky M E and Nikuni T 1998 *Phys. Rev.* B **57** 5013
[9] Gluzman S 1993 *Z. Phys.* B **90** 313
[10] Popov V N 1972 *Theor. Math. Phys.* **11** 565
[11] Fisher D S and Hohenberg P C 1988 *Phys. Rev.* B **37** 4936

Critical properties of generalized four-state clock model on square lattices

Takafumi Suzuki[1], **Yusuke Tomita**[1], **Naoki Kawashima**[1] **and Pinaki Sengupta**[2]

[1]Institute for Solid State Physics, University of Tokyo, Chiba Kashiwa 5-1-5, JP
[2]School of Physical and Mathematical Sciences, Nanyang Technological University, Singapore 21 Nanyang Link, SG

E-mail: takafumi-s@issp.u-tokyo.ac.jp

Abstract. We study the finite-temperature transition to the 1/2 magnetization plateau in a model of interacting $S = 1/2$ spins with longer range interactions and strong exchange anisotropy on the geometrically frustrated Shastry-Sutherland lattice. In previous studies, it was obtained from Monte Carlo calculations that the transition to the plateau state occurs via two successive transitions with the two-dimensional Ising universality class, when the quantum exchange interactions are finite, while a single phase transition takes place in the purely Ising limit[1]. To understand these behaviors, we introduce the generalized four-state chiral clock model and perform Monte Carlo calculations for this model. By comparing the phase diagrams of the two models, we find that the topology of the thermal phase diagram is the same each other - the criticality of the thermal phase transition in the original model can be well explained by that of the generalized four-state chiral clock model.

1. Introduction

Geometrically frustrated interactions inhibits the stabilization of classical orderings. It sometimes become a trigger for the emergence of several exotic orders in magnetic fields. Fractional magnetic plateaus have recently been discovered at low temperatures in several rare-earth tetraborides RB_4 [R is a rare-earth element][2, 3, 4, 5, 6]. The magnetic moments from R^{3+} ions are arranged in a Shastry-Sutherland lattice (SSL) in the ab-plane. Therefore, these compounds propose a good stage to study the frustration effect on the magnetic properties in the SSL model. In TmB_4 case, an extended magnetization plateau at $m = 1/2$ was confirmed for $H_{c1} \sim 1.9$[T]$< H < H_{c2} \sim 3.6$[T] when a magnetic field is applied along the c-axis[4, 5]. Here m is the normalized value by the saturation magnetization. Specific heat results[5] suggested that the degeneracy of the lowest J=6 multiplet of Tm^{3+} is lifted by a strong anisotropy along the c-axis direction; the lowest energy state for a single ion is the non-Kramers doublet with J_z=$\pm$6 and a large energy gap to the first excited doublet exists. When perturbative treatment to the lowest energy doublets is applied to describe the effective model for the low-energy magnetic part, a S=1/2 XXZ model with Ising-like exchange anisotropy is obtained. This effective model has a characteristic point that the transverse coupling is ferromagnetic.

In our previous studies[1], we discussed that the longer range interactions, such as ferromagnetic J_4 and antiferromagnetic J_3 couplings shown in Figure 1 (e), are important to stabilize the 1/2 plateau state. We also investigated the finite-temperature phase transition to the 1/2 plateau state. The finite-size scaling analysis indicated that a two-step second-order transition takes place and the universality class at both critical points is explained by the critical exponents of the two-dimensional(2d) Ising model.

For the finite-temperature transition, nontrivial behaviors remain uncleared. In the 1/2 plateau phase, the lowest energy state is four-fold degenerate. Therefore, it is naively expected that the universality class should be the same as that of the four-state Potts model. As we show the phase diagram in Figure 1 (a), the critical behavior can indeed be explained by the four-state Potts universality in the Ising limit, while the finite quantum fluctuation causes a two-step transition with both transitions belonging to the 2d Ising universality. Since the same symmetry breaking takes place at a very low temperature, the schematic thermal phase diagram is also expected to be the same. However, the results obtained are different from this expectation. To understand these, therefore, is valuable from the view point of statistical mechanics.

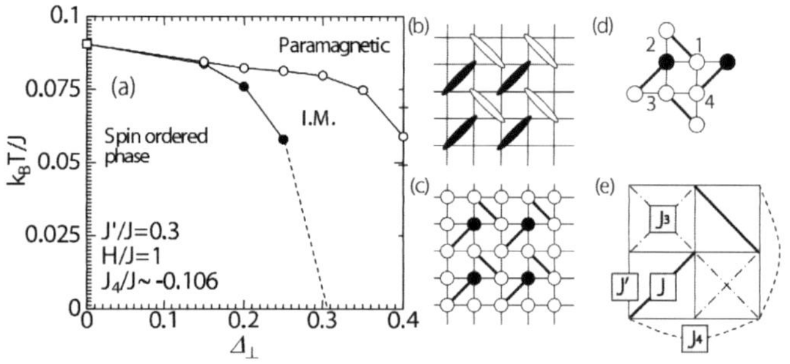

Figure 1. (a)Phase diagram for the 1/2 plateau state in the S=1/2 Ising-like XXZ model on the SSL. Open and solid circles indicate the second order transition with the 2d Ising universality. All lines are guide to the eyes. (b)Spin configuration in the intermediate (I.M.) phase. Solid (open) ellipsoids denote spin pairs carrying $S^z \sim 1$ ($S^z \sim 0$). (c)Spin configuration in the spin ordered phase. Solid (open) circles are down (up) spins. (d)Sublattice indecies for the cluster spin. (e)Bond configuration in the effective model for TmB_4.

2. Model and method

To explain the thermal transition to the 1/2 plateau phase in the S=1/2 Ising-like XXZ model, we consider the simplified Hamiltonian for the low-energy part. Before the obtained Hamiltonian is introduced, we briefly summarize the previous results[1] again. The spin configuration in the 1/2 plateau state can be described by Figure 1 (c). Such spin configuration is characterized by the 90° (C_4)-rotation symmetry breaking around the center of the plaquette without diagonal coupling. Therefore there are two possible candidates for the symmetry-breaking scenarios at a finite temperature. One is that the system shows a single phase transition belonging to the four-state Potts universality class. The other is that the two-step phase transition takes place. In the former case, the C_4 symmetry breaks down to trivial group at a critical point. When the system exhibits a two-step phase transition, the C_4 symmetry breaks down to C_2 group (180° rotational symmetric state) at the higher critical temperature T_{c1}, and the remaining one breaks down to the trivial group at the lower critical temperature T_{c2}. The resulting universality class at both critical temperatures belongs to the 2d Ising model.

In Figure 1 (a), the Ising anisotropy ($\Delta_\perp$) dependence of the critical temperatures for fixed $J'/J = 0.3$, $H/J = 1$ and $J_4/J \sim -0.106$ is summarized. At the Ising-spin model limit ($\Delta_\perp$=0), the system shows a single phase transition, and the two-step phase transition takes place for $\Delta_\perp > 0$ clearly. The critical temperature associated with the 180°-rotation symmetry breaking decreases as $\Delta_\perp$ increases toward the isotropic (Heisenberg) point, $\Delta_\perp = 1$. In the intermediate phase, the Néel order of spin pairs with total $S^z \sim 1$ and $S^z \sim 0$ on diagonal bond J realizes (see Figure 1 (b)). The dimerization of antiparallel spin pairs, in other words, the quantum fluctuation thus seems to play a key role in the thermal transition.

For the simplicity, we begin by introducing eight-spin clusters in the 1/2 plateau state as shown in Figure 2 (a). In the 1/2 plateau phase, a down spin occupies one of four sites on a plaquette without the diagonal bond and up spins occupy the remaining three sites. If a down spin occupies the site labeled by '2' in Figure 1(d), the nearest down spin prefers to locate at the third-neighbor distance from it as long as the ferromagnetic J_4 coupling. This indicates that the down spins in the 1/2 plateau state occupies the same edge of diagonal bond connecting each plaquette. Since the ferromagnetic J_4 coupling is essential for stabilizing the 1/2 plateau state as discussed in ref. [1], it is reasonable to describe the 1/2 plateau state by an arrangement of such clusters. This description becomes exact in the limit $|J_4/J| \gg 1$ and $\Delta_\perp = 0$. Since there are four kinds of clusters corresponding to the 90°-rotation symmetry breaking in the 1/2 plateau, we assign clock spins pointing at an angle to each clusters. Here we ignore clusters having $m \neq 1/2$ for simplicity. Thus we obtain the simplified classical Hamiltonian that can be regarded as the generalized four-state clock model on a square lattice with the nearest and next-nearest neighbor interactions,

$$\begin{aligned}\mathcal{H}_{sim} &= \mathcal{H}_{\mathrm{N.N.}} + \mathcal{H}_{\mathrm{N.N.N}} \\ \mathcal{H}_{\mathrm{N.N.}} &= \sum_i \mathcal{J}_x(\theta_{i,j}, \theta_{i+1,j}) + \sum_j \mathcal{J}_y(\theta_{i,j}, \theta_{i,j+1}) \\ \mathcal{H}_{\mathrm{N.N.N}} &= \sum_i \mathcal{J}'_x(\theta_{i,j}, \theta_{i+1,j+1}) + \sum_j \mathcal{J}'_y(\theta_{i,j}, \theta_{i-1,j+1}), \end{aligned} \quad (1)$$

where $\theta_{i,j}$ denotes an angle of clock spins and takes $0, \pi/2, \pi$, or $3\pi/2$. The value of $\theta_{i,j}$ denotes the position of a down spin in the clusters (see Figure 2 (a)). $\mathcal{J}_x$ and $\mathcal{J}_y$ are interactions between two clock spins, and their values are listed in Table I. The interactions, $\mathcal{J}_x$ and $\mathcal{J}_y$, are calculated from the sums of coupling energy between spins located in different clusters. When $\Delta_\perp = 0$, the values, A, B, C, and D, in $\mathcal{J}_x$ and $\mathcal{J}_y$ (see Table I) are evaluated as A=$4J_4$, B=$2J'$, C=0, D=$4J'$, respectively. The resulting classical Hamiltonian retains C_4 rotational symmetry conditionally; the system requires a lattice rotation when a global angle shift of all clock spins is executed. Such requirement arises from the geometric characteristics of the SSL. Since the symmetry of the Hamiltonian $\mathcal{H}_{\mathrm{N.N.}}$ is lower than that of $\mathcal{H}_{\mathrm{N.N.N.}}$, we focus on the critical properties of $\mathcal{H}_{\mathrm{N.N.}}$ in the followings.

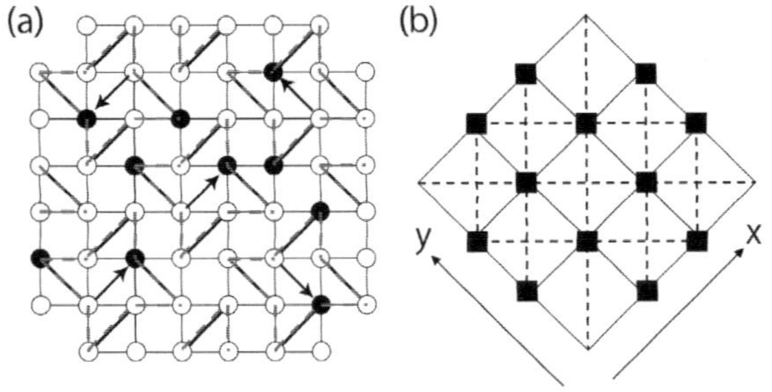

Figure 2. (a) Spin clusters in the 1/2 plateau state and clock spins in the simplified classical model. A cluster is constructed by eight spins. Red dashed lines drawing windmills correspond to the clusters and arrows in the windmills denote angles of clock spins that represent states of the clusters. (b) Generalized four-state clock model on a square lattice. Solid squares correspond to each clusters described in (a). Solid (dashed) lines denote the nearest (next-nearest) neighbor interaction $\mathcal{J}_x$ and $\mathcal{J}_y$ ($\mathcal{J}'_x$ and $\mathcal{J}'_y$).

3. Results and Discussion

The simplified Hamiltonian $\mathcal{H}_{\mathrm{N.N.}}$ is identical to the conventional four-state Potts model at B=C=D=0. Therefore, the system shows the trivial criticality: the finite-temperature phase transition is a single second-order transition belonging to the four-state Potts universality. However, the universality class becomes nontrivial, when B≠D≠0. Performing Monte Carlo simulations, we numerically investigated the finite-temperature transition of the Hamiltonian

Table 1. Coupling constants of the simplified classical Hamiltonian. Note that (A, B, C, D)=($4J_4$,$2J'$,0, $4J'$) corresponds to the parameter set we have treated in the section III. Upper (lower) side tables are matrices of the nearest neighbor (next-nearest neighbor) interaction.

$\mathcal{J}_x(\theta_{i,j}, \theta_{i+1,j})$

$\theta_{i,j} \backslash \theta_{i',j'}$	0	$\pi/2$	π	$3\pi/2$
0	A	B	D	B
$\pi/2$	B	A	B	C
π	-D	B	A	B
$3\pi/2$	B	C	B	A

$\mathcal{J}_y(\theta_{i,j}, \theta_{i,j+1})$

$\theta_{i,j} \backslash \theta_{i',j'}$	0	$\pi/2$	π	$3\pi/2$
0	A	B	D	B
$\pi/2$	B	A	B	C
π	-D	B	A	B
$3\pi/2$	B	C	B	A

$\mathcal{J}'_x(\theta_{i,j}, \theta_{i+1,j})$

$\theta_{i,j} \backslash \theta_{i',j'}$	0	$\pi/2$	π	$3\pi/2$
0	A	C	A	C
$\pi/2$	C	A	C	-A
π	A	C	A	C
$3\pi/2$	C	-A	C	A

$\mathcal{J}'_y(\theta_{i,j}, \theta_{i,j+1})$

$\theta_{i,j} \backslash \theta_{i',j'}$	0	$\pi/2$	π	$3\pi/2$
0	A	C	-A	C
$\pi/2$	C	A	C	A
π -A	C	A	C	
$3\pi/2$	C	A	C	A

$\mathcal{H}_{\text{N.N.}}$ with D=2B and A=-4 up to L=72. We summarize the obtained critical temperatures in Figure 3 and find that the system shows a two-steps transition for -B/A $\gg$ 0.1 and C/A $\gg$ 1. These critical temperatures were estimated from the correlation ratio $\Gamma_{\cos n\theta}(L/2,0)/\Gamma_{\cos n\theta}(L/4,0)$, where $\Gamma_{\cos n\theta}(x,y) = \langle \cos n\theta_{x,y} \cos n\theta_{0,0} \rangle$ and n=1,2. The correlation ratio is a excellent estimator for the critical point because its temperature dependence disappears at the transition temperature.

To discuss the criticality, we perform the finite-size analysis for the forth-order cumulant of the order parameters, $\langle \cos\theta \rangle$ and $\langle \cos 2\theta \rangle$ and show the results in Figure 4[11]. Note that we employed critical temperatures that were estimated from the correlation ratio. At -B/A=0.0125 and C/A=0, the data collapse is obtained by a fixed critical exponent $\nu = 2/3$ and a single critical temperature $T_c \sim 3.688(2)$. This strongly suggests that the universality class belongs to the four-state Potts model. However, we confirm a clear evidence of a two-steps phase transition that is characterized by the critical exponent of the two dimensional Ising model as |-B/A| or C/A increases. Figure 4 (b1) and (b2) are the results of the finite-size scaling analysis at B/A=-0.075 and C/A=0.75. When the two-step phase transition takes place, the 90°-rotation symmetry breaks at the higher critical temperature and the 180°-rotation symmetry survives in the intermediate phase. As the difference between A and C decreases, the lower critical temperature decreases and the system undergoes a single phase transition for C/A $>$ 1.

Further confirmation of the universality class is obtained from phenomenological renormalization group analysis for the correlation functions. The exponent η in several two dimensional models with discrete symmetry, such as the Ising model, the Ashkin-Teller model, and the eight-vertex model, takes a value $\eta = 1/4$[8, 9, 10]. Therefore, it is expected that the simplified model $\mathcal{H}_{\text{N.N.}}$ has also $\eta = 1/4$ at critical points. In the analysis, we assume the scaling form, $\log\left[\frac{\Gamma_{\cos n\theta}(L/2)}{\Gamma_{\cos n\theta}(L'/2)}\right] \sim (2-d-\eta)\log\left(\frac{L}{L'}\right) + \log\left[F(tL^{1/\nu})/F(tL'^{1/\nu})\right]$, where F is a scaling function. Figure 3 shows the value of η when B/A is varied at C/A=0 and C/A is varied at -B/A=0.05. The scaling results allow us to conclude that the critical exponent satisfies $\eta \sim 0.25$ at the both critical temperatures.

By comparing the phase diagrams shown in Figure 1 (a) and Figure 3, we find that the topology of the phase diagram is the same each other. In both simplified classical model

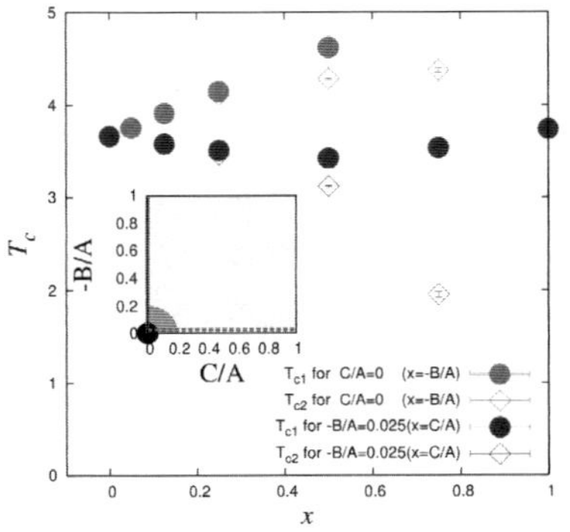

Figure 3. Critical temperatures of the generalized four-state clock model when we fix at B/A=-0.05 and at C/A=0. Schematic phase diagram of the present simplified classical model is shown in the inset. Red and blue symbols correspond to the cross sections on red solid and dashed lines drawn in the phase diagram, respectively. In the phase diagram, an area painted by green indicates the region where the two-step phase transition takes place clearly. A blue quarter circle means the area where the system seems to undergo a single phase transition. Note that increase of x fixing the value -B/A corresponds to increase of $\Delta_\perp$ in Figure 1 (see in text).

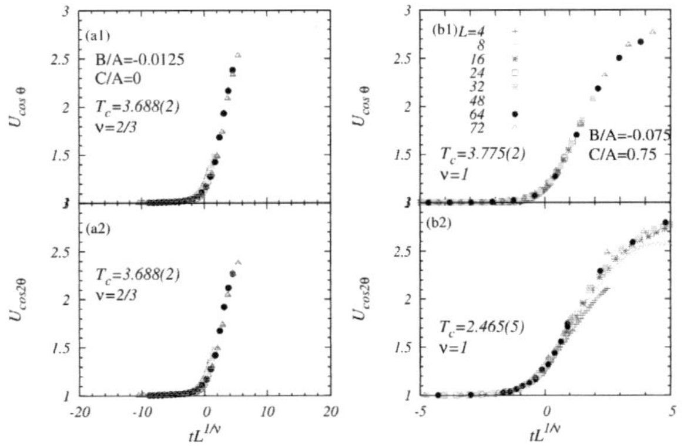

Figure 4. Finite size scaling analysis for the fourth-order cumulant of $\langle\cos 2\theta\rangle$ and $\langle\cos\theta\rangle$. In these analysis, we estimated all critical temperatures from crossing points of the correlation ratios as not shown here. The data collapse in (a1) and (a2) ((b1) and (b2)) was obtained by assuming the fixed critical exponents $\nu = 2/3$ ($\nu = 1$). The parameter set for (a1) and (a2) ((b1) and (b2)) corresponds to that in the blue circle (outside of the blue circle) shown in the inset of Figure 3.

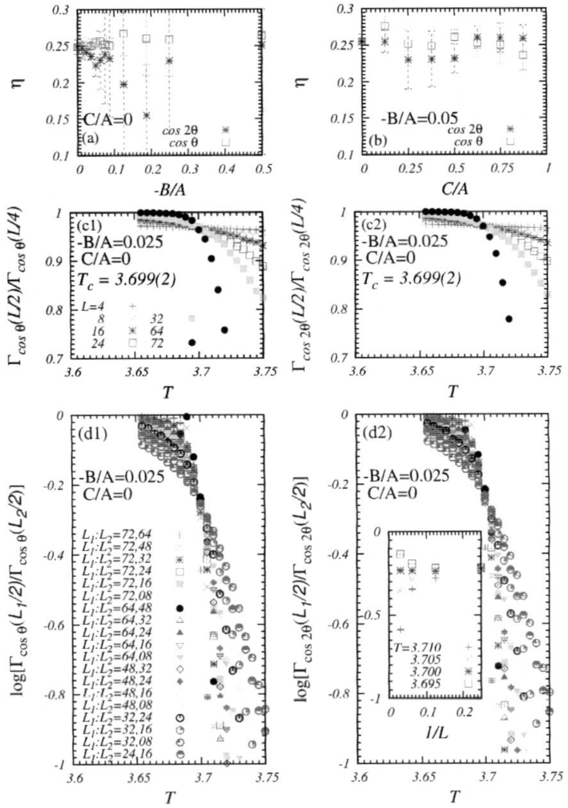

Figure 5. (a) and (b) are estimated critical exponents η at each critical point. Red (blue) symbols are related to the critical point of 90° (180°) rotational symmetry breaking. (c1) and (c2) are the results of temperature dependence for correlation ratio when B/A=0.025 and C/A=0. (d1) and (d2) are the results of phenomenological renormalization-group analysis for correlation functions $\Gamma_{\cos\theta}(r)$ and $\Gamma_{\cos 2\theta}(r)$, respectively. Correlation ratios for different system size also show a cross at a critical temperature giving the critical exponent η, because a correlation function behaves as $\Gamma(t, L) \sim L^{2-d-\eta}\Phi(tL^{1/\nu})$ around $t \sim 0$. Inset of (d2) is the result of the size dependence of $\log[\Gamma(2L)/\Gamma(L)]$.

and Ising-like XXZ model cases, the region of the intermediate phase, where only the 90° rotational symmetry is broken, shrinks rapidly as the system approaches the four-state-Potts universality point. The quantum effect in the original XXZ model corresponds to fluctuation of antiferromagnetic spin pairs on the diagonal bond. Therefore it is enhanced as $\Delta_\perp$ increases. Then, the sublattice magnetization of antiparallel spin pair closes to zero. In the case, where the antiparallel spin pair becomes a quantum-triplet or classical dimer with $S^z = 0$, the clock spins pointing toward $\theta = 0$ and π (or $\pi/2$ and $3\pi/2$) direction are identical each other. Such dimerization effect is captured as a reduction in the difference between A and C. Consequently, the increase of $\Delta_\perp$ means that C/A approaches to 1 and then, the lower transition temperature T_{c2} associated with the 180° rotational symmetry breaking shifts toward $T = 0$ as $\Delta_\perp$ increases in the original XXZ model.

As for the universality class around -B/A=C/A=0, there are two possible scenarios. One is that the system show the four-state Potts universality class only at -B/A=C/A=0 or $\Delta_\perp = 0$. The other is that there exists the *area* showing the same universality class as that the four-state Potts model. Further computations are undergoing to clarify this point.

4. Conclusion

We have studied the thermal transition to the 1/2 plateau phase for the S=1/2 Ising-like XXZ model on the geometrically frustrated Shastry-Sutherland lattice. The previous results from the Monte Carlo computations suggested that the transition to the 1/2 plateau state occurs via two successive transitions with the 2d Ising universality class, when the quantum exchange interactions are finite, whereas a single phase transition takes place in the purely Ising limit. To understand these behaviors, we have performed Monte Carlo simulations for the classical generalized four-state chiral-clock model and compared the phase diagrams of the two models. From the results, we have found that the above discrepancy is successfully explained by the critical properties of the generalized four-state chiral-clock model.

Acknowledgment

The present research subject was suggested by C. D. Batista and we would like to thank him. The computation was executed on computers at the Supercomputer Center, ISSP, University of Tokyo. This work is financially supported by Grant-in-Aid for Young Scientists (B) (21740245), for Scientific Research (B) (19340109), and for Scientific Research on Priority Areas(19052004).

References

[1] Suzuki T, Tomita Y, Kawashima N 2009 *Phys. Rev. B* **80** 180405(R)
[2] Yoshii S, Yamamoto T, Hagiwara M, Takeuchi T, Shigekawa A, Michimura S, Iga F, Takabatake T and Kindo K 2007 *J. Magn. Magn. Mater.* **310** 1282-1284
[3] Michimura S, Shigekawa A, Iga F, Sera M, Takabatake T, Ohyama K and Okabe Y 2006 *Physica B* **378-380** 596-597
[4] Yoshii S, Yamamoto T, Hagiwara M, Shigekawa A, Michimura S, Iga F, Takabatake T and Kindo K 2006 *J. Phys.: Conf. Ser.* **51** 59-62
[5] Gabáni S, Matas S, Priputen P, Flachbart K, Siemensmeyer K, Wulf E, Evdokimova A and N. Shitsevalova 2008 *Acta. Phys. Pol. A* **116** 227
[6] Siemensmeyer K, Wulf E, Mikeska H J, Flachbart K, Gabáni S, Matas S, Priputen P, Efdokimova A and Shitsevalova N 2008 *Phys. Rev. Lett.* **101** 177201
[7] Niehuis B 1987 *Phase transition and Critical Phenomena* vol 11, ed Domb C and Lebowitz J L (New York: Adademic Press)
[8] Ashkin J and Teller E 1943 *Phys. Rev.* **64** 178
[9] Baxter R J 1982 *Exactly Solved Models in Statistical Mechanics* (New York: DOVER PUBLICATIONS, INC.)
[10] Cardy J 1980 *J. Phys. A: Math. Gen.* **13** 1507
[11] In both analysis, the logarithmic corrections discussed in Salas J and Sokal A D 1997 *J. Stat. Phys.* **88** 567 exists, but it looks small because the resolution of the data points seems to be low.

International Conference on Frustration in Condensed Matter (ICFCM) IOP Publishing
Journal of Physics: Conference Series **320** (2011) 012013 doi:10.1088/1742-6596/320/1/012013

Double-q Order in a Frustrated Random Spin System

Ryo Tamura and Naoki Kawashima

Institute for Solid State Physics, University of Tokyo, 5-1-5 Kashiwa-no-ha, Kashiwa, Chiba 277-8581, Japan

E-mail: `r.tamura@issp.u-tokyo.ac.jp`

Abstract. We use the three-dimensional Heisenberg model with site randomness as an effective model of the compound $Sr(Fe_{1-x}Mn_x)O_2$. The model consists of two types of ions that correspond to Fe and Mn ions. The nearest-neighbor interactions in the *ab*-plane are antiferromagnetic. The nearest-neighbor interactions along the *c*-axis between Fe ions are assumed to be antiferromagnetic, whereas other interactions are assumed to be ferromagnetic. From Monte Carlo simulations, we confirm the existence of the double-$\boldsymbol{q}$ ordered phase characterized by two wave numbers, $(\pi\pi\pi)$ and $(\pi\pi 0)$. We also identify the spin ordering pattern in the double-$\boldsymbol{q}$ ordered phase.

1. Introduction

Infinite-layer iron oxide $SrFeO_2$ exhibits a Néel transition to the G-type antiferromagnetic ordered phase at $T_N = 473\text{K}$[1, 2]. The transition temperature decreases as Mn^{2+} ions are substituted for Fe^{2+} ions[3]. In the compound $Sr(Fe_{1-x}Mn_x)O_2$, Fe and Mn ions are magnetic and placed randomly on a tetragonal lattice. In neutron scattering measurements on $Sr(Fe_{0.7}Mn_{0.3})O_2$, magnetic peaks have been observed to develop at two distinct wave vectors $\boldsymbol{q} = (\pi\pi\pi)$ and $(\pi\pi 0)$. These orders coexist at low temperatures[3]. However, the spin ordering pattern in the low-temperature phase is not known.

The site-random model has been used as an effective model of the random magnets. The site-random model consists of two types of ions labeled A and B. The ions are placed randomly on the lattice. The interactions of each bond are set depending on the combination of ions. In previous works that employed the site-random model[4, 5, 6, 7, 8], all interactions depended on the arrangement of A and B ions: the interactions between A ions were ferromagnetic, whereas other interactions were antiferromagnetic. In the case of the Ising spin model on a simple cubic lattice[4], it was found through Monte Carlo simulations that the phase diagram lacks both a mixed ordered phase and spin-glass phase. In the Heisenberg spin model[5, 6, 7, 8], as in the Ising spin case, there is also no spin-glass phase in the phase diagram. However, the ferromagnetic and antiferromagnetic mixed ordered phase exists at low temperatures. In this mixed ordered phase, the peaks of the structure factor develop at wave vectors $\boldsymbol{q} = (000)$ and $(\pi\pi\pi)$. Moreover, the ferromagnetic and antiferromagnetic ordering vectors are mutually perpendicular. For this model, the occurrence of a two-step phase transition has been reported: one is the transition from the paramagnetic phase to the ferromagnetic (antiferromagnetic) ordered phase, the other is the transition from the ferromagnetic (antiferromagnetic) ordered phase to the mixed ordered phase. The universality class at each phase transition has been suggested to belong to the three-dimensional Heisenberg universality class[8].

Published under licence by IOP Publishing Ltd

In the previous works that used the site-random Heisenberg model[5, 6, 7, 8], $\boldsymbol{q} = (000)$ and $(\pi\pi\pi)$ orderings have been found. However, since the magnetic peak positions are $\boldsymbol{q} = (\pi\pi\pi)$ and $(\pi\pi 0)$ in $Sr(Fe_{0.7}Mn_{0.3})O_2$, the spin ordering pattern is different from that observed previously. Thus, our current aim is to clarify the spin ordering pattern in the low-temperature phase of $Sr(Fe_{1-x}Mn_x)O_2$.

2. Model

To introduce suitable randomness into $Sr(Fe_{1-x}Mn_x)O_2$, we investigate a classical Heisenberg model with site-random interlayer couplings on a simple cubic lattice:

$$\mathcal{H} = -J \sum_{\langle i,j\rangle_{ab\text{-plane}}} \boldsymbol{s}_i \cdot \boldsymbol{s}_j - \sum_{\langle i,j\rangle_{c\text{-axis}}} J_{ij} \boldsymbol{s}_i \cdot \boldsymbol{s}_j, \qquad (|J_{ij}| = |J|). \tag{1}$$

The first term of the Hamiltonian denotes the uniform antiferromagnetic nearest-neighbor interaction ($J < 0$) in the ab-plane. The sign of the nearest-neighbor coupling J_{ij} in the second term depends on the arrangement of two ions along the c-axis; the interactions between Fe ions are assumed to be antiferromagnetic ($J_{ij} < 0$), whereas other interactions are assumed to be ferromagnetic ($J_{ij} > 0$). For simplicity, we further assume that the absolute values of all interactions are the same and that the spin lengths of Fe and Mn ions are the same. Hereinafter, we set $|J| = 1$ and denote the concentration of Mn ions by p_{Mn}. From the above assumptions regarding the interactions, at $p_{\mathrm{Mn}} = 0$ where all magnetic ions are Fe ions, the system exhibits a Néel transition to the G-type antiferromagnetic ordered phase characterized by $\boldsymbol{q} = (\pi\pi\pi)$, because all interactions along the c-axis are antiferromagnetic. In contrast, at $p_{\mathrm{Mn}} = 1$ where all magnetic ions are Mn ions, the C-type antiferromagnetic ordered phase characterized by $\boldsymbol{q} = (\pi\pi 0)$ becomes stable at low temperatures, because all interactions along the c-axis are ferromagnetic. For $0 < p_{\mathrm{Mn}} < 1$, the frustration is caused by the random ferromagnetic and antiferromagnetic interactions along the c-axis direction.

3. Simulation Results

We use Monte Carlo simulations based on the standard heat-bath method to the model on an $N = L \times L \times L$ simple cubic lattice with a periodic boundary condition. Before starting the simulations, we set p_{Mn} so that the number of Mn ions is an integer, and we set random configurations of Fe and Mn ions according to p_{Mn}. In this paper, p_{Mn} is set to 0.1875. For this value of p_{Mn}, there are more antiferromagnetic couplings than ferromagnetic couplings along the c-axis. We prepare 64 samples, and each run contains 10^6–10^7 Monte Carlo steps per spin at each temperature.

The temperature dependence of the structure factor $S(\boldsymbol{q})$ at $\boldsymbol{q} = (\pi\pi\pi)$ and $(\pi\pi 0)$ is shown in Figure 1. The structure factor $S(\boldsymbol{q})$ is defined by

$$S(\boldsymbol{q}) = \frac{1}{N} \sum_{i,j} \langle \boldsymbol{s}_i \cdot \boldsymbol{s}_j \rangle e^{i\boldsymbol{q}\cdot(\boldsymbol{r}_i - \boldsymbol{r}_j)}, \tag{2}$$

where $\langle \cdots \rangle$ indicates the thermal average. $S(\pi\pi\pi)$ and $S(\pi\pi 0)$ increase at $T \cong 1.09$ and 0.46, respectively. From Figure 1, it is clear that there are three types of phases in this model. In the low-temperature phase (phase I), the peaks of the structure factor develop at two distinct wave vectors $\boldsymbol{q} = (\pi\pi\pi)$ and $(\pi\pi 0)$. Accordingly, this phase is called a double-$\boldsymbol{q}$ ordered phase. In the intermediate-temperature phase (phase II), since the structure factor has a peak at $\boldsymbol{q} = (\pi\pi\pi)$, this phase is the G-type antiferromagnetic ordered phase. The high-temperature phase (phase III) is the paramagnetic phase. The transition temperature from the G-type antiferromagnetic ordered phase to the paramagnetic phase is lower than the Néel temperature of the pure system

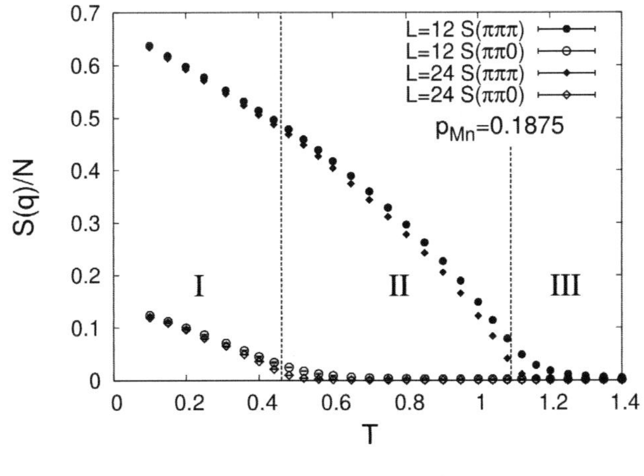

Figure 1. Temperature dependence of the structure factor at $\boldsymbol{q} = (\pi\pi\pi)$ and $(\pi\pi 0)$ when the lattice size is $L = 12$ and 24. Phases denoted by I, II, and III are the double-$\boldsymbol{q}$ ordered phase, the G-type antiferromagnetic ordered phase, and the paramagnetic phase, respectively.

$T_N \cong 1.44$[9]. From the results of the structure factor, we conclude that the double-$\boldsymbol{q}$ ordered phase characterized by the wave vectors $\boldsymbol{q} = (\pi\pi\pi)$ and $(\pi\pi 0)$ exists in the model. The positions of the magnetic peaks for the model are the same as those observed for $Sr(Fe_{1-x}Mn_x)O_2$.

To identify the spin ordering pattern in the double-$\boldsymbol{q}$ ordered phase, we calculate the correlation function along the c-axis between two spins. The correlation function along the c-axis is given by

$$G^c(r_z) = \frac{1}{N}\sum_i \langle \boldsymbol{s}(\boldsymbol{r}_i) \cdot \boldsymbol{s}(\boldsymbol{r}_i + r_z \boldsymbol{e}_z)\rangle, \tag{3}$$

where r_z is the distance between two spins, and $\boldsymbol{e}_z$ is unit vector of the c-axis direction. Figure 2 shows dependence of the correlation function on distance r_z for $L = 24$ at $T = 0.1$. Since the spin correlation is ferromagnetic for an even number of r_z, all spins in odd (even)-numbered layers are co-linear; The Néel-like configuration of each layer is stacked in parallel to other layers. Furthermore, the spin correlation is weakly coupled antiferromagnetic for an odd-number of r_z. This suggests that the angle between nearest-neighbor spins along the c-axis is close to $2\pi/3$ at $T = 0.1$.

To investigate the temperature dependence of spin ordering pattern, we calculate the angle θ between the staggered magnetization of an odd-numbered layer, $\boldsymbol{M}_o$, and that of an even-numbered layer, $\boldsymbol{M}_e$. $\boldsymbol{M}_o$ and $\boldsymbol{M}_e$ are given by

$$\boldsymbol{M}_o = \frac{2}{N} \sum_{\substack{\boldsymbol{r}\in\text{odd-numbered}\\ \text{layer}}} e^{i\boldsymbol{q}\cdot\boldsymbol{r}} \boldsymbol{s}(\boldsymbol{r}), \tag{4}$$

$$\boldsymbol{M}_e = \frac{2}{N} \sum_{\substack{\boldsymbol{r}\in\text{even-numbered}\\ \text{layer}}} e^{i\boldsymbol{q}\cdot\boldsymbol{r}} \boldsymbol{s}(\boldsymbol{r}), \tag{5}$$

where $\boldsymbol{q} = (\pi\pi 0)$. We define the angle θ between $\boldsymbol{M}_o$ and $\boldsymbol{M}_e$ as follows:

$$\cos\theta = \left\langle \frac{\boldsymbol{M}_o \cdot \boldsymbol{M}_e}{|\boldsymbol{M}_o||\boldsymbol{M}_e|} \right\rangle. \tag{6}$$

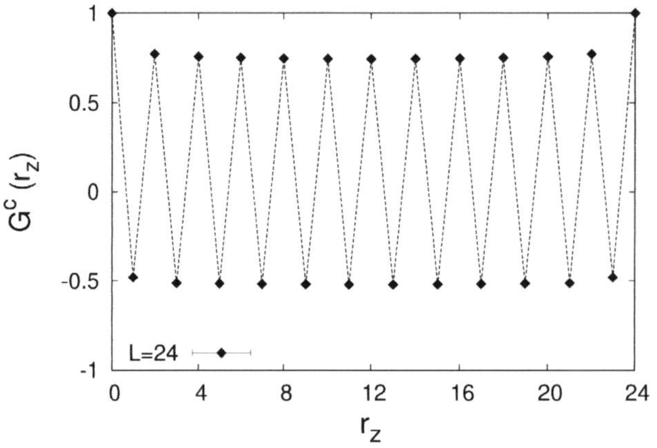

Figure 2. Correlation function along the c-axis for $L = 24$ at $T = 0.1$.

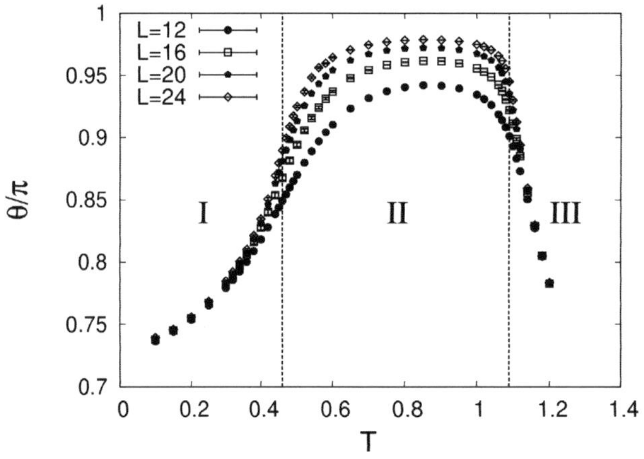

Figure 3. Temperature dependence of the angle θ between the staggered magnetization of an odd-numbered layer and that of an even-numbered layer.

Figure 3 shows the temperature dependence of θ. At the higher-temperature transition point, the value of θ/π approaches 1 and the $(\pi\pi\pi)$ order develops. At the lower-temperature transition point, the value of θ/π deviates from 1. In phase I, the value of θ depends on the temperature and monotonically decreases as the temperature decreases. Thus, the spin configuration characterized by θ changes depending on the temperature in phase I. At a temperature of zero, the value of θ/π appears to converge to 0.73. From the results, we find that the lower-temperature phase transition is caused by θ/π deviating from 1.

Let us consider the translational symmetry and the spin rotation symmetry in each phase. The intermediate-temperature phase is translationally symmetric with the O(3) spin rotation symmetry broken down to the U(1). In the double-$\boldsymbol{q}$ ordered phase, both the translational symmetry and the U(1) spin rotation symmetry are broken. Thus, we conclude that the transition to the double-$\boldsymbol{q}$ ordered phase is characterized by the breaking of the U(1) spin rotation symmetry.

4. Discussion

We discuss dependence of physical properties on the concentration of Mn ions $p_{\rm Mn}$. In this paper, we have fixed $p_{\rm Mn} = 0.1875$. In this case, there are more antiferromagnetic couplings than ferromagnetic couplings along the c-axis, and the intermediate G-type antiferromagnetic ordered phase exists in a broad temperature range ($0.46 \lesssim T \lesssim 1.09$). At $p_{\rm Mn} \cong 0.3$, the number of antiferromagnetic couplings and ferromagnetic couplings along the c-axis become nearly the same. Then, $(\pi\pi\pi)$ order and $(\pi\pi 0)$ order develop at nearly the same temperature, and the intermediate-temperature phase exists in a very narrow temperature range. Furthermore, for $p_{\rm Mn} > 0.3$, the intermediate C-type antiferromagnetic ordered phase exists, and the transition to the double-$\boldsymbol{q}$ ordered phase occurs at low temperature. In future work, we will attempt to clarify the details of the phase diagram of the temperature versus the concentration of Mn ions.

5. Conclusion

We have used the three-dimensional Heisenberg model with site-random interlayer couplings as an effective model of the compound $\mathrm{Sr(Fe_{1-x}Mn_x)O_2}$. The model consists of two types of ions that correspond to Fe and Mn ions. The interactions have been assumed to be antiferromagnetic in the ab-plane. The interactions along the c-axis between Fe ions have been assumed to be antiferromagnetic, whereas other interactions have been assumed to be ferromagnetic. From Monte Carlo simulations, we have confirmed the existence of the double-$\boldsymbol{q}$ ordered phase characterized by the wave vectors $\boldsymbol{q} = (\pi\pi\pi)$ and $(\pi\pi 0)$. The positions of the magnetic peaks for the model are the same as those observed for $\mathrm{Sr(Fe_{1-x}Mn_x)O_2}$. We have also identified the spin ordering pattern in the double-$\boldsymbol{q}$ ordered phase. In the double-$\boldsymbol{q}$ ordered phase, all spins in odd (even)-numbered layers are co-linear. Furthermore, the staggered magnetization of an odd-numbered layer and that of an even-numbered layer are not parallel, and the angle between them changes depending on the temperature. We conclude that the transition to the double-$\boldsymbol{q}$ ordered phase is characterized by the breaking of the U(1) spin rotation symmetry.

Acknowledgments

We would like to thank Takafumi Suzuki, Shu Tanaka, Liis Seinberg, Takafumi Yamamoto, and Hiroshi Kageyama for useful comments and discussions. R.T. acknowledges support from the Global COE Program "The Physical Sciences Frontier" of the Ministry of Education, Culture, Sports, Science and Technology (MEXT), Japan. This work is financially supported by Grants-in-Aid for Scientific Research (B) (22340111) and for Scientific Research on Priority Areas "Novel States of Matter Induced by Frustration" (19052004) from MEXT, Japan, and by Next Generation Supercomputing Project, Nanoscience Program, MEXT, Japan. The computation in this work is executed on computers at the Supercomputer Center, Institute for Solid State Physics, University of Tokyo.

References

[1] Tsujimoto Y, Hayashi C, Watanabe T, Kageyama H, Yoshimura K, Takano M, Ceretti M, Ritter C and Paulus W 2007 *Nature* **450** 1062
[2] Xiang H J and Wei Su-Huai 2008 *Phys. Rev. Lett.* **100** 167207
[3] Kageyama H, private communications
[4] Hukushima K, Nonomura Y, Ozeki Y and Takayama H 1997 *J. Phys. Soc. Jpn.* **66** 215
[5] Nielsen M, Ryan D H, Guo H and Zuckermann M 1996 *Phys. Rev.* B **53** 343
[6] Matsubara F, Tamiya T and Shirakura T 1996 *Phys. Rev. Lett.* **77** 378
[7] Bekhechi S and Southern B W 2004 *Phys. Rev.* B **70** 020405(R)
[8] Beath A D and Ryan D H 2006 *Phys. Rev.* B **73** 214445
[9] Chen K, Ferrenberg A M and Landau D P 1993 *Phys. Rev.* B **48** 3249

NMR relaxation rate in the field-induced octupolar liquid phase of spin-$\frac{1}{2}$ J_1-J_2 frustrated chains

Masahiro Sato[1], **Toshiya Hikihara**[2] **and Tsutomu Momoi**[3]

[1]Department of Physics and Mathematics, Aoyama Gakuin University, Sagamihara, Kanagawa 252-5258, Japan
[2]Department of Physics, Hokkaido University, Sapporo 060-0810, Japan
[3]Condensed Matter Theory Laboratory, RIKEN, Wako Saitama 351-0198, Japan

E-mail: `sato@phys.aoyama.ac.jp`

Abstract. In the spin-$\frac{1}{2}$ frustrated chain with nearest-neighbor ferromagnetic exchange J_1 and next-nearest-neighbor antiferromagnetic exchange J_2 under magnetic eld, magnetic multipolar-liquid (quadrupolar, octupolar, and hexadecapolar) phases are widely expanded from the saturation down to a low- eld regime. Recently, we have clari ed characteristic temperature and eld dependence of the NMR relaxation rate $1/T_1$ in the quadrupolar phase. In this paper, we examine those of $1/T_1$ in the octupolar phase combining eld theoretical method with numerical data. The relevance of the results to quasi one-dimensional J_1-J_2 magnets such as $PbCuSO_4(OH)_2$, $Rb_2Cu_2Mo_3O_{12}$ and Li_2ZrCuO_4 is shortly discussed.

1. Introduction

The spin-$\frac{1}{2}$ chain with nearest-neighbor exchange coupling J_1 and next-nearest-neighbor coupling J_2 is one of the representatives of frustrated spin systems [1]. If J_2 is antiferromagnetic (AF) i.e., $J_2 > 0$, the system has magnetic frustration regardless of the sign of J_1. Recently, several quasi one-dimensional (1D) edge-sharing cuprates [2, 3, 4, 5, 6, 7, 8, 9] such as $LiCu_2O_2$, $LiCuVO_4$, $Rb_2Cu_2Mo_3O_{12}$, $PbCuSO_4(OH)_2$ and Li_2ZrCuO_4, have been investigated as quantum multiferroic materials or frustrated magnets, and many of them are believed to be approximately described by a J_1-J_2 spin chain with ferromagnetic (FM) J_1 and AF J_2 [10, 11] at least in the high temperature regime above their ordering temperatures.

Recent theoretical studies have elucidated e ects of external magnetic eld H on the J_1-J_2 chain [12, 13]. It was found that the eld induces three multipolar Tomonaga-Luttinger (TL) liquid phases, i.e., quadrupolar, octupolar, and hexadecapolar TL-liquid phases, in the wide region of $(J_1/J_2, H)$. In these three phases, multipolar operators $\prod_{n=1}^{p} S_{j+n}$ with $p = 2$, 3, and 4 are respectively shown to exhibit a quasi long-range order. In addition, in these phases, the longitudinal spin correlator also decays in an algebraic form, while the transverse spin correlator obeys an exponential-decay law.

It is generally di cult to experimentally distinguish magnetic multipolar phases in spin systems. There is no established method of measuring multipolar correlations, and static quantities such as speci c heat and susceptibility exhibit no large di erence between the multipolar and standard TL liquids. Recently, we have shown [14] for these multipolar TL liquids that (i) in contrast to usual TL-liquid phase, nuclear-magnetic-resonance (NMR) relaxation rate

$1/T_1$ of the multipolar liquid phases decreases with lowering temperature near the saturation, and (ii) gapless points and peak positions of dynamical structure factors are clearly different from those of usual AF spin chains. Furthermore, in Ref. [15] we have numerically evaluated $1/T_1$ of the quadrupolar phase as a function of temperature and magnetic field, combining field theoretical approach with numerical data. The obtained result demonstrated that field dependence of $1/T_1$ is also characteristic: $1/T_1$ decreases with increasing the field except for the vicinity of saturation. These features of dynamical quantities would be helpful to detect multipolar TL liquids in experiments for the above quasi-1D magnets. In this paper, using the same method as Ref. [15], we determine the field and temperature dependence of $1/T_1$ in the octupolar phase of spin-$\frac{1}{2}$ J_1-J_2 chain. The result is expected to be relevant for quasi-1D J_1-J_2 magnets with large FM $|J_1|$. For instance, $PbCuSO_4(OH)_2$ might be a possible candidate for the realization of octupolar TL liquid [16].

2. Properties of multipolar phases

Here we shortly review the low-energy properties of multipolar TL-liquid phases of spin-$\frac{1}{2}$ J_1-J_2 chain [12, 13]. The Hamiltonian of J_1-J_2 chain under magnetic field H is defined as

$$\mathcal{H} = \sum_{n=1,2}\sum_j J_n \boldsymbol{S}_j \cdot \boldsymbol{S}_{j+n} - \sum_j H S_j^z, \quad (1)$$

where we focus on the case with $J_1 < 0$ and $J_2 > 0$. For the saturated state in high-field regime, two-, three- and four-magnon bound states with momentum π, respectively, become the lowest-energy excitation in the regions $0 < |J_1|/J_2 < 2.699$ $(0.371 < J_2/|J_1| < \infty)$, $2.720 < |J_1|/J_2 < 3.514$ $(0.285 < J_2/|J_1| < 0.368)$, and $3.514 < |J_1|/J_2 < 3.764$ $(0.266 < J_2/|J_1| < 0.285)$. With decreasing field H below the saturation field, these bound magnons are condensed, and the system can be regarded as a hard-core Bose gas of bound magnons. We call these gases of two-, three-, and four-magnon bound states, respectively, the second, third, and fourth order multipolar (quadrupolar, octupolar, and hexadecapolar) TL liquids [17]. These multipolar phases expand from the saturation down to the low-field regime and turn into the low-field vector-chiral phase with $\langle(\boldsymbol{S}_j \times \boldsymbol{S}_{j+1})^z\rangle \neq 0$. In the p-th order multipolar phase, correlation functions of both the longitudinal spin operator S_j^z and the p-th multipolar operators $\prod_{n=1}^{p} S_{j+n}^{\pm}$ (i.e., creation/annihilation operator of bound magnons) can be calculated from the effective TL-liquid theory for the hard-core Bose gas of multi-magnon bound states [12]. They are shown to decay algebraically as follows:

$$\langle S_0^z S_j^z\rangle = M^2 + C_0 \cos\Big[(1-2M)\pi j/p\Big]|j|^{-2\kappa} + \frac{p^2\kappa}{2\pi^2}|j|^{-2} + \cdots, \quad (2)$$

$$\langle\prod_{n=1}^{p} S_n^+ \prod_{n=1}^{p} S_{j+n}^-\rangle = (-1)^j C_1 |j|^{-1/(2\kappa)} + \cdots, \quad (3)$$

where $M = \langle S_j^z\rangle$ is the uniform magnetization per site, C_m are nonuniversal constants, and κ is the TL-liquid parameter. On the other hand, correlators of $q(< p)$-th multipolar operators including transverse spin operator $S_j^{\pm}$ obey an exponential-decay form $\sim e^{-|r|/\xi}$ [14] since any q-magnon excitation requires a finite excitation energy to break p-magnon bound states. The correlation length ξ must be proportional to the inverse of the binding energy of magnons. The absence of the quasi-long-range order in the transverse-spin correlation plays a key role in the following discussion of NMR relaxation rate.

3. NMR relaxation rate

For spin systems in solids, NMR relaxation time T_1 is usually determined through weak hyperfine interactions between electron and nuclear spins. Therefore, we can obtain information about electron-spin correlations from measurement of T_1. The formula for T_1 [18, 19, 20] is given by

$$1/T_1 \;=\; \frac{1}{2}A_\perp^2[\mathcal{S}^{\pm}(\omega)+\mathcal{S}^{\mp}(\omega)]+A_\parallel^2\mathcal{S}^{zz}(\omega), \tag{4}$$

where $A_{\perp,\parallel}$ are the hyper ne coupling constants, and $\mathcal{S}^{\mu\nu}(\omega)$ is the local dynamical structure factor

$$\mathcal{S}^{\mu\nu}(\omega) \;=\; \int dte^{i\omega t}\langle S_j^{\mu}(t)S_j^{\nu}(0)\rangle. \tag{5}$$

The strength of $A_{\perp,\parallel}$ depends on kinds of nuclei, crystal structures, and direction of eld H. The value of ω is xed to a nearly resonating frequency of nuclear-spin precession motion, and it is generally much smaller than the energy scale of electron spin systems. Therefore, we take the limit $\omega/(k_BT) = \omega\beta \to 0$ in the following discussions. Utilizing these formulas, we will discuss the temperature and eld dependence of $1/T_1$ in the octupolar liquid phase.

4. Temperature and field dependence of $1/T_1$

From Eqs. (2), (4) and (5) and eld-theoretical method [18, 19, 20], $1/T_1$ in the multipolar liquid phases is obtained as

$$\begin{aligned} 1/T_1 \;=\; & A_\parallel^2 C_0\frac{2a}{u}\cos(\kappa\pi)B(\kappa,1\quad 2\kappa)\left(\frac{2\pi a}{\beta u}\right)^{2\kappa\quad 1} \\ & +A_\parallel^2\frac{p^2\kappa}{\pi}\left(\frac{a}{u}\right)^2\beta^{\quad 1}+\quad\quad . \end{aligned} \tag{6}$$

Here, $\beta = 1/(k_BT)$, u is the excitation velocity, a is lattice spacing, and $B(x, y)$ is the Beta function. We note that Eq. (6) is valid for the temperature regime $k_BT \quad \min(u/a, \quad)$, where the low-energy e ective theory of the bound-magnon Bose gas is reliable. If spatial dependence of hyper ne couplings is taken account of beyond the formula (4), the prefactors $A_\parallel^2$ in the rst and second term of Eq. (6) are generally modi ed and become di erent.

The form of $1/T_1$ in Eq. (6) consists of contributions from the longitudinal spin correlation; The rst and second terms in Eq. (6) originate from the second and third terms in Eq. (2), respectively. Remarkably, both terms are power-law type functions of temperature. On the other hand, since the transverse spin correlation exhibits a gapped behavior, its contribution to $1/T_1$ is a thermal-activation form $\quad e^{\ \beta\Delta}$. Therefore, if we consider the temperature region lower than the binding energy $\ /k_B$, the contribution from transverse spin correlation is negligible. The disappearance of transverse-spin contribution is a characteristic feature of multipolar TL liquids and makes the multipolar liquids distinct from the usual TL liquid in various quasi-1D AF magnets as discussed below.

Parameters κ, u, and C_0 can be evaluated as a function of eld H (or magnetization M) and J_1/J_2 at zero temperature by using density-matrix renormalization group method [12, 21]. Substituting them into Eq. (6), we obtain the temperature and eld dependence of $1/T_1$. In particular, the temperature dependence of each term is determined by the TL-liquid parameter κ. For the multipolar phases, κ is known to increase almost monotonically with increasing H (or M) and approach unity at the saturation [12, 13]. We can thereby nd that $1/T_1$ decreases with lowering temperature in the high- eld region with $\kappa > 1/2$, while it diverges in the low- eld region with $\kappa < 1/2$ [14]. This behavior highly contrasts to the usual TL liquid such as that of the spin-$\frac{1}{2}$ AF chain, in which one of the longitudinal- or transverse-spin contributions diverges at $T \to 0$ and $1/T_1$ is a monotonically increasing function of temperature in su ciently low-temperature regime. Furthermore, for the quadrupolar phase ($p = 2$), the eld dependence

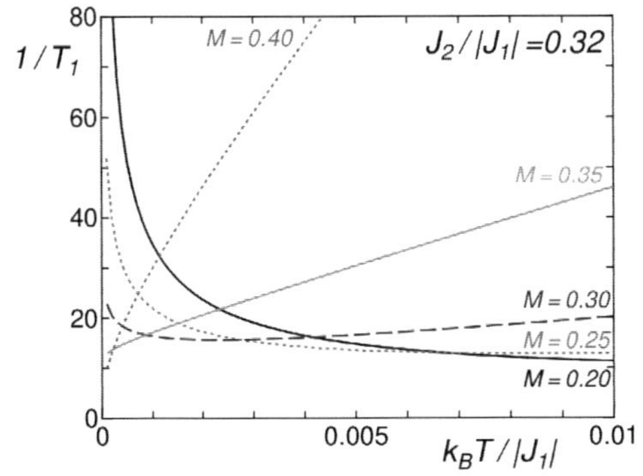

Figure 1. Temperature dependence of NMR relaxation rate $1/T_1$ in the octupolar TL-liquid phase for $J_2/|J_1| = 0.32$ ($J_1/J_2 =\ \ 3.125$). We simply set the coupling constant $A_{\parallel} = 1$.

of $1/T_1$ has also been determined rather accurately [15]. It has then been shown that with increasing H, $1/T_1$ decreases for low eld regime and increases in the vicinity of the saturation eld (see Fig. 1 of Ref. [15]). This peculiar non-monotonic eld dependence of $1/T_1$ is expected to be a common property of all the multipolar TL-liquid phases in the J_1-J_2 chain and would be also a hallmark of them.

Let us now examine the temperature and eld dependence of $1/T_1$ in the octupolar TL-liquid phase ($p = 3$). We evaluate $1/T_1$ combining the formula (6) with numerical estimates of κ, u, and C_0 which are obtained in the same manner as Refs. [12, 21, 22]. The coupling constant $A_{\parallel}$ is simply set to unity. Hereafter, we concentrate on the point of $J_2/|J_1| = 0.32$ ($J_1/J_2 =\ \ 3.125$), at which the octupolar phase widely exists for $0.07 \lesssim M < 1/2$ ($0.022|J_1| \lesssim H < H_c = 0.05|J_1|$) [12]. We note that in the octupolar phase, the excitation velocity u, i.e., the width of low-energy excitation band, is rather small compared to those of quadrupolar phase [15]. For instance, we evaluate $u/a \simeq 0.06|J_1| \simeq 0.19J_2$ for $M = 0.2$ and $u/a \simeq 0.01|J_1| \simeq 0.03J_2$ for $M = 0.4$ at the point of $J_2/|J_1| = 0.32$. This is a natural result because the parameter region of the octupolar liquid phase is closer to the ferromagnetic instability point $J_2/|J_1| = 0.25$ at zero eld. The magnon binding energy is also evaluated as $\ \simeq 0.025|J_1| \simeq 0.078J_2$ for $M = 0.2$ and $\ \simeq 0.041|J_1| \simeq 0.13J_2$ for $M = 0.4$ at $J_2/|J_1| = 0.32$ [23]. Therefore, when we apply the formula (6) at this point, temperature must be low enough, say, $k_BT < 0.01|J_1|$. We also note that the numerical estimates of C_0 in the present calculation are subject to a rather large nite-size e ect and the evaluated $1/T_1$ is less accurate than that of the quadrupolar liquid in Ref. [15]. However, we have checked that the result below is stable and semi-quantitatively unchanged against such an error of C_0.

We show in Fig. 1 the temperature dependence of $1/T_1$. The result con rms the previous prediction that, as temperature decreases, $1/T_1$ increases divergently for the low- eld regime ($\kappa < 1/2$) and decreases for the high- eld regime ($\kappa > 1/2$). Figure 2 shows the eld dependence of $1/T_1$ for several xed temperatures. It is remarkable that, for the su ciently low temperature $k_BT = 0.001|J_1|$, $1/T_1$ decreases with increasing H (or M) except for the vicinity of the saturation eld. The contribution from each term in Eq. (6) is plotted in Fig. 3 separately. It is clear that the rst term in Eq. (6) proportional to $\beta^{\ (2\kappa\ \ 1)}$ contributes dominantly to a peculiar behavior of $1/T_1$, decrease with increasing H, for a wide eld range. This feature becomes less signi cant as temperature increases and the eld dependence of $1/T_1$ becomes nearly at for a rather high temperature, $k_BT = 0.01|J_1|$. The T and H dependence of $1/T_1$ found above is qualitatively the same as that of the quadrupolar TL liquid [15], as expected. Even in the neighboring hexadecapolar liquid phase, similar properties must be realized (although the formula (6) would be valid only in extremely low temperature regime).

Quasi-1D J_1-J_2 magnets with high crystallographic symmetry often have principal axes for hyper ne coupling tensor. For such a case, we can eliminate all the contributions from the

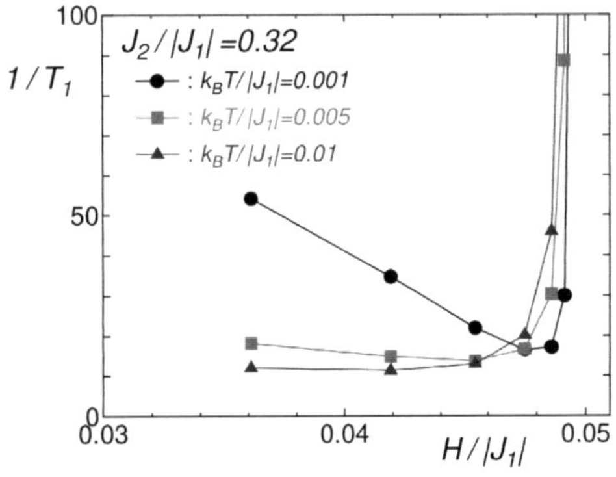

Figure 2. Field dependence of NMR relaxation rate $1/T_1$ in the octupolar TL-liquid phase for $J_2/|J_1| = 0.32$ ($J_1/J_2 = 3.125$). Symbols represent the data obtained from Eq. (6) with numerical values of κ, u, and C_0. Lines are guide for the eyes.

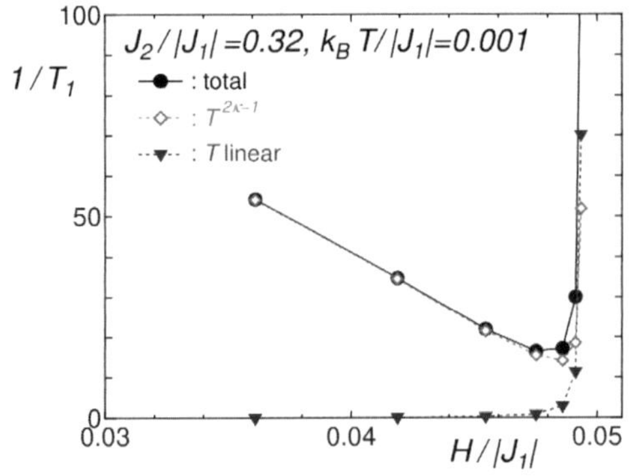

Figure 3. Field dependence of each contribution in NMR relaxation rate $1/T_1$ [Eq. (6)] in the octupolar TL-liquid phase for $J_2/|J_1| = 0.32$ ($J_1/J_2 = 3.125$) and $k_BT/|J_1| = 0.001$.

longitudinal spin correlations, i.e., $A_\parallel$ terms, in $1/T_1$, setting the eld H parallel to a principal axis. As a result, thermal activated contribution $e^{\beta\Delta}$ proportional to $A_\perp^2$, which is the hallmark of the multipolar liquids, becomes detectable in $1/T_1$. In this case, therefore, we can easily distinguish the multipolar TL liquid and usual TL liquid phases from temperature dependence of $1/T_1$.

5. Discussions

We have studied the NMR relaxation rate $1/T_1$ in eld-induced octupolar TL liquid of the spin-$\frac{1}{2}$ J_1-J_2 chain. Substituting numerical estimates of parameters κ, u, and C_0 into the eld-theoretical result, we have evaluated $1/T_1$ as a function of temperature and magnetic eld. We have then found that, similarly in the quadrupolar liquid, $1/T_1$ in the octupolar TL liquid also exhibits characteristic features which are qualitatively di erent from those in the usual TL liquid; In the octupolar liquid, (i) $1/T_1$ decreases (increases) with increasing temperature T for the low- (high-) eld regime, and (ii) for su ciently low temperatures, $1/T_1$ decreases with increasing magnetic eld H for a wide eld range except the vicinity of the saturation eld. These features can be understood from the absence of the quasi-long-range order in the transverse-spin correlation, which is a natural consequence of the formation of magnon bound states, and therefore be generally expected in all the multipolar TL liquids in J_1-J_2 chain. We note that the above features of $1/T_1$ in the octupolar phase is observed only in su ciently low temperature regime ($k_BT < 0.01|J_1|$ at $J_2/|J_1| = 0.32$). This is a quantitative di erence from the case of the quadrupolar phase, which has relatively small $|J_1|$.

The result in this paper indicates that NMR relaxation rate $1/T_1$ could be useful to detect a signature of the multipolar TL-liquid phases. However, only the temperature dependence of $1/T_1$ cannot be a su cient evidence for them. For instance, $1/T_1$ also decreases with lowering temperature in high- eld spin-gapped states such as magnetization plateaus. Therefore, we propose that observations of both characteristic behavior in $1/T_1$ and gapless behavior of bulk quantities (speci c heat, susceptibilities, etc.) present an indirect but strong evidence for multipolar TL liquids.

It has been reported [5, 6, 7, 8, 9] that three quasi-1D cuprates $PbCuSO_4(OH)_2$,

$Rb_2Cu_2Mo_3O_{12}$, and Li_2ZrCuO_4 have relatively large value of $|J_1|$ ($2 \lesssim |J_1|/J_2 \lesssim 4$). We expect that our result will be helpful for searching for the appearance of octupolar phase in these compounds under high magnetic eld.

Acknowledgments

This work was supported by Grants-in-Aid for Scienti c Research from MEXT, Japan (Grants No. 21740277, No. 21740295, and No. 22014016).

References

[1] See, for a review, 2004 *Frustrated spin systems*, edited by Diep T H (Singapore World Scienti c)
[2] Masuda T, Zheludev A, Roessli B, Bush A, Markina M and Vasiliev A 2005 *Phys. Rev.* B **72** 014405
[3] Enderle M, Mukherjee C, Fåk B, Kremer R K, Broto J -M, Rosner H, Drechsler S -L, Richter J, Malek J, Proko ev A, Assmus W, Pujol S, Raggazzoni J -L, Rakoto H, Rheinstädter M and Rønnow H M 2005 *Europhys. Lett.* **70** 237
[4] Svistov L E, Fujita T, Yamaguchi H, Kimura S, Omura K, Proko ev A, Smirnov A I, Honda Z and Hagiwara M 2011 *JETP Lett.* **93** 24
[5] Hase M, Kuroe H, Ozawa K, Suzuki O, Kitazawa H, Kido G and Sekine T 2004 *Phys. Rev.* B **70** 104426
[6] Kamieniarz G, Bieliński M, Szukowski G, Szymczak R, Dyeyev S and Renard J -P 2002 *Comp. Phys. Comm.* **147** 716
[7] Baran M, Jedrzejczak A, Szymczak H, Maltsev V, Kamieniarz G, Szukowski G, Loison C, Ormeci A, Drechsler S -L and Rosner H 2006 *Phys. Stat. Sol.* (c) **3** 220
[8] Wolter A U B, Vogel R, Skourski Y, Rule K C, Süllow S, Heide G and Buechner B 2010 International conference "Magnetic resonance in highly frustrated magnetic systems" (Slovenia Feb. 2010)
[9] Drechsler S -L, Volkova O, Vasiliev A N, Tristan N, Richter J, Schmitt M, Rosner H, Málek J, Klingeler R, Zvyagin A A and Büchner B 2007 *Phys. Rev. Lett.* **98** 077202
[10] Furukawa S, Sato M and Onoda S, 2010 *Phys. Rev. Lett.* **105** 257205
[11] Sato M, Furukawa S, Onoda S and Furusaki A 2011 arXiv:1101.1374; submitted to *Mod. Phys. Lett.* B as Proceedings of StatPhys24 Satellite Meeting, Hanoi, Jul. 2010.
[12] Hikihara T, Kecke L, Momoi T and Furusaki A, 2008 *Phys. Rev.* B **78** 144404
[13] Sudan J, Luscher A and Läuchli A M 2009 *Phys. Rev.* B **80** 140402(R)
[14] Sato M, Momoi T and Furusaki A 2009 *Phys. Rev.* B **79** 060406(R)
[15] Sato M, Hikihara T and Momoi T 2011 *Phys. Rev.* B **83** 064405
[16] Wolter A U B and Drechsler S -L private communication
[17] In the narrow range $-2.720 < J_1/J_2 < -2.669$, the soft two-magnon bound state has an incommensurate momentum $k \neq \pi$ to realize a two-component TL liquid. In this paper, we do not consider this incommensurate case and focus on the one-component multipolar TL liquids of the multimagnon bound states with momentum $k = \pi$. However, the qualitative behavior of $1/T_1$ in the incommensurate multipolar phase is expected to be similar to those of the one-component multipolar phase since it mainly results from the formation of multimagnon bound states.
[18] See, for example, Giamarchi T 2004 *Quantum Physics in One Dimension* (New York Oxford University Press).
[19] Chitra R and Giamarchi T 1997 *Phys. Rev.* B **55** 5816
[20] Giamarchi T and Tsvelik A M 1999 *Phys. Rev.* B **59** 11398
[21] Hikihara T and Furusaki A 2004 *Phys. Rev.* B **69** 064427
[22] Hikihara T, Momoi T, Furusaki A and Kawamura H 2010 2010 *Phys. Rev.* B **81** 224433
[23] We determine the magnon binding energy Δ in the octupolar TL liquid as follows. First we prepare $\Delta_n(\mathcal{M}) = E_0(\mathcal{M}+n) - E_0(\mathcal{M})$, where $E_0(\mathcal{M})$ is the lowest energy eigenvalue in the subspace with xed $\sum_j S_j^z = \mathcal{M} = ML$ (L is total number of sites). Then, using Δ_n we de ne four kinds of energy gaps $E_{b1+} = 3\Delta_1 - \Delta_3$, $E_{b2+} = \Delta_1 + \Delta_2 - \Delta_3$, $E_{b1-} = 3\Delta_{-1} - \Delta_{-3}$, and $E_{b2-} = \Delta_{-1} + \Delta_{-2} - \Delta_{-3}$. Finally the true binding energy is de ned as the minimum value of them, $\Delta = \min(E_{b1+}, E_{b2+}, E_{b1-}, E_{b2-})$.

Numerical Diagonalization Study on the S=1/2 Frustrated Three-Leg Quantum Spin Ladder Systems

Shohei Abe[1], Toru Sakai[1,2], Kiyomi Okamoto[3] and Kenji Tsutsui[2]

[1] Department of Material Science, University of Hyogo, Kamigori, Hyogo 678-1297, Japan
[2] Japan Atomic Energy Agency, SPring-8, Sayo, Hyogo 679-5148, Japan
[3] Department of Physics, Tokyo Institute of Technology, Meguro-ku, Tokyo 152-8551, Japan

E-mail: abe.shohei@jaea.go.jp

Abstract. The spin ladder systems have attracted a lot of interest in the field of low-dimensional physics. It is well known that the $S = 1/2$ two-leg spin ladder has a spin gap, while the three-leg one has a gapless spin liquid ground state. Some spin liquid behaviors were observed in the recent experiments of some frustrated systems; the distorted triangular lattice and kagome lattice antiferromagnets etc.. Thus we consider the $S = 1/2$ three-leg spin ladder system with the next-nearest-neighbor interaction to investigate the effect of frustration on the spin liquids. Using the exact numerical diagonalization of finite-cluster systems, we study the ground state of this system. The phase diagram with respect to some exchange interaction constants is presented. It is revealed that sufficiently large next-nearest-neighbor coupling leads to a ferrimagnetic phase, which has a frustration-induced spontaneous magnetization. In addition we find two spin liquid phases, one of which corresponds to a Néel-like spin configuration and the other a collinear spin configuration. As a result of the numerical study, we find a new phase which is different from three phases.

1. Introduction

The spin liquid has attracted a lot of interest, since the resonating valence bond was proposed as a mechanism of the high temperature superconductivity[1, 2]. Recently a spin liquid-like behavior was observed in the organic triangular-lattice quantum spin system[3]. In the system the spin frustration and the quantum fluctuation interplay to create the spin liquid state. The $S = 1/2$ three-leg spin ladder[4] is one of typical systems with the gapless spin liquid ground states. In this paper, we investigate the $S = 1/2$ three-leg spin ladder model with the next-nearest-neighbor interaction, to consider a role of the frustration in the spin liquid. Using the exact numerical diagonalization, we calculate the low-lying energy levels to present the ground-state phase diagram.

2. Model

We consider the $S = 1/2$ frustrated three-leg spin ladder, as shown in Fig. 1, described by the Hamiltonian,

$$H = J_l \sum_{i=1}^{3} \sum_{j=1}^{L} \vec{S}_{i,j}\vec{S}_{i,j+1} + J_r \sum_{i=1}^{2} \sum_{j=1}^{L} \vec{S}_{i,j}\vec{S}_{i+1,j} + J_d \sum_{i=1}^{2} \sum_{j=1}^{L} \vec{S}_{i+1,j}\vec{S}_{i,j+1},$$

Published under licence by IOP Publishing Ltd

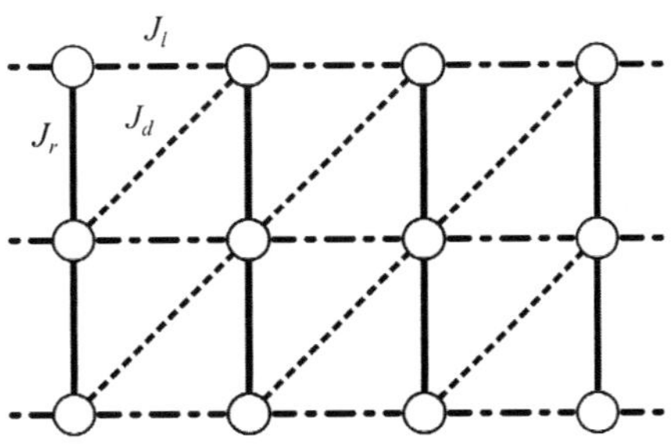

Figure 1. The structure of the $S = 1/2$ frustrated three-leg spin ladder. Open circles denote the locations of $S = 1/2$ spins. The AF exchange interaction constants J_r, J_l and J_d are for the neighboring spin pairs along rung, legs and diagonals viewed as the solid, broken dotted and dashed lines, respectively.

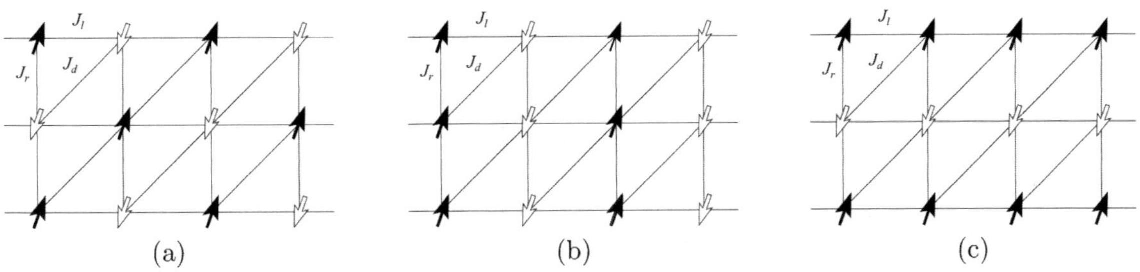

Figure 2. Schematic ground states of the $S = 1/2$ frustrated three-leg spin ladder expected for some extreme values of AF exchange interaction constants. Up arrows and down arrows are the $S^z = 1/2$ up spins and down spins in the figure. (a) $J_d \ll J_l, J_r$. The ground-state is the Néel-like spin liquid. (b) $J_r \ll J_l, J_d$. The ground state is the collinear spin liquid. (c) $J_l \ll J_r, J_d$. The ground-state is the ferrimagnetic.

where $\vec{S}_{i,j}$ is the spin-1/2 oprator and L is the length of the ladder in the leg direction. The Antiferromagnetic(AF) exchange interaction constant, J_l, is for the neighboring spin pairs along the legs, while J_r and J_d are the rung and diagonal coupling constant, respectively.

3. Method

Using exact numerical diagonalization, we calculate the low-lying energy levels for finite-size systems. ($L = 4, 6, 8$.) When we calculate the energy levels for various AF-exchange parameters (J_r, J_l, J_d), we use the spin quantum number S and wave number k to identify each states. We evaluate the value of S from the degree of degeneracy with respect to S^z. We decide a critical point of quantum phase transition by comparing some energy eigenvalues. Finally, we use all of numerical result to make a phase diagram.

4. Results

4.1. Expected phases

One can expect several different phases in the ground state for some extreme values of the coupling constants. In the limit that $J_d \ll J_l, J_r$, the ground state is the Néel-like spin liquid as shown in Fig. 2(a). In the Néel-like spin liquid phase, the neighboring spin pairs along the legs and rungs tend to be antiparallel to each other. In the limit that $J_r \ll J_l, J_d$, the ground state is the collinear spin liquid as shown in Fig. 2(b). In the collinear spin liquid phase, the

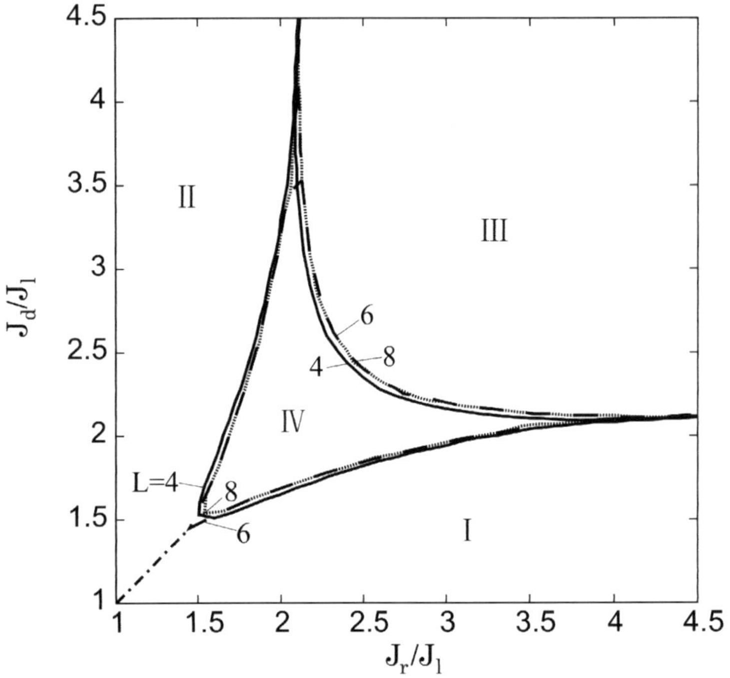

Figure 3. Phase diagram of the $S = 1/2$ frustrated three-leg spin ladder with three AF-exchange interaction constants.
(I) ~ (IV) show the Néel-like spin liquid phase, collinear spin liquid phase, ferrimagnetic phase and new phase. The solid, dashed, dotted lines show the phase boundary lines of finite-size result for $L = 4, 6, 8$. The broken dotted line shows energy crossover line.

neighboring spin pairs along the legs and diagonals tend to be antiparallel to each other. In the Néel-like and collinear spin liquid phases, $S = 0$. In the limit that $J_l \ll J_r$,J_d, the ground state is the ferrimagnetic as shown in Fig. 2(c). In the ferrimagnetic phase, the neighboring spin pairs along the rungs and diagonals tend to be antiparallel to each other. In the ferrimagnetic phase $S = M_s/3$. (M_s is the saturated magnetization. e.g. $M_s = 6$ when $L = 4$.)

4.2. Numerical results

As a result from the numerical calculation, we make the phase diagram of $S = 1/2$ frustrated three-leg spin ladder with three AF exchange interaction constants, as shown in Fig. 3. All of the phase boundaries correspond to first-order transitions, detected as level crosses in the ground state. Three phases discussed in Sec. 4.1. are shown as (I) the Néel-like spin liquid phase, (II) the collinear spin liquid phase and (III) the ferrimagnetic phase in Fig. 3. In our current study, we find the new phase shown as (IV) in Fig. 3. The size dependence of phase boundaries are so small that we can neglect them to present the phase diagram. The broken dotted line shown in Fig. 3 is the energy crossover line. On the line, a level cross between the first- and second-excited states exists.

To make sure of the validity of our numerical results, we try an analytical approach to calculate phase boundaries between ferrimagnetic phase and two spin liquid phases. We use the first-order perturbation expansion from the limit that $J_r \gg J_l$,J_d or $J_d \gg J_l$,J_r. In case that $J_r \gg J_l$,J_d, we diagonalize the Hamiltonian of the three-spin unit cell along the rung and take the lowest two states. We introduce a pseudospin to describe the two states as follows:

$$|\Uparrow\rangle = \frac{1}{\sqrt{6}} \left(|\uparrow\uparrow\downarrow\rangle - 2\,|\uparrow\downarrow\uparrow\rangle + |\downarrow\uparrow\uparrow\rangle \right).$$

$$|\Downarrow\rangle = -\frac{1}{\sqrt{6}} \left(|\downarrow\downarrow\uparrow\rangle - 2\,|\downarrow\uparrow\downarrow\rangle + |\uparrow\downarrow\downarrow\rangle \right).$$

Within the pseudospin effective theory, the nearest neighbor pairs should be $|\Uparrow\rangle\otimes|\Uparrow\rangle$ and $|\Uparrow\rangle\otimes|\Downarrow\rangle$ in the ferrimagnetic and Néel-like spin liquid phases, respectively. For each phase we calculate the expectation value of the following Hamiltonian:

$$H_{J_l,J_d} = J_l \sum_{i=1}^{3}\sum_{j=1}^{2} \vec{S}_{i,j}\vec{S}_{i,j+1} + J_d \sum_{i=1}^{2}\sum_{j=1}^{2} \vec{S}_{i+1,j}\vec{S}_{i,j+1}$$

For ferrimagnetic phase, the value E_{ferri} is:

$$E_{\text{ferri}} = \langle\Uparrow|\otimes\langle\Uparrow|\, H_{J_l,J_d}\, |\Uparrow\rangle\otimes|\Uparrow\rangle = \frac{1}{4}J_l - \frac{1}{9}J_d.$$

For Néel-like spin liquid phase, the value E_{SL} is:

$$E_{\text{SL}} = \langle\Uparrow|\otimes\langle\Downarrow|\, H_{J_l,J_d}\, |\Uparrow\rangle\otimes|\Downarrow\rangle = -\frac{1}{4}J_l + \frac{1}{9}J_d.$$

The phase boundary can be determined by $E_{\text{ferri}} = E_{\text{SL}}$. Therefore the analytical phase boundary between ferrimagnetic and Néel-like spin liquid phase is obtained as $J_d/J_l = 9/4$. In the same way, the analytical phase boundary between ferrimagnetic and collinear spin liquid phase is $J_r/J_l = 9/4$ in the case that $J_d \gg J_l, J_r$. Numerical calculation results approach asymptotically to perturbation calculation results.

In the present study we found a new phase between the spin liquid and ferrimagnetic phases, as shown in Fig. 3. However, it is quite difficult to specify the spin structure of this phase with the numerical diagonalization, because many size-dependent level crosses appear. Now we can just conclude that the ground state is a singlet in this phase. Thus at the boudary of the ferrimagnetic phase the magnetization has a jump from $S = 0$ to $S = M_s/3$. In constrast, another three-leg quantum spin model which exhibits a partial ferrimagnetism was proposed[5].

5. Conclusions

We have investigated the $S = 1/2$ three-leg spin ladder model with the next-nearest-neighbor interaction, to consider a role of the frustration in the spin liquid. Using the exact numerical diagonalization, we have calculated the low-lying energy levels to present the ground-state phase diagram. We have found four ground-state phases; (I) the Néel-like spin liquid phase, (II) the collinear spin liquid phase, (III) the ferrimagnetic phase and (IV) the new phase.

6. Acknowledge

We wish to thank Profs. T. Tonegawa, H. Nakano and K. Nomura for fruitful discussions. This work was partly supported by Grants-in-Aid (No.20340096), Grants-in-Aid for Scientific Research and Priority Area "Novel States of Matter Induced by Frustration" from the Ministry of Education, Culture, Sports, Science and Technology of Japan. A part of the computations was performed using facilities of the Supercomputer Center, Institute for Solid State Physics, University of Tokyo.

References

[1] Anderson P W 1973 *Mater. Res. Bull.* **8** 153.
[2] Anderson P W 1987 *Science* **235** 1196.
[3] Shimizu Y, Miyagawa K, Kanoda K, Maesato M and Saito G 2003 *Phys. Rev. Lett.* **91** 107001.
[4] Azuma M, Hiroi Z, Takano M, Ishida K and Kitaoka Y 1994 *Phys. Rev. Lett.* **73** 3463.
[5] Shimokawa T and Nakano H 2011 submitted to J. Phys.: Conf. Series, proceedings of ICFCM.

Journal of Physics: Conference Series **320** (2011) 012016 doi:10.1088/1742-6596/320/1/012016

Quantum critical phenomena in magnetization process of the Kagome and triangular lattice antiferromagnets

Tôru Sakai^{A,B,C} **and Hiroki Nakano**^{B}

^{A}Japan Atomic Energy Agency, SPring-8, Sayo, Hyogo 679-5148, Japan
^{B}Graduate School of Material Science, University of Hyogo, Kamigori, Hyogo 678-1297, Japan
^{C}TRIP JST, Saitama 332-0012, Japan

E-mail: sakai@spring8.or.jp

Abstract. Using the numerical exact diagonalization of finite-size clusters up to 39 spins, we investigate the magnetization process of the Kagome lattice antiferromagnet, as well as the triangular lattice antiferromagnet. A finite-size scaling analysis at 1/3 of the saturation magnetization indicates that an unconventional quantum phase transition, so called a magnetization ramp, occurs for the Kagome lattice antiferromagnet. In contrast, the triangular lattice antiferromagnet is revealed to exhibit a conventional 1/3 magnetization plateau. These conclusions are based on the estimated critical exponet δ defined as $|m-m_c| \sim |H-H_c|^{1/\delta}$. The exponents at the lower and higher field sides of $m = 1/3$ δ_- and δ_+ are estimated as $\delta_- = 1.9 \pm 1.0$ and $\delta_+ = 0.5 \pm 0.2$ for the Kagome kattice, while $\delta_- = 1.0 \pm 0.2$ and $\delta_+ = 0.8 \pm 0.2$ for the triangular lattice.

1. Introduction

The $S = 1/2$ Kagome[1] and triangular lattice antiferromagnets are ones of the most popular frustrated quantum spin systems. The previous theoretical studies indicated that the former system is disordered in the ground state [5, 6, 7, 8, 9, 10, 11, 12, 13, 14, 15, 16], while the latter has the 120° long-range order[2, 3, 4]. Experimental studies to observe a novel spin liquid phase have been accelarated since discoveries of several realistic materials; the herbertsmithite[18, 19], the volborthite[20, 21] and the vesignieite[22] for the Kagome lattice, the organic compound κ-(BEDT-TTF)$_2$Cu$_2$(CN)$_3$ for the triangular lattice[17]. Since the quantum Monte Carlo simulation and the DMRG calculation are useless for these systems, the numerical exact diagonalization is one of the best numerical method for them. The numerical diagonalization studies suggested that both systems have the 1/3 magnetization plateau [23, 24, 25, 26], although the classical spin systems have no plateau on both lattices in the ground state[28, 29]. In our recent numerical diagonalization study on the $S = 1/2$ Kagome lattice antiferromagnet up to $N = 36$, the calculated field derivatives revealed an anomalous behavior at 1/3 of the saturation magnetization[30]. Namely, the field derivative is diversing at the low-field side of the critical field H_c, while almost zero at the high-field side. This critical behavior is quite different from conventional magnetization plateaux in two-dimensional systems where the field derivative is finite at both sides of H_c. To distinguish such an anomalous property at the 1/3 magnetization of the Kagome lattice from conventional plateaux, we called it a "magnetization ramp". However, its mechanism is still an open problem. In this paper, to clarify such an unconventional behavior around the 1/3 magnetization of the $S = 1/2$ Kagome lattice antiferromagnet, comparing with the triangular one, we applied the numerical diagonalization for both systems up to $N = 39$ which is the largest cluster at present. In addition we estimated the critical exponent δ by the finite-size scaling proposed by the previous work[32], to investigate the quantum critical behevior more quantitatively.

2. Model and calculation

The magnetization processes of the $S = 1/2$ kagome and triangular lattice antiferromagnets are described by the Hamiltonian

$$\begin{aligned} \mathcal{H} &= \mathcal{H}_0 + \mathcal{H}_Z, \\ \mathcal{H}_0 &= \sum_{\langle i,j \rangle} \mathbf{S}_i \cdot \mathbf{S}_j, \\ \mathcal{H}_Z &= -H \sum_j^N S_j^z, \end{aligned} \tag{1}$$

where $\langle i, j \rangle$ means all the nearest neighbor pairs on each lattice. Throughout we use the unit such that $g\mu_B = 1$. For N-site systems, the lowest energy of $\mathcal{H}_0$ in the subspace where $\sum_j S_j^z = M$ (the macroscopic magnetization is $m = M/N$) is denoted as $E(N, M)$. We restrict us to the rhombic cluster under the periodic boundary condition to keep the 120° rotational symmetry for a systematic finite-size scaling. Using the numerical exact diagonalization, we can calculate all the values of $E(N, M)$ available for the rhombic clusters with N =12, 21, 27, 36 and 39, to obtain the ground state magnetization curves. In our previous work[30] the analysis of the field derivative calculated as the form

$$\chi^{-1} = \frac{E(N, M+1) - 2E(N, M) + E(N, M-1)}{1/M_s}, \tag{2}$$

revealed an exotic behavior of the Kagome lattice antiferromagnet at $m = 1/3$, called a magnetization ramp, using the result up to $N = 36$. This phenomenon is characterized by a difference between the lower and higher field sides of $m = 1/3$; χ is diverging at the lower side like a plateau in one-dimensional systems, while is very small (possibly zero) at the higher one. The magnetization and the field derivative curves of the Kagome lattice antiferromagnet for $N = 39$ also supported such a ramp-like behavior[31] at least qualitatively.

3. Critical exponent

In order to characterize the unconvetional magnetization behavior at $m = 1/3$ of the Kagome lattice antiferromagnet in a quantitative way, we introduce the critical exponent δ. It is defined by the form

$$|m - m_c| \sim |H - H_c|^{1/\delta}, \tag{3}$$

which is an important index to specify the universality class of the field induced quantum phase transition. The previous theoretical works indicated $\delta = 2$ for some typical one-dimensional gapped systems[33, 34], while $\delta = 1$ for two-dimensional systems[35]. In order to clarify the anomalous critical behavior at $m = 1/3$ of the Kagome lattice antiferromagnet, we estimate δ by the finite-size scaling developed by the previous work[32], and compare it with the triangular lattice antiferromagnet. Although the method was proposed for one-dimensional systems, it can be easily generalized for two dimensions. We assume the asymptotic form of the size dependence of the energy as

$$\frac{1}{N}E(N,M) \sim \epsilon(m) + C(m)\frac{1}{N^\theta} \qquad (N \to \infty), \tag{4}$$

where $\epsilon(m)$ is the bulk energy and the second term describes the leading size correction. We also assume that $C(m)$ is an analytic function of m. The lowest and highest magnetic field corresponding to $m = 1/3$ in the thermodynamic limit are defined as H_{c1} and H_{c2}, respectively, as the form

$$E(N,\frac{N}{3}) - E(N,\frac{N}{3}-1) \to H_{c1} \quad (N \to \infty), \tag{5}$$

$$E(N,\frac{N}{3}+1) - E(N,\frac{N}{3}) \to H_{c2} \quad (N \to \infty). \tag{6}$$

In order to consider the critical magnetization behaviors for $m < 1/3$ and $m > 1/3$ independently, we define the critical exponents δ_{c-} and δ_{c+} by the forms

$$m - \frac{1}{3} \sim (H - H_{c2})^{1/\delta_+}, \tag{7}$$

$$\frac{1}{3} - m \sim (H_{c1} - H)^{1/\delta_-}. \tag{8}$$

If we define the quantities $f_+(N)$ and $f_-(N)$ by the forms

$$f_\pm(N) \equiv \pm[E(N,\frac{N}{3}\pm 2) + E(N,\frac{N}{3}) - 2E(N,\frac{N}{3}\pm 1)], \tag{9}$$

the asymptotic forms of them are expected to be

$$f_\pm(N) \sim \frac{1}{N^{\delta_\pm}} + O(\frac{1}{N^{\theta+1}}) \qquad (N \to \infty), \tag{10}$$

as far as we assume the form (5). Thus the exponents δ_- and $\delta+$ can be estimated from the slope of the $\log f_\pm$-$\log N$ plot, respectively, under the condition $\theta > \delta_\pm - 1$. In order to avoid an oscilation of the finite-size correction due to the cluster shape dependence, we just use the rhombic clusters under the periodic boundary condition with N =12, 21, 27, 36, and 39. The plots of $\log f_\pm$ versus $\log N$ for the triangular and Kagome lattice antiferromagnets are shown in Figures. 1 and 2, respectively. Figure 1 suggests that the calculated points are well fitted to a line for each of f_- and f_+ in the case of the triangular lattice. Thus applying the standard least square fitting to lines (Dashed and long-dashed lines are used to obtain δ_+ and δ_-, respectively.) for N =12, 21, 27, 36, 39, δ_- and δ_+ are estimated as follows:

$$\delta_- = 1.0 \pm 0.2, \qquad \delta_+ = 0.8 \pm 0.2,$$

for the triangular lattice. The errors are estimated from the deviation of points from the fitted lines. It would be reasonable to conclude $\delta_- = \delta_+ = 1$ at $m = 1/3$ of the triangular lattice antiferromagnet, as expected for conventional magnetization plateau in two dimensions. On the other hand, Figure 2 indicates quite different feature of the Kagome lattice antiferromagnet. The same least square fitting yields the following estimates:

$$\delta_- = 1.9 \pm 1.0, \qquad \delta_+ = 0.5 \pm 0.2,$$

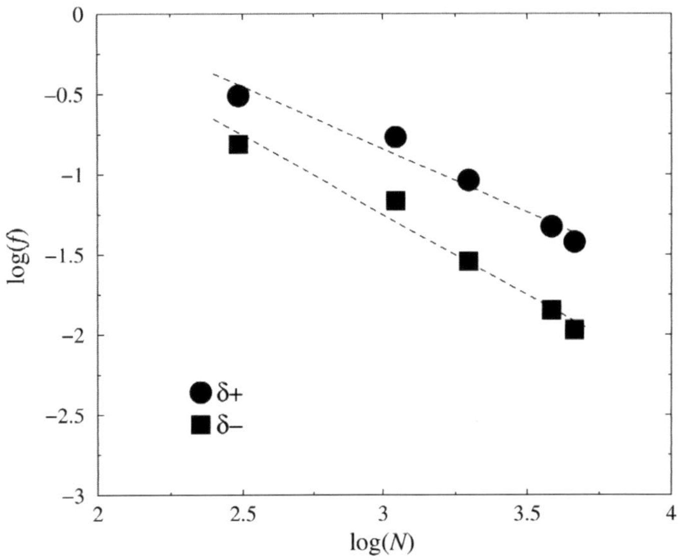

Figure 1. $\log(f)$ is plotted versus $\log(N)$ for the triangular lattice antiferromagnet. Solid circles and squares are useful to estimate the critical exponents $\delta+$ and $\delta-$, respectively.

for the Kagome lattice antiferromagnet. Exponent δ_- has a large error because the line fitting is not good. It does not converge with respect to the system size well, but seems to still increasing with N. The same line fitting to the points for N =27, 36 and 39 yields the estimation $\delta_- = 4.6 \pm 0.3$. Thus we can just conclude $\delta_- \geq 2$ at most. It means that the diversing behavior of the field derivative at H_{c1} is stronger than one-dimensional systems. It is led to two possibilities. One is a jump (a first-order transition) in the magnetization curve. A magnetization jump which also appears near the saturation was proved[36]. The other is an anomalous continuous transition. A similar phenomenon was reported in the metal-insulator transition of the Hubbard chain with next-nearest neighbor hopping[?]. In comparison with δ_-, δ_+ is more conclusive, because the fitting error is much smaller. According to the above result of the line fitting, we conclude $\delta+$ is smaller than unity. Thus the field derivative χ should be zero at the higher field side of H_{c2} It also justifies a property of the magnetization ramp.

4. Ramp versus plateau

We consider whether a flat part of the magnetization curve at $m = 1/3$ exists or not for the triangular and Kagome lattice antiferromagnets. Namely, we examine whether each system has no plateau ($H_{c1} = H_{c2}$) or a finite plateau ($H_{c1} \neq H_{c2}$) at $m = 1/3$ in the thermodynamic limit. We evaluate the length of the flat part $H_{c2} - H_{c1}$ corresponding to the plateau width of the finite-size clusters with N =12, 21, 27, 36 and 39 for both systems. If the system has a gapless excitation like a spin wave from some ordered states, the low-lying energy spectrum is expected to be proportional to the wave vector k in the long wave lenght limit. Thus the excitation energy gap of the finite-size systems should have the asymptotic form $\sim 1/N^{1/2}$ in two-dimensional gapless systems. On the other hand, in gapped systems the gap is expected to converge to the thermodynamic limit with exponentially decaying (faster than $1/N^{1/2}$) finite-size correction, as the system size increases. Thus if the extrapolation by fitting the gap versus $1/N^{1/2}$ leads to a finite gap in the thermodynamic limit, it would be a strong evidence to confirm the gapped ground state. The length of a flat part $H_{c2} - H_{c1}$ is plotted versus $1/N^{1/2}$ in Figure 3, where open triangles and solid circles are for the triangular and Kagome lattice antiferromagnets,

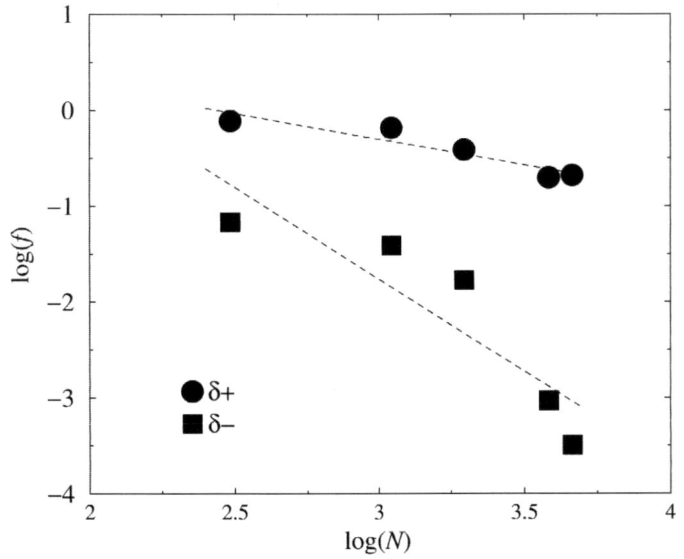

Figure 2. log(f) is plotted versus log(N) for the Kagome lattice antiferromagnet. Solid circles and squares are useful to estimate the critical exponents $\delta+$ and $\delta-$, respectively.

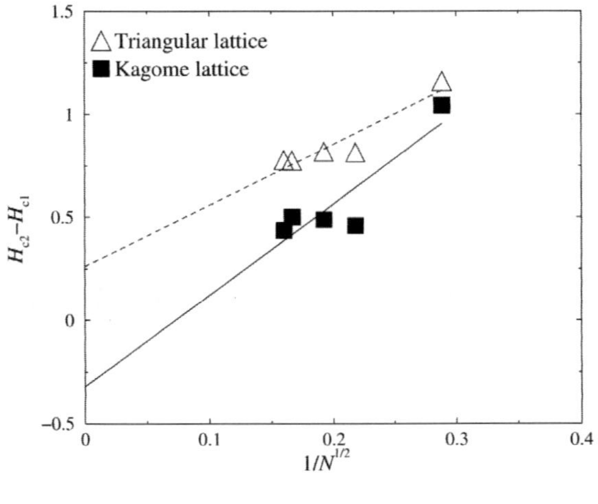

Figure 3. Plateau width $H_{c2} - H_{c1}$ is plotted versus $1/N^{1/2}$. Open triangles and solid squares are for the triangular and Kagome lattice antiferromagnets, respectively. Fitted lines are used for the extrapolation to the thermodynamic limit.

respectively. The least square fitting to a line leads to the following results: $H_{c2} - H_{c1} = 0.3 \pm 0.2$ for the triangular lattice and $H_{c2} - H_{c1} = -0.3 \pm 0.5$ for the Kagome lattice. Obviously we can conclude that the triangular lattice antiferromagnet has the 1/3 magnetization plateau. In contrast, the result for the Kagome lattice suggests that it possibly has a single critical field $H_c = H_{c1} = H_{c2}$. Thus we conclude that the Kagome lattice antiferromagnet exhibits a 1/3 magnetization ramp as a single quantum phase transition at H_c.

In the recent magnetization measurement[20] on a candidate of the Kagome lattice antiferromagnet Volborthite several step-like behaviors were observed, but it has not reached $m = 1/3$ yet. The same measurement is sitll going on to observe an anomaly at $m = 1/3$, which is expected to be about 60T. It would be interesitng to detect some unconventional features.

5. Summary

We have investigated critical magnetization behaviors at $m = 1/3$ for the $S = 1/2$ triangular and Kagome lattice quantum antiferromagnets, using the numerical exact diagonalization of rhombic clusters up to $N = 39$. The triangular lattice is revealed to have the critical exponets $\delta_- = \delta_+ = 1$ and a finite plateau, which are consistent with a conventional magnetization plateau in two-dimensional systems. On the other hand, the Kagome lattice is revealed to exhibit unconventional critical properties; $\delta_- < 1 < \delta_+$, namely the field derivative χ is diverging at the lower field side, while zero at the higher one of a single critical field $H_c = H_{c1} = H_{c2}$. The conclusion supports the magnetization ramp behavior at $m = 1/3$ of the Kagome lattice antiferromagnet.

References

[1] A brief history of the study on the kagome-lattice magnetism was given in M. Mekata, Phys. Today Feb., 12 (2003).
[2] D. A. Huse and V. Elser, Phys. Rev. Lett. **60**, 2531 (1988).
[3] B. Bernu, P. Lecheminant, C. Lhuillier and L. Pierre, Phys. Rev. B **50**, 10048 (1994).
[4] L. Capriotti, A. E. Trumper and S. Sorella, Phys. Rev. Lett. **82**, 3899 (1999).
[5] S. Sachdev, Phys. Rev. B **45**, 12377 (1992).
[6] T. Nakamura and S. Miyashita, Phys. Rev. B **52**, 9174 (1995).
[7] P. Lecheminant, B. Bernu, C. Lhuillier, L. Pierre, and P. Sindzingre, Phys. Rev. B **56**, 2521 (1997).
[8] Ch. Waldtmann, H.-U. Everts, B. Bernu, C. Lhuillier, P. Sindzingre, P. Lecheminant and L. Pierre, Eur. Phys. J. B **2**, 501 (1998).
[9] F. Mila, Phys. Rev. Lett. **81**, 2356 (1998).
[10] M. Hermele, T. Senthil and M. P. A. Fisher, Phys. Rev. B **72**, 104404 (2005).
[11] R. R. P. Singh and D. A. Huse, Phys. Rev. B **76**, 180407(R) (2007).
[12] Y. Ran, M. Hermele, P. A. Lee and X. -G. Wen, Phys. Rev. Lett. **98**, 117205 (2007).
[13] H. C. Jiang, Z. Y. Weng and D. N. Sheng, Phys. Rev. Lett. **98**, 117203 (2008).
[14] O. Cepas, C. M. Fong, P. W. Leung, and C. Lhuillier, Phys. Rev. B **78**, 140405(R) (2008).
[15] P. Sindzingre and C. Lhuillier, Europhys. Lett. **88**, 27009 (2009).
[16] G. Evenbly and G. Vidal, Phys. Rev. Lett. **104**, 187203 (2010).
[17] Y. Shimizu, K. Miyagawa, K. Kanoda, M. Maesato and G. Saito, Phys. Rev. Lett. **91**, 107001 (2003).
[18] M. P. Shores, E. A. Nytko, B. M. Barlett, and D. G. Nocera, J. Am. Chem. Soc. **127**, 13462 (2005).
[19] P. Mendels and F. Bert, J. Phys. Soc. Jpn. **79**, 011001 (2010).
[20] H. Yoshida, Y. Okamoto, T. Tayama, T.Sakakibara, M.Tokunaga, A. Matsuo, Y. Narumi, K. Kindo, M. Yoshida, M. Takigawa, and Z. Hiroi, J. Phys. Soc. Jpn. **78**, 043704 (2009).
[21] M. Yoshida, M. Takigawa, H. Yoshida, Y. Okamoto, and Z. Hiroi, Phys. Phys. Lett. **103**, 077207 (2009).
[22] Y. Okamoto, H. Yoshida, and Z. Hiroi, J. Phys. Soc. Jpn. **78**, 033701 (2009).
[23] K. Hida, J. Phys. Soc. Jpn. **70**, 3673 (2001).
[24] D. C. Cabra, M. D. Grynberg, P. C. W. Holdsworth, and P. Pujol, Phys. Rev. B **65**, 094418 (2002).
[25] J. Schulenberg, A. Honecker, J. Schnack, J. Richter, and H.-J. Schmidt, Phys. Rev. Lett. **88**, 016207 (2002).
[26] A. Honecker, J. Schulenberg, and J. Richter, J. Phys.: Condens. Matter **16**, S749 (2004).
[27] D. C. Cabra, M. D. Grynberg, P. C. W. Holsworth, A. Honecker, P. Pujol, J. Schmalfus and J. Schulenburg, Phys. Rev. B **71**, 144420 (2005).
[28] H. Kawamura and S. Miyashita, J. Phys. Soc. Jpn. **54**, 4530 (1985).
[29] M. E. Zhitomirsky, Phys. Rev. Lett. **88**, 057204.
[30] H. Nakano and T. Sakai, J. Phys. Soc. Jpn. **79**, 053707 (2010).
[31] Nakano H and Sakai T submitted to *J. Phys.: Conf. Series*, proceedings of ICFCM 2011
[32] T. Sakai and M. Takahashi, Phys. Rev. B **57**, R8091 (1998).
[33] I. Affleck, Phys. Rev. B **43**, 3215 (1991).
[34] T. Sakai and M. Takahashi, Phys. Rev. B **43**, 13383 (1991); J. Phys. Soc. Jpn. **60**, 3615 (1991).
[35] N. Katoh and M. Imada, J. Phys. Soc. Jpn. **63**, 4529 (1994).
[36] M. E. Zhitomirsky and H. Tsunetsugu, Phys. Rev. B. **70**, 100403 (2004).

International Conference on Frustration in Condensed Matter (ICFCM) IOP Publishing
Journal of Physics: Conference Series **320** (2011) 012017 doi:10.1088/1742-6596/320/1/012017

Magnetization ramp of the kagome lattice antiferromagnet

Hiroki Nakano and Toru Sakai[A]

Univ of Hyogo. Hyogo, Japan
[A]SPring8, JAEA, Hyogo, Japan

E-mail: `hnakano@sci.u-hyogo.ac.jp`

Abstract. The magnetization process of the Heisenberg antiferromagnet on the kagome lattice is studied by numerical-diagonalization method. Though the magnetization plateau was considered to appear at the one-third of the magnetization saturation, recent reexamination of the magnetization process of the same model from another viewpoint of the eld-derivative of the magnetization in nite-size data from the numerical-diagonalization calculation up to 36 sites suggests that the behavior at around one-third height of the saturation is clearly di erent from typical magnetization plateaux, which we should call the magnetiztaion ramp. The present study reports our new numerical-diagonalization result up to 39 sites, which reveals well the characteristics of the magnetization ramp. The behavior of the kagome lattice antiferromagnet is compared with the typical magnetization plateux of the Heisenberg antiferromagnet on the triangular lattice.

1. Introduction

Kagome lattice is one of the typical cases where frustrations play an important role to the properties of magnetic materials; the extensive studies have been carried out over 50 years. The kagome-lattice antiferromagnet has become a hot topic due to the discovery of new materials[1, 2, 3, 4, 5].

The magnetization process of the Heisenberg antiferromagnet on the kagome lattice was reported by numerical-diagonalization studies[6, 7]. These studies concluded that the magnetization plateau appears at the one-third of the magnetization saturation. Recently, reexamination from another viewpoint of the eld-derivative of the magnetization in nite-size data from the numerical-diagonalization calculation up to 36 sites indicates that the behavior at around one-third height of the saturation is clearly di erent from typical magnetization plateaux[8]. The new phenomenon is called the *magnetiztaion ramp*. In order to con rm that the behavior of the magnetization ramp is not a nite-size e ect but an intrinsic phenomenon, the present study reports our new numerical-diagonalization result of the cluster including 39 sites. Our result of 39 sites reveals the characteristics of the magnetization ramp markedly. We also present the magnetization process of the Heisenberg antiferromagnet on the triangular lattice up to 39 sites and perform the same analysis of the eld-derivative of the magnetization. It is clari ed that the behavior of the kagome lattice antiferromagnet is quite di ernt from the typical magnetization plateu of the Heisenberg antiferromagnet on the triangular lattice.

This paper is organized as follows. In the next section, the model Hamiltonian and the method of calculation will be explained. In §3, we present numerical results of the kagome lattice antiferromagnet and the triangular lattice antiferromagnet. The nal section is devoted to the summary of this study and some discussions.

2. Hamiltonian and method

In this study, we examine the model whose Hamiltonian is given by

$$\mathcal{H} = \sum_{\langle i,j\rangle} J \boldsymbol{S}_i \quad \boldsymbol{S}_j + h \sum_i S_i^z, \tag{1}$$

where $\boldsymbol{S}_i$ is an $S = 1/2$ spin operator at site i. Spin sites are supposed to be on the kagome lattice or the triangular lattice. We consider the case when the interaction in the rst term is isotropic. Energies are measured in units of J; thus we assume $J = 1$ hereafter. The second term in eq. (1) is the Zeemann term; h is the external magnetic eld. For a system with size N, we can obtain the lowest energy $E(N, M)$ in each of the subspace of a given value of $S_z^{\rm tot}$ denoted by M by the numerical diagonalization of the Lanczos algorithm and/or the householder one, where $S_z^{\rm tot}$ represents the z-component of the total spin. The lowest energy $E(N, M)$ enable us to obtain magnetization process; at the eld given by $h = E(N, M + 1) \quad E(N, M)$, the magnetization increases from M to $M + 1$. The eld-derivative of the magnetization is evaluated by the expression

$$\chi^{\;\,1} = \frac{E(N, M+1) \quad 2E(N, M) + E(N, M \quad 1)}{1/M_{\rm sat}}, \tag{2}$$

where $M_{\rm sat}$ denotes the saturation of the magnetization, namely $M_{\rm sat} = N/2$.

It is known that numerical diagonalization method can treat only small clusters because the dimension of the matrix grows exponentially with respect to the number of spin sites. The maximum of available system sizes is determined by the quantity of resources in computers that one uses. The largest system sizes for the $S = 1/2$ spin model on the kagome lattice so far was $N = 36$[9, 10, 11, 12, 8]. In our numerical diagonalization, only the conservation of $S_z^{\rm tot}$ is used for the dividision into subspaces. Thus, we can perform diagonalizations of the systems irrespective of the shapes of clusters. The shapes of $N = 39$ clusters for the kagome lattice and the triangular lattice are depicted in Fig.1 and Fig.2, respectively.

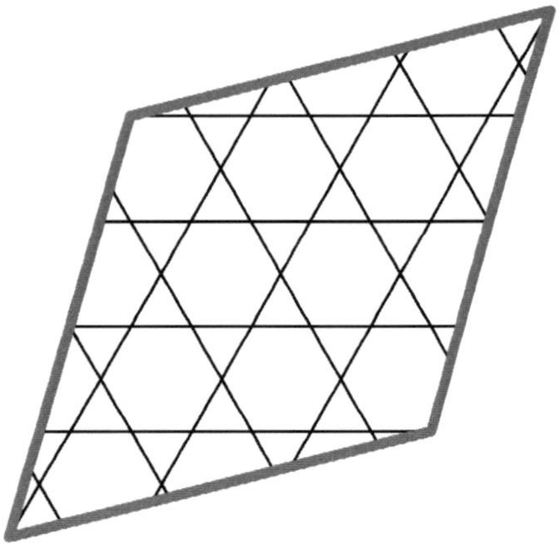

Figure 1. The shape of $N = 39$ cluster on the kagome lattice.

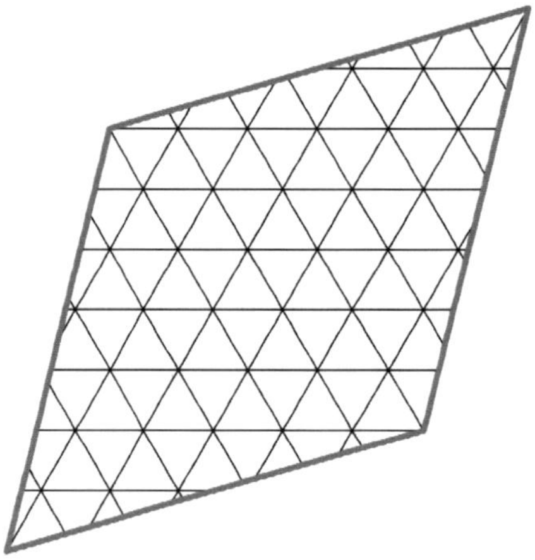

Figure 2. The shape of $N = 39$ cluster on the triangular lattice.

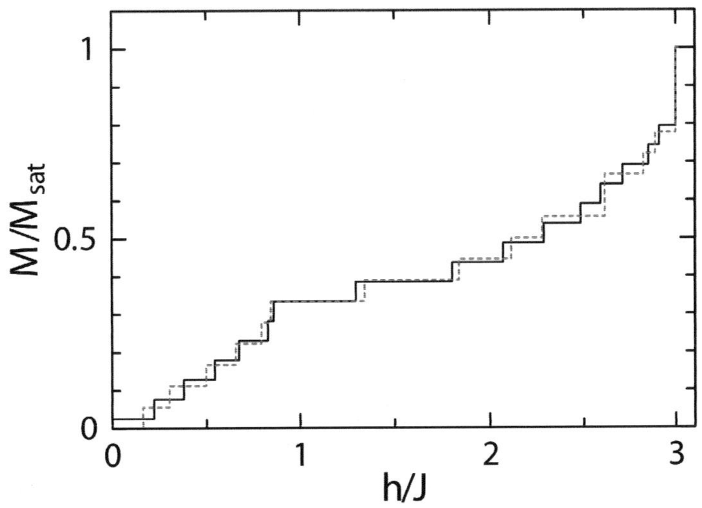

Figure 3. Magnetization process of the Heisenberg antiferromagnet on the kagome lattice. Results for $N = 39$ and 36 are represented by a black solid line and a red dotted line.

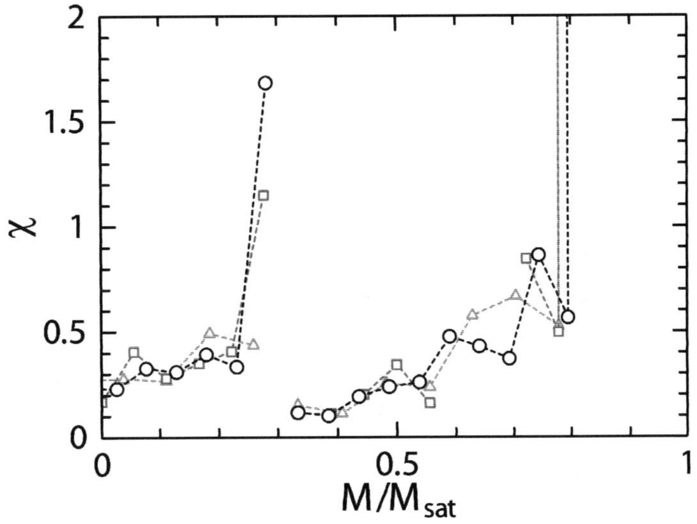

Figure 4. Field derivative of the magnetization of the Heisenberg antiferromagnet on the kagome lattice. Results for $N = 39$, 36, and 27 are denoted by black circles, red squares, and green triangles.

Note here that both the clusters of $N = 39$ are rhombic. Since the clusters show a high symmetry in thier shapes, it is reasonalbe to consider that two-dimensionality is included in our calculations for $N = 39$. All the results in our present study are based on the calculations for the case of the rhombic clusters. The dimension of the subspace divided by $S_z^{\rm tot}$ is the largest when $S_z^{\rm tot} = 0$ ($S_z^{\rm tot} = 1/2$) for a system of even (odd) N. The largest dimension of the $N = 39$ system is 68 923 264 410. In order to treat such huge matrices in computers, we have carried out parallel calculations using the MPI-parallelized code that we developed[13]. We have succeeded in calculations of $N = 39$ in Hitachi SR16000 at National Institute for Fusion Science. Note here that 8192 logical cores have been used in the largest case of our calculations.

3. Numerical results

3.1. Kagome lattice antiferromagnet

First, we present our new result of the magnetization process of the kagome lattice antiferromagnet for $N = 39$ depicted in Figure 3; for comparison, the magnetization process for $N = 36$ [7, 8] are accompanied. One observes that the two magnetization processes for $N = 36$ and $N = 39$ in Figure 3 seem to agree with each other. Near the 1/3 magnetization that is the primary interest, a at part of the processes certainly exists in both the cases. However, the width of $N = 39$ is getting slightly narrower than that of $N = 36$. It is not so clear whether the width survives or not in the thermodynamic limit. Every nite-size step of the magnetization process of $N = 39$ appears from $M = 0$ to the saturation whereas a jump at $M/M_{\rm sat} = 11/18$ appears for $N = 36$. This suggests two possibilities. One is that the jump is a nite-size e ect that is absent in the thermodynamic limit. The other is that the anomalous behavior occurs at the value of $M/M_{\rm sat}$ that is not included in the system of $N = 39$.

In order to observe carefully the behavior near the 1/3 magnetization, we evaluate the gradient χ depicted in Figure 4. One nds that the characteristics of *magnetization ramp* appear more clearly in our new result of $N = 39$. In the smaller-M side of 1/3 magnetization, the divergent behavior occurs more remarkably. In the larger-M side of 1/3 magnetization, on the other hand, no discontinuous behavior exists. These two characteristics will be compared with the case in the triangular lattice antiferromagnet which reveals the behavior of a typical magnetization plateau in a two-dimensional magnet.

3.2. Triangle lattice antiferromagnet

Numerical-diagonalization studies were done from the results up to $N = 36$[7, 14]. Reference 15 reported that diagonalization calculations of $N = 39$ were carried out; unfortunately we were not able to nd printed research papers.

We here present the magnetization process of the triangular lattice; our results are depicted in Figure 5. One observes that the behavior of a plateau appears at 1/3 magnetization irrespect to N. No signi cant change can be seen in the width at 1/3 magnetization with increasing N. No anomalous behavior seems present in other parts of $M/M_{\rm sat}$ than $M/M_{\rm sat} = 1/3$.

Next, let us examine the behavior of the magnetization process by evaluating the gradient χ. One cannot nd a signi cant system-size dependence in the region of $M/M_{\rm sat} < 0.5$ although a small oscilating behavior appears in the region of $M/M_{\rm sat} > 0.5$. One observes clearly that $\lim_{M/M_{\rm sat}\to\frac{1}{3}+0}\chi$ does not meet χ just at $M/M_{\rm sat} = 1/3$ and that $\lim_{M/M_{\rm sat}\to\frac{1}{3}\;\;0}\chi$ does not meet χ just at $M/M_{\rm sat} = 1/3$; the nite-size gradient χ just at $M/M_{\rm sat} = 1/3$ takes a small value that is getting smaller with increasing N. Note that no diverging behavior is observed in the smaller-M side of $M/M_{\rm sat} = 1/3$ although χ gradually increases when $M/M_{\rm sat}$ is increases. This behavior of χ is the one of a typical magnetization plateau in a two-dimensional magnet.

4. Summary and discussions

We have studied the magnetization process of the kagome lattice Heisenberg antiferromagnet by numerical diagonalization of nite-size clusters. We have presented new results of the system size $N = 39$. The magnetization process of the kagome lattice model for $N = 39$ shows the characteristics of the *magnetization ramp* more clearly, which are di erent from the behavior of the typical magnetization plateau observed in the triangular lattice antiferromagnet. The diverging behavior in the smaller-M side of $M/M_{\rm sat} = 1/3$ suggests that a one-dimensional spin structure forms in the two-dimensional lattice structure, which is an interesting issue in

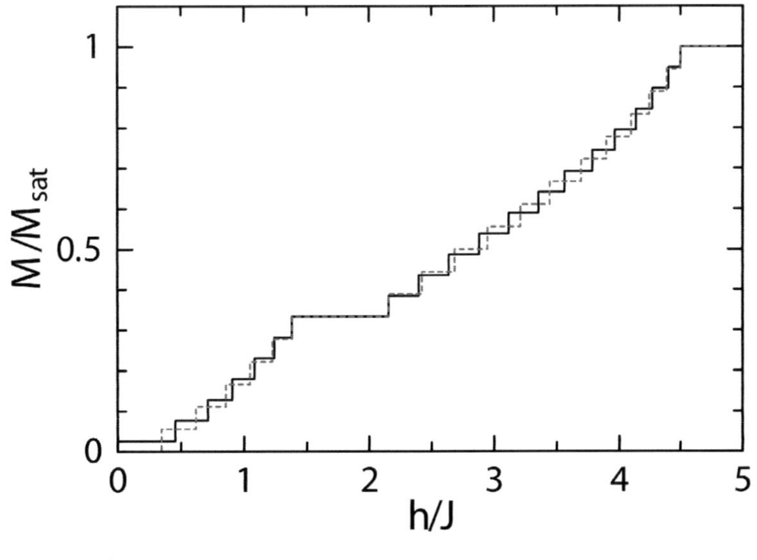

Figure 5. Magnetization process of the Heisenberg antiferromagnet on the trianglar lattice. Results for $N = 39$ and 36 are represented by a black solid line and a red dotted line.

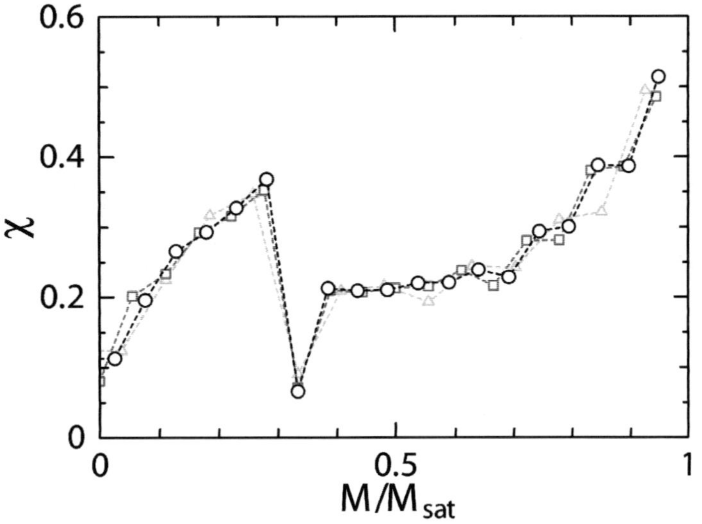

Figure 6. Field derivative of the magnetization of the Heisenberg antiferromagnet on the triangular lattice. Results for $N = 39$, 36, and 27 are denoted by black circles, red squares, and green triangles.

future studies. The critical behavior around the magnetization ramp of the kagome lattice antiferromagnet is discussed based on the results up to $N = 39$ in Reference 16.

It was pointed out that the ground state at $M/M_{\mathrm{sat}} = 1/3$ in the magnetic eld in the isotropic case is related to the state of the system of the XXZ-type anisotropic interactions[17]. Our study is now in progress concerning with the relationship between the magnetization ramp of the isotropic case and the magnetization process of the anisotropic case; results will be published elsewhere.

Acknowledgments

We wish to thank Prof. K. Hida and Prof. T. Tonegawa for fruitful discussions. This work was partly supported by Grant-in-Aid (No. 20340096) from the Ministry of Education, Culture, Sports, Science and Technology of Japan. This work was partly supported by Grant-in-Aid (No. 22014012) for Scienti c Research and Priority Areas "Novel States of Matter Induced by Frustration" from the Ministry of Education, Culture, Sports, Science and Technology of Japan. Non-hybrid thread-parallel calculations in numerical diagonalizations were based on the TITPACK ver.2 coded by H. Nishimori. A part of the computations was performed using facilities of the Information Technology Center, Nagoya University, Department of Simulation Science, National Institute for Fusion Science, and the Supercomputer Center, Institute for Solid State Physics, University of Tokyo.

References

[1] Shores M P, Nytko E A, Barlett B M and Nocera D G 2005 *J. Am. Chem. Soc.* **127** 13462
[2] Mendels P and Bert F 2010 *J. Phys. Soc. Jpn.* **79** 011001
[3] Okamoto Y, Yoshida H and Hiroi Z 2009 *J. Phys. Soc. Jpn.* **78** 033701
[4] Yoshida H, Okamoto Y, Tayama T, Sakakibara T, Tokunaga M, Matsuo A, Narumi Y, Kindo K, Yoshida M, Takigawa M and Hiroi Z 2009 *J. Phys. Soc. Jpn.* **78** 043704.
[5] Yoshida M, Takigawa M, Yoshida H, Okamoto Y and Hiroi Z 2009 *Phys. Phys. Lett.* **103** 077207.
[6] Hida K 2001 *J. Phys. Soc. Jpn.* **70** 3673
[7] Honecker A, Schulenberg J and Richter J 2004 *J. Phys. Conden. Matter* **16** S749
[8] Nakano H and Sakai T 2010 *J. Phys. Soc. Jpn.* **79** 053707
[9] Lecheminant P, Bernu B, Lhuillier C, Pierre L and Sindzingre P 1997 *Phys. Rev. B* **56** 2521
[10] Waldtmann Ch, Everts H-U, Bernu B, Lhuillier C, Sindzingre P, Lecheminant P and Pierre L 1998 *Eur. Phys. J. B* **2** 501
[11] Cepas O, Fong C M, Leung P W and Lhuillier C 2008 *Phys. Rev. B* **78** 140405(R)
[12] Sindzingre P and Lhuillier C 2009 *Europhys. Lett.* **88** 27009
[13] Nakano H and Terai A 2009 *J. Phys. Soc. Jpn.* **78** 14003
[14] Bernu B, Lecheminant P, Lhuillier C and Pierre L 1994 *Phys. Rev. B* **50** 10048
[15] Noack R M and Manmana S R 2005 *AIP Conf. Proc.* **789** 93
[16] Sakai T and Nakano H submitted to *J. Phys.: Conf. Series*, proceedings of ICFCM 2011
[17] Cabra D C, Grynberg M D, Holdsworth P C W, Honecker A, Pujol P, Richter J, Schmalfus D and Schulenburg J 2005 *Phys. Rev. B* **71** 144420

How to distinguish the Haldane/Large-D state and the intermediate-D state in an $S = 2$ quantum spin chain with the XXZ and on-site anisotropies

Kiyomi Okamoto[1], Takashi Tonegawa[2], Hiroki Nakano[3], Tôru Sakai[3,4,5], Kiyohide Nomura[6] and Makoto Kaburagi[2]

[1]Department of Physics, Tokyo Institute of Technology, Meguro-ku, Tokyo 152-8551, Japan
[2]Professor Emeritus, Kobe University, Kobe 657-8501, Japan
[3]Graduate School of Material Science, University of Hyogo, Hyogo 678-1297, Japan
[4]Japan Atomic Energy Agency, SPring-8, Hyogo 679-5148, Japan
[5]Transformative Research-Project on Iron Pnictide, JST, Saitama 332-0012, Japan
[6]Department of Physics, Kyushu University, Fukuoka 812-8581, Japan

E-mail: `kokamoto@phys.titech.ac.jp`

Abstract. We numerically investigate the ground-state phase diagram of an $S = 2$ quantum spin chain with the XXZ and on-site anisotropies described by $\mathcal{H} = \sum_j (S_j^x S_{j+1}^x + S_j^y S_{j+1}^y + \Delta S_j^z S_{j+1}^z) + D \sum_j (S_j^z)^2$, where Δ denotes the XXZ anisotropy parameter of the nearest-neighbor interactions and D the on-site anisotropy parameter. We restrict ourselves to the $\Delta > 0$ and $D > 0$ case for simplicity. Our main purpose is to obtain the definite conclusion whether there exists or not the intermediate-D (ID) phase, which was proposed by Oshikawa in 1992 and has been believed to be absent since the DMRG studies in the latter half of 1990's. In the phase diagram with $\Delta > 0$ and $D > 0$ there appear the XY state, the Haldane state, the ID state, the large-D (LD) state and the Néel state. In the analysis of the numerical data it is important to distinguish three gapped states; the Haldane state, the ID state and the LD state. We give a physical and intuitive explanation for our level spectroscopy method how to distinguish these three phases.

1. Introduction

In this paper, using mainly numerical methods, we [1] investigate the ground-state phase diagram of the $S = 2$ quantum spin chain described by the Hamiltonian

$$\mathcal{H} = \sum_j (S_j^x S_{j+1}^x + S_j^y S_{j+1}^y + \Delta S_j^z S_{j+1}^z) + D \sum_j (S_j^z)^2 , \qquad (1)$$

where Δ and D are, respectively, the XXZ anisotropy parameter of the nn interactions and the on-site anisotropy parameter. We restrict ourselves to the $\Delta > 0$ and $D > 0$ case for simplicity.

The ground-state phase diagram of this model was first discussed by Schulz [2]. In his phase diagram for $\Delta > 0$ and $D > 0$, there appear the Néel phase, the XY phase, the Haldane phase and the large-D (LD) phase. In 1992 Oshikawa [3] predicted, for $S \geq 2$ integer quantum spin cases, the existence of the intermediate-D (ID) phase, the valence bond picture of which is depicted in Fig.1(b). After that, however, by use of the density-matrix renormalization-group

Published under licence by IOP Publishing Ltd

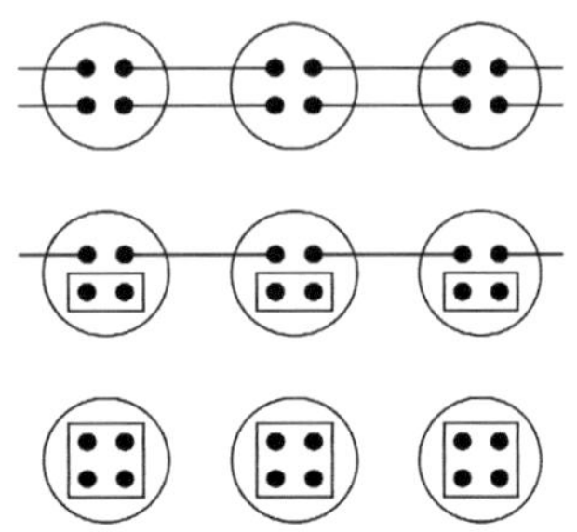

Figure 1. Valence bond pictures for (a) the Haldane state, (b) the ID state and (c) the LD state. Big circles denote $S=2$ spins and dots $S=1/2$ spins. Solid lines represent valence bonds (singlet pairs of two $S=1/2$ spins, $(1/\sqrt{2})(\uparrow\downarrow - \downarrow\uparrow)$). Two $S=1/2$ spins in rectangles are in the $(S_{\rm tot}, S^z_{\rm tot})=(1,0)$ state and similarly four $S=1/2$ spins in squares are in the $(S_{\rm tot}, S^z_{\rm tot})=(2,0)$ state.

(DMRG) calculation, Schollwöck et al. [4, 5] and Aschauer and Schollwöck [6] concluded the absence of the ID phase. By use of the level spectroscopy (LS) analysis of numerical results of exact-diagonalization calculations, Nomura and Kitazawa [7] showed in the case of $\Delta = 1$ that, with the increase of D from zero, the ground state changes as Haldane state $\Rightarrow$ XY state $\Rightarrow$ LD state. The first and second transitions occur at $D_{\rm c1} = 0.043$ and $D_{\rm c2} = 2.39$, respectively. Since these works it has been believed for a long time that the ID phase does not exist in the phase diagram of the present $S = 2$ model.

The DMRG is a very powerful method for spin chains, especially when the magnitude of the spin gap is targeted. However, it is difficult to deal with the phase transition in some cases, because the phase transition point is determined by the extrapolated values of the finite-size being equal to zero. Thus, it is somewhat hard to obtain accurate results if the magnitude of the gap is very small (i.e. the correlation length is very long). This difficulty of zero-or-finite problem is not special to the DMRG method but is common to almost all numerical methods. On the other hand, the LS method is conspicuous on this point because the phase transition point is determined from the crossing point of two related excitations, which is free from the zero-or-finite problem.

The LS method [7, 8] is firmly based on the effective Hamiltonian, renormalization group method and conformal field theory. In this paper we give a physical and intuitive explanation for our LS method how to distinguish the Haldane, ID and LD states sketched in Fig.1.

2. A very simple example: $S = 1/2$ chain with bond alternation

Before discussing $S = 2$ chain problem, let us visit a very simple example of an $S = 1/2$ chain with bond alternation described by

$$\mathcal{H}_{\rm ba}(\delta) = \sum_j [1 + (-1)^{j+1}\delta] \boldsymbol{S}_j \cdot \boldsymbol{S}_{j+1}, \qquad S = 1/2 \tag{2}$$

where δ is the bond-alternation parameter ($-1 \le \delta \le 1$). For $\delta = 0$ the ground state of the Hamiltonian (2) is the Tomonaga-Luttinger liquid state. We treat an $N = 8$ system for a while. When $\delta = \pm 1$ the ground states of the Hamiltonian (2) under the periodic boundary condition (PBC) are trivial

$$\psi_0^{\rm PBC}(\delta = +1) = [1,2]\,[3,4]\,[5,6]\,[7,8] \tag{3}$$
$$\psi_0^{\rm PBC}(\delta = -1) = [2,3]\,[4,5]\,[6,7]\,[8,1] \tag{4}$$

where $[i,j] \equiv (1/\sqrt{2})(\uparrow_i\downarrow_j - \downarrow_i\uparrow_j) = -[j,i]$. It is well known that the ground state for $\delta > 0$, $\psi_0^{\rm PBC}(\delta > 0)$, is similar to $\psi_0^{\rm PBC}(\delta = +1)$ and $\psi_0^{\rm PBC}(\delta < 0)$ is similar to $\psi_0^{\rm PBC}(\delta = -1)$. The quantum phase transition occurs at $\delta = 0$. The state $\psi_0^{\rm PBC}(\delta = \pm 1)$ is the highest energy states at $\delta = \mp 1$, respectively. However, since the crossing of the ground state energy does not occur when δ is swept, as depicted in Fig.2, we cannot determine the transition point from the crossing. Because $\psi_0^{\rm PBC}(\delta=\pm1)$ are not distinguished by the eigenvalue of the discrete symmetry, there occurs the mixing of these two states resulting in the level repulsion. In other words, the degeneracy is lifted by the "perturbation".

Let us impose the twisted boundary condition (TBC)

$$\boldsymbol{S}_8 \cdot \boldsymbol{S}_1 \Rightarrow -S_8^x S_1^x - S_8^y S_1^y + S_8^z S_1^z. \tag{5}$$

In this case the ground state of Hamiltonian (2) for $\delta=\pm1$ cases are, respectively

$$\psi_0^{\rm TBC}(\delta=+1) = [1,2]\,[3,4]\,[5,6]\,[7,8] \tag{6}$$
$$\psi_0^{\rm TBC}(\delta=-1) = [2,3]\,[4,5]\,[6,7]\,\{8,1\} \tag{7}$$

where $\{i,i\} \equiv (1/\sqrt{2})(\uparrow_i\downarrow_j + \downarrow_j\uparrow_i) = \{j,i\}$. Here we introduce the lattice inversion operator $\mathcal{I}$ about the bond center of the boundary, which works $1 \leftrightarrow 8$, $2 \leftrightarrow 7$, $3 \leftrightarrow 6$, $4 \leftrightarrow 5$. From

$$\mathcal{I}\psi_0^{\rm TBC}(\delta=+1) = [8,7]\,[6,5]\,[4,3]\,[2,1] = +\psi_0^{\rm TBC}(\delta=+1) \tag{8}$$
$$\mathcal{I}\psi_0^{\rm TBC}(\delta=-1) = [7,6]\,[5,4]\,[3,2]\,\{1,8\} = -\psi_0^{\rm TBC}(\delta=-1) \tag{9}$$

we see that $\psi_0^{\rm TBC}(\delta=\pm1)$ have different parity eigenvalues $P=\pm1$. Then the states $\psi_0^{\rm TBC}(\delta>0)$ and $\psi_0^{\rm TBC}(\delta<0)$ have different parities P since they are adiabatically connected to $\psi_0^{\rm TBC}(\delta=+1)$ and $\psi_0^{\rm TBC}(\delta=+1)$, respectively. The Hamiltonian (2) is P-invariant, which means there is no mixing between the states with different P. Thus the lowest $P=+1$ state and $P=-1$ state crosses with each other at the quantum phase transition point $\delta=0$, as demonstrated in Fig.3. We note that $\psi_0(\delta>0)$ and $\psi_0(\delta<0)$ have same eigenvalues P for the PBC case.

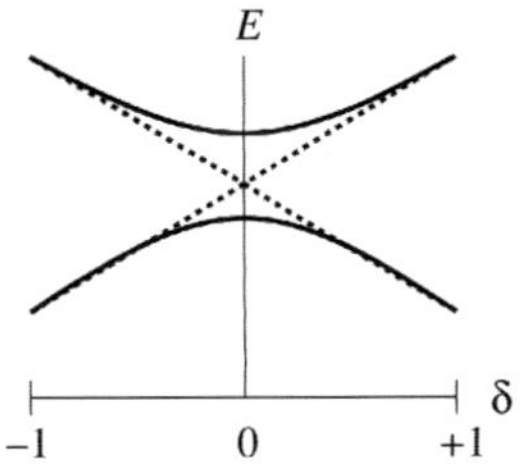

Figure 2. Schematic behaviors of $\psi_0^{\rm PBC}(\delta>0)$ and $\psi_0^{\rm PBC}(\delta<0)$ under the PBC. They do not cross with each other (not broken lines). They smoothly changes with each other because the mixing of these two states occurs.

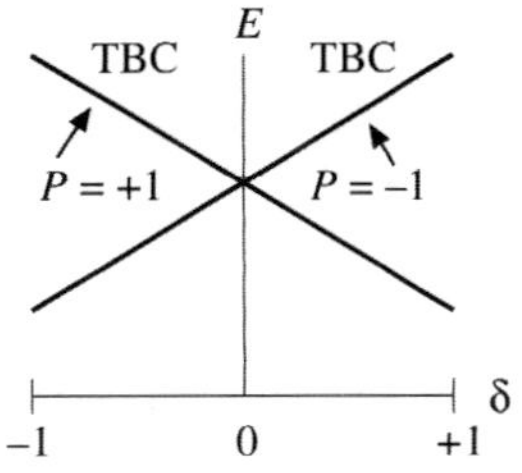

Figure 3. Schematic behaviors of $\psi_0^{\rm PBC}(\delta>0)$ and $\psi_0^{\rm PBC}(\delta<0)$ under the TBC. They cross with each other at the transition point $\delta=0$. The mixing of these two states does not occur because they are protected by different P.

3. $S=2$ case

Here we explain the $S=2$ case, considering $N=4$ system for simplicity. Under the PBC the Haldane state and the ID state corresponding to the valence bond pictures are, respectively

$$\psi_{\rm H}^{\rm PBC} = [1,2]^2\,[2,3]^2\,[3,4]^2\,[4,1]^2 \tag{10}$$
$$\psi_{\rm ID}^{\rm PBC} = [1,2]\,[2,3]\,[3,4]\,[4,1] \tag{11}$$

where two valence bonds between the ith and jth spins are abbreviated as $[i,j]^2$. Although these wave functions are not exact ones in general, the exact wave functions for the Haldane and ID states are adiabatically connected to above two wave functions respectively. When we operate $\mathcal{I}$ on these states, we obtain

$$\mathcal{I}\psi_{\rm H}^{\rm PBC} = [4,3]^2\,[3,2]^2\,[2,1]^2\,[1,4]^2 = \psi_{\rm H}^{\rm PBC} \tag{12}$$
$$\mathcal{I}\psi_{\rm ID}^{\rm PBC} = [4,3]\,[3,2]\,[2,1]\,[1,4] = \psi_{\rm ID}^{\rm PBC} \tag{13}$$

Then, $P=+1$ for both the Haldane state and the ID state. Since there are no valence bonds in the LD state, it is clear $P=+1$ for the LD state. We cannot distinguish these three states under the PBC.

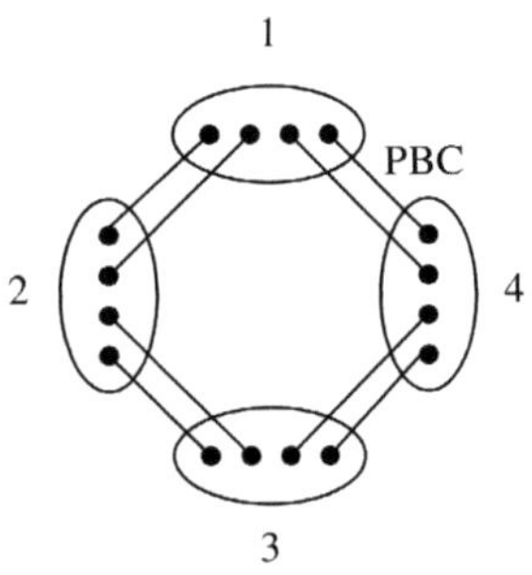

Figure 4. Haldane state under the PBC.

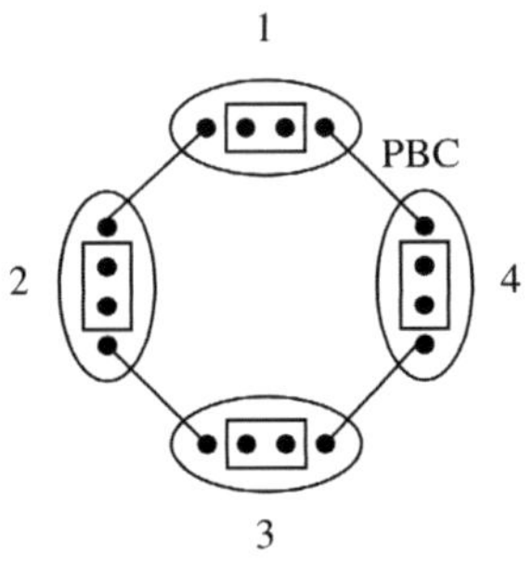

Figure 5. ID state under the PBC.

Under the TBC the Haldane state and the ID state corresponding to the valence bond picture are, respectively

$$\psi_{\rm H}^{\rm TBC} = [1,2]^2\,[2,3]^2\,[3,4]^2\,\{4,1\}^2 \tag{14}$$
$$\psi_{\rm ID}^{\rm TBC} = [1,2]\,[2,3]\,[3,4]\,\{4,1\} \tag{15}$$

When we operate $\mathcal{I}$ on these states, we obtain

$$\mathcal{I}\psi_{\rm H}^{\rm TBC} = [4,3]^2\,[3,2]^2\,[2,1]^2\,\{1,4\}^2 = \psi_{\rm H}^{\rm TBC} \tag{16}$$
$$\mathcal{I}\psi_{\rm ID}^{\rm TBC} = [4,3]\,[3,2]\,[2,1]\,\{1,4\} = -\psi_{\rm ID}^{\rm TBC} \tag{17}$$

Then, $P=+1$ for the Haldane state and $P=-1$ for the ID state. Since there are no valence bonds in the LD state, it is clear $P=+1$ for the LD state. Thus we can distinguish the ID state from the Haldane and LD states by the eigenvalue P. Namely, under the TBC, if the lowest eigenstate is $P=-1$, the ground state is the ID state. On the other hand, if the lowest eigenstate is $P=+1$, the ground state is the Haldane state or the LD state.

There may arise a question how to distinguish the Haldane state and the LD state. Recently Pollmann, Berg, Turner and Oshikawa [9] stated that the Haldane state is essentially indistinguishable from the LD state. They also constructed a one-parameter matrix product state which interpolates the Haldane state and the LD state without any quantum phase transition. Our results [1] also indicate that the Haldane state and the LD state belong to the same phase, as shown later in Figs. 10 and 11.

4. Ground state phase diagram of an $S=2$ chain

From the discussion of the previous section, we should compare the two energy levels $E_0^{\rm TBC}(M=0,P=+1)$ and $E_0^{\rm TBC}(M=0,P=-1)$, where $E_0^{\rm TBC}(M,P)$ is the lowest energy in the

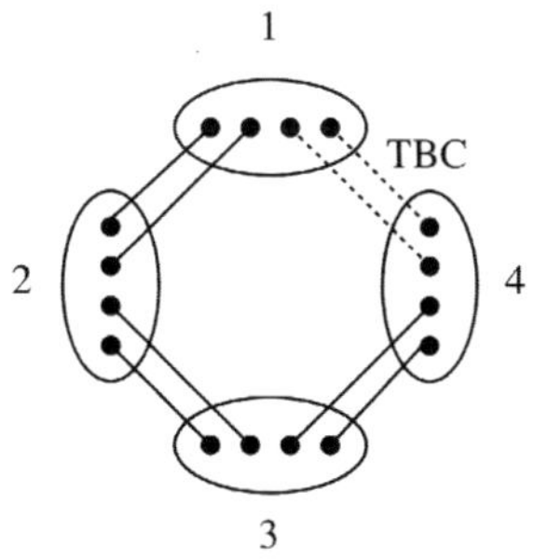

Figure 6. Haldane state under the TBC. Broken lines denote $\{i, j\}$.

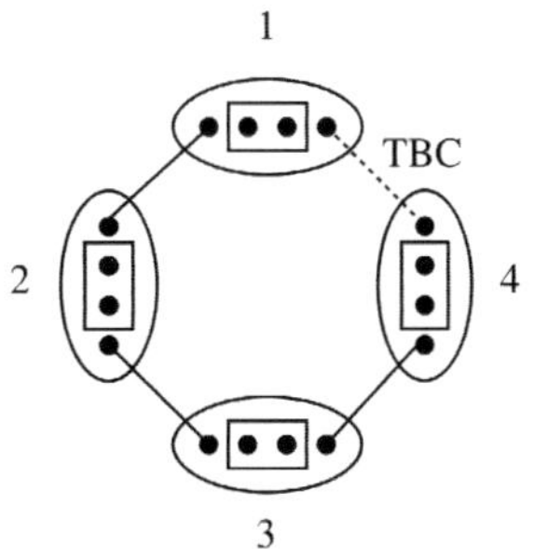

Figure 7. ID state under the TBC. Broken line denotes $\{i, j\}$.

subspace of magnetization $M = \sum_j S_j^z$ and parity P under the TBC. To obtain the phase diagram, we further check whether the ground state is gapless (XY) state or gapped state (Haldane, ID or LD). Nomura and Kitazawa [7] showed that the ground state is gapless when $E_0^{\rm PBC}(M=2) < E_0^{\rm TBC}(M=0,P)$, where $E_0^{\rm PBC}(M=2)$ is the lowest energy in the subspace $M = 2$ under the PBC. This condition was obtained through the effective Hamiltonian and the renormalization group method, unfortunately for which we have no physical and intuitive explanation. Anyway, we have to compare three energy levels, $E_0^{\rm TBC}(M=0,P=+1)$, $E_0^{\rm TBC}(M=0,P=-1)$ and $E_0^{\rm PBC}(M=2)$. Namely, the ground state is the Haldane/LD (H/LD) state, the ID state or the XY state according as $E_0^{\rm TBC}(M=0,P=+1)$, $E_0^{\rm TBC}(M=0,P=-1)$ or $E_0^{\rm PBC}(M=2)$ is the lowest.

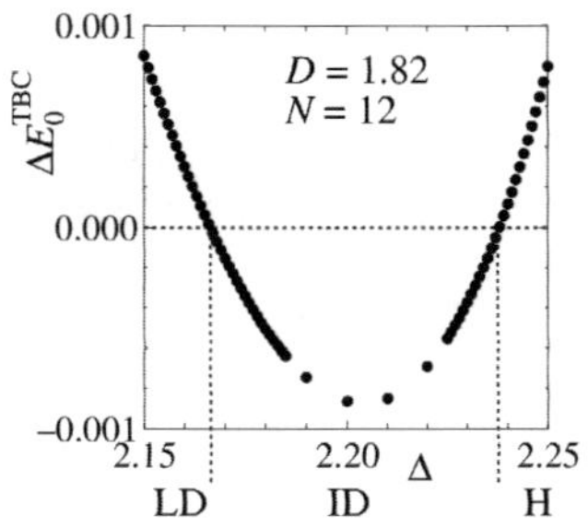

Figure 8. Behavior of $\Delta E_0^{\rm TBC} \equiv E_0^{\rm TBC}(M=0,P=+1) - E_0^{\rm TBC}(M=0,P=-1)$.

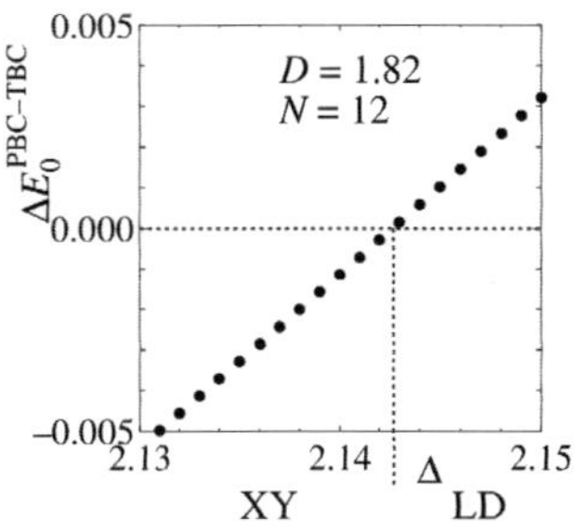

Figure 9. Behavior of $\Delta E_0^{\rm PBC-TBC} \equiv E_0^{\rm PBC}(M=2) - E_0^{\rm TBC}(M=0,P=+1)$.

Figure 8 shows $\Delta E_0^{\rm TBC} \equiv E_0^{\rm TBC}(M=0,P=+1) - E_0^{\rm TBC}(M=0,P=-1)$ as a function of Δ when $D = 1.82$ obtained by the numerical diagonalization of 12 spin systems. From this figure we can determine the phase boundary between the Haldane/LD phase and the ID phase. We note that $E_0^{\rm PBC}(M=2)$ is higher than $E_0^{\rm TBC}(M=0,P=\pm 1)$ in this region. The behavior of $\Delta E_0^{\rm PBC-TBC} \equiv E_0^{\rm PBC}(M=2) - E_0^{\rm TBC}(M=0,P=+1)$ is shown in Fig. 9, from which the phase boundary between the XY phase and Haldane/LD phases is determined. The energy $E_0^{\rm TBC}(M=0,P=-1)$ is higher in this region. Examples of the size dependence of the critical values of Δ are summarized in Table 1, from which we see that the finite-size effects are not so serious in our LS anayses. Since the transition between the Haldane/LD phase and the Nèel phase is expected to of the Ising type, we use the phenomenological renormalization group method for the numerical analysis [1].

Our final phase diagram is shown in Figs. 10 and 11. The remarkable natures of the phase

Table 1. Examples of critical values of Δ in the case of $D = 1.82$ obtained by the LS method.

N	$\Delta_c^{(H,ID)}(N)$	$\Delta_c^{(ID,LD)}(N)$	$\Delta_c^{(XY,LD)}(N)$
6	2.17687	2.14971	2.08931
8	2.22262	2.16106	2.12246
10	2.23529	2.16527	2.13607
12	2.23793	2.16639	2.14265
∞	2.241 ± 0.001	2.167 ± 0.001	2.156 ± 0.001

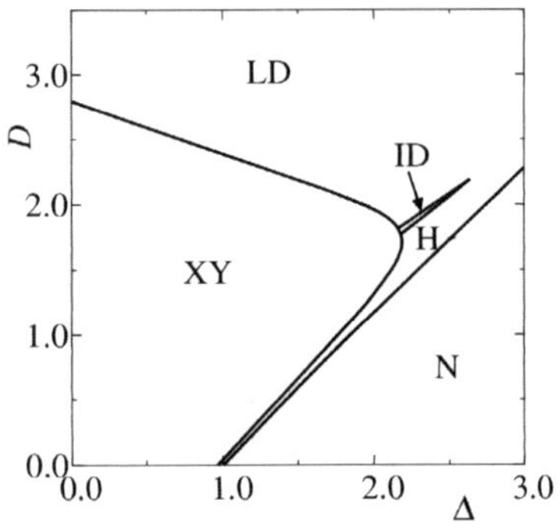

Figure 10. Phase diagram of an $S = 2$ chain. H and N are the abbreviations of Haldane and Néel, respectively.

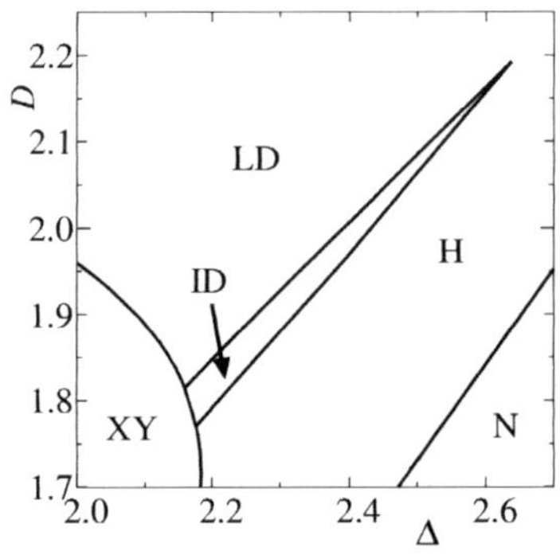

Figure 11. Enlarged phase diagram near the ID phase.

diagram are: (1) there exists the ID phase which was predicted by Oshikawa in 1992 and has been believed to be absent for a long time; (2) the Haldane state and the LD state belong to the same phase.

Acknowledgments
We would like to express our appreciation to Professor Masaki Oshikawa and Dr. Frank Pollmann for their invaluable discussions and comments. We also thank the Supercomputer Center, Institute for Solid State Physics, University of Tokyo, the Computer Room, Yukawa Institute for Theoretical Physics, Kyoto University and the Cyberscience Center, Tohoku University for computational facilities. The present work has been supported in part by a Grant-in-Aid for Scientific Research (B) (No. 20340096), and a Grant-in-Aid for Scientific Researches on Priority Areas "Novel states of matter induced by frustration" from the Ministry of Education, Culture, Sports, Science and Technology.

References
[1] Tonegawa T, Okamoto K, Nakano H, Sakai T, Nomura K and Kaburagi M 2011 *J. Phys. Soc. Jpn* **80** No.4
[2] Schulz H J 1986 *Phys. Rev. B* **34** 6372
[3] Oshikawa M 1992 *J. Phys.: Cond. Matter* **4** 7469
[4] Schollwöck U and Jolicœur Th 1995 *Europhys. Lett.* **30** 493
[5] Schollwöck U, Golinelli O and Jolicœur Th 1996 *Phys. Rev. B* **54** 4038
[6] Aschauer H and Schollwöck U 1998 *Phys. Rev. B* **58** 359
[7] Nomura K and Kitazawa A 1998 *J. Phys. A: Math. Gen.* **31** 7341
[8] Kitazawa A 1997 *J. Phys. A: Math. Gen.* **30** L285
[9] Pollmann F, Berg E, Turner A M and Oshikawa M 2009 arXiv:0909.4059

International Conference on Frustration in Condensed Matter (ICFCM) IOP Publishing
Journal of Physics: Conference Series **320** (2011) 012019 doi:10.1088/1742-6596/320/1/012019

Z_2 topological number of local quantum clusters in the orthogonal dimer model

Isao Maruyama[1], Sho Tanaya[2], Mitsuhiro Arikawa[3,4], Yasuhiro Hatsugai[2]

[1] Graduate School of Engineering Science, Osaka University, Toyonaka, Osaka 560-8531, Japan
[2] Institute of Physics, Univ. of Tsukuba 1-1-1 Tennodai, Tsukuba Ibaraki 305-8571, Japan
[3] Center for Computational Sciences, Univ. of Tsukuba 1-1-1 Tennodai, Tsukuba Ibaraki 305-8571, Japan
[4] CREST, Japan Science and Technology Agency, Chiyoda-ku, Tokyo, 102-0075, Japan

E-mail: `maru@mp.es.osaka-u.ac.jp`

Abstract. We have studied the Z_2 topological number defined by the Berry phase for the gapped frustrated systems including the orthogonal dimer model which has a direct product state of local quantum clusters as the exact ground state. The Z_2 topological number can clarify what kind of the local quantum clusters is formed to lift the macroscopic degeneracy due to frustration, even when the exact ground state is unknown. As a demonstration, the dimer-singlet and plaquette-singlet phase are identified by two kinds of Z_2 topological numbers in the Shastry-Sutherland model and its generalization realized experimentally as $SrCu_2(BO_3)_2$ and CaV_4O_9.

1. Introduction

Macroscopic degeneracy that survives at low temperature is a key feature of frustrated materials. Approaching the absolute zero temperature, the degeneracy will be lifted to meet the third law of thermodynamics. How is the degeneracy resolved? Its answer has a rich diversity, because frustrated systems have many exotic phases that are sensitive to small change. A standard scenario is the symmetry breaking mechanism that tells us an order parameter to characterize how the degeneracy is resolved. However, recent development in condensed matter physics reveals that there are topological phases where usual order parameters do not play any fundamental roles. Instead of usual order parameters such as a magnetization, topological invariants are used to characterize the topological phases and they are linked to the presence or absence of gapless edge modes. A historical example is the quantum Hall system where the topological invariant is the Chern number that is an integer defined by the Berry connection theoretically and observed as the Hall conductance experimentally. In addition to the Chern number, Z_2 topological values, i.e., the integers modulo two, have been focused in recent studies on quantum spin Hall systems[1, 2]. In such topological phases, many of them are gapped.

In some frustrated materials, gap formation is another possible scenario for lifting the macroscopic degeneracy as in the two-dimensional gapped spin system, $SrCu_2(BO_3)_2$[3, 4]. A discovery of $SrCu_2(BO_3)_2$ shed a light on theoretical construction of the model Hamiltonian with an exact ground state of local spin-singlets, i.e., the Shastry-Sutherland model[5, 6]. Inspired by Majumdar's one-dimensional model[7], Shastry and Sutherland constructed the two-dimensional

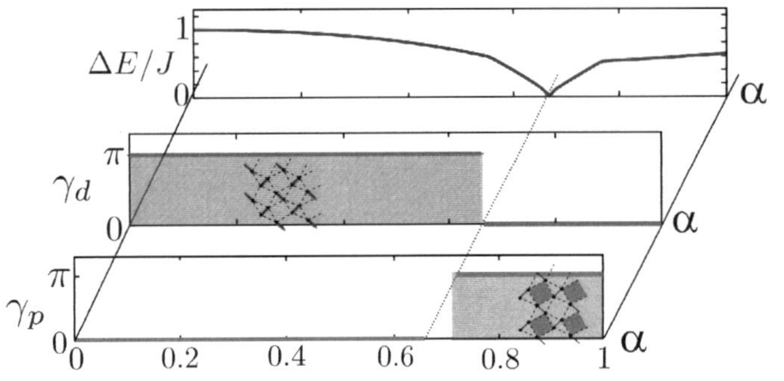

Figure 1. Gap ΔE and Berry phases of the Shastry-Sutherland model for system size $N = 16$ with the adiabatic parameter $\alpha = J'/J$. J (and J') is the nearest neighbor (next nearest neighbor) interaction. γ_d and γ_p are Berry phases for the DS and PS, respectively. The detail is discussed in §3.

geometrically frustrated spin system in which the exact ground state is the dimer-singlet(DS) state which is a direct product state of local spin-singlets[5]. Unless the spin gap closes, the DS state is still the exact ground state against the particular kind of perturbation due to the orthogonality of the dimers. Such construction is common in orthogonal dimer models including three-dimensional generalization[8].

The Shastry-Sutherland model also has an gapped phase, the plaquette-singlet(PS) phase[9–11], in addition to the DS phase. Unlike the DS phase, it is difficult to obtain its exact ground state in the PS phase. In fact, Weihong *et.al.*[8] have raised a possibility that the PS phase is unstable. To clarify it, one can consider an adiabatic modification of the Hamiltonian connected to the decoupled PS Hamiltonian which has a direct product state of PSs as the exact ground state. Actually, the ground state in the PS phase is adiabatically connected to that of the decoupled PS Hamiltonian[9]. The generalized theoretical model discussed in the adiabatic connection includes the 1/5-depleted square lattice that is realized as CaV_4O_9[12, 13]. In this sense, the direct product state of local quantum objects, such as DSs and PSs, and the adiabatic modification toward decoupled Hamiltonians are not just theoretical toys.

Recently, a method to obtain a decoupled Hamiltonian is proposed, that is, the Z_2 topological number defined by the Berry phase[14–18]. Unless the spin gap closes, the Z_2 topological number remains invariant, i.e., topologically protected against perturbation. These situations are quite similar to the case of the Chern number. The difference is that the Chern number is integer while the Z_2 topological number is quantized to the following two values: a trivial value 0 or nontrivial value (mod 2) due to the time reversal symmetry of the spin system. However, since the Berry phase is generally defined through the local perturbation, we can consider many kinds of the Z_2 topological numbers defined by the Berry phase. The Z_2 topological numbers have successfully identified the spin-singlets, plaquette-singlets, and Haldane state in the spin ladders. It can be applied to not only the spin system but also the electron system including the BCS system[18]. In addition, the itinerant singlets and the Kondo singlets are identified. Recently, we have succeeded in generalizing the Z_2 Berry phase into Z_Q Berry phase with any integer Q[19].

In this paper, we apply the Z_2 topological number to the Shastry-Sutherland model. Our numerical result in Fig. 1 shows that the DS and PS phase are clearly identified by two kinds of Z_2 topological numbers, γ_d and γ_p, respectively. Numerical detail will be explained in §3, but here we note that the ground state in the PS phase is not a simple direct product state, while there is the simple exact ground state in the DS phase. Generally speaking, although the direct product state is not an exact ground state when small long range Heisenberg exchange interactions are considered as a perturbation for real material of $SrCu_2(BO_3)_2$, it will be widely accepted that the ground state in the perturbation is similar to the direct product state. The similarity is naively based on the adiabatic connection of two ground states. By using the Z_2 topological classification, it is possible to discuss clear equivalence of ground states via the adiabatic connection, as we will explain in §2. This approach is demonstrated for the generalized theoretical model including $SrCu_2(BO_3)_2$ and CaV_4O_9 in §3.

2. De nition of the $\mathbf{Z}_2$ topological number for orthogonal dimer models

To describe our approach, let us start with local quantum clusters. A local quantum state $|\psi_i\rangle$ is supposed to be the singlet ground state of a quantum local Hamiltonian h_i of an ith cluster, such as a spin-singlet, plaquette-singlet, local Kondo-singlet and dimerized electrons state. The decoupled Hamiltonian is given by $H_0 = \sum_i h_i$ with $[h_i, h_j] = 0$. Due to small system-size of h_i, the energy gap Δ generally opens above the exact ground state $|\Psi_0\rangle = \prod_i |\psi_i\rangle$. Although the local singlet-cluster $|\psi_i\rangle$ is a pure quantum object and does not have a classical analogue, the direct product state $|\Psi_0\rangle$ is a classical state in the sense that there is no quantum entanglement between the clusters. The decoupled Hamiltonian is a start point to consider both construction of the orthogonal dimer model and definition of the Z_2 topological number.

To construct an orthogonal dimer model generically, we connect the decoupled Hamiltonian H_0 with an interaction H_1. Introducing an adiabatic parameter α, we define $H_\alpha = (1-\alpha)H_0 + \alpha H_1$ or $H_\alpha = H_0 + \alpha H_1$. To keep the direct product state $|\Psi_0\rangle$ as the exact ground state for $\alpha > 0$, a condition $H_1|\Psi_0\rangle = 0$ is required. Since we supposed $|\psi_i\rangle$ is a singlet-state, H_1 can be constructed by a total spin operator for ith cluster $T_i \ = \sum_l S_{il}$ ($= x, y, z$) with $T_i \ |\psi_i\rangle = 0$, where il indicates the lth site in the ith cluster. The condition guarantees that the ground state of H_α is the direct product state around $\alpha = 0$ to some extent as far as the gap is non-zero.

In the orthogonal dimer model, we set the local spin 1/2 cluster $h_i = J\mathbf{S}_{i1} \cdot \mathbf{S}_{i2}$, with the local singlet ground state $|\psi_i\rangle = \frac{1}{\sqrt{2}}(|\uparrow\rangle_{i1}|\downarrow\rangle_{i2} - |\downarrow\rangle_{i1}|\uparrow\rangle_{i2})$. Using the total spin operator $T_i \ = S_{i1} + S_{i2}$, we can construct $H_1 = \sum_{i,i' \neq i,l'} J'_{ii'l'} \mathbf{S}_{i'l'} \cdot (\mathbf{S}_{i1} + \mathbf{S}_{i2})$. The coefficients $J'_{ii'l'}$ are determined by the lattice geometry. In this construction, the orthogonality of the dimer bonds is always existing, that is, the interaction between $i1$ and $i'l'$ is identical to that between $i2$ and $i'l'$ always. H_1 can be generalized to a scalar product type $H_1 = \sum J'_{ij\ j'\ '} (\mathbf{S}_j \ \times \mathbf{S}_{j'\ '}) \cdot (\mathbf{S}_{i1} + \mathbf{S}_{i2})$ and more complicated types. In addition, the local quantum objects can be generalized to a plaquette singlet when we consider $h_i = J(\mathbf{S}_{i1} \cdot \mathbf{S}_{i2} + \mathbf{S}_{i2} \cdot \mathbf{S}_{i3} + \mathbf{S}_{i3} \cdot \mathbf{S}_{i4} + \mathbf{S}_{i4} \cdot \mathbf{S}_{i1})$ with $T_i \ = \sum_{l=1}^{4} S_{il} \ = T_i^{\ \dagger}$. In the same manner, one can consider larger cluster singlets and generalizations to local quantum objects of bosonic or fermionic systems. Note that the Hermite condition $H_1 = H_1^\dagger$ imposes some limitation on H_1. In the above construction, the gap Δ of H_0, which comes from a finite-size gap of the local quantum cluster, plays an important role for the stability of the exact ground state against the adiabatic parameter α.

If the local ground state $|\psi_i\rangle$ is not a singlet but a multiplet, we must deal with the macroscopic degeneracy. The gap will open above the macroscopically degenerate states and is identified by the Z_2 topological number defined by non-Abelian Berry phases. Of-course, the perturbation H_1 may lift the degeneracy and we need further discussion if we focus on the states below newly opened gap. In the following, we limit ourselves to a non-degenerate ground state and Abelian Berry phases in the definition of the Z_2 topological number.

The Berry's geometrical phases[20] are used to calculate the macroscopic polarization in the solid with surfaces. We emphasize that an edge or surface property such as the macroscopic polarization can be obtained via the Berry phase which is calculated in the periodic bulk-system without surfaces. This is bulk-edge correspondence, which is originally proposed as the one-to-one correspondence between the Chern number defined by the Berry connections and the number of the edge states in the quantum Hall systems[21]. Such local degrees of freedom on the boundaries are also characteristic in the topological insulators, and quantum spin Hall systems[1, 2]. The bulk-edge correspondence is important for the Z_2 topological number used in this paper. For example, the non-trivial Z_2 topological number guarantees the existence of the Kennedy triplet for an open chain[17].

To illustrate the definition of the Z_2 topological number for H_α, let us consider the decoupled Hamiltonian H_0 first. In our previous studies, it have been clarified that the gapped local quantum cluster, such as the spin-singlet, plaquette-singlet, and Kondo singlet, gives the non-trivial Berry phase. The key ingredient to obtain the Berry phase for the local cluster h_i is the unitary gauge transformation. Let us explain it with detailed definition of the Berry phase[21]. The Abelian Berry phase is defined by a local $SU(2)$ spin twist on a specified site. When we choose a local cluster h_I and a lth site in h_I, the local spin twist is introduced by the substitution $h_I \to h_I(\phi) = U^\dagger(\phi) h_I U(\phi)$ with $U(\phi) = \exp\left[i\phi(S - S^z_{Il})\right]$ for a spin $S = 1/2$, where Il indicates the chosen site. With using $h_I(\phi)$, the decoupled Hamiltonian has one-parameter dependence as $H_0(\phi) = h_I(\phi) + \sum_{i' \neq I} h_{i'}$. With the adiabatic parameter α we define $H_\alpha(\phi) = (1-\alpha)H_0(\phi) + \alpha H_1$ or $H_\alpha(\phi) = H_0(\phi) + \alpha H_1$.

For an one-parameter dependent Hamiltonian $H_\alpha(\phi)$, the Berry phase γ is defined as $\gamma = \int_0^{2\pi} \mathrm{d}\phi \langle \Psi_\alpha(\phi) | \frac{\mathrm{d}}{\mathrm{d}\phi} | \Psi_\alpha(\phi) \rangle$, where $|\Psi_\alpha(\phi)\rangle$ is the ground state of $H_\alpha(\phi)$. The numerical integration is given by discretizing the parameter space of ϕ into finite number of points[22], which is enough up to 60. The Berry phase here is quantized into 0, or if the ground state is invariant under an anti-unitary operator. The anti-unitary operator is the time-reversal operator for the spin system. For the decoupled Hamiltonian $H_0(\phi)$, the ground state is a direct product state and has one-parameter dependence through the unitary transformation $|\Psi_0(\phi)\rangle = U(\phi)|\Psi_0(0)\rangle$, which gives the Berry phase, because $\gamma = 2\ (S - \langle \Psi(0) | S^z_{Il} | \Psi(0) \rangle) =$ in the spin $S = 1/2$ case. This result does not depend on the detail of the local Hamiltonian h_i. When the Hamiltonian is $H_\alpha(\phi) = (1-\alpha)H_0(\phi) + \alpha H_1$, there should be the change of the Z_2 topological number. At $\alpha = 1$, since the Hamiltonian $H_{\alpha=1}(\phi) = H_1$ does not depend on ϕ, the Berry phase γ cannot become the non-trivial value , that is, γ is zero if the gap opens or undefined if the gap closes. Then, there is inevitably the change of γ in the adiabatic modification via α, which signals whether the ground state is equivalent to the direct product states or not. It should be noted that since $H_\alpha(\phi) \neq U^\dagger(\phi) H_\alpha U(\phi)$ due to H_1, the Hamiltonian $H_\alpha(\phi)$ is able to have a local flux, which cannot be gauged out. Then, the problem is non-trivial for generic H_1.

Typically, there are two types of the transitions for the Z_2 Berry phase in the adiabatic modification α. To discuss it, let us show an example of a local cluster, the $N = 4$ spin $S = 1/2$ system with the decoupled Hamiltonian $H_0(\phi) = U^\dagger(\phi) h_1 U(\phi) + h_2$ of the dimer singlets $h_i = \mathbf{S}_{i1} \cdot \mathbf{S}_{i2}$ and the local spin twist $U(\phi) = \exp\left[i\phi(S - S^z_{11})\right]$. Here we consider two kinds of Hamiltonians $H_\alpha(\phi) = (1-\alpha)H_0(\phi) + \alpha H_1$: (a) the non-orthogonal interaction $H_1 = \mathbf{S}_{12} \cdot \mathbf{S}_{21} + \mathbf{S}_{11} \cdot \mathbf{S}_{22}$ and (b) the orthogonal interaction $H_1 = (\mathbf{S}_{11} + \mathbf{S}_{12}) \cdot (\mathbf{S}_{21} + \mathbf{S}_{22})$. Due to the lattice symmetry, the Berry phase does not depend on the site where $U(\phi)$ is introduced. In the case (b), the exact ground state of $H_\alpha(0)$ is a direct product state of the dimer singlets for $\alpha < \alpha_c$ and there is a level cross at $\alpha = \alpha_c = 1/2$. On the contrary, in the case (a), the direct product state is not an exact ground state for $\alpha > 0$ and there is finite gap without any level cross in the range of $\alpha \in [0, 1]$ for $H_\alpha(0)$. We expect that in the latter adiabatic modification two DSs delocalize asymmetrically except for the symmetric point $\alpha = \alpha_c$. The Z_2 Berry phase detects the asymmetry of the DSs. In Fig. 2, the result of the Z_2 Berry phase γ is shown in both cases (a) and (b). In the case (a), the Z_2 Berry phase clearly identified the two phases connected to the two decoupled models, $\alpha = 0$ and 1, respectively. At the phase boundary $\alpha = \alpha_c$, the Z_2 Berry phase becomes undefined because the level cross occurs at $\phi =$. In other words, the diabolic point of the Dirac cone generates the Berry phase. Here, we emphasize that the Hamiltonian $H_\alpha(0)$ at $\phi = 0$ has no gap closing in the adiabatic modification via α and there is not a transition but a cross-over[18]. In the case (b) as shown in Fig. 2(b), the level cross at $\alpha = \alpha_c$ without twist ϕ induces a finite range of the undefined region of the Z_2 Berry phase.

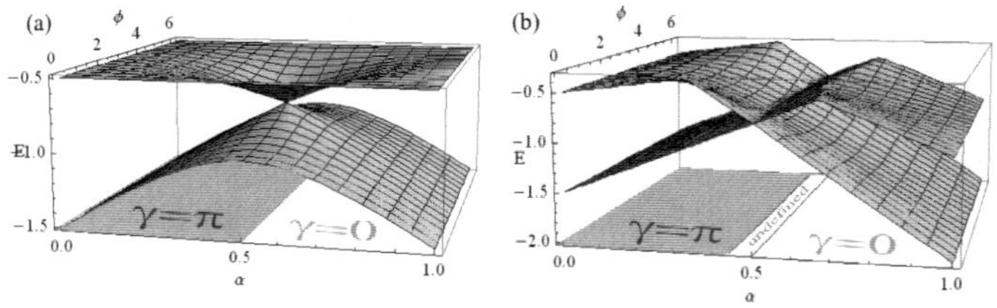

Figure 2. Z_2 Berry phase γ versus α, and two lowest eigenenerges as a function of α and ϕ for $H_\alpha(\phi) = (1-\alpha)H_0(\phi) + \alpha H_1$ with (a) the non-orthogonal interaction $H_1 = \mathbf{S}_{12} \cdot \mathbf{S}_{21} + \mathbf{S}_{11} \cdot \mathbf{S}_{22}$ and (b) the orthogonal interaction $H_1 = (\mathbf{S}_{11} + \mathbf{S}_{12}) \cdot (\mathbf{S}_{21} + \mathbf{S}_{22})$.

3. Results on the Shastry-Sutherland model

Here we discuss adiabatic modifications of the Shastry-Sutherland model as shown in Fig. 1. The Shastry-Sutherland model with J and $J' = \alpha J$ can be defined as the case $J_1 = J$ and $J_2 = J_3 = J'$ of the generalized model as shown in Fig. 3. As mentioned in previous sections, the definition of the Berry phase depends on the local spin twist and on the choice of the decoupled Hamiltonian for a given Hamiltonian. For this Shastry-Sutherland model, $\gamma_{\rm d}$ (and $\gamma_{\rm p}$) is defined by the local spin twist $\phi_{\rm d}$ ($\phi_{\rm p}$) shown in Fig. 3. The decoupled DS (and PS) Hamiltonian only with J_1 (J_2) corresponds to $\gamma_{\rm d}$ ($\gamma_{\rm p}$). Since the Shastry-Sutherland model has the exact ground state of the DSs, the level cross at $\alpha = \alpha_c$ is expected as in Fig. 2(b). Actually, in our calculation with the small system size $N = 16$, there is the first-order quantum phase transition, that is, a gap closing $\Delta = 0$ occurs at $\alpha_c = 0.67$ as shown in Fig. 1. Moreover, $\gamma_{\rm d}$ shows a clear transition at $\alpha = \alpha_c$ in Fig. 1. This is quite natural because we know the exact DS ground state for $\alpha = J'/J < \alpha_c$. The question arising here is how the ground state for $\alpha > \alpha_c$ is identified. To answer this question we introduce another Z_2 topological number, $\gamma_{\rm p}$, which corresponds to the decoupled PS Hamiltonian having the direct product state of local PSs as the exact ground state. Figure 1 shows that the two Z_2 topological numbers can identify the DS phase for $\alpha < \alpha_c$ and the PS phase for $\alpha > \alpha_c$. However, since there is the first-order phase transition at α_c, the Z_2 Berry phase will show instability just around the level cross point as shown in Fig. 2(b). In fact, $\gamma_{\rm p}$ shows an undefined region around $\alpha = \alpha_c$ as shown in Fig. 1. Although the system size is too small and the transition point α_c is not comparable to previous studies, the Z_2 Berry phase clarifies a PS-type character of the ground state for $\alpha > \alpha_c$.

In addition, we consider $J_2/J_1 = 1.82$ and $J_3/J_1 = 0.95$ in the parameter space J_1, J_2, and J_3 of the generalized model in Fig. 3, which corresponds to CaV_4O_9 and is in the PS phase[10]. To clarify the property of the ground state by the Z_2 topological number, we calculated $\gamma_{\rm p}$ and we obtained $\gamma_{\rm p} =$ for $N = 16$ sites. To illustrate the result, we show the gap ΔE as a function of twist $\phi = \phi_{\rm p}$ and the adiabatic parameter α from the decoupled PS Hamiltonian at $\alpha = 0$ to CaV_4O_9 at $\alpha = 1$. The decoupled PS Hamiltonian has $\gamma_{\rm p} =$ by definition. Figure 4 shows that there is no gap closing in the adiabatic modification not only at $\phi = 0$ but in whole region of $\phi \in [0, 2\ ]$. To change the Z_2 topological number, the gap closing is required as discussed in §2. In this sense, $\gamma_{\rm p} =$ means that the state is adiabatically connected to the direct product state in the two dimensional parameter space including the gauge twist angle $\phi \in [0, 2\ ]$.

4. Summary

In summary, we have studied the Z_2 topological number of the orthogonal dimer model as the gapped frustrated spin system. The construction of the orthogonal dimer model with the exact ground state and the Z_2 topological number $\gamma_{\rm d}$ are based on the same decoupled DS Hamiltonian. As shown in Fig. 1, the DS phase has the non-trivial Z_2 topological number, $\gamma_{\rm d} =$, corresponding the direct product state of DSs. The orthogonal dimer model shows the first-order transition with a level crossing, as shown in Fig. 2(b), in the finite-size calculation.

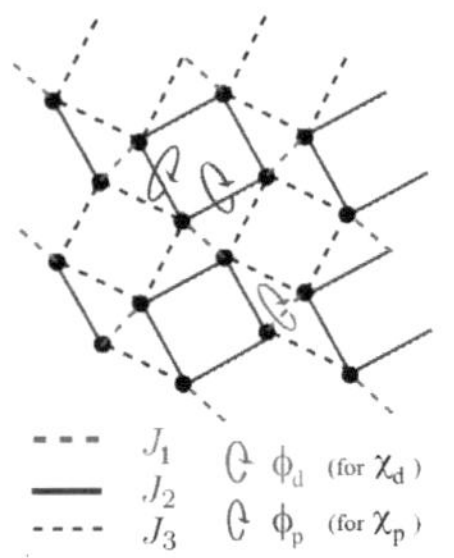

Figure 3. Hamiltonian with J_1, J_2, and J_2.

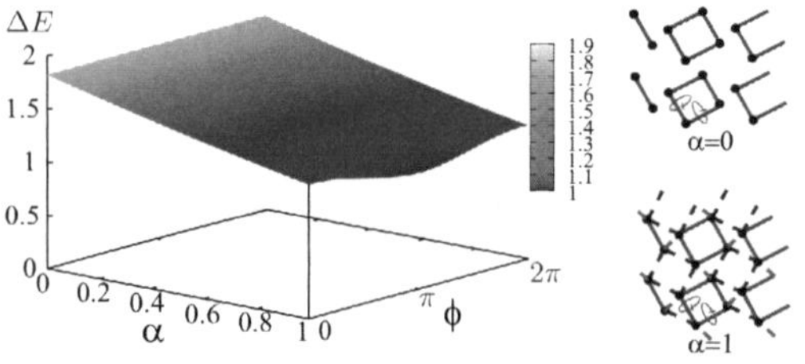

Figure 4. The gap ΔE for the Z_2 topological number $\gamma_{\rm p}$. The twist $\phi = \phi_{\rm p}$ is defined for the decoupled PS Hamiltonian with the bonds $J_2 = 1.82$. The adiabatic parameter α is introduced as $J_1 = \alpha, J_3 = 0.95\alpha$.

Although the exact ground state after the transition is unknown, the Z_2 topological number which corresponds to the decoupled PS Hamiltonian shows $\gamma_{\rm p} =$ for $\alpha > \alpha_c$ as shown in Fig. 1 and for the specified parameter corresponding to CaV_4O_9 as clarified in Fig. 4. Even though our analysis using the exact diagonalization is limited for the small system-size, this result agrees with existence of the PS phase discussed in the previous studies[9, 10].

Acknowledgments
The authors are grateful to A. Koga for his valuable discussion. This work was supported by a Grant-in-Aid from the Ministry of Education, No. 20340098 from JSPS, No.22014002 on Priority Areas from MEXT. Some numerical calculations were carried out on Altix3700BX2 at YITP in Kyoto University and the facilities of the Supercomputer Center, Institute for Solid State Physics, University of Tokyo.

References

[1] Hasan M Z and Kane C L 2010 *Rev. Mod. Phys.* **82** 3045
[2] Moore J 2009 *Nature Physics* **5** 378
[3] Kageyama H, Nishi M, Aso N, Onizuka K, Yoshihama T, Nukui K, Kodama K, Kakurai K and Ueda Y 2000 *Phys. Rev. Lett.* **84** 5876
[4] Takigawa M, Waki T, Horvati M, and Berthier C 2010 *J. Phys. Soc. Jpn.* **79** 011005
[5] Shastry B S and Sutherland B 1981 *Physica B* **108** 1069
[6] Miyahara S and Ueda K 1999 *Phys. Rev. Lett.* **82** 3701
[7] Majumdar C K 1979 *L735*
[8] Miyahara S and Ueda K 2003 *J. Phys: cond. Matt.* **15** 327
[9] Koga A and Kawakami N 1999 *Phys. Rev. Lett.* **84** 4461
[10] Takushima Y, Koga A, and Kawakami N 2001 *J. Phys. Soc. Jpn.* **70** 1369
[11] Lauchili A, Wessel S and Sigrist M 2002 *Phys. Rev. B* **66** 014401
[12] Ueda K, Kontani H, Sigrist M and Lee P A 1996 *Phys. Rev. Lett.* **76** 1932
[13] Troyer M, Kontani H, and Ueda K 1996 *Phys. Rev. Lett.* **76** 3822
[14] Hatsugai Y 2006 *J. Phys. Soc. Jpn.* **75** 123601
[15] Hirano T, Katsura H and Hatsugai Y 2008 *Phys. Rev. B* **78** 054431
[16] Maruyama I, Hirano T and Hatsugai Y 2009 *Phys. Rev. B* **79** 115107
[17] Arikawa M, Tanaya S, Maruyama I and Hatsugai Y 2009 *Phys. Rev. B* **79** 205107
[18] Arikawa M, Maruyama I and Hatsugai Y 2010 *Phys. Rev. B* **82** 073105
[19] Hatsugai Y and Maruyama I 2010 *ArXiv: 1009.3792*
[20] Resta R 2000 *Berry's Phase and Geometric Quantum Distance: Macroscopic Polarization and Electron Localization* Lecture Notes for the "Troisième Cycle de la Physique en Suisse Romande" also available online at http://www-dft.ts.infn.it/ resta/publ/notes2000.ps.gz. (Ecole Polytechnique Federale de Lausanne, Lausanne)
[21] Hatsugai Y 1993 *Phys. Rev. Lett.* **71** 3697
[22] King-Smith R D and Vanderbilt D 1993 *Phys. Rev. B* **47** 1651

Journal of Physics: Conference Series **320** (2011) 012020 doi:10.1088/1742-6596/320/1/012020

Duality among competing orders in antiferromagnets and topological insulators: nonlinear sigma model approach

Akihiro Tanaka

National Institute for Materials Science, 1-2-1 Sengen, Tsukuba, Ibarakai, Japan 305-0047

E-mail: TANAKA.Akihiro@nims.go.jp

Abstract. We revisit our framework for studying competing orders in quantum antiferromagnets (A. Tanaka and X. Hu, Phy. Rev. Lett. **95** (2005) 036402; Phys. Rev. B 74 (2006) 140407(R)), in which we showed that when two or more ordering tendencies are organized into a single composite order parameter, the corresponding effective sigma model action can generally host a new topological term. It has since been argued by several authors that such terms are indicative of novel critical behaviors. Here we reinforce this assertion by searching for nonvanishing fermionic bilinears (Golstone-Wilczek-type currents) associated with the same models. We arrive at a duality relation among the competing orders (e.g. a direct relation between the topological charge of the antiferromagnetic order sector and a Noether current associated with the rotational symmetry in the valence bond solid order sector), which becomes relevant upon approaching criticality. We also discuss the physical implications of this duality in the context of topological insulators.

1. Introduction

Several years ago[1, 2] the author showed how a new nonlinear sigma model description of competing orders within a two-dimensional antiferromagnet can emerge, when one chooses to start from massive Dirac fermions which accomodates a manifold of chiral-symmetry breaking states (each realization of chiral symmetry breaking corresponds to a different ordering pattern in the antiferromagnet). The possible connection of this theory to the problem of deconfined quantum criticality[3] has since been taken up by various authors[4]. Our sigma model framework, which builds on a general field theory prescription laid out by Abanov and Wiegmann[5], features a new Wess-Zumino-type topological term, and (1) can be generalized to arbitrary dimensions (2) and applies as well to other combinations of competing orders, e.g. to the competition between a quantum spin Hall state and s-wave superconductivity[6]. More recently, a general mathematical correspondence between this term and momentum space topological invariants (Chern numbers) employed in the study of topological insulators was laid out, which adds a further motivation for investigating the properties of this theory[7].

Unfortunately, the effective action itself is difficult to analyze, and has so far lead to little information on the detailed nature of the criticality which might arise out of the competition among multiple ordering tendencies. The Abanov-Wiegmann machinery, however, can also be used to obtain various Goldstone-Wilczek-type currents/fermion numbers[8] which are induced by order parameter configurations with nontrivial topology. Here we utilize this aspect of the theory to derive nontrivial duality relations which hold between currents/charges associated with the symmetry of one order, and the topological charge of an opposing order. In the original context[1] of the 2d quantum antiferromagnet, for instance, we can show that when the valence-bond-solid (VBS) sector of the model possesses a rotational symmetry (which implies the existence of a conserved Noether current) –such as had been observed in several numerical studies on frustrated spin systems, the Skyrmion number of the Néel vector field configuration is conserved. Since the conservation of Skyrmion number is a central feature of the theory of deconfined quantum criticality[3], its appearance here suggests that further investigations along this direction can contribute some fresh insights into this intriguing phenomenon. The duality relation, when applied to the case of topological insulators can be given a more direct physical interpretation: it suggests that the Skyrmion number of the spin Hall effect order parameter is to be identified with a preformed Cooper pair, as anticipated in Ref. [6].

2. Competing orders as patterns of chiral symmetry breaking

We begin by summarizing the salient points of the forementioned framework[1] employed in our following discussions. In our original settings[1], our point of departure was the Affleck-Marston π-flux state of an antiferromagnet residing on a 2d square lattice. This state accomodates two inequivalent "valleys" of massless Dirac cones within the Brioullin zone, which can be lumped together into a 4-component Dirac spinor with spin indices. The latter would also describe (with suitable readjustment of the gamma matrix representation) the continuum limit of tight-binding electrons hopping on a 2d honeycomb lattice, i.e. graphene.

Let us now open up a mass gap at the Dirac points. Due to the enlarged 4-by-4 gamma matrix representation, there are two generators of chiral transformations, denoted below as γ_3 and γ_5, which together span a manifold of chiral symmetry breaking states. This chiral structure offers an interesting setting (which readilly generalizes to other spatial dimensions) for analyzing the competition among various orders, each of which corresponds to a different manifestation of broken chiral symmetry. An example of particular interest is the case of competing antiferromagnetic (AF) and VBS orders, which can be incorprated into the fermionic model by setting the mass term to be [1]:

$$\mathcal{L}_{mass} = \bar{\psi}\hat{m}_{\mathrm{AF-VBS}}\psi = m\bar{\psi}(\mathbf{N}\cdot\sigma + iN_{\mathrm{xVBS}}\gamma_3 + iN_{\mathrm{yVBS}}\gamma_5)\psi. \tag{1}$$

Here the vector $\mathbf{N}$ is the AF order parameter, while N_{xVBS} and N_{yVBS} are the horizontal and vertical compenents of the VBS order parameter. Together they comprise a five-component composite order parameter $\vec{N} \equiv (\mathbf{N}, N_{\mathrm{xVBS}}, N_{\mathrm{yVBS}})$ whose length we set to be unity. The effective action for $\vec{N}$, valid below the energy scale m, can be obtained[1] by the Abanov-Wiegmann gradient expansion scheme[5], which systematically salvages contributions which may arise from anomalies. This leads us to the effective action $\mathcal{S}_{eff}[\vec{N}] = \mathcal{S}_{NL\sigma} + \mathcal{S}_{\mathrm{WZ}}$, where $\mathcal{S}_{NL\sigma}$ is an O(5) nonlinear sigma model, and $\mathcal{S}_{\mathrm{WZ}}$ the Wess-Zumino term (written in Euclidean space-time),

$$\mathcal{S}_{\mathrm{WZ}} = ik\int_{S_3\times[0,1]} \tilde{\vec{N}}^*\omega. \tag{2}$$

In the above, the map $\vec{N} : S_3 \to S_4$ has been extended to $\tilde{\vec{N}} : S_3 \times [0,1] \to S_4$ according to standard prescriptions for constructing a Wess-Zumino (WZ) term[9], ω is the normalized volume form $\sim \vec{N}\wedge d\vec{N}\wedge d\vec{N}\wedge d\vec{N}\wedge d\vec{N}$ on S_4, and finally, $\tilde{\vec{N}}^*$ is the pullback from S_4 to the base manifold $S_3 \times [0,1]$. The level k for our case is unity.

It is clear that this method can be generalized to different choices of competing orders, provided they can all be expressed as chiral symmetry breaking terms of an underlying Dirac fermion system. A particularly interesting situation arises when one of the orders involve pairing channels. Refs. [6, 11, 12], for example, consider 2d systems where the competition between a quantum spin Hall state (i.e. a 2d topological insulator) and s-wave superconductivity is present. Here again the two orders comprise a 5 component vectorial order parameter with a unit norm. The authors of these papers find that the corresponding effective action reduces to $\mathcal{S}_{eff}[\vec{N}]$, which we had found for the AF-VBS competition. It is also straightforward to work out lower/higher dimensional analogues. Hence the present work, which primarily deals with 2d quantum antiferromagnets, also should offer some insights into similar problems in different dimensionalities, as well as in the field of topological insulators.

3. Fermion currents

It has been suggested [1, 4] that the presence of the term $\mathcal{S}_{\mathrm{WZ}}$ accounts for the highly unconventional phenomenology of *deconfined quantum criticality* (DQC) proposed by Senthil et al[3]. However, an emergent U(1) symmetry corresponding to the conservation of Skyrmion number, which is a hallmark of the original DQC scenario, is not manifest in the O(5) model as it stands. This observation has motivated the search for a different fixed point theory[10]. Furthermore, as mentioned above, the form of the action $\mathcal{S}_{eff}[\vec{N}] = \mathcal{S}_{NL\sigma} + \mathcal{S}_{\mathrm{WZ}}$, while strongly suggesting the exotic nature of the system, is not suited for submitting to a detailed analysis.

To gain further insight into the symmetry properties of the model $\mathcal{L} = \bar{\psi}(i\partial\!\!\!/ + \hat{m}_{\mathrm{AF-VBS}})\psi$ (hereafter the Feynman slash will represent, as usual, the contraction with the space-time components of the gamma matrices, e.g. $\partial\!\!\!/ \equiv \gamma_\mu\partial_\mu$), we recall that in addition to deriving the effective action, the Abanov-Wiegman gradient expansion also enables us to evaluate fermionic quantum numbers[8], or in other words the expectation values of various fermionic bilinears (Goldstone-Wilczek currents) which arise when the background order parameters form a topologically nontrivial configuration.

Let us first illustrate the general strategy using the simpler case of a two-component Dirac fermion interacting with a three-component unit vector field $\mathbf{n}$, $\mathcal{L}_F = \bar{\psi}(i\partial\!\!\!/ + m\mathbf{n}\cdot\sigma)\psi \equiv \bar{\psi}\mathcal{D}\psi$. (For this case there are only three gamma matrices, which can conveniently be identified with the Pauli matrices.) Couple the fermions to a source gauge field a_μ via the prescription $i\partial\!\!\!/ \to i\partial\!\!\!/ + a\!\!\!/$. The current $j_\mu \equiv \langle\bar{\psi}\gamma_\mu\psi\rangle$ is then easily calculated by noting that $\frac{\delta\mathcal{S}_F}{\delta a_\mu} = \mathrm{tr}[(\mathcal{D}^\dagger\mathcal{D})^{-1}\mathcal{D}^\dagger\frac{\delta\mathcal{D}}{\delta a_\mu}]|_{a_\mu=0}$ and by treating the operator $(\mathcal{D}^\dagger\mathcal{D})^{-1}$ in a gradient expansion. We find that

$$\langle j_\mu\rangle = \epsilon_{\mu\nu\lambda}\frac{1}{4\pi}\mathbf{n}\cdot\partial_\nu\mathbf{n}\times\partial_\lambda\mathbf{n}. \tag{3}$$

Repeating this procedure for our fermionic action $\mathcal{L} = \bar{\psi}(i\partial\!\!\!/ + \hat{m}_{\mathrm{AF-VBS}})\psi$, we immediately find that $\langle j_\mu\rangle = 0$. A quick inspection reveals that an exact sign cancellation between the contributions coming from each Dirac "valley" is responsible for this null result. We thus seek a nonvanishing current by introducing a gauge field $\tilde{a}_\mu$ whose gauge charge has *opposite* signs at the two valleys. [We note some interesting parallels to stategies taken in graphene physics. The Haldane model[13], which exhibits a zero-field integer quantum Hall effect, is in essence a graphene system perturbed by a mass term which has opposite signs at the two valleys. The unusual mass term was so designed to prevent precisely this same intervalley cancellation. In addition, the unconventional gauge field described above (referred to in the graphene context as the ripple gauge field) can be utilized to yield a quantized valley Hall effect.] It can be shown that the valley degree of freedom is labled by the eigenvalue of the operator $\gamma_{35} \equiv \gamma_3\gamma_5$. Thus, inserting the gauge coupling $i\partial\!\!\!/ \to i\partial\!\!\!/ + \gamma_{35}\tilde{a}\!\!\!/$, and carrying out the gradient expansion as before, we arrive at

$$\langle j^{35}_\mu\rangle \quad = \quad \langle\bar{\psi}\gamma_{35}\psi\rangle = \left\langle\frac{\delta S_{\mathrm{F}}(i\partial\!\!\!/ \to i\partial\!\!\!/ + \tilde{a}\!\!\!/\gamma_{35})}{\delta\tilde{a}_\mu}\right\rangle_{\vec{N}(x)}|_{\tilde{a}_\mu=0}$$

$$= \epsilon_{\mu\nu\lambda}\frac{1}{4\pi}\mathbf{N}\cdot\partial_\nu\mathbf{N}\times\partial_\lambda\mathbf{N}, \tag{4}$$

which bears an apparent similarity to eq.(3).

Noting that the set of matrices $\{\gamma_3, \gamma_5, i\gamma_{35}\}$ forms an SU(2) algebra, and recalling[1] that the pseudoscalars $\bar{\psi}\gamma_3\psi$ and $\bar{\psi}\gamma_5\psi$ are the fermionic expressions for the VBS order parameters, we can see that γ_{35} generates rotation within the VBS order parameter space, and that j^{35}_μ is the Noether current associated with rotational symmetry (when such symmetry is present) within this space. In other words, when the VBS sector exhibits a U(1) rotational symmetry, the right hand side of the above equation, which is a quantity associated with the AF sector, is a conserved current. In the case where there is no amplitude fluctuation between the two sectors, the latter is, modulo a constant prefactor, the topological current of AF Skyrmion events. It is interesting to interpret this duality relation in light of numerical results related to the DQC[14, 15], where the U(1) rotational symmtery in the VBS sector was indeed observed when the system was (presumably) in proximity to the critical point.

Let us now apply the same argument to the case where the competition is between quantum spin Hall and superconducting orders[6]. Here the quantum spin Hall state is assumed to be induced as a result of strong electron correlations (i.e. a topological Mott insulator), and the spin quantization axis is treated as a dynamically fluctuating entity. The duality relation now states that the conserved charge associated with a U(1) symmetry within the superconducting sector is given by the Skyrmion topological charge of the opposing (quantum spin Hall) sector. When combined with the conjugation relation between the phase and the amplitude of the superconducting order parameter, we find that this is in accord with the assertion[6] that the Skyrmions of this system are actually preformed Cooper pairs, which when condensed lead to superconductivity. Duality relations similar to eq.(4) are readily obtained for other spatial dimensions, and will be reported elsewhere.

Finally we note that while preparing this report, we came across a preprint by Fu, Sachdev, and Xu[16] which also addresses in part some similar aspects of competing orders in 2d. Instead of assuming a Dirac-like fermionic system at the outset, these authors begin with a Hubbard model, and incorporate the band structure of the electron system. They also find a nonvanishing Berry phase-induced electron biinear in the background of a finite Skyrmion density, but point out that the results depend on the band structure. It is interesting, in view of these results, to attempt to incorporate nonrelativistic electron dispersions into the approach that we have described in this article.

In summary, we have studied the Goldstone-Wilczek currents associated with Dirac fermion systems which describe the competition between different ordering tendencies in antiferromagnets and topological insulators. We found that the low effective actions of such systems, nonlinear sigma models with a topological term, have a nontrivial built-in duality between the competing orders. We expect this approach to be useful in gaining new insights into competing orders in quantum magnets and topological insulators.

Acknowledgments

We thank Y. Kato, N. Kawashima, K. Totsuka, A. Furusaki, M. Kohno, M. Sato and R. Yamashita for helpful discussions. This work is supported in part by Grant-in -Aid for Scientific Research (C) 20540386 from MEXT, Japan.

[1] Tanaka A and Hu X, Phy.Rev.Lett. **95** (2005) 036402.
[2] Tanaka A and Hu X, Phys. Rev. B 74 (2006) 140407(R).
[3] Senthil T, Vishwanath A, Balents L, Sachdev S and Fisher M P A, Science **303** (2004) 1490.
[4] Senthil T and Fisher M P A, Phys. Rev. B **74** (2006) 064405.
[5] Abanov A G and Wiegmann P, Nucl. Phys. B **570** (2000) 685.
[6] Grover T and Senthil T, Phys. Rev. Lett. **100** (2008) 156804.

[7] H. Yao and D.-H. Lee, Phys. Rev. B **82** (2010) 245117.
[8] J. Goldstone and F. Wilczek, Phys. Rev. Lett. **47** (1981) 986.
[9] Altland A and Simons B 2010, *Condensed Matter Field Theory* (2nd ed.) (Cambridge UK, Cambridge Univ. Press) Chap.9.
[10] Xu C K and Sachdev S, Phys. Rev. Lett. **100** (2008) 137201.
[11] Ghaemi P, Ryu S and Lee D H, Phys. Rev B **81** (2010) 081403(R).
[12] Hosur P, Ryu S and Vishwanath A, Phys. Rev. B **81** (2010) 045120.
[13] Haldane F D M, Phys. Rev. Lett. **61** (1988) 2015.
[14] Tanabe Y and Kawashima N, Phys. Rev. Lett **98** (2007) 057202.
[15] Sandvik A W, Phys. Rev. Lett. **98** (2007) 227202.
[16] Fu L, Sachdev S and Xu C, arXiv:1010.3745.

International Conference on Frustration in Condensed Matter (ICFCM) IOP Publishing
Journal of Physics: Conference Series **320** (2011) 012021 doi:10.1088/1742-6596/320/1/012021

Relation between high-energy quasiparticles of quasi-one-dimensional antiferromagnets in a magnetic field and a doublon of a Hubbard chain

Masanori Kohno

International Center for Materials Nanoarchitectonics (MANA), National Institute for Materials Science (NIMS), Tsukuba 305-0044, Japan

E-mail: KOHNO.Masanori@nims.go.jp

Abstract. In spin-1/2 one-dimensional Heisenberg antiferromagnets and anisotropic triangular Heisenberg antiferromagnets, high-energy states carrying considerable spectral weights have been observed in a magnetic eld using inelastic neutron scattering. Such high-energy properties cannot be explained in terms of either Nambu-Goldstone bosons due to spontaneous breaking of continuous symmetries or quasiparticles in a Tomonaga-Luttinger liquid. Here, we show that the mechanism causing the high-energy states is analogous to that of the upper Hubbard band in the one-dimensional Hubbard model, by theoretically tracing the origin of the high-energy states back to string solutions of the Bethe ansatz.

1. Introduction

In conventional magnets with magnetic long-range order, dynamical properties are explained in terms of magnons in spin-wave theory [1, 2]. In a magnetic eld, the dispersion relation of the magnon deforms according to the change of the ground-state spin con guration [3]. However, for systems without magnetic long-range order, the spin-wave theory based on classical spin con gurations is not justi ed, and the dynamical properties are not necessarily explained in terms of the magnon. In fact, in a one-dimensional (1D) spin-1/2 antiferromagnetic Heisenberg chain, whose ground state has no magnetic long-range order, the dynamical properties in zero eld are explained in terms of spinons [4, 5]. Also, in spatially anisotropic two-dimensional (2D) frustrated Heisenberg antiferromagnets, bound states of spinons, called triplons, dominate the spectra in zero eld [6]. Although the triplon might appear similar to the magnon, its stabilization mechanism di ers from that of the magnon: the triplon is induced by spin exchange, i.e., two-spinon hopping, between neighboring chains [6], while the magnon is stabilized by the classical spin con guration [1, 2].

The di erence between such quasiparticles (QP's) without ordered ground states and the magnon created from magnetic long-range order becomes clearer in a magnetic eld. For example, in an antiferromagnetic Heisenberg chain in a magnetic eld, low-energy properties are primarily explained in terms of QP's called psinons ('s) and anti-psinons (*'s) [7, 8], which have gapless points at incommensurate momenta di erent from those of the magnon, in addition to the gapless points at commensurate momenta. In anisotropic-2D frustrated Heisenberg antiferromagnets, bound states of 's and *'s dominate the low-energy spectra, leading to incommensurate magnetic long-range order whose momentum shifts significantly with the magnetic field and multi-particle crossover in a magnetic field [9]. These features cannot be explained in terms of the magnon in linear spin-wave theory.

Furthermore, in an antiferromagnetic Heisenberg chain, a continuum is separated from other low-energy continua in a magnetic field, and shifts to higher energies as the magnetic field increases [8]. In fact, signatures of the high-energy continuum have been observed in inelastic neutron scattering experiments on the 1D Heisenberg antiferromagnet $CuCl_2$ $2N(C_5D_5)$ in a magnetic field [10]. This high-energy continuum cannot be explained in terms of either magnons in linear spin-wave theory or 's and *'s. By using the Bethe ansatz [11], the continuum has been identified [8] as that of 2-string solutions [12], and a QP representing the 2-string has been introduced [8]. Such a high-energy feature is not special to 1D chains, but is also relevant to frustrated antiferromagnets in higher dimensions. In anisotropic-2D frustrated Heisenberg antiferromagnets, a high-energy mode, which can be regarded as bound states of and the QP for the 2-string, emerges from the continuum of 1D chains [9]. Indeed, signatures of the high-energy mode have been observed in inelastic neutron scattering experiments on the anisotropic triangular Heisenberg antiferromagnet Cs_2CuCl_4 in a magnetic field [13]. For intuitive understanding of the nature of such high-energy states, it would be helpful to indicate relations to well-known features in other systems. In this paper, we argue that the high-energy continuum of antiferromagnetic Heisenberg chains and the high-energy mode of frustrated Heisenberg antiferromagnets in a magnetic field are caused by the mechanism similar to that of the upper Hubbard band in the Hubbard model, by using the Bethe ansatz.

2. Upper Hubbard band and Doublon

The Hubbard model is defined by the following Hamiltonian:

$$\mathcal{H}_{\text{Hub}} = -t \sum_{\langle i,j\rangle\sigma} \left(c^{\dagger}_{i\sigma} c_{j\sigma} + \text{H.c.} \right) + U \sum_{i} n_{i\uparrow} n_{i\downarrow} - \mu \sum_{i\sigma} n_{i\sigma}, \tag{1}$$

where $c_{i\sigma}$ and $n_{i\sigma}$ denote annihilation and number operators of an electron with spin σ at site i, respectively. Here, $\langle i,j\rangle$ means that i and j are nearest neighbors. In the repulsive ($U > 0$) Hubbard model, the ground state at half-filling becomes insulating when the on-site Coulomb repulsion is strong enough, because addition of an electron to the ground state costs energy of order U. This implies that the band splits into two bands, called Hubbard bands, by the repulsive interaction. One might naively think that double occupancy would characterize the upper Hubbard band and that it may be regarded as a QP, doublon. Here, we show that this naive picture is not accurate, by using the 1D Hubbard model. In the Bethe ansatz, exact solutions of the 1D Hubbard model are parameterized by quasimomenta $\{k_j\}$ and rapidities $\{\Lambda_\alpha\}$ that satisfy the following Bethe equations [14]:

$$Lk_j = 2\pi I_j + 2\sum_{\alpha=1}^{M} \tan^{-1}\frac{4t(\Lambda_\alpha - \sin k_j)}{U} \quad \text{for} \quad j = 1 \cdots N,$$

$$\sum_{j=1}^{N} \tan^{-1}\frac{4t(\Lambda_\alpha - \sin k_j)}{U} = \pi J_\alpha + \sum_{\beta=1}^{M} \tan^{-1}\frac{2t(\Lambda_\alpha - \Lambda_\beta)}{U} \quad \text{for} \quad \alpha = 1 \cdots M,$$

where L, N, and M denote the number of sites, electrons, and down spins, respectively. Here, I_j and J_α are called Bethe quantum numbers, which are integers or half-odd integers constrained by L, N, and M under the (anti-)periodic boundary condition. The energy and momentum of an eigenstate are expressed as $E = -2t\sum_j \cos k_j - \mu N$ and $K = \sum_j k_j$, respectively. Since $\{k_j\}$ and $\{\Lambda_\alpha\}$ are obtained through the above Bethe equations once $\{I_j\}$ and $\{J_\alpha\}$ are given, eigenstates are characterized by $\{I_j\}$ and $\{J \ \}$. Essential spectral features in the lower Hubbard band have been well explained in terms of QP's de ned by the distributions of $\{I_j\}$ and $\{J \ \}$ for real $\{k_j\}$ and real $\{\Lambda \ \}$ solutions [15, 16]. Here, the QP's de ned by $\{I_j\}$ are called holons and anti-holons, and those de ned by $\{J \ \}$ are called spinons. In the lower Hubbard band, there are also continua of 2-Λ-string solutions [16, 17], for which the QP representing the 2-Λ-string [8] is relevant. However, for the upper Hubbard band, we need to consider other solutions.

It is known that there are solutions with complex $\{k_j\}$ [17]. Recently, single-particle properties in the upper Hubbard band have been explained by using k-Λ string solutions [16] where the k-Λ string is a pair of quasimomenta k^+ and k complex conjugate to each other, i.e., $k^{\pm} = \pi \quad \sin^{\ 1}\ \Lambda' \pm \frac{\imath U}{4t}$ with real Λ' within the accuracy of $O(\mathrm{e}^{\ cL})$ with $c > 0$ [17]. In the k-Λ string solutions, $\{k_j\}$, $\{\Lambda \ \}$, and Λ' satisfy the following Bethe-Takahashi equations [17]:

$$L\mathrm{Re}\left[\sin^{\ 1}\left(\Lambda' + \frac{\imath U}{4t}\right)\right] = \pi J_1' + \sum_{j=1}^{N\ 2} \tan^{\ 1}\frac{4t(\Lambda' \quad \sin k_j)}{U},$$

$$Lk_j = 2\pi I_j + 2\tan^{\ 1}\frac{4t(\Lambda' \quad \sin k_j)}{U} + 2\sum_{\ =1}^{M\ 1}\tan^{\ 1}\frac{4t(\Lambda \quad \sin k_j)}{U} \quad \text{for } j = 1 \quad N \quad 2,$$

$$\sum_{j=1}^{N\ 2}\tan^{\ 1}\frac{4t(\Lambda \quad \sin k_j)}{U} = \pi J \ + \sum_{\beta=1}^{M\ 1}\tan^{\ 1}\frac{2t(\Lambda \quad \Lambda_\beta)}{U} \quad \text{for } \ = 1 \quad M \quad 1.$$

To explain the properties in the upper Hubbard band, a QP representing the k-Λ string has been de ned by using the Bethe quantum number for the k-Λ string, J_1' [16]. Noting that the k-Λ string has energy of order U in the large U/t regime, i.e., $E_{k\ \Lambda} = \mathrm{Re}\sqrt{16t^2 \quad (4t\Lambda' \quad \imath U)^2}$ [17] and that it represents a bound state of electrons [17], we can regard the QP as a doublon [16]. This doublon characterizes the upper Hubbard band: while properties in the lower Hubbard band are explained by using solutions without k-Λ strings, those in the upper Hubbard band are explained by using solutions with the k-Λ string (Fig. 1(c)) [16, 18]. Also, it should be noted that the number of the doublons is an integer in each eigenstate. These features contrast with conventional doublons de ned as double occupancies. Namely, the number of double occupancies is not an integer in eigenstates, and double occupancy exists even in the ground state at half- lling, i.e., in the lower Hubbard band, which implies that double occupancies do not behave as QP's for the upper Hubbard band. In terms of the doublon de ned as the k-Λ string, the upper Hubbard band can be well distinguished from the lower Hubbard band by the presence of the doublon (k-Λ string) in eigenstates. Or, we can de ne the upper and lower Hubbard bands as eigenstates with and without the doublon, respectively. From the above argument, we can readily conclude that properties in the lower Hubbard band have no relation to the doublon. For example, pseudo-gap behaviors in the lower Hubbard band can be essentially explained in terms of spinon, holons, and anti-holons without the doublon [16], which contrasts with the picture of doublon (double occupancy)-holon (empty site) binding for pseudo-gap behaviors near the Mott transition proposed in the literature [19, 20].

3. High-energy continuum of spin chains in a magnetic field

The spin-1/2 Heisenberg model is de ned by the following Hamiltonian:

$$\mathcal{H}_{\mathrm{Heis}} = J\sum_{\langle i,j\rangle}\boldsymbol{S}_i \quad \boldsymbol{S}_j \quad H\sum_i S_i^z, \tag{2}$$

where $\boldsymbol{S}_i$ and S_i^z denote the spin-1/2 operator at site i and its z-component, respectively. In 1D, exact solutions have been obtained by using the Bethe ansatz [11]. In the Bethe ansatz, exact

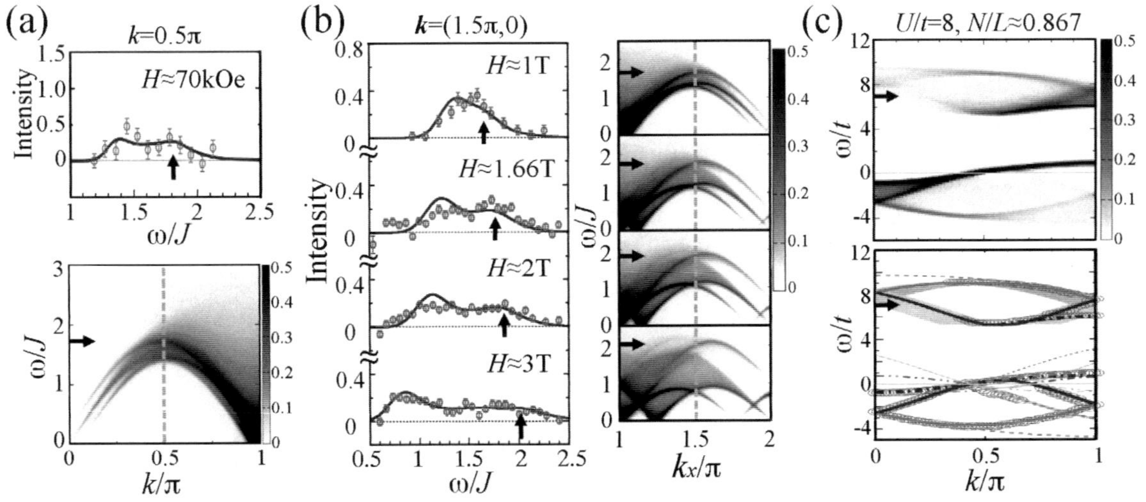

Figure 1. (a) Dynamical structure factor $S(k,\omega) = \left[S^{\ +}(k,\omega) + S^{+\ }(k,\omega) + 4S^{zz}(k,\omega)\right]/6$ of a spin-1/2 antiferromagnetic Heisenberg chain in a magnetic eld. The blue solid line and intensity plot show the results calculated using the Bethe ansatz in $L = 320$ in Ref. [8]. Red open circles denote the experimental results on the 1D Heisenberg antiferromagnet $CuCl_2\,2N(C_5D_5)$ in Ref. [10]. Arrows indicate the high-energy states of the 2-string solutions. (b) $S(\boldsymbol{k},\omega)$ of a spin-1/2 anisotropic triangular Heisenberg antiferromagnet in a magnetic eld. Blue solid lines and intensity plots show the results calculated using a weak-interchain coupling approximation combined with Bethe-ansatz results of 1D chains in Ref. [9]. The ratio of the interchain coupling constant J' to the intrachain coupling constant J was chosen as $J'/J = 0.34$, corresponding to that of the anisotropic triangular Heisenberg antiferromagnet Cs_2CuCl_4 ($J = 0.374$ meV and $J' = 0.128$ meV [24]). Red open circles denote the experimental results on Cs_2CuCl_4 in Ref. [13]. Arrows indicate the high-energy mode originating from the 2-string solutions. (c) The upper panel shows the single-particle spectral function $A(k,\omega)$ of a repulsive Hubbard chain for $U/t = 8$ at $N/L \simeq 0.867$ calculated using the dynamical density-matrix renormalization group method in $L = 60$ in Ref. [16]. The lower panel shows dispersion relations calculated using the Bethe ansatz in $L = 240$ in Ref. [16]. Solid lines show holon modes and anti-holon modes. Blue dashed-dotted lines denote spinon modes. Purple dashed double-dotted lines indicate modes of 2-Λ-string solutions. Dashed lines denote edges of continua. The lines in the upper Hubbard band are those in the k-Λ string solutions. Pink and yellow regions represent doublon-holon continua. Open circles indicate peaks of dominant modes of the upper panel. Arrows indicate the upper Hubbard band, i.e., the high-energy states of the k-Λ string solutions.

solutions of the 1D antiferromagnetic ($J > 0$) Heisenberg model are parameterized by rapidities $\{\Lambda\ \}$ that satisfy the following Bethe equations [21, 11]:

$$L \tan^{\ 1} \Lambda\ = \pi J\ + \sum_{\beta=1}^{M} \tan^{\ 1} \frac{\Lambda\ \ \ \Lambda_\beta}{2} \quad \text{for} \quad\ = 1 \quad M,$$

where J are Bethe quantum numbers having integers or half-odd integers constrained by L and M under the periodic boundary condition. In a magnetic eld, dominant low-energy properties have been explained in terms of 's and *'s [7, 8] which are de ned as holes and particles created from the consecutive distribution of $\{J\ \}$ in the ground state in a magnetic eld, respectively [7]. The 's reduce to spinons in the zero- eld limit. In addition to the low-energy continua of 's and *'s, a high-energy continuum appears in a magnetic eld [8] in the dynamical structure factor $S^{+}\ (k,\omega)$. Here, dynamical structure factors are de ned as $S^{\ }\ (k,\omega)=\sum_i|\langle i|S_k\ |\mathrm{GS}\rangle|^2\delta(\omega\ \ \epsilon_i)$ for $\ =+,\ \ $, and z, where $|i\rangle$ and $|\mathrm{GS}\rangle$ denote an excited state with excitation energy ϵ_i and the ground state, respectively. The high-energy continuum has been identi ed as that of 2-string solutions [8] where the 2-string is a pair of rapidities Λ^{+} and Λ complex conjugate to each other, i.e., $\Lambda^{\pm}=\bar{\Lambda}\pm\imath$ with real $\bar{\Lambda}$ within the accuracy of $O(\mathrm{e}^{\ cL})$ with $c>0$ [12]. In the 2-string solutions, $\{\Lambda\ \}$ and $\bar{\Lambda}$ satisfy the following Bethe-Takahashi equations [12]:

$$L\tan^{\ 1}\frac{\bar{\Lambda}}{2}=\pi\bar{J}_1+\sum_{\beta=1}^{M\ 2}\left[\tan^{\ 1}\frac{\bar{\Lambda}\ \ \Lambda_\beta}{3}+\tan^{\ 1}\ \bar{\Lambda}\ \ \Lambda_\beta\ \right],$$

$$L\tan^{\ 1}\Lambda\ \ =\pi J\ \ +\tan^{\ 1}\frac{\Lambda\ \ \bar{\Lambda}}{3}+\tan^{\ 1}\ \Lambda\ \ \ \bar{\Lambda}\ +\sum_{\beta=1}^{M\ 2}\tan^{\ 1}\frac{\Lambda\ \ \ \Lambda_\beta}{2},$$

for $\ =1\ \ \ M\ \ 2$. By using the Bethe quantum number for the 2-string, $\bar{J}_1$, a QP representing the 2-string has been de ned [8], which can be regarded as a bound state of two *'s [8]. Thus, the 2-string in the Heisenberg model is analogous to the k-Λ string in the Hubbard model: the high-energy states in the Heisenberg model in a magnetic eld and those in the Hubbard model are both due to string solutions with string length greater than one. In other words, the QP responsible for the high-energy continuum in the Heisenberg model in a magnetic eld is related in origin with the doublon in the Hubbard model through the Bethe ansatz.

The physical picture of this correspondence relation will be more intuitively understood by mapping the spin-1/2 antiferromagnetic Heisenberg model to the hard-core boson model with nearest neighbor repulsion:

$$\mathcal{H}_{\mathrm{hcb}}=\ \frac{J_{xy}}{2}\sum_{\langle i,j\rangle}\ b_i^{\dagger}b_j+\mathrm{H.c.}\ +J_z\sum_{\langle i,j\rangle}\left(n_{\mathrm{b}i}\ \ \frac{1}{2}\right)\left(n_{\mathrm{b}j}\ \ \frac{1}{2}\right)+H\sum_i\left(n_{\mathrm{b}i}\ \ \frac{1}{2}\right),\qquad(3)$$

where b_i and $n_{\mathrm{b}i}$ denote annihilation and number operators of a hard-core boson at site i. For the Heisenberg model, $J_{xy}=J_z=J$. As in the Hubbard model, the repulsive interaction (J_z-term) raises energy, when two bosons are located at neighboring sites. Thus, it is naturally expected that the band will split due to the repulsion, if the repulsive interaction is strong enough ($J_z\gg J_{xy}$). For the Heisenberg model ($J_{xy}=J_z$), although the band does not split in zero eld, it splits in a magnetic eld [8]. The continuum of 2-string solutions shifts to higher energies as the magnetic eld increases [8], as the upper Hubbard band does with hole doping [16]. Likewise, a corresponding continuum appears in the spinless fermion model with nearest neighbor repulsion on a 1D lattice [22, 23], since the hard-core boson model can be mapped to the spinless fermion model through the Jordan-Wigner transformation.

4. High-energy mode of frustrated antiferromagnets in a magnetic field

In anisotropic-2D frustrated Heisenberg antiferromagnets, interchain spin exchange induces bound states of 1D QP's from continua of 1D chains [6, 9]. The mechanism can be intuitively understood by noting that spin exchange corresponds to two-spinon hopping in zero eld [6]: the kinetic energy in the direction perpendicular to chains gains by forming bound states of spinons due to the pair hopping of spinons, i.e., spin exchange, between neighboring chains. The modes of such bound states dominate the spectra and behave as QP's in anisotropic-2D frustrated Heisenberg antiferromagnets [6, 9]. This implies that the origin of the dominant modes can be traced back to the 1D QP's. Namely, the high-energy mode in the frustrated Heisenberg antiferromagnets in a magnetic eld originates from and the QP for the 2-string [9]. Thus, the high-energy mode observed in the anisotropic triangular Heisenberg antiferromagnet Cs_2CuCl_4 in a magnetic eld [13] can be interpreted as the mode originating from the mechanism similar to that of the upper Hubbard band in a Hubbard chain.

5. Summary

In spin-1/2 antiferromagnetic Heisenberg chains and anisotropic-2D frustrated Heisenberg antiferromagnets, high-energy states carry considerable spectral weights in a magnetic eld [8, 9, 10, 13]. Such high-energy properties cannot be explained in terms of either Nambu-Goldstone bosons due to spontaneous breaking of continuous symmetries or gapless QPs in a Tomonaga-Luttinger liquid. In this paper, the origin of the high-energy states in the spin-1/2 Heisenberg antiferromagnets in a magnetic eld was shown to be analogous to that of the upper Hubbard band in the 1D repulsive Hubbard model, by using the Bethe ansatz: the high-energy states in both the Heisenberg antiferromagnets and the Hubbard model originate from string solutions with string length greater than one, i.e., the 2-string solutions for the Heisenberg antiferromagnets and the k-Λ string solutions for the Hubbard model. Thus, we can interpret the high-energy states in spin-1/2 1D Heisenberg antiferromagnets (Fig. 1(a)) and anisotropic-2D frustrated Heisenberg antiferromagnets (Fig. 1(b)) observed in inelastic neutron scattering experiments in a magnetic eld [10, 13] as those caused by the mechanism similar to that of the upper Hubbard band in the Hubbard model (Fig. 1(c)). In other words, the QP's responsible for the high-energy states in the Heisenberg antiferromagnets in a magnetic eld are related in origin with the doublon in the Hubbard model through the Bethe ansatz. It should be noted that, in contrast to an anti-bonding band due to multi-site unit cells, the high-energy states are induced by repulsive interactions without sublattices. Thus, the high-energy properties of the Heisenberg antiferromagnets in a magnetic eld can be understood in the context of the Mott physics. I expect that such high-energy states will generally appear in the one-particle addition spectral function in repulsively interacting systems if the number of particles is small enough.

Acknowledgments

This work was supported by World Premier International Research Center Initiative on Materials Nanoarchitectonics, MEXT, Japan, and KAKENHI 22014015.

References

[1] Bloch F 1930 *Z. Phys.* **61** 206
[2] Anderson P W 1952 *Phys. Rev.* **86** 694
[3] Nagamiya T 1967 *Solid State Physics* vol. 2, ed F Seitz *et al.* (Academic, New York) p 305
[4] Faddeev L D and Takhtajan L A 1981 *Phys. Lett. A* **85** 375
[5] Haldane F D M 1991 *Phys. Rev. Lett.* **66** 1529
[6] Kohno M, Starykh O A and Balents L 2007 *Nature Phys.* **3** 790
[7] Karbach M, Biegel D and Müller G 2002 *Phys. Rev.* B **66** 054405
[8] Kohno M 2009 *Phys. Rev. Lett.* **102** 037203
[9] Kohno M 2009 *Phys. Rev. Lett.* **103** 197203
[10] Heilmann I U *et al.* 1978 *Phys. Rev.* B **18** 3530
[11] Bethe H 1931 *Z. Phys.* **71** 205
[12] Takahashi M 1971 *Prog. Theor. Phys.* **46** 401
[13] Coldea R *et al.* 1997 *Phys. Rev. Lett.* **79** 151
[14] Lieb E and Wu F Y 1968 *Phys. Rev. Lett.* **20** 1445
[15] Benthien H, Gebhard F and Jeckelmann E 2004 *Phys. Rev. Lett.* **92** 256401
[16] Kohno M 2010 *Phys. Rev. Lett.* **105** 106402
[17] Takahashi M 1972 *Prog. Theor. Phys.* **47** 69
[18] There are solutions with multiple k-Λ strings and longer k-Λ strings [17]. But, they have much higher energies [17] and much smaller single-particle spectral weights [16] than the solutions with the k-Λ string de ned as a pair of quasimomenta in the large U/t regime.
[19] Phillips P 2010 *Rev. Mod. Phys.* **82** 1719
[20] Sakai S, Motome Y and Imada M 2009 *Phys. Rev. Lett.* **102** 056404
[21] Yang C N and Yang C P 1966 *Phys. Rev.* **150** 327
[22] Kohno M, Arikawa M, Sato J and Sakai K 2010 *J. Phys. Soc. Jpn.* **79** 043707
[23] Pereira R G, White S R and Affleck I 2009 *Phys. Rev.* B **79** 165113
[24] Coldea R *et al.* 2002 *Phys. Rev. Lett.* **88** 137203

Journal of Physics: Conference Series **320** (2011) 012022 doi:10.1088/1742-6596/320/1/012022

Exotic quantum phase transitions in the spin nanotube

T Sakai[1,2], M Sato[3], K Okunishi[4], K Okamoto[5], C Itoi[6]

[1]*Japan Atomic Energy Agency, SPring-8, Hyogo 679-5148, Japan*
[2]*Graduate School of Material Science, Univ. of Hyogo, Hyogo 678-1297, Japan*
[3]*Department of Physics and Mathematics, Aoyama Gakuin Univ., Kanagawa 251-5258, Japan*
[4]*Department of Physics, Niigata University, Niigata 9502181, Japan*
[5]*Department of Physics, Tokyo Institute of Technology, Tokyo 152-8551, Japan*
[6]*Department of Physics, Nihon Univ., Tokyo 101-8308, Japan*

E-mail: sakai@spring8.or.jp

Abstract. We investigate the S=1/2 three-leg spin tube, which is the simplest spin nanotube, using the density matrix renormalization group (DMRG) and the numerical exact diagonalization (ED), conbined with a finite-size scaling analysis. Particularly, we intruduce the lattice distortion from the regular triangle unit to the isosceles one. In our previous works, we revealed that the S=1/2 isosceles triangle spin tube exhibits following exotic quantum phenomena: the quantum phase transition between the spin gapped and gapless spin liquid phases, the 1/3 magnetization plateau due to two different mechanisms, and a new phase between the two plateau phases where the translational symmetry is spontaneously broken. We briefly review these results and show our new calculation of the expectation value of the magnetization at each site to specify the spin structure in the new field induced phase.

1. Introduction

Recently some quantum spin systems on tube lattices, so called spin nanotubes, have been synthesized. They are expected to be interesting low-dimensional systems like the carbon nanotubes. As the first step of theoretical study on the spin nanotube, we investigate the S=1/2 three-leg spin tube, which is the simplest one, using the density matrix renormalization group (DMRG) and the numerical exact diagonalization (ED), conbined with a finite-size scaling analysis. Particularly, we introduce the lattice distortion from the regular triangle unit to the isosceles one. In our previous works[1-4], we revealed that the S=1/2 isosceles triangle spin tube exhibits following exotic quantum phenomena:

(1) the quantum phase transition between the spin gapped and gapless spin liquid phases[1,2];
(2) the 1/3 magnetization plateau due to two different mechanisms[1, 4];
(3) the field-induced symmetry breaking between the two plateau phases[1-4].

In the present paper, we briefly review these results and present the calculated spin excitation values in the new phase between the two 1/3 magnetization plateau phases, to confirm the symmetry breaking and specify the spin structure.

2. Model

We consider the S=1/2 asymmetric three-leg spin tube, shown in Fig. 1, described by the Hamiltonian

$$\hat{H} = J_1 \sum_{i=1}^{3} \sum_{j=1}^{L} \vec{S}_{i,j} \cdot \vec{S}_{i,j+1} + J_r \sum_{i=1}^{2} \sum_{j=1}^{L} \vec{S}_{i,j} \cdot \vec{S}_{i+1,j} + J'_r \sum_{j=1}^{L} \vec{S}_{3,j} \cdot \vec{S}_{1,j} \quad (1)$$

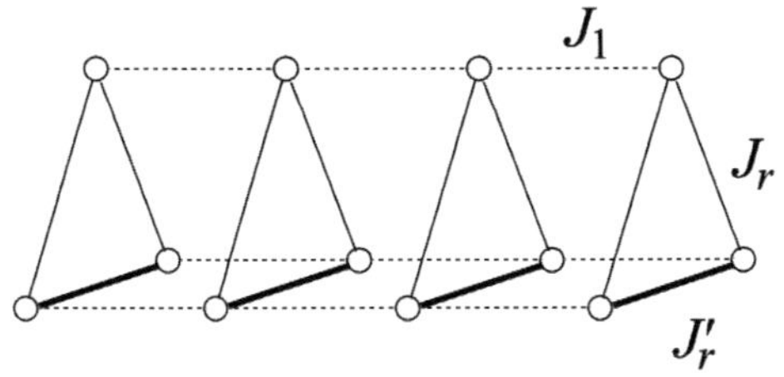

Fig. 1 Structure of the three-leg asymmetric spin tube (1).

where $\vec{S}_{i,j}$ is the spin-1/2 operator and L is the length of the tube in the leg direction. The exchange interaction constant J_1 is for the neighbouring spin pairs along the legs, while J_r and J'_r are the rung interaction constants. All the exchange interactions are supposed to be antiferromagnetic (namely, positive).

The ratio $\alpha = J'_r/J_r$ stands for the degree of the asymmetry of the rung interactions. We will vary α and J_1 to investigate the quantum phase transitions. Throughout this paper, we fix J_r to one. The effect of the magnetic fields is incorporated by adding the Zeeman energy

$$\hat{H}_Z = -H \sum_{i=1}^{3} \sum_{j=1}^{L} S_{i,j}^z \quad (2)$$

The present model includes three typical models as limiting cases; (a) $\alpha=0$: the three-leg spin ladder, (b) $\alpha=1$: the symmetric spin tube, and (c) $\alpha \to \infty$: the single chain plus rung dimers. Since the system is gapless in the cases (a) and (c), while gapful in the case (b), at least two quantum phase transitions should occur with increasing α from 0 to infinity. The one-site translational symmetry along the leg ($\vec{S}_{i,j} \to \vec{S}_{i,j+1}$) was revealed to be spontaneously broken in the symmetric spin tube at least in the strong-rung-coupling regime in the previous works[5,6].

3. Spin gap

According to the Lieb-Schultz-Mattis theorem, the spin gap phase of the S=1/2 three-leg spin tube should have two-fold degeneracy in the ground state, due to the translational symmetry break down. A schematic picture of the gapped phase for the symmetric spin tube was proposed as shown in Fig. 2, where the three equivalent singlet dimer covering patterns are resonating at each two unit cell triangles.

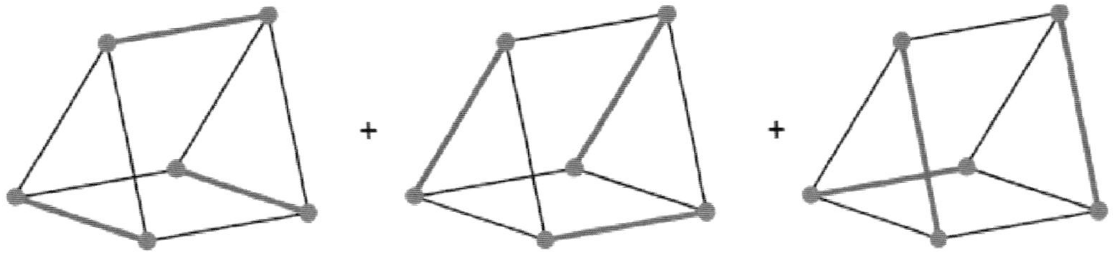

Fig. 2. A schematic picture of the spin gap formation mechanism of the S=1/2 three-leg spin tube. Blue circles and thick red lines are spins and singlet bonds, respectively. The three equivalent patterns at each two unit cells are resonating.

If this picture is valid, a double periodicity in the chain direction should be realized in the ground state. In order to confirm it, we calculate the singlet excitation energy with the momentum $k=\pi$ and test the degeneracy of it with the ground state in the thermodynamic limit. The phenomenological renormalization is a good method to determine the parameter region where the two states are degenerated in the infinite length limit. For each value of J_1, the critical point α_c is determined by the fixed point equation

$$L_1 \Delta_s(\alpha_c, L_1) = L_2 \Delta_s(\alpha_c, L_2),$$

for two different system sizes L_1 and L_2, where Δ_s is the energy difference between the singlet excitation and ground states. The calculated Δ_s is plotted versus α for L=4, 6 and 8 with fixed J_1=0.5 in Fig. 3. Extrapolating the size-dependent fixed point α_c (4,6) and α_c (6,8) with respect to the system size L (we assume the size correction is proportional to $1/L^2$.), we determine α_c in the thermodynamic limit. The result is shown as solid circles in the α-J_1 plane in Fig. 4. The phase boundary between the spin gap and gapless phases determined by the phenomenological renormalization applied for the singlet-triplet gap in the previous work[1,2] is also plotted as a solid line in Fig. 4. Although there is some difference between the two boundaries due to finite size effects at larger J_1, Fig. 4 suggests that the two singlet states are degenerated in the thermodynamic limit in the spin gap phase. It is consistent with that the translational symmetry is broken and double periodicity is realized in the gapped phase.

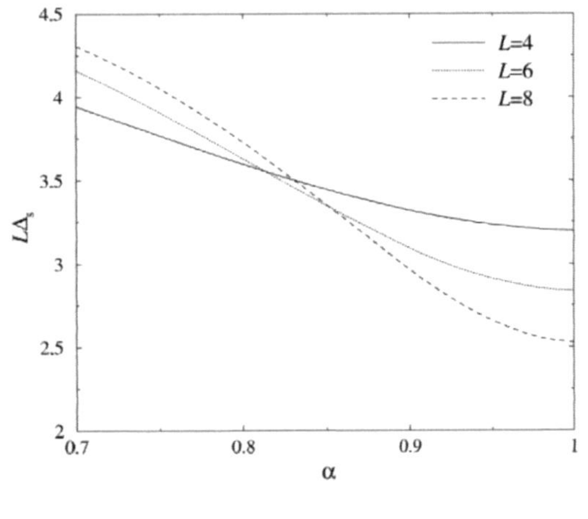

Fig. 3 Scaled singlet-singlet excitation gap $L\Delta_s$ plotted versus α with fixed J_1=0.5.

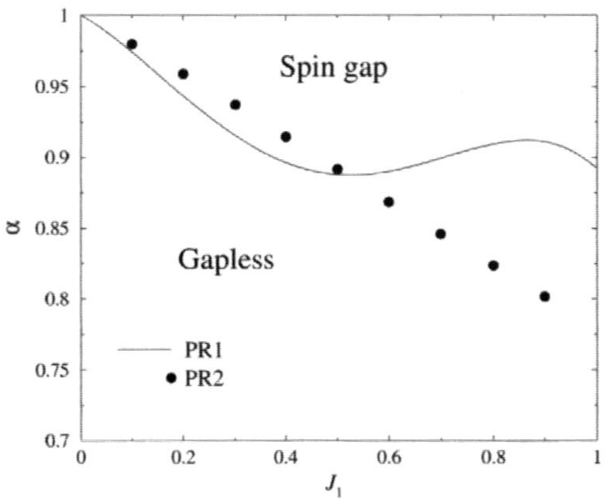

Fig. 4 Phase boundaries between the spin gap and gapless phases. Solid circles are determined by the phenomenological renormalization of the singlet-singlet excitation gap, a solid line is determined by the singlet-triplet excitation gap from Ref. [2].

4. 1/3 magnetization plateau

In order to study the 1/3 magnetization plateau, we calculated the spin excitation gap at m/m_s=1/3 (m is the magnetization and m_s is the saturation magnetization), namely

$$\Delta = E(M+1) + E(M-1) - 2E(M) \qquad (3)$$

where $E(M)$ is the lowest energy in the subspace $\sum S_j^z = M$ and $M=L/2$, using the numerical exact diagonalization. The scaled gap $L\Delta$ for α=1 is plotted versus J_1 for L=4,6 and 8 in Fig. 5. From the cross point of L=6, 8 curves, the quantum critical point is estimated as J_1~0.5 between the plateau and gapless (no plateau) phases.

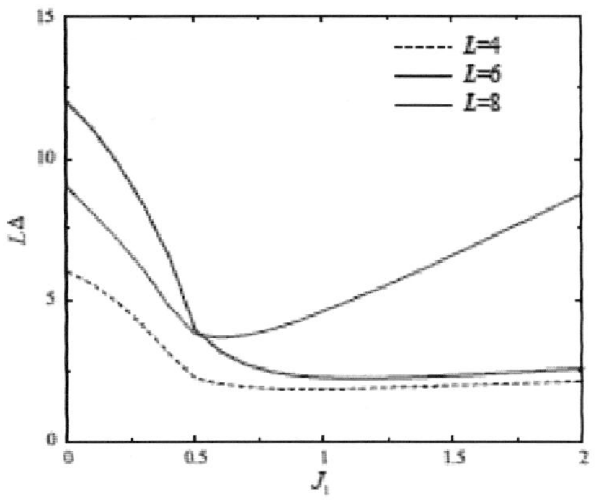

Fig. 5 Scaled gaps for L=4, 6 and 8 plotted versus J_1 with fixed J'_r/J_r=1.

In comparison with the spin gap at H=0, the 1/3 plateau is expected to be stable against the asymmetry (α≠1), because it was also predicted to exist even for the three-leg ladder (α =0).

At the 1/3 plateau phase close to α=0, the state |↑ ↓ ↑› is expected to be stabilized at each triangle unit, while |singlet, ↑› for the other limit α→∞. We call the two mechanisms of the plateau formation as the udu mechanism and dimer-monomer mechanism, respectively. Thus the three phases; the udu plateau, the dimer-monomer plateau and gapless phases are expected to occur at $m/m_s=1/3$.

Next we consider the boundary between the two different plateau phases. It is expected to appear around $\alpha=1$ and $J_l < 0.5$. The α dependences of the plateau width estimated by the DMRG calculation for L=128 for J_l=0.1, 0.3, and 0.5 are shown in Fig. 6. Figure 6 indicates clear anomalous increases of the plateau width for J_l=0.1 and 0.3. The anomalous increase region is supposed to be a new phase between the two plateau phases. According to the perturbation study in the strong J_r region in our previous work [1], the translational symmetry in the leg direction is broken and two different low-lying states at each triangle unit alternate like the Neel order in the new phase. The lowest excitation gap within the $M=M_s/3$ space, Δ_0, with $k=\pi$ and the odd parity for the inversion with respect to the center line of the isosceles triangle, is a useful order parameter to distinguish the new phase with the other two plateau phases. The gap closes in the new phase, while it is open in the other ones. Thus the phase boundary of the new plateau phase can be determined by the same phenomenological renormalization as the one used to obtain the ground state phase diagram in Fig. 4, applied for Δ_0. We just show the phase diagram at $m/m_s=1/3$ obtained by the method in Fig. 7, where the plateau 1, 2, and 3 phases are the udu, dimer-monomer, and new plateau phases, respectively. The boundaries between different plateau phases belong to the Ising universality class, while the one between the plateau and gapless phases the Berezinskii-Kosterlitz-Thouless (BKT) universality class[1].

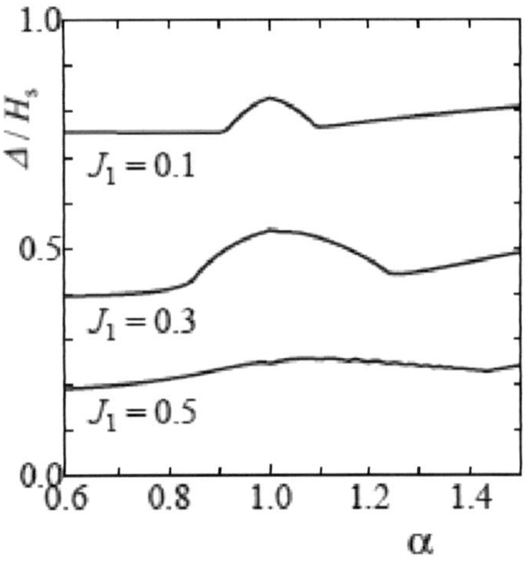

Fig. 6 α dependence of the plateau width normalized by the saturation magnetic field H_s calculated by DMRG for J_l=0.1, 0.3 and 0.5.

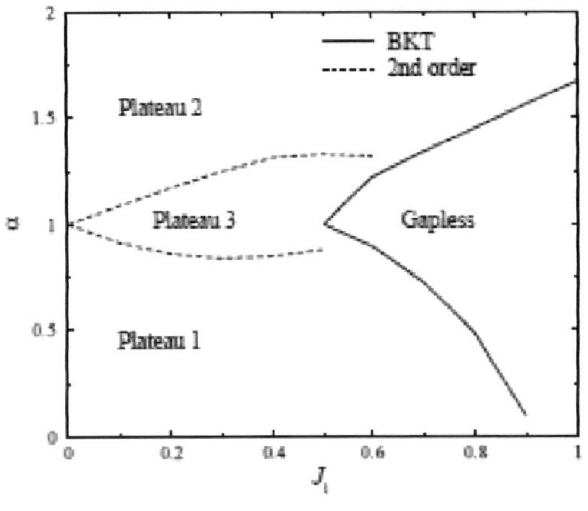

Fig. 7 Phase diagram at m=1/3.

5. Spin structure of the new phase

Finally we show the calculated spin structure in the new phase by the DMRG. The expectation value of S_j^z at each site calculated by DMRG is plotted versus α in Fig. 8, where blue stars, green pluses and red crosses correspond to the apical, the left, and right bottom sites of the triangle unit, respectively for J_1=0.3. The spins have different values between the two bottom sites in the new phase (0.84< α <1.24) and they are switched at the adjacent triangle. It implies that the long-range Neel order is realized at the two bottom sites along the leg. Thus the present calculation justifies the existence of the new phase characterized by the staggered long-range order at the two bottom sites.

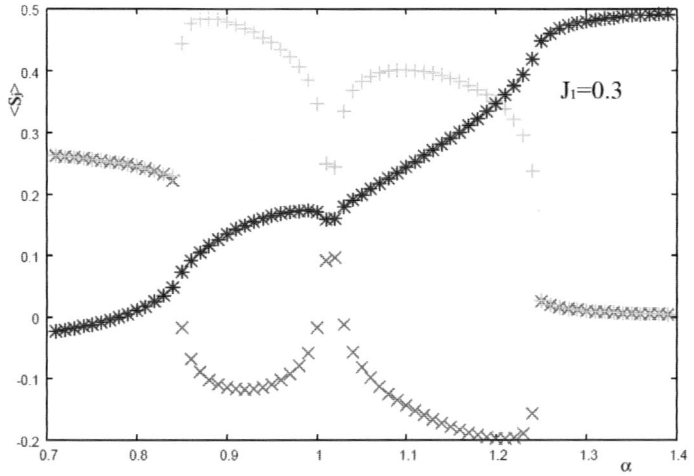

Fig. 8 Calculated S_j^z at the apical (blue stars), left (green pluses), and right (red crosses) bottom sites of the triangular unit for J_1=0.3 by DMRG (L=48).

6. Summary

The S=1/2 isosceles triangle spin tube was investigated by the DMRG and numerical diagonalization. The present study indicated that two singlet states are degenerated in the m=0 ground state, which is consistent with the translational symmetry breaking. In addition in the new phase between the two different 1/3 magnetization plateau phases, the staggered long-range order was revealed to realized.

Acknowledgment

This work was partly supported by Grants-in-Aid (No.20340096) for Scientific Research and Priority Area "Novel States of Matter Induced by Frustration" from the Ministry of Education, Culture, Sports, Science and Technology of Japan and the Computer Center ISSP Univ. of Tokyo.

References

[1]T. Sakai et al., *J. Phys.: Condens. Matter* **22**, 403201 (2010).
[2]T. Sakai et al., *Phys. Rev. B*, **78**, 184415 (2008).
[3] M. Sato and T. Sakai, *Phys. Rev. B*, **75**, 014411 (2007).
[4] K. Okamoto et al., *Physica E*, **43**, 769 (2011).
[5] K. Kawano and M. Takahashi, *J. Phys. Soc. Jpn.*, **66**, 4001 (1997).
[6] S. Nishimoto and M. Arikawa, *Phys. Rev. B*, **78**, 054421 (2008).

Journal of Physics: Conference Series **320** (2011) 012023 doi:10.1088/1742-6596/320/1/012023

Dimensional crossover in the quasi-two-dimensional Ising-O(3) model

Y Kamiya[1], **N Kawashima**[1] **and C D Batista**[2]

[1] Institute for Solid State Physics, University of Tokyo, Kashiwa, Chiba 227-8581, Japan
[2] Theoretical Division, Los Alamos National Laboratory, Los Alamos, New Mexico 87545, USA

E-mail: kamiya@issp.u-tokyo.ac.jp

Abstract. We study the Ising-O(3) model on two-dimensional (2D) and quasi-2D lattices. This is an unfrustrated effective model for the stacked square-lattice J_1-J_2 Heisenberg model where the nearest neighbor (J_1) and next-nearest neighbor (J_2, $J_2 \gtrsim J_1/2$) exchange interactions are frustrated. By means of Monte Carlo simulation we show that this model has an Ising ordered phase where the O(3) spins remain disordered in a moderate quasi-2D region. We obtain a first-order transition in a region with a sufficiently large inter-layer coupling. Our results provide a qualitative explanation of the experiments on ferropnictides, namely the observed sequence and the orders of the structural and magnetic transitions as a function of the ratio between the inter-layer and intra-layer couplings.

1. Introduction

The square-lattice J_1-J_2 Heisenberg model is a prototypical example of a frustrated spin system. This model is believed to have a nontrivial ground state and finite temperature properties due to the combination of frustration and low-dimensionality [1]. The corresponding single-layer Hamiltonian is

$$H = J_1 \sum_{\text{n.n.}} \mathbf{S}_i \cdot \mathbf{S}_j + J_2 \sum_{\text{n.n.n.}} \mathbf{S}_i \cdot \mathbf{S}_j, \tag{1}$$

where the nearest neighbor (J_1) and next-nearest neighbor (J_2) couplings are frustrated. Here we consider a stacked square lattice by introducing a finite inter-layer coupling. It is known that the ground state of the H (1) has Néel ordering for $J_2/J_1 \lesssim 0.4$ while it exhibits a two-sublattice collinear antiferromagnetic (CAF) ordering for $J_2/J_1 \gtrsim 0.65$ (Fig. 1). The intervening region $J_2 \sim J_1/2$ is still controversial [2].

Both theoretical and experimental efforts have been extensively devoted to understanding the phase diagram of this Hamiltonian [1]. Recently, the discovery of high-T_c superconductivity in ferropnictides has revived this interest [3]. The local moment picture is most likely inadequate for a microscopic description of the metallic ferropnictides. However, in this work we show that this simple model provides a qualitative understanding of almost universal features in the structural and magnetic transitions observed in ferropnictides. Particularly, although lattice degrees of freedom are not included in the model, the electronically driven Ising-like ordering shown in Fig. 1 will necessarily trigger the observed structural transition (C_4 symmetry breaking) [4, 5].

The model may be related to ferropnictides in the CAF region. More specifically, we are interested in the 1111-family of quasi-2D compounds RFeAsO, where R is a rare earth ion [6, 7]

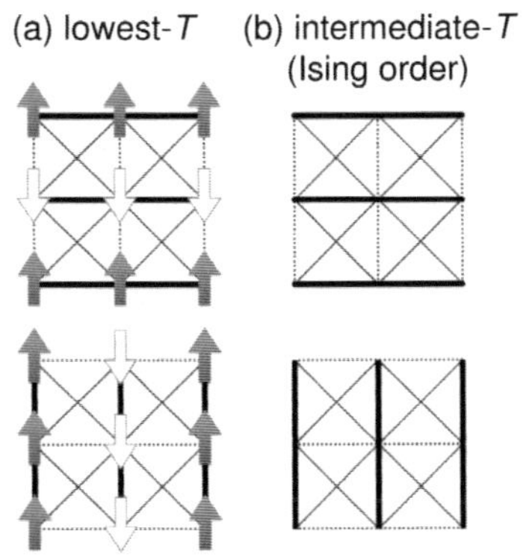

Figure 1. Schematic pictures of (a) the lowest-T and (b) the Ising ordered phases (spins remain disordered) of the J_1-J_2 Heisenberg model ($J_2 \gtrsim J_1/2$). The solid (dashed) lines on the non-diagonal edges represent the parallel (antiparallel) local spin alignment. The horizontal (top panels) or vertical (bottom panel) stripe can be distinguished by an Ising variable $\sim$ $\phi_A \cdot \phi_B$ with $\phi_{A,B}$ being the Néel order parameters of the $\sqrt{2} \times \sqrt{2}$ sublattices.

and 122-type 3D compounds $A\mathrm{Fe_2As_2}$ with alkaline earth elements A [8]. Different experiments have confirmed the following features for the undoped ground states of these ferropnictides [3]: (i) spin-density-wave (SDW) order at (,0) and (ii) orthorhombic lattice structure, distorted from high-T tetragonal structure. Interestingly, the tetragonal-orthorhombic structural transition occurs at a temperature (T_{c1}) that is either the same as the Néel temperature (T_{c2}), as in the 3D 122-type undoped compounds [8], or slightly higher than T_{c2}, as in the more quasi-2D 1111-type compounds [6, 7]. Experiments also confirmed first-order features such as hysteresis in the undoped 122 compounds [8], while the separate transitions in the 1111 compounds seem to be of second order or very weakly first order [7]. The proximity of these transitions is evidence in favor of a relevant magneto-elastic coupling. This was also confirmed by density functional calculations in Ref. [9]. As it was suggested in Ref. [10], it seems that there is a correlation between the sequence of the transition(s) and the magnitude of the inter-layer coupling. Finally, $T_{c1} \geq T_{c2}$ seems to be a universal feature among ferropnictides.

Several authors have argued that these almost universal characteristics of ferropnictides can be qualitatively explained by using an oversimplified J_1-J_2 Heisenberg model in the CAF region [4, 5]. In this picture, the structural transition is identified with the Ising-like order that emerges from an order-by-disorder effect [11], while the SDW order is identified with the magnetic transition that breaks O(3) spin rotational symmetry as well as the translational symmetry. The ordering wave-vectors are (,0) or (0,) [4, 5]. The inequality $T_{c1} \geq T_{c2}$ immediately follows from this picture because the Ising order is described by a composite order parameter in terms of spins [4, 5]. Furthermore, according to Mermin-Wagner theorem [12, 13], only the Ising order can survive at $T > 0$ in $d = 2$: $T_{c1}^{2D} > T_{c2}^{2D} = 0$. This result was confirmed by a Monte Carlo study [14]. This also implies that the Ising order without magnetic order will be present in some quasi-2D region of the phase diagram even in the presence of a finite inter-layer coupling [5]. However, the precise form of the phase diagram of the J_1-J_2 Heisenberg model as a function of the inter-layer coupling is still unknown. This is the main motivation of our numerical study.

2. Model

We consider an unfrustrated classical effective model, called the Ising-O(3) model [5], adequate for describing the J_1-J_2 Heisenberg model in the CAF region. The Hamiltonian of this model is defined by

$$H_{I\text{-}O(3)} = -\sum_{\langle i,j \rangle} J_{ij} \left(1 + \quad _i \quad _j\right) \mathbf{S}_i \cdot \mathbf{S}_j, \tag{2}$$

where ${}_i$ and $\mathbf{S}_i$ denote the classical Ising and Heisenberg spins respectively. We consider a 2D square lattice and a quasi-2D cubic lattice. For the latter we assume that the coupling constant J_{ij} is spatially anisotropic in such a way that:

$$J_{ij} = \begin{cases} J & \text{if } (i,j) \text{ are on the same layer} \\ J_z & \text{if } (i,j) \text{ are on the nearest neighbor layers,} \end{cases} \tag{3}$$

where $0 < J_z \lesssim J$. We use the unit $J \quad 1$ below. The following effective Hamiltonian can be derived by introducing auxiliary O(3) fields $\boldsymbol{\phi}_A \sim \mathbf{S}$ and $\boldsymbol{\phi}_B \sim \quad \mathbf{S}$:

$$\mathcal{H}_{\mathit{eff}} = \int \mathrm{d}^d x \left[\sum_{a=A,B} \quad \frac{1}{2}|\nabla\boldsymbol{\phi}_a|^2 + r|\boldsymbol{\phi}_a|^2 + u|\boldsymbol{\phi}_a|^4 \quad + u_{AB}|\boldsymbol{\phi}_A|^2|\boldsymbol{\phi}_B|^2 + \lambda\left(\boldsymbol{\phi}_A\cdot\boldsymbol{\phi}_B\right)^2 \right], \tag{4}$$

where $\lambda < 0$. It was argued that an effective field theory of the J_1-J_2 Heisenberg model can also be written in this form when $J_2 \gtrsim J_1/2$ [5]. In that case, $\boldsymbol{\phi}_A$ and $\boldsymbol{\phi}_B$ are Néel order parameters of the $\sqrt{2}\times\sqrt{2}$ sublattices. Therefore, a comparison between $H_{I\text{-}O(3)}$ and the effective Hamiltonian (4) suggests that ${}_i$ in the Ising-O(3) model represents a degree of freedom that specifies whether the local spin configuration is "vertical" or "horizontal" (see Fig. 1).

3. Numerical results

We simulated the Ising-O(3) model (2) on the square and the anisotropic cubic lattices by employing the cluster Monte Carlo method where the Ising and O(3) spins are updated alternatively. When $J_z \neq 0$, we combine the cluster update with a local update. In this case, we also use a semi-global cluster update where the clusters are allowed to expand only within a given layer.

3.1. 2D case

First, we show the results in $d = 2$. The non-negativity of $J_{ij}\,(1 + \quad {}_i \quad {}_j)$ favours a parallel alignment of the O(3) spins at any temperature. This implies that the effective coupling between the Ising spins is also ferromagnetic because, at the mean-field level, it is given by $J_{ij}\,\langle \mathbf{S}_i \cdot \mathbf{S}_j \rangle$. This cooperative relation leads to a finite-T transition of the 2D system where only the Ising spins become critical. A standard finite-size scaling analysis strongly indicates that the transition belongs to the 2D Ising universality class [15].

3.2. Quasi-2D case

Here we show the results for the quasi-2D region ($J_z \neq 0$). The inter-layer coupling stabilizes the O(3) spin order at $T > 0$. It also makes the Ising order 3D. In particular, we are interested in a region where the Ising and O(3)-like transitions occur at different temperatures because this may be related to the separate transitions observed in the 1111 compounds [6]. We also note that previous analytical studies suggested different scenarios when these transitions are close to each other: a large-N approximation predicted always two separate second-order transitions [4] whereas a mean-field theory predicted a phase diagram with richer structure [16].

We obtained two transitions for $J_z = 0.01$ (Fig. 2). Naturally, the separation between both transitions increases for a smaller inter-layer coupling [5, 15]. The lattice that we simulated has a tetragonal shape $L \times L \times rL$ with aspect ratio $r = 1/12$. The value of r was determined in such a way that $G^m(L_x/2,0,0) \approx G^m(0,0,L_z/2)$ and $G \ (L_x/2,0,0) \approx G \ (0,0,L_z/2)$, where $G^m(r)$ ($G \ (r)$) is the correlation function of the O(3) (Ising) order parameter. We studied system sizes that range from $L = 48$ to $L = 120$.

Figure 2 shows the specific heat $C = N^{-1} \quad {}^2\left[\left\langle (H_{I\text{-}O(3)})^2 \right\rangle - \left\langle H_{I\text{-}O(3)} \right\rangle^2\right]$ where $N \quad rL^3$. The double peak structure that develops for $L = 120$ suggests two separate transitions. This

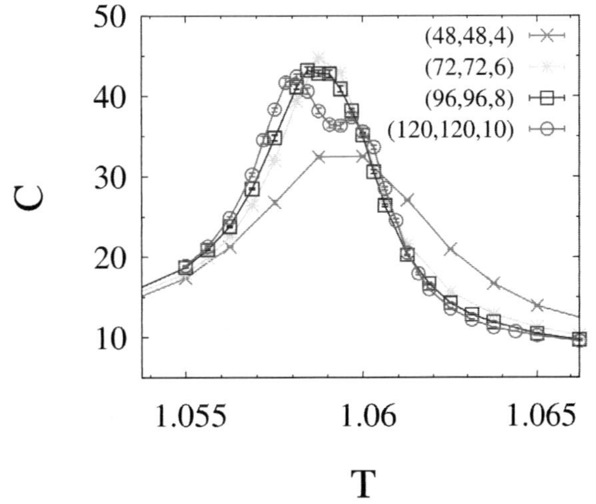

Figure 2. Specific heat of the quasi-2D Ising-O(3) model ($J_z = 0.01$). The lines are guides to the eye.

observation is also supported by the behavior of Binder parameters $U_m = \langle \mathbf{m}^4\rangle/\langle \mathbf{m}^2\rangle^2$ with $\mathbf{m} = N^{-1}\sum_i \mathbf{S}_i$ and $U \ = \langle \ ^4\rangle/\langle \ ^2\rangle^2$ ($= N^{-1}\sum_i \ _i$). Figure 3 shows their finite-size scaling plots. Based on symmetry arguments we assume that the higher-T (lower-T) transition is in the 3D Ising (Heisenberg) universality class and we use critical exponents available in the literature: $\eta_{\rm Is} = 0.03639(15)$ and $_{\rm Is} = 0.63012(16)$ for the 3D Ising universality class [17] and $\eta_{\rm H} = 0.0375(5)$ and $_{\rm H} = 0.7112(5)$ for the 3D Heisenberg universality class [17]. Although there are large finite-size effects, there is an asymptotic tendency towards data collapse that supports our assumptions for the universality classes. From these results we obtain $T_{c1} = 1.0610(8)$ and $T_{c2} = 1.0575(5)$. As for the sizable sub-leading corrections observed in these scaling plots, it appears that they are largely due to (i) the proximity of the two transitions and (ii) the spatially anisotropic correlations [15]. A more comprehensive finite-size scaling study including other dimensionless parameters and the correlation functions will be discussed elsewhere [15].

3.3. 3D case with weak spatial anisotropy

Finally, we move on to the 3D region with a larger inter-layer coupling. We simulated the system for $J_z = 0.1$, 0.0625, 0.04, 0.0278 and 0.0204, and aspect ratios of the lattice that turned out to be $r = 1/3$, 1/4, 1/5, 1/6 and 1/7, respectively. We obtained a bimodal structure of the internal energy distribution for all of these parameters [15], which is an unambiguous sign of a *rst-order* transition. We also found that the first-order feature is stronger for a larger inter-layer coupling [15]. We also confirmed the diverging behavior in the Binder parameters of both of order parameters for all the inter-layer couplings listed above except for $J_z = 0.0204$ [15]. This means that the first-order transition takes place between the paramagnetic phase and the lowest-T phase, where $\langle \mathbf{m}\rangle \neq 0$ and $\langle \ \rangle \neq 0$. The intervening region including $J_z = 0.0204$ is currently under investigation. Figure 4 shows the phase diagram that summarizes our results.

4. Conclusion

In summary, we studied the Ising-O(3) model on 2D and quasi-2D lattices. This is an effective model of the J_1-J_2 Heisenberg model with $J_2 \gtrsim J_1/2$. By solving this effective model we

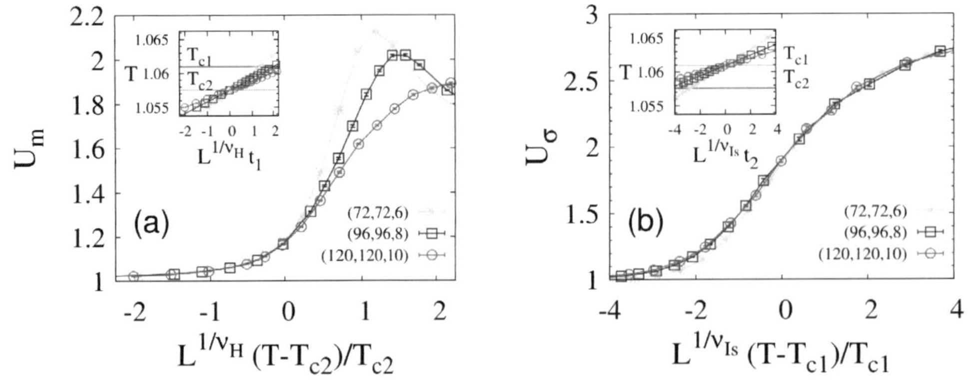

Figure 3. Finite-size scaling plots of the Binder parameters (a) U_m and (b) U for $J_z = 0.01$. The insets show explicit relations between the scaled temperatures $t_{1,2}$ $(T - T_{c1,2})/T_{c1,2}$ and the real temperature T. The lines are guides to the eye.

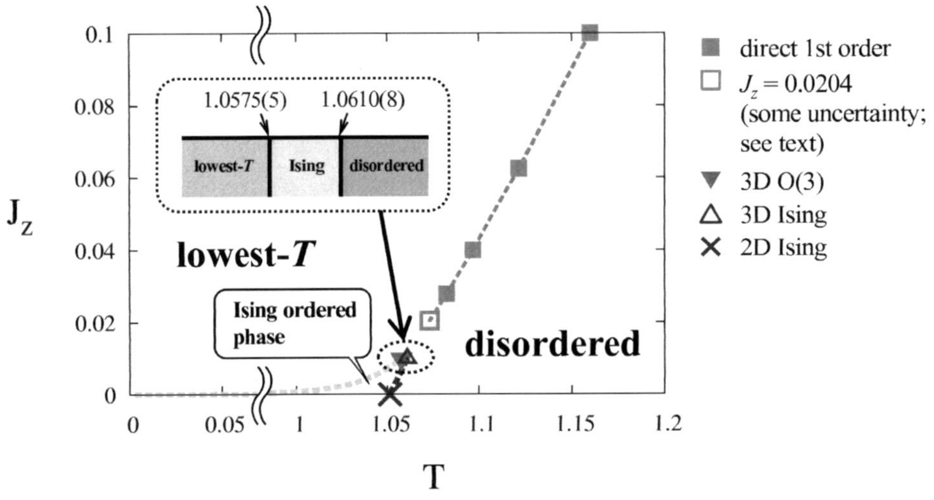

Figure 4. Phase diagram of the quasi-2D Ising-O(3) model. The Ising ordered phase retaining full O(3) symmetry exists in a region with a weak inter-layer coupling $J_z \lesssim 0.01$. The line that separates the Ising ordered phase and the lowest-T phase is a schematic one. The other lines are guides to the eye.

identified the region where the O(3)-symmetric Ising ordered phase exists. For sufficiently large inter-layer coupling, we found that a single first-order transition occurs between the paramagnetic and lowest-T ordered phase. The other subtle features discussed in the context of a mean-field theory [16] have not been reproduced but may need further investigation.

Our results provide a qualitative explanation for the sequence and the orders of transitions observed in ferropnictides as a function of the ratio between the inter-layer and intra-layer exchange couplings. This suggests that the separate structural and SDW transitions observed in the quasi-2D 1111 compounds [6, 7] are due to the fragility of the SDW order against fluctuations, which makes it more sensitive to the magnitude of the inter-layer coupling. The first-order nature of the simultaneous structural and SDW transition observed in the more 3D 122 parent compounds [8] is also consistent with our results. Despite the oversimplified nature of our local-moment model for the microscopic description of ferropnictides, we obtained a qualitatively correct phase diagram. This result indicates that the J_1-J_2 Heisenberg model and the related Ising-O(3) model are good starting points for describing universal properties of these compounds.

Acknowledgments
The computation in the present work was executed at the Supercomputer Center, Institute for Solid State Physics, University of Tokyo. This work is financially supported by GCOE for Phys. Sci. Frontier, MEXT, Japan, the MEXT Grand-in-Aid for Scientific Research (B) (22340111), the MEXT Grand-in-Aid for Scientific Research on Priority Areas "Novel States of Matter Induced by Frustration" (19052004), and the Next Generation Supercomputing Project, Nanoscience Program, MEXT, Japan.

References

[1] Misguich G and Lhuillier C 2004 *Frustrated Spin Systems* ed Diep H (Singapore: World Scientific) pp 229–306
[2] Reuther J and Wölfle P 2010 *Phys. Rev.* B **81** 144410
[3] Paglione J and Greene R 2010 *Nature Physics* **6** 645–658
[4] Fang C, Yao H, Tsai W F, Hu J and Kivelson S A 2008 *Phys. Rev.* B **77** 224509
[5] Xu C, Müller M and Sachdev S 2008 *Phys. Rev.* B **78** 020501(R)
[6] de La Cruz C *et al.* 2008 *Nature* **453** 899–902
[7] McGuire M A *et al.* 2008 *Phys. Rev.* B **78** 094517
[8] Krellner C, Caroca-Canales N, Jesche A, Rosner H, Ormeci A and Geibel C 2008 *Phys. Rev.* B **78** 100504(R)
[9] Yildirim T 2008 *Phys. Rev. Lett.* **101** 057010
[10] Qi Y and Xu C 2009 *Phys. Rev.* B **80** 094402
[11] Chandra P, Coleman P and Larkin A I 1990 *Phys. Rev. Lett.* **64** 88–91
[12] Mermin N D and Wagner H 1966 *Phys. Rev. Lett.* **17** 1133–1136
[13] Hohenberg P C 1967 *Phys. Rev.* **158** 383–386
[14] Weber C, Capriotti L, Misguich G, Becca F, Elhajal M and Mila F 2003 *Phys. Rev. Lett.* **91** 177202
[15] Kamiya Y, Kawashima N and Batista C D (unpublished)
[16] Cano A, Civelli M, Eremin I and Paul I 2010 *Phys. Rev.* B **82** 020408(R)
[17] Pelissetto A and Vicari E 2002 *Phys. Rep.* **368** 549–727

Sliding of a vortex solid with self-generated randomness in a frustrated Josephson junction array

Hajime Yoshino, Tomoaki Nogawa[A], Bongsoo Kim[B]
Osaka Univ. Osaka, Japan, [A]U. of Tokyo, Tokyo, Japan, [B]Changwon Univ., Changwon, Korea
E-mail: yoshino@ess.sci.osaka-u.ac.jp

Abstract. A class of Josephson junction arrays under external magnetic field with anisotropic Josephson couplings realizes a gapless band of metastable vortex solid states with *self-generated* randomness. Each of the metastable states is characterized by a frozen pattern of transverse undulation of vortex stripes which run along the direction of weaker Josephson coupling. The vortex stripes can slide into the direction of stronger Josephson coupling even at zero temperature, eliminating the pinning effects of the underlying lattice structure of the array, but blocked into the direction of weaker coupling. Because of the sliding of the vortex matter, the phase of superconducting order parameter remain disordered at macroscopic level even at zero temperature. This is an extremely strong version of the *spin-chirality decoupling* which should be manifested in electric transport properties.

1. Introduction

Josephson junction arrays (JJA) under the magnetic field [1, 2] provide a very interesting play ground to study consequences of frustration. JJA is a network of superconducting islands connected by Josephson junctions in the form, say, of a square lattice of size $N = L \times L$ as shown in Fig. 1. At low enough temperatures, the amplitude of the superconducting order parameter on each of the island becomes finite and take the same value. However the phases θ_i of the order parameter on the islands $i = 1, 2, \ldots N$ would take any values independently if the Josephson coupling is absent. With the (ferromagnetic) Josephson coupling, the difference of the phases $\theta_i - \theta_j$ between the neighbouring islands i and j become minimized leading to ordering of the phases across the network just like a ferromagnetic XY model. Now if one switches on an external magnetic field of strength B, the JJA becomes a kind of *frustrated crystal* - a crystal which is forced to host a given number density $f = Ba^2/\phi_0$ of dislocations (vorticies in the phase of the superconducting order parameter). Here a^2 is the area of a plaquette and ϕ_0 is the flux quantum. Because of the injected dislocations the crystal is now strongly frustrated.

Recently we found a novel state of matter in a class of frustrated JJA on a square lattice with a) *anisotropic* Josephson coupling into x and y directions (see Fig. 1) and with b) *irrational* number density f of the vortices per plaquette [3, 4]. Here we review the results with particular emphasis on the following two prominent features. Firstly we found a band of metastable vortex solid states with self-generated randomness *in the absence of quenched disorder*. Whether geometrical frustration alone can realize disordered, glassy states in the absence of extrinsic quenched disorder is a long-standing issue[6, 7]. Such a question is relevant in the context of frustrated magnets. Secondly we found an extremely strong realization of an analogue of the *spin-chirality decoupling* [8] found in frustrated magnets. Here the ordering of the "chirality" is translated to a formation of a vortex solid in the JJA. We found that the vortex solid can slide into the direction of stronger Josephson coupling. This should lead to destruction of the long-ranged ordering of the phase of the superconducting order parameter even at zero temperature. In this paper we also present some numerical results of Monte Carlo simulations performed to observe the spin-chirality decoupling at finite temperatures.

In the limit $f \to 1/2$, the system is called fully frustrated XY (FFXY) model [10]. In this case the charge of the vortices take half-integer values $\cdots -3/2, -1/2, 1/2, 3/2, \ldots \cdots$ (see below) and the vortices become literally equivalent to the *chiralities* which play important roles in frustrated magnets (see [8] for a recent review). Recent extensive numerical studies on the FFXY model [11, 12, 13] as well as vectorial spin-glasses [14, 15] show convincingly that the chirality exhibit long-ranged order at a transition temperature T_c above the spin transition temperature $T_s > 0$. Quite remarkably the decoupling can become much more stronger in the present system with *irrational* f in the sense that $T_s = 0$ due to the free sliding of the vortex solid at zero temperature while the vortex solid itself appear to be developed at a finite temperature $T_c > 0$. (See sec.4)

In the next section we define our model. In sec. 3 we review our resent work [4]. In sec. 4 we present numerical results on the "spin-chirality decoupling" at finite temperatures. Finally we conclude this paper in sec. 5 with additional discussions on the possibility of experiments.

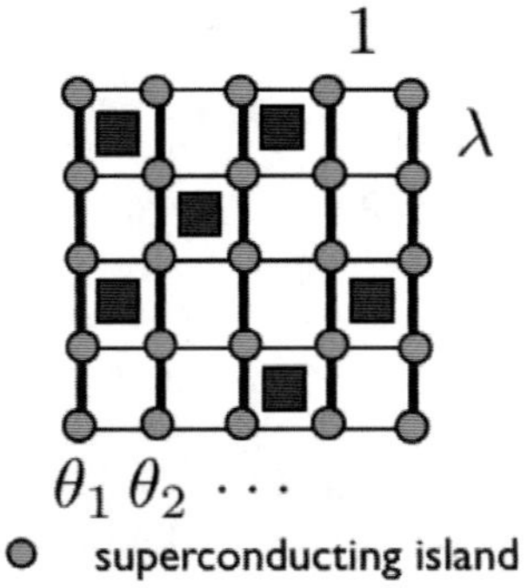

Figure 1. A schematic picture of a Josephson junction array (JJA). The superconducting islands (circles) are connected via Josephson junctions (links). Anisotropic JJA with different strength of the Josephson coupling into x and y directions with the ratio λ can be created in laboratories by lithography techniques with which the spacings between the superconducting islands can be controlled [9]. The squares in the plaquettes represent vortices. (Figure taken from [4])

2. Model

The effective Hamiltonian of the JJA is given by [1],

$$H = -\sum_{<i,j>\|x\text{-axis}} \cos(\psi_{ij}) - \lambda \sum_{<i,j>\|y\text{-axis}} \cos(\psi_{ij}) \tag{1}$$

with the gauge-invariant phase difference, $\psi_{ij} \equiv \theta_i - \theta_j - A_{ij}$. We measure temperature T in a unit with $k_B = 1$. Here λ is the ratio of the strength of the Josephson couplings along x and y-axis. In the following we only examine $\lambda \geq 1$, which is obviously sufficient by symmetry. The vector potential $A_{ij}(= -A_{ji})$ is defined such that directed sum of them around each cell is $2\pi f$.

Vortex charge v_i of the vortex at the plaquette associated with the i-th vertex is defined by taking directed sum of $(\psi_{ij}/2\pi - [\psi_{ij}/2\pi]_n)$ on the junctions around the plaquette. Here $[x]_n$ denotes the nearest integer of the real variable x. It takes values $\ldots, -1-f, -f, 1-f, \ldots$.

Any irrational number f can be approximated by a series of rational numbers p/q with integer p and q. For instance we used a series of rational numbers $p/q = 5/13$, $8/21$, $13/34$, $21/55$, $34/89$, $55/144$, $89/233$ which approximate $(3-\sqrt{5})/2 = 0.38196601\ldots$. This allows us to use the periodic boundary conditions for both x and y directions in systems of size $L \times L$. With this convention the ratio $f = p/q$ converges to the target irrational number in $L \to \infty$ limit.

Journal of Physics: Conference Series **320** (2011) 012024 doi:10.1088/1742-6596/320/1/012024

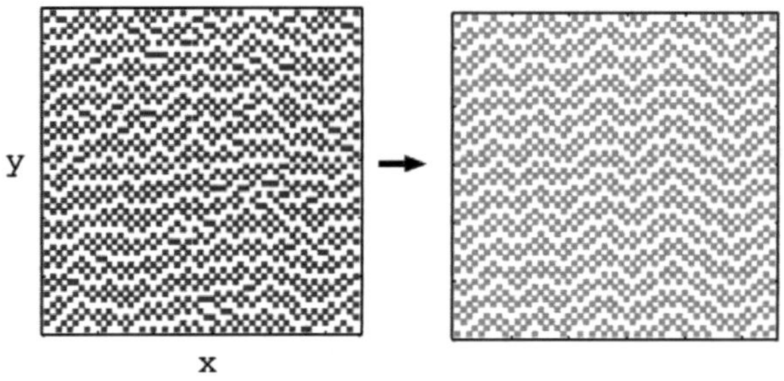

Figure 2. Vortex patterns in an irrationally frustrated JJA under external magnetic field with anisotropic coupling. Here $\lambda = 1.5$ so that the coupling is stronger along y direction. The system size is $L = 55$. A fraction $f = 21/55$, which approximates an irrational number $(3-\sqrt{5})/2 = 0.381966..$, of the plaquettes are occupied by vortices with charge $1-f$ represented by filled squares. The left panel shows an equilibrium vortex pattern at $T = 0.2$ and the right panel displays that at a nearby energy minimum. (Figure taken from [4])

3. Vortex stripes in the strongly anisotropic limit

Let us now review our recent theoretical results [4] focusing on the formation of vortex stripe states with frozen transverse undulations in strongly anisotropic case $\lambda \gg 1$.

In Fig. 2 we show an example of the vortex pattern observed numerically in an equilibrium at finite temperature (left panel) and in a nearly energy minimum (right panel). The equilibrium configuration is generated during a Monte Carlo simulation (see next section for the details). The energy minimum is obtained by an energy descent algorithm starting from the thermalized configuration. Apparently the vortices exhibit stripes running parallel to the direction of weaker coupling. The formation of the stripes itself may be not so surprising since the effective repulsive interactions between the vortices becomes anisotropic with the anisotropic Josephson coupling. Remarkable features are that I) the stripes are disordered by transverse undulations (in the absense of quenched disorder) but that II) the undulated patterns are stacked regularly along the stronger coupling. Starting from different thermalized configurations, different energy minima with different realizations of the stacked undulations can be sampled [4].

In [4], we obtained explicit analytical forms which describe the ground state (vortex stripes without the undulation) and the metastable states (vortex stripes with the undulation). The strategy is to start from an ansatz that the gauge invariant phase differences ψ across the Josephson junctions along the x and y directions connecting neighbouring islands (vertexes) can be written as,

$$\psi_{(x,y)(x+1,y)} = \phi_x[y + \alpha(x)] \qquad \psi_{(x,y)(x,y+1)} = \phi_y[y + \alpha(x)] \tag{2}$$

Here $\phi_x[y]$ and $\phi_y[y]$ are unknown functions defined on the "folded coordinate" $[y] = fy - \text{int}(fy)$ where int(x) is the floor function. The folded coordinate takes values limited in the range $0 \leq [y] < 1$. The crucial point is that if f is *irrational*, the vertexes of the JJA *uniformly* fill the entire range of the folded coordinate $[y]$ in the limit $N \to \infty$ (even with, e. g., $f = 1/2 + \epsilon$ with *infinitesimal irrational* ϵ). Thus we can treat $[y]$ as a *continuous* variable. Quite amusingly the set of functions can be regarded as a two dimensional generalization of the so called hull function which play central roles in the analysis of the sliding/jamming in the one-dimensional Frenkel-Kotorova model extensively studied by S. Aubry and coworkers [16] and related one-dimensional friction models such as the Matsukawa-Fukuyama model [17, 18]. (See [5] for a review on the relation among the seemingly unrelated problems.)

The functions $\phi_x[y]$ and $\phi_y[y]$ obey two equations(see [4] for the details). One follows from the constraint that directed sum over ψ_{ij} around each plaquette must be $-2\pi f$. The other follows from the conservation of the Josephson currents at each vertex (force balance condition) in each energy minimum. By a perturbative analysis in series of $1/\lambda$ starting from infinite anisotropy limit $\lambda = \infty$, we solved the equations and computed the functions $\phi_x[y]$ and $\phi_y[y]$ explicitly.

Now let us focus on the parameter $\alpha(x)$ which appears in Eq. (2). Physically it describes the transverse undulation of vortices such as the one shown in the right panel of Fig. 2. In the ground state we find $\alpha(x)$ does not depend on x, i.e. $\alpha(x) = \text{const}$ and the solution for the hull functions $\phi_x[y]$ and $\phi_y[y]$ describe a simple horizontal stripe of vortices running along the direction of weaker coupling.

Quite remarkably we found a continuous spectrum of solutions with non-zero transverse undulation parametrized by an *arbitrary* continuous function $\alpha(x)$. By computing their energies we find that they constitute a continuous, gapless band of the energy. Although they have higher energies than the ground state, they must be regarded as *metastable* states since they themselves satisfy the force balance (current conservation) condition. Thus the irrationally frustrated JJA admits a continuous, gapless spectrum metastable states with self-generated randomness.

A very important consequence of the existence of the analytic hull functions $\phi_x[y]$ and $\phi_y[y]$ is the *sliding*. Starting from any metastable state parametrized by an arbitrary continuous function $\alpha(x)$ (including the ground state $\alpha(x) = \text{const}$), there is a flat band of infinitely many states with exactly the same energy which can be obtained by a uniform translation $\alpha(x) \to \alpha(x) + \Delta\alpha$ with arbitrary shift $\Delta\alpha$. Physically this means sliding of the vortices into the direction of stronger coupling. This would not be surprising at all in the case of clean bulk superconductors without any pinning centers [1]. However it must be recalled that there is an underlying lattice structure in the present system due to the array itself which naturally gives rise to pinning effects. Indeed there are non-zero critical currents in the cases of *rational* f [19]. The presence of the sliding mode across the flat band of the states mean that this pinning effect is effectively cancelled out in the case of *irrational* f. Moreover we must note that the sliding mode is present only along the direction of stronger coupling and absent along the weaker coupling. Thus the vortex solid can flow without energy dissipation along the stronger coupling but completely blocked along the weaker coupling. This strong anisotropy of the mobility of the vortex solid should be directly reflected in the transport properties of the system as we discuss in the end of this paper.

4. "Spin-chirality decoupling"

Sliding of the vortices discussed at the end of the preceding section immediately means extremely strong spin-chirality decoupling. The point is that configuration of the phase (spin) θ_i is not fixed only by the structure of the vortices. The uniform translation of the vortices along one axis changes the phase differences across the system along the orthogonal axis. The presence of the sliding mode along the stronger coupling means the phase differences along the weaker coupling can be changed by an arbitrary amount. The phase cannot establish long-ranged order even at zero temperature.

Now let us present some numerical results to look for the signature of the spin-chirality decoupling at finite temperatures. We performed Monte Carlo (MC) simulations on systems with $L = 13 - 89$ using $20 - 120$ temperatures in the temperature range $T = 0.2 - 0.4$ by the exchange MC method [20]. More precisely each MC step consists of one sweep over the system by the Metropolis updates of each phase θ_i, one step of the over-relaxation [21] and one step of the exchange MC. We used $10^5 - 10^6$ MC steps for the equilibration and observations.

In Fig. 3 we display the scaled correlation length ξ_V/L of the vortices measured by the same method as in [22]. (See [22] for the details of the method, which is a standard technique for spin-glasses [14, 15].) Here the correlation length along the direction of stronger coupling is shown.

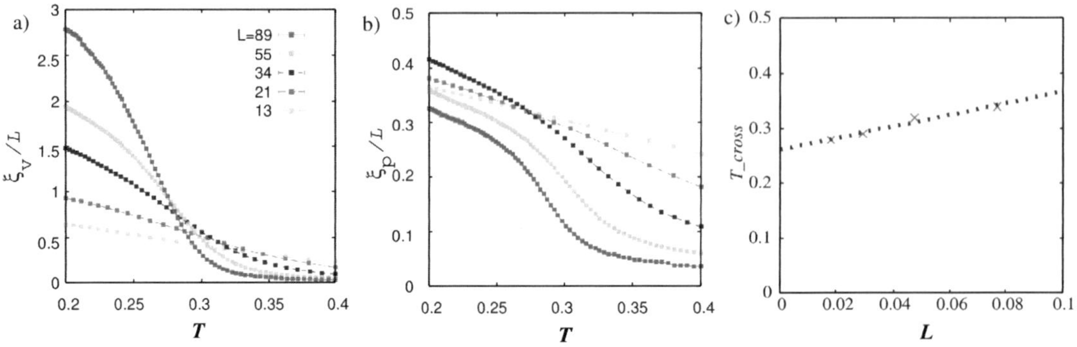

Figure 3. Scaled correlation lengths in equilibrium obserbed by Monte Carlo simulations. a) Correlation length of the vortex ξ_V and b) phase ξ_P. Here $\lambda = 1.75$ for which we estimate the critical temperature of the vortices $T_c \sim 0.26$ by the linear fit shown in c).

The scaled correlation length of vortices ξ_V/L of different sizes exhibit crossings suggesting a 2nd order phase transition at some finite critical temperature T_c. To estimate the critical temperature T_c, we extracted the crossing temperature $T_{\rm cross}(L)$ of the scaled correlation length of the vortices ξ_V between two system sizes $(L, L') = (13, 21), (21, 34), (34, 55)$ and $(55, 89)$. By fitting the crossing temperatures $T_{\rm cross}(L)$ to a fitting formula $T_{\rm cross}(L) = T_c + {\rm const}/L$ we estimated the critical temperature $T_c(\lambda)$. As shown in the panel c) of Fig. 3, we find non-zero critical temperature $T_c(\lambda = 1.75) \sim 2.6$. This is at marked variance with the isotropic $\lambda = 1$ case [22] in which $\xi_V \sim T^{-2.2}$ suggesting $T_c(\lambda = 1) = 0$.

On the other hand the scaled correlation length ξ_p of the phases exhibit no hint of phase transition at finite temperatures, as expected from the analysis on the low lying states discussed in the preceding section. To sum up, we find clear numerical evidences that the vortex (chiral) solid state appears at a finite temperature $T_c(\lambda) > 0$ while the phase (spin) remains disordered at finite temperatures, i.e. $T_s = 0$.

5. Discussions

To conclude we presented a review on our recent theoretical results on the irrationally frustrated anisotropic Josephson junction arrays. In particular we forcused on the nature of the low lying metastable states which exhibit several novel features, i.e. self-generated randomness without quenched disorder, sliding and spin-chirality decoupling. We also presented a set of numerical results which strongly suggests formation of the vortex (chiral) solid state a a finite temperature $T_c > 0$ while the phase (spin) remains disordered at finite temperature $T_s = 0$. This is an extremely strong form of the spin-chirality decoupling. We plan to present more details of the numerical simulations at finite temperatures elsewhere.

Finally let us discuss possible experimental measurements concerning our problem. In [3] we reported numerical simulations of the transport properties of the present system by the method of RCSJ (Resistively and Capacitively Shunted Junction) model performed at zero temperature. We found clear evidences of sliding/jamming of the vortex solid along the direction of stronger/weaker coupling. This agrees well with the analysis of the low lying states discussed in sec. 3. Thus the standard transport measurement, i.e characterization of the current-voltage relation, should yield superconducting/normal behavior with respect to the electric current injected along the direction of stronger/weaker coupling. The coexistence of the macroscopic superconducting/normal behaviour would be a proof the spin (phase) - chirality (vortex) decoupling. Let us recal again that anisotropic JJAs can be created in laboratory by standard lithography techniques [9].

Similar *frustrated crystals* in which the frustration is brought into the system in the form of gauge field would be found in also other systems. An example is the charge density wave system on ring crystal [23, 24]. Another example is oxide high-Tc superconductors under external magnetic field paralle to the ab -plane [25]. It would be very interesting to investigate these systems from the view point similar to our works.

Acknowledgement We gratefully acknowledge Hikaru Kawamura and Hiroshi Matsukawa for interesting discussions over many years on related problems which inspired our works. We thank Toshihito Osada for useful discussions on the experiments on JJAs. We thank the Supercomputer Center, ISSP, University of Tokyo for the use of the facilities. This work is supported by Grant-in-Aid for Scientific Research on Priority Areas "Novel States of Matter Induced by Frustration" (1905200*) and Grant-in-Aid for Scientific Research (C) (21540386).

References

[1] Tinkham M. 2004 *Introduction to Superconductivity*, Courier Dover Publications.
[2] van der Zant H. S. J. , Rijken H. A. , and Mooij J. E. 1991 J. Low. Temp. Phys., **82** 67.
[3] Yoshino H., Nogawa T. and Kim Bongsoo 2009 *New J. Phys.* **11** 013010.
[4] Yoshino H., Nogawa T. and Kim Bongsoo 2010 *Phys. Rev. Lett.* **105** 257004.
[5] Yoshino H., Nogawa T. and Kim Bongsoo 2010 *Prog. Theor. Phys. Supplement* **184** 153.
[6] Sadoc J. F. Sadoc, Mosseri R. 1999 *Geometrical frustration*, Cambridge University Press.
[7] Tarjus G., Kivelson S. A., Nussinov Z. and Viot P. 2005 *J. Phys. Condens. Matters* **17** R 1143.
[8] Kawamura H. 2010 *J. of Phys. Soc. Jpn.* **79** 011007.
[9] Saito S. and Osada T. 2000, *Physica B: Condensed Matter* **284-288** 614.
[10] Teitel S. and Jayaprakash C. 1983, *Phys. Rev.* **B** 27, 598.
[11] Hasenbusch M., Pelissetto A. and Vicari E. 2005, *Phys. Rev.* **B** 72, 184502.
[12] Hasenbusch M., Pelissetto A. and Vicari E. 2005, *J. Stat. Mech.*, P12002.
[13] Okumura S. and Yoshino H. and Kawamura H., arXiv:1008.0026.
[14] Viet D. X. and Kawamura H. 2009 Phys. Rev. Lett. **102** 027202.
[15] Viet D. X. and Kawamura H. 2010 Phys. Rev. Lett. **05** 097206.
[16] Peyard M. and Aubry S. 1983 *J. Phys. C* **16** 1593.
[17] Matsukawa H. and Fukuyama H. 1994 *Phys. Rev B.* **49**, 17286.
[18] Kawaguchi T. and Matsukawa H. 1997 *Phys. Rev. B* **56** 13932.
[19] Teitel S. and Jayaprakash C. 1983 *Phys. Rev. Lett.* **51** 1999.
[20] Hukushima K. and Nemoto K. 1996 *J. Phys. Soc. Jpn* **65** 1604.
[21] Creutz M. 1987, *Phys. Rev. D* **36**, 515.
[22] Granato E. 2008 *Phys. Rev. Lett.* **101** 027004.
[23] Tanda S., Tsuneta T., Okajima Y., Inagaki K., Ya-aya K. and Hatakenaka N. 2002 *Nature* **417** 397.
[24] Nogawa T. and Nemoto K. 2006, *Phys. Rev. B* **73** 184504.
[25] Hu X. and Tachiki M. 1998 *Phys. Rev. Lett.* **80** 4044.

Dynamical Properties of Potts Model with Invisible States

Shu Tanaka

Research Center for Quantum Computing, Interdisciplinary Graduate School of Science and Engineering, Kinki University. 3-4-1, Kowakae, Higashi-Osaka, Osaka 577-8502, Japan

E-mail: shu-t@alice.math.kindai.ac.jp

Ryo Tamura

Institute for Solid State Physics, University of Tokyo. 5-1-5, Kashiwanoha Kashiwa, Chiba 277-8581, Japan

Abstract. We study dynamic behavior of Potts model with invisible states near the rst-order phase transition temperature. This model is regarded as a standard model to analyse nature of phase transition. We can control the energy barrier between the ordered state and disordered state without changing the symmetry which breaks at the transition point. We focus on melting process starting from the perfect ordered state. We calculate time-dependency of the order parameter, density of invisible state, and internal energy. All of them show two-step relaxation behavior. We also analyze the relationship between the characteristic melting time and characteristic scale of the energy barrier by changing the number of invisible states. We nd that characteristic melting time increases as the energy barrier enlarges in this model. This model is regarded as a fundamental model to analyze dynamic behavior near the rst-order phase transition point.

1. Introduction

Frustration causes many interesting static and dynamic behavior which are not observed in unfrustrated systems because of peculiar density of states [1–10]. In two-dimensional frustrated systems, there have been found many nontrivial phase transitions such as order by disorder [11, 12], reentrant phase transition [13–17], topological phase transition [18], and novel type of rst-order phase transition. Recently, strange rst-order phase transitions have been found in two-dimensional frustrated continuous spin systems [19–22].

In [19], the authors studied equilibrium properties of the classical Heisenberg model on triangular lattice with nearest neighbor ferromagnetic interaction J_1 and third-nearest neighbor antiferromagnetic interaction J_3. They found that a rst-order phase transition with threefold symmetry breaking occurs at nite temperature. This looks a strange phase transition, since phase transition with threefold symmetry breaking is often second-order phase transition on two-dimensional lattice *e.g.* the three-state ferromagnetic Potts model [23]. After this study, similar nature of rst-order phase transition have been found by a number of researchers [20–22]. Stoudenmire *et al.* found a rst-order phase transition with breaking of threefold symmetry in J_1 J_3 model with biquadratic interaction on triangular lattice. Okumura *et al.* also found similar rst-order phase transition in frustrated J_1 J_2 model on hexagonal lattice. Such a rst-order phase transition sometimes takes place in two-dimensional frustrated systems with a number of competed short-range interactions. It is an open problem why such a rst-order phase transition appears in some two-dimensional frustrated systems.

To consider this problem, we constructed a model which exhibits a rst-order phase transition with threefold symmetry breaking by introducing a new kind of degree of freedom into the standard ferromagnetic three-state Potts model [24, 25]. We introduced the invisible states which does not contribute to the internal energy. The ground state of the Potts model with invisible state and its degeneracy are the same as that of the standard Potts model. We found that the rst-order phase transition with threefold symmetry breaking occurs by just adding the invisible states into the standard three-state ferromagnetic Potts model. The standard Potts model has been regarded as a fundamental model to consider the properties of phase transition in statistical physics [23,26]. It is believed that the Potts model with invisible states is a potential model which can clarify inherent nature of rst-order phase transition with breaking of threefold symmetry on two-dimensional lattice.

Our purpose of the present study is to clarify dynamic properties of Potts model with invisible states. It is an interesting topic in statistical physics to study dynamical nature of the systems which exhibit a rst-order phase transition, since some similar points between these systems and glassy systems have been pointed out [27,28]. It is expected that some dynamic nature in glassy systems can be explained in terms of dynamic behavior of the systems in which a rst-order phase transition occurs. Potts model with invisible states can be considered a potential model to study dynamical nature of rst-order phase transition. This is because we can control the energy barrier between the ordered state and disordered state without changing the symmetry which breaks at the transition temperature by modulating the number of invisible states. In this paper, we focus on melting process, then, we study dynamics of the Potts model with invisible states at xed temperatures which are above the transition temperature.

2. Model

We consider the ferromagnetic Potts model with invisible states on square lattice. The Hamiltonian of this model is

$$\mathcal{H} = \ J \sum_{\langle i,j\rangle} \delta_{\sigma_i,\sigma_j} \sum_{=1}^{q} \delta_{\sigma_i,}\ , \quad (J>0), \quad \sigma_i = 1, \quad ,q,q+1, \quad ,q+r, \tag{1}$$

where $\langle i,j\rangle$ denotes the nearest neighbor pairs on square lattice. In this paper we take J as an energy unit. If and only if $1 \le \sigma_i = \sigma_j \le q$, the interaction works. Obviously this model for $r=0$ corresponds to the standard q-state ferromagnetic Potts model. We call the state in $1 \le \sigma_i \le q$ "colored state" and the state in $q+1 \le \sigma_i \le q+r$ "invisible state". Hereafter we call this model (q,r)-state Potts model. The invisible state does not a ect the internal energy. It should be noted that the number of ground states of the (q,r)-state Potts model is the same as that of the standard ferromagnetic q-state Potts model. Then the (q,r)-state Potts model exhibits a phase transition with q-fold symmetry breaking. To change the number of invisible states r corresponds to changing form of the density of states. The invisible states contribute the entropy as $\log r$. Then, it is expected that the order of phase transition can be changed by adding the invisible states. Actually, a rst-order phase transition with q-fold symmetry breaking occurs for large enough the number of invisible states r even for $q = 2, 3$, and 4 on two-dimensional lattice [24]. As the number of the invisible states r increases, the latent heat increases and the transition temperature decreases [25].

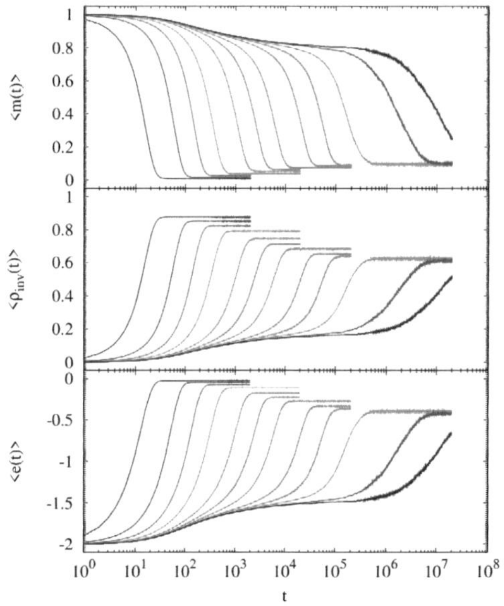

Figure 1. Time-evolution of the order parameter $\langle m(t)\rangle$, density of the invisible states $\langle \rho_{\rm inv}(t)\rangle$, and the internal energy $\langle e(t)\rangle$ of the (3,27)-state Potts model starting from the completely ordered state $\sigma_i(0) = 1$ (for all i) for xed di erent temperatures. From left to right, the temperatures are $T = T_{\rm c} + 0.5$, $T_{\rm c} + 0.2$, $T_{\rm c} + 0.1$, $T_{\rm c} + 0.05$, $T_{\rm c} + 0.02$, $T_{\rm c} + 0.01$, $T_{\rm c} + 0.005$, $T_{\rm c} + 0.002$, $T_{\rm c} + 0.001$, $T_{\rm c} + 0.0005$, $T_{\rm c} + 0.0002$, and $T_{\rm c} + 0.0001$, where $T_{\rm c} = 0.58513$ [25].

3. Result

In this paper, we study melting dynamics of the (q,r)-state Potts model on square lattice whose size is $N = 128 \quad 128$ at xed temperatures which are above the transition temperature following [28]. We impose periodic boundary condition. We prepare independent 1024 samples for obtaining data. It should be noted that since the width of error bars obtained from 1024 samples is half the size of that obtained from 256 samples, it is enough to consider a statistic error by using 1024 samples. The initial state is set to be perfect ordered state such as $\sigma_i = 1$ for all i. We adopt single-spin- ip Metropolis type of Monte Carlo method as the time-evolution rule.

To consider dynamic behavior of the (q,r)-state Potts model, we de ne the time-dependent order parameter $m(t)$, density of invisible states $\rho_{\rm inv}(t)$, and internal energy $e(t)$ as follows:

$$m(t) = \frac{(q+r)\sum_{i=1}^{N}\left[\delta_{\sigma_i(t),1} \quad \frac{1}{q+r}\right]}{N(q+r \quad 1)}, \tag{2}$$

$$\rho_{\rm inv}(t) = \frac{1}{N}\sum_{i=1}^{N}\sum_{=q+1}^{q+r}\delta_{\sigma_i(t),}, \tag{3}$$

$$e(t) = \quad \frac{J}{N}\sum_{\langle i,j\rangle}\delta_{\sigma_i(t),\sigma_j(t)}\sum_{=1}^{q}\delta_{\sigma_i(t),}, \tag{4}$$

where t denotes Monte Carlo step and $\sigma_i(t)$ is the state at time t. Figure 1 shows the time-evolution of the order parameter $\langle m(t)\rangle$, density of the invisible states $\langle \rho_{\rm inv}(t)\rangle$, and the internal

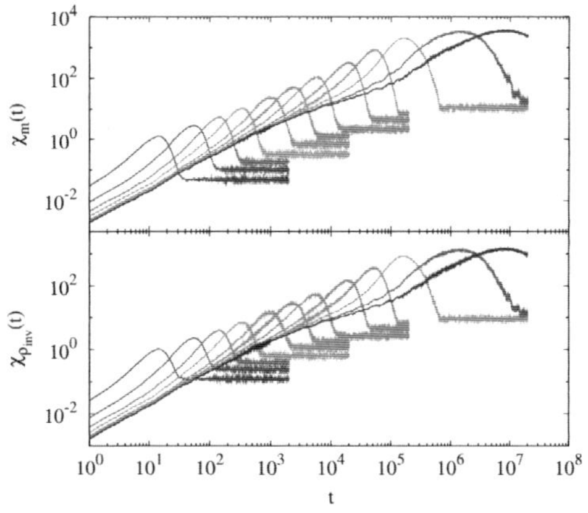

Figure 2. Dynamical susceptibility of the order parameter $\chi_m(t)$ and density of invisible states $\chi_{\rho_{\rm inv}}(t)$ of the (3,27)-state Potts model starting from the completely ordered state $\sigma_i(0) = 1$ (for all i) for xed di erent temperatures. From left to right, the temperatures are $T = T_{\rm c} + 0.5$, $T_{\rm c} + 0.2$, $T_{\rm c} + 0.1$, $T_{\rm c} + 0.05$, $T_{\rm c} + 0.02$, $T_{\rm c} + 0.01$, $T_{\rm c} + 0.005$, $T_{\rm c} + 0.002$, $T_{\rm c} + 0.001$, $T_{\rm c} + 0.0005$, $T_{\rm c} + 0.0002$, and $T_{\rm c} + 0.0001$, where $T_{\rm c} = 0.58513$ [25].

energy $\langle e(t) \rangle$ of the (3,27)-state Potts model starting from the completely ordered state $\sigma_i(0) = 1$ (for all i) for xed di erent temperatures. Here $\langle \ \rangle$ denotes an ensemble average. In Fig. 1, there are obvious two-step relaxations near the transition temperature. This nature comes from a rst-order phase transition. Two-step relaxation often appears in systems where a rst-order phase transition takes place and also in glassy systems [27–29]. Melting occurs after stabilization in the ordered state which corresponds to plateau region, in other words, rst-step relaxation. As the temperature approaches to the transition point, the plateau region enlarges and as a result, the characteristic melting time becomes long.

Next we consider the dynamical susceptibility to consider the characteristic melting time systematically. The dynamical susceptibility of the time-dependent physical quantity $A(t)$ is de ned as

$$\chi_A(t) = N\beta(\langle A(t)^2 \rangle \quad \langle A(t) \rangle^2), \tag{5}$$

where β denotes the inverse temperature. We calculate the dynamical susceptibility of the order parameter $\chi_m(t)$ and density of the invisible states $\chi_{\rho_{\rm inv}}(t)$. Figure 2 shows the dynamical susceptibility $\chi_m(t)$ and $\chi_{\rho_{\rm inv}}(t)$ starting from the completely ordered state $\sigma_i(0) = 1$ (for all i) for xed di erent temperatures of the (3,27)-state Potts model. In Fig. 2, the dynamical susceptibilities have a peak, and the peak position indicates characteristic melting time. Characteristic melting time de ned from $\chi_m(t)$ and $\chi_{\rho_{\rm inv}}(t)$ are almost same as shown in Fig. 2. Then we de ne $\tau_{\rm max}$ as the peak position of dynamical susceptibility of the order parameter $\chi_m(t)$. As the temperature decreases, the peak height increases. This result means that characteristic length scale becomes large as the temperature approaches to the transition point, since the peak height relates to characteristic length scale. The characteristic melting time $\tau_{\rm max}$ also increases as the temperature decreases. The behavior of the dynamical susceptibilities is quantitatively similar with that of χ_4 which is often used in analysis of glassy systems [28].

We also study temperature-dependency of $\tau_{\rm max}$ for the (3,25)-state Potts model and (3,27)-state Potts model to consider the e ect of the number of invisible states. Figure 3 shows the

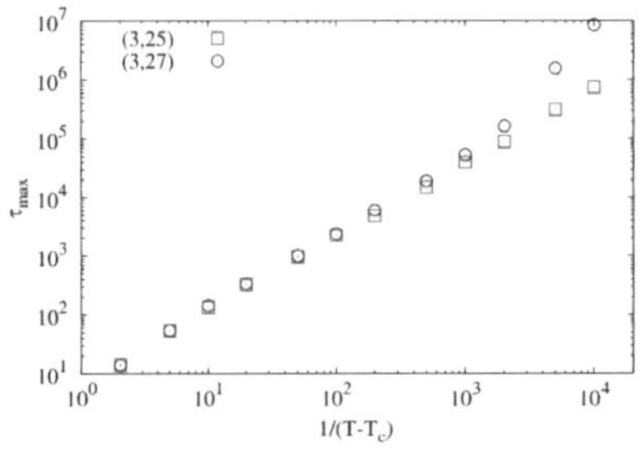

Figure 3. Peak position of the dynamical susceptibility of the order parameter $\tau_{\rm max}$ as a function of $1/(T \quad T_{\rm c})$, where $T_{\rm c}$ is the transition temperature. The red squares and blue circles indicate the case for the (3,25)-state Potts model and the (3,27)-state Potts model, respectively. Transition temperatures for the (3,25)-state Potts model and the (3,27)-state Potts model are 0.59630 and 0.58513, respectively [25].

characteristic melting time as a function of $1/(T \quad T_{\rm c})$, where $T_{\rm c}$ denotes transition temperature. The transition temperatures for (q,r)=(3,25) and (3,27) are 0.59630 and 0.58513, respectively [25]. In high temperature region, behavior of $\tau_{\rm max}$ is almost same both for (3,25)-state Potts model and for (3,27)-state Potts model. However, there is a di erence between the behavior of $\tau_{\rm max}$ for (3,25)-state Potts model and that of (3,27)-state Potts model near the transition temperature. The characteristic time $\tau_{\rm max}$ for the (3,27)-state Potts model is larger than that for the (3,25)-state Potts model, since the energy barrier between the ordered state and paramagnetic state for the (3,27)-state Potts model is larger than that for the (3,25)-state Potts model [24, 25]. From this result, we expect that $\tau_{\rm max}$ enlarges as the number of invisible state r increases. Thus, in the the Potts model with invisible states, we can change the time-scale and length-scale of rst-order phase transition without changing the symmetry which breaks at the transition point. Since the time-scale and length-scale are very important physical properties for dynamic nature, the Potts model with invisible states can be regarded as a standard model to analyze dynamic behavior of systems which exhibits a rst-order phase transition.

Before the conclusion, we should mention the size-dependency of melting dynamics. At high temperature, there is almost no size-dependency. On the other hand, size-dependence clearly appears in the second-step relaxation near the transition temperature whereas the rst-step relaxation does not depend on system size. From this result, the rst-step relaxation is just a local event, however, the second-step relaxation comes from cooperative phenomena. Note that it is de nitely consistent that the characteristic melting time $\tau_{\rm max}$ becomes large, as the number of invisible states r increases for xed lattice sizes. This size-dependency is very complicated and it will be reported elsewhere [30].

4. Conclusion and Future Perspective

We studied dynamic properties of the Potts model with invisible states at xed temperatures which are above the transition temperature. In this paper, we focused on melting process which is an important nature of rst-order phase transitions. First, we considered the dynamical properties of the order parameter, density of invisible states, and the internal energy. We found that there is a two-step relaxation which is typical behavior in systems which exhibit a rst-order phase transition and also glassy systems. Next we calculated the dynamical susceptibility of the order parameter and density of invisible states. As the temperature approaches the transition temperature, the peak height of the dynamical susceptibility grows. Furthermore, we studied the relation between the peak position of the dynamical susceptibility of the order parameter

$\tau_{\max}$ and temperature. We found that $\tau_{\max}$ depends on the energy barrier between the ordered state and paramagnetic state. We also found that in high temperature region, r-independent nature of $\tau_{\max}$, whereas $\tau_{\max}$ obviously depends on r near the transition temperature.

In the Potts model with invisible states, we can control the energy barrier between the ordered state and paramagnetic state without changing the symmetry which breaks at the transition point. We should study the relation between the energy barrier and the time-scale and length-scale in this model more carefully. Mori and Nakada have studied dynamical properties in this model systematically to clarify whether there is new *universal* behavior near the rst-order phase transition point [31]. It is also an important topic to clarify whether the Potts model with invisible states relates to strange type of rst-order phase transition in two-dimensional frustrated systems. It will be reported elsewhere [30].

Acknowledgments

The authors are grateful to Naoki Kawashima, Jie Lou, Yoshiki Matsuda, Seiji Miyashita, Takashi Mori, Yohsuke Murase, Taro Nakada, Masayuki Ohzeki, and Eric Vincent for their valuable comments. S.T. is partly supported by Grant-in-Aid for Young Scientists Start-up (21840021) from the JSPS, Grant-in-Aid for Scienti c Research (B) (22340111), and the "Open Research Center" Project for Private Universities: matching fund subsidy from MEXT. R.T. is partly supported by Global COE Program "the Physical Sciences Frontier", MEXT, Japan. The computation in the present work was performed on computers at the Supercomputer Center, Institute for Solid State Physics, University of Tokyo.

References

[1] Toulouse G 1977 *Commun. Phys.* (London) **2** 115
[2] Liebmann R 1986 *Statistical Mechanics of Periodic Frustrated Ising Systems* (Springer-Verlag, Berlin/Heidelberg GmbH, Heidelberg)
[3] Kawamura H 1998 *J. Phys.: Condens. Matter* **10** 4707
[4] Diep H T 2005 *Frustrated Spin Systems* (World Scienti c, Singapore)
[5] Moessner R, Sondhi S L and Cahndra P 2000 *Phys. Rev. Lett.* **84** 4457
[6] Tanaka S and Miyashita S 2009 *J. Phys. Soc. Jpn.* **78** 084002
[7] Matsuda Y, Nishimori H and Katzgraber H G 2009 *New. J. Phys.* **11** 073021
[8] Tanaka S, Hirano M and Miyashita S 2010 *Lecture Note in Physics* **802** 215
[9] Tanaka S and Miyashita S 2010 *Phys. Rev.* E **81** 051138
[10] Tanaka S, Hirano M and Miyashita S 2011 *Physica* E **43** 766
[11] Villain J, Bidaux R, Carton J and Conte R 1980 *J. Phys.* (Paris) **41** 1263
[12] Henley C L 1989 *Phys. Rev. Lett.* **62** 2056
[13] Nakano H 1968 *Prog. Theor. Phys.* **39** 1121
[14] Syozi I 1968 *Prog. Theory. Phys.* **39** 1367
[15] Fradkin E H and Eggarter T P 1976 *Phys. Rev.* A **14** 495
[16] Tanaka S and Miyashita S 2005 *Prog. Theor. Phys. Suppl.* **157** 34
[17] Miyashita S, Tanaka S and Hirano M 2007 *J. Phys. Soc. Jpn.* **76** 083001
[18] Kawamura H and Miyashita S 1984 *J. Phys. Soc. Jpn.* **53** 4138
[19] Tamura R and Kawashima N 2008 *J. Phys. Soc. Jpn.* **77** 103002
[20] Stoudenmire E M, Trebst S and Balents L 2009 *Phys. Rev.* B **79** 214436
[21] Okumura S, Kawamura H, Okubo T and Motome Y 2010 *J. Phys. Soc. Jpn.* **79** 114706
[22] Tamura R and Kawashima N in preparation.
[23] Wu F Y 1982 *Rev. Mod. Phys.* **54** 235
[24] Tamura R, Tanaka S and Kawashima N 2010 *Prog. Theor. Phys.* **124** 381
[25] Tanaka S, Tamura R and Kawashima N in preparation.
[26] Potts R B 1952 *Proc. Cambridge Philos. Soc.* **48** 106
[27] Krzakala F and Zdeborová L arXiv:1006.2479.
[28] Krzakala F and Zdeborová L arXiv:1006.2480.
[29] Tanaka S and Miyashita S 2007 *J. Phys. Soc. Jpn.* **76** 103001
[30] Tanaka S and Tamura R in preparation.
[31] Mori T and Nakada T in preparation.

Journal of Physics: Conference Series **320** (2011) 012026 doi:10.1088/1742-6596/320/1/012026

High field ESR measurements of S=1/2 low dimensional antiferromagnet (2,3-dmpyH)$_2$CuBr$_4$

Hitoshi Ohta[1,2], Tatsuya Yamasaki[2], Susumu Okubo[1], Takahiro Sakurai[3], Masashi Fujisawa[4], Hikomitsu Kikuchi[4]

[1]Molecular Photoscience Research Center, Kobe University, Kobe, Japan
[2]Graduate School of Science, Kobe University, Kobe, Japan
[3]Center for Supports to Research and Education Activities, Kobe University, Kobe, Japan
[4]Department of Applied Physics, University of Fukui, Fukui, Japan

E-mail: hohta@kobe-u.ac.jp

Abstract. High field ESR measurements of (2,3-dmpyH)$_2$CuBr$_4$ powder sample have been performed in the frequency range from 60 to 315 GHz using the pulsed magnetic field up to 16 T. Observed temperature range is 1.8 to 265 K. ESR spectrum observed at 265 K shows a typical powder pattern spectrum of Cu^{2+} ion, and obtained g-values are $g_{//}$=2.307 and $g_{\perp}$=2.053. Temperature dependence measurements are also performed at 80, 160 and 315 GHz, and almost no temperature dependence of g-value is observed for all frequencies down to 1.8 K. These results are rather unusual for the low dimensional strongly correlated spin system, and possible mechanisms by spin frustration is suggested.

1. Introduction

Spin dynamics in low dimensional antiferromagnet have attracted much interest and have been studied intensively by electron spin resonance (ESR). Spin dynamics of one-dimensional (1D) Heisenberg antiferromagnet are well studied and they typically show the g-shift [1] and the broadening of linewidth [2] at low temperature due to the development of short-range order. However, low dimensional antiferromagnets with frustration show rich variety of spin dynamics. The S=1/2 diamond chain antiferromagnet $Cu_3(CO_3)_2(OH)_2$ (azurite) shows a large g-shift with small anomaly of linewidth at low temperature [3], while almost no g-shift is observed in S=1/2 zig-zag chain antiferromagnets Cu(ampy)Br_2 [4] and (VO)(SO_4)(2,2-bpy) [5]. In case of S=1/2 kagome lattice antiferromagnets, $ZnCu_3(OH)_6Cl_2$ (herbertsmithite) with some site exchanges between Cu^{2+} and Zn^{2+} ions shows almost no temperature dependence of g-value and linewidth [6,7], and $Cu_3V_2O_7(OH)_2 2H_2O$ (volborthite) with slightly distorted kagome lattice shows complicated temperature dependence of g-value and linewidth [8-10], while $BaCu_3V_2O_8(OH)_2$ (vesigniete) with nearly ideal kagome lattice shows g-shift and broadening of linewidth at low temperature [10,11]. In this paper we will deal with S=1/2 two-leg ladder antiferromagnet (2,3-dmpyH)$_2$CuBr$_4$, which has a dimensionality in between 1D and 2D, and a possibility of frustration in the system will be suggested. Fig. 1 shows the crystal structure of (2,3-dmpyH)$_2$CuBr$_4$ which has a monoclinic symmetry with the space group $P2(1)/n$ [12]. Lattice parameters are a = 7.5037 Å, b = 31.613 Å, c = 8.2016 Å, β = 98.972°. Magnetic ions Cu^{2+} ($3d^9$, S =

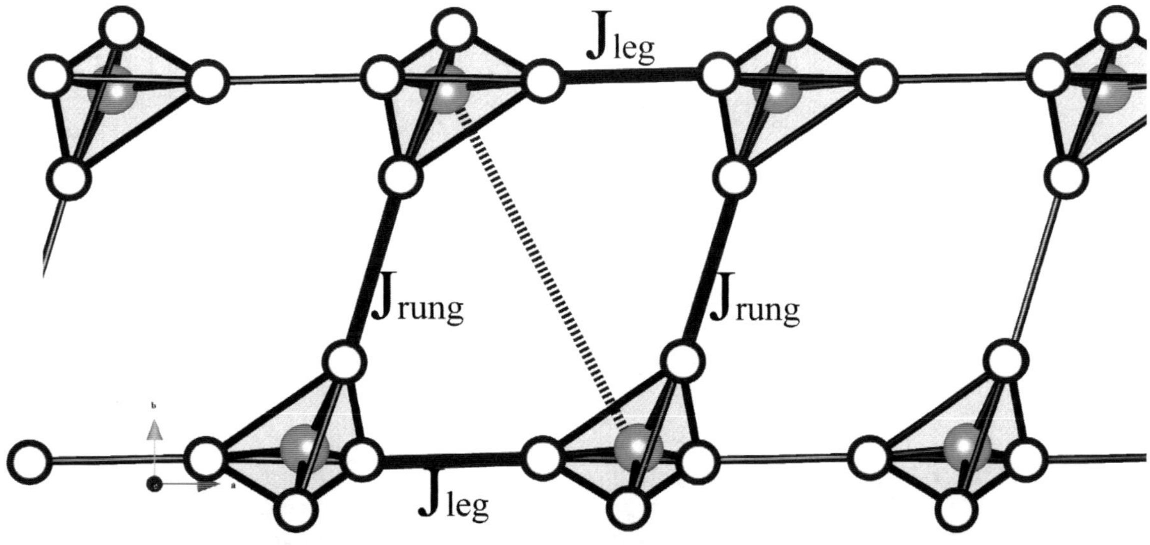

Figure 1. Crystal structure of $(2,3\text{-dmpyH})_2CuBr_4$. Cu2+ ions are shown by solid circles. The leg direction is the a-axis. The dashed line shows the possible antiferromagnetic exchange interaction which will cause the spin frustration.

1/2) form the spin-ladder along the a-axis as shown in Fig.1. Cu^{2+} ion is situated in the tetrahedral ligand, which consists of four Br^- ions, and it is elongated along the *z*-axis as shown in Fig. 2. The magnetic susceptibility as a function of temperature shows a maximum at 11 K, which is a characteristic of a low dimensional Heisenberg antiferromagnet, and decreases rapidly to zero below 11 K, suggesting a singlet ground state [12]. The data analyses by a phenomenological strong-leg exchange model suggest antiferrmagnetic exchange interactions J_{leg}=-8.42 K and J_{rung}=-4.34 K. From first principles theoretical calculations using the crystal structure at 88 K, Shapiro *et al.* suggest that other exchange interactions are negligible [12] but the coincidence between the calculated magnetic susceptibility by J_{leg} and J_{rung} obtained by this first principles calculations and the experimental result is rather poor. Therefore, we think that the existence of other exchange interactions remains.

In order to obtain information about the spin dynamics of $(2,3\text{-dmpyH})_2CuBr_4$, we have performed high field ESR measurements.

2. Experimental

High-field ESR measurements have been performed using pulsed magnetic fields up to 16T in the

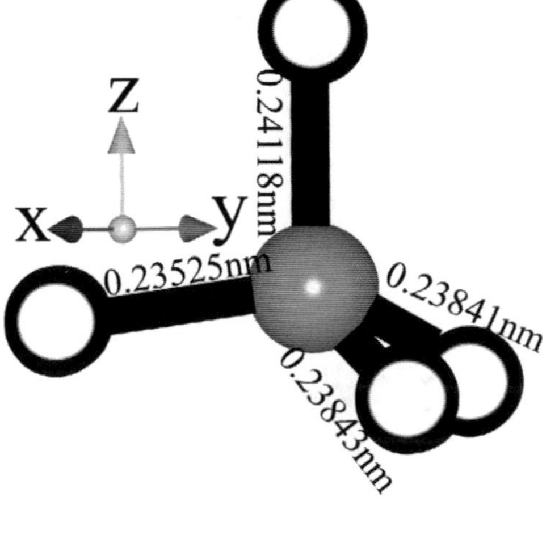

Figure 2. Ligand structure of Cu^{2+} site. Solid and open circles correspond to Cu^{2+} ion and Br^- ion, respectively.

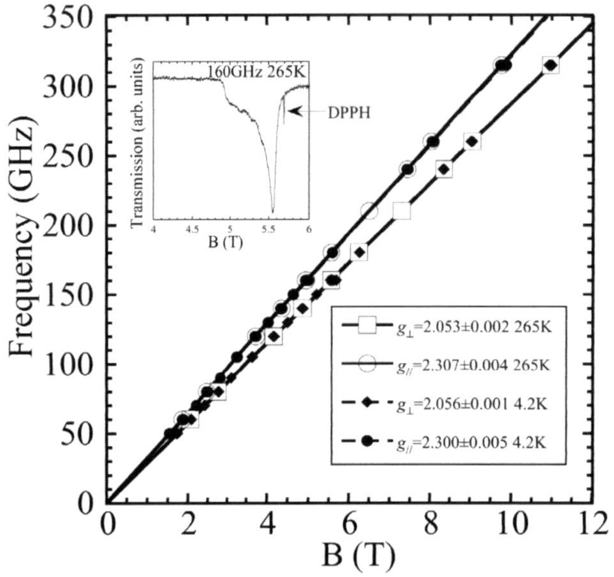

Figure 3. Frequency-field relations obtained at 265 and 4.2 K. The inset shows the typical spectrum observed at 160 GHz and 265 K. DPPH is a field marker corresponding to g=2.

frequency region from 60 to 315GHz. Gunn Oscillators and multipliers are used as light sources. The temperature region is from 1.8 to 265K. Details of our experimental setup can be found in refs [13-15]. The powder sample of (2,3-dmpyH)$_2$CuBr$_4$ is obtained by the crystallization from aqua solutions as written in the reference [12] at the University of Fukui.

3. Results and discussion

Frequency-field relation obtained at 265 K is shown in Fig. 3. The inset, which shows a typical spectrum observed at 160 GHz and 265 K, suggests that it is a typical powder pattern spectrum of Cu^{2+} ion [16]. From the intensity distribution, we can attribute the lower field shoulder and the higher field

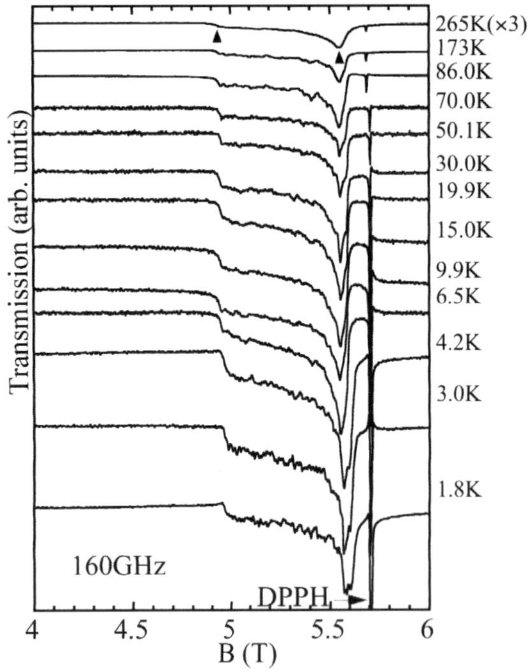

Figure 4. Temperature dependence of ESR spectra observed at 160 GHz.

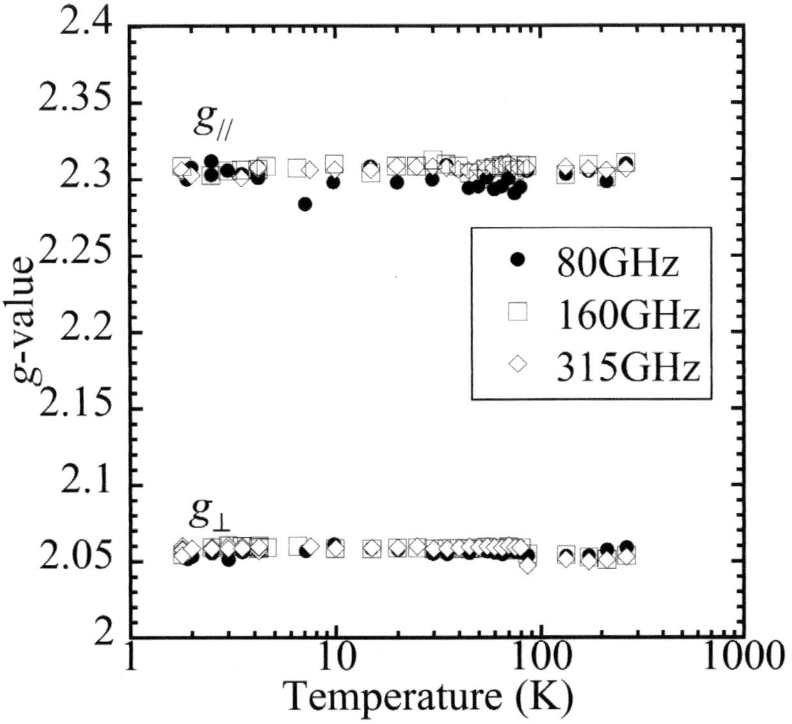

Figure 5. Temperature dependence of g-values observed at 80, 160 and 315 GHz.

peak to $g_{//}$ and $g_{\perp}$, respectively. Obtained g-values at 265 K are $g_{//} = 2.307 \pm 0.004$ and $g_{\perp} = 2.053 \pm 0.002$ as shown in Fig. 3. These g-values look typical for Cu^{2+} ion [16]. However, Cu^{2+} ion is situated in the tetrahedral ligand as shown in Fig. 2 in $(2,3\text{-dmpyH})_2CuBr_4$. Therefore, Γ_5 triplet orbital states has lower energy than Γ_3 doublet orbital states, while it is opposite in the octahedral ligand [17]. Three Γ_5 orbital states can be considered as fictitious angular momentum components $|\pm 1>$ and $|0>$, and g-values of Cu^{2+} ion depend strongly on the ground orbital state [17]. For instance, they will show very anisotropic g-values when $|\pm1>$ states are the ground state because the contribution of the fictitious angular momentum should be considered, while they will show isotropic g-values when $|0>$ state is the ground state. In the case of elongated z-axis as shown in Fig. 2, $|0>$ state is the orbital ground state and the obtained g-values are rather reasonable, where $g_{//}$ direction corresponds to the z-axis. Using a simple spin Hamiltonian approximation, g-values can be expressed as follows;

$$g_{//} = g_e - 8\lambda/(E_4 - E_0) \qquad (1)$$
$$g_{\perp} = g_e - 2\lambda/(E_{\pm1} - E_0) \qquad (2)$$

where g_e is a g-factor of free electron, λ is a spin-orbit coupling, and E_4 is the energy of the lower orbital state from Γ_3 doublet. Using the observed g-values, $E_4-E_0=-26\lambda$ and $E_{\pm1}-E_0=-37\lambda$ can be obtained. This result suggests comparable energies for E_4 and $E_{\pm1}$. However, this is a very simple model and precise orbital energy levels require further consideration.

Figure 4 shows the temperature dependence of ESR spectra observed at 160 GHz. We can see almost no temperature dependence of resonance fields. Temperature independent g-values observed in $(2,3\text{-dmpyH})_2CuBr_4$ are summarized in Fig. 5. It shows clearly that g-values observed at 80, 160 and 315 GHz for all temperatures are almost identical within the error, and it is also supported by the nearly identical g-values obtained at 265 K and 4.2 K in Fig. 3. These results are rather surprising because the temperature dependence of magnetic susceptibility shows a maximum at 11 K [12], which suggests the development of short-range order below 11 K. In case of 1D antiferromagnet it is well known that g-values tend to shift at low temperature due to a short-range order [1]. However, our recent ESR results show that zig-zag chain systems, which have geometrical frustration, show no g-shift at low temperature [4,5] while their magnetic susceptibilities show a maximum at low temperature. As it is suggested that J_{leg} is stronger than J_{rung} in $(2,3\text{-dmpyH})_2CuBr_4$ [12], the system is a spin ladder antiferromagnet closer to quasi-1D chain. Moreover, as the S=1/2 quasi-1D antiferromagnet $Cu_2Cl_4H_8C_4SO_2$ with couplings between chains but no frustration shows g-shift below the temperature of magnetic susceptibility maximum [18], we suggest that the temperature independent g-values observed in $(2,3\text{-dmpyH})_2CuBr_4$ are due to the frustration in the system. From the crystal structure, we also suggest that the frustration can be expected by the antiferromagnetic interaction (dashed line) in Fig. 1. Further detailed theoretical study remains as a future issue.

4. Conclusion

High field ESR measurements of $(2,3\text{-dmpyH})_2CuBr_4$ powder sample have been performed in the temperature region from 1.8 to 265 K using the pulsed magnetic field up to 16 T. The observed frequency range is from 60 to 315 GHz using Gunn oscillators and multipliers. The *g*-values at 265 K are determined to be $g_{//}$=2.307 and $g_{\perp}$=2.053. Almost no temperature dependence of g-value is observed for 80, 160 and 315 GHz down to 1.8 K. We suggest that the observed temperature independent g-values down to 1.8 K may be due to the existence of frustration in the system.

Acknowlegments

One of the authors (HO) would like to acknowledge Prof. Shojiro Kimura of IMR, Tohoku University for valuable discussions. This work was partly supported by a Grant-in-Aid for Scientific Research on Priority Area (No. 19052005 “Novel States of Matter Induced by Frustration”) from the Ministry of Education, Culture, Sports, Science and Technology of Japan.

References

[1] K. Nagata, Y. Tazuke, *J. Phys. Soc. Jpn.* **32**, 337-345 (1972)
[2] Y. Ajiro, S. Matsukawa, T. Yamada, T. Haseda, *J. Phys. Soc. Jpn.* **39**, 259-260 (1975)
[3] H. Ohta, S. Okubo, T. Kamikawa, T. Kunimoto, Y. Inagaki, H. Kikuchi, T. Saito, M. Azuma, M. Takano, J. Phys. Soc. Jpn. **72**, 2464 (2003)
[4] H. Ohta, S. Takano, S. Okubo, M. Yoshida, H. Kikuchi, J. Magn. Magn. Mater. **310**, 1234-1235 (2007)
[5] N. Takahashi1, S. Okubo, H. Ohta, T. Sakurai, Y. Ishikawa, M. Fujisawa, H. Kikuchi, submitted to Appl. Magn. Reson.
[6] A. Zorko, S. Nellutla, J. van Tol, L.C. Brunel, F. Bert, F. Duc, J.-C. Trombe, M.A. de Vries, A. Harrison, P. Mendels, Phys. Rev. Lett. **101**, 026405 (2008)
[7] S. Okubo, M. Tomoo, H. Ohta, H. Kikuchi, J. Phys.: Conf. Series **145**, 012011 (2009)
[8] S. Okubo, H. Ohta, K. Hazuki, T. Sakurai, N. Kobayashi, Z. Hiroi, Physica B **294-295**, 75 (2001).
[9] H. Ohta, W-M Zhang, S. Okubo, M. Tomoo, M. Fujisawa, H. Yoshida, Y. Okamoto, Z. Hiroi, J. Phys.: Conf. Series **145**, 012010 (2009).
[10] H. Ohta, W-M Zhang, S. Okubo, M. Fujisawa, T. Sakurai, Y. Okamoto, H. Yoshida, Z. Hiroi, Phys. Status Solidi B **247**, 679 (2010).
[11] W-M Zhang, H. Ohta, S. Okubo, M. Fujisawa, T. Sakurai, Y. Okamoto, H. Yoshida, Z. Hiroi, J. Phys. Soc. Jpn. **79**, 023708 (2010).
[12] A. Shapiro, C.P. Landee, M.M. Turnbull, J. Jornet, M. Deumal, J.J. Novoa, M.A. Robb, W. Lewis, J. Am. Chem. Soc. 129, 952 (2007)
[13] M. Motokawa, H. Ohta, N. Makita, Int. J. Infrared MMW **12**, 149 (1991)
[14] S. Kimura, H. Ohta, M. Motokawa, S. Mitsuda, W.-J. Jang, M. Hasegawa, H. Takei, Int. Int. J. Infrared MMW **17**, 833 (1996)

[15] H. Ohta, S. Okubo, K. Kamikawa, D. Fukuoka, Y. Inagaki, T. Kunimoto, Z. Hiroi, J. Phys. Soc. Jpn. **72**, Suppl. B 26 (2003)
[16] H. Ohta, N. Yamauchi, T. Nanba, M. Motokawa, S. Kawamata, K. Okuda: J. Phys. Soc. Jpn. **62**, 785 (1993)
[17] A. Abragam, B. Bleaney, *Electron Paramagnetic Resonance of Transition Ions* (Clarendon Press, Oxford, 1970)
[18] M. Fujisawa, K. Shiraki, S. Okubo, H. Ohta, M. Yoshida, H. Tanaka, T. Sakai, Phys. Rev. B 80, 012408 (2009)

Journal of Physics: Conference Series **320** (2011) 012027 doi:10.1088/1742-6596/320/1/012027

AC heat capacities of κ-$(BEDT\text{-}TTF)_2Cu_2(CN)_3$ measured by microchip calorimeter

Y Muraoka[1], S Yamashita[2], T Yamamoto[1] and Y Nakazawa[1,*]

[1]*Department of Chemistry, Graduate School of Science, Osaka University, Machikaneyama 1-1, Toyonaka, Osaka 560-0043, JAPAN*
[2]*RIKEN, Hirosawa 2-1, Wako, Saitama 351-0198, JAPAN*

*E-mail: nakazawa@chem.sci.osaka-u.ac.jp

Abstract. Thermodynamic measurements of an organic spin liquid compound of κ-$(BEDT\text{-}TTF)_2Cu_2(CN)_3$ where BEDT-TTF is bis(ethylenedithio)tetrathiafulvalene were performed by the ac calorimetry technique using a microchip device of TCG3880. This technique is effective to measure relative temperature and magnetic-field dependences of heat capacity for tiny single crystal samples less than 1 μg. Broad hump structures in C_p vs T which are known as so-called 6 K anomaly were observed in κ-$(BEDT\text{-}TTF)_2Cu_2(CN)_3$ and its deuterated compound. The hump temperatures are evaluated as 5.7 K in both compounds. This result demonstrates that the TCG3880 is useful for performing thermodynamic investigations of such kind of organic charge transfer complexes with much reduced sample quantity than the conventional techniques and that the existence of hump structure is intrinsic for κ-$(BEDT\text{-}TTF)_2Cu_2(CN)_3$. The in-plane angular dependence of the magnetic field of 7 T applied parallel to the two dimensional layer is also studied and absence of in-plane anisotropy of the hump structure is discussed in both pristine and deuterated compounds.

1. Introduction

There are increasing interests on low-temperature magnetic properties of organic triangular lattice compounds, since realization of quantum liquid state and appearance of novel physical properties peculiar for molecule-based materials are expected in them [1-2]. Up to now, two compounds are known as candidates of the organic spin liquid. The first compound is κ-$(BEDT\text{-}TTF)_2Cu_2(CN)_3$ of which absence of long range ordering nor glass-like freezing down to 30 mK was reported by Shimizu *et al.* by magnetic susceptibility and ^{13}C-NMR experiments [3]. This compound has a similar molecular arrangement as the famous 10 K class superconductors consisting of BEDT-TTF donor molecules. The other is an acceptor and cation compound of $EtMe_3Sb[Pd(dmit)_2]_2$ whose frustrated character was suggested by magnetic susceptibility measurement by Tamura *et al.* in 2002 [4]. Quite recently, detail ^{13}C-NMR experiments [5] revealed that this compound also do not show any long-range orderings down to 19 mK and the ground state is a spin-liquid state which are also confirmed by thermal conductivity measurement [6]. The possibility of electronic structure change around 1 K was suggested by ^{13}C-NMR experiments [7]. The two compounds are dimer based Mott insulators in which one electron/hole localizes on each dimer site owing to strong electron correlations in two-dimensional (2D) layer.

Concerning on κ-$(BEDT\text{-}TTF)_2Cu_2(CN)_3$, we performed a low temperature heat capacity measurement down to dilution temperature region and reported absence of thermal anomalies related to a long range ordering. Instead of the formation of magnetically ordered structures, the fluctuating spins give a distinct T-linear contribution in heat capacity with a finite electronic heat capacity coefficient γ at low temperatures[8]. The γ value is reported as 12.6 mJK^{-2}mol^{-1} from the measurement in the dilution temperature range, which scales to the magnetic susceptibility $\chi(0)$ extrapolated down to T=0. This fact indicates that the formation of spin-liquid ground state in 2D S-1/2 triangular system is really established in this material and the excitations from the ground state are gapless as is suggested by theories [9,10] for example long-range resonation valence bond (RVB) theory which predicts Fermionic excitations. However, a possibility of small gap opening at extremely low temperature region below 0.5 K was reported by thermal transport measurement [11]. In addition to the gapless or tiny gapped excitations, this compound shows a broad anomaly around 6 K, which is insensitive to external magnetic fields up to 8 T applied perpendicular to the 2D layer. The thermal expansion measurements by Manna *et al.* observed a clear kink in lattice parameters at this temperature [12]. Furthermore, a curious dielectric property which may be related to the internal degrees of freedom in each dimer has been observed by Abdel-Jawad *et al.* [13]. The origin of the hump structure in the thermodynamic properties is still an unsolved problem and probably gives profound physics which have not been discussed within spin-liquid theories.

κ-(BEDT-TTF)$_2$Cu$_2$(CN)$_3$ sample whose ethylene groups in BEDT-TTF molecule are all substituted by deuterons is known to show almost the same spin-liquid behaviors as pristine compound, although some κ-(BEDT-TTF)$_2$X compounds are known to show drastic isotope effects by deuteration of ethylene groups. We reported that Ŧ the heat capacity of deuterated sample shows similar γ value and temperature dependence of heat capacity below 5K is almost the same as pristine compound [14]. Since the deutrated samples used for the previous heat capacity measurements are thin plates, to unveil precise temperature dependence of heat capacity around the hump temperature was difficult in the previous work in Ref. [14]. We have recently developed a microchip calorimetry system by which we can analyze relative change of heat capacity using tiny piece of crystal less than 1μg [15]. It is an aim of the present work to report capability of the calorimeter for the study of organic charge transfer salts and discuss the magnetic fields dependence of the broad hump of hydrogenated and deuterated samples. We also discuss angular dependence of external magnetic fields in the in-plane configuration.

2. Experimental

2.1 Experimental apparatus for ac heat capacity measurements

For the measurements of thermodynamic properties of tiny single crystal, we used a micro-chip device of TCG3880 supplied by Xensor company [16]. This chip is utilized for high-speed thermal analysis with a temperature scanning rate of 10^4-10^5 K s^{-1} in order to study melting, recrystallization, and glass formation for polymer substances in inequilibrium condition by H. Huth and C. Schick and their collaborators [17-18]. A photo picture and a schematic view of the chip is displayed in Fig.1.

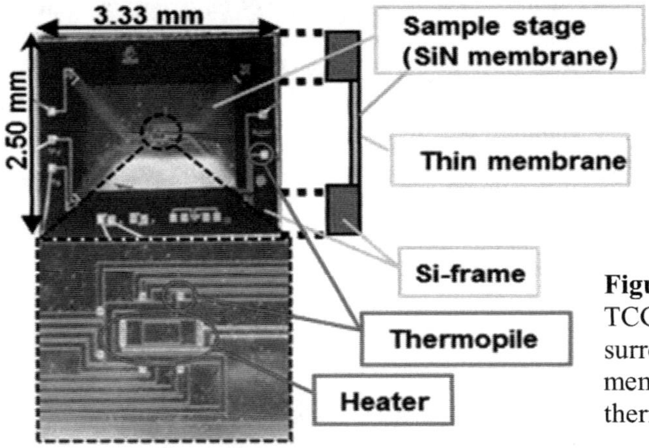

Figure 1. Schematic view of the microchip device TCG3880 .The chip consist of SiN membrane surrounded by Si frame. In the centre of the membrane a small heater and hot junctions of thermopiles are fabricated.

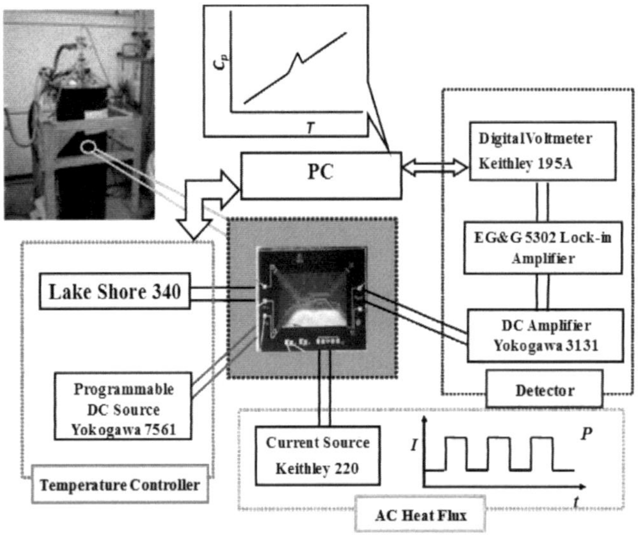

Figure 2. The block diagram of the ac calorimetry system using TCG3880.The system consisting of AC current unit, temperature control unit and a detection unit controlled by personal computer (PC).

This chip consists of a Si frame and a SiN membrane with nearly 20 μm thickness. In the centre of the membrane, small resistance heater with approximately 600 Ω at room temperature is fabricated. The resistance of the heater decreases about 20% at liquid helium temperature, but the temperature dependence is moderate enough to be fitted by a simple polynomial formula. Surrounding this heater part, six hot junctions of a thermopile sensor are fabricated. The cold junction of the thermopile is on the Si frame and therefore detection of the temperature difference between sample part and Si frame is possible by measuring the output voltage of the thermopile. The Si frame is thermally linked to the heat sink made of Cu block on which calibrated thermometers are mounted. The sensitivity (S) of the thermopile is reported as 0.75 mV K^{-1} at room temperature and decreases down to about 5% of the room temperature value at 15 K. Since the sensitivity decreases with the decrease of temperatures especially below 100 K, we have constructed a signal-amplify unit consisting of low-noise DC amplifiers (YOKOGAWA 3132) and low pass filter. By this unit, we can amplify small signal efficiency up to 100 times and low temperature experiments down to liquid helium temperature are possible. The sample was adhered on the resistance heater and the hot junctions by small amount of Apiezon N grease. The temperature of the Cu block with a diameter of 25 mm was controlled accurately by a temperature controller (LakeShore 340). We used a calibrated Pt (PT-111 Lake Shore) sensor and a cernox (Cx1030 Lake Shore) sensor to determine temperature of the chip. The temperature ranges of 30-350 K and 0.5-30 K were covered by Pt sensor and cernox sensor, respectively. The block diagram of the measurement system is shown in Fig.2.

The Cu block with TCG3880 was mounted on a ^{3}He cryostat which is available in a VTI (variable temperature insert) system. By a 7T split pair magnet, we can apply magnetic field just parallel to the donor layers and therefore angular dependence of in-plane heat capacity is measureable. In order to detect small signal efficiently, we used the ac calorimetric technique to perform thermodynamic experiments. By studying the frequency dependence of the temperature oscillation signal, we have determined a proper frequency of 3.3Hz at which T_{ac}/ω gives the largest peak in the high vacuum condition.

2.2 Sample preparation

The hydrogenated sample used for this study is selected from the same batch as those used in previous reports. The deuterated samples were synthesized by electrochemical oxidation technique using Pt electrode according to the procedure reported by Komatsu *et al.* [19]. Small dc current with 1 μA was applied by a constant current source to oxidize the donor molecule solved in organic solvent. Since it was very difficult to measure sample mass of tiny crystals used in this study, we roughly estimated the magnitude of sample mass from the dimension of the crystals estimated by microscope and density of the sample. The mass of the pristine sample was about 1μg and that of deuterated sample was about 20-30% of the pristine sample.

3. Results and Discussion

In Fig.3, we show ac heat capacity data of κ-(BEDT-TTF)$_2$Cu$_2$(CN)$_3$ in a C_p vs T plot. In general, the ac calorimetry is known as a high-resolution technique to detect relative change of heat capacity. Thermodynamic nature of small thermal anomalies and their frequency dependences are studied widely by this technique. However, it is difficult to determine absolute values of heat capacity especially in the case of tiny sample measurement as in the present case. Therefore, the vertical axis in Fig.3 is shown

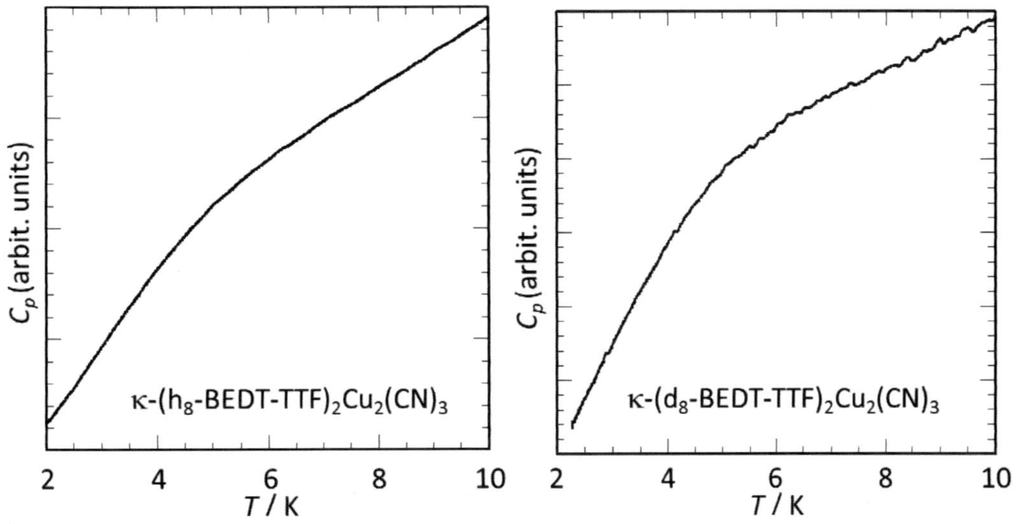

Figure 3. Temperature dependences of heat capacity of κ-(BEDT-TTF)$_2$Cu$_2$(CN)$_3$ and its deuterated compound.

in arbitrary unit to discuss only relative change of heat capacity as a function of temperature. It is noteworthy from the figure that a broad hump structure ranging from 3 to 7 K in C_p vs T plot exists in this compound [8]. The peak of the hump is estimated to be 5.7 K and this temperature coincides well with the previous work, although the experimental technique in previous reports was relaxation calorimetry technique and was different from the present method. It is emphasized that the broadness and peak position are almost the same as that reported in the previous work obtained by the relaxation technique which usually use 10^2-10^3μg samples. This fact demonstrates that micro-chip device TCG3880 is effective tool to detect thermal anomaly of organic compound. By this chip calorimetry technique, we have also succeeded to detect thermal anomaly related to superconductive transition and glass freezing of ethylene groups in BEDT-TTF molecules in κ-(BEDT-TTF)$_2$Cu[N(CN)$_2$]Br [20], but the observation of such kind of broad hump structure can expand further possibility of the chip devices. The data of the deuterated sample is also shown in Fig.3. Since the deuterated sample was very thin plate (with the thickness of about 20 μm), the resolution of the data was worse than that of hydrogenated sample. As is observed in Fig.3, the deuterated sample also shows the similar hump structure with exactly the same temperature as pristine sample. The thermodynamic properties including both γ and the hump structure are almost the same in both compounds and therefore the chemical pressure effect is confirmed to be not so serious in κ-(BEDT-TTF)$_2$Cu$_2$(CN)$_3$.

Our next interest on this broad thermal anomaly of κ-(BEDT-TTF)$_2$Cu$_2$(CN)$_3$ is how the magnetic field affects on the hump structure. In our first heat capacity report of this compound in Ref.[8], thermodynamic data up to 8 T with $H\perp$ plane configuration were presented and negligibly small magnetic field dependence was reported. The ratio of two kind of transfer integrals between neighbouring dimers is evaluated as t'/t=1.06 by extended Hückel calculation which claims that 6% distortion from regular triangular structure exists in this compound [3]. Therefore, small in-plane anisotropy against external magnetic fields may appear in thermodynamic properties. In order to examine whether the in-plane anisotropy exists or not, especially between of H //a-axis and $H\perp$ a-axis configurations, we have performed heat capacity measurements under magnetic fields applied parallel to the plane. Using a split pair magnet, we applied magnetic field of 7 T from three different directions shown schematically in Fig. 4. The application of magnetic field in all directions did not show any difference in both hydrogenated and deuterated compounds which means that the hump structure is insensitive to magnetic fields. The clear kink in temperature dependence of thermal expansion and existence of dielectric anomaly related to the intradimer charge fluctuations reported in Ref. [12, 13] demonstrates that the 6 K anomaly contains the entropy related to the charge and lattice degrees of freedom. The absence of anisotropic behaviour even in the in-plane direction is consistent with these experiments. At present, it is not clear how these degrees of freedom affect on the liquid-like magnetic property appears at the lowest temperature region. In discussing the spin-liquid behaviour of organic dimer systems, the role of these freedoms should be taken into account in addition to spin degrees of freedom.

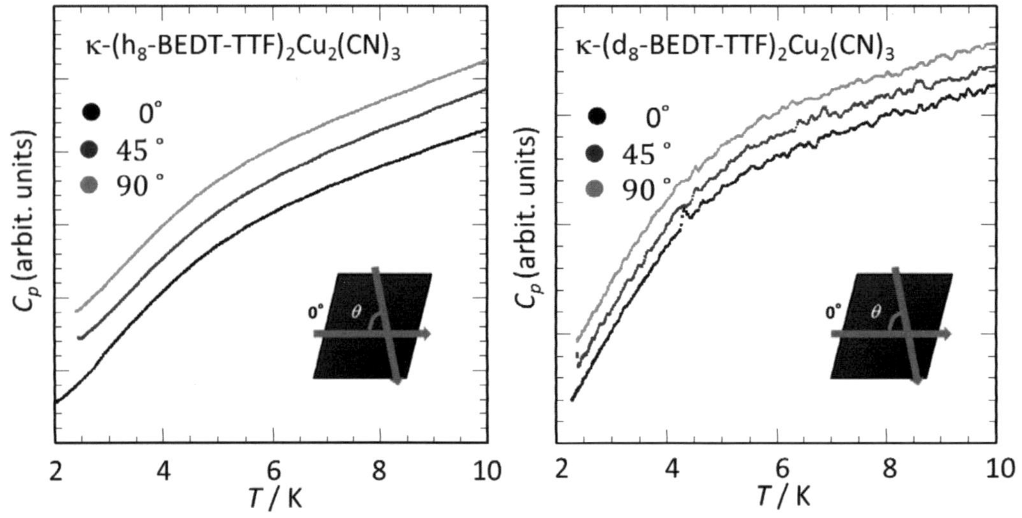

Figure 4. Temperature dependences of heat capacity of κ-(BEDT-TTF)$_2$Cu$_2$(CN)$_3$ and its deuterated compound obtained under magnetic fields of 7 T applied from three different directions.

4. Summary

AC heat capacity measurements using tiny single piece of crystal were performed for the spin liquid compound of κ-(BEDT-TTF)$_2$Cu$_2$(CN)$_3$. The thermodynamic hump structure around 5.7 K corresponding to so-called 6 K anomaly was observed. Substitution of deuterons for eight hydrogen atoms in an ethylene group of BEDT-TTF molecule did not show any systematic change for the formation of hump structure. The effects of magnetic fields applied parallel to the two-dimensional layer did not show any angular dependence in heat capacity, which means that the magnetic fields dependence of the hump structure is very small due to the strong magnetic interactions dominated by large J/k_B. The results presented here demonstrate that the present chip calorimeter is effective to detect small thermodynamic anomaly of molecule based conducting materials even though their single crystals are small.

Acknowledgements
This work was supported by a Grant-in-Aid on Priority Area "Novel States of Matter Induced by Frustration (22014007)" and a Grant-in-Aid for Scientific Research (No. 20654033) from JSPS.

References

[1] Lee P A 2008 *Science* **321** 1306-1307
[2] Balents L 2010 *Nature* **464** 199-208
[3] Shimizu Y, Miyagawa K, Kanoda K, Maesato M and Saito G 2003 *Phys. Rev. Lett.* **91** 1007001
[4] Tamura M and Kato R 2002 *J. Phys. Condens. Matter.* **7** L729-734
[5] Itou T, Oyamada A, Maegawa S, Tamura M and Kato R 2009, *J. Phys. Conference Series* **145** 012039
[6] Yamashita M, Nakata N, Senshu Y, Nagata M, Yamamoto H M, Kato R, Shidauchi T and Matsuda Y 2010 *Science* **328** 1246 -1248
[7] Itou T, Oyamada A, Maegawa S and Kato R 2010 *Nature Phys.* **6** 673-676
[8] Yamashita S, Nakazawa Y, Oguni M, Oshima Y, Nojiri H, Shimizu Y, Miyagawa K and Kanoda, K 2008, *Nature Phys.* **4** 459-462
[9] Anderson P W 1973 *Mater. Res. Bull.* **8** 153-160
[10] Grover T, Trivedi N, Senthil T and Lee P A 2010 *Phys. Rev.* B **81** 245121
[11] Yamashita M, Nakata N, Kasahara Y, Sasaki T, Yoneyama N, Kobayashi N, Fujimoto S, Shibauchi T and Matsuda Y 2009 *Nature Phys.* **5** 44–47
[12] Manna R S, de Souza M, Brühl A, Schlueter J A and Lang M 2010 *Phys. Rev. Lett.* **104** 016403
[13] Abdel-Jawad M, Terasaki I, Sasaki T, Yoneyama N, Kobayashi N, Uesu Y and Hotta C 2010 *Phys. Rev.* B **82** 125119
[14] Yamashita S, Yamamoto T and Nakazawa Y 2010 *Physica B* **405** S240-243.
[15] Inoue Y, Nakazawa Y 2009 *Thermochimica Acta*, **492** 85-88
[16] http://www.xensor.nl/pdffiles/sheets/xen-tcg3880.pdf
[17] Minakov A A, Roy S B, Bugoslavsky Y V, Cohen L F 2005 *Rev. Sci. Instrum.* **76** 043906
[18] Huth H, Minakov A A, Schick C 2006, *J. Polym. Sci.* Part B *Polymer Physics* **44** 2996-3005
[19] Komatsu T, Matsukawa N, Inoue T and Saito G 1996 *J. Phys. Soc. Jpn.* **65** 1340-1354
[20] Muraoka Y, Yamashita S, Yamamoto T and Nakazawa Y 2011 *Thermochimica Acta in press*

NMR study of magnetic excitation in $LiVX_2$ (X = O, S)

Yu Kawasaki, Yutaka Kishimoto, Tetsu Tanaka, Takashi Ohno, Seiji Niitaka[A], Naoyuki Katayama[B] and Hidenori Takagi[A]

Department of Quantum Materials Science, University of Tokushima, Tokushima 770-8506, Japan
[A]RIKEN (The Institute of Physical and Chemical Research), Wako 351-0198, Japan
[B]Department of Physics, University of Virginia, 382 McCormick Road, Charlottesville, VA 22904, USA

E-mail: yu@pm.tokushima-u.ac.jp

Abstract. Magnetic excitation in the two-dimensional triangular lattice $LiVO_2$ and its itinerant analogue $LiVS_2$ are investigated by ^{51}V- and ^{7}Li-NMR experiments. The nuclear spin-lattice relaxation rate $1/T_1$ on ^{51}V and ^{7}Li nuclei shows thermally activated behavior with different size of spin gap, $\Delta/k_B \sim 3400$ K for $LiVO_2$ and ~ 1900 K for $LiVS_2$, below the transition temperature T_c from magnetic to non-magnetic ground state. In $^{51}(1/T_1(T))/^{51}(1/T_1(T_c))$ vs. T/T_c plot, the normalized relaxation rate for both materials seems to fall on the same curve. This scaling behavior indicates that the ground states of $LiVO_2$ and $LiVS_2$ are the same, *i.e.*, V^{3+} ($3d^2, S = 1$) spins form spin singlet with the formation of triangular vanadium clusters, although the replacement of oxygen by sulfur may broaden the band width and largely suppresses T_c from ~ 500 K in $LiVO_2$ to ~ 310 K in $LiVS_2$. However, the magnetic properties in the high temperature phase are different between $LiVO_2$ and $LiVS_2$. The value of $^7(1/T_1T)$ in $LiVS_2$ becomes smaller with decreasing temperature, which suggests the development of pseudogap above T_c, in qualitatively consistent with the results from the magnetic susceptibility and the small change in entropy at T_c. This is a contrast to the case in $LiVO_2$ where $^{51}(1/T_1T)$ is proportional to $1/T$ and it is explained by a localized spin model well above T_c.

1. Introduction

Two-dimensional triangular lattice has attracted much attention concerning the unusual frustration effects and the exotic character of the ground state. $LiVO_2$ is regarded as one of the ideal model compounds of such a system and, therefore, many experimental and theoretical works have been devoted to this material for several decades [1, 2, 3, 4, 5, 6]. $LiVO_2$, in which the magnetic V^{3+} ions ($3d^2$, $S = 1$) form a two-dimensional triangular lattice, is a paramagnetic insulator with strong antiferromagnetic interaction among well localized $3d$ electron spins at high temperature. Upon cooling, at $T_c \sim 500$ K, $LiVO_2$ exhibits a first order phase transition to a spin singlet state accompanied by the formation of triangular vanadium clusters (trimers). This exotic magnetic ground state have been discussed in terms of novel interplay between spin and orbital degrees of freedom in the two-dimensional triangular lattice [7].

Recently, its itinerant analogues $LiVS_2$ and $LiVSe_2$ have been synthesized and systematically investigated by Katayama et al. [8]. The replacement of oxygen by lager anions such as S and

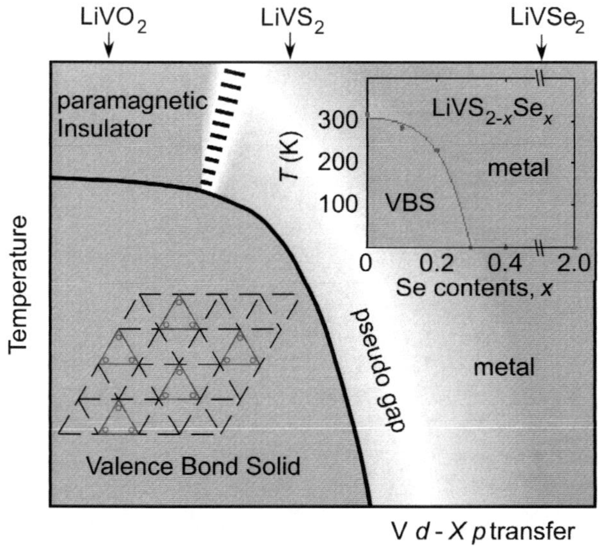

Figure 1. Representation of schematic phase diagram in $LiVX_2$ (X = O, S and Se) system from Ref.[8].

Se is considered to increase the overlap between $3d$ of V and p electrons of O and S (Se) and, as a result, broaden the band width. Actually, $T_c \sim 310$ K in $LiVS_2$ is markedly suppressed by the replacement from $T_c \sim 500$ K in $LiVO_2$. The ground state is, however, robust, *i.e.*, V^{3+} spins in $LiVS_2$ form a spin singlet with the formation of vanadium trimer below T_c as realized in $LiVO_2$. The formation of vanadium trimer is evidenced by the electron diffraction and the EXFS measurements. More itinerant analogue $LiVSe_2$ is a paramagnetic metal down to 2 K, *i.e.*, the spin singlet state disappears. The schematic phase diagram in $LiVX_2$ (X = O, S and Se) system is represented in Fig. 1 from Ref.[8]. The phase diagram indicates that $LiVS_2$ is located near the border between paramagnetic metal and spin singlet insulating phases.

Interestingly, in the high temperature phase of $LiVS_2$, the magnetic susceptibility χ exhibits strong temperature dependence, which is similar to the pseudogap behavior found in underdoped superconducting cuprates [9]. The fact that the change in entropy of $LiVS_2$ at T_c is about one-half as small as that of $LiVO_2$ also supports the existence of pseudogap. To investigate these exotic magnetic properties in $LiVO_2$ and $LiVS_2$ from a microscopic view point, we have measured nuclear spin-lattice relaxation rate $^{51}(1/T_1)$ by ^{51}V-NMR and $^{7}(1/T_1)$ by ^{7}Li-NMR experiments. Here, we stress that NMR may be one of the most powerful tools to investigate a spin singlet and a pseudogap, because $1/T_1$ probes low energy magnetic excitation and the pseudogap in underdoped superconducting cuprates was found by NMR for the first time [9].

2. Experimental

The detailed preparation method of the powder samples of $LiVO_2$ and $LiVS_2$ has been already reported elsewhere [8]. The synthesized samples were characterized crystallographically with X-ray diffractometer, and magnetically with SQUID magnetometer. The temperature dependence of χ indicates that the transition from the non-magnetic ground state to the magnetic one is of a first order and occurs at $T_c \sim 310$ K for $LiVS_2$ and at $T_c \sim 500$ K for $LiVO_2$ in heating process. The obtained data were in accordance with those reported previously [6, 8, 10].

For the present NMR measurements, the powder samples were enclosed in a quartz tube with He gas as a heat exchanger to avoid possible degeneration of the samples in atmosphere. The ^{51}V- and ^{7}Li-NMR signals were observed by the spin-echo technique and nuclear spin-lattice relaxation rates $^{51}(1/T_1)$ and $^{7}(1/T_1)$ were measured by the saturation recovery method. A longitudinal recovery of nuclear magnetization $m(t)$, t sec after a saturating pulse, is theoretically

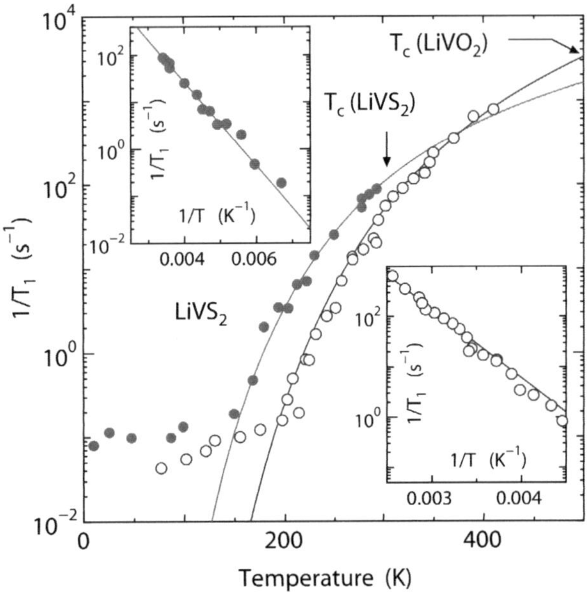

Figure 2. Temperature dependence of $^{51}(1/T_1)$ in $LiVO_2$ (open circles) and $LiVS_2$ (filled circles) measured by ^{51}V-NMR. Insets: Arrhenius plot of $^{51}(1/T_1)$ for $LiVO_2$ (bottom right) and $^{51}(1/T_1)$ of $LiVS_2$ (upper left). The solid lines represent the fits by the activation-type curves to the data.

expressed by

$$\frac{m(\infty)-m(t)}{m(\infty)} = 0.0119\exp\left(-\frac{t}{T_1}\right)+0.0682\exp\left(-\frac{6t}{T_1}\right)+0.206\exp\left(-\frac{15t}{T_1}\right)+0.714\exp\left(-\frac{28t}{T_1}\right)$$

for the transition between $I_z = -1/2 \leftrightarrow +1/2$ of the nuclear spin $I = 7/2$ [11]. This is the case for our measurements of $^{51}(1/T_1)$ by ^{51}V-NMR . $^{51}(1/T_1)$ is determined by the fits of the theoretical curve to the measured $m(t)$ at each temperature. The high quality of the fits guarantees validity of the measurements [12]. In the measurements of $^{7}(1/T_1)$ by ^{7}Li-NMR, the measured $m(t)$ follows $(m(\infty) - m(t))/m(\infty) = \exp(-t/T_1)$. Such a single exponential relaxation is expected if all the satellite lines overlap the central line in NMR spectrum [11]. This overlap is, generally, likely to occur in ^{7}Li-NMR experiments, partly because the nuclear quadrupole moment of ^{7}Li nuclei ^{7}Q is small as $^{7}Q = -4 \times 10^{-26}$ cm^2. Actually, ^{7}Li-NMR spectrum in $LiVO_2$ and $LiVS_2$ consists of single line at all the measured temperature [13].

3. Results and Discussion

3.1. ^{51}V-NMR

Figure 2 shows the temperature dependence of nuclear spin-lattice relaxation rate $^{51}(1/T_1)$ for $LiVO_2$ and $LiVS_2$ by open and filled circles, respectively. The overall temperature dependence of $^{51}(1/T_1)$ in $LiVX_2$ (X = O, S) is quite similar each other and show thermally activated behavior below T_c except the data at low temperature. We show the Arrhenius plot of the data in the bottom right and the upper left of the figure for $LiVO_2$ and $LiVS_2$, respectively. The linear relation between $\log^{51}(1/T_1)$ and $1/T$ clearly confirms the thermally activated behavior. These results indicate that the electronic density of states or magnetic spin excitation has an energy gap at the Fermi surface in consistent with the formation of spin singlet in V^{3+} ions. The formation of spin singlet is also indicated by non-magnetic natures inferred from temperature independent Knight shift below T_c (not shown) in agreement with the previous reports [2, 4, 13].

The activation-type relaxation rate varies following

$$\frac{1}{T_1} \propto \exp\left(-\frac{\Delta}{k_B T}\right), \quad (1)$$

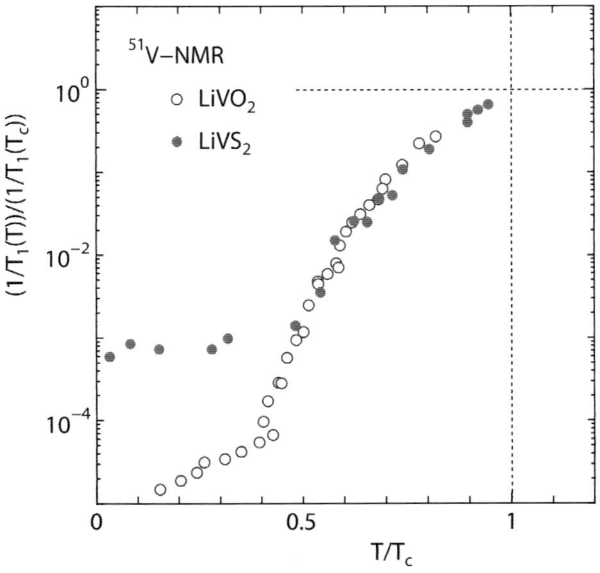

Figure 3. The scaling behavior in $^{51}(1/T_1(T))/^{51}(1/T_1(T_c))$ vs. T/T_c plot for $LiVO_2$ (open circles) and $LiVS_2$ (filled circles).

where Δ/k_B is the size of energy gap in the unit of Kelvin and k_B is the Boltzmann constant. The best fit to the data is represented by solid lines in the main panel of Fig. 2 and its insets. In the fits, we adopt the data in the temperature range 180 K $< T < T_c$ for $LiVO_2$ and 150 K $< T < T_c$ for $LiVS_2$. The size of energy gap Δ/k_B is estimated to be about 3400 K for $LiVO_2$ and 1900 K for $LiVS_2$ from the fits.

At low temperature, $^{51}(1/T_1)$ becomes weakly temperature dependent and deviates from the thermally activated behavior below 180 K for $LiVO_2$ and 150 K for $LiVS_2$. The deviation is most probably due to possible fluctuations of paramagnetic impurity spins and, therefore, extrinsic, because a relaxation rate is generally expected to be independent of temperature if independent fluctuations of local magnetic moments dominate a relaxation process of nuclear spin [14, 15]. Actually, there exist paramagnetic impurity spins and an amount of them is estimated to be as small as $\sim$ 1 % from a Curie tail in χ at low temperature [8]. The weakly temperature dependent values of $^{51}(1/T_1) \sim 0.1$ s^{-1} in $LiVO_2$ is the same order of magnitude as that in $LiVS_2$. This result suggests that hyperfine coupling constants on ^{51}V nuclei in $LiVO_2$ and $LiVS_2$ are comparable, if we assume that spin-fluctuation properties of paramagnetic impurity are similar for both materials.

The estimated sizes of the spin gap are in good agreement with those from $^7(1/T_1)$ by ^{7}Li-NMR as discussed later for both materials, but the size of it for $LiVO_2$ is larger than $\sim$ 1600 K reported in the literature [3]. The difference may originate from the different temperature range adopted for the fits, where the wider temperature range gives the smaller size of the spin gap due to the weakly temperature dependent region at low temperature. Although the estimation of the spin-gap size somewhat depends on fits, one could compare the temperature dependence of $^{51}(1/T_1)$ in $LiVO_2$ and $LiVS_2$ as follows. Fig. 3 shows $^{51}(1/T_1(T))$ normalized by the value at T_c against T/T_c for $LiVO_2$ (open circles) and $LiVS_2$ (filled circles). In this plot, the normalized relaxation rate for both materials seems to fall on the same curve below T_c except at low temperature. This scaling behavior indicates that the difference in T_c of $LiVO_2$ and $LiVS_2$ is closely associated with the difference in the size of spin gap and that the ground state of them are essentially same. In other words, although the itinerancy of $3d$ electrons in $LiVS_2$ are increased from that in $LiVO_2$, V^{3+} spins in $LiVS_2$ form the spin singlet with the formation of vanadium trimer in the ground state as realized in $LiVO_2$ in the ground state.

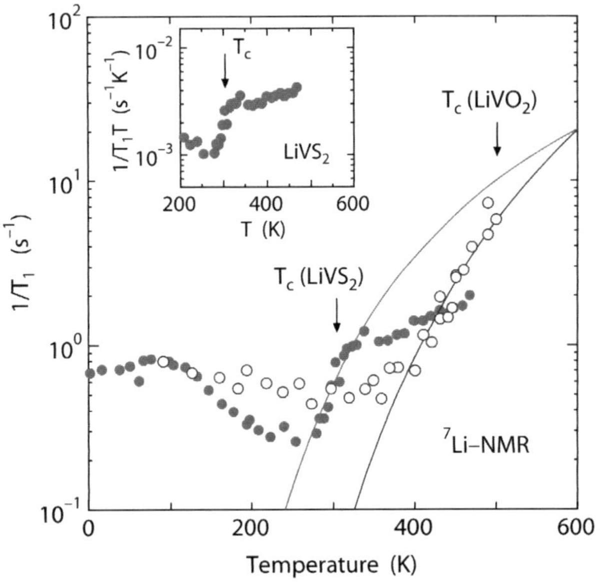

Figure 4. Temperature dependence of $^7(1/T_1)$ for $LiVO_2$ (open circles) and $LiVS_2$ (filled circles) measured by ^{7}Li-NMR. The solid lines represent the fits of the activation-type curve to the data. Inset shows the temperature dependence of $^7(1/T_1T)$ for $LiVS_2$.

The intensity of the ^{51}V-NMR signal decreases rapidly on approaching T_c and the signal was not measurable above T_c in our experimental condition for both materials. The rapid decrease in the signal is most probably due to the decrease in spin-spin relaxation time T_2 near the transition temperature, *i.e.*, a critical slowing down [13]. Nonetheless, one could estimate the value of $^{51}(1/T_1(T))$ divided by temperature at T_c, $^{51}(1/T_1(T_c)T_c)$ to be 4.8 $s^{-1}K^{-1}$ and 0.44 $s^{-1}K^{-1}$ for $LiVO_2$ and $LiVS_2$, respectively, from the extrapolation of the data below T_c. $^{51}(1/T_1(T_c)T_c)$ in $LiVS_2$ is one order of magnitude smaller than that in $LiVO_2$, although the hyperfine coupling constants are comparable for both materials. This result suggests that the magnetic property is different between $LiVS_2$ and $LiVO_2$ in the high temperature phase. The existence of pseudogap above T_c, suggested from the temperature dependence of χ and the small change in the entropy at T_c [8], may explain the much smaller value of $^{51}(1/T_1(T_c)T_c)$ in $LiVS_2$.

3.2. ^{7}Li-NMR

The intensity of ^{7}Li-NMR signal is larger than that of ^{51}V-NMR one. This is partly because two satellite lines overlap the central line in the ^{7}Li-NMR spectrum and spin-spin relaxation in ^{7}Li nuclei is much slower than that in ^{51}V one. The larger signal intensity enables us to measure the nuclear spin-lattice relaxation rate $^7(1/T_1(T))$ up to the higher temperature, $\sim$ 500 K. In Fig. 4, we show the temperature dependence of $^7(1/T_1)$ in $LiVO_2$ and $LiVS_2$ by open and filled circles, respectively. In a limited temperature range, 320 K $< T < T_c$ for $LiVO_2$ and 260 K $< T < T_c$ for $LiVS_2$, we find the thermally activated behavior characterized by Eq. (1). The best fits to the data, represented by solid lines in Fig. 4, yield the size of spin gap $\Delta/k_B \sim$ 3100 K for $LiVO_2$ and $\sim$ 2000 K for $LiVS_2$. These values are in good agreement with those obtained by the ^{51}V-NMR measurements and, therefore, support the discussion in the previous subsection. At low temperature, $^7(1/T_1)$ shows weak temperature dependence and deviates from the thermally activated behavior for both materials. The deviation may be extrinsic, because it most probably originates from the independent fluctuations of paramagnetic impurity spins as discussed in the previous subsection.

To focus on the high temperature phase of $LiVS_2$, we show the temperature dependence of $^7(1/T_1T)$ above 200 K in the inset of Fig. 4. The value of $^7(1/T_1T)$ in $LiVS_2$ becomes smaller with decreasing temperature in the high temperature phase above T_c. This result is a contrast to the case in $LiVO_2$ where $^{51}(1/T_1T)$ is proportional to $1/T$ well above T_c and it is explained by the localized spin model with $S = 1$ [2]. The decrease in $^7(1/T_1T)$ of $LiVS_2$ indicates the existence of gap in magnetic excitation or Li^+ diffusion. The latter is, however, ruled out, because no motional narrowing is observed in the ^{7}Li-NMR spectrum up to 500 K, *i.e.*, the line width of ^{7}Li-NMR spectrum, $\sim$ 22 Gauss, is independent of temperature in the measured temperature range. The temperature dependence of $^7(1/T_1T)$, therefore, suggests the opening of pseudogap in the high temperature phase, where the short-range ordering of the spin singlet may develop.

4. Conclusions

We have measured the nuclear spin-lattice relaxation rate $1/T_1$ on ^{51}V and ^{7}Li nuclei in the two-dimensional triangular lattice $LiVO_2$ and its itinerant analogue $LiVS_2$. In the low temperature phase below T_c, the temperature dependence of $1/T_1$ on both ^{51}V and ^{7}Li nuclei consistently follows the thermally activated behavior, $1/T_1 \propto \exp(-\Delta/k_BT)$, and evaluates the magnitudes of spin gap; $\Delta/k_B \sim 3400$ K for $LiVO_2$ and ~ 1900 K for $LiVS_2$. In $^{51}(1/T_1(T))/^{51}(1/T_1(T_c))$ vs. T/T_c plot, the normalized relaxation rate for both materials seems to fall on the same curve. These results indicate that the ground state of $LiVS_2$ is essentially the same as that in $LiVO_2$, where V^{3+} $(3d^2, S = 1)$ spins form the spin singlet with the formation of vanadium trimer, although the replacement of oxygen by sulfur broadens the band width and markedly suppresses T_c from $\sim$ 500 K in $LiVO_2$ to $\sim$ 310 K in $LiVS_2$. In the high temperature phase of $LiVS_2$ above T_c, the nuclear spin-lattice relaxation rate divided by temperature $^7(1/T_1T)$ decreases on cooling in $LiVS_2$, exhibiting a pseudogap-like behavior. This is a contrast to the case in $LiVO_2$ where $^{51}(1/T_1T)$ is proportional to $1/T$ well above T_c and it is explained by a localized spin model.

References

[1] J.B. Goodenough, G. Dutta and A. Manthiram, Phys. Rev. B, **43** (1991) 10170.
[2] M. Onoda, T. Naka and H. Nagasawa, J. Phys. Soc. Jpn., **60** (1991) 2550.
[3] M. Onoda and T. Inabe, J. Phys. Soc. Jpn., **62** (1993) 2216.
[4] J. Kikuchi, S. Kambe, H. Yasuoka, Y. Ueda, K. Tomimoto and J. Akimitsu, J. Phys. Soc. Jpn., **60** (1991) 3620.
[5] K. Imai, H. Sawa, M. Koike, M. Hasegawa and H. Takei, J. Solid State Chem., **114** (1995) 184.
[6] W. Tian, M.F. Chisholm, P.G. Khalifah, R. Jin, B.C. Sales, S.E. Nagler and D. Mandrus, Mater. Res. Bull., **39** (2004) 1319.
[7] H.F. Pen, H. van den Brink, D.I. Khomskii and G.A. Sawatzky, Phys. Rev. Lett., **78** (1997) 1323.
[8] N. Katayama et al., Phys. Rev. Lett., **103** (2009) 146405.
[9] H. Yasuoka, T. Imai and T. Shimizu, *Strong Correlation and Superconductivity*, ed. H. Fukuyama, S. Maekawa and A.P. Malozemoff, p.254 (Springer Verlag, 1989)
[10] B. van Laar and D.J.W. Jido, J. Solid State Chem., **3** (1971) 590.
[11] A. Narath, Phys. Rev., **162** (1967) 320.
[12] T. Tanaka et al., J. Phys. Chem. Solid, **69** (2008) 3142.
[13] T. Tanaka et al., J. Phys. Soc. Jpn., **78** (2009) 054709.
[14] T. Moriya, Prog. Theort. Phys., **16** (1956) 320.
[15] T. Moriya, Prog. Theort. Phys., **16** (1956) 641.

International Conference on Frustration in Condensed Matter (ICFCM) IOP Publishing
Journal of Physics: Conference Series **320** (2011) 012029 doi:10.1088/1742-6596/320/1/012029

An exotic ordered phase in the pyrochlore-like antiferromagnet $Ni_2(OH)_3Cl$ studied by NMR

S Sato[1,*], A Oyamada[1], M Nishiyama[1], T Itou[1], S Maegawa[1], X G Zheng[2], M Hagihala[2]

[1]Graduate School of Human and Environmental Studies, Kyoto University, Kyoto 606-8501, Japan
[2]Faculty of Science and Engineering, Saga University, Saga 804-8502, Japan

E-mail[*] : manochori@baka.mbox.media.kyoto-u.ac.jp

Abstract. We investigated a three-dimensional frustrated 3d-electron magnet on a pyrochlore-like lattice $Ni_2(OH)_3Cl$ with $s = 1$ by means of ^{1}H-NMR measurements. Previous specific heat measurements reported that this material shows a phase transition at 4.0 K and also has an anomaly at around 20 K. Our ^{1}H-NMR measurement in 0.73 T has demonstrated that both the spin-lattice relaxation rates T_1^{-1} and the spectra show clear evidences of a magnetic long-range ordering below 4.0 K. However, they do not show any anomalies at 20 K. By contrast the width of the ^{1}H-NMR spectra in 4.3 T starts to deviate from linearity to the magnetization below 20 K, which suggests that the magnetic anisotropy is induced by the magnetic field. In addition, T_1^{-1} starts to decrease below 20 K, which indicates the suppression of spin fluctuation. We discuss the unusual spin state below 20K.

1. Introduction

The pyrochlore lattice is one of the most fascinating stages of three-dimensional geometrical frustration. Magnets on the pyrochlore lattice raise many interesting topics. Neither classical nor quantum Heisenberg antiferromagnets with nearest-neighbour interaction on the pyrochlore lattice have any magnetic ordering, and spin liquid states emerge [1, 2]. Our interest is rather in the novel ordering state induced by the additional perturbative interactions. For instance, it is predicted that further neighbouring exchange interactions will produce a spin nematic state or an unusual state with an ordered spin Fourier component and fluctuating phase degree of freedom [3, 4]. Biquadratic exchange interactions induce the spin nematic state and vector multipolar state [5]. Another perturbation such as anisotropies, lattice deformations or external magnetic field can also lead to novel magnetic ordered states.

The 3d electron compound $Ni_2(OH)_3Cl$ is a Heisenberg antiferromagnet on a pyrochlore-like lattice, where the magnetic ions Ni^{2+} with spin $s = 1$ are located on the vertices of distorted tetrahedra. Magnetic susceptibility and neutron scattering measurements have been reported to show an antiferromagnetic ordering at 4.0 K, however the Weiss temperature is 19 K, which indicates the existence of ferromagnetic interactions [6, 7]. Thus, it is considered that this material has both antiferromagnetic and ferromagnetic interactions. Specific heat measurements performed on this material have revealed a successive phase transition at 20 and 4.0 K [8]. A large peak of the specific heat at 4.0 K in zero field shifts to lower temperatures and diminishes when the magnetic field is increased. The peak almost vanishes in 3 T [8]. The other transition at 20 K was observed as a kink in the temperature dependence of the specific heat, whose anomaly is smaller than that at 4.0 K in zero magnetic field. A large broad peak of the specific heat was observed at 20 K in a field of 3 T, in contrast to the diminution of the peak at 4.0 K. No obvious anomaly was observed around 20 K by magnetic susceptibility and neutron scattering measurements in zero or nonzero magnetic fields. The nature of the intermediate phase of this frustrated magnet remains open question.

We performed magnetization and ^{1}H-NMR measurements on $Ni_2(OH)_3Cl$ in order to reveal the nature of the intermediate phase from a microscopic view point.

2. Experimental Methods

A powder sample of $Ni_2(OH)_3Cl$ was prepared by hydrothermal reactions [7]. ^{1}H-NMR measurements were conducted at frequencies of 31.0 and 178.9 MHz, whose resonance fields are about 0.73 and 4.3 T, respectively, by a coherent pulsed NMR method from 1.5 to 300 K. The NMR spectra were obtained from the integrated intensities of spin echo signals measured by sweeping the magnetic field at a fixed frequency. The spin-lattice relaxation rates T_1^{-1} were determined from the recovery of the integrated spin–echo intensity $M(t)$ after a time delay t from saturation comb pulses. The magnetization measurements were performed by a SQUID magnetometer between 1.7 and 300 K in a magnetic field of 0.73 and 4.3 T.

3. Results and Discussion

Figure 1 shows the temperature dependence of the magnetization divided by the external field M / H, obtained in 0.73 and 4.3 T. These magnetic fields correspond to the 31.0 and 178.9 MHz of the resonant frequencies used in our ^{1}H-NMR experiment. The values M / H in both magnetic fields increase with decreasing temperature in the paramagnetic phase. The peak of M / H at 4.0 K in 0.73 T indicates the antiferromagnetic transition, while the value in 4.3 T increases monotonically over the entire temperature region, and no antiferromagnetic ordering is observed down to 2.0 K. It is thought that antiferromagnetic ordering is suppressed in 4.3 T, which is consistent with the specific heat measurements. Meanwhile, there is no anomaly near 20 K in the temperature dependence of the magnetization, in contrast to the specific heat measurement results, which show an anomaly around 20 K in 3 T [8].

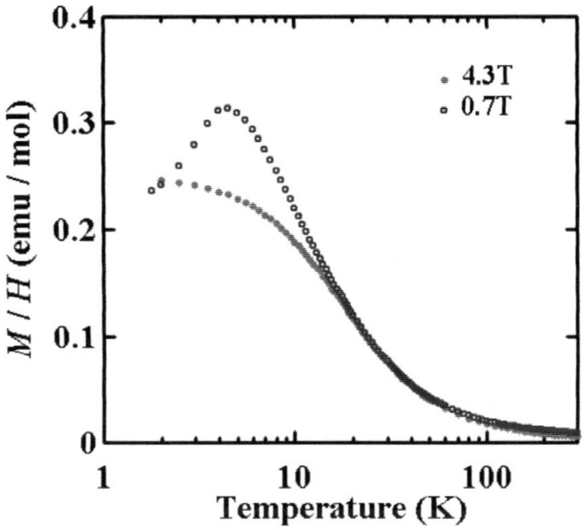

Figure 1. Temperature dependence of M / H, measured under the fields of 0.73 and 4.3 T.

Journal of Physics: Conference Series **320** (2011) 012029 doi:10.1088/1742-6596/320/1/012029

Figure 2(a) shows the temperature variation of the ^{1}H-NMR spectra of $Ni_2(OH)_3Cl$ at 31.0 MHz. Peaks of ^{1}H resonance in the paramagnetic phase were observed around 0.73 T for 31.0 MHz. The spectra above 4.0 K have asymmetric profiles characteristic of powder patterns, which are caused by the anisotropic hyperfine interaction that is dipolar tensor. For the powder sample, the angles between grains in the sample and the external magnetic field distribute randomly. This random distribution of the angle causes the width of the spectra, because the internal fields at ^{1}H sites produced by Ni^{2+} moments varies with the angle owing to the anisotropy of the dipolar interaction. We estimated the width as the full width at half maximum (FWHM) of the ^{1}H-NMR spectra. The width increases gradually as the temperature decreases down to 4.0 K. It should be noted that this linewidth is proportional to the macroscopic magnetization in usual paramagnets where field-induced magnetic moments are homogeneous and parallel to the applied field. This proportionality is clearly seen above 4 K in figure 2(b), which is the plot of the width as a function of the macroscopic magnetization measured in a field of 0.73 T. If a conventional magnetic ordering occurs, the additional large broadening of the NMR spectra is observed as the additional internal fields at ^{1}H sites. Actually, below 4.0 K the width increases steeply and the shape of the spectra changes into a pattern with three steps, which indicates the emergence of the large internal field due to the conventional magnetic ordering. On the other hand, no additional broadening is observed around 20 K, which indicates there is no conventional ordering at 20 K.

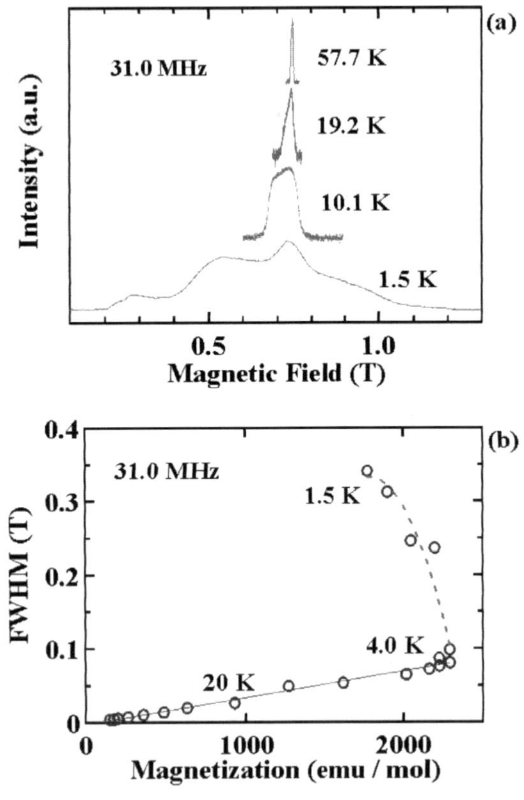

Figure 2. (a) Temperature variation of ^{1}H-NMR spectra at 31.0 MHz. **(b)** FWHM vs. magnetization plot. Magnetization is measured in 0.73 T.

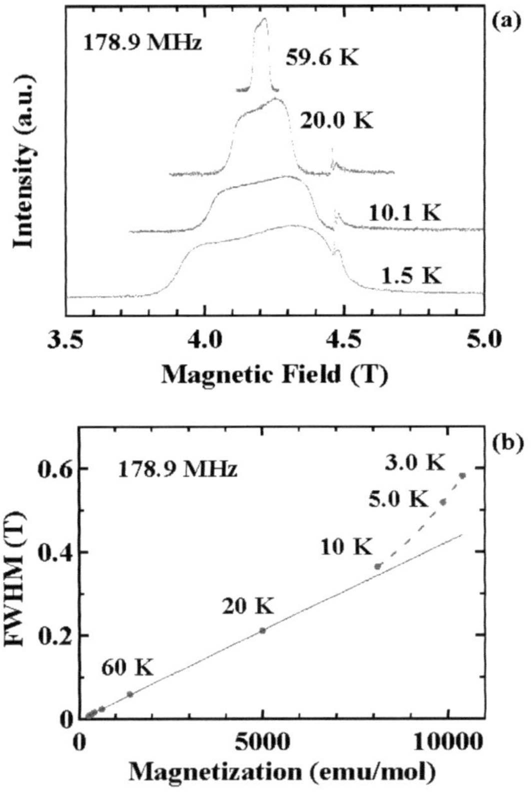

Figure 3. (a) Temperature variation of ^{1}H-NMR spectra at 178.9 MHz. **(b)** FWHM vs. magnetization plot. Magnetization is measured in 4.3 T.

Figure 3(a) shows that the width of the spectra at 178.9 MHz gradually increases with decreasing temperature, while the shape of the spectra remains almost the same down to 1.5 K. The sharp peaks at 4.5 T are due to the resonance of ^{19}F nuclei contained in the Teflon coating of the resonance coil. It seems that magnetic long-range ordering does not occur down to 1.5 K in this magnetic field. However, a noticeable change appears in the width of the spectra. The width FWHM at 178.9 MHz is shown as a function of magnetization in 4.3 T in figure 3(b). The width of the spectra starts to deviate gradually from linearity to the magnetization below 20 K. This deviation shows the additional broadening of the spectra with no obvious change in the shape of the spectra. As the hyperfine interaction has no temperature dependence, this additional broadening indicates that field-induced magnetic moments become anisotropic or inhomogeneous below 20 K.

Figure 4 shows the temperature dependence of the ^{1}H spin-lattice relaxation rates T_1^{-1}, which were measured at 31.0 and 178.9 MHz. All the recoveries of the nuclear magnetization at 31.0 MHz are fitted by double-exponential functions

$$\frac{M(\infty) - M(t)}{M(\infty)} = A\exp\left(-\frac{t}{T_1}\right) + (1-A)\exp\left(-\frac{t}{T_{1L}}\right), \qquad (1)$$

where $M(\infty)$ is the nuclear magnetization at thermal equilibrium and $M(t)$ is the recovered nuclear magnetization after a time delay t from the saturation pulses. Single-exponential recoveries that correspond to A = 1 were observed above 20 K, and the longer relaxation time T_{1L} appeared below around 20 K. The rate T_1^{-1} increases with decreasing temperature down to 4.0 K for 31.0 MHz. The temperature dependence of T_1^{-1} is expected to be proportional to χT, where χ is magnetic susceptibility and T is temperature, when spins fluctuate randomly without spin correlation in the paramagnetic phase. The temperature dependence of χT is shown by a solid curve in figure 4. The experimental data of T_1^{-1} agrees well down to 40 K with the curve calculated using the experimental susceptibility, but deviates below this temperature, which indicates that spin correlations with $\boldsymbol{q} \neq 0$ develop below 40 K.

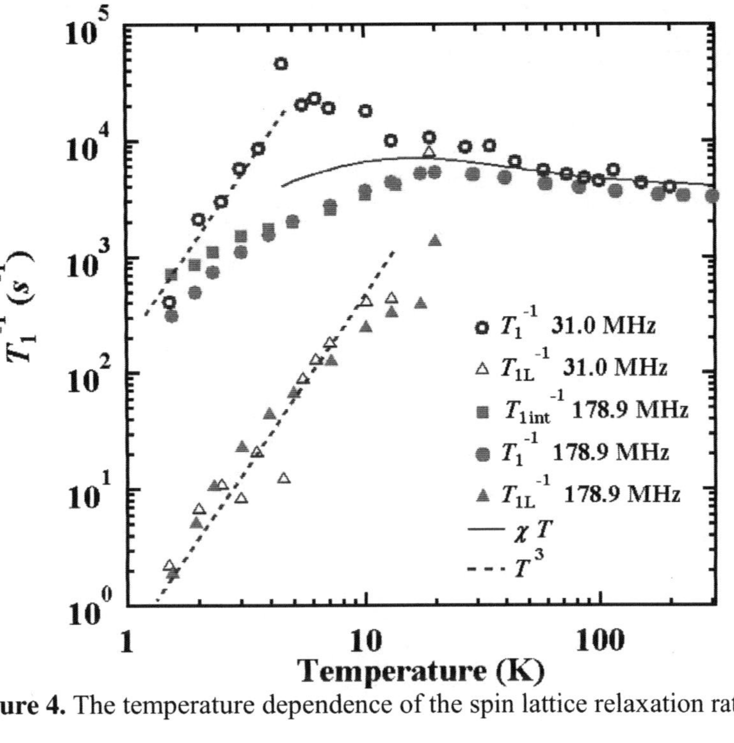

Figure 4. The temperature dependence of the spin lattice relaxation rate T_1^{-1}.

Journal of Physics: Conference Series **320** (2011) 012029 doi:10.1088/1742-6596/320/1/012029

Then the rate T_1^{-1} exhibits a sharp peak at 4.0 K, which clearly indicates the phase transition to the long-range ordering. The rate decreases sharply with decreasing temperature proportionally to T^3 below 4.0 K. It should be noted that no anomaly was observed in the temperature dependence of T_1^{-1} around the higher transition temperature of 20 K. Meanwhile, the relaxation rate T_{1L}^{-1} has the temperature dependence proportional to T^3. However, we are not yet certain whether the T_{1L} component is intrinsic, because the value of the coefficient (1-A) is only about 0.1, which might be due to impurities. We need to examine the sample dependence of the T_{1L} component.

The recovery curves at 178.9 MHz in the higher field of 4.3 T could be fitted by a single-exponential function above 20 K as well as those at 31.0 MHz, as shown in figure 5. On the other hand, another characteristic behavior at 178.9 MHz was observed below about 20 K. Relaxation curves below 20 K deviate from the single-exponential decay as shown in figure 5, and could be fitted by the following stretched exponential function:

$$\frac{M(\infty)-M(t)}{M(\infty)} = B\exp\left[-\left(\frac{t}{T_1}\right)^{\beta}\right] + (1-B)\exp\left(-\frac{t}{T_{1L}}\right). \tag{2}$$

This behavior was not seen at 31.0 MHz, while the slow relaxation component T_{1L} was also observed below 20 K at 178.9 MHz as well as at 31.0 MHz, and the relaxation rate T_{1L}^{-1} is almost the same as that of 31.0 MHz. The stretching coefficient β denotes the degree of inhomogeneity of the spin system, while the nuclear magnetization in the homogeneous system relaxes exponentially with $\beta = 1$. The temperature dependence of β is shown in figure 6. The value β decreases as temperature decreases below 20 K, which indicates the growth of the inhomogeneity with decreasing temperature. The magnetic field of 4.3 T gives rise to some inhomogeneity in the material. The rate T_1^{-1} in the high-temperature phase increases as temperature decreases, and starts to decrease gradually around 20 K. This temperature dependence is essentially the same as the alternative estimation of relaxation rate $T_{1\mathrm{int}}^{-1}$, which is obtained from the initial decay of the relaxation curves. The difference between T_1^{-1} and $T_{1\mathrm{int}}^{-1}$, which indicates an inhomogeneity of the system, becomes clear in low temperatures. The

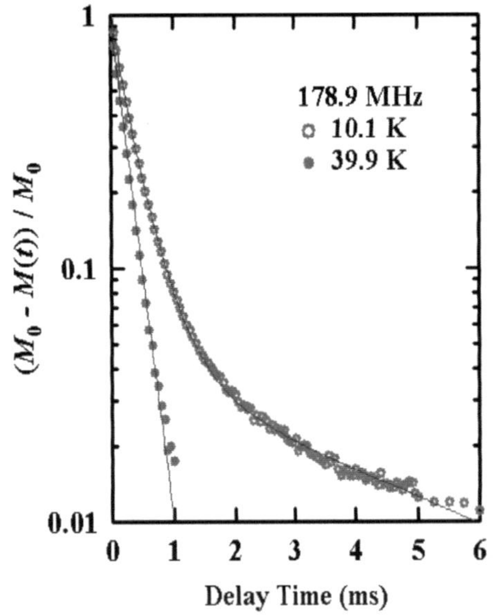

Figure 5. Nuclear magnetization recoveries at 10.1 and 39.9 K for 178.9 MHz.

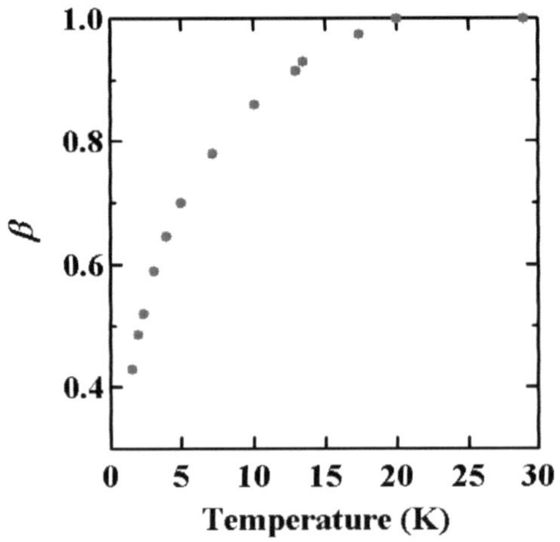

Figure 6. The temperature dependence of β below 30 K, observed at 178.9 MHz (4.3 T).

decreases of T_1^{-1} and $T_{1\mathrm{int}}^{-1}$ with lowering temperature below 20 K suggest that the spin fluctuation is suppressed by the magnetic field below this temperature. This result may correspond to the results of specific heat measurements, in which the large broad peak around 20 K was observed in 3 T [8].

No obvious anomaly like critical phenomenon observed in a conventional magnetic ordering is observed around 20 K in our NMR experiment. The temperature dependences of the spectra and T_1^{-1} between 4.0 K and 20 K are quite paramagnetic-like in the low field of 0.73 T. However, we observed the appearance of an additional broadening in the ^{1}H-NMR spectra and the suppression of spin fluctuation in the high field of 4.3 T. These unusual behaviors may be closely related to the transition at 20 K, which was observed as a large broad peak of specific heat measurement in 3 T [8].

We comment on the nematic state which has been proposed theoretically [3, 4, 9]. The simplest order parameter of the spin quadrupolar moment among nematic orderings can be expressed as

$$Q_{\mu\mu'} = \frac{1}{2}\langle S^{\mu}S^{\mu'} + S^{\mu'}S^{\mu}\rangle - \frac{1}{3}S(S+1)\delta_{\mu\mu'}, \tag{3}$$

where μ is the spin index. This spin quadrupolar moment, which denotes the anisotropy of spin fluctuation, has a finite value in the spin nematic ordered state and exhibits anisotropic spin susceptibility. This anisotropic susceptibility is compatible with our observation that the NMR spectra show the additional broadening below 20 K. The anisotropic spin fluctuation in nematic states can also cause the inhomogeneity observed in the distribution of T_1^{-1}. The rate T_1^{-1} depends on the angle between a grain and the applied field directions in such a case, since the rate reflects the transverse component of the fluctuating local fields with respect to the applied field.

The unusual behaviors, however, can be explained also by other scenarios. For example, the Ni^{2+} ions have more or less inherent anisotropic susceptibility like single-ionic anisotropy. Moreover, the Ni^{2+} ions have two different crystallographical sites. Both of these can also cause the additional broadening, which indicates the anisotropy or inhomogeneity in the spin system.

Further study is required to elucidate the spin state of the frustrated magnet on the pyrochlore lattice Ni_2OH_3Cl.

Acknowledgements
This work was supported by Grants-in-Aid for Scientific Research on Priority Areas ''Novel States of Matter Induced by Frustration" (19052005) from the Ministry of Education, Culture, Sports, Science and Technology, Japan.

References

[1] Moessner R and Chalker J T 1998 *Phys. Rev. Lett.* **80** 2929
[2] Canals B and Lacroix C 2000 *Phys. Rev.* B **61** 1149
[3] Chern G W, Moessner R and Tchernyshyov O 2008 *Phys. Rev.* B **78** 144418
[4] Tsuneishi D, Ioki M and Kawamura H 2007 *J. Phys. Cond. Matt.* **19** 145273
[5] Shannon N, Penc K and Motome Y 2010 *Phys. Rev.* B **81** 184409
[6] Maegawa S, Oyamada A and Sato S 2010 *J. Phys. Soc. Jpn.* **79** 011002
[7] Zheng X G, Hagihala M, Nishiyama K and Kawae T 2009 *Physica* B **404**, 677
[8] Zheng X G, Hagihala M and Kawae T private communication.
[9] Tsunetsugu H and Arikawa M 2007 *J. Phys. Cond. Matt.* **19** 145248

NMR study of pyrochlore lattice antiferromagnet, melanothallite Cu_2OCl_2

Masahide Nishiyama[1], Akira Oyamada[1], Tetsuaki Itou[1], Satoru Maegawa[1], Hirotaka Okabe[2] and Jun Akimitsu[2]

[1]Graduate School of Human and Environmental Studies, Kyoto University, Kyoto 606-8501, JAPAN

[2]Department of Physics and Mathematics, Aoyama Gakuin University, Kanagawa 229-8558, JAPAN

E-mail: `nishiyama.masahide.2r@kyoto-u.ac.jp`

Abstract. The melanothallite Cu_2OCl_2 is a new example of pyrochlore-like antiferromagnet, which is composed of 3d transition metal electrons. We performed Cu- and Cl-NMR experiments on powder samples of Cu_2OCl_2 below transition temperature $T_N = 70\,K$ and we observed six resonant peaks of Cu nuclei, which are composed of three symmetric peaks corresponding to ^{63}Cu and three corresponding to ^{65}Cu. The Cu nuclei feel the strong hyperfine fields because of ordered magnetic moments and the electric field gradients. We determined the spin structure by analyzing the Cu-NMR spectra. The melanothallite has an all-in-all-out spin structure. The spin lattice relaxation rates T_1^{-1} of Cu- and Cl-NMR in the ordered phase are proportional to the temperature; This suggests that although long-range ordering occurs at rather high temperature, the large spin fluctuations caused by the geometrical frustration still remain.

1. Introduction

The pyrochlore lattice is composed of corner-sharing tetrahedra. The antiferromagnet on the pyrochlore lattice with nearest neighbour Heisenberg interactions has the macroscopic number of degeneracy and has no long-range magnetic ordering even at zero temperature. This magnetic tetrahedron system has attracted much interest because of three-dimensional geometrical frustration. A well-known example is an $A_2B_2O_7$ system, in which A is a rare earth element and B is a 3d-transition metal element. In these compounds 4f-electrons dominate their magnetic properties and their main interaction is a long-range dipolar interaction. In contrast, the interaction in the 3d-electron system is short-range exchange interaction. Thus, the 3d-electron antiferromagnets on the pyrochlore lattice are considered to be simpler than 4f-electron pyrochlore antiferromagnets. Recently, clinoatacamite $Cu_2(OH)_3Cl$ with $s = 1/2$, atacamite $Ni_2(OH)_3Cl$ with $s = 1$, Cu_3O_4 with $s = 1/2$, and FeF_3 with $s = 5/2$ have been reported to be pyrochlore-like antiferromagnets with 3d-electron system [1–3]. These compounds were reported to show interesting behaviours, such as successive transitions, exotic critical behavior, large spin fluctuation in the spin-freezing phase, and novel spin structure in the ordered phase. These behaviors are considered to be due to additional perturbations, such as further neighbouring interactions, anisotropies, or a lattice distortions [1–5]. The melanothallite is a new example of the pyrochlore-like antiferromagnet with the 3d-electron system.

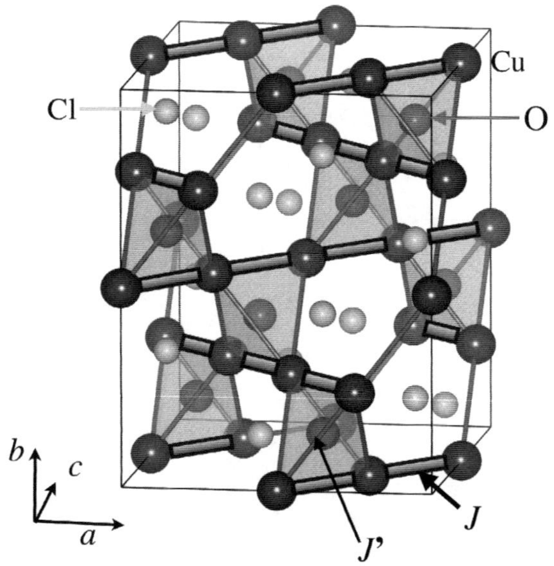

Figure 1. Crystal structure of melanothallite. The blue, red, and green balls represent copper, oxygen, and chlorine, respectively. The thick blue bonds between Cu^{2+} represent the exchange interaction J, and the thin ones represent the exchange interaction J'. The tetrahedra composed of Cu are shown in yellow. The figures was prepared using the VESTA program [6].

The crystal structure of melanothallite is orthorhombic, as shown in figure 1, and its space group is $Fddd$. The lattice constants are $a = 9.542$ Å, $b = 9.712$ Å, and $c = 7.396$ Å. All Cu^{2+}, O^{2-} and Cl^{-} belongs crystallographically to corresponding identical sites. The magnetic ion Cu^{2+} has a spin of 1/2. The oxygens are located at the centre of the tetrahedra composed of Cu^{2+}. These tetrahedra form the pyrochlore lattice by the corner-sharing network of Cu^{2+}. On the other hand, the crystal structure could be considered to be composed of the vertical cross-stacking of one-dimensional chains of Cu^{2+}, as shown in figure 1 by the thick blue bonds. The super-exchange path between Cu moment that consists Cu–O–Cu and Cu–Cl–Cu is represented by thick blue line. Thin blue line represents the super-exchange Cu–O–Cu path. Controversy had existed on whether this compound should magnetically be considered a three-dimensional pyrochlore lattice or a one-dimensional chain.

The temperature dependence of susceptibility shows a broad peak at about 140 K and an antiferromagnetic ordering at $T_N = 70$ K [7]. The broad maximum cannot be explained by a one-dimensional antiferromagnetic model with only intrachain interactions J. If a modified one-dimensional model with intrachain interactions J and interchain interactions J', as shown in figure 1, is assumed, the fitting can be improved. However, the obtained J and J' are 113 K and 108 K, respectively, and thus they are nearly the same [7]. Therefore the one-dimensional model is not appropriate, and it is suggests that the melanothallite, Cu_2OCl_2, is a slightly distorted pyrochlore lattice antiferomagnet.

The specific heat measurement also indicates that the long-range magnetic ordering occurs below T_N =70 K [7]. At temperature below T_N, the clear muon spin precession signals were observed in the μSR measurement in the long-range ordered phase [8]. However, the time spectra of the μSR measurement show exponential damping even below T_N, which suggests that spin fluctuations still exist even below T_N because of the geometrical frustration [8]. The spin structure and the origin of large spin fluctuation in the ordered phase have not been determined.

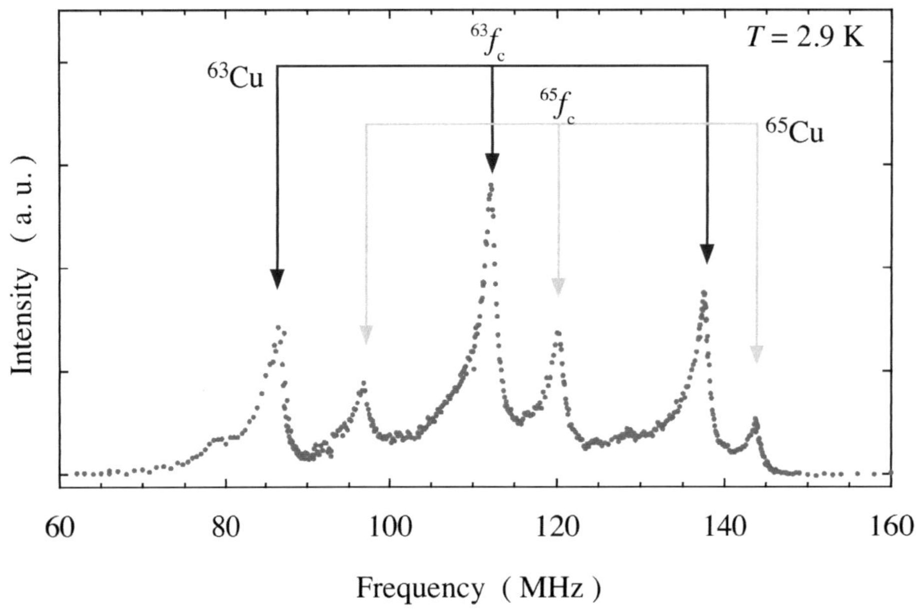

Figure 2. Cu-NMR spectrum at $T = 2.9\,\mathrm{K}$ below T_N with no external magnetic field. The blue and green solid arrows indicate ^{63}Cu, and ^{65}Cu peaks, respectively.

Therefore we have investigated the spin structure and spin fluctuations of the new 3d-electron pyrochlore antiferromagnet, Cu_2OCl_2, microscopically by nuclear magnetic resonance (NMR) measurement.

2. Experiment

We have performed Cu- and Cl-NMR experiments on polycrystalline samples of Cu_2OCl_2. The sample was synthesized by a conventional solid-state reaction of high-purity $CuCl_2$ and CuO powders. The stoichiometric powder mixtures were pressed into pellets and sintered at 623 K for 18 h in CO_2 flow.

The spin-echo spectra of ^{63}Cu ($I = 3/2$), ^{65}Cu ($I = 3/2$)A and ^{35}Cl ($I = 3/2$)@was obtained and the spin-lattice relaxation rates $^{63}T_1^{-1}$ of ^{63}Cu and $^{35}T_1^{-1}$ of ^{35}Cl were obtained using a coherent phase spectrometer. The spectra of Cu nucleus were obtained by a frequency sweep without an external magnetic field. In the case of the Cl-NMR, the spectra were obtained by a field sweep at an operating frequency of 24.08 MHz. No magnetic field was applied for the measurements of the relaxation rates of the Cu nuclei.

3. Experimental Results and Discussions

The Cu spin-echo spectra at $T = 2.9\,\mathrm{K}$ below T_N without external magnetic field is shown in figure 2. Six sharp resonant peaks were clearly observed; three equally-spaced peaks corresponding to ^{63}Cu nucleus and three equally-spaced peaks corresponding to ^{65}Cu nucleus. The three symmetric peaks for each Cu nucleus are due to the internal magnetic field and the electric field gradient (EFG). These resonant peaks can be analyzed on the basis of the Zeeman interaction and electric quadrupole interactions. The Hamiltonian of Cu nucleus is expressed as

$$\mathcal{H} = \frac{eQV_{zz}}{4I(2I-1)}\left[(3I_z^2 - I^2) + \frac{\eta}{2}\;I_+^2 + I_-^2\;\right] + \gamma\hbar H_{\mathrm{local}} I_{z'},$$

Table 1. Values of center-peak frequencies f_c, local fields at Cu site H_{local}, and NQR frequencies ν_Q, and modified EFG ν_Q/Q.

nucleus	f_c (MHz)	H_{local} (T)	ν_Q (MHz)	ν_Q/Q (10^{24} MHz/cm^2)
^{63}Cu	112.2	9.94	25.2	119.4
^{65}Cu	120.2	9.94	23.5	120.5

^{63}Cu; γ_{63}=11.285 (MHz/T) and Q_{63}=−0.211 (10^{24} cm^2).
^{65}Cu; γ_{65}=12.089 (MHz/T) and Q_{65}=−0.195 (10^{24} cm^2).

where I_z, I_+ and I_- are the nuclear spin operators; Q is the nuclear quadrupole moment; γ is the gyromagnetic ratio; and H_{local} is the magnetic local field at the Cu nucleus position. The second-order derivatives of the electric potential $V_{\alpha\alpha} = \partial^2 V/\partial\alpha^2$ ($\alpha = x, y, z$) denote components of the EFG tensor in the principal coordinate system with $|V_{zz}| \geq |V_{yy}| \geq |V_{xx}|$, and $\eta = (V_{xx} - V_{yy})/V_{zz}$ is the asymmetric parameter. The principal axis z of the EFG is generally not along the direction of the localized magnetic field z'. The nuclear quadrupole resonant (NQR) frequency $\nu_Q = eQV_{zz}/4I(2I-1)h$ is a measure of size of the EFG, and the value of ν_Q is experimentally determined by measuring half the distance between the two satellite peaks. The values of ν_Q are obtained from the spectra of ^{63}Cu and ^{65}Cu in figure 2, and these values are presented in table 1. The value of the centre peak frequencies f_c are obtained from the spectra in figure 2, and these values are listed in table 1.

The NQR frequency ν_Q is a quarter of the f_c, as is seen in table 1. Thus the electric quadrupole interaction is too large to be treated as a perturbation for Zeeman interaction. In this case, the shift of the spectra becomes asymmetrical depending on the angle between the principal axis z of the EFG and the principal axis z' of the local magnetic field. The obtained Cu-NMR spectra are symmetric, which implies that the axis z and axis z' are parallel. The local magnetic fields H_{local} are obtained from the centre-peak frequencies f_c by $H_{\mathrm{local}} = f_c/\gamma$, and are listed in table 1. For convenience, the modified EFG are determined as $\nu_Q/Q = eV_{zz}/4I(2I-1)h$, and these values are listed in table 1. The values of $\nu_Q/Q = eV_{zz}/4I(2I-1)h$ and H_{local} depend only on the environment around the Cu nucleus and not on kinds of isotope nucleus. The modified EFG ν_Q/Q and the local field H_{local} at all Cu nucleus positions are found to be identical, as seen in from table 1. In addition, since the obtained spectra are very sharp, H_{local}, ν_Q/Q, and the angle between the principal axes z and z' have no distribution. We find that the EFG, the magnetic moment of Cu spin, and the angle between the principal axis z of the EFG and the principal axis z' of the local magnetic field are uniform. This is suggested that the magnetic and crystallographic unit cells are the same.

We calculated the EFG at the Cu site by assuming a point charge on O^{2-} and on Cl^{1-}. The direction of V_{zz} at the Cu site is found to be the direction almost from Cu to O that locates the center of tetrahedron. The spin structures that the spin directs toward the center of tetrahedron are, so-called, the 2-in-2-out, 3-in-1-out, and all-in-al-out spin structures. Since the nearest neighbour interaction in Cu_2OCl_2 is determined to be antiferromagnetic by the susceptibility measurement, the all-in-all-out spin structure is energetically stable. Therefore, the magnetic structures of Cu_2OCl_2 are the all-in-all-out type, as shown in figure 3.

The ground state of Heisenberg pyrochlore antiferromagnet with only nearest-neighbour interaction J has the macroscopic number of degeneracy even at zero temperature, and has no magnetic long-range ordering. Thus, the observed spin ordering in Cu_2OCl_2 should be caused by additional perturbative effects. Further neighboring interactions, distorted nearest-neighbour interactions, and the anisotropy are considered to be the additional perturbative effects in Cu_2OCl_2. First, we consider further neighbouring interactions and distorted nearest-

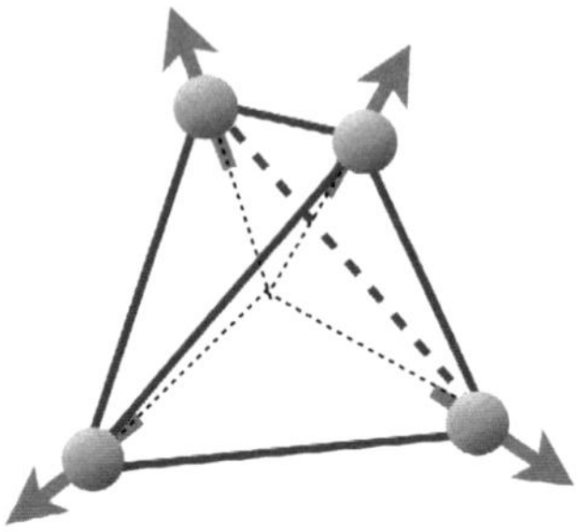

Figure 3. Illustration of all-in-all-out spin arrangement.

neighbouring interactions. The case with the additional exchanges of the second-neighbouring interaction J_2 and the third-neighbouring interaction J_3 were studied by Monte Carlo simulations [4]. The simulations show that the $q = 0$ states are stable. The $q = 0$ states contain the all-in-all-out spin structure and collinear spin arrangement. The collinear spin arrangement is thermally selected due to the entropy effect. In Cu_2OCl_2, the pyrochlore lattice is slightly distorted with $|J| = 113\,\mathrm{K} > |J'| = 108\,\mathrm{K}$. In this case, the collinear ordered spin arrangement is also energetically favored. Thus, models with $|J| > |J'|$ and further interactions cannot explain the obtained spin structure of Cu_2OCl_2 in the ordered phase.

Next we consider the effect of the anisotropy. Because the pyrochlore lattice has no inversion centre between the lattice points, it is possible that Dzyaloshinsky-Moriya interactions (DMI) exist in Cu_2OCl_2. Therefore it is reasonable to discuss the DMI. The DMI are written as $\mathcal{H}_{\mathrm{DMI}} = \sum_{<ij>} \mathbf{D}_{ij} \cdot (\mathbf{S}_i \times \mathbf{S}_j)$, where $\mathbf{D}_{ij}$ is a DMI vector. The ground state of Heisenberg pyrochlore antiferromagnet with DMI has been studied by Monte Carlo simulations [5]. In reference 5, it has been considered to have two types of DMI configurations — *direct* and *indirect* DMI. In the pyrochlore lattice, the $\mathbf{D}$ vector exists in the vertical plane at the middle point of the edge and is parallel to the opposite side edge. The difference between *direct* and *indirect* is the sign of $\mathbf{D}$. The *direct* and *indirect* DMI lead to the all-in-all-out structure and a collinear ordered spin structure, respectively. Thus the DMI resolve the degeneracy between the all-in-all-out state and collinear ordered states. Since Cu_2OCl_2 is not an ideal pyrochlore structure and DMI configuration depend on the symmetry of the crystallographic structure, we cannot simply apply this DMI configuration to Cu_2OCl_2. It is, however, thought that DMI can exist in Cu_2OCl_2, because Cu_2OCl_2 has no inversion centre between the Cu^{2+} ions and because the perpendicular bisector at the middle point of the J bond is a mirror plane. It is likely that the DMI in J bond is similar to the *direct* DMI configuration. Thus, we conclude that the DMI play the important roles on the magnetic ordering and structure in Cu_2OCl_2.

The rates ${}^{63}T_1^{-1}$ of ^{63}Cu were measured at the centre peak that corresponds to $-1/2 \leftrightarrow +1/2$ transition. The temperature dependences of ${}^{63}T_1^{-1}$ and ${}^{35}T_1^{-1}$ of ^{35}Cl are shown in figure 4. The divergent behaviour of the temperature dependence of ${}^{35}T_1^{-1}$ is observed near T_{N} as shown in figure 4. This implies that a critical slowing down occurs around T_{N}. We found that both rates in the ordered phase are proportional to the temperature, $T_1^{-1} \propto T$. When the spin-wave excitation is dominant in the long-range ordered spin system, the T_1^{-1} exhibits an approximately a power-like-law dependence on temperature, T^{α} where $\alpha > 2$. The results for T_1^{-1} may suggest that although the long-range ordering occurs at rather high temperature, the large spin fluctuations caused by the geometrical frustration still remain. These results are consistent with the results of the μSR results.

In summary, we performed the NMR study of the 3d-electron pyrochlore antiferromagnet Cu_2OCl_2. Based on the analysis of the Cu-NMR spectra, Cu_2OCl_2 has long-range ordering

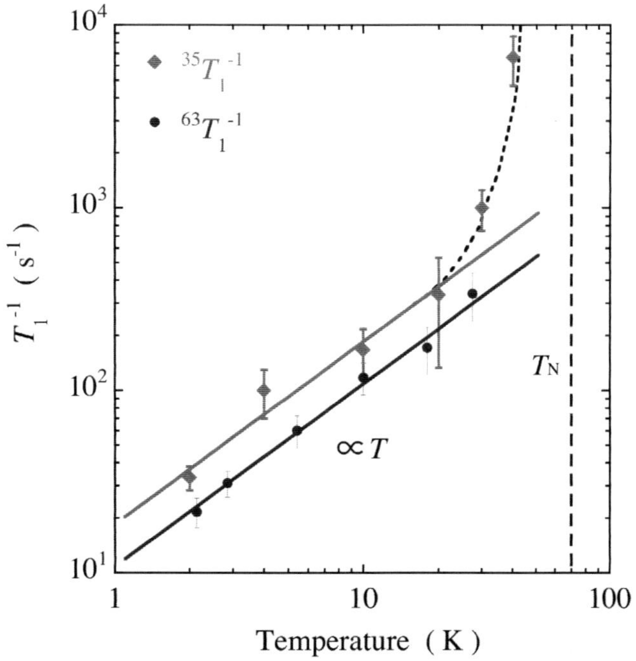

Figure 4. Temperature dependence of spin-lattice relaxation rates, $^{35}T_1^{-1}$ of ^{35}Cl and $^{63}T_1^{-1}$ of ^{63}Cu. Red squares represent $^{35}T_1^{-1}$, and blue circles represent $^{63}T_1^{-1}$. The dashed line indicates T_N. The rate $^{35}T_1^{-1}$ was measured under the external field of 57.7 kOe at an operating frequency of 24.08 MHz. The rate $^{63}T_1^{-1}$ was measured without the external field at the centre peak. The red and blue solid lines represent $T_1^{-1} \propto T$, and the dotted line is a guide for the eye.

below T_N. The spin structure of Cu_2OCl_2 in ordered phase is all-in-all-out type spin structure, and the DMI is a plausible cause for this spin ordering. The temperature dependence of T_1^{-1} suggests that large spin fluctuations exist in the ordered phase.

Acknowledgments
This study was supported by a Grant-in-Aid for Scientific Research on Priority Areas "Novel States of Matter Induced by Frustration" (19052005) and a Grant-in-Aid for Scientific Research(C) (20540360) from MEXT, Japan.

References
[1] Maegawa S, Sato S and Oyamada A 2010 *J. Phys. Soc. Jpn.* **79** 011002.
[2] De Pape R and Ferey G 1986 *Mater. Res. Bull.* **21** 971.
[3] Pinsard-Gaudart L, Rodríguez-Carvajal, Gukasov A and Monod P 2004 *Pys. Rev.* B **69** 104408.
[4] Reimers JN, Greedan JE and Björgvinsson M 1992 *Pys. Rev.* B **45** 7295.
[5] Elhajal M,Canals B, Sunyer R and Lacroix C 2005 *Pys. Rev.* B **71** 094420.
[6] Komma K and Izumi F 2008 *J. Appl. Crystallogr.* **41** 653.
[7] Okabe H, Suzuki H, Kawashima K and Akimitsu J 2006 *J. Phys. Soc. Jpn.* **75** 123705.
[8] Kawashima K, Okabe H, Suzuki K, Kuroiwa S, Akimitsu J, Sato KH, Koba A and Kadono R 2007 *J. Phys.: Cond. Matt.* **19** 145275.

Journal of Physics: Conference Series **320** (2011) 012031 doi:10.1088/1742-6596/320/1/012031

New category of the frustrated quantum magnets composed of spin-1/2 triple-chains

Masashi Fujisawa[1], **Hikomitsu Kikuchi**[A], **Yutaka Fujii**[B], **Seitaro Mitsudo**[B], **Akira Matsuo**[C], **Koichi Kindo**[C]

Headquarters for the Advancement of High Priority Research, University of Fukui, Fukui 910-8507, Japan

[A]Department of Applied Physics, University of Fukui, Fukui 910-8507, Japan

[B]Research Center for Development of Far-Infrared Region, University of Fukui, Fukui 910-8507, Japan

[C]Institute for Solid State Physics, Univ. of Tokyo, Kashiwa, Chiba 277-8581, Japan

E-mail: fujisawa.m.ad@m.titech.ac.jp

Abstract. $Cu_3(OH)_4MO_4$, which is a natural mineral szenicsite, has been studied by the magnetic susceptibility, the specific heat, and the high field magnetization measurement. This compound consists of S =1/2 triple-chain of the edge-sharing octahedra CuO_6. On the basis of the crystal structure in szenicsite, spin frustration can strongly affect the ground state. The magnetic susceptibility has a broad maximum around 80 K, which is characteristic of the quasi-one-dimensional antiferromagnet. There are no sign of magnetic phase transitions from 300 K down to 2 K in our measurements. The magnetization curve measurement up to 48 T at 1.3 K indicates that neither magnetic order nor spin gap is present. Assuming the triple-chains as the isolated uniform chains, we estimated the dominant exchange interaction $J/k_B =$ 66 K.

1. Introduction

Spin frustration is one of big issues in the solid state physics. Recently much attention has been paid to peculiar magnetic properties of a new category of frustrated quantum magnets, spin-1/2 triple-chains. Antlerite ($Cu_3(OH)_4SO_4$) [1, 2, 3, 4] and szenicsite ($Cu_3(OH)_4MoO_4$)[5] are candidates for spin-1/2 triple-chains. Figure 1(a) and (b) show the exchange interactions of triple-chains which represent antlerite and szenicsite, respectively. Because triple-chains include competing interactions, spin frustration can strongly affect their ground state.

In antlerite, the magnetic susceptibility above 40 K follows the Curie-Weiss law with the Weiss temperature θ = 1.69 K. The magnetic susceptibility has broad maximum at 6.5 K, which is characteristic of quasi-one-dimensional antiferromagnet. Antlerite has a magnetic phase transition at T_N = 5 K observed by magnetic susceptibility and specific heat measurement[1, 2]. The anomalous spin state, which is called "idle-spin", is realized at low temperature in antlerite [1, 3].

Natural mineral szenicsite is also a model substance of an S = 1/2 triple-chain system. The crystal structure of szenicsite belongs to orthorhomobic, space group $Pnnm$ with a= 8.5201(8) Å, b = 12.545(1) Å, and c = 6.0794(6) Å. Figure 2(a) shows the crystal structure of szenicsite.

[1] Present address: Research Center for Low Temperature Physics, Tokyo Institute of Technology, Tokyo

Journal of Physics: Conference Series **320** (2011) 012031 doi:10.1088/1742-6596/320/1/012031

The Cu^{2+} triple-chains are separated from each chains by MoO_4 tetrahedra. Edge-sharing CuO_6 octahedra form one-dimensional chain of Cu^{2+} ions, and these three chains form $S = 1/2$ triple-chains along c-axis. From the crystal structure of szenicsite, we can expect than the hole orbitals $d(x^2 - y^2)$ lie in a shaded square, and are linked together along the chain, as shown in Fig. 2 (b).

No magnetic properties of szenicsite have yet been reported. As shown in Figs. 1 (a) and (b), the symmetry of exchange interactions in szenicsite is different from that of antlerite. Therefore novel magnetic properties are expected. In order to investigate the magnetic properties of szenicsite, we have performed the magnetic susceptibility, the high field magnetization and the specific heat measurements.

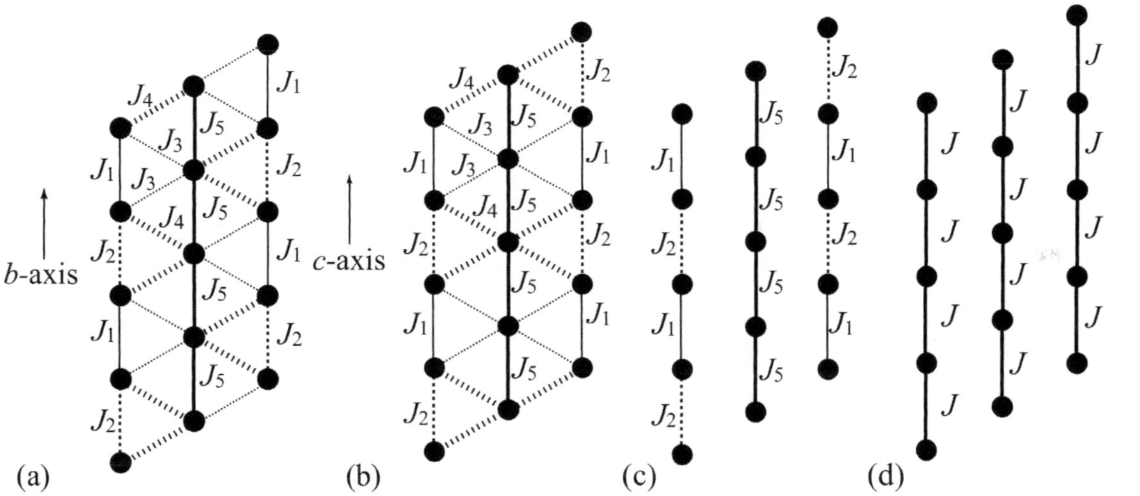

Figure 1. Exchange interactions for (a) antlerite, (b) szenicsite, (c) a uniform chain and alternating chains, and (d) three uniform chains.

2. Experiment

The natural mineral of szenicsite, which was purchased at a stone shop, was used for the measurements because it is difficult to synthesize this compound. The magnetic susceptibility were measured down to 1.9 K at $H = 0.1$ T using a SQUID magnetometer (Quantum Design MPMS XL). The high-field magnetization measurement was performed using an induction method with a multilayer pulse magnet at the Ultra-High Magnetic Field Laboratory, Institute for Solid State Physics, the University of Tokyo. The specific heat measurement was carried out down to 1.9 K using a Quantum Design PPMS by the relaxation method.

3. Results and Discussion

Figure 3 shows the temperature dependence of magnetic susceptibility of szenicsite measured at $H = 0.1$ T. Above 200 K, the susceptibility can be fitted to the Curie-Weiss law $(T) = C/(T - \theta)$, where C is Curie constant and θ is Weiss temperature. The Curie-Weiss fit leads to Weiss temperature $\theta = -70$ 5 K. Therefore the main exchange interactions are antiferromagnetic in szenicsite.

The magnetic susceptibility has a broad maximum around $T_{max} = 80$ K, which indicates that szenicsite can be classified as quasi-one-dimensional antiferromagnets. The rapid increase of the magnetic susceptibility below 30 K results from a paramagnetic impurity contribution

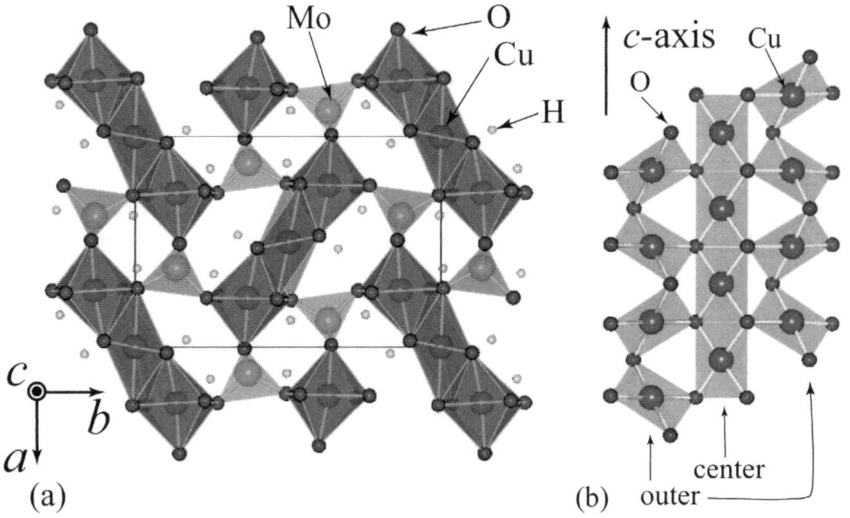

Figure 2. (a) Crystal Structure viewed along the c-axis in szenicsite. The triple-chains are located at the corner and the center of the chemical unit cell in the $a - b$ plane. (b) A triple-chain structure of Cu^{2+} in szenicsite. The hole orbitals $d(x^2 - y^2)$ of Cu^{2+} ions lie in shaded square.

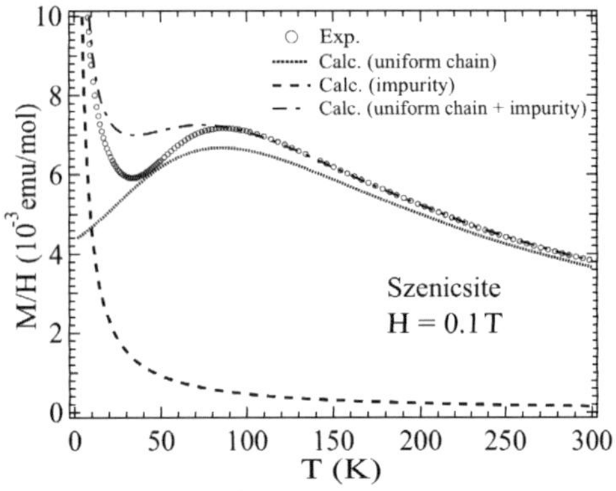

Figure 3. Temperature dependence of magnetic susceptibility in szenicsite measured at $H = 0.1$ T. Open circles, solid, dashed, and dashed-and-dotted lines denote the experimental data, the calculation of uniform chain model ($J/k_{\rm B} = -66$ K), the calculation of paramagnetic impurity, and the calculation of uniform chain and paramagnetic impurity, respectively.

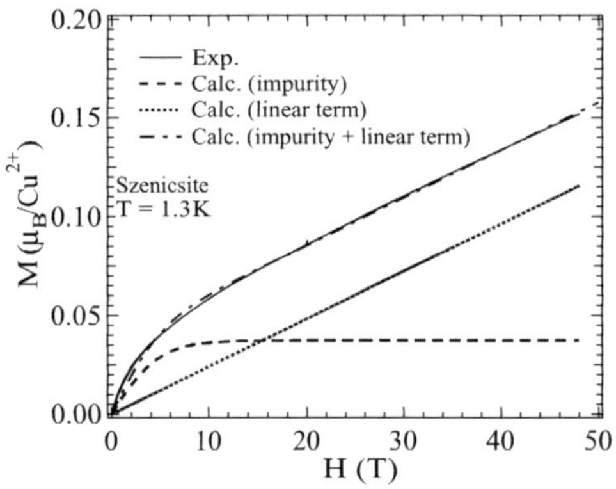

Figure 4. Magnetization curve in szenicsite measured at $T = 1.3$ K. Solid line, dashed line, dotted line, and dashed-and-dotted line denote the experimental data, the calculation of paramagnetic impurity, the calculation of linear term, and the calculation of paramagnetic impurity and linear term, respectively.

because the magnetic susceptibility below 20 K is described by $_{\rm imp.} \propto 1/T$. Magnetic phase transition is not observed from $T = 300$ K to 2.0 K. The absence of magnetic order and the broad maximum of magnetic susceptibility can prove the low dimensionality of szenicsite.

Figure 4 shows the high-field magnetization process measured at 1.3 K in magnetic field

Journal of Physics: Conference Series **320** (2011) 012031 doi:10.1088/1742-6596/320/1/012031

of up to 48 T. The magnetization smoothly increased as the field was increased. Below 7 T, the magnetization due to paramagnetic impurities, which is described by Brillouin function, is dominant. The intrinsic magnetization, which is given by subtraction of the impurity contribution from the raw data, has linear dependence on the magnetic field. From these results, the experimental magnetization can be reproduced by,

$$M(H) \propto \frac{N_{\mathrm{imp}} g \ \ _{\mathrm{B}}}{2} \tanh \frac{g \ \ _{\mathrm{B}} H}{2k_{\mathrm{B}} T} + aH, \tag{1}$$

where a and N_{imp} are constant and the number of paramagnetic impurity, respectively. The experimental data agree well the calculation as shown in Fig. 4. Even if this system has spin gap, the magnitude of the gap is negligible.

Figure 5 shows the total specific heat in szenicsite. There are no anomalies in the measured range. Magnetic phase transition is not observed above 2.0 K. This result is consistent with that of magnetic susceptibility. On the other hand, the specific heat of antlerite has the λ like anomaly, which indicates magnetic phase transition[2].

Even though the crystal structure of szenicsite is similar to that of antlerite, the magnetic properties of szenicsite are considerably different compared to those of antlerite. From the experimental results, the exchange interactions between the triple-chains via MoO_4 octahedra is negligible and the absolute value of the dominant exchange interactions in szenicsite are expected to be large compared to that in antlerite.

It is difficult to analyze the magnetic properties of szenicsite without simplification because the triple-chains of szenicsite has the five different exchange interactions. We tried to reproduce the temperature dependence of magnetic susceptibility by employing a simple and reasonable model. In order to analyze the magnetic susceptibility, we let us take assumptions on the basis of crystal structure.

Table 1 shows the distance between Cu-Cu and the angle between Cu-O-Cu in the exchange interactions J_i (i = 1, 2, 3, 4, 5) in szenicsite. The Cu-O-Cu angles in szenicsite are from 99.4° to 104.9°. Hase *et al.* reported that the exchange interaction between the coppers is antiferromagnetic if the Cu-O-Cu angles are larger than 97° in the exchange pass of Cu-O-Cu[6]. Therefore the exchange interactions J_i (i = 1, 2, 3, 4, 5) of szenicsite are supposed to be antiferromagnetic.

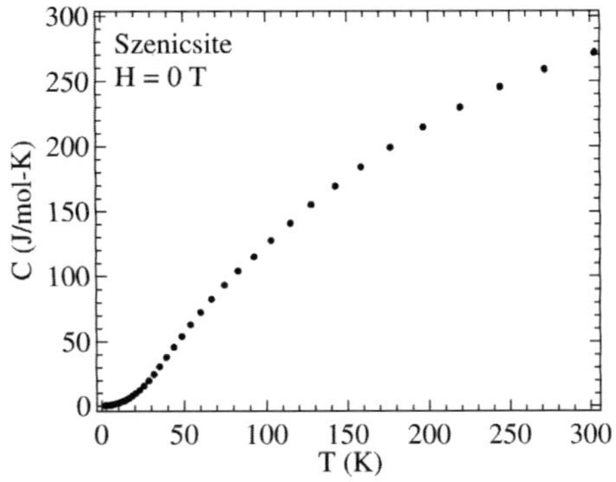

Figure 5. Total specific heat in szenicsite measured at zero field.

Table 1. Cu-Cu distance and Cu-O-Cu angle for J_i.

J_i	Cu-Cu distance (Å)	Cu-O-Cu angle (°)
J_1	3.034	103.6
J_2	3.046	104.9
J_3	3.211	103.8
J_4	3.213	104.2
J_5	3.040	99.4

The magnitude of the exchange interaction J_3 can be close to that of J_4 because the Cu-O-Cu angle and the Cu-Cu distance in J_3 are close to those in J_4. And we suppose the magnitude of J_3 and J_4 is smaller than J_1,J_2, and J_5 because the distances of Cu-Cu in J_3 or J_4 are larger than those of J_1, J_2, and J_5.

In addition, geometrical spin frustration can affect the magnetic properties of szenicsite because the network of the exchange interactions in the triple-chains are composed of triangle units and each of the exchange interactions is antiferromagnetic. Therefore the effect of J_3 and J_4 can be canceled out. If we ignore the effect of zig-zag rung exchange interactions J_3 and J_4, we can regard the network of the exchange interactions in szenicsite (as shown in Fig. 1 (b)) as a simple network composed of both a uniform chain and two alternating chains (as shown in Fig.1 (c)).

Furthermore the magnitude of J_1 can be close to J_2 because the Cu-O-Cu angle and the Cu-Cu distance in J_1 are close to those in J_2. Generally a ground state of alternating spin chain is singlet. However if the alternation parameter $\alpha = J'/J$ is close to 1, the temperature dependence of the magnetic susceptibility is similar to that of the uniform chain except for very low temperatures ($T << J/k_B$)[7]. Because the distances between Cu-Cu of J_1, J_2 and J_5 are close to each other, we would assume $J_1 = J_2 = J_5 = J$ as shown in Fig. 1(d).

The magnetic susceptibility of $S = 1/2$ antiferromagnetic uniform chain is calculated[8, 9]. Figure 3 shows the reproduction of temperature dependence of magnetic susceptibility by using uniform chain term and impurity term. By comparing the calculation and the experimental data, we estimated the exchange interaction $J/k_B = -66$ K. The discrepancy between experimental data and calculation below 70 K is attributed to oversimplification of our model.

4. Conclusion

We have performed the magnetic susceptibility, the specific heat, and the high field magnetization measurements on the $S = 1/2$ triple-chain antiferromagnet szenicsite ($Cu_3(OH)_4MO_4$). The magnetic susceptibility exhibits broad maximum around 80 K, which is characteristic of the one-dimensional antiferromagnet. Specific heat data from $T = 2.0$ K to 300 K indicate no magnetic phase transition. High field magnetization process up to 48 T at 1.3 K has revealed that szenicsite has neither magnetic phase transition and nor spin gap. Even though the framework of exchange interactions in szenicsite is similar to that of antlerite, the magnetic properties of szenicsite are fairly different from those of antlerite. The magnetic properties of szenicsite can be described by the $S = 1/2$ triple-chain model with the exchange interactions along the leg (J_1, J_2 in outer alternating chains and J_5 in center uniform chain) and the zig-zag diagonal exchange interactions (J_3 and J_4). We reproduced the temperature dependence of magnetic susceptibility by using the simple model, in which the exchange interactions J_i were satisfied with the following relations from the crystal structural point of view; $J_1 = J_2 = J_5 = J \neq 0$ and $J_3 = J_4 = 0$. We estimated the dominant exchange interaction $J/k_B = -66$ K.

Acknowledgement
This work was partly supported by a Grant-in-Aid for Scientific Research (C) No. 18540343 from the Japan Society fro the promotion of Science and a Grant in Aid for Scientific Research on Priority Area (No. 19052005 "Novel states of Matter Induced by Frustration") from the Ministry of Education, Culture, Sports, Science and Technology of Japan.

References
[1] Vilminot S, Richard-Plouet M, Andre G, Swierczynski D, Guillot M, Bouree-Vigneron F, and Drillon M, 2003 *J. Solid State Chem.* **170** 255
[2] Vilminot S, Andre G, Bouree-Vigneron F, Richard-Plouet M, and Kurmoo M 2007 *Inorg. Chem.*, **46** 10079.
[3] Fujii Y, Azuma Y, Kikuchi H, and Yamamoto Y, 2009 *J. Phys. Conf. Series* **145** 012061.
[4] Okubo S, Yamamoto H, Fujisawa M, Ohta H, Nakamura T, and Kikuchi H 2009 *J. Phys. Conf. Series* **150** 042156.
[5] Burns P C, 1998 *Mineral. Mag.* **62** 461.
[6] Hase M, Kohno M, Kitazawa H, Tsujii N, Suzuki O, Ozawa K, Kido G, Imai M, and Hu X 2006 *Phys. Rev. B* **73** 104419.
[7] Johnston D C, Kremer R K, Troyer M, Wang X, Klumper A, Bud'ko S L, Panchula A F, and C. Canfield P 2000 *Phys. Rev. B* **61** 9558.
[8] Bonner J C, and Fisher M E 1964 *Phys. Rev.* **135** A640.
[9] Hatfield W E 1981 *J. Appl. Phys.* **52** 1985.

NMR study of quantum spin liquid and its phase transition in the organic spin-1/2 triangular lattice antiferromagnet $EtMe_3Sb[Pd(dmit)_2]_2$

Satoru Maegawa[1,*], Tetsuaki Itou[1], Akira Oyamada[1] and Reizo Kato[2]

[1] Graduate School of Human and Environmental Studies, Kyoto University, Kyoto 606-8501, Japan

[2] Condensed Molecular Materials Laboratory, RIKEN, Wako-shi, Saitama 351-0198, Japan.

E-mail[*] : maegawa.satoru.3w@kyoto-u.ac.jp

Abstract. We found by ^{13}C-NMR experiment that a quasi-two-dimensional organic antiferromagnet with spin 1/2 on the triangular lattice, $EtMe_3Sb[Pd(dmit)_2]_2$, has no classical magnetic ordering down to 19.4mK, although the magnetic exchange interaction J is 220-250K. The quantum spin liquid state is realized in this material due to the significant spin frustration and quantum fluctuation. The ^{13}C nuclear spin-lattice relaxation rate T_1^{-1} under an external magnetic field of 7.65T indicates that the excitations have no spin gap in this spin liquid phase above 1.0K. Then the relaxation rate exhibits a clear kink at 1.0K, and shows a steep decrease below this temperature proportional to T^2. This clearly shows a continuous phase transition at 1.0K and the steep decrease of the spin fluctuation could show the appearance of the excitation gap in the lower spin liquid phase. The power law but not exponential law of the temperature dependence may imply a nodal gap rather than a full gap.

1. Introduction

An antiferromagnet on the triangular lattice is the most fundamental geometrically frustrated magnet. The most simple example is an Ising type antiferromagnet on the triangular lattice, which has a macroscopic number of degeneracy in the ground state, and has no magnetic ordering even at zero temperature. For the Heisenberg type case the exotic resonating valence bond (RVB) state composed of a quantum mechanical superposition of singlet pairs has been proposed [1], and attracted much attention in the field of condensed matter physics also because of its relation to the possible origin for high temperature superconductivity. Now it is recognized theoretically, however, that the nearest neighbor Heisenberg antiferromagnet on the triangular lattice has a 120° spin ordered structure in the ground state for both classical and quantum spin systems. On these theoretical interests the real model materials for the triangular lattice antiferromagnet have been sought. Especially the spin 1/2 system is more attractive, as the quantum fluctuation would destroy the conventional magnetic ordering and is expected to give rise to a new quantum spin state. Recently we discovered that the organic compound $EtMe_3Sb[Pd(dmit)_2]_2$ (Et = C_2H_5, Me = CH_3, dmit = 1,3-dithiole-2-thione-4,5-dithiolate) is an ideal model material for the-two dimensional antiferromagnet on the triangular lattice and realizes a quantum spin liquid state [2], as well as the organic compound of κ- $(BEDT\text{-}TTF)_2Cu_2(CN)_3$ (BEDT-TTF = bis(ethylenedithio)-tetrathiafulvalene) [3].

In this paper, we report the discovery of a quantum spin-liquid state and its instability found in $EtMe_3Sb[Pd(dmit)_2]_2$ by means of ^{13}C-NMR experiment down to 19.4mK [2] [4] [5].

A layered organic compound $EtMe_3Sb[Pd(dmit)_2]_2$ is a good candidate material for the quantum antiferromagnet on the triangular lattice, as shown in Fig. 1. The $Pd(dmit)_2$ molecules are strongly dimerized, and these dimers $[Pd(dmit)_2]_2$ form the nearly regular-triangular lattice and have spin s = 1/2. The magnetic $[Pd(dmit)_2]_2$ layers are separated by nonmagnetic layers of the closed-shell monovalent cation $EtMe_3Sb$, and thus the compound is considered to be a quasi two-dimensional magnet. The localized spins on the dimers interact antiferromagnetically with each other. The transfer integrals were calculated using the extended Hückel method to be 28.3, 27.7 and 25.8meV between the $[Pd(dmit)_2]_2$ dimers, while 453.5meV for intradimer. Thus the three exchange interactions are nonequivalent but close to each other [6]. The exchange interactions are estimated to be 220–250K by the static susceptibility measurement and the high temperature expansion of the regular-triangular spin-1/2 Heisenberg antiferromagnet [7] [2] with a g-value of 2.038 obtained by ESR measurement [8]. The ambiguity of the value of exchange interaction may be attributed to the approximation of a regular-triangular lattice and/or an existence of higher-order exchange interactions.

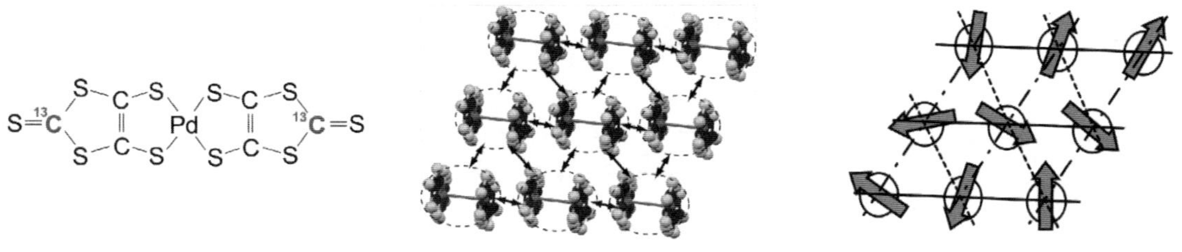

(a) (b) (c)

Figure 1. (a) $Pd(dmit)_2$ molecule with enriched ^{13}C isotope at both ends of the molecule. (b) Crystal structure of a $[Pd(dmit)_2]_2$ layer. Two $Pd(dmit)_2$ molecules are dimerized. (c) Schematic of the spin system of $EtMe_3Sb[Pd(dmit)_2]_2$, where circles represent $[Pd(dmit)_2]_2$ dimers with localized 1/2 spins.

2. Experimental method

Fine high-quality single crystals of $EtMe_3Sb[Pd(dmit)_2]_2$ were synthesized by an aerial oxidation method. The typical sizes were of 0.3mm×0.3mm×0.01mm. For the ^{13}C-NMR measurements, the carbon atoms at both ends of the $Pd(dmit)_2$ molecules were enriched with ^{13}C isotopes, as shown in figure 1(a). A large number of the single crystals were packed into a glass tube with random orientation. The ^{13}C-NMR measurements were performed from 300K to 19.4mK under a magnetic field of 7.65T. The NMR spectra were obtained by Fourier transformation of spin echo signals refocused after the $\pi/2$ - π pulse sequence. The typical pulse widths of π /2 and π were 3 and 6μs, respectively. The inverse of these values were sufficiently larger than the present spectral widths, and thus the pulses could cover the whole NMR spectra. The spin-lattice relaxation rate T_1^{-1} was obtained from the recovery curve of the integrated spin-echo intensity $M(t)$ after time delay t from saturation comb pulses.

3. Experimental results and discussions

Figure 2 shows the ^{13}C-NMR spectra of $EtMe_3Sb[Pd(dmit)_2]_2$. The shape of the spectra scarcely changes, and the critical broadening of the spectrum, which is characteristic of spin freezing, was not observed in the whole temperature range from 272K to 19.4mK. The observed local static field, which is estimated from the width of the spectrum, is too small even at lowest temperature to be understood as a magnetic long range ordering or a spin glass state. The spectral tail is at most within

50kHz, which is much smaller than the scale of the hyperfine coupling constant discussed later. This result indicates that $EtMe_3Sb[Pd(dmit)_2]_2$ does not undergo either spin ordering or freezing down to 19.4mK, which is 1/10,000 of the exchange interaction J. Therefore we can conclude that a quantum spin liquid state is realized in this material. A very small broadening of the spectra at low temperatures is attributed to an inhomogeneous broadening due to slight static local fields [2]. This small inhomogeneous broadening is also observed in the other spin liquid material κ-(BEDT-TTF)$_2Cu_2(CN)_3$ [3], and thus is considered to be a universal nature of these spin liquids, which are expected to have considerably long-range spatial spin correlation.

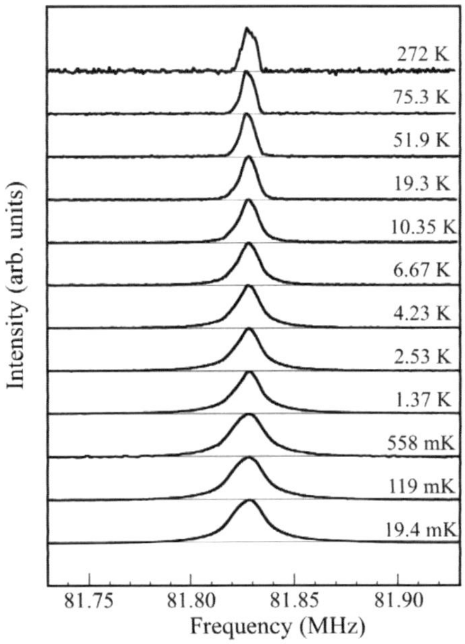

Figure 2. Temperature variation of ^{13}C-NMR spectra of $EtMe_3Sb[Pd(dmit)_2]_2$ at 7.65T.

The temperature dependence of the ^{13}C nuclear spin-lattice relaxation rates T_1^{-1} at an external magnetic field of 7.65T is shown in figure 3. The recoveries of the ^{13}C nuclear magnetization $M(t)$ were not single exponential at some temperature region, but could be expressed by the stretched exponential function in the whole experimental temperatures as,

$$\frac{M(\infty)-M(t)}{M(\infty)}=\exp\left\{-\left(\frac{t}{T_1}\right)^{\beta}\right\} \quad , \tag{1}$$

where β is the stretching exponent. The stretching coefficient β generally denotes the degree of inhomogeneity of the spin system, while the nuclear magnetization in the homogeneous system relaxes exponentially with $\beta=1$. The temperature dependence of the obtained coefficient β is shown in figure 4. In figure 3, we also show the relaxation rate determined from an initial decay of the recovery curve. The relaxation rate derived from the initial decay gives the average value among different relaxation rates. The difference between the two relaxation rates shows the degree of inhomogeneity.

The rate T_1^{-1} gradually decreases as temperature decreases from 300K. The rate T_1^{-1} stops decreasing and has a finite value between 5 and 1K. Then a steep decrease is observed below 1.0K with a kink.

First, we consider the spin dynamics at high temperatures. We can evaluate the antiferromagnetic spin correlations by comparing T_1^{-1} with the static spin susceptibility χ. The spin-lattice relaxation rate is the NMR frequency ω_n component of Fourier transformation of the time-autocorrelation function of the fluctuating transverse local field $\delta H^{\pm}(t)$ at nuclear sites as,

$$T_1^{-1} = \frac{\gamma_n^{\ 2}}{2}\int_{-\infty}^{\infty} dt \left\langle \{\delta H^{-}(t)\delta H^{+}(0)\} \right\rangle \exp(-i\omega_n t) \ , \qquad (2)$$

where γ_n is the gyromagnetic ratio of nuclear spin. The fluctuating local field at nuclear sites comes from the fluctuation of electron spins through the hyperfine interaction, and the contributing spin correlation function is related to the static susceptibility $\chi'(q,\ \omega = 0)$ owing to the fluctuation-dissipation theorem [9]. For the high-temperature paramagnetic state where the spin autocorrelation

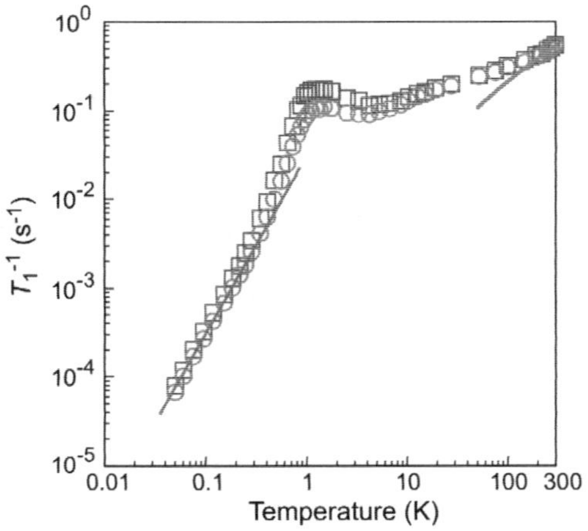

(a)

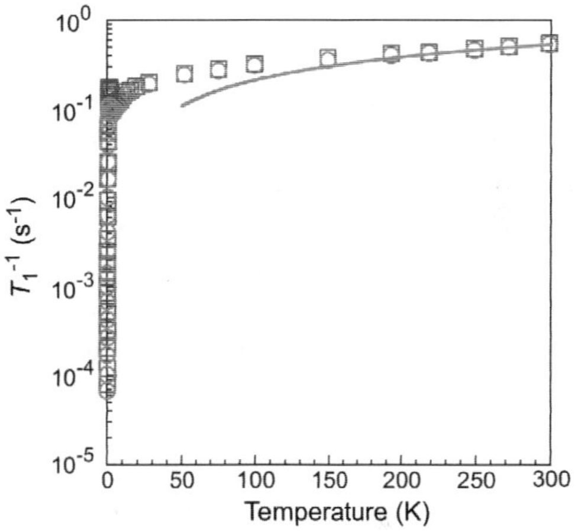

(b)

Figure 3. Temperature dependence of relaxation rate T_1^{-1}. The horizontal axis is logarithmic in figure (a), while it is linear in figure (b). The blue circles indicate the values determined from the stretched-exponential analysis, and the red squares denote the values determined from the initial decay of recovery curves. An orange solid curve shows values calculated using 3.6 (mol emu^{-1} K^{-1} s^{-1})$\times\chi T$ for the relaxation rate in the paramagnetic fluctuation at high temperatures. The green solid line denotes the temperature dependence proportional to T^2 in the lower quantum spin liquid phase.

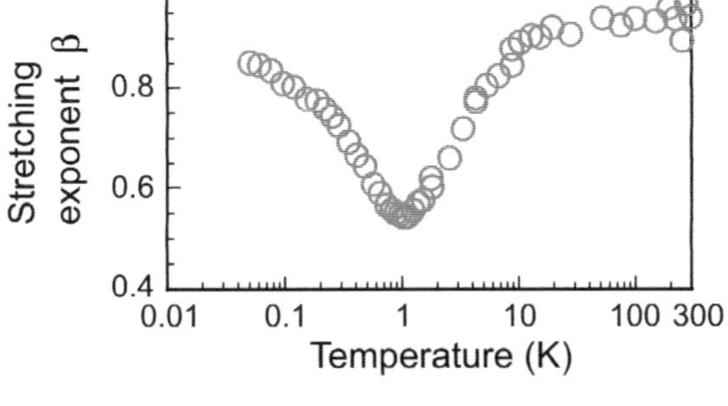

Figure 4. Temperature dependence of stretching exponent β in the recovery curve fitted to equation (1).

Journal of Physics: Conference Series **320** (2011) 012032 doi:10.1088/1742-6596/320/1/012032

function decays exponentially with time, the rate is derived as

$$T_1^{-1} \propto T \sum_q \frac{\chi'(q,0)}{\Gamma_q} \quad , \tag{3}$$

where $\chi(q,\omega) = \chi'(q,\omega) + i\chi''(q,\omega)$ is the dynamic spin susceptibility and Γ_q is the decay rate of the autocorrelation function. At sufficiently high temperatures ($T >> J$) we can also assume that the autocorrelation function has no q-dependence due to random fluctuation, and is equal to the static susceptibility, i.e. $\chi'(q, 0) = \chi$. So the temperature dependence of T_1^{-1} in the paramagnetic region can be described as [10]

$$T_1^{-1} \propto \chi T \ . \tag{4}$$

The calculated values using the experimental susceptibility are shown by the orange solid lines in figure 3(a) and (b). The calculated curve agrees well with the experimental data above 180K, as is clearly found in figure 3(b). Thus, antiferromagnetic correlations are almost absent around room temperature. The proportionality coefficient in equation (4) is 3.6 mol emu^{-1} K^{-1} s^{-1}, which gives a hyperfine coupling constant to be $9{\times}10^2$ kHz/μ_B by assuming Gaussian fluctuation [10] and by using an exchange frequency $\omega_{ex} = 5{\times}10^{13}$ rad/ s, which is calculated from the exchange J = 220–250K and the neighboring number z = 6. This hyperfine value is consistent with our previous estimation of several hundreds of kHz/μ_B [11], which was calculated on an analysis of the spectral width.

The experimental data is, however, enhanced from the χT line below 180K, which corresponds to the energy of J. The substantial deviation of the experimental T_1^{-1} data suggests the development of the antiferromagnetic spin correlations, because it indicates an enhancement of $\chi'(q, 0) /\chi'(0, 0)$ and a slowing down of Γ_q at a specific q.

The relaxation rate T_1^{-1} smoothly decreases down to 1K without any features of the spin ordering or freezing, and has a finite value between 5 and 1K, as is shown in figure 3(a). This temperature dependence indicates that this system does not have an appreciable spin gap. The paramagnetic state remains in spite of the development of antiferromagnetic correlation from much higher temperatures around 180K. Since 1K is lower than 1% of J, thermal fluctuations are negligible in this temperature region. Thus, the absence of spin ordering or freezing is attributed not to thermal fluctuations but to quantum fluctuations. Considering the absence of an appreciable spin gap, this state is clearly distinct from the VBS state accompanied by the spin dimerization. Therefore, we can conclude that this state is regarded as the quantum spin-liquid state with no spin gap above the temperature of 1.0K.

The temperature dependence of the relaxation rate T_1^{-1} shows an obvious kink at around 1.0K, as is found in figure 3(a). This strongly suggests that a phase transition occurs at this temperature. This transition is not a first-order but a continuous transition, as no discontinuous jump is observed in the temperature dependence of T_1^{-1}. The continuous phase transitions always accompany changes of states, that is symmetry breaking and/or topological ordering. Therefore, our result indicates that the gapless spin liquid changes to a different spin state accompanying symmetry breaking and/or topological ordering. This transition is not a classical magnetic ordering or freezing, because the anomaly in T_1^{-1} is not a critical divergence and also the NMR spectrum shown in figure 2 has no broadening below this temperature.

The relaxation rate T_1^{-1} in the lower temperature phase below 1.0K shows a steep decrease, as is found in figure 3(a). The stretching coefficient β decreases to 0.5 around the transition temperature, and returns gradually to 1.0 at lower temperatures, as shown in figure 4, that is, the recovery curves of nuclear magnetization return to single exponential functions at sufficiently low temperatures. This temperature dependence suggests that the distribution of T_1^{-1} is amplified by the slight imperfections in the sample and the increase of the correlation length in the vicinity of the phase transition of this quantum spin liquid.

The rate T_1^{-1} in the lower temperature phase is clearly proportional to T^2. This is in contrast to the nature of the fully gapless spin liquid with the spinon Fermi surface, where $T_1^{-1} \propto T$ for a Fermi-liquid case or diverges for a non-Fermi-liquid case. Thus, the low temperature phase is not fully gapless, and has a spin gap at least in some portion of q-space as a nodal gap, which is similar to superconducting gaps in anisotropic superconductors.

4. Conclusion

In conclusion, our ^{13}C-NMRexperiments show that the s = 1/2 Heisenberg antiferromagnet on the triangular lattice $EtMe_3Sb[Pd(dmit)_2]_2$ does not undergo classical spin ordering or freezing down to 19.4mK, which is less than 1/10,000 of the exchange interaction J. While this quantum spin liquid has a gapless spin excitation above 1.0K, we found clear evidence that the spin system under 7.65T shows an instability other than classical ordering at around 1.0K and has a spin gap below this temperature. This gap may be nodal similar to that in anisotropic superconductors.

Quantum liquids of ^{3}He, ^{4}He and electrons in metals exhibit exotic phase transitions with the symmetry breaking as superfluidity and superconductivity. Various attractive predictions for instability or transition in the quantum spin liquid have been proposed theoretically, such as the spinon pairing [12][13][14], chiral RVB [15] [16], Z_2-vortex [17] and spin nematic [18]. Our discovery of the transition in $EtMe_3Sb[Pd(dmit)_2]_2$ is expected to provide important information on a new instability of the quantum spin liquid.

Acknowledgements

This work was supported by Grants-in-Aid for Scientific Research (Grants Nos. 19052005, 18043024, 20110003, 18740199 and 21740255) from the Ministry of Education, Culture, Sports, Science and Technology, Japan.

References

[1] Anderson P W 1973 *Mater. Res. Bull.* 8 153.
[2] Itou T, Oyamada A, Maegawa S, Tamura M and Kato R 2008 *Phys. Rev. B* **77** 104413.
[3] Shimizu Y, Miyagawa K, Kanoda K, Maesato M and Saito G 2003 *Phys. Rev. Lett.* **91** 107001.
[4] Itou T, Oyamada A, Maegawa S, Tamura M and Kato R 2009 *J. Phys.: Conf. Ser.* **145** 012039.
[5] Itou T, Oyamada A, Maegawa S and Kato R 2010 *Nature Phys.* **6** 673.
[6] Kato R 2004 *Chem. Rev.* **104** 5319.
[7] Tamura M and Kato R 2002 *J. Phys. Condens. Matter* **14** L729.
[8] Nakamura T, Takahashi T, Aonuma S and Kato R 2001 *J. Mater. Chem.* **11** 2159.
[9] Kubo R 1957 *J. Phys. Soc. Jpn.* **12** 570.
[10] Moriya T 1956 *Prog. Theor. Phys.* **16** 23.
[11] Itou T, Oyamada A, Maegawa S, Tamura M and Kato R 2007 *J. Phys. Condens. Matter* **19** 145247.
[12] Lee S-S, Lee P A and Senthil T 2007 *Phys. Rev. Lett.* **98** 067006.
[13] Galitski V and Kim Y B 2007 *Phys. Rev. Lett.* **99** 266403.
[14] Grover T, Trivedi N, Senthil T and Lee P A 2010 *Phys. Rev. B* **81** 245121.
[15] Kalmeyer V and Laughlin R B 1987 *Phys. Rev. Lett.* **59** 2095.
[16] Baskaran G 1989 *Phys. Rev. Lett.* **63** 2524.
[17] Kawamura H, Yamamoto A and Okubo T 2010 *J. of Phys. Soc. Jpn*, **79** 023701.
[18] Ritchey L and Coleman P 1990 *J.Phys. Condens. Matter* **2** 9227.

Journal of Physics: Conference Series **320** (2011) 012033 doi:10.1088/1742-6596/320/1/012033

Ground state of the organic spin-liquid material $EtMe_3Sb[Pd(dmit)_2]_2$

Tetsuaki Itou[1,*], Akira Oyamada[1], Satoru Maegawa[1] and Reizo Kato[2]

[1] Graduate School of Human and Environmental Studies, Kyoto University, Kyoto 606-8501, Japan

[2] Condensed Molecular Materials Laboratory, RIKEN, Wako-shi, Saitama 351-0198, Japan

E-mail: *tetsuaki@staff.mbox.media.kyoto-u.ac.jp

Abstract. The organic Mott insulator $EtMe_3Sb[Pd(dmit)_2]_2$ is a 1/2-spin system on a nearly-regular triangular lattice. It does not show classical antiferromagnetic ordering in spite of strong antiferromagnetic exchange interactions J = 220-250 K, and thus it is regarded as a spin liquid material. We have found that $(T_1T)^{-1}$ of this spin liquid state shows a peak at around 1 K and after that $(T_1T)^{-1}$ decreases on cooling. This is not attributed to slowing down of the spin dynamics because T_2^{-1} does not show any enhancement down to low temperature limit. Therefore, the decrease of $(T_1T)^{-1}$ is attributed to disappearance of spin excitations, suggesting that magnetic excitations acquire an energy gap at least in some portion of q-space. This indicates that the ground state of the present spin-liquid material is not a state with spinon Fermi surface, where magnetic excitations are fully gapless and thus $(T_1T)^{-1}$ does not decrease down to low temperature limit.

1. Introduction

Quantum spin liquid state, where spins are vital but does not show classical magnetic ordering even at absolute zero temperature due to quantum uctuations, has been a long-sought state of matter. Although it was theoretically proposed early on that this state may be realized in 1/2-spin systems on frustrated lattices [1], this state was not experimentally found for many years. This is because real model materials of ideal frustrated 1/2-spin systems were limited. Recently, however, some organics are recognized as ideal frustrated 1/2-spin systems, and the spin-liquid state has been found in two triangular-lattice antiferromagnets, κ-$(BEDT\text{-}TTF)_2Cu_2(CN)_3$ (BEDT-TTF = bis(ethylenedithio)-tetrathiafulvalene) [2] and $EtMe_3Sb[Pd(dmit)_2]_2$ (dmit = 1,3-dithiole-2-thione-4,5-dithiolate, Me = CH_3, Et = C_2H_5) [3].

Stimulated by these experimental ndings, theorists have discussed the nature of the quantum spin liquid on triangular lattices. This is a profound topic, and there are many di erent ideas and theories which have proposed so far. Under this circumstance, further experimental studies of the spin liquid state on triangular lattices are highly desired, to develop understanding of the spin liquid physics on the triangular lattice. One of the important points to understand the spin-liquid state is what kind of uctuations and excitations the state has. We have been performing experiments for the spin liquid material $EtMe_3Sb[Pd(dmit)_2]_2$, aiming to understand this point.

The family of layered organic salts X[Pd(dmit)$_2$]$_2$ (X : nonmagnetic monovalent counter cation) have scalene-triangular lattices of [Pd(dmit)$_2$]$_2$ dimers, as shown in Figs. 1 and 2. They are Mott insulators and thus have a localized 1/2 spin on each dimer [4]. Therefore, they are triangular-lattice spin systems. Among them, EtMe$_3$Sb[Pd(dmit)$_2$]$_2$ is a nearly regular-triangular spin system, because the interdimer transfer integrals between the localized orbitals along the three directions of the triangles are close to each other [3].

The exchange interactions of EtMe$_3$Sb[Pd(dmit)$_2$]$_2$ between neighboring spins along the three directions were estimated to be 220-250 K from magnetic susceptibility data [3]. In spite of the strong exchange interactions, we found from ^{13}C-NMR (nuclear magnetic resonance) measurements that this material does not undergo classical antiferromagnetic ordering down to at least down to 19.4 mK, which is the lowest stable temperature achieved using our dilution refrigerator and is less than 0.01 % of the exchange interactions [5, 6].

Importantly, we have also discovered that temperature dependence of the spin-lattice relaxation rate of ^{13}C-NMR shows an obvious kink at around 1 K and the rate shows a rapid decrease below this temperature.

We discuss the low-temperature phase of this spin system in this report.

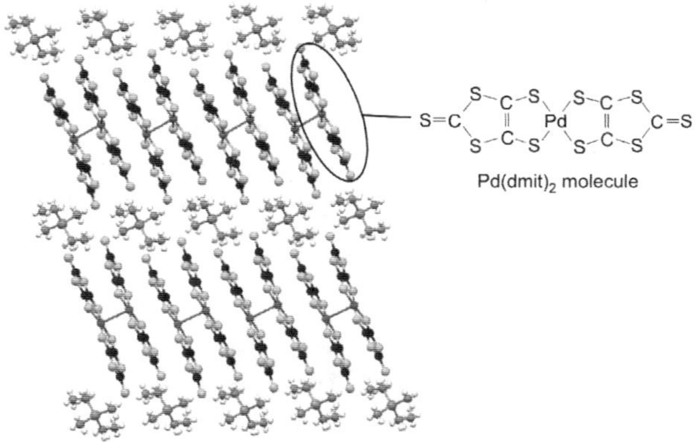

Figure 1. Side view of the crystal structure of EtMe$_3$Sb[Pd(dmit)$_2$]$_2$. Two dimensional magnetic Pd(dmit)$_2$ layers are separated by non-magnetic layers of the closed-shell monovalent cation EtMe$_3$Sb.

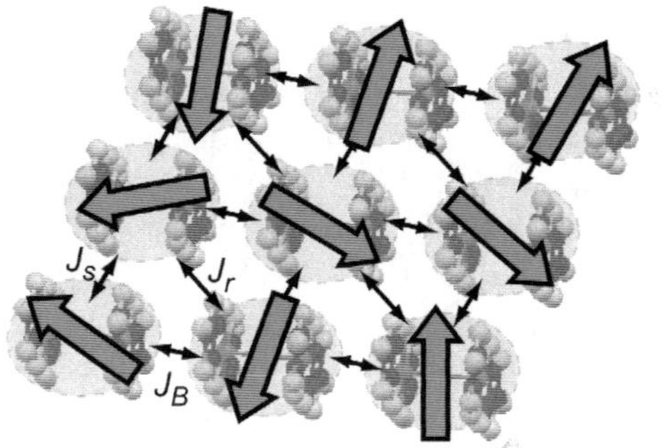

Figure 2. Top view of the crystal structure of the Pd(dmit)$_2$ layer. The Pd(dmit)$_2$ molecules are strongly dimerized. One electron with a 1/2-spin is localized on each [Pd(dmit)$_2$]$_2$ dimer. The exchange interactions, J_s, J_r, J_B, are close to each other and estimated to be 220-250 K.

2. Experimental method

We prepared ne high-quality single crystals of EtMe$_3$Sb[Pd(dmit)$_2$]$_2$ with typical sizes of 0.3 mm 0.3 mm 0.01 mm, using an aerial oxidation method. For the ^{13}C-NMR measurements, the carbon atoms at both ends of the Pd(dmit)$_2$ molecules were enriched with ^{13}C isotopes, as shown in Fig. 3. The NMR measurements were carried out at a eld of 7.65 T using many single crystals with no particular orientation packed into a glass tube. The NMR spectra were obtained by Fourier transformation of spin-echo signals refocused after the pulse sequence of $\pi/2$ -τ-π, where τ was the rf pulse interval. The typical pulse widths of $\pi/2$ and π were 3 and 6 μs, respectively. These values were su ciently smaller than the inverse of the spectral widths, and thus the pulses could cover the whole NMR spectra. The spin-spin relaxation ($T_2\ ^{1}$) curves were obtained from the integrated spin-echo spectral intensity M(2τ) with varying τ. The spin-lattice relaxation ($T_1\ ^{1}$) curves were obtained from the spin-echo spectral intensity $M(t)$ after time delay t following saturation comb pulses. The spin-lattice relaxation time T_1 in this report is determined from the stretched-exponential analysis [6]. The spin-spin relaxation time T_2 is de ned as the time when the recovery curve of M(2τ) reaches 1/e

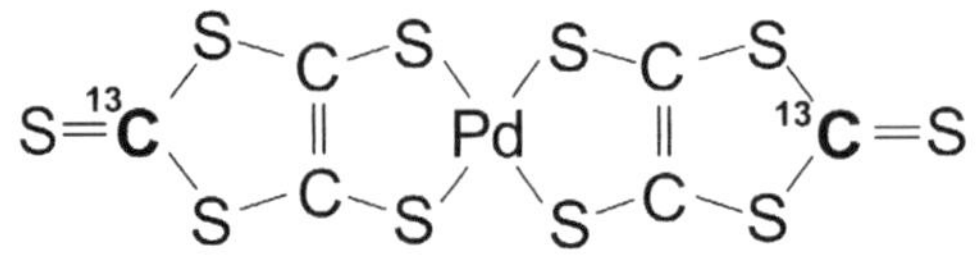

Figure 3. Pd(dmit)$_2$ molecule with selective substitution of ^{13}C isotope. The carbon atoms at both ends of the molecule are enriched.

3. Result and Discussion

Figure 4 shows the temperature dependence of the spin-lattice relaxation rate multiplied by the inverse of temperature (T). It shows a clear kink at around 1 K, below which $(T_1T)^{\;1}$ decreases on cooling. The value $(T_1T)^{\;1}$ measures the imaginary part of the spin susceptibility $\sum_q \chi''(q, \omega_\mathrm{N})/\omega_\mathrm{N}$, where q is wavenumber and ω_N is the NMR frequency (in the present case, $\omega_\mathrm{N} = 81.8$ MHz). We explained in a previous report [6] that this decrease of $(T_1T)^{\;1}$ is due to a decrease of spin excitations and thus due to a spin gap formation.

Some people may think that the present decrease of $(T_1T)^{\;1}$ might be attributed to the other scenario, that is, *slowing-down scenario* as follows. If the characteristic frequency of the spin dynamics Γ would decrease beyond NMR frequency on cooling, $(T_1T)^{\;1}$ would be expected to decrease on cooling after showing a peak. The present material is a quasi-two-dimensional system, and thus, they may think that, the present material might show slowing down of Γ toward around zero temperature, because the Mermin-Wagner theorem guarantees that pure-two-dimensional Heisenberg spin systems can undergo magnetic ordering only at absolute zero temperature, even though the ground state is a magnetically-ordered state.

However, this slowing-down scenario is not the case in the present material. We explain this point.

The linewidth of NMR spectrum re ects the sum of the three elds which nuclear spins experience: static inhomogeneous eld, slowly uctuating (particularly, on the order of kHz) eld from electron spins, and eld from other nuclear spins. If Γ decreased far beyond the NMR frequency of the order of MHz and reached around the order of kHz, the ^{13}C-NMR linewidth would increase toward the scale of the hyper ne coupling constant, which is about $9 \quad 10^2$ kHz/μ_B [3, 7], due to growth of slowly uctuating eld. However, experimental results reported in Ref. [6] show that the ^{13}C-NMR linewidth does not broaden (remains within 50 kHz) down to 19.4 mK, which is much lower than the peak temperature of 1 K. That is, the ^{13}C-NMR linewidth shows no sign of slowing down of spin dynamics. This is one evidence that can deny the slowing-down scenario.

Spin-spin relaxation rate, $T_2^{\;1}$, is a quantity which is more sensitive than the spectral linewidth to detect slowly uctuating eld, because $T_2^{\;1}$ is not a ected by static inhomogeneous eld in contrast to the spectral linewidth. This value probes only the two elds: slowly uctuating eld from electron spins and eld from other nuclear spins. Figure 5 shows the temperature dependence of $T_2^{\;1}$ of EtMe$_3$Sb[Pd(dmit)$_2$]$_2$. The observed values of $T_2^{\;1}$ correspond to about 1 kHz, which is much smaller than the spectral linewidth of about 50 kHz. This means that the spectral linewidth is mainly dominated by static inhomogeneous eld. The rate $T_2^{\;1}$ is not a ected by it and thus is a much better quantity to check for the slowly uctuating eld. Importantly, we can see that $T_2^{\;1}$, as well as the spectral linewidth, shows no enhancement down to low temperatures, retaining an almost constant value. This is a strong

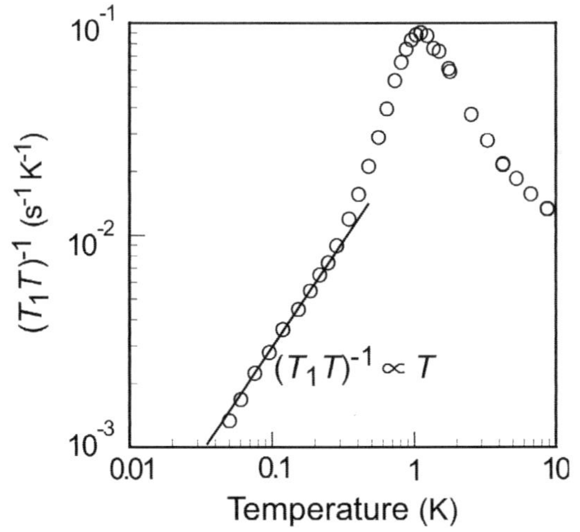

Figure 4. Temperature dependence of $(T_1T)^{\ 1}$ for many single crystals of EtMe$_3$Sb[Pd(dmit)$_2$]$_2$ with no particular orientation. The NMR measurements were performed at 7.65 T on the ^{13}C nuclear spins at both ends of the Pd(dmit)$_2$ molecules.

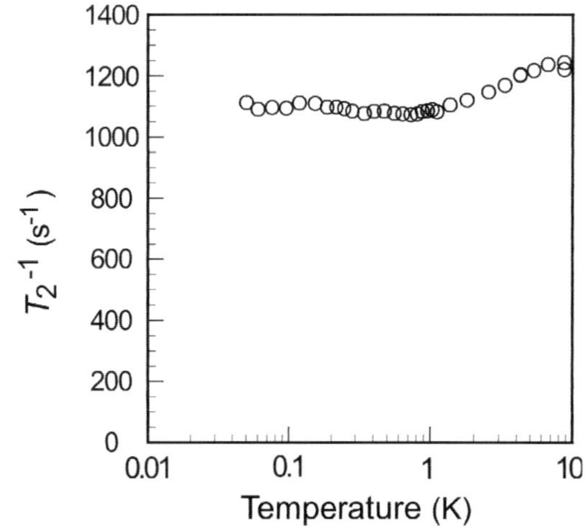

Figure 5. Temperature dependence of $T_2^{\ 1}$ for many single crystals of EtMe$_3$Sb[Pd(dmit)$_2$]$_2$ with no particular orientation.

evidence for absence of slowly uctuating eld and thus we conclude that the slowing-down scenario is not the case in the present material. The constant value of $T_2^{\ 1}$ is well accord with contribution of the dipole eld from other nuclear spins. The nuclear dipole eld makes the dominant contribution to $T_2^{\ 1}$ in the present material because strong slowly- uctuating eld is absent.

Therefore, we can conclude that the decrease of $(T_1T)^{\ 1}$ below 1 K does not mean the slowing down of the spin dynamics. The behaviors of the spectral linewidth and $T_2^{\ 1}$ guarantee that the spin dynamic does not undergo serious slowing down, and thus the decrease of $(T_1T)^{\ 1}$ simply indicates a decrease of the spin excitations, as discussed in the previous report [6].

Our nding, the decrease of the spin excitations, denies the picture of the fully gapless spin-liquid state with spinon Fermi surface, where $(T_1T)^{\ 1}$ remains constant (Fermi-liquid case) or diverges (non-Fermi-liquid case) on cooling. Figure 4 means that the imaginary part of the spin susceptibility vanishes on approaching zero temperature and thus the ground state has a spin gap at least in some portion of q-space. In addition, the decrease of $(T_1T)^{\ 1}$ does not follow an exponential law but a power law in temperature. This result implies the possibility that the spin gap is a nodal one.

Here we comment on a few concerns about our NMR data, which some people might have. One concern is that the NMR spin-lattice relaxation might be a ected by impurities. Such impurities, if any, are few and thus generally only slightly in uence the relaxation. However, in a spin-gapped system, a spin gap causes the intrinsic relaxation to be so small at low temperatures that the slight relaxation caused by the impurities might become detectable. In such cases, $(T_1T)^{\ 1}$ might not re ect the intrinsic spin dynamics. Another concern is that, in the present system, the counter nonmagnetic cation EtMe$_3$Sb has low-energy molecular motion. The ethyl group and methyl group of the cation EtMe$_3$Sb shows quantum rotation, which survives even at low temperatures. This molecular motion does not a ect the spin state because the motion exists not on the magnetic layer but on the nonmagnetic counter cation layer. However, it might a ect physical quantities such as the NMR relaxation rates and thermodynamic quantities. In this context, it is important to con rm the reliability and reproducibility of the NMR data, and to obtain a guarantee that the present behavior of $(T_1T)^{\ 1}$ shown in Fig. 4 de nitely re ects the intrinsic spin dynamics of the present system. We have performed several experiments to check these points and succeeded in getting good reliability and reproducibility. We will report it elsewhere in the near future.

Lastly, we discuss the recently published data on the thermal conductivity κ of the present material by M. Yamashita *et al* [8]. They measured κ down to about 100 mK and found that κ/T seems to retain a nonzero value in the zero temperature limit. This reminds us of metal-like excitations. In contrast, our results deny the fully gapless spin-liquid state with spinon Fermi surface. We do not have a conclusive idea to reconcile this seeming contradiction, but here comment on one point. One clue may be the fact that, NMR relaxation rates probe only *magnetic* excitations, while thermal quantities probe both *magnetic* and *nonmagnetic* excitations. This di erence may be the reason why observations look seemingly di erent between our NMR data and their thermal-conductivity data. Our NMR data does not provide any information on *nonmagnetic* excitations, which do not contribute to the dynamic spin susceptibility. The thing that can be de nitively determined from our NMR data is that *magnetic* excitations have an energy gap at least in some portion of q-space in the ground state of $EtMe_3Sb[Pd(dmit)_2]_2$.

4. Conclusion

We measured the spin-lattice relaxation rate $T_1^{\ 1}$ and spin-spin relaxation rate $T_2^{\ 1}$ of the organic spin-liquid material $EtMe_3Sb[Pd(dmit)_2]_2$ at ultra-low temperatures. We found that $(T_1T)^{\ 1}$ shows a peak at around 1 K and after that it decreases on cooling. This is not attributed to slowing down of the spin dynamics, because $T_2^{\ 1}$ does not show any enhancement at low temperatures. The decrease of $(T_1T)^{\ 1}$ indicates the disappearance of the spin excitations and thus indicates the appearance of a gap structure (at least in some portion of q-space) in magnetic excitations.

Acknowledgments

This work was supported by Grant-In-Aid for Scienti c Research from MEXT, Japan (numbers 18740199, 19052005, 20110003 and 21740255).

References

[1] Anderson P W 1973 *Mater. Res. Bull.* **8** 153.
[2] Shimizu Y, Miyagawa K, Kanoda K, Maesato M and Saito G 2003 *Phys. Rev. Lett.* **91** 107001.
[3] Itou T, Oyamada A, Maegawa S, Tamura M and Kato R 2008 *Phys. Rev. B* **77** 104413.
[4] Kato R 2004 *Chem. Rev.* **104** 5319.
[5] Itou T, Oyamada A, Maegawa S, Tamura M and Kato R 2009 *J. Phys.: Conf. Ser.* **145** 012039.
[6] Itou T, Oyamada A, Maegawa S and Kato R 2010 *Nature Phys.* **6** 673.
[7] Itou T, Oyamada A, Maegawa S, Tamura M and Kato R 2007 *J. Phys.: Condens. Matter* **19** 145247.
[8] Yamashita M, Nakata N, Senshu Y, Nagata M, Yamamoto H M, Kato R, Shibauchi T and Matsuda Y 2010 *Science* **328** 1246.

High-frequency ESR study on a frustrated mixed spin system BIPNNBNO

T Kanzawa[1], Y Hosokoshi1[*1], K. Katoh[2], K. Inoue[3] and H Nojiri[4]

[1]Department of Physical Science, Osaka Prefecture University, Sakai, Osaka 599-8531, Japan
[2]Department of Chemistry, Tohoku University, Aramaki, Sendai 980-8578, Japan
[3]Department of Chemistry, Hiroshima University, Hiroshima 739-8526, Japan
[4]Institute of Materials Research, Tohoku University, Katahira, Sendai 980-8577, Japan

*E-mail: yhoso@p.s.osakafu-u.ac.jp

Abstract. High-frequency ESR measurements were made for single crystals of an organic triradical BIPNNBNO having $S = 1/2$ and $S = 1$ species within a molecule. [BIPNNBNO=3,5-bis(*N-tert*-butylaminoxyl)-4'-(1-oxyl-3-oxide-4,4,5,5-tetramethylimidazolin-2-yl)biphenyl.] The temperature and field dependences of the linewidth and resonance field were studied. The dominant magnetic exchange coupling is discussed and the frustrated spin structure of this compound is suggested.

1. Introduction

There has been considerable interest in the study of low-dimensional quantum spin systems with spin frustration. Since organic radicals consisting of only light elements show very small spin-orbit coupling, they are regarded as ideal Heisenberg spin systems and are suitable to study quantum nature of spin systems. Recently we have developed a frustrated spin system by using an organic triradical BIPNNBNO, where BIPNNBNO = 3,5-bis(*N-tert*-butylaminoxyl)-4'-(1-oxyl-3-oxide-4,4,5,5-tetramethylimidazolin-2-yl)biphenyl. Since two radical units of $S = 1/2$ ferromagnetically couple strongly to form an $S = 1$ species, BIPNNBNO is regarded as a mixed spin system with $S = 1$ and $S = 1/2$. The study of the mixed spin systems with two kinds of antiferromagnetically exchange coupled spins, in particular, the mixed spin systems with integer and half-integer spins is much attractive [1]. The ground state properties depend on the periodic arrangement of the different spins. For example, an alignment of elementary cells with spins 1, 1, 1/2, 1/2 in this order results in a singlet ground state with an excitation gap [2]. Properties of alternating-spin ladder were also theoretically studied and the quantum fluctuations play a crucial role in the nonmagnetic ground state [3]. We have developed some mixed spin systems using organic triradicals and different ground states were observed depending on the intermolecular networks. The crystals of PNNBNO, where PNNBNO = 2-[3',5'-bis(*N-tert*-butylaminoxyl)phenyl]-4,4,5,5-tetramethyl-4,5-dihydro-1*H* imidazol-1-oxyl 3-oxide, has the intermolecular antiferomagnetic coupling between the $S = 1$ and $S = 1/2$ in three dimension and is the first example of organic ferrimagnets [4]. On the other hand, the crystals of PIMBNO and BIPNNBNO involve the intermolecular contacts between $S = 1$ units and/or the ones between $S = 1/2$ units, where PIMBNO = 2-[3',5'-bis(*N-tert*-butylaminoxyl)phenyl]-4,4,5,5-tetramethyl-4,5-dihydro-1*H* imidazol-1-oxyl. Only BIPNNBNO shows the singlet ground state with an excitation gap, whereas

Published under licence by IOP Publishing Ltd

PIMBNO shows a gapless ground state [5]. The crystal structural analysis has revealed the complex spin network in BIPNNBNO crystals. The flexibility of the molecule gives BIPNNBNO competing magnetic exchange couplings, which induces frustrated spin structure.

The two dimensionality in BIPNNBNO crystals is confirmed by the angular dependence of the peak-to-peak linewidth of X-band (9.5 GHz) ESR spectra at room temperature [6]. However, the spin correlation by the magnetic exchange interaction is not clear yet. In an organic radical, the spin densities distribute on the whole molecule due to π–conjugation. Since the magnetic exchange is related to the overlap between the molecular orbitals, the treatment of an organic spin as a point dipole is not always good approximation. The largest exchange interaction does not always occur between spins with the shortest distance which show the largest magnetic dipole-dipole interaction. For example, the temperature dependence of the ESR *g*-factor in an organic spin-ladder compound BIP-TENO shows opposite tendency to the theory proposed to one-dimensional antiferromagnetic chain [7] and importance of the consideration of the spin density distribution within a molecule was pointed out [8]. The spin density distribution in organic molecules also makes it difficult to understand the spin correlation. For the elucidation of the magnetic exchange interaction, we have started to study the ESR measurements by using single crystals. In this paper, we report the temperature and frequency dependences of the linewidth and resonant field of BIPNNBNO crystals.

2. Experimental

BIPNNBNO was synthesized by following the conventional procedure [4,9] and crystallized to rhombus shaped dark green single crystals. The BIPNNBNO crystal belongs to an orthorhombic system, *Pbcn*, a = 9.0126(7), b = 17.578(1), c = 33.924(3) Å, V = 5374.3(6) Å^3, Z = 8 [5]. The rhombus plane corresponds to the crystallographic *bc* plane and the longer and shorter diagonal is parallel to the *b*- and *c*-axis, respectively. Submillimeter wave ESR measurements have been performed at frequencies of up to 716 GHz and in pulsed magnetic fields of up to 29 T, with the field direction parallel to the crystallographic axes. Three pieces of single crystals were used. An optical pumped far-infrared laser, backwardtravelling wave tubes and Gunn oscillators have been employed as radiation sources. We have employed a simple transmission method with Faraday configuration where the propagation vector of the incident radiation is aligned parallel to the external magnetic field [10]. The polarization of light is random in the plane normal to the propagation vector. An InSb is used as the detector. For the field calibration, DPPH or ruby was used.

3. Summary of the magnetic properties of BIPNNBNO

The crystal structure of BIPNNBNO is schematically shown in Figure 1 (a). Molecules are connected by three kinds of radical-radical contacts [5,6], which always yield antiferromagnetic interactions [4,11]. The two dimensional sheet is spread on the *ab*-plane. Along the *a*-axis, a ladder-like structure is formed by two kinds the radical-radical contacts, which will induce a frustrated spin structure. On the other hand, ferrimagnetic chains are formed along the *b*-axis.

The temperature dependence of the paramagnetic susceptibility (χ_p) of BIPNNBNO crystals were compared with the one for isolated molecules diluted in diamagnetic polyvinylenechloride matrix [12]. Below 30 K, the χ_p values of crystals become smaller than the ones of isolated molecules, reflecting the intermolecular antiferromagnetic interactions. The double-peak structure was observed at 4.2 and 15 K in the temperature dependence of χ_p of BIPNNBNO crystals, which suggests the existence of at least two kinds of intermolecular antiferromagnetic interactions. The value of χ_p of BIPNNBNO crystals decreases to zero as temperature decreases below 4.2 K, which suggests the singlet ground state of this material. The magnetization isotherm of BIPNNBNO measured at 0.4 K is shown in Figure 1(b) [5]. The finite magnetization is not observed below B_{c0} = 4.5 T. Above B_{c0}, magnetization increases rapidly and reaches 1μ_B/f.u., which corresponds to the value of 1/3 of the full saturation magnetization (3μ_B/f.u.). Between B_{c1} = 6 T and B_{c2} = 23 T, the magnetization takes a constant value of 1μ_B/f.u., which is called 1/3 magnetization plateau. Above B_{c2}, it increases again and saturates to 3μ_B/f.u. at B_{sat} = 29 T. A very narrow magnetization plateau of 2/3 is also appears around B_{c3} = 26 T.

We categorize the magnetic field in three regions: gapped, gapless, and 1/3 magnetization plateau regions are defined as the one below B_{c0}, between B_{c0} and B_{c1}, and between B_{c1} and B_{c2}, respectively.

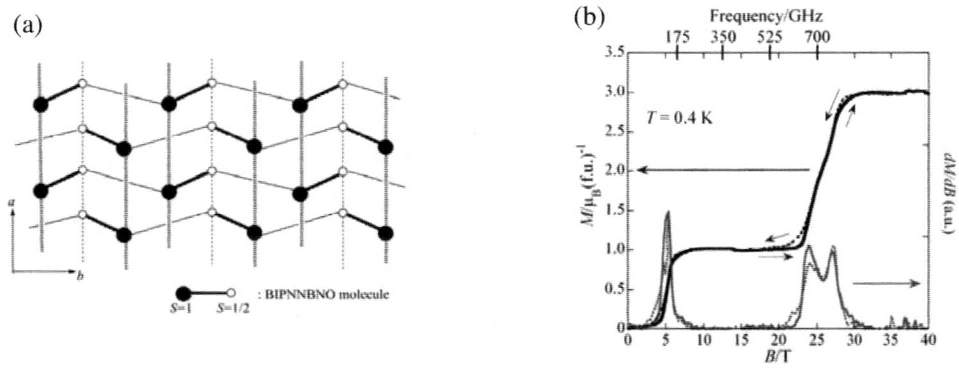

Figure 1. (a) Schematic diagram of the crystal structure of BIPNNBNO. Open and closed circles and bold solid lines represent $S = 1/2$ and $S = 1$ species and intramolecular coupling, respectively. Bold and thin dotted lines correspond to the intermolecular contacts along the *a*-axis, whereas thin solid lines along the *b*-axis.
(b) Magnetization isotherm of BIPNNBNO at 0.4 K. Derivative curve is also plotted.

4. Results and discussions

4.1. X-band ESR

Figure 2 shows the temperature dependence of the X-band ESR *g*-factors and peak-to-peak linewidth (ΔB_{pp}) when the magnetic field is applied parallel to the orthogonal crystallographic axes ($B \parallel a, b, c$). The small anisotropy of the principal values of *g*-factors at room temperature ($g_a = 2.0066$, $g_b = 2.0068$, $g_c = 2.0051$) means that this compound is an almost ideal Heisenberg spin system. The temperature dependence of ΔB_{pp} shows a broad maximum at about 100 K. The temperature dependence of χ_p of crystalline states above 30 K is identical with the one of isolated molecules and the $\chi_p T$ value of isolated molecules starts to decrease below 100 K. Thus, the broad maximum in ΔB_{pp} at about 100 K is related to the spin-spin correlation between $S = 1$ and $S = 1/2$ within a molecule. Below 20 K, where χ_p takes a broad maximum, the broadening of ΔB_{pp} is remarkable at any field direction. It is related to

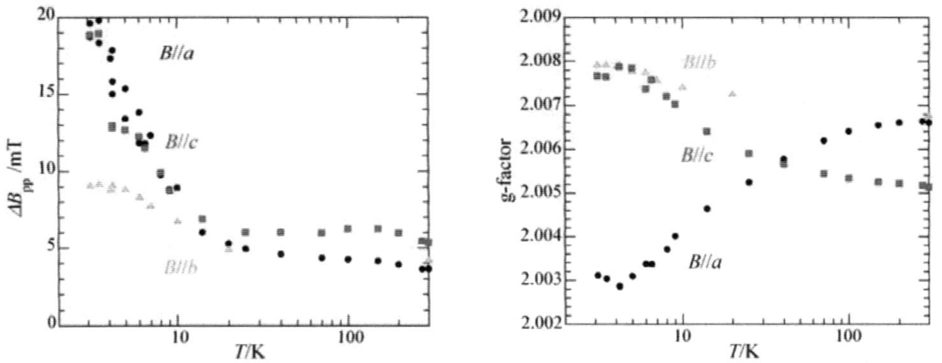

Figure 2. Temperature dependences of (a) ΔB_{pp} and (b) *g*-factor in X-band ESR measurements. Black, green, and red circles represent observed data for *B//a*, *b*, and *c*, respectively.

the spin-spin correlation due to the intermolecular antiferromagnetic interactions. In one and two dimensional antiferromagnets, broadening of the linewidth due to the short range order was reported [13,14]. The *g*-factor shows positive and negative shift upon cooling for $B \parallel b$, c and $B \parallel a$, respectively. The *a*-axis is perpendicular to the thin rhombus plane and the largest demagnetizing field is expected along this direction. By using the approximation of ellipsoid [15], the contribution of the demagnetizing field is estimated to be at most 50 % of the observed *g*-shift. Thus the observed *g*-shift is intrinsic to the spin network of this material. The observed *g*-shift does not agree with the reports on the one and two dimensional antiferromagnets [7,16], probably due to the spin density distribution [8].

4.2. High-frequency ESR

4.2.1. *Temperature dependence of linewidth*

In the high-frequency ESR measurements, a single absorption signal was observed in the whole frequency and temperature range. Figure 3 shows the temperature dependences of the full width at the half maximum ($\Delta B_{1/2}$) measured at 190, 357, 716 GHz. The measurements at 95 and 135 GHz were also made but the data were not shown. The signal became broad as increasing frequency. The value of $\Delta B_{1/2}$ for $B \parallel b$ is smaller than the ones for other two field directions. As decreasing temperatures, the value of $\Delta B_{1/2}$ gradually increases below 20 K similarly to the X-band result. The $\Delta B_{1/2}$ below 10 K behaves differently depending on the field region. In the gapped (95 GHz) and plateau region (357, 716 GHz), the value of $\Delta B_{1/2}$ approaches a maximum below about 10 K, whereas in gapless region (135 and 190 GHz), the monotonous increase with decreasing temperature was observed.

At 190 GHz corresponding to the end of the gapless region, the rapid increase of $\Delta B_{1/2}$ was remarkable below 4 K for B // a, b. For B // c, only the subtle change of $\Delta B_{1/2}$ was observed, which is probably due to the considerably small magnetic correlation along this direction.

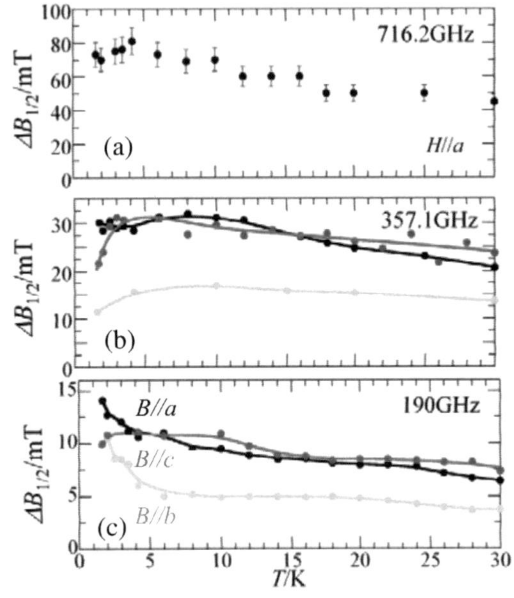

Figure 3. Temperature dependence of $\Delta B_{1/2}$ of high-frequency ESR measured at (a) 716 GHz, (b) 357 GHz, and (c) 190 GHz. Black, green, red circles represent the observed data for *B*//*a*, *b*, and *c*, respectively. The solid curves are guide for eyes.

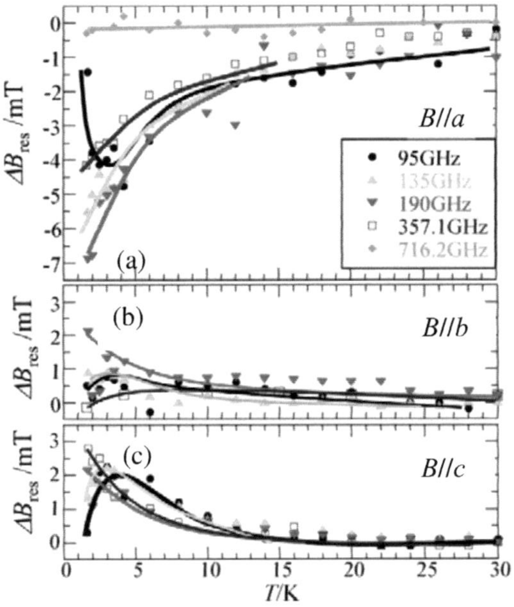

Figure 4. Temperature dependence of ΔB_{res} of high-frequency ESR for (a) *B*//*a*, (b) *B*//*b*, and (c) *B*//*c*. Black, green, red, blue, and light blue marks represent the data measured at 95, 135, 190, 357, and 716 GHz, respectively. Solid curves are guide for eyes.

4.2.2. *Temperature dependence of the shift of resonant fields*

Figure 4 shows the temperature dependences of the shift of the resonant field (B_{res}) for B // a, b, c. The plotted symbols represent the difference of the resonance field from that at 50 K, $\Delta B_{res} = B_{res}(T) - B_{res}(50\text{ K})$. The positive and negative shift was observed for B // b, c and B // a, respectively. At 95 GHz in the gapped region, the maximum shift was observed at 4 K in any field directions, which is the same tendency observed at X-band. At 716 GHz in the 2/3 plateau region, no shift was observed for *B//a*, which suggests the saturated magnetic behaviour. The behaviour in gapless (135, 190 GHz) and 1/3-plateau (357 GHz) region depends on the field direction. For *B // a*, monotonous shift was observed. The maximum shift was observed at 190 GHz in the end of the gapless region. The shift in the 1/3 magnetization plateau region is moderate as compared with that in the gapless region. For *B // b*, similarly to *B//a*, the largest shift was observed at 190 GHz. In the gapless region, the value of B_{res} was almost constant. At 357 GHz in the 1/3 magnetization plateau region, the temperature dependence of ΔB_{res} shows a broad maximum at about 4 K. The behaviour for *B // c* is somewhat different from other two axes. The temperature dependence of ΔB_{res} at 135 GHz shows a maximum at about 3 K, whereas monotonous shift of ΔB_{res} was observed at 190 GHz. Peculiar feature is the moderate shift at 190 GHz. The shift at 190 GHz was smaller than that at 135 GHz (gapless region) and at 357 GHz (1/3 magnetization plateau region). Along the *c*-axis, considerably small intermolecular magnetic exchange coupling is expected as compared with that for other two axes. Therefore, the behaviour for *B // c* is possibly related to the spin-spin correlation within a molecule.

4.2.3. *Field dependence of the resonant field*

We also examined the field dependence of the resonant field (B_{res}) at 1.6 K for *B // a, b, c*. Tiny deviations from the paramagnetic line was observed. The resonant field of paramagnetic state (B_{para}) was calculated by $B_{para} = h\nu/g_i\mu_B$, where g_i ($i = a, b, c$) is the principal value estimated by X-band ESR at room temperature. The observed shift of the resonant field was calculated by

$$\delta B_{res} = B_{res} - B_{para} - N_i M, \quad (I = a, b, c) \qquad (1)$$

where N_i is the demagnetizing factor by using the approximation of ellipsoid[15] and *M* is the observed magnetization value. Figure 5 (a) shows the field dependence of δB_{res}. The anisotropic behaviour was observed. Above 10 T, positive and negative slope was observed for *B // b, c* and for

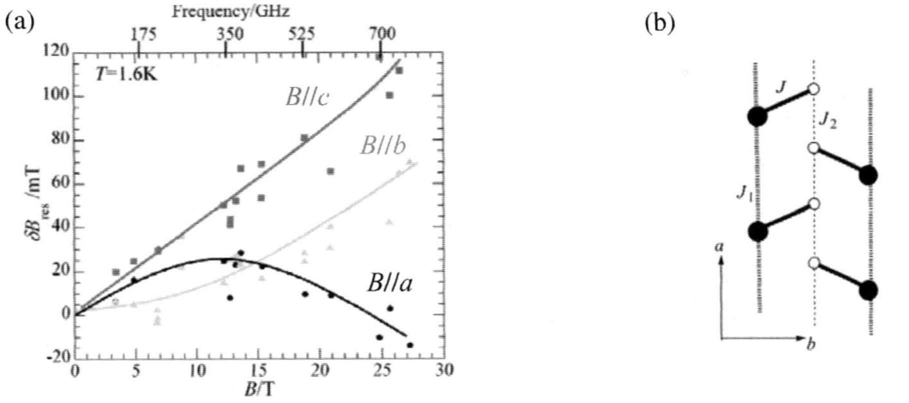

Figure 5. (a) Field dependence of resonant fields in BIPNNBNO crystals. The deviation from the paramagnetic line is plotted. Black, green, and red circles represent the observed data for *B//a, b*, and *c*, respectively. Solid curves are guide for eyes. (b) Schematic illustration of the magnetic model of BIPNNBNO. The same definition as figure 1(a) is used. Two dimensional structure shown in figure 1 (a) is reduced to a ladder-like structure in consideration of the dominant magnetic interactrions.

B // a, respectively. The largest demagnetizing factor is expected for *B // a*, and the contribution is less than 20 %. The correction of demagnetizing field does not change the sign of the slope of δB_{res} against *B*. The homogeneity of the pulsed field is about 0.01 % of the applied field for 1mm discrepancy. At 25 T, the inhomogeneity is 2.5 mT, which corresponds to about 4 % of $\Delta B_{1/2}$ and a quarter of the observed shift of δB_{res}. The field dependence at any field direction involves the contribution showing positive field dependence, which is considered as the effect of the thermal excitations. The anisotropic change of the slope above 10 T should be related to the anisotropic magnetic exchange coupling. Then we can conclude that the *a*-axis is the direction along which dominant magnetic exchange couplings work between spins.

5. Conclusion

We have performed high-frequency ESR measurements of single crystals of BIPNNBNO. The temperature dependence of $\Delta B_{1/2}$ and ΔB_{res} differs between the gapped, gapless, and plateau region. In the field dependence of B_{res}, small deviation from B_{para} was observed. Because of the very small anisotropy of organic radical, the efficiency of analysing δB_{res} has been demonstrated. From the results of the temperature and field dependences of B_{res}, the different behaviour was observed for $B \mathbin{/\!/} a$ from that for $B \mathbin{/\!/} b, c$. It can be related to the magnetic exchange coupling. Although two-dimensional network is spread within the *ab* plane, the larger molecular overlap is expected along the *a*-axis, which means that the dominant magnetic exchange couplings work along the *a*-axis. The schematic magnetic model of BIPNNBNO is shown in figure 5 (b). We can see two kinds of intermolecular antiferromagnetic exchange couplings, which will yield frustrated spin structure. Recent study of the specific heat in magnetic field also suggests that the antiferromagnetic coupling along the *a*-axis is crucial, especially the coupling between $S = 1$ units [17].

Acknowledgements

Submillimeter ESR measurements wert performed under the interuniversity cooperative research program of the Institute for Materials Research, Tohoku University. This work was supported in part by Grant-in-Aid for Scientific Research on Priority Areas ``Novel States of Matter Induced by Frustration" from the Ministry of Education, Culture, Sports, Science and Technology (MEXT) of Japan. Support from The Asahi Glass Foundation is also acknowledged. HN acknowledges Kakenhi No.20244052 and GCOE Program:Material Integration.

References

[1] Yamamoto S 1999 Phys. Rev. B **59** 1024
[2] Tonegawa T, Hikihara T, Kaburagi M, Nishino T, Miyashita S, Mikesuka H.-J 1998 J. Phys. Soc. Jpn. **67** 1000.
[3] Fukui T, Kawakami N 1998 Phys. Rev. B **57** 398
[4] Hosokoshi Y, Katoh K, Nakazawa Y, Nakano H, Inoue K 2001 J. Am. Chem. Soc., **123** 7921
[5] Goto T, Mushnikov N V, Hoskoshi Y, Katoh K, Inoue K 2003 Physica B, **329-333** 1160
[6] Kanzawa T, Hosokoshi Y, Katoh K, Nishihara S, Inoue K, Nojiri H 2006 J. Phys. Conference Series, **51** 91
[7] Nagata K, Tazuke Y 1972 J. Phys. Soc. Jpn. **32** 337
[8] Ohta H, Kirita K, Kunimoto T, Okubo S, Hosokoshi Y, Katoh K, Inoue K, Ogasawara A, Miyashita S 2002 J. Phys. Soc. Jpn., **71** 2640
[9] Hosokoshi Y, Tamura M, Kinoshita M, Sawa H, Kato R, Fujiwara Y, Ueda Y 1994 J. Mater.Chem, **4** 1219
[10] Nojiri H, Ajiro Y, Asano T, Boucher J P 2006 New J. Phys. **8** 218
[11] Hosokoshi Y, Nakazawa Y, Inoue K, Takizawa K, Nakano H, Takahashi M, and Goto T 1999 Phys. Rev. B **60** 12924
[12] The magnetic behavior of an isolated BIPNNBNO molecule is well described by the antiferromagnetic coupling between $S = 1$ and $S = 1/2$ with $2J/k_B = -26$ K
[13] Ajiro Y, Matsuoka S, Yamada T, Haseda T 1975 J. Phys. Soc. Jpn, **39** 259
[14] de Wijn H W, Davis L R,. Guggenheim H J 1972 Solid State Commun., **11** 803
[15] Osborn J A 1945 Phys. Rev. **67** 351
[16] Yokozawa Y. 1971 J. Phys. Soc. Jpn, **31** 1590
[17] Imazawa T, Hosokoshi Y, Kanzawa T, Katoh K, Inoue K, Takano Y, in preparation

Two-dimensional Frustrated Antiferromagnets $(MCl)LaNb_2O_7$ (M = Mn, Co, Cr)

A Kitada[1, 2], Y Tsujimoto[1, 2, 3], T Yajima[1], K Yoshimura[2], Y Ajiro[1], Y Kobayashi[1] and H Kageyama[1, 2, 4]
[1] Department of Energy and Hydrocarbon Chemistry, Faculty of Engineering, Kyoto University, Nishikyo, Kyoto 615-8510, Japan
[2] Department of Chemistry, Graduate School of Science, Kyoto University, Kyoto 606-8502, Japan
[3] International Center for Young Scientists (ICYS), International Center for Materials Nanoarchitechtonics (MANA), National Institute for Materials Science (NIMS), 1-1 Namiki, Tsukuba 305-0044, Japan
[4] Institute for Integrated Cell-Material Sciences (iCeMS), Kyoto University, Sakyo, Kyoto 606-8501, Japan

E-mail: kage@scl.kyoto-u.ac.jp

Abstract. Magnetic susceptibility and specific heat measurements have been performed on two-dimensional spin systems $(M^{2+}Cl)LaNb_2O_7$ (M = Mn (S = 5/2), Cr (S = 2), Co (S = 3/2)), prepared via a topotactic ion-exchange reaction. All three compounds establish antiferromagnetic order at T_N = 53 K, 61 K and 52 K, respectively for M = Mn, Co, Cr. Together with T_N = 78 K for M = Fe (S = 2), this result indicates that the T_N is not simply scaled by the magnitude of spin. In particular, the presence of strong spin-orbit interactions is suggested for $(CoCl)LaNb_2O_7$.

1. Introduction

Geometrically frustrated magnets have been received considerable interests over the last several decades. Among the models comprising of squares with some diagonal interactions are the J_1-J_2 model $Li_2VO(Si, Ge)O_4$ [1], the checkerboard model $A_2F_2Fe_2OQ_2$ (A = Sr and Ba; Q = S and Se) [2] and the Shastry-Sutherland $SrCu_2(BO_3)_2$ [3]. These materials show intriguing magnetic properties such as spin-disordered state and quantized magnetization plateaus to name only a few [3, 4].

Soft chemical methods offer promising routes to explore low-dimensional magnetic compounds, as thermodynamically metastable phases. Recently S = 1/2 layered copper oxyhalides $(CuX)A_{n-1}B_nO_{3n+1}$ (where X = Cl^-, Br^-; A = La^{3+}, Ca^{2+}, Na^+,...; B = Nb^{5+}, Ta^{5+}, Ti^{4+},...; n = 2, 3) were obtained by topotactic ion-exchange reactions, where the magnetic $CuCl_4O_2$ octahedral layers are separated by magnetically inert perovskite blocks [5-7]. Various exotic quantum magnetic phenomena have emerged from these quasi-two-dimensional magnets. The properties range from spin-singlet ground states based on quantum dimers in $(CuCl)LaNb_2O_7$ [8-12], stripe magnetic order with reduced magnetic moments in $(CuBr)LaNb_2O_7$ and $(CuCl)LaTa_2O_7$ [13, 14], quantum phase separation in the solid solution system $(CuCl)La(Nb, Ta)_2O_7$ [14, 15], to successive phase transitions and 1/3 magnetization plateaus in $(CuBr)A_2B_3O_{10}$ [16]. These interesting magnetic behaviors were interpreted as due to the presence of competing antiferromagnetic and ferromagnetic interactions within the CuX layer.

This ion-exchange strategy was expanded by replacing the copper by other transition-metal elements, yielding (MCl)$LaNb_2O_7$ (S = 3/2 for V^{2+} and Co^{2+}, S = 2 for Cr^{2+} and Fe^{2+}, and S = 5/2 for Mn^{2+}) [17-19], (FeCl)$Ca_2Ta_3O_{10}$ (S = 2) [7] and (NiCl)$Sr_2Ta_3O_{10}$ (S = 1) [20]. However, only a little is still known about the spin number dependence of the magnetic properties of $(MX)A_{n-1}B_nO_{3n+1}$. In (NiCl)$Sr_2Ta_3O_{10}$ magnetic long-range-order is significantly suppressed down to 50 K despite a large Weiss temperature θ of −125 K. Additionally, μSR measurements of (NiCl)$Sr_2Ta_3O_{10}$ revealed the coexistence of the magnetically ordered and disordered states for 20 K < T < 50 K [20]. The antiferromagnetic transition temperatures T_N of (FeCl)$LaNb_2O_7$ and (MnCl)$LaNb_2O_7$ are suggested to be 78 K and 54 K [18, 19]. In this paper we report on the magnetic properties of (MCl)$LaNb_2O_7$ (M = Mn, Co, Cr).

2. Experimental

2.1. Synthesis

The synthesis of (MCl)$LaNb_2O_7$ (M = Mn, Co, Cr) is expressed as the following two-step ion-exchange reactions [17]:

$$RbLaNb_2O_7 + LiNO_3 \rightarrow LiLaNb_2O_7 + RbNO_3 \quad (1)$$

$$LiLaNb_2O_7 + MCl_2 \rightarrow (MCl)LaNb_2O_7 + LiCl \quad (2)$$

First, $RbLaNb_2O_7$ was prepared via a conventional high-temperature route, using stoichiometric amount of La_2O_3 (99.99% purity) and Nb_2O_5 (99.99%) and 25% molar excess of Rb_2CO_3 (99.9%). Second, $LiLaNb_2O_7$ was obtained from $LiNO_3$ and $RbLaNb_2O_7$ in 10:1 molar ratio through the ion-exchange reaction (1) at 300 °C for 24 hours in air. The product was washed with warm water and then dried at 120 °C overnight. Third, $LiLaNb_2O_7$ was mixed with two-fold molar excess of ultradry MCl_2 (M = Mn, Co, Cr; 99.9%) and pressed into pellets in an Ar-filled glove box (<1ppm O_2/H_2O). The ion-exchange reaction (2) was carried out in sealed, evacuated (<10^{-3} Torr) Pyrex tubes at 390 ~ 400 °C for 7 days, followed by washing with distilled water for M = Mn, Co and with ethanol for M = Cr to eliminate the excess MCl_2 (M = Mn, Co, Cr) and LiCl, and dried at 120 °C overnight. The schematic structure is represented in figure 1.

2.2. Characterization

In-house X-ray diffraction study at room temperature confirmed the tetragonal symmetry with room-temperature cell constants: (a, c) = (3.899 Å, 12.04 Å), (3.908 Å, 11.63 Å), and (3.899 Å, 11.97 Å) for M = Mn, Co and Cr, respectively, in good agreement with those previously reported [17]. Magnetic susceptibilities were studied using a superconducting quantum interference device (SQUID) magnetometer (Quantum Design, MPMS) for a temperature range T = 2 − 300 K in a magnetic field H = 0.1 T. Specific heat measurements were performed by the thermal relaxation method in a T range between 4 K and 100 K in the absence of magnetic field using Physical Property Measurement System (PPMS, Quantum Design) at Institute for Solid State Physics, University of Tokyo. Hand-pressed pellets were attached to an alumina platform with a small amount of Apiezon N grease.

3. Results and Discussions

Figure 2(a) shows the temperature dependence of magnetic susceptibility χ_{raw} for (MnCl)$LaNb_2O_7$. A Curie-tail is seen below 30 K, which is most likely due to impurities and/or defects of Mn^{2+} ions in (MnCl)$LaNb_2O_7$, as was also present in the previous report [19]. Using the Curie equation $\chi_{imp} = C_{imp}/T$, we fitted the raw data χ_{raw} below 30 K and the best fit gave a small C_{imp} = 0.12 emu K^{-1} mol^{-1}, corresponding to about 2.7% of noninteracting S = 5/2 Mn^{2+} ions. After subtracting this upturn, one obtains the intrinsic susceptibility χ_{spin}, where a broad maximum characteristic of low-dimensional magnet was centered at T^{χ}_{max} = 65 K. Above 160 K, the inverse susceptibility χ_{spin}^{-1} (figure 2(b)) obey a Curie-Weiss law, and the fitting gave the Curie constant C = 4.38 emu K^{-1} mol^{-1} together with θ = −131 K with a slight temperature-independent impurity term of χ_0 = −4.0 × 10^{-4} emu K^{-1} mol^{-1}. The value of T^{χ}_{max} is comparable with that previously reported (63 K) [19]. Moreover, the value of C obtained in this study is in excellent agreement with the theoretical value for 1 mol of S = 5/2 Mn^{2+} ions (4.375 emu K^{-1} mol^{-1}), while the one obtained in the previous study was a little larger (4.467 emu K^{-1} mol^{-1}) [19]. However, the value of θ obtained in this study is somewhat smaller than the one obtained in the previous study (−145.7 K) [19]. This is probably because the value of C largely depends on the fitting range and the value of χ_0. It is also to be noted that C and θ are influenced by the way in which the Curie-tail at low temperature is subtracted from the raw data. The specific heat C_p at zero field is shown in figure 3. A tiny anomaly at around 53 K being located 15 K below T^{χ}_{max} strongly indicates the occurrence of the antiferromagnetic phase transition. Indeed, this temperature is in excellent agreement with T_N = 54 K estimated from $d\chi/dT$ [19]. The anomaly in the specific heat is not so obvious probably because of the use of polycrystalline sample. We could not estimate magnetic specific heat C_m by subtracting lattice contribution βT^3 because this approximation should be valid up to around 30 K. The large difference between θ and T_N manifests two-dimensionality of the magnetic system and also certain frustrated interactions.

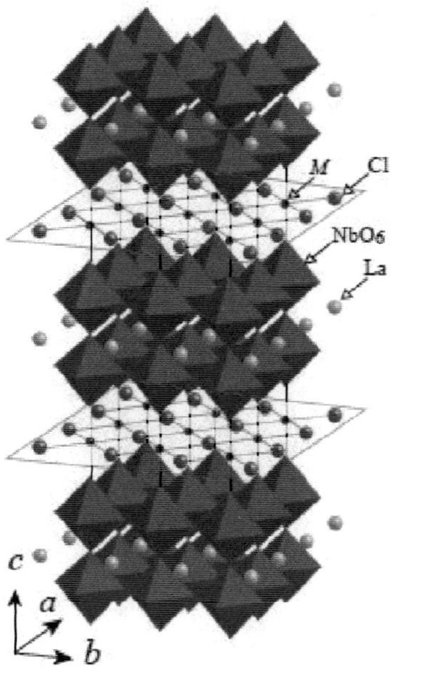

Figure 1. Schematic structure of (MCl)$LaNb_2O_7$ (M = Mn, Co, Cr).

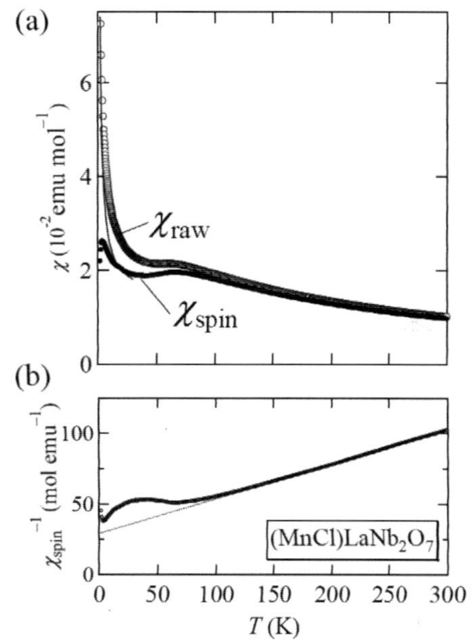

Figure 2. (a) Temperature dependence of the susceptibility for M = Mn. The open circles, closed circles and the solid line represent χ_{raw} , χ_{spin} and χ_{imp}, respectively. (b) Inverse susceptibility χ_{spin}^{-1}. The dotted line represents the theoretical curve.

The specific heat C_p of M = Co and Cr at zero field (figures 4 and 5) has a slight anomaly at 61 K and 52 K, respectively. As shown in figures 6(a) and 7(a), each magnetic susceptibility has, as in the case of M = Mn, a broad maximum at a slightly higher temperature of 67 K for M = Co and 55 K for M = Cr, indicating that the anomaly in the specific heat is due to the magnetic phase transition. As is the case with M = Mn, we could not estimate magnetic specific heat C_m by subtracting lattice contribution βT^3. In the lower temperature region below this transition, the susceptibility grows considerably with decreasing temperature. However, in contrast to the case of M = Mn as demonstrated above, both materials exhibit another anomaly in a different manner: in (CoCl)$LaNb_2O_7$, a hysteresis behavior is observed below about 6 K, while in (CrCl)$LaNb_2O_7$ a cusp without hysteresis is observed at 3 K (see the insets of figures 6(a) and 7(a)). The low-temperature anomaly featured by the hysteresis between zero-field and field cooling processes is also seen in (FeCl)$LaNb_2O_7$ and was attributed to the second magnetic transition [18]. We would consider, however, that these low-temperature anomalies likely come from the defect of the magnetic ions in the crystal (and thus being extrinsic to the pure system) because the magnitude of the Curie-like tail has sample dependence. The estimation of the defect amount is not straightforward because of the presence of low-temperature anomalies. However, since the Co-, Cr- and Mn-samples have a similar size of the Curie-like tail, the amount of the magnetic defect should be roughly the same.

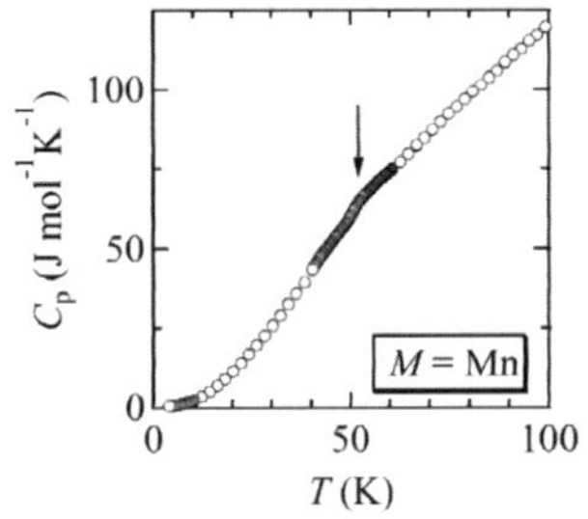

Figure 3. Temperature dependence of C_p for M = Mn showing magnetic order at 53 K.

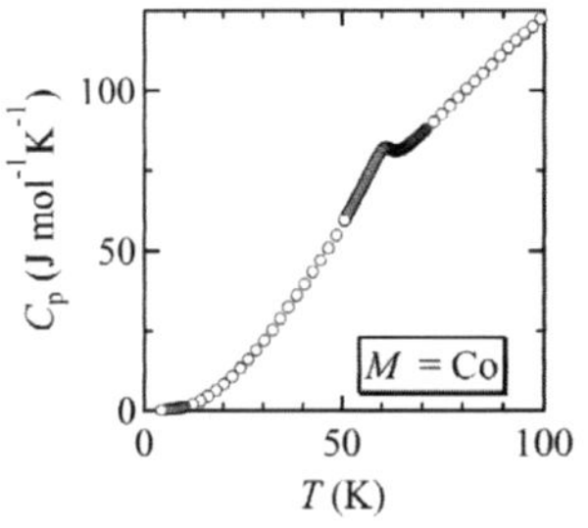

Figure 4. Temperature dependence of C_p for M = Co showing magnetic order at 61 K.

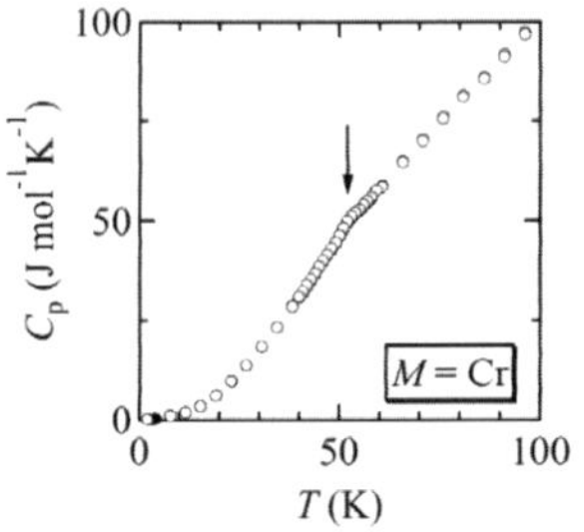

Figure 5. Temperature dependence of C_p for M = Cr showing magnetic order at 52 K.

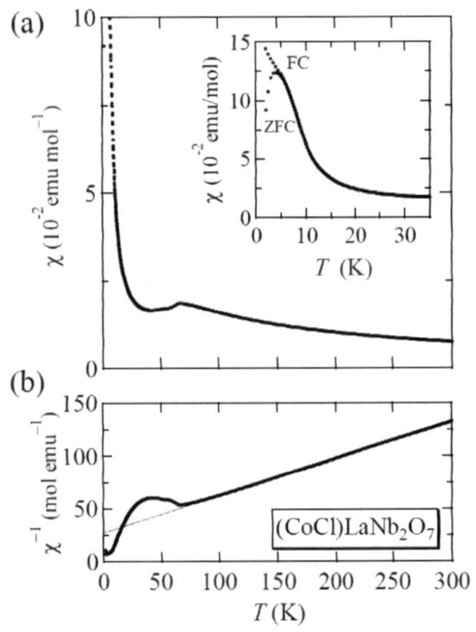

Figure 6. (a) Temperature dependence of the susceptibility for M = Co. Inset: enlarged plot of the susceptibility below 35 K. (b) Inverse susceptibility. The dotted line represents the Curie-Weiss fit.

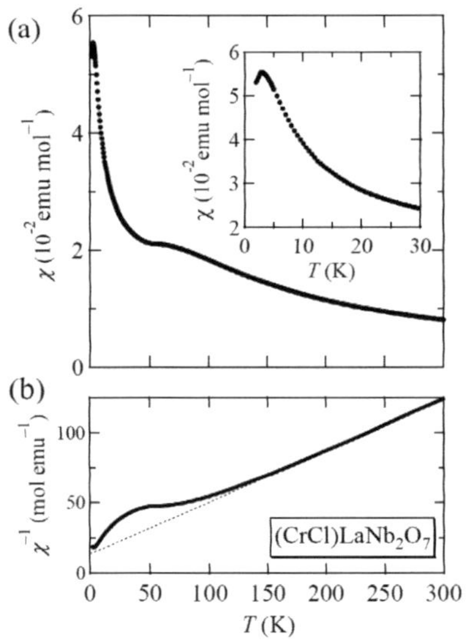

Figure 7. (a) Temperature dependence of the susceptibility for M = Cr. Inset: enlarged plot of the susceptibility below 30 K. (b) Inverse susceptibility. The dotted line represents the Curie-Weiss fit.

Given the lower temperature anomaly in these two materials, subtracting the Curie-like tail from the raw data would not be appropriate. Accordingly, the raw data were fitted to the Curie-Weiss formula in the temperature range above 90 K (Co) and 175 K (Cr). We obtained, for M = Cr, C = 3.24 emu K^{-1} mol^{-1}, which agrees reasonably with the theoretical value for 1 mol of S = 2 Cr^{2+} ions (3.0 emu K^{-1} mol^{-1}), indicating the completion of the ion-exchange reaction. Interestingly, the obtained value of C for M = Co is 2.775 emu K^{-1} mol^{-1}, which is significantly larger than the theoretical value for 1 mol of S = 3/2 Co^{2+} ions (1.875 emu K^{-1} mol^{-1}), where g = 2 is assumed. Thus orbital angular momentum should sizably contribute to the g-factor, resulting in a strong anisotropy in the Co moment due to spin-orbit interactions. Such a strong anisotropy has been observed in the Co^{2+}-containing compounds [21, 22]. The values of θ for M = Co and Cr were, respectively, −77.95 K, −61 K, the magnitude of which is only slightly higher than T_N (61 K and 52 K). Here we would like to stress that the proximity between $|\theta|$ and T_N does not mean that the frustration effect is negligible because when competing antiferromagnetic and ferromagnetic interactions are present (which is the case of the related systems), $|\theta|/T_N$ cannot be a measure of frustration. Notably, recent reinvestigations of the structure of $(CuCl)LaNb_2O_7$ using a single crystal X-ray diffraction and the state-of-the-art structural analysis revealed the superstructure in the space group *Pbam*, which accounts for the spin-singlet formation [12]. We suppose that there will be also a possibility that M and Cl atoms in the present compounds displace from the vertices of square lattice in an ordered manner, which might result in complex relationship between $|\theta|$ and T_N.

4. Conclusion

The two-dimensional antiferromagnets $(M\mathrm{Cl})LaNb_2O_7$ (M = Mn (S = 5/2), Co (S = 3/2) and Cr (S = 2)) obtained via a topotactic ion-exchange reaction are found to exhibit antiferromagnetic long range ordering at 53 K, 61 K and 52 K, respectively, indicating that the T_N is not simply scaled by the spin quantum number but affected by competing magnetic interactions and a possible formation of superstructure. A strong anisotropy was indicated in $(CoCl)LaNb_2O_7$, where the orbital degrees of freedom may affect the magnetic property. Further magnetic study is needed such as powder neutron diffraction.

Acknowledgments

This work was supported by Grants-in-Aid for Science Research (No. 19052004) from the Ministry of Education, Culture, Sports, Science and Technology (MEXT) of Japan. This work was also supported by the Global COE Program "Integrated Material Science" (No. B-024) from MEXT of Japan. One of the authors (A. K.) was supported by JSPS for Young Scientists.

References

[1] Melzi R, Carretta P, Lascialfari A, Mambrini M, Troyer M, Mittlet P and Mila F 2000 *Phys. Rev. Lett.* **85** 1318

[2] Kabbour H, Janod E, Corraze B, Danot M, Lee C, Whangbo M-H and Cario L 2008 *J. Am. Chem. Soc.* **130** 8261

[3] Kageyama H, Nishi M, Aso N, Onizuka K, Yoshihama T, Nukui K, Kodama K, Kakurai K and Ueda Y 2000 *Phys. Rev. Lett.* **84** 5876

[4] Onizuka K, Kageyama H, Narumi Y, Kindo K, Ueda Y and Goto T 2000 *J. Phys. Soc. Jpn.* **69** 1016

[5] Kodenkandath T A, Lalena J N, Zhou W L, Carpenter E E, Sangregorio C, Falster A U, Simmons W B, O'Connor C J, Wiley J B 1999 *J. Am. Chem. Soc.* **121** 10743

[6] Kodenkandath T A, Kumbhar A S, Zhou W L and Wiley J B 2001 *Inorg. Chem.* **40** 710

[7] Kageyama H, Viciu L, Caruntu G, Ueda Y and Wiley J B 2004 *J. Phys.: Condens. Matter* **16** S585

[8] Kageyama H, Kitano T, Oba N, Nishi M, Nagai S, Hirota K, Viciu L, Wiley J B, Yasuda J, Baba Y, Ajiro Y and Yoshimura K 2005 *J. Phys. Soc. Jpn.* **74** 1702

[9] Kageyama H, Yasuda J, Kitano T, Totsuka K, Narumi Y, Hagiwara M, Kindo K, Baba Y, Oba

N, Ajiro Y and Yoshimura K 2005 *J. Phys. Soc. Jpn.* **74** 1702
[10] Kitada A, Hiroi Z, Tsujimoto Y, Kitano T, Kageyama H, Ajiro Y and Yoshimura K 2007 *J. Phys. Soc. Jpn.* **76** 15523
[11] Yoshida M, Ogata N, Takigawa M, Yamaura J, Oba N, Kitano T, Kageyama H, Ajiro Y, and Yoshimura K 2007 *J. Phys. Soc. Jpn.* **76** 104703
[12] Tassel C, Kang J, Lee C, Hernandez O, Qiu Y, Paulus W, Collet E, Lake B, Guidi T, Whangbo M-H, Ritter C, Kageyama H and Lee S-H 2010 *Phys. Rev. Lett.* **105** 167205
[13] Oba N, Kageyama H, Kitano T, Yasuda J, Baba Y, Nishi M, Hirota K, Narumi Y, Hagiwara M, Kindo K, Saito T, Ajiro Y and Yoshimura K 2006 *J. Phys. Soc. Jpn.* **75** 113601
[14] Kitada A, Tsujimoto Y, Kageyama H, Ajiro Y, Nishi M, Narumi Y, Kindo K, Ichihara M, Ueda Y, Uemura Y J and Yoshimura K 2009 *Phys. Rev. B* **80** 174409
[15] Uemura Y J, Aczel A A, Ajiro Y, Carlo J P, Goko T, Goldfeld D A, Kitada A, Luke G M, MacDougall G J, Mihailescu I G, Rodriguez J A, Russo P L, Tsujimoto Y, Wiebe C R, Williams T J, Yamamoto T, Yoshimura K and Kageyama H 2009 *Phys. Rev. B* **80** 174408
[16] Tsujimoto Y, Kageyama H, Baba Y, Kitada A, Yamamoto T, Narumi Y, Kindo K, Nishi M, Carlo J P, Aczel A A, Williams T J, Goko T, Luke G M, Uemura Y J, Ueda Y, Ajiro Y and Yoshimura K 2008 *Phys. Rev. B* **78** 214410
[17] Viciu L, Caruntu G, Royant N, Koenig J, Zhou W L, Kodenkandath T A and Wiley J B 2002 *Inorg, Chem.* **41** 3385
[18] Viciu L, Koenig J, Spinu L, Zhou W L and Wiley J B 2003 *Chem. Mater.* **15** 1480
[19] Viciu L, Golub V O and Wiley J B 2003 *J. Solid State Chem.* **175** 88
[20] Tsujimoto Y, Kitada A, Uemura Y J, Goko T, Aczel A A, Williams T J, Luke G M, Narumi Y, Kindo K, Nishi M, Ajiro Y, Yoshimura K and Kageyama H 2010 *Chem. Mater.* **22** 4625
[21] Breed D J, Gilijamse K and Miedema A R 1969 *Physica* **45** 205
[22] Manaka H, Miyashita Y, Watanabe Y and Masuda T 2007 *J. Phys. Soc. Jpn.* **76** 085003

O^{2-}-to-F^{-} substitution on the quasi-two-dimensional quantum antiferromagnet $(CuCl)LaNb_2O_7$

C Tassel, Y Kobayashi, S Mitsuoka, F Takeiri, Y Ajiro, H Kageyama

Department of Energy and Hydrocarbon Chemistry, Graduate School of Engineering, Kyoto University, Nishikyo, Kyoto 615-8510, Japan

E-mail: tassel.cedricpascaleric.6x@kyoto-u.ac.jp

Abstract. We present the preparation of the electron-doped quasi-two dimensional quantum antiferromagnet $(CuCl)LaNb_2O_7$ by two step-wise topochemical reactions. The first step involves a reductive fluorination of the insulating layered perovskite $RbLaNb_2O_7$ using polytetrafluoroethylene that allows electric conduction in the perovskite blocks. The product $RbLaNb_2O_6F$ was then ion-exchanged with copper dichloride to yield $(CuCl)LaNb_2O_6F$. The synchrotron x-ray diffraction study indicates that the structure of the final compound is different from that of the pure $(CuCl)LaNb_2O_7$.

1. Introduction

The Dion-Jacobson perovskite phases $A'A_{n-1}B_nO_{3n+1}$ (A' = Li^+, Na^+, K^+, Rb^+, Cs^+,...; A = La^{3+}, Nd^{3+}, Eu^{3+}, Sr^{2+}, Ca^{2+}, Ba^{2+},...; B = Nb^{5+}, Ta^{5+}, Ti^{4+}, n = 2, 3, 4,...) consist of negatively charged, non-magnetic perovskite slabs $[A_{n-1}B_nO_{3n+1}]^-$ with the B site being occupied by a d^0 transition metal that is separated by a monovalent cationic (A'^+) unit.[1-3] The interlayer site, the A' site, can be modified using a variety of low-temperature reactions such as intercalation and ion-exchange to give various new chemical compositions.[4]

Wiley and coworkers modified the A'^+ site with the metal-halide array $(MX)^+$ (M = Co^{2+}, Cu^{2+}, Fe^{2+}, Mn^{2+}... and X = Cl^-, Br^-) through ion-exchange reaction using corresponding metal dihalides.[5-10] The resulting structure consists of the magnetic MO_2X_4 octahedra layers separated by the nonmagnetic AB_2O_7 arrays with the B cation remaining in the d^0 electronic configuration. The original structural analysis showed that the M^{2+} ions form an ideal two-dimensional square lattice. These compounds exhibit interesting magnetic features due to competing ferromagnetic (FM) and antiferromagnetic (AFM) exchange interactions within the CuX layer. For instance, the magnetism of $(CuCl)LaNb_2O_7$ is characterized by a spin singlet ground state with a spin gap of 2.3 meV.[11] Under magnetic fields, the Bose-Einstein condensation of triplet magnons occurs at around 10 T.[12] Recently, its structure was refined using powder and single crystal neutron diffraction and x-ray diffraction (XRD) revealing that this material corresponds to the otherwise solely theoretical ferromagnetically coupled Shastry-

Published under licence by IOP Publishing Ltd

Sutherland model, where the four next-nearest neighbour Cu^{2+} ions forming spin-singlets are arranged *orthogonally* in the *ab* plane (J_4) and ferromagnetically coupled with respect to each other.[13]
How itinerant electrons behave in the frustrated lattice has recently been attracting considerable attention. Examples include the anomalous Hall effect related to chiral magnetic structures,[14,15] heavy fermion behaviour in the spinel oxide LiV_2O_4 [16] and anisotropic superconductivity in the triangular lattice $Na_{0.35}CoO_2 \cdot 1.3H_2O$.[17] Up till now, however, most experimental studies were performed on triangle-based frustrated lattices. Although several theoretical investigations of the Shastry-Sutherland lattice demonstrated anisotropic superconductivity when moderate carriers are introduced, no experimental report on metallization exists on a Shastry-Sutherland system either in the ferromagnetic case, or even the antiferromagnetic Shastry-Sutherland $SrCu_2(BO_3)_2$.[18-21] In light of the fact that a ferromagnetically coupled Shastry-Sutherland compound has now been realized, it would be very interesting to inject electrons or holes into $(CuCl)LaNb_2O_7$. Recently, Kobayashi *et al.*[22] performed the reaction of $RbLaNb_2O_7$ with polytetrafluoroethylene (PTFE) and succeeded to reduce partially the valence of the niobium centre from Nb^{5+} to Nb^{4+}, yielding $RbLaNb_2O_{7-x}F_x$ ($0 < x < 1$). Remarkably, this reductive fluorination induces electronic conductivity in the Nb_2O_7 blocks while preserving the ion-exchangeable Rb^+ layer. The present study combines the aforementioned ion-exchange reaction with the reductive fluorination of $RbLaNb_2O_7$ to finally obtain $(CuCl)LaNb_2O_7F$.

2. Experimental

$RbLaNb_2O_7$ was synthesized using high temperature synthesis. Powders of Rb_2CO_3 (99.99%), La_2O_3 (99.99%) and Nb_2O_5 were mixed in an agate mortar. The mixture was heated in an alumina crucible at 800 °C for 12 hours. Next, the powder was ground and sintered at 1050 °C for 24 hours. Finally, the obtained sample was washed using water and heated at 1050 ºC for 24 hours. The quality of the sample was checked using a Bruker AXS (D8 Advance) XRD apparatus.

Fluorination was performed by mixing $RbLaNb_2O_7$ with PTFE (99.9%). A 1:0.5 mol mixture was ground thoroughly in an agate mortar and pelletized. The pellets were inserted in a Pyrex tube, sealed under dynamic vacuum (1.3×10^{-8} MPa) and heated at 400 °C for 24 hours. Following this, fluorinated samples were mixed with $CuCl_2$ (99.999%) in a 1:3 molar ratio in an Ar-filled glove box, pelletized, inserted in pyrex tubes and sealed under dynamic vacuum ($< 1.3\times10^{-8}$ MPa). The tubes were heated at 325 °C for one week. Finally the powders were washed using methanol to remove RbCl and excess $CuCl_2$ byproducts. Chemical composition was determined by combustion ion chromatography.

The synchrotron powder XRD experiments were performed at 293 K on the large Debye–Scherrer camera installed at the Japan Synchrotron Radiation Research Institute SPring-8 BL02B2 by using an imaging plate as a detector. Incident beams from a bending magnet were monochromatized to 0.77709 Å. The synchrotron XRD data were collected in a 2θ range from 0° to 75° with a step interval of 0.01°.

The crystal structure was refined using the Rietveld method (RIETAN-2000).[23] The peak shape model used was a modified split-pseudo-Voigt function. The agreement indices are *R*-weighted pattern, $R_{wp} = [\sum w_i(y_{io} - y_{ic})^2/\sum w_i(y_{io})^2]^{1/2}$, *R*-pattern, $R_p = \sum|y_{io} - y_{ic}|/\sum(y_{io})$, and goodness of fit (GOF), $\chi^2 = [R_{wp}/R_{exp}]^2$, where $R_{exp} = [(N-P)/\ \sum w_i y_{io}^2]^{1/2}$, y_{io} and y_{io} are the observed and calculated intensities, w_i is the weighting factor, $I_o(h_k)$ and $I(h_k)$ are the observed and calculated integrated intensities, N is the total number of y_{io} data when the background is refined, and P is the number of adjusted parameters.

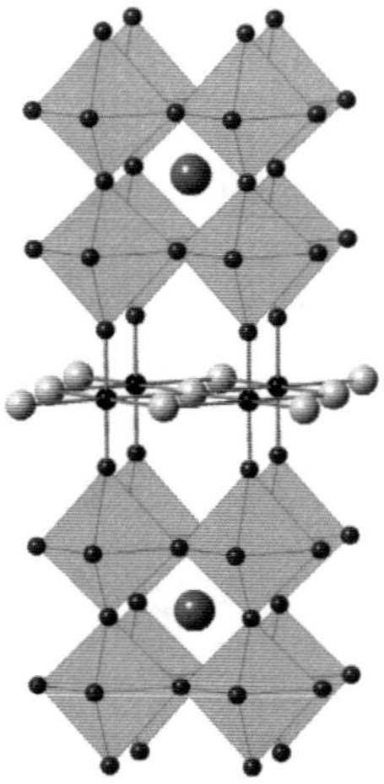

Figure 1. Schematic structure of (CuCl)$LaNb_2O_7$. Grey polyhedra are NbO_6 octahedra. White, black and grey balls represent Cl^-, Cu^{2+} and La^{3+}, respectively.

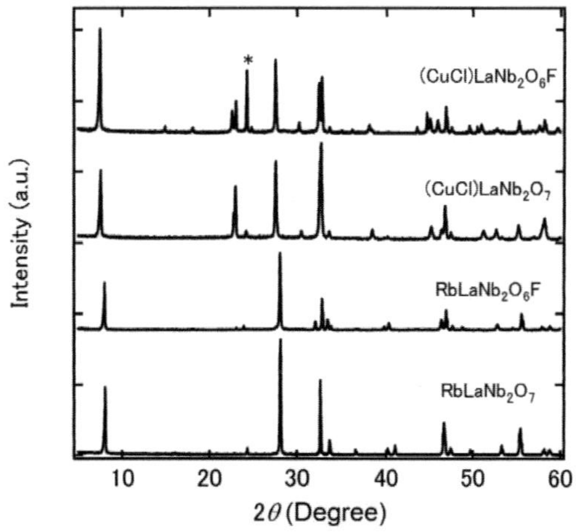

Figure 2. Comparison of x-ray diffraction patterns of $RbLaNb_2O_7$, $RbLaNb_2O_6F$, (CuCl)$LaNb_2O_7$ and (CuCl)$LaNb_2O_6F$. Asterisks indicate the position of Si.

3. Results and discussion

The synthesis of (CuCl)$LaNb_2O_6F$ occurs in two steps: the reductive fluorination of the niobium layer of $RbLaNb_2O_7$ and the ion-exchange of the interlayer cation Rb^+ with $CuCl^+$. The colour of the sample changes from white to black in the first reaction. After the fluorination, we observed a larger change of the lattice parameters, i.e. the increase of the c axis (~0.26 Å) is greater than ~0.21 Å reported earlier [22] (see Table 1). This indicates that a larger amount of fluorine was inserted in the sample, despite the fact that less PTFE was used in our preparation compared to ref. 22. This difference may be due to the difference in the fluorination method. The fluorination was performed previously under flowing N_2, while in the present study the $RbLaNb_2O_7$:PTFE mixture was inserted in a vacuum sealed pyrex tube. The a axis decreases only slightly, which comes from the preferential occupation of the fluorine atoms at the apical site as pointed out in.[22]

Table 1. Lattice parameters of the reduced unit cell of $RbLaNb_2O_7$, $RbLaNb_2O_6F$, (CuCl)$LaNb_2O_7$ and (CuCl)$LaNb_2O_6F$.

	Space Group	a (Å)	b (Å)	c (Å)	V (Å)
$RbLaNb_2O_7$[a]	*P4/mmm*	3.89170(9)	3.89170(9)	11.0072(4)	166.71
$RbLaNb_2O_6F$[a]	*P4/mmm*	3.86575(16)	3.86575(16)	11.2135(7)	167.58
$RbLaNb_2O_7$	*P4/mmm*	3.8894(8)	3.8894(8)	10.9870(2)	166.20
$RbLaNb_2O_6F$	*P4/mmm*	3.8685(4)	3.8685(4)	11.243(8)	168.27
(CuCl)$LaNb_2O_7$	*P4/mmm*	3.8807(4)	3.8807(4)	11.7337(9)	176.70
(CuCl)$LaNb_2O_6F$	*Pbam*	3.8711(8)	3.8711(8)	11.863(5)	177.77

[a] ref. 22

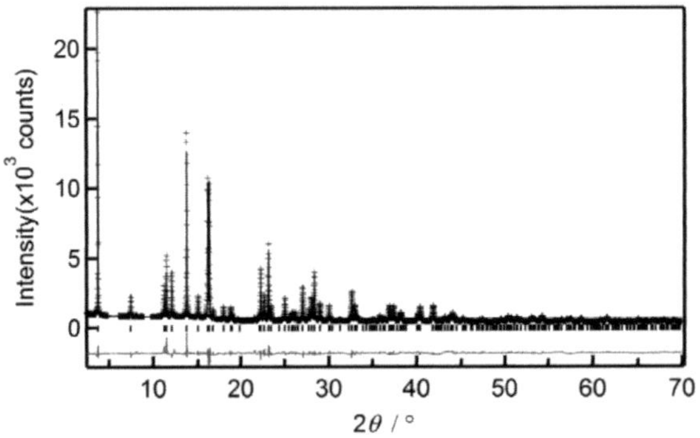

Figure 3. Rietveld refinement of $(CuCl)LaNb_2O_6F$. Dark crosses, the black line, black ticks and the bottom line indicate the observed pattern, calculated pattern, peak positions and difference between observed and calculated pattern, respectively.

The second reaction led to a colour change from black to dark green. Elemental analysis showed that the average content of fluorine is ~1.1(7) in $RbLaNb_2O_6F$ and ~1.4(4) in $(CuCl)LaNb_2O_6F$, indicating no loss of fluorine during the ion-exchange reaction. Unfortunately, the electric conductivity was lost. The unit cell of the final product was calculated to be tetragonal with a = 3.8711(8) Å and c = 11.863(5) Å. The c axis increase of ~ 0.13 Å with respect to the previously reported non-fluorinated $(CuCl)LaNb_2O_7$ indicates both the successful intercalation of CuCl and the preservation of the fluorine. We first refined the structure using the simple tetragonal unit cell and the space group *P*4/*mmm*. The fluorine atom was placed at the 1*c* site bridging the two octahedra within the layers. The refinement gave reasonable agreement values of R_{wp} = 7.09% and χ^2 = 3.43. However, several peaks were not fitted well and high anisotropic displacement parameters (ADP) values were observed for both Cu (3.64×10^{-2} Å^2) and Cl (8.56×10^{-2} Å^2). From this, we tried a disordered model with copper and chlorine shifting from their ideal (0, 0, 0.5) and (0.5, 0, 0.5) positions, respectively. The agreement values consequently decreased to R_{wp} = 5.97% and χ^2 = 2.48 (Figure 3); the position of copper remained within the ideal position but chlorine drifted to (0.094, 0.056, 0.4999) (Table 2). The refinement of the ADP led to similar values for copper and better values for chlorine. The occupation factor of copper and chlorine was refined and gave reasonable values g_{Cu} = 0.98(2) for copper and g_{Cl} = 1.00(6) for chlorine indicating that the ion exchange was complete.

The genuine structure of $(CuCl)LaNb_2O_7$ is orthorhombic *Pbam* 2*a* × 2*b* × *c* with a pseudo tetragonal unit cell (a = 7.7556(5) Å and b = 7.7507(5) Å).[13,24] When we applied this distorted model to the obtained data, however, the refinement did not lead to better values (R_{wp} = 6.33% and χ^2 = 2.72). While the ADP parameters for copper decrease significantly from the disordered model (Table 3), the occupancy factors for both elements are not satisfactory (g_{Cu} = 0.95(2) and g_{Cl}=0.90(2)). The refinement results based on the non-distorted structure (*P*4/*mmm*) and the superstructure (*Pbam*) indicate that although the structure of $(CuCl)LaNb_2O_6F$ is possibly distorted, the structure is different from that of $(CuCl)LaNb_2O_7$ and the FM Shastry-Sutherland model. Complementary analysis of these materials using other techniques (electron and neutron diffraction) will be necessary to understand their genuine structure.

The valence state of the transition metal elements (i.e., copper and niobium) in the final product is not clear yet. A possible reason for the loss of the electronic conductivity after the ion-exchange reaction is that mobile carriers in the perovskite blocks in $RbLaNb_2O_6F$ become frozen out as a consequence of the CuCl layer insertion. The insertion of such a covalent network may induce the deformation of oxygen octahedral in NbO_6 and the tilting and rotation of octahedra, both of which would strongly modulate the transfer integral between Nb 3*d* and O 2*p* orbitals. Another plausible reason is the case, where all conducting electrons in the perovskite blocks are transferred to the CuCl layer, leaving no partially occupied *d*-electrons for both Cu (d^{10}) and Nb (d^0). Then an interesting challenge toward metallization would be controlling F/O ratio. It was already demonstrated that the F/O ratio in $RLaNb_2(O_{6+x}F_x)$ can be tuned in the range of $0 \leq x \leq 1$.[22] It would be thus possible to obtain $(CuCl)LaNb_2(O_{6+x}F_x)$. In any case, band calculations would be helpful to know whether the chemical potential is located closer to the Cu 3*d* band or to the Nb 4*d* band.

Table 2. Structural parameters obtained from Rietveld refinement CuCl)$LaNb_2O_6F$.

Atom	site	*sof*	*x*	*Y*	*z*	$100B_{iso}$ (Å^2)
Cu	1d	0.97(7)	0.5	0.5	0.5	3.64(3)
Cl	8	0.125(6)	0.094(3)	0.056(5)	0.5	3.96(8)
La	1*a*	1	0	0	0	1.29(2)
Nb	2*h*	1	0.5	0.5	0.193(6)	0.28(1)
O(1)	4*i*	1	0	0.5	0.152(3)	1.13(9)
O(2)	2*h*	1	0.5	0.5	0.333(2)	1.27(4)
O/F(3)	1*c*	1	0.5	0.5	0	2.67(2)

P4/mmm, a = 3.8711(8) Å, c = 11.863(5) Å, R_{wp} = 5.97 %, R_p = 4.07 %, and χ^2 = 2.48.

Table 3. Structural parameters obtained from Rietveld refinement of (CuCl)$LaNb_2O_6F$, distorted model.

Atom	site	*Sof*	*x*	*y*	*z*	$100B_{iso}$ (Å^2)
Cu	4*h*	0.95(2)	0.492(7)	−0.220(7)	0.5	2.09(8)
Cl	4*h*	0.90(2)	0.737(2)	−0.057(8)	0.5	4.83(6)
La	4*g*	1	0.750(5)	−0.505(2)	0	1.24(8)
Nb	8*i*	1	0.500(8)	−0.248(3)	0.192(5)	0.44(8)
O(1)	4*f*	1	0.5	0	0.152(3)	1.48(5)
O(2)	4*e*	1	0.5	0.160(2)	0.374(6)	0.37(4)
O(3)	8*i*	1	0.761(8)	−0.251(6)	0.148(7)	0.87(3)
O/F(4)	4*g*	1	0.519(7)	−0.265(9)	0	1.82(5)
O(5)	8*i*	1	0.500(1)	−0.257(1)	0.322(7)	1.09(2)

Pbam, a = 7.7419(5) Å, b = 7.7426(2) Å, c = 11.863(5) Å, R_{wp} = 6.33 %, R_p = 4.25%, and χ^2 = 2.72.

4. Conclusion

We prepared (CuCl)$LaNb_2O_6F$ by two successive topotactic reactions. The first consisted in the substitution of O^{2-} by F^- and the second the ion-exchange of Rb^+ with $(CuCl)^+$ to yield (CuCl)$LaNb_2O_6F$ from $RbLaNb_2O_7$. By using synchrotron powder diffraction, we were able to show that the structure of the fluorinated (CuCl)$LaNb_2O_6F$ appears to be different from (CuCl)$LaNb_2O_7$. Distortions from the ideal model are common in these layered perovskites and have been observed in

(FeCl)$LaNb_2O_7$, having a $\sqrt{2}a \times \sqrt{2}b \times c$ monoclinic cell [25], and (CuCl)$LaNb_2O_7$ having a $2a \times 2b \times c$ orthorhombic cell.[13,24] In the same way, the detailed structures of the fluorinated phases $RbLaNb_2O_6F$ and (CuCl)$LaNb_2O_6F$ are unknown; i.e. $RbLaNb_2OF$ has been demonstrated to have a $4a \times 4b \times c$ superstructure, but the details are not clear yet.[22] Although the electronic conductivity was lost after the ion-exchange reaction, further synthetic efforts such as controlling the fluorine content may eventually change the unfortunate situation. The synthesis pathway and structure analysis developed here will allow the systematic study of new fluorinated ion-exchanged perovskite $(MX)A_{n-1}B_nO_{3n+1}$ which we hope will provide crucial hints to understand and realize new structures with novel magneto-electric properties.

Acknowledgements

The authors acknowledge Science Research on Priority Area (19052004) from MEXT of Japan.

References

[1] Dion, M.; Ganne, M.; Tournoux, M. 1981 *Mater. Res. Bull.* **16** 1429.
[2] Dion, M.; Ganne, M.; Tournoux, M.; Ravez, J. 1984 *Rev. Chim. Miner.* **21** 92.
[3] Jacobson, A. J.; Johnson, J. W.; Lewandowski, J. T. 1985 *Inorg. Chem.* **24** 3729.
[4] Schaak, R. E.; Mallouk, T. E. 2002 *Chem. Mater.* **14** 1455, and references therein.
[5] Kodenkandath, T. A. *et al.* 1999 *J. Am. Chem. Soc.* **121** 10743.
[6] Kodenkandath, T. A.; Kumbhar, A. S.; Zhou W. L.; Wiley, J. B. 2001 *Inorg. Chem.* **40** 710.
[7] Viciu, L. *et al.* 2002 *Inorg. Chem.* **41** 3385.
[8] Viciu, L. *et al.* 2003 *Chem. Mater.* **15** 1480.
[9] Viciu, L. *et al.* 2003 *J. Solid State Chem.* **175** 88.
[10] Tsujimoto, Y. *et al.* 2010 *Chem. Mater.* **22** 4625.
[11] Kageyama, H. *et al.* 2005 *J. Phys. Soc. Jpn.* **74** 1702.
[12] Kitada, A. *et al.* 2007 *J. Phys. Soc. Jpn.* **76** 093706-1-4.
[13] Tassel, C. *et al.* 2010 *Phys. Rev. Lett.* **105** 167205.
[14] Taguchi, Y. *et al.* 2001 *Science* **291** 2573.
[15] Nakatsuji, S. *et al.* 2006 *Phys. Rev. Lett.* **96** 087204.
[16] Kondo, S. *et al.* 1997 *Phys. Rev. Lett.* **78** 3729.
[17] Takada, K. *et al.* 2003 *Nature* **422** 53.
[18] Shastry, B S.; Kumar, B. 2002 *Prog. Theor. Phys. Suppl.* **145** 1.
[19] Chung, C-H.; Kim, Y. B. 2004 *Phys. Rev. Lett.* **93** 207004.
[20] Kimura, T.; Kuroki, K.; Arita, R.; Aoki, H. 2004 *Phys. Rev. B* **69** 054501.
[21] Liu, J.; Trivedi N.; Lee, Y.; Harmon, B. N.; Schmalian, J. 2007 *Phys. Rev. Lett.* **99** 227003.
[22] Kobayashi, Y.; Tian M.; Eguchi M.; Mallouk, T. E. 2009 *J. Am. Chem. Soc.* **131** 9849.
[23] Izumi, F.; Ikeda, T. 2000 *Mater. Sci. Forum* **198** 321.
[24] Yoshida, M. *et al.* 2007 *J. Phys. Soc. Jpn.* **76** 104703-1-9.
[25] Oba, N. *et al.* 2007 *J. Magn. & Magn. Mater.* **310** 1337.

ToF inelastic neutron scattering studies on quantum spin systems (CuCl)LaB_2O_7 (B = Nb, Ta)

S Ohira-Kawamura[1], K Nakajima[1], Y Inamura[1], Y Tsujimoto[2], A Kitada[3], F Takeiri[3], H Kageyama[3], Y Ajiro[3], M Nishi[4] and K Kakurai[5]
[1]J-PARC Center, Japan Atomic Energy Agency, Tokai, Ibaraki 319-1195, Japan
[2]National Institute for Materials Science, Tsukuba, Ibaraki 305-0047, Japan
[3]Graduate School of Engineering, Kyoto University, Kyoto 615-8510, Japan
[4]Institute for Solid State Physics, University of Tokyo, Tokai, Ibaraki 319-1106, Japan
[5]Quantum Beam Science Directorate, Japan Atomic Energy Agency, Tokai, Ibaraki 319-1195, Japan

E-mail: seiko.kawamura@j-parc.jp

Abstract. Magnetic excitations in quantum spin system (CuCl)LaB_2O_7 (B = Nb, Ta) have been studied by the time-of-flight (ToF) inelastic neutron scattering technique. (CuCl)LaNb_2O_7 exhibits a spin-singlet ground state with an excitation gap of ΔE ~ 2.3 meV, while (CuCl)LaTa_2O_7 exhibits an antiferromagnetic (AFM) ordering below T_N = 7 K. In (CuCl)LaNb_2O_7, a band-like excitation with an energy width of ~1 meV was observed at ΔE ~ 2 meV. The magnitude of scattering vector Q dependence of the integrated intensity of the magnetic excitation around ΔE ~ 2 meV for this system well reproduces that observed by a triple-axis spectrometer, indicating a spin-singlet feature. The excitation spectra show modulations for both the intensity of excitation and the peak position in ΔE along the Q direction, which indicates that spin pairs forming a singlet state are not simply isolated but have some interaction. Using the isolated dimer model as well as the previous study, the modulation is characterized by a length ~ 8.7 Å, which corresponds to the fourth-neighbor distance. (CuCl)LaTa_2O_7 in the AFM state also shows modulating behaviour in the excitation spectra. However, the integrated intensity exhibits a different Q dependence, reflecting a different ground state.

1. Introduction

Low-dimensional magnetic systems where quantum fluctuations play an important role for their magnetic properties have been of great interest in recent years. In particular, spin frustration often brings about some exotic ground state. While some quantum phenomena have been reported for geometrically frustrated systems such as triangular lattice and Kagomé lattice, two-dimensional (2D) square lattice can also exhibit interesting quantum spin behavior by competing magnetic interactions. A series of compounds (CuCl)LaB_2O_7 (B = Nb or Ta) have the CuCl and LaB_2O_7 alternative layers, and these systems are candidates of such novel frustrated 2D spin systems [1-3]. The magnetic Cu ions form a square lattice within the layer. It has been expected that the system is described by J_1-J_2 model [4-6], namely, the nearest and next-nearest exchange interactions, J_1 and J_2, compete within the CuCl layer. The magnetic ground state depends on the ratio of J_1 to J_2. The ground state of (CuCl)LaNb_2O_7 (B = Nb) has been suggested to be a spin-singlet state [1]. A gap energy is estimated to be ΔE ~ 2.3 meV (~27 K) assuming an S = 1/2 isolated dimer model. A neutron scattering measurement on this

Journal of Physics: Conference Series **320** (2011) 012037 doi:10.1088/1742-6596/320/1/012037

system by using a triple-axis spectrometer by Kageyama *et al.* showed magnetic excitation at 2.2 meV, where modulation of the intensity along the Q direction was observed (Q is the magnitude of scattering vector). Furthermore, the results of the structure analysis on the single crystal by using x-ray and neutron diffraction measurements were recently reported by Tassel *et al.* [7]. There band structure calculations reveal that the magnetism of this system is described as antiferromagnetic (AFM) spin pairs between fourth neighbors coupled by frustrated ferromagnetic interactions in a Shastry-Sutherland-type arrangement. On the other hand, (CuCl)$LaTa_2O_7$ (B = Ta) exhibits an antiferromagnetically ordered state with collinear spin structure below $T_N = 7$ K [2,3].

The time-of-flight (ToF) inelastic neutron scattering technique using a chopper spectrometer is useful to investigate a whole aspect in magnetic excitation in such systems. Especially, new-generation chopper spectrometers installed recently at new neutron sources such as J-PARC can give detailed information for dynamics in a wide Q-E range, on which measurements give us alternative opportunities to those on the conventional triple-axis spectrometers. We have performed ToF inelastic neutron scattering measurements on (CuCl)LaB_2O_7 (B = Nb and Ta) to investigate their magnetic excitation. We report here the characteristic magnetic excitation in these systems. A band-like excitation was observed at $\Delta E \sim 2$ meV in both systems. The Q dependence for (CuCl)$LaNb_2O_7$ well reproduces that observed at a triple-axis spectrometer, which is indicative of a spin-singlet feature with a characteristic length ~ 8.7 Å assuming the isolated dimer model. On the other hand, (CuCl)$LaTa_2O_7$ in the AFM state exhibits a different Q dependence. In both systems, characteristic modulations for both the intensity of magnetic excitation and its peak position were observed along the Q direction.

2. Experimental procedure

Powder samples of (CuCl)$LaNb_2O_7$ and (CuCl)$LaTa_2O_7$ were prepared by the method reported elsewhere [1,2]. About 10g of the (CuCl)$LaNb_2O_7$ sample and 17g of (CuCl)$LaTa_2O_7$ sample were respectively packed into an Al foil to shape a cylinder with 10mm of diameter and 50mm of height. The packed sample was mounted into an Al sample can with the He gas, and set to a closed-cycle refrigerator. The inelastic neutron scattering measurements were carried out using a cold-neutron disk-chopper spectrometer AMATERAS installed at BL14 in Materials and Life science Experimental Facility at J-PARC, Japan [8,9]. We chose a chopper condition in which neutrons with the incident energies E_i = 41.9, 15.1, 7.7, 4.7 and 3.1 meV were extracted simultaneously. All the obtained data were analyzed with the software suite Utsusemi [10].

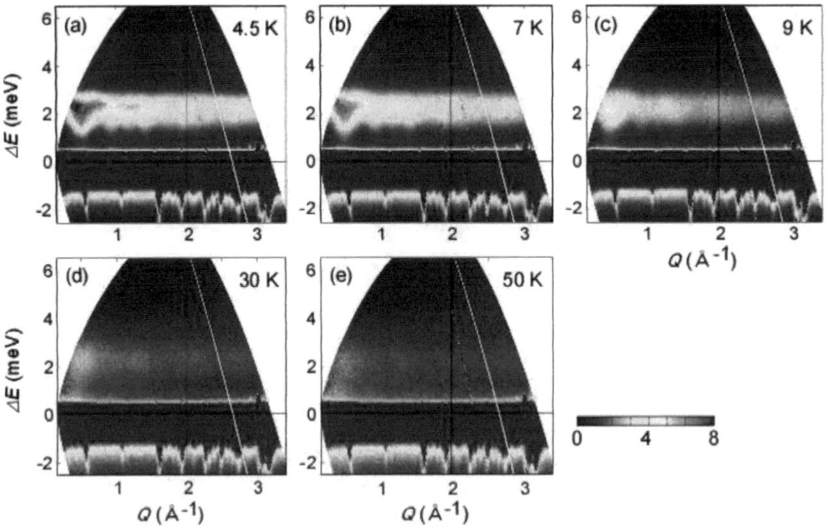

Figure 1. Excitation spectra in Q-E maps for (CuCl)$LaNb_2O_7$ with E_i = 7.7 meV obtained at (a) 4.5 K, (b) 7 K, (c) 9 K, (d) 30 K and (e) 50 K.

3. Results and discussion

Figure 1 shows the excitation spectra observed in *Q-E* maps for (CuCl)$LaNb_2O_7$ with E_i = 7.7 meV at various temperatures. The neutron count is normalized by proton beam current. A band-like excitation with an energy width of ~1 meV is clearly observed at around ΔE ~ 2 meV, indicating a singlet-triplet excitation. This excitation signal becomes ambiguous with increasing temperature, but survives even at 50 K. The peak position for the excitation along the ΔE direction does not change with temperature. The *Q* dependence of the integrated intensity of the magnetic excitation, which is obtained from a *Q*-scan at ΔE = 2.1 ± 1.2 meV, is displayed in figure 2. A clear oscillation of the intensity is observed here, and its amplitude decreases as temperature increases. However, the *Q* dependence does not change up to 50 K. Four peaks are seen at a glance. Its *Q* dependence well reproduces that obtained from the neutron scattering study by a triple-axis spectrometer, indicating the spin-singlet feature [1]. We introduce the isolated dimer model to analyze the present data as well as the previous neutron scattering study. In this model, the intensity as a function of *Q* can be described by the product of the squared form factor $f^2(Q)$ [11] and the interference term $(1 - \sin QR / QR)$ parameterized by the intradimer separation distance *R* [12,13]. Using this model, *R* is estimated to be ~ 8.7 Å. This value corresponds to the distance for fourth neighbors, 8.67 Å. In figure 2, the obtained *R* value seems to reasonably fit the observed oscillation frequency, which clearly indicates the fourth-neighbor interaction is dominant.

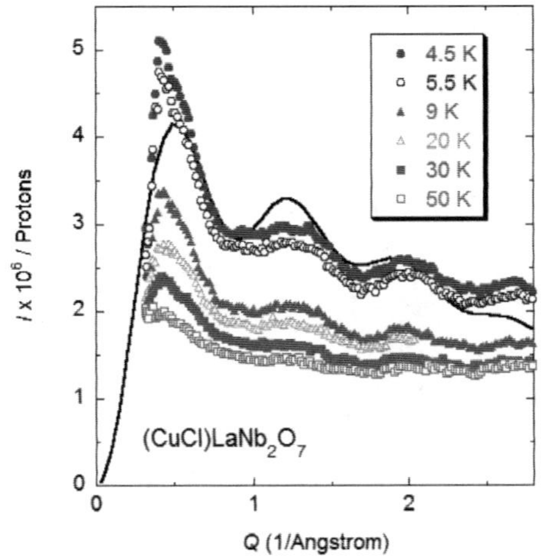

Figure 2. *Q* dependence of the integrated intensity of the magnetic excitation in (CuCl)$LaNb_2O_7$, obtained from a *Q*-scan at ΔE = 2.1 ± 1.2 meV. Solid line is the fit to the 4.5 K result with the function described in the text.

Looking at the *Q-E* maps carefully, we notice remarkable behavior: not only the intensity but the ΔE value of the peak position slightly modulates along the *Q* direction. This fact implies that the singlet spin pairs are not simply isolated ones but have some interaction. The spin-singlet state in this system neighbors long-range AFM states appearing for the Ta-for-Nb or Cl-for-Br substitution and application of magnetic field [4,5]. Therefore, it is likely that such correlation exists between the coupled spin pairs. We also note that the recent report shows that this system is described as a ferromagnetically coupled Shastry-Sutherland quantum spin system with the spin pairs between fourth neighbors [7]. The observed structure might be a precursor to induce the long-range ordering. In a constant-*Q* scan at *Q* = 0.8 ± 0.1 $Å^{-1}$ with E_i = 15.1 meV, another excitation was observed at around 5 meV (not shown), showing consistency with the result obtained by a triple-axis spectrometer [1].

Q-E maps for (CuCl)$LaTa_2O_7$ with E_i = 7.7 meV are shown in figure 3. The neutron count is normalized by proton beam current as well as in (CuCl)$LaNb_2O_7$. The collinear AFM ordering in this system is characterized by a propagation vector $\boldsymbol{q}$ = (1/2, 0, 1/2), which corresponds to *Q* = 0.84 $Å^{-1}$[2]. Clear dispersive behavior is not seen there. Instead, at 4.5 K, a band-like magnetic excitation was observed similarly to that in (CuCl)$LaNb_2O_7$ at ΔE ~ 2 meV. The excitation spectra in (CuCl)$LaTa_2O_7$ also exhibits modulations for both the intensity and the peak position in ΔE along the *Q* direction.

However, the modulation pattern is different from (CuCl)$LaNb_2O_7$. This difference in dispersion is presumably attributed to spin dynamics in this three-dimensionally ordered system being different from those in (CuCl)$LaNb_2O_7$. In contrast to (CuCl)$LaNb_2O_7$, the excitation energy decreases with increasing temperature. Its intensity also decreases and then the signal disappears at 30 K.

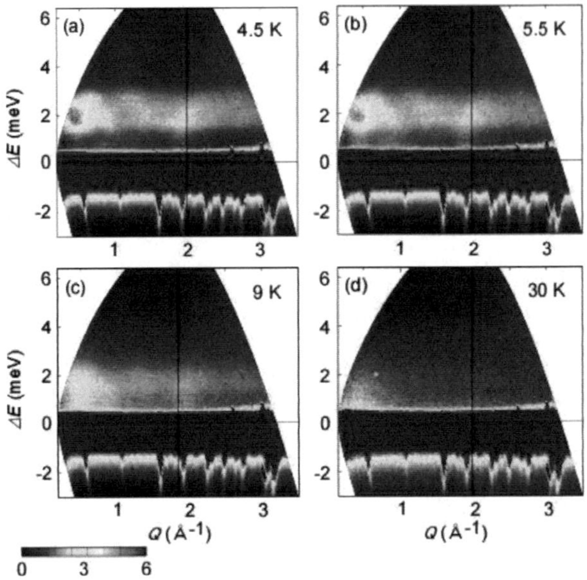

Figure 3. Excitation spectra in *Q-E* maps for (CuCl)$LaTa_2O_7$ with E_i = 7.7 meV obtained at (a) 4.5 K, (b) 5.5 K, (c) 9 K and (d) 30 K.

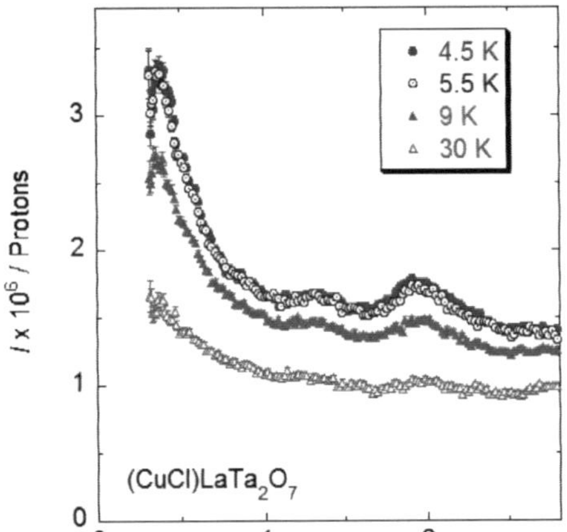

Figure 4. *Q* dependence of the integrated intensity of the magnetic excitation in (CuCl)$LaTa_2O_7$, obtained from a *Q*-scan at $\Delta E = 2.1 \pm 1.2$ meV.

The *Q* dependence of the integrated intensity of magnetic excitation, obtained from a *Q*-scan at $\Delta E = 2.1 \pm 1.2$ meV, is shown in figure 4. This system also shows oscillating behavior. However, the *Q* dependence at 4.5 K is different from that in (CuCl)$LaNb_2O_7$ reflecting the different ground state. Below T_N, peaks are seen at around 0.5, 0.85, 1.4 and 1.9 $Å^{-1}$. The second peak appears approximately at a position for $\boldsymbol{q} = (1/2, 0, 1/2)$. Above T_N, the second peak is ambiguous and the third peak appears to be slightly shifted to the lower *Q* side. The *Q* dependence at 9 K resembles that in (CuCl)$LaNb_2O_7$, rather than (CuCl)$LaTa_2O_7$ below T_N. The correlating coupled-spin feature lying in (CuCl)$LaNb_2O_7$ may remain even in (CuCl)$LaTa_2O_7$. Comparing the results in both systems to each other, it may be a common feature that a band-like excitation exists around 2 meV. Though the excitation observed in (CuCl)$LaTa_2O_7$ cannot be explained at the present stage, the obtained Q dependence of the intensity will give an important hint to understand the magnetic excitation. We need some model to describe the Q dependence of the intensity of the magnetic excitation in (CuCl)$LaTa_2O_7$.

4. Summary

ToF inelastic neutron scattering measurements using a chopper spectrometer were performed on (CuCl)LaB_2O_7 (B = Nb and Ta) to study their magnetic excitation. In a spin-singlet system (CuCl)$LaNb_2O_7$, a band-like excitation with an energy width of ~1 meV was observed at ΔE ~ 2 meV. The Q dependence of the integrated intensity of magnetic excitation for this system shows good agreement with the previous result obtained by using a triple-axis spectrometer, indicating the characteristics of a singlet-triplet excitation. The excitation spectra further shows modulations for not only its intensity but the peak position in ΔE along the Q direction. This implies that the singlet spin pairs are not simply isolated ones but have some interaction. (CuCl)$LaTa_2O_7$ in the antiferromagnetically ordered state exhibits similar excitation spectra with a band-like excitation at ΔE ~ 2 meV and peculiar modulations. However, the Q dependence of its intensity differs from that for (CuCl)$LaNb_2O_7$, reflecting the different ground state.

Acknowledgments

The authors acknowledge Science Research on Priority Area (No. 19052004) from the Ministry of Education, Culture, Sports, Science and Technology (MEXT) of Japan. One of the authors (K. N.) acknowledges supports by a Grant-in-Aid for Scientific Research (B) (No. 20340097) from MEXT of Japan.

References

[1] Kageyama H, Kitano T, Oba N, Nishi M, Nagai S, Hirota K, Viciu L, Wiley J B, Yasuda J, Baba Y, AjiroY and Yoshimura K 2005 *J. Phys. Soc. Jpn.* **74** 1702
[2] Kitada A, Tsujimoto Y, Kageyama H, Ajiro Y, Nishi M, Narumi Y, Kindo K, Ichihara M, Ueda Y, Uemura Y J and Yoshimura K 2009 *Phys. Rev. B* **80** 174409
[3] Uemura Y J, Aczel A A, Ajiro Y, Cario J P, Goko T, Goldfeld D A, Kitada A, Luke G M, MacDougall G J, Mihailescu J G, Rodriguez J A, Russo P L, Tsujimoto Y, Wiebe C R, Williams T J, Tamamoto Y, Yoshimura K and Kageyama H 2009 *Phys. Rev. B* **80** 174408
[4] Read N and Sachdev S 1989 *Phys.Rev. Lett.* **62** 1694
[5] Gelfand M P *et al.* 1989 *Phys. Rev. B* **40** 10801
[6] Becca F and Mila F 2002 *Phys. Rev. Lett.* **89** 037204
[7] Tassel C, Kang J, Lee C, Hernandez O, Qiu Y, Paulus W, Collet E, Lake B, Guidi T, Whangbo M -H, Ritter C, Kageyama H and Lee S -H 2010 *Phys. Rev. Lett.* **105** 167205
[8] Nakajima K, Nakamura M, Kajimoto R, Osakabe T, Kakurai K, Matsuda M, Metoki N, Wakimoto S, Sato T J, Itoh S, Arai M, Yoshida K and Niita K 2007 *J. Neutron Res.* **15** 13–21
[9] Nakajima K and Ohira-Kawamura S 2010 *Hamon* **20** 49–53 (in Japanese)
[10] Inamura Y, Nakajima K, Kajimoto R, Takatani N, Arai M, Otomo T, Suzuki J, So J Y and Park J G *Proc. 19th meeting on Collaboration of Advanced Neutron Sources*, to be published
[11] International Tables for Crystallography, ed. A. J. C. Wilson, Vol. C.
[12] Furrer A and Güdel H U 1977 *Phys. Rev. Lett.* **39** 657
[13] Güdel H U *et al.* 1979 *Inorg. Chem.* **18** 1021

Spin-orbit coupling inactivity of Co^{2+} ion in geometrically frustrated magnet $GeCo_2O_4$

Keisuke Tomiyasu[1], Ayaka Tominaga[2], Shigeo Hara[2], Hirohiko Sato[2], Tadataka Watanabe[3], Shin-Ichi Ikeda[4], Haruhiro Hiraka[5], Kazuaki Iwasa[1] and Kazuyoshi Yamada[6]

[1]Department of Physics, Tohoku University, Aoba, Sendai 980-8578, Japan
[2]Department of Physics, Chuo University, Bunkyou, Tokyo 112-8551, Japan
[3]Department of Physics, CST, Nihon University, Chiyoda, Tokyo 101-8308, Japan
[4]Nanoelectronics Research Institute, AIST, Tsukuba, Ibaraki 305-8568, Japan
[5]Institute for Materials Research, Tohoku University, Aoba, Sendai 980-8577, Japan
[6]WPI Advanced Institute for Materials Research, Tohoku University, Aoba, Sendai 980-8577, Japan

E-mail: tomiyasu@m.tohoku.ac.jp

Abstract. We report single-crystal neutron di raction studies on a spinel antiferromagnet $GeCo_2O_4$, which exhibits magnetic order with a trigonal propagation vector and tetragonal lattice expansion ($c/a \simeq 1.001$) below $T_N = 21$ K. For this inconsistency between spin and lattice in symmetry, magnetic Bragg re ections with a tetragonal propagation vector were discovered below T_N. We discuss spin and orbital states of Co^{2+} ion underlying the new magnetic component.

1. Introduction

Geometrical frustration causes mysterious spin-orbital-lattice states. In spinel-type oxides AB_2O_4, where A and B are metal ions, the B site octahedrally surrounded by anions forms a network of corner-sharing tetrahedra, also known as the pyrochlore lattice, which is geometrically frustrated. Recently, exotic phenomena in spinel-type magnets with orbital degree of freedom have attracted much attention. For example, AV_2O_4 (A = Zn, Mn, Li, Al) with t_{2g} orbital degree of freedom exhibit spin and orbital chain, orbital ordering, heavy fermion behavior and heptamer spin singlet state [1–4].

A spinel-type antiferromagnet $GeCo_2O_4$ (Co^{2+}: d^7, $S = 3/2$) is also expected to have degree of freedom in t_{2g} orbital states [5]. The cobaltite simultaneously exhibits magnetic long-range order described by a trigonal propagation vector $\boldsymbol{Q}_{\rm II} = (1/2, 1/2, 1/2)$ and cubic-tetragonal lattice expansion at $T_N = 21$ K ($c/a \simeq 1.001$) [6; 7]. The spin and lattice symmetries contradict each other.

A NaCl-type antiferromagnet CoO (Co^{2+}: d^7, $S = 3/2$) once had the same problem: this material also exhibits magnetic order with $\boldsymbol{Q}_{\rm II}$ and tetragonal lattice contraction below $T_N = 290$ K ($c/a \simeq 0.99$) [8–10]. However, careful neutron di raction experiments later revealed the existence of additional antiferromagnetic order with a tetragonal propagation vector $\boldsymbol{Q}_{\rm I}$=(0,0,1) [11]. Therefore, $GeCo_2O_4$ likely hides an additional magnetic order with tetragonal symmetry as well.

In this paper, we report our discovery of magnetic Bragg re ections with a tetragonal propagation vector by single-crystal neutron di raction on $GeCo_2O_4$. The spin and orbital state of Co^{2+} ion is discussed in comparison with that in CoO.

2. Experiments

Neutron di raction (elastic scattering) experiments were carried out on a triple-axis spectrometer, AKANE (T1-2) of Institute of Materials Research, Tohoku University, installed at the thermal guide tube of JRR-3 in Japan Atomic Energy Agency (JAEA) [12] and TOPAN (6G) of the same University, installed at the same reactor. AKANE equips a Ge 311 monochromator and a PG 002 analyzer, and e ectively eliminates higher-order contamination without a lter because of a Ge 622 forbidden re ection. Incident energy was xed at E_i = 19.4 meV with horizontal collimation sequence of guide-open-60′-60′. On TOPAN, E_i was xed at 13.5 meV and 30.5 meV with that of blank-60′-60′-blank. Usually one pyrolytic graphite lter is used to eliminate the half-lambda contamination, but we used two lters to completely eliminate it down to the order of 10 ^{6}. A single-crystal rod of $GeCo_2O_4$ was enclosed in an aluminum container with ^{4}He exchange gas, which was mounted on the cold nger of a ^{4}He closed-cycle refrigerator. The crystal rod was grown by a oating zone method. Detail of the crystal growth was summarized in Ref. [13]. The size of rod is about 4 mm diameter and 20 mm height.

Since our crystal has multiple tetragonal domains and the $|1 \quad c/a|$ value is only 0.1%, we cannot distinguish 001 and 100 (parallel and perpendicular to the tetragonal axis) in the present experiments. Hereafter we use the cubic notation for arbitrary *hkl* indices and propagation vectors.

3. Results

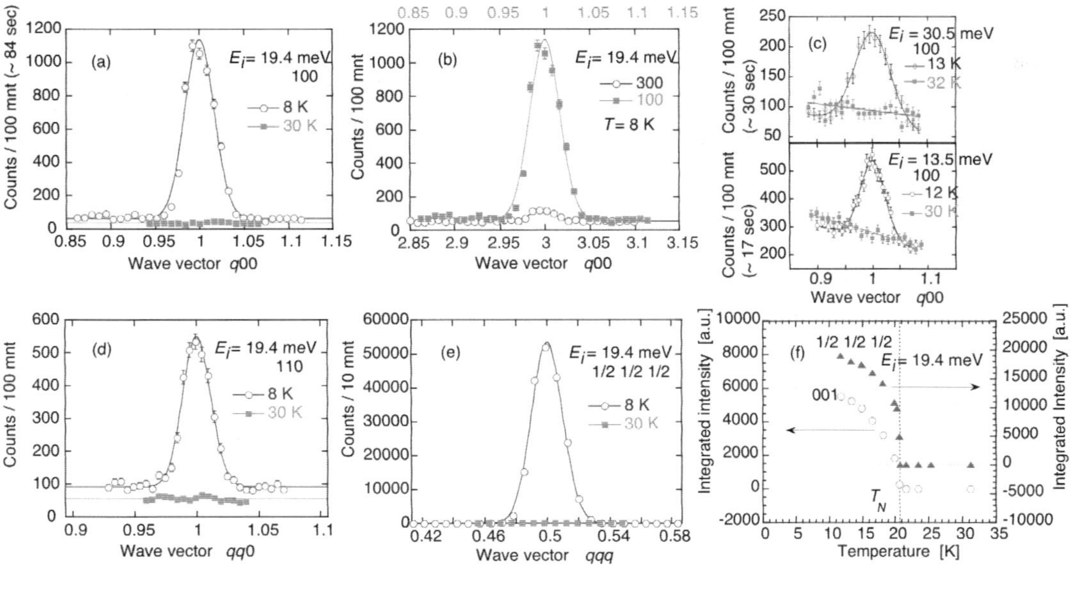

Figure 1. (Color online) Neutron di raction data, measured on a single-crystal specimen of $GeCo_2O_4$. The 100 re ection (a), comparison between 100 and 300 re ections (b), 100 re ections with di erent E_i (c), 110 re ection (d), 1/2 1/2 1/2 re ection (e) and temperature dependence of integrated intensity of 100 and 1/2 1/2 1/2 re ections (f).

Figure 1(a) shows the neutron di raction data, obtained by scanning along the ⟨100⟩ direction

Journal of Physics: Conference Series **320** (2011) 012038 doi:10.1088/1742-6596/320/1/012038

below and above T_N with E_i = 19.4 meV. The unreported re ection is observed at the 100 reciprocal lattice point only below T_N. Figure 1(b) shows the data of 100 re ection and equivalent 300 one below T_N. The former intensity is much stronger than the latter one, indicating that these re ections are magnetic.

The integrated intensity ratio of 100 re ection to conventional 1/2 1/2 1/2 magnetic Bragg re ection is only less than 1/500. Therefore, in order to completely eliminate the doubt of double re ection (multiple scattering), we carried out the same scans with di erent E_i = 13.5 and 30.5 meV. Since multiple scattering is very sensitive to the re ection angles, it should instantly disappear for di erent E_i. As shown in Fig. 1(c), however, the 100 re ection was observed for the di erent E_i again. Thus the 100 magnetic re ection undoubtedly exists.

We also con rmed the existence of temperature-dependent re ections at 110, 112 and 221. The 110 data is shown in Fig. 1(d) as the representative. All the new re ections are described by $\boldsymbol{Q}_{\mathrm{I}} = (1, 0, 0)$.

Figure 1(e) shows the data of conventional 1/2 1/2 1/2 magnetic Bragg re ection. The linewidth of the 100 re ection is as sharp as that of the 1/2 1/2 1/2 one, meaning that the correlation length of $\boldsymbol{Q}_{\mathrm{I}}$ order is longer than the instrumental resolution limit of 130 A (full width at half maximum). Figure 1(f) shows the temperature dependence of integrated intensities of 100 and 1/2 1/2 1/2 re ections. Both the magnetic re ections disappear at T_N = 21 K with increasing the temperature.

In this way, we discovered the additional $\boldsymbol{Q}_{\mathrm{I}}$ magnetic long-range order, which simultaneously appears below T_N in addition to the dominant $\boldsymbol{Q}_{\mathrm{II}}$ one.

4. Discussion

We discuss the magnetic structure described by $\boldsymbol{Q}_{\mathrm{I}}$. The present experiments cannot distinguish (1,0,0) and (0,0,1) for $\boldsymbol{Q}_{\mathrm{I}}$. In the tetragonal lattice expansion [7], however, spin interaction is expected to be more dominant in the c plane than along the c axis; (1,0,0) is more plausible than (0,0,1). In fact, $\boldsymbol{Q}_{\mathrm{I}}$ is equal to (0,0,1) in CoO with lattice contraction [11].

Furthermore, recent ultrasound measurements reported that, as temperature decreases, longitudinal elastic moduli not only along the ⟨111⟩ direction but also the ⟨110⟩ direction remarkably soften towards T_N [5]. The softening along the ⟨111⟩ direction is most likely connected to the trigonal magnetic ordering by strong spin-lattice coupling [5]. Although an origin of the other softening has been unresolved, this suggests that the $\boldsymbol{Q}_{\mathrm{I}}$ order is composed of spin chains along the [110] and [$1\bar{1}0$] directions in the c plane, as shown in Fig. 2(a).

Determination of more precise magnetic structure is di cult owing to the complex structure of pyrochlore lattice, the multi-domains and the neutron absorption e ect of Co nuclei. Even for an isomorphic cubic material $GeNi_2O_4$ with no strong absorption, single-crystal polarized neutron di raction experiments have recently determined its magnetic structure at last [14].

The $\boldsymbol{Q}_{\mathrm{I}}$ magnetic order in $GeCo_2O_4$ can be understood in the standard orbital picture under tetragonal lattice expansion described as $(d_{yz})^2$, $(d_{zx})^2$, $(d_{xy})^1$, $(d_{3z^2\ r^2})^1$ and $(d_{x^2\ y^2})^1$. An unpaired electron occupies d_{xy} without orbital degree of freedom in t_{2g} states below T_N. This electron will lead to kinetic direct exchange between the nearest neighbor Co^{2+} ions [5], and will promote the formation of spin chains shown in Fig. 2. If so, however, it is interesting that a Co^{2+} ion would exhibit the character of d_{xy} orbital, normally vanished by spin-orbit coupling like in CoO ($S = 3/2$, $L = 1$, m 4 μ_B) [15; 16]. In fact, the Curie-Weiss tting to magnetic susceptibility data gives us magnetic moment of about 4.9 μ_B for $GeCo_2O_4$, which is much larger than the spin-only value 3 μ_B [17]. Therefore, it seems that the LS orbital state (L 1 μ_B) with $\boldsymbol{Q}_{\mathrm{II}}$ component is dominant, and that the non-LS (d_{xy}) character with $\boldsymbol{Q}_{\mathrm{I}}$ component supplementarily adds to the LS state.

We consider why the spin-orbit coupling tends to be inactive in $GeCo_2O_4$ compared to in CoO. The Co^{2+} ions with $S = 3/2$ have a hole in t_{2g} orbital states, which generates $\boldsymbol{L}$ characters [15].

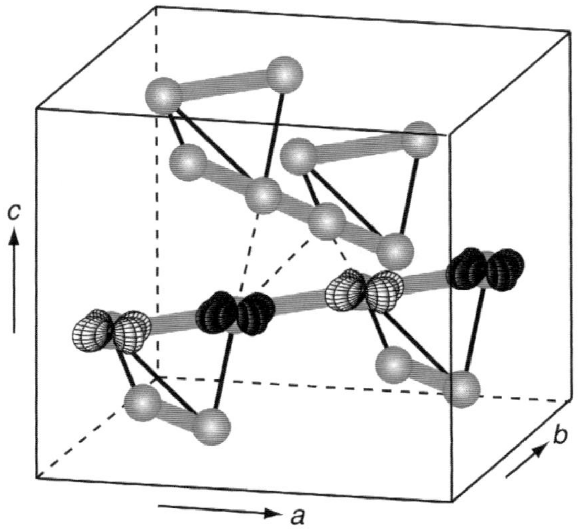

Figure 2. (Color online) Spin and orbital model for $\boldsymbol{Q}_{\rm I}$ component in $GeCo_2O_4$. The orange bold lines show the antiferromagnetic spin chains connected by d_{xy} orbital states. The orbitals with up and down spins (black and white polygons) on one chain are depicted as the representative.

Since the former and latter materials undergo tetragonal lattice *expansion* and *contraction* below $T_{\rm N}$, respectively [7; 9; 10], the hole will occupy d_{xy} in the former one and degenerated d_{yz} or d_{zx} in the latter one. Furthermore, the d_{xy}, d_{yz}, d_{zx}, $d_{x^2\ y^2}$ and $d_{3z^2\ r^2}$ states are connected to $\boldsymbol{L}$ states by the following equations, $d_{xy} = (|2\rangle \quad |\ 2\rangle)/(\sqrt{2}i)$, $d_{yz} = (|1\rangle + |\ 1\rangle)/(\ \sqrt{2}i)$, $d_{zx} = (|1\rangle \quad |\ 1\rangle)/(\ \sqrt{2})$, $d_{x^2\ y^2} = (|2\rangle + |\ 2\rangle)/\sqrt{2}$ and $d_{3z^2\ r^2} = |0\rangle$, where the kets mean $|L_z\rangle$ states with $L = 2$ [15]. Therefore, unless $d_{x^2\ y^2}$ is degenerated with d_{xy}, the spin-orbit coupling cannot split d_{xy} to the pure $\boldsymbol{L}$ states of $|2\rangle$ and $|\ \ 2\rangle$, which is consistent with the inactivity. In contrast, the degenerated d_{yz} and d_{zx} states can transform to $|1\rangle$ and $|\ \ 1\rangle$ by linear combination between themselves.

5. Summary

By single-crystal neutron di raction on $GeCo_2O_4$, we discovered an additional $\boldsymbol{Q}_{\rm I}$ magnetic long-range order, which simultaneously emerges at $T_{\rm N}$ with the dominant $\boldsymbol{Q}_{\rm II}$ one. The $\boldsymbol{Q}_{\rm I}$ vector matches the tetragonal lattice deformation below $T_{\rm N}$ in symmetry. This discovery suggests the appearance of t_{2g} character in Co^{2+}, normally vanished by spin-orbit coupling like in CoO.

Acknowledgements

We thank Mr. M. Ohkawara, Mr. K. Nemoto, Mr. M. Onodera and Mr. T. Asami for their supports at JAEA and Tohoku University. The neutron experiments were performed under User Programs conducted by ISSP, University of Tokyo. This work was supported by the MEXT of Japan, Grants in Aid for Young Scientists (B) (22740209), Priority Areas (22014001), Scienti c Researches (S) (21224008) and (A) (22244039) and Innovative Areas (20102005), and by Tohoku University, Inter-university Cooperative Research Program of the Institute for Materials Research.

References

[1] Lee S H, Louca D, Ueda H, Park S, Sato T J, Isobe M, Ueda Y, Rosenkranz S, Zschack P, Iniguez J, Qiu Y and Osborn R 2004 *Phys. Rev. B* **93** 156407

[2] Suzuki T, Katsumura M, Taniguchi K, Arima T and Katsufuji T 2007 *Phys. Rev. Lett* **98** 127203

[3] Matsushita Y, Ueda H and Ueda Y 2005 *Nature* **4** 845

[4] Horibe Y, Shingu M, Kurushima K, Ishibashi H, Ikeda N, Kato K, Motome Y, Furukawa N, Mori S and Katsufuji T 2006 *Phys. Rev. Lett* **96** 086406

[5] Watanabe T, Hara S and Ikeda S 2008 *Phys. Rev. B* **78** 094420

[6] Hubsch J and Gavoille G 1987 *J. Magn. Magn. Mater.* **66** 17

[7] Hoshi T, Katori H, Kosaka M and Takagi H 2007 *J. Magn. Magn. Mater.* **310** 448

[8] Shull C G, Strauser W A and Wollan O 1951 *Phys. Rev.* **83** 333

[9] Tombs N C and Rooksby H P 1950 *Nature* **165** 442

[10] Germann K H, Maier K and Strauss E 1974 *Phys. Status Solidi B* **61** 449

[11] Tomiyasu K, Inami T and Ikeda N 2004 *Phys. Rev. B* **70** 184411

[12] Hiraka H, Ohoyama K, Yamaguchi Y and Yamada K 2007 *J. Magn. Magn. Mat.* **310** e971

[13] Hara S, Yoshida Y, Ikeda S, Shirakawa N, Crawford M K, Takase K, Takano Y and Sekizawa K 2005 *J. Cryst. Growth* **283** 185

[14] Matsuda M, Chung J H, Park S, Sato T J, Matsuno K, Katori H A, Takagi H, Kakurai K, Kamazawa K, Tsunoda Y, Kagomiya I, Henley C L and Lee S H 2008 *Euro. Phys. Lett.* **82** 37006

[15] Kanamori J 1957 *Prog. Theor. Phys.* **17** 177 and 197

[16] Tomiyasu K and Itoh S 2006 *J. Phys. Soc. Jpn* **75** 084708

[17] Lashley J C, Stevens R, Crawford M K, Boerio-Goates J, Wood eld B F, Qiu Y, Lynn J W, Goddard P A and Fisher R A 2008 *Phys. Rev. B* **78** 104406

Neutron scattering study of spiral-type spin correlations in the frustrated spinel $Mn_{0.07}Mg_{0.93}Cr_2O_4$

T. Yokobori[1], K. Tomiyasu[2], Y. Kousaka[1], H. Matsui[1], H. Hiraka[3], K. Iwasa[2], K. Yamada[4] and J. Akimitsu[1,*]
[1] Dept. of Physics and Mathematics, Aoyama-Gakuin Univ. Kanagawa, Japan
[2] Dept. of Physics, Tohoku Univ., Miyagi, Japan
[3] IMR, Tohoku Univ., Miyagi, Japan
[4] WPI-AIMR, Tohoku Univ., Miyagi, Japan
[*] E-mail: jun@phys.aoyama.ac.jp.

Abstract. We studied spiral-type spin correlations in $Mn_{0.07}Mg_{0.93}Cr_2O_4$ by single-crystalline neutron scattering. The correlation lengths are evaluated to be 12.7 Å and 4.4 Å along parallel and perpendicular to the propagation vector, respectively. The spiral rotational plane is estimated to be the proper-screw type rather than the cycloid-screw one. This fact suggests that Mn moments stabilize the cycloidal structure, generating the ferroelectricity in the parent material $MnCr_2O_4$.

1. Introduction

A series of the spinel-type chromates ACr_2O_4 is a well-known geometrically frustrated system, in which Cr^{3+} ions (spin S = 3/2) construct a pyrochlore lattice, and A^{2+} ions construct a non-frustrated diamond lattice. $ZnCr_2O_4$ and $MgCr_2O_4$ with nonmagnetic Zn^{2+} and Mg^{2+} exhibit molecular spin fluctuations [1-3] and excitations [4]. $CoCr_2O_4$ and $MnCr_2O_4$ with magnetic Co^{2+} (S = 3/2) and Mn^{2+} (S = 5/2) exhibit both ferromagnetic and ferroelectric polarizations called multiferroics, induced by conical spin structures [5; 6]. The spiral components rapidly grow below T_S = 25 K and 14 K, but cannot form long-range order because of the geometrical frustration among the Cr sites, whereas the axial ferrimagnetic components form long-range order by non-frustrated A-Cr interactions below T_C = 93 K and 51 K [7-9].

Recently, a further frustration effect on the spiral correlation was discovered in lowest temperature phases in the solid solution materials $Co_{1-x}Zn_xCr_2O_4$ by neutron scattering and magnetic susceptibility measurements [10]. As x increases from 0, the ferrimagnetic component disappears at $x \sim 0.35$, the remaining spiral clusters interestingly become not only small in size but also anisotropic in shape, and antiferromagnetic order with multiple propagation vectors appears above $x \sim 0.95$ [10]. However, no quantitative information of the anisotropic spiral clusters was given owing to lack of a large single crystal.

In this paper, we study the anisotropic spiral clusters using $Mn_{1-x}Mg_xCr_2O_4$, of which large single crystals are available above $x \sim 0.9$ by our current technique. The spin system is also expected to be almost the same as in $Co_{1-x}Zn_xCr_2O_4$. First, an unreported x-T magnetic phase diagram is presented with polycrystalline magnetic susceptibility data. Next, single-crystalline neutron scattering data measured on $Mn_{0.07}Mg_{0.93}Cr_2O_4$ are given. Then, the anisotropic spiral clusters are semi-quantitatively discussed.

2. Experiments

Polycrystalline samples of $Mn_{1-x}Mg_xCr_2O_4$ were synthesized by a solid-state reaction method in air from mixture of MnO, MgO and Cr_2O_3. All the samples were confirmed to be a single phase of a spinel structure by x-ray diffraction at room temperature. A single-crystalline sample of $Mn_{0.07}Mg_{0.93}Cr_2O_4$ with 4.4 mm diameter and 27 mm long was grown by a floating zone method under 6 atm Ar atmosphere.

Magnetic susceptibility was measured by a superconducting quantum interference device magnetometer (SQUID: Quantum Design MPMS-5). Neutron scattering was carried out on triple-axis spectrometers AKANE and TOPAN, installed at the thermal guide tube at JRR-3 and at the thermal beam hall of the same reactor in Japan Atomic Energy Agency (JAEA), Tokai, Japan, respectively. On AKANE, the incident energy was fixed at E_i = 19.4 meV with horizontal collimation sequence of guide-60'-60'-blank. On TOPAN, the final energy was fixed at E_f = 14.7 meV with horizontal collimation sequence of blank-100'-100'-blank. Each sample was enclosed with He exchange gas in an aluminum container, which was placed under the cold head of a closed-cycle He refrigerator.

3. Results

Figures 1(a) and 1(b) show the zero-field-cooled (ZFC) and field-cooled (FC) magnetic susceptibility data measured on the polycrystalline samples with several compositions. Both T_C and T_S are observed in Fig. 1(a), and only T_S is observed in Fig. 1(b). The two curves do not trace each other below T_S. Figure 1(c) shows the resultant phase diagram obtained from the anomaly temperatures. As x increases from 0, T_C rapidly decreases and disappears at $x \sim 0.35$. The value of T_S slowly decreases towards 9 K at $x = 0.90$ and increases again above $x = 0.90$. All the features are almost the same as in $Co_{1-x}Zn_xCr_2O_4$ [10].

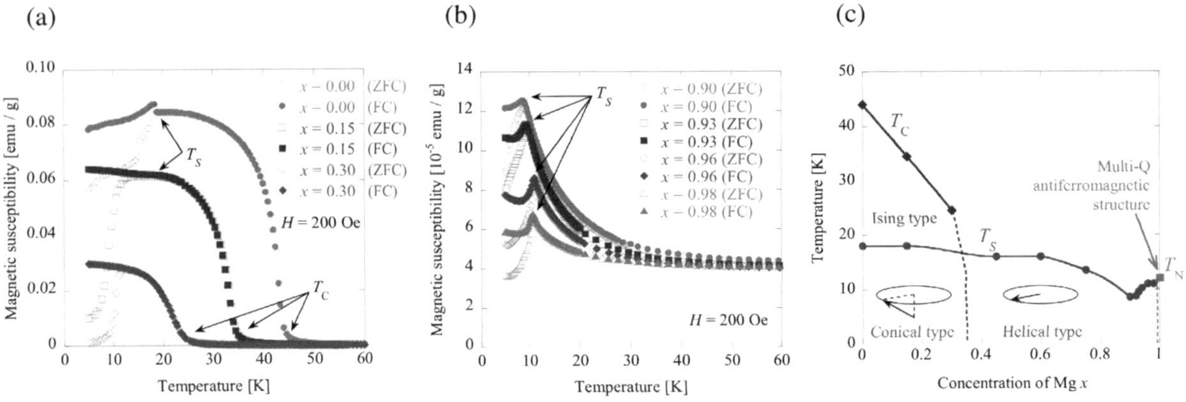

Figure 1. (Color online) (a)(b) Temperature dependence of zero-field-cooled (ZFC) and field-cooled (FC) magnetic susceptibility measured on polycrystalline $Mn_{1-x}Mg_xCr_2O_4$ under a magnetic field of H = 200 Oe. (c) Magnetic phase diagram of $Mn_{1-x}Mg_xCr_2O_4$ obtained from the susceptibility data. The solid lines connect the experimental plots, and the dotted lines are guide for eyes.

Figure 2(a) shows the elastic neutron scattering intensity distribution in the $hk0$ zone measured at 4 K ($< T_S = 11$ K) on the single-crystalline sample $Mn_{0.07}Mg_{0.93}Cr_2O_4$. Very diffusive and anisotropic magnetic satellite reflections were observed around (1.32, 1.32, 0), (2.68, 1.32, 0) and (1.32, 2.68, 0), which are described by the magnetic propagation vectors $\boldsymbol{Q}_s$ = (-0.68, -0.68, 0), (0.68, -0.68, 0) and (-0.68, 0.68, 0) measured from (2, 2, 0), respectively. The ratio of the peak intensity at (1.32, 1.32, 0) to those at (2.68, 1.32, 0) is estimated to be about 4.0(3). The diffuse scattering profile is well fitted by a Lorentzian like in $MnCr_2O_4$ [7] both along parallel and perpendicular to each $\boldsymbol{Q}_s$. The full width at half maximum (FWHM) is estimated to be about 0.22 $Å^{-1}$ along $\boldsymbol{Q}_s$ and about 0.63 $Å^{-1}$ perpendicular to $\boldsymbol{Q}_s$.

These values are much larger than the resolution limit 0.03 Å^{-1}, estimated from the nuclear Bragg reflection (2, 2, 0). Thus the real-space correlation lengths (FWHM) are estimated to be 12.7 Å and 4.4 Å, respectively, corresponding to about 4.3 and 1.5 times of Cr-Cr distance by using the value of 2.95 Å in $MgCr_2O_4$ [11].

Figure 2(b) shows the constant-$\boldsymbol{Q}$ scan data measured at 4 K at (1.25, 1.25, 0) as the representative point in the diffuse scattering. A quasielastic feature is observed, indicating that the spiral correlation dynamically fluctuates.

Figure 2(c) shows the temperature dependence of the peak intensity at Q = 1.5 Å^{-1} ~ |(1.32, 1.32, 0)| measured on the polycrystalline $Mn_{0.07}Mg_{0.93}Cr_2O_4$. The magnetic scattering intensity gradually decreases and almost disappears at 90 K.

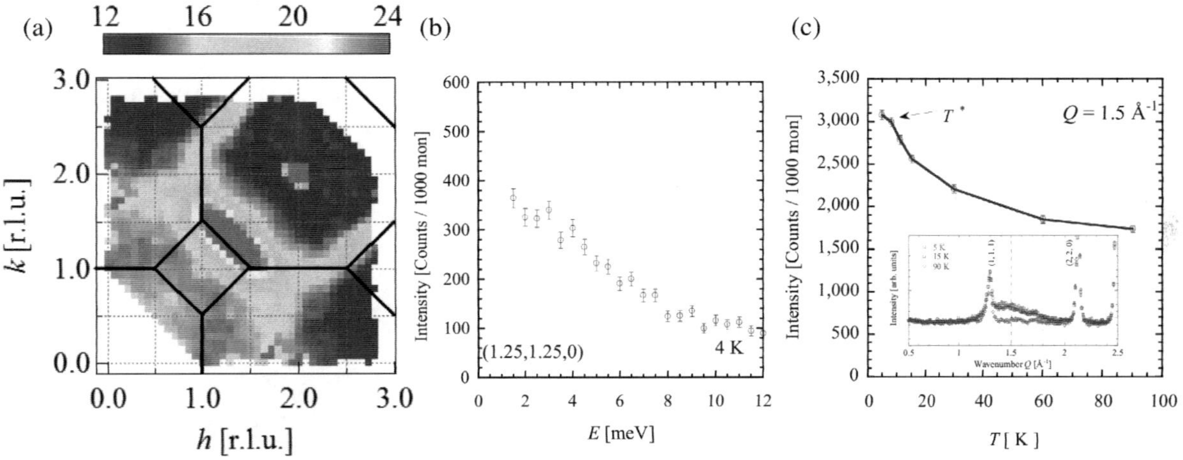

Figure 2. (Color online) Neutron scattering data measured on $Mn_{0.07}Mg_{0.93}Cr_2O_4$. (a) Single-crystalline elastic scattering intensity distribution measured in the *hk*0 zone at 4 K below T_S. The solid lines represent the Brilloun zone boundaries of the spinel structure. The horizontal tone indicates intensity in arbitrary units. We also observed the similar pattern at 15 K above T_S at the elastic condition. (b) Constant-$\boldsymbol{Q}$ scan data measured at (1.25, 1.25, 0). (c) Temperature dependence of elastic peak intensity at Q = 1.5 Å^{-1} measured on the polycrystalline sample. The inset shows the Q dependence of elastic scattering intensity measured at several temperatures.

4. Discussion

We roughly estimated the rotational plane of the spiral correlation in $Mn_{0.07}Mg_{0.93}Cr_2O_4$ from the observed intensity ratio. The spiral order is described by

$$\vec{m}_{j,k} = m_j^{(0)}\{\vec{e}_x \cos(\vec{Q}_s \cdot \vec{r}_{j,k} + \gamma_j) + \vec{e}_y \sin(\vec{Q}_s \cdot \vec{r}_{j,k} + \gamma_j)\}, \quad (1)$$

where $\vec{m}$ is the magnetic moment, j labels the magnetic sublattice from 1 to 6 (two Mn sites and four Cr sites) in the kth unitcell, $m_j^{(0)}$ is magnitude of the moment, $\vec{r}_{j,k}$ is the position, $\vec{Q}_s$ is identical to the propagation vector $\boldsymbol{Q}_s$, $\vec{e}_x$ and $\vec{e}_y$ are unit vectors defining the rotational plane, and γ_j is the relative phase angle in the plane [8]. Then the magnetic scattering intensity is represented by

$$S(\vec{Q}) = C\left|F(\vec{Q})\right|^2 L(\vec{Q})(1+\cos^2\theta)\left|\sum_{j=1}^{6} \exp\{\mathrm{i}(\vec{G}\cdot\vec{r}_j + \gamma_j)\}\right|^2, \quad (2)$$

where C is a constant, F is the magnetic form factor, for which we used the values in Ref. [12], L is the Lorentz factor, θ is an angle between the scattering vector $\vec{Q}$ and the normal vector of the rotational plane, and $\vec{G}$ is the reciprocal lattice vector corresponding to (2, 2, 0) in the present estimation [8]. Here we ignore the 7% Mn spins as the first approximation and reduce the j range from 1 to 4. In addition, the four Cr sites are resolved into two sublattices $B1$ and $B2$, as shown in Fig. 3, and the phase angles are assumed to be $\gamma_{B1} = -\gamma_{B2}$, which are the same as for $CoCr_2O_4$ and $MnCr_2O_4$ [7-9].

Table I summarizes the intensity ratios calculated for a proper-screw model and two cycloid-screw models: *ab*-plane cycloid ($CoCr_2O_4$ type), $\boldsymbol{Q}_s c$-plane cycloid ($MnCr_2O_4$ type) [7-9]. Among the three models the proper-screw type is closest to the observed ratio. Figure 3 depicts the spatial correlation of the clusters, taking into account the anisotropic correlation lengths as well.

The proper-screw formation suggests disappearance of the ferroelectricity observed in $MnCr_2O_4$ with inversed Dzyaloshinskii-Moriya interactions. The origin is unclear at this stage. Further researches of the *A*-site substituted system are needed to understand and control the spiral clusters.

Table I. Calculated and observed ratios of the peak intensity at (1.32, 1.32, 0) to that at (2.68, 1.32, 0). The $CoCr_2O_4$ and $MnCr_2O_4$ types mean *ab*-plane and $\boldsymbol{Q}_s c$-plane cycloid models.

Calculations			Observation
Proper-screw type	$CoCr_2O_4$ type	$MnCr_2O_4$ type	
4.0	2.2	2.0	4.0(3)

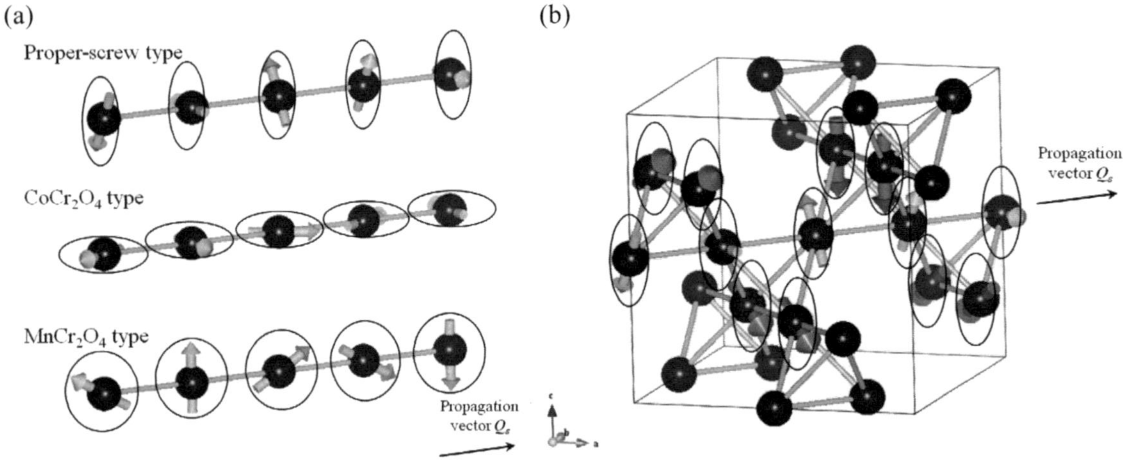

Figure 3. (Color online) (a) Spiral ordering models for magnetic scattering intensity calculation. (b) A schematic picture of the anisotropic spiral clusters in $Mn_{0.07}Mg_{0.93}Cr_2O_4$. The balls represent Cr sites. The thick red and thin green arrows represent the spins on the two sublattices $B1$ and $B2$.

5. Conclusions

We studied the spiral short-range correlations in $Mn_{0.07}Mg_{0.93}Cr_2O_4$ by single-crystalline neutron scattering. The anisotropic correlation lengths (FWHM) were estimated to be about 4.3 and 1.5 times of Cr-Cr distance along parallel and perpendicular to the propagation vector $\boldsymbol{Q}_s$, respectively. The spatial spin correlation was suggested to be the proper-screw type rather than the cycloid one.

Acknowledgement

This work was supported by Grant-in-Aid for Scientific Research on Priority Areas "Novel States of Matter Induced by Frustration" (No. 22014001) from the Ministry of Education, Culture, Sports, Science and Technology (MEXT) of the Japanese Government. K. T. was supported by Grants-in-Aid for Young Scientists (B) (No. 22740209), and Scientific Researches (S) (No. 21224008) and (A) (No.

22244039) from the MEXT. Y. K. is supported by Iketani Science and Technology Foundation and by Grant-in-Aid for Research Activity Start-up (No. 22840038) from the MEXT. J. A. is supported by a High-Tech Research Center project for private universities with a matching fund subsidy and Grant-in-Aid for Scientific Research (A) (No. 22245023) from the MEXT.

References

[1] S-H Lee, Broholm C, Kim T H, Ratclllif II W and Cheong S-W 2000 *Phys. Rev. Lett.* **84** 3718
[2] S-H Lee, Broholm C, Ratcllif W, Gasparovic G, Huang Q, Kim T H and Cheong S-W 2002 *Nature (London)* **418** 856
[3] Suzuki H and Tsunoda Y 2007 *J. Phys. Chem. Solids* **68** 2060
[4] Tomiyasu K, Suzuki H, Toki M, Itoh S, Matsuura M, Aso N and Yamada K 2008 *Phys. Rev. Lett.* **101** 177401
[5] Yamasaki Y, Miyasaka S, Kaneko Y, He J-P, Arima T and Tokura Y 2006 *Phys. Rev. Lett.* **96** 207204
[6] Mufti N, Blake G R and Palstra T T M 2009 *J. Mag. Mag. Mater.* **321** 1767
[7] Menyuk N, Dwight K and Wold A 1964 *J. Phys. (Paris)* **25** 528
[8] Hastings J M and Corliss L M 1962 *Phys. Rev.* **126**, 556
[9] Tomiyasu K, Fukunaga J and Suzuki H 2004 *Phys. Rev. B* **70** 214434
[10] Tomiyasu K, Horigane K, Yokobori T, Kousaka Y, Takahasi M, Hiraka H, Akimitsu J and Yamada K 2009 *J. Phys. Soc. Jpn.* **78** 084704
[11] Ortega-San-Martin L, Williams A J, Gordon C D, Klemme S and Attfield J P 2008 *J. Phys., Condens. Matt.* **20** 104238.
[12] Sears V F 1999 *International Tables for Crystallography, Vol. C 2nd Edition* ed Wilson A J C and Prince E (Kluwer Academic Publishers, Dordrecht-Boston-London)

Journal of Physics: Conference Series **320** (2011) 012040 doi:10.1088/1742-6596/320/1/012040

Stability of hexamer-type spin excitations in the frustrated spinel $Mg_{1-x}Cr_2O_{4-x}$

Yusuke Kousaka[1], **Keisuke Tomiyasu**[2], **Toshio Yokobori**[1], **Kazumasa Horigane**[3], **Haruhiro Hiraka**[4], **Kazuyoshi Yamada**[3] **and Jun Akimitsu**[1]

[1] Department of Physics and Mathematics, Aoyama-Gakuin University, Sagamihara 252-5258, Japan
[2] Department of Physics, Tohoku University, Sendai 980-8578, Japan
[3] WPI-AIMR, Tohoku University, Sendai 980-8577, Japan
[4] IMR, Tohoku University, Sendai 980-8577, Japan

E-mail: `koyu@phys.aoyama.ac.jp`

Abstract. We performed magnetic susceptibility and inelastic neutron scattering measurements on the frustrated spinel $Mg_{1\;x}Cr_2O_{4\;x}$ $(0.02 \leq x \leq 0.11)$. Susceptibility shows that Curie-Weiss temperature is almost independent of x, meaning that exchange interactions sustain. Inelastic neutron scattering reveals that the spin-hexamer excitations also persist with almost the same energy around 4 meV in a lowest temperature phase, whereas the resonance width increases with increasing x. These results suggest that nearest-neighboring Cr$\cdots$Cr direct exchange predominantly forms the hexamers regardless of lattice defect and superexchange via Mg and O sites.

1. Introduction

A series of spinel chromates ACr_2O_4, in which Cr^{3+} ions with spin $S = 3/2$ form a pyrochlore structure, is well known as a geometrically frustrated system. $ZnCr_2O_4$ and $MgCr_2O_4$, in which non-magnetic ions Zn^{2+} and Mg^{2+} occupy the A sites, exhibit characteristic dynamical short-range spin clusters in paramagnetic and magnetically ordered phases, called spin molecules [1, 2, 3, 4]. In an ordered phase, the molecules are observed as resonance-like excitations with discrete energies $E = 4.5$ and 9.0 meV, which are identified to be hexamer and heptamer, respectively [4].

The hexamer-type resonance-like excitations, albeit broadened in energy, interestingly persist even in a spin glass phase in the A-site-substituted $Zn_{0.95}Cd_{0.05}Cr_2O_4$, in which static magnetic long-range order is suppressed [5]. This fact suggests that the hexamer formation does not rely on the magnetic order [5].

In this paper, we further approach an origin of the hexamers by magnetization measurements and inelastic neutron scattering on the A- and O-site-defected $Mg_{1-x}Cr_2O_{4-x}$ (0.02 x 0.11). The defect with the wider x range is expected to more effectively deform lattice, modify superexchange, and destroy the hexamers than the substitution. However, it is revealed that the hexamers still persist in all the x range.

2. Experiments

Single-crystalline samples of $Mg_{1-x}Cr_2O_{4-x}$ ($x = 0.02, 0.06, 0.11$) were grown by a floating zone method (4-Xe lamp design, Crystal Systems Corporation) for the first time. The polycrystalline feed rods were synthesized from mixtures of MgO and Cr_2O_3 under Ar atmosphere by a solid-phase reaction method. Since MgO was inevitably evaporated in our crystal growth, the molar ratio was adjusted to be $(1-x) \times 1.08 : 1$ with excess MgO. The rods were crystallized with 20 mm/h growth and seeding speeds under 6 atm Ar atmosphere. Then the single crystals with 4 mm diameter and 20 to 30 mm long were obtained. The final molar ratios of Mg to Cr and the resultant x values were confirmed by inductively coupled plasma atomic emission spectroscopy (ICP). The degree of the oxygen defect was assumed to keep the valence of Cr at 3+, since Cr normally takes only 3+ or 6+.

Single-crystalline magnetic susceptibility was measured by a superconducting quantum interference device magnetometer (SQUID: Quantum Design MPMS-5). Single-crystalline neutron scattering experiments were performed on a triple axis spectrometer, AKANE (T1-2), installed at the thermal guide tube of JRR-3, Japan Atomic Energy Agency (JAEA), Tokai, Japan. The incident energy was fixed at $E_i = 19.4$ meV with a horizontal collimation sequence of guide$-60'-60'-$blank. Each sample was mounted to use the $(hk0)$ scattering plane, and was cooled in an Al can in a closed-cycle He refrigerator.

3. Results

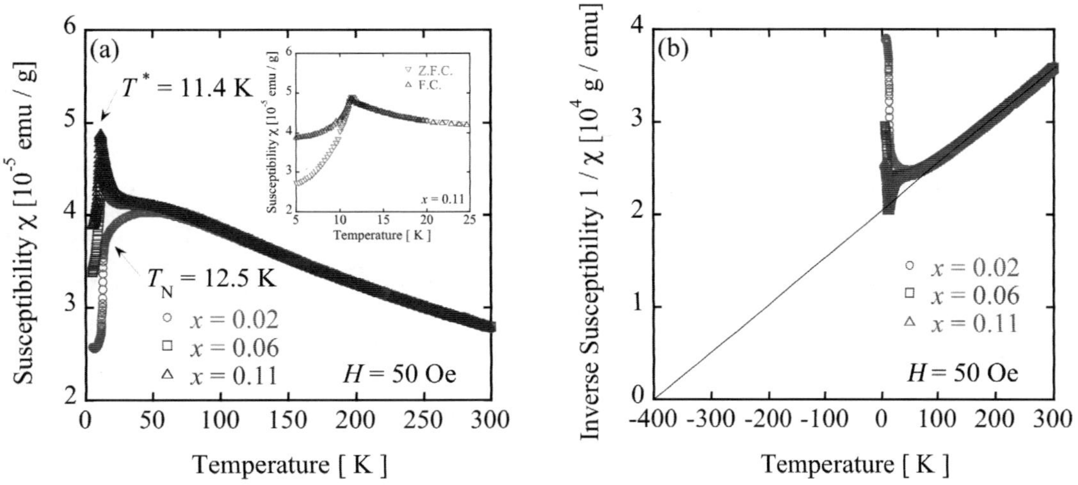

Figure 1. Temperature dependence of field-cooled (FC) magnetic susceptibility (a) and inverse one (b) measured under a magnetic field of 50 Oe. Open circles, open squares and open triangles represent the data for $Mg_{0.98}Cr_2O_{3.98}$ ($x = 0.02$), $Mg_{0.94}Cr_2O_{3.94}$ ($x = 0.06$) and $Mg_{0.89}Cr_2O_{3.89}$ ($x = 0.11$), respectively. The inset in (a) shows temperature dependence data of zero-field-cooled (ZFC) and FC magnetic susceptibility for $Mg_{0.89}Cr_2O_{3.89}$ ($x = 0.11$).

Figure 1(a) shows the temperature dependence of the field-cooled (FC) magnetic susceptibility measured on $Mg_{1-x}Cr_2O_{4-x}$ in a magnetic field of 50 Oe. The $x = 0.02$ susceptibility decreases and abruptly drops at Neel temperature $T_N = 12.5$ K with no difference between the zero-field-cooled (ZFC) and FC curves, which is the same as in $MgCr_2O_4$. Meanwhile, the $x = 0.06$ and 0.11 susceptibility exhibits a sharp cusp at T = 11.4 K. The ZFC-FC difference appears below T , as the $x = 0.11$ curves are shown in the inset in Fig. 1(a) as the representative, suggesting appearance of a spin-glass-like state. There seems to be also another anomaly at 10 K.

Figure 1(b) shows the temperature dependence of the inverse FC susceptibility. The data between 150 and 300 K is linear, giving the Curie-Weiss temperatures $\Theta_{CW} = -408$, -396 and -401 K for $x = 0.02$, 0.06 and 0.11, respectively. Assuming that nearest-neighboring direct exchange interactions J_1 are dominant as for $x = 0$ ($MgCr_2O_4$), these values correspond to -4.7, -4.6 and -4.6 meV. All the values obtained above coincide with those for $MgCr_2O_4$ ($\Theta_{CW} = -400$ K [6] and $J_1 = -4.6$ meV).

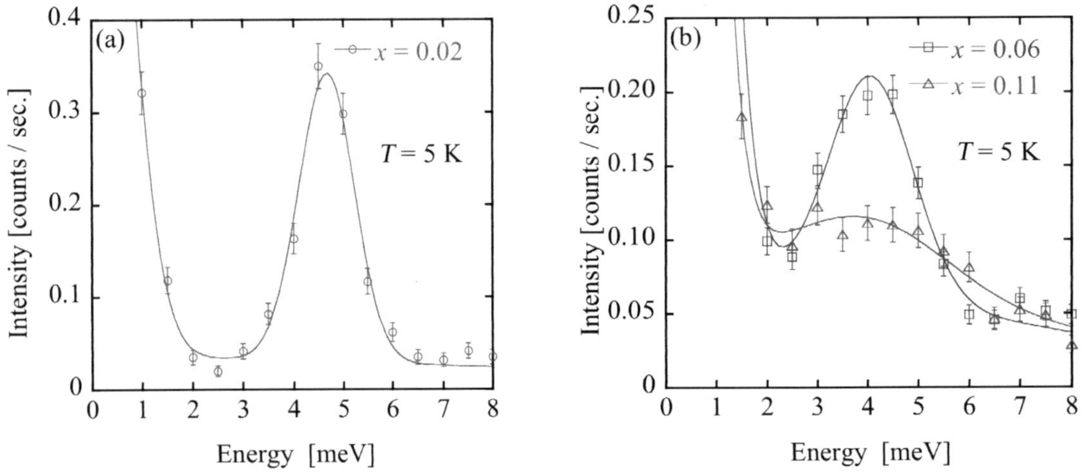

Figure 2. Constant-Q scan data measured at (1.25,1.25,0) in $Mg_{0.98}Cr_2O_{3.98}$ ($x = 0.02$) (a), $Mg_{0.94}Cr_2O_{3.94}$ ($x = 0.06$) and $Mg_{0.89}Cr_2O_{3.89}$ ($x = 0.11$) (b). Open circles, open squares and open triangles correspond to $x = 0.02$, 0.06 and 0.11, respectively. The solid lines are obtained by double Gaussian fitting.

Next, we measured the resonance energy by inelastic neutron scattering. Figures 2(a) and 2(b) show the constant-Q scan data at (1.25,1.25,0) reciprocal lattice unit, where strong diffusive magnetic scattering appears from the hexamers [4]. The resonance-like peak of the hexamers is robustly observed at all the x values, though the broadening is severe at $x = 0.11$. By fitting the data with double Gaussians, the resonance energies (the center positions of Gaussian) are estimated to be 4.7, 4.1 and 4.1 meV, and the resonance widths (full widths at half maximum) are 1.3, 1.8 and 3.3 meV for $x = 0.02$, 0.06 and 0.11, respectively.

4. Discussion

Thus we revealed that x as well as static magnetic ordering/freezing does not influence both the exchange interactions (Θ_{CW}) and the resonance energy in $Mg_{1-x}Cr_2O_{4-x}$. The defect at A and O sites is expected to substantially deform lattice and many bond angles on superexchange paths via A and O sites. Therefore, this x independency suggests that nearest-neighboring direct exchange interactions predominantly form the hexamers rather than lattice deformation and superexchange. On the other hand, the resonance width increases with increasing x (Figs. 2(a) and 2(b)), suggesting that the hexamers gradually become unstable probably because the lattice deformation and superexchange enhance inter-hexamer correlations.

Acknowledgments

This work was supported by Grant-in-Aid for Scientific Research on Priority Areas "Novel States of Matter Induced by Frustration"(No. 22014001) from the Ministry of Education, Culture,

Sports, Science and Technology (MEXT) of the Japanese Government. Y. K. is supported by Iketani Science and Technology Foundation and Grant-in-Aid for Research Activity Start-up (No. 22840038) from the MEXT. K. T. was supported by the MEXT of Japan, Grants in Aid for Young Scientists (B) (No. 22740209), and Scientific Researches (S) (No. 21224008) and (A) (No. 22244039). J. A. is supported by a High-Tech Research Center project for private universities with a matching fund subsidy from the MEXT and Grant-in-Aid for Scientific Research (A) (No. 22245023) from the MEXT.

References

[1] S.-H. Lee, C. Broholm, T. H. Kim, W. Ratchli, and S.-W. Cheong, Phys. Rev. Lett. **84**, 3718 (2000).
[2] S.-H. Lee, C. Broholm, W. Ratchli, G. Gasparovic, Q. Huang, T. H. Kim, and S.-W. Cheong, Nature **418**, 856 (2002).
[3] H. Suzuki, Y. Tsunoda, J. Chem. Solids **68**, 2060 (2007).
[4] K. Tomiyasu, H. Suzuki, M. Toki, S. Itoh, M. Matsuura, N. Aso, and K. Yamada, Phys. Rev. Lett. **101**, 177401 (2008).
[5] W. Ratcliff II, S.-H. Lee, C. Broholm, S.-W. Cheong, and Q. Huang, Phys. Rev. B **65**, 220406(R) (2002).
[6] M. T. Rovers, P. P. Kyriakou, H. A. Dabkowska, G. M. Luke, M. I. Larkin and A. T. Savici, Phys. Rev. B **66**, 174434 (2002).

Evidence for spin-glass state in nonmagnetic atom disorder compound Pr_2AgIn_3

D. X. Li [1], T. Yamamura [1], K. Yubuta [1], S. Nimori [2], Y. Haga [3], T. Shikama [1]

[1] *Institute for Materials Research, Tohoku University, Oarai, Ibaraki, 311-1313 Japan*
[2] *Tsukuba Magnet Laboratory, National Research Institute for Metals, Tsukuba, 305-0003 Japan*
[3]*Advanced Science Research Centre, Japan Atomic Energy Research Institute, Tokai, Ibaraki 319-1195, Japan*

E-mail: dxli@imr.tohoku.ac.jp

Abstract. A systematic investigation of magnetic properties has been performed on intermetallic compound Pr_2AgIn_3, which crystallizes in a disordered hexagonal $CaIn_2$-type structure. We present the experimental results of ac and dc susceptibility, magnetization, magnetic relaxation, specific heat and electrical resistivity measurements. The results provide clear evidence for the formation of spin-glass state in Pr_2AgIn_3 below a spin freezing temperature T_f~3.7 K accompanied with a short-range magnetic order. The spin-glass behavior of Pr_2AgIn_2 is discussed and compared with that observed for the isostructural compounds Ce_2AgIn_3 and Nd_2AgIn_3 as well as other 2:1:3 spin-glass materials. The spin-glass state in Pr_2AgIn_3 can be considered to originate from the continued site randomness of the non-magnetic elements, Ag and In, which introduce the random distribution of the RKKY interactions.

1. Introduction

Ternary intermetallic compounds with a general formula R_2TX_3 (R=U or rare-earth element, T=transition 3d, 4d and 5d element, X = Ga, In, Si, Ge) have been the subjects of intensive studies during the past years because of their intriguing physical properties. Many members of these materials are reported to crystallize in the "binary-type" structure, e.g., the AlB_2-type structure (space group *P6/mmm*), the $CaIn_2$-type structure (space group $P6_3/mmc$), or derivatives of those hexagonal structures [1]. The interest in these compounds stems from the distribution of the R, T and X elements on the crystallographic sites with a complete order of magnetic R atoms on the Al/Ca site and a random distribution of the transition metal T and X atoms on the B/In site of the AlB_2-/$CaIn_2$-type structure. Because the d/p electronic states of ligands are considered to affect the magnetic interactions of R_2TX_3 compounds strongly, the random distribution of T and X elements can be expected to cause some anomalous magnetic properties such as spin-glass (SG) behavior. This has indeed been confirmed by recently experimental results, e.g., the SG behavior in Ce_2CuX_3 (X=Si [2] and Ge [3]) and U_2TSi_3 (T = Pd [4], Pt [5], Au [6] and Rh [7]); the coexistence of long range magnetic order state and SG state in Tb_2PdSi_3 [8] and Tb_2CuIn_3 [9]; and the complex magnetic ordering in Nd_2CuSi_3 [10] and R_2PdSi_3 (R=Eu [11], Dy, Ho, Er [8], Gd [12]), etc. However, in many cases, a clear consensus regarding the magnetic nature is still elusive. In particular, it is a common belief that competing between ferromagnetic and antiferromagnetic near-neighbour interactions caused by randomly varying bond angles and interatomic distances are mainly responsible for the formation of SG state in amorphous or diluted metallic SG materials. Intermetallic compounds R_2TX_3, however, are crystallographically ordered substances. In such systems, the origin of SG behavior is evidently different from that in amorphous or diluted metallic SG systems, and is not elucidated theoretically so far. To clarify the mechanism of magnetic interactions systematic experimental data on other R_2TX_3 compounds are necessary. Recently, we have been paying special attention to a new family of the 2:1:3 compounds, namely R_2AgIn_3 (R=rare-earth element). According to neutron diffraction, the Tb compound is the only example in this series to show long-range magnetic order [13] with the Néel temperature T_N=42 K. On the other hand, experimental measurements of thermal, magnetic and transport properties reported for R_2AgIn_3 reveal the canonical SG behavior for R=Ce (T_f=1.8 K) [14], SG-like behavior for R=Tm (T_f=3.2 K) and SG behavior with short-range antiferromagnetic order for R=Nd (T_f=12.6 K) [15]. In this paper, we present the results of ac and dc susceptibility, magnetization, magnetic relaxation, electrical resistivity and specific heat measurements of polycrystalline Pr_2AgIn_3. Our results provide clear evidence for the existence of SG state in Pr_2AgIn_3, and suggest that the occurrence of SG behavior is accompanied with the formation of short-range magnetic order.

2. Experimental

The polycrystalline sample of Pr_2AgIn_3 was prepared by melting appropriate amounts of the constituent elements in an arc furnace under purified argon atmosphere. The sample was then wrapped into tantalum foil and annealed in evacuated silica tube at 750 °C for two weeks. X-ray powder diffraction was performed at room temperature with Cu$K\alpha$ radiation. The samples used in the experiments are small pieces cut from the annealed button. The ac and dc susceptibility, low-field magnetization and magnetic relaxation were measured between 1.8 and 300 K in magnetic fields up to 10 kOe using a SQUID magnetometer. High-field magnetization experiments at 2.7 K were carried out using an Oxford Instruments VSM 12 T magnetometer in magnetic fields up to 110 kOe. The adiabatic heat pulse method was employed for specific heat measurements over the temperature range between 1.7 and 20 K. Electrical resistivity measurements were performed between 3.2 and 284 K using a standard four-terminal DC method.

3. Results

From the results of powder x-ray diffraction, the annealed Pr_2AgIn_3 sample was found to be almost single-phased. The diffraction lines can be indexed based on the disordered hexagonal $CaIn_2$-type structure model (space group $P6_3/mmc$) with Pr atoms on 2b sites (0, 0, 1/4) and Ag and In atoms statistically distributed over the 4f sites (1/3, 2/3, z) with z=0.465 [13]. The determined room-temperature lattice constants are a=4.872(3) Å and c=7.594(5) Å.

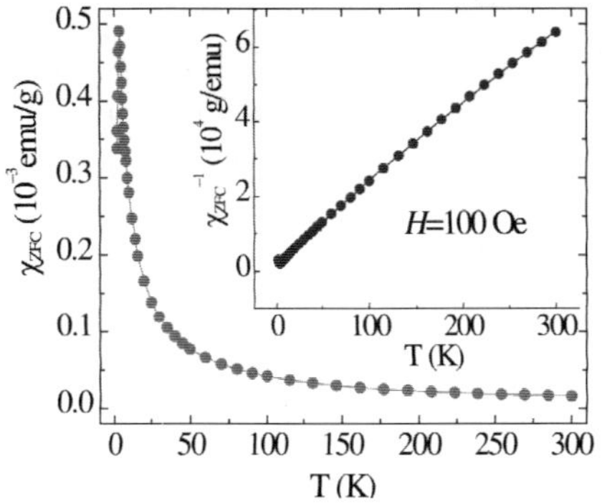

Figure 1. Temperature dependence of ZFC susceptibility ($\chi=M/H$) of Pr_2AgIn_3 between 2 and 300 K in a magnetic field of 100 Oe. The inset shows the data of reciprocal ZFC susceptibility.

The temperature dependence of the zero field-cooled dc susceptibility χ_{ZFC} ($=M_{ZFC}/H$) of Nd_2AgIn_3 measured in a field of H=100 Oe is shown in Fig. 1. Above 20 K the observed $\chi_{ZFC}(T)$ behavior can be well described by the modified Curie-Weiss law $\chi=\chi_0+C/(T-\theta_p)$ (solid line in the insets of Fig. 1). Where C is the Curie constant, θ_p the paramagnetic Curie temperature and χ_0 denotes temperature-independent contributions such as core diamagnetism, Landau diamagnetism and Pauli paramagnetism. From the fit, we obtain the values of μ_{eff}=3.49 μ_B/Pr, θ_p=–4.9 K, and $\chi_0=2.08\times10^{-6}$ emu/g. The value of effective magnetic moment μ_{eff} is close to the value of 3.58 μ_B expected for free-ion Pr^{+3} in the J=4 Hund's rule ground state indicating the 4f electrons are almost localized within the

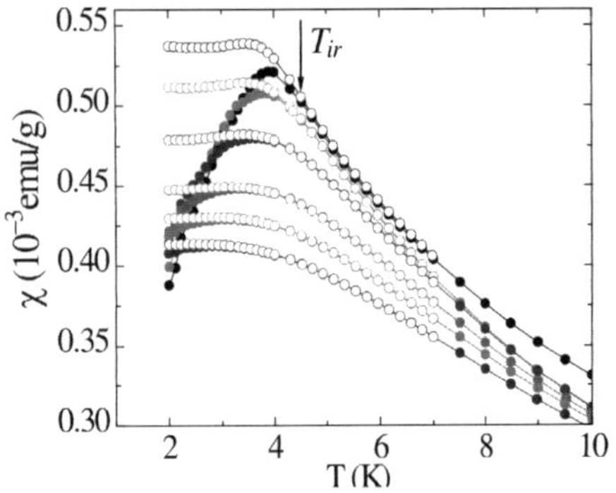

Figure 2. Temperature dependence of FC (○) and ZFC (•) susceptibility of Pr_2AgIn_3 between 2 and 10 K measured in magnetic fields of 0.1, 1.0, 2.5, 5.0, 7.5 and 10 kOe (from top to bottom).

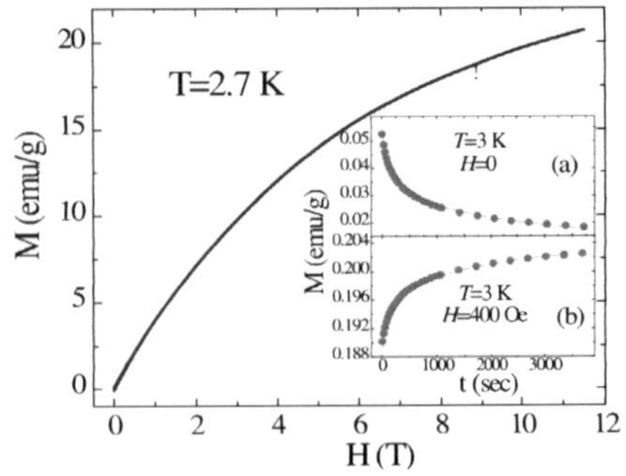

Figure 3. High-field magnetization $M(H)$ of Pr_2AgIn_3 measured by VSM at 2.7 K up to 110 kOe. The inset show the time dependence of magnetization measured at 3 K in zero field (a) and 400 Oe (b).

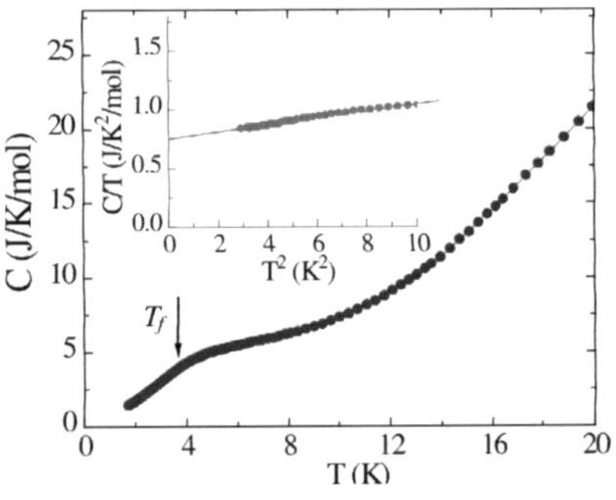

Figure 4. Temperature dependence of specific heat of Pr_2AgIn_3 between 1.7 and 20 K. The inset shows the plot of C/T vs. T^2.

Pr atoms. At low temperatures, a sharp peak is observed near T_f=3.7 K indicating some kind of magnetic phase transition near this temperature.

In order to character the nature of this transition, the field-cooled (FC) and zero field-cooled (ZFC) susceptibilities in various applied fields were measured around T_f. As shown in Fig. 2, there is no difference between the FC and ZFC curves in the paramagnetic state, and the magnetic transition peak can be observed only in the $\chi_{ZFC}(T)$ curve at the strongly field dependent temperature $T_f(H)$ that closes to 3.7 K in low field. Moreover, irreversible magnetism manifesting as the bifurcation between the FC and ZFC curves appears below a characteristic temperature $T_{ir}(H)$. With increasing H, the peak in $\chi_{ZFC}(T)$ becomes broader and its height decreases, while both $T_{ir}(H)$ and $T_f(H)$ shift toward lower temperatures. The strong field dependence of the peak position T_f in $\chi_{ZFC}(T)$ curve and the evident irreversible magnetism below T_{ir} are the typical features of SG system. Note that T_{ir} for a canonical SG is equals to T_f even in low magnetic field. For Pr_2AgIn_3, however, T_{ir} is clearly higher than T_f in applied field below 1 kOe suggesting the observed features are not simple SG behaviors.

The irreversible magnetism described above indicates the nonequilibrium feature of the low-temperature magnetic states in Pr_2AgIn_3. Thermodynamically, such nonequilibrium states directly relate to slow dynamics and could also lead to the remanence phenomenon in field dependence of magnetization and long-time magnetic relaxation effect. High-field magnetization $M(H)$ of Pr_2AgIn_3 measured at 2.7 K is illustrated in Fig. 3. It is clear that $M(H)$ curve does not saturate up to 110 kOe, and when H is returned from 110 kOe to zero, remanent magnetization of about 0.13 emu/g is detected. From the hysteresis loop we determine the coercive field H_C of Pr_2AgIn_3 to be 230 Oe at 2.7 K. Magnetic relaxation effect of Pr_2AgIn_3 was studied at 3 K by measured the magnetization M as a function of time t in zero and 400 Oe applied fields, the results are displayed in the insets of Fig. 3. Before the measurement, the sample was first ZFC from 50 K to 3 K, then, in the case of inset of Fig. 3(a) a magnetic field of 5 kOe was applied for 5 min and switched off at t=0, and in the case of inset of Fig. 3(b) a field of 400 Oe was applied at t=0. In both cases, the decay of $M(t)$ is remarkably slow, nonzero remanence in zero field [inset of Fig.3(a)] and unsaturated magnetization in 400 Oe [inset of Fig.3(b)] can be observed even after waiting for 1h. Using a logarithmic function, $M(t) = M_0 - S\ln(t + t_0)$, the obtained relaxation behaviors shown in the insets of Fig. 3 can been fitted very well over

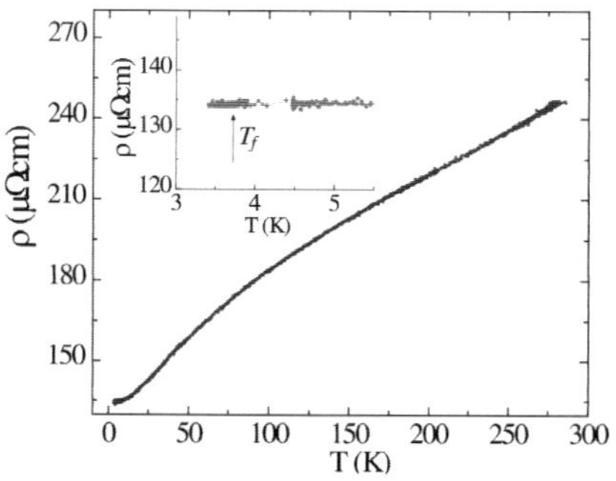

Figure 5. Temperature dependence of electrical resistivity of Pr_2AgIn_3. The inset shows the low temperature part in an expanded scale.

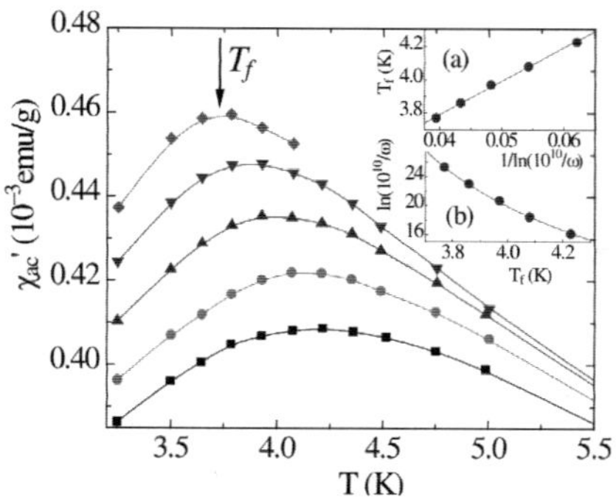

Figure 6. Real component $\chi_{ac}'(T)$ of the ac susceptibility of Pr_2AgIn_2 versus temperature between 3.2 and 5.5 K at various frequencies 0.1, 1, 10, 100 and 1000 Hz (from top to bottom). The insets show frequency dependence of the dynamic spin freezing temperature T_f of plotted as (a) T_f vs. $1/\ln(10^{10}/\omega)$, and (b) $\ln(10^{10}/\omega)$ vs. T_f. The solid line in (a) and (b) represents the fit to the Vogel-Fulcher law and to the expression of critical slowing down, respectively, as explained in the text.

the full time range studied with three H- and T-dependent fitting parameters: initial zero-field magnetization M_0, magnetic viscosity S and characteristic time t_0. The best fitting results obtained by using the least-squares method are shown by the solid lines in the insets of Fig. 3 with positive and negative S values in zero field and 400 Oe, respectively.

It is well known that the characteristic features of irreversible magnetism and magnetic relaxation effect can be observed not only for SG material, but also for long-range magnetic order system with high magnetic anisotropy [16]. For Pr_2AgIn_3, the existence of long-range spatial magnetic ordering can be excluded by our specific heat and electronic resistivity measurements. Figure 4 illustrates the temperature dependence of specific heat $C(T)$ of Pr_2AgIn_3. Although it is not easy to separate the magnetic contribution C_m from the total specific heat (consisting of vibrational, electronic and magnetic parts) accurately, the absence of dramatic change in $C(T)$ around T_f ~3.7 K indicates no sharp peak in $C_m(T)$ and thus excludes the existence of long-range magnetic ordering at T_f. At low temperatures, the C/T vs. T^2 plot shown in the inset of Fig. 4 yields for $T\to0$ K a γ value (the specific heat coefficient of T-linear term) of 375 mJ (mole-Pr)$^{-1}$ K^{-2} for Pr_2AgIn_3 much larger than that of a normal metal. Note that $C(T)$ curve shows a broad bend around 5 K just above T_f similar to that observed for Ce_2AgIn_3 [14] and Nd_2AgIn_3 [15], and different from that observed for U_2PdSi_3 [4]. This broad bend may be due to the formation of magnetic cluster (short-range magnetic order). Thus in addition to the influence of random distribution of Ag and In atoms (which usually originates the large the γ value in nonmagnetic atom disorder system [17]), the short-range magnetic correlation could also contribute to the observed large γ value considering the magnetic cluster formed at sufficiently low temperature.

Figure 5 gives the results of electrical resistivity [$\rho(T)$] measurement for Pr_2AgIn_3 between 3.2 and 284 K. $\rho(T)$ shows metallic conductivity with some negative curvature between 50 and 150 K, which may be resulted from crystal field interaction and/or s-d interband scattering of conduction electrons [18]. As clearly shown in the inset of Fig. 5 in an expanded scale, no anomaly can be detected at T_f~3.7 K, which provides a new evidence for the absence of long-range magnetic ordering in Pr_2AgIn_3.

In order to confirm the SG effects and characterize the spin freezing state, we have also performed an ac susceptibility measurement on the Pr_2AgIn_3 sample at various frequencies. Figure 6 shows the temperature dependence of the real component (χ_{ac}') of the ac susceptibility of Pr_2AgIn_3 between 3.2 and 5.5 K at the frequency range $0.1\le\omega/2\pi\le1000$ Hz. χ'_{ac} exhibits a pronounced maximum with amplitude and position depending on the frequency of the applied magnetic field. At a frequency of 0.1 Hz, the cusp in χ_{ac}' appears at T_f=3.7 K, which shifts to 4.2 K at $\omega/2\pi$=1000 Hz. Although the value of the imaginary component (χ_{ac}'', not shown here) is much smaller than that of χ_{ac}' and has relatively large error, a cusp in χ_{ac}'' curve, which shifts to higher temperatures with frequency, is also observed clearly. These results provide important evidence for the formation of SG state in Pr_2AgIn_3. For convenience, in this paper we define the spin freezing temperature $T_f\,(\omega)$ as the peak temperature in $\chi_{ac}'(T)$ curve. Similar frequency shift of the ac susceptibility maximum is also observed for Ce_2AgIn_3 [14] and Nd_2AgIn_3 [15]. The initial frequency shift δT_f calculated as $\delta T_f = \Delta T_f /(T_f \Delta \log \omega)$ is 0.031 comparable to the typical values (from a few thousandths to a few hundredths) for most SG materials [19]. In order to further identify the nature of spin freezing state, the Vogel and Fulcher empirical law [20], $\omega = \omega_0 \exp[-E_a / k_B (T_f - T_0)]$, and the conventional critical slowing down model [21], $\tau_{max} = \tau_0[(T_f - T_S)/T_S]^{-z\nu}$, were used to analyze the frequency dependence of ac susceptibility. Assuming $\omega_0/2\pi = 10^{10}$ Hz, typically taken in the SG systems [19, 22], we obtained the best fitting results as shown by the solid lines in the insets of Fig. 6 with the values of fitting parameters: activation energy E_a/k_B=20.25 K, characteristic temperature T_0=2.98 K, static freezing temperature T_S=3.38 K and critical (dynamical) exponent $z\nu$=11.75.

4. Discussion and conclusions

It is well known that some typical properties for SG material are (1) frozen-in magnetic moments below a freezing temperature T_f and hence a cusp-like anomaly in the frequency-dependent ac susceptibility around T_f; (2) irreversible magnetism behaving as the difference between FC and ZFC magnetization below T_f; (3) lack of periodic long-range magnetic order and (4) remanence and magnetic relaxation on macroscopic time scales after changing the applied magnetic field below T_f [23]. All these features are observed in Pr_2AgIn_3 in the present work, which can be considered as the evidences for the formation of SG state in this system.

The existence of frustration and randomness are the necessary conditions for SG state [23]. It means that there must be a competition between ferromagnetic and antiferromagnetic interactions so that no single configuration of the spins is uniquely favored by all the interactions, and the distribution of ferromagnetic and antiferromagnetic interactions between magnetic atoms must be at least partially random. Ternary intermetallic compound Pr_2AgIn_3 crystallizes in the $CaIn_2$-type structure, which consists of layers of magnetic Pr atoms alternating with non-magnetic Ag-In layers along the *c*-axis. Within one magnetic layer Pr atoms form triangles of nearest neighbours. The statistical disorder of the Ag and In atoms within the Ag-In layers could vary the electronic environment around the Pr ions, lead to the random distribution of the RKKY interactions, and thus induce the formation of frustrated magnetic moments of Pr ions and /or Pr clusters. Below a critical temperature T_f (~3.7 K, freezing temperature) the frustrated magnetic moments could be frozen in random directions forming the SG state. In fact, the neutron diffraction study for Tb_2AgIn_3 has confirmed the existence of competition between ferromagnetic and antiferromagnetic interactions within Tb layers [24]. Similar competing magnetic interactions and thus frustrated magnetic moments are naturally expected in Pr_2AgIn_3 due to the isostructuralism.

As described above, the observed magnetic behavior of Pr_2AgIn_3 suggests the formation of SG state with short-range magnetic order in this compound similar to that in Ce_2AgIn_3 and Nd_2AgIn_3. For Nd_2AgIn_3, the FC susceptibility (χ_{FC}) does not show tendency of saturation as T is lowed below T_f (~13 K), in contrast, slow decrease of χ_{FC} can be observed down to 7 K. In addition, the $\rho(T)$ curve of Nd_2AgIn_3 illustrates an upturn below about 25 K (not published). These properties indicate the antiferromagnetic nature of the magnetic coupling within magnetic cluster. Thus Nd_2AgIn_3 is categorized as a SG with short range *antiferromagnetic* order. In the present case, however, the $\chi_{FC}\,(T)$ curve of Pr_2AgIn_3 does not decrease below T_f even in a very low field (see Fig. 2), while the low field $\chi_{FC}(T)$ and $\chi_{ZFC}(T)$ curves separate from each other at the temperature $T_{ir} > T_f$ as usually observed in ferromagnetic cluster-glass systems. Moreover, there is no upward tendency can be observed in $\rho(T)$ curve down to the lowest temperature measured (see Fig.5). These features are significantly different with that observed for Nd_2AgIn_3, suggesting the *ferromagnetic-like* nature of the short-range magnetic order within the magnetic cluster in Pr_2AgIn_3 in spite of the negative paramagnetic Curie temperature.

Of course, to confirm the formation of SG state in Pr_2AgIn_3 from different point of view, further experimental researches including non-linear susceptibility measurement (which is expected to show the negative divergence at T_f) are necessary.

In conclusion, ac and dc susceptibility, magnetization, magnetic relaxation, specific heat and electrical resistivity measurements were preformed for Pr_2AgIn_3. This compound was found to undergo a SG transition with a freezing temperature T_f~3.7 K. The lack of anomaly in both specific heat and electrical resistivity around T_f, the up-shift of the ac susceptibility peak with increasing frequency, the down-shift of the dc susceptibility peak with increasing field, the long-time magnetic relaxation effect and the low temperature irreversible magnetism can be considered as the evidence for the formation of SG state. To characterize the spin freezing state of this system, the static spin freezing temperature T_S, critical exponent $z\nu$ and activation energy E_a are determined from dynamical analyses of the ac susceptibility data. In addition, FC-ZFC susceptibilities and specific heat measurements suggest the existence of magnetic clusters with evident influence on the magnetic properties. Thus it seems to be appropriate to consider Pr_2AgIn_3 as a SG with short-range magnetic order.

References

[1] Gladyshevskii R E, Cenzual K and Parthé E 1992 *J. Alloys Compd.* **189** 221.
[2] Hwang J S , Lin K J and Tien C 1996 *Solid State Commun.* **100** 169.
[3] Tien C, Feng C H, Shui C and Lu J J 2000 *Phys. Rev. B* **61** 12151.
[4] Li D X, Shiokawa Y, Homma Y, Uesawa A, Dönni A, Suzuki T, Haga Y, Yamamoto E, Honma T and Onuki Y 1998 *Phys. Rev. B* **57** 7434.
[5] Li D X, Shiokawa Y, Homma Y, Uesawa A and Suzuki T 1997 *J. Magn. Magn. Mater.* **176** 261.
[6] Li D X, Kimura A, Homma Y, Shiokawa Y, Uesawa A and Suzuki T 1998 *Solid State Commun.* **108** 863.
[7] Li D X, Dönni A, Kimura Y, Shiokawa Y, Homma Y, Haga Y, Yamamoto E, Honma T and Onuki Y 1999 *J. Phys.: Condensed Matter.* **11** 8263.
[8] Szytuła A, Hofmann M, Penc B, Ślaski M, Majumdar S, Sampathkumaran E V and Zygmunt A 1999 *J. Magn. Magn. Mater.* **202** 365.
[9] Siouris I M, Semitelou I P, Yakinthos J K, Schäfer W and Arons R R 2001 *J. Alloys Compd.* **314** 1.
[10] Tien C and Luo L 1998 *Solid State Commum.* **107** 295.
[11] Mallik R, Sampathkumaran E V, Strecker M, Wortmann G, Paulose P L and Ueda Y 1998 *J. Magn. Magn. Mater.* **185** L135.
[12] Saha S R, Sugawara H, Matsuda T D and Sato H 1999 *Phys. Rev. B* **60** 12162.
[13] Siouris J, Semitelou J P, Yakinthos J K and Schäfer W 2000 *Physica B* **276-278** 582.
[14] Nishioka T, Tabata Y, Taniguchi T and Miyako Y 2000 *J. Phys. Soc. Jpn.* **69** 1012.
[15] Li D X, Nimori S, Shiokawa Y, Tobo A, Onodera H, Haga Y, Yamamoto E and Onuki Y 2001 *Appl. Phys. Lett.* **79** 4183.
[16] Street R and Brown S D 1994 *J. Appl. Phys.* **76** 6386.
[17] Gschneidner K A, Tang J, Dhar S K andGoldman A 1990 *Physica B* **163** 507.
[18] Kaczorowski D, Czopnik A, Suski W, Nikiforov V N and Gribanov A V 2001 *Solid State Commun.* **117** 373.
[19] Mydosh J. A *Spin Glass: An Experimental Introduction* (Taylar & Francis, London, 1993).
[20] Vogel H 1921 *Phys. Z.* **22** 645; Fulcher G S 1925 *J. Am. Ceram. Soc.* **8** 339.
[21] Hohenberg P C and Halperin B I 1977 *Rev. Mod. Phys.* **49** 435; Gunnarsson K, Svedlindh P, Nordblad P, Lundgren L, Aruga H and Ito A 1988 *Phys. Rev. Lett.* **61** 754.
[22] Tholence J L 1980 *Solid State Commun.* **35** 113; Prejean J J 1978 *J. de Phys.* **39** C6-907; Dho J, Kim W. S and Hur N H 2002 *Phys. Rev. Lett.* **89** 27202.
[23] Binder K and Young A P 1986 *Rev. Mod. Phys.* **58** 801.
[24] Semitelou J P, Siouris J, Yakinthos J K, Schäfer W and Schmitt D 1999 *J. Alloys Compd.* **283** 12.

Interfacial disorder and magnetic frustration in epitaxial Fe/Cr/Fe trilayers studied by thermoremanent magnetization

Masahiro Nomura[1], Tomoya Koba[1], Ryuichi Nakamura[1], Chikahide Kanadani[2], Daisuke Akahoshi[1] and Toshiaki Saito[1,2]

[1]Department of Physics, Faculty of Science, Toho University, Chiba 274-8510, Japan
[2]Research Center for Materials with Integrated Properties, Toho University, Chiba 274-8510, Japan

E-mail: `saito@ph.sci.toho-u.ac.jp`

Abstract. In order to investigate the relaxation phenomena in Fe/Cr(001) system with various degrees of interfacial magnetic frustration, we have epitaxially grown Fe/Cr/Fe(001) trilayers on MgO(001) substrate under several thermal conditions by using molecular beam epitaxy (MBE) technique, and measured the time dependence of the thermoremanent magnetization (TRM) in the absence of magnetic eld after eld cooling process from 300 K to a measuring temperature 250 K under a magnetic eld of 100 Oe after a waiting time $t_w = 0 \sim 100$ min. The re ection high energy electron di raction (RHEED) was employed for con rming the epitaxial growth and for evaluating degree of the interfacial atness in each sample. From the experimental results, we found that 1) all the present epitaxially grown Fe/Cr/Fe(001) trilayers are ferromagnetic at 300 K, 2) the TRM shows the logarithmic time dependence with larger magnetic viscosity coe cient for the samples which have more disordered interface relative to the other samples, and 3) the most disordered sample shows clearly t_w dependence as observed in spin glasses. These results indicate that the interfacial magnetic frustration in the epitaxial Fe/Cr(001) system increases with increasing the interfacial disorder.

1. Introduction

Fe/Cr multilayer is a well-known system to exhibit the giant magnetoresistance (GMR) [1] and the oscillatory interlayer exchange coupling through Cr spacer [2]. One of recent subjects which have attracted much attention is the interfacial magnetic frustration. It has been reported in the measurements of magneto-optical Kerr e ect for epitaxial Fe/Cr [3] and Mossbauer spectroscopy for epitaxial Fe/Cr/Sn/Cr [4] that the interfacial frustration reduces the magnetic moments of Cr. The model proposed in ref. [4] is schematically shown in gure 1 for the spin arrangement near the epitaxial Fe/Cr(001) interface. Exchange interaction is ferromagnetic for Fe-Fe, and antiferromagnetic for Fe-Cr and Cr-Cr spin pairs. For the ideally at interface, there is no frustration because all the spin pairs satisfy all the interactions (a). When there are Fe steps in Cr layer in the Fe/Cr interface, both directions of Fe spin in the step cannot satisfy all the interactions as shown in gures 1(b) and 1(c), causing the interfacial frustration. The interfacial frustration also occurs in the same way when there are Cr steps in Fe layer. The interfacial frustration in this model, therefore, increases with increasing the interfacial

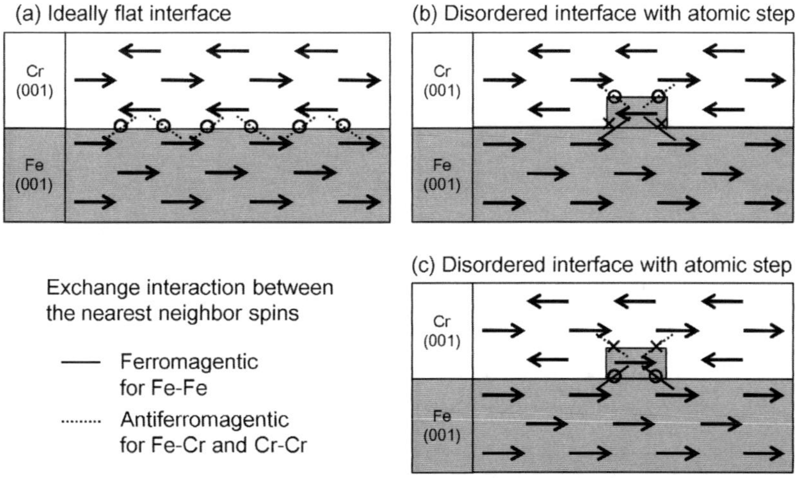

Figure 1. Schematic model of the spin arrangement near the epitaxial Fe/Cr(001) interface. The arrows indicate the Fe and Cr spins. Exchange interaction between the nearest neighbor spins is ferromagnetic for Fe-Fe and antiferromagnetic for Fe-Cr and Cr-Cr, respectively. In the ideally at interface, there is no frustration (a). The spin at the step site can not satisfy all the interactions for both spin directions (b) and (c).

disorder (atomic step). It is reasonable that such an interfacial frustration a ects the magnetic relaxation mechanism and more or less causes magnetic viscosity as observed in spin glasses [5]. The direct observation of the slow dynamics is possible by measuring time dependence of the thermoremanent magnetization (TRM). As far as we know, time decay of the TRM has been measured only in a polycrystalline Fe/Cr system [6] or an epitaxial Fe/Cr with considerable interdi usion [7] in which dominant origin of the frustration is not the interfacial disorder itself but that including substantial internal disorder. In fact, the magnetic transitions in such systems occur well below room temperature due to the bulk-like disorder [6, 7]. In this study, in order to investigate the relaxation phenomena of Fe/Cr(001) system in which the interfacial disorder is essential, we prepared the epitaxial Fe/Cr/Fe(001) trilayers with various degrees of the interfacial atness by using the molecular beam epitaxy (MBE) technique and measured the time dependence of the TRM.

2. Experimental

Fe/Cr/Fe(001) trilayers were prepared by using MBE technique. The residual pressure in the MBE chamber was in the order of 10 ^{9} Torr. MgO(001) single-crystal substrate was used to obtain the epitaxial Fe and Cr layers with (001) orientation [8]. The sample layer structure was MgO(001)/Cr(50 A)/Fe(20 A)/Cr(18 A)/Fe(20 A)/Au(10 A). Cr(50 A) was evaporated as a bu er layer prior to the Fe/Cr/Fe trilayer evaporation. Au(10 A) was capped to prevent the oxidation. The thickness of the Cr spacer, 18 A, was selected so that in our samples the rst and second Fe layers are coupled ferromagnetically [2] to ensure su ciently large magnetization for the relaxation measurement even in a small applied eld. In order to obtain the samples with several degrees of the interfacial atness, heat treatment condition was varied as below: 1) the growth temperatures of T_s = 300 C and 400 C were selected based on the fact that the optimal temperature to obtain the at Fe(001) surface on the MgO(001) substrate is about 300 C [9], 2) after evaporating each layer, the sample was annealed at 500 C for 15 min, and 3) the samples without annealing process for each T_s were also prepared. The re ection high

Journal of Physics: Conference Series **320** (2011) 012042 doi:10.1088/1742-6596/320/1/012042

energy electron di raction (RHEED) was employed for con rming the epitaxial growth and for evaluating of the interfacial atness in each sample.

Magnetic properties were measured by using a SQUID magnetometer (Quantum Design MPMS). The procedure of the relaxation measurement of the TRM was as follows: rst, a sample was rapidly cooled (-10 K/min) in the eld H_{FC} (= 100 Oe) from room temperature to a measuring temperature T_{m} (= 250 K), and after a waiting time t_{w} (= 0, 5, 15, and 100 min), the measurement was started in the absence of eld and continued for about 10, 000 sec.

3. Results and discussion

3.1. Interfacial structure observed by the RHEED

From *in situ* RHEED observations, we con rmed that all the layers in Cr/Fe/Cr/Fe/Au grew epitaxially with the (001) basal plane. Figure 2 shows the RHEED patterns at the surface of the rst Fe layer in the sample of T_{s} = 300 C before annealing (a) and after annealing (b), respectively. The so-called $c(2\quad 2)$ reconstructed structure was observed as shown in gure 2 (a). It has been reported, in the study of the at surface structure for Fe(001) epitaxially grown on MgO(001) substrates, that the $c(2\quad 2)$ reconstruction indicates two fold periodicity in the Fe[110] direction of the arrangement of the topmost Fe atoms and the surface with atomically at and wide terraces separated by monoatomic steps [10, 11]. On the other hand, in the sample of T_{s} = 400 C, no reconstructed structure was observed before annealing as shown in the gure 3 (a). By annealing at 500 C for 15 min, $c(3\sqrt{2}\quad \sqrt{2})$ reconstructed structure with three fold periodicity in the Fe[110] direction emerged in both samples as shown in gure 2 (b) and gure 3 (b). Such a $c(3\sqrt{2}\quad \sqrt{2})$ reconstructed structure was also observed for Cr layer surface after annealing as shown in gure 2 (c). It has been reported that $c(3\sqrt{2}\quad \sqrt{2})$ reconstruction is formed by C atoms segregated from the MgO(001) substrate [12]. Although it remains controversial what kind of element forms the reconstructed structure, an atomically at surface at least equivalent to the $c(2\quad 2)$ case is needed to induce the $c(3\sqrt{2}\quad \sqrt{2})$ reconstructed structure.

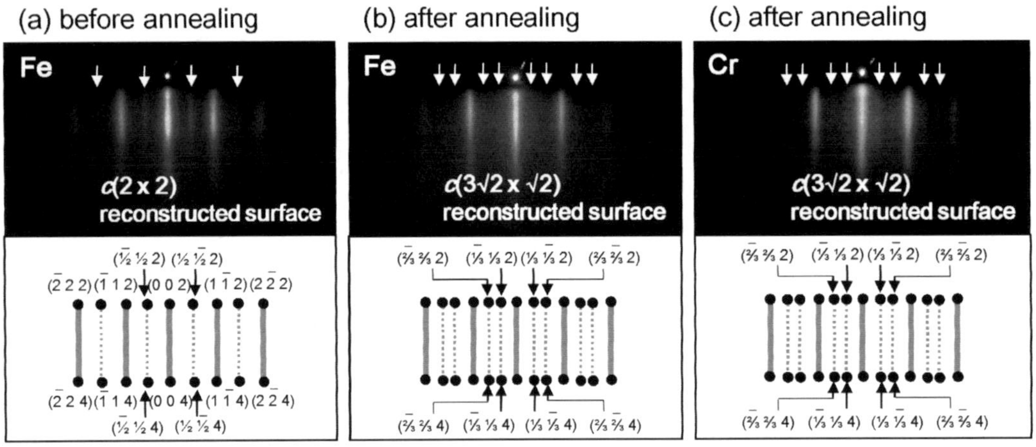

Figure 2. (Color online) RHEED patterns for the sample of T_{s} = 300 C at the surface of the rst Fe (a), (b) and Cr (c) layers obtained in the Fe[110] and Cr[110] directions. Arrows in pattern (a) indicate the $c(2\quad 2)$ reconstructed structure, and arrows in pattern (b) and (c) indicate the $c(3\sqrt{2}\quad \sqrt{2})$ reconstructed one. By annealing at 500 C for 15 min, the RHEED pattern changes from (a) into (b). The lower panels show the Miller's indices of the RHEED patterns. The dotted lines indicate the Miller's indices for $c(2\quad 2)$ and $c(3\sqrt{2}\quad \sqrt{2})$ reconstructed surfaces.

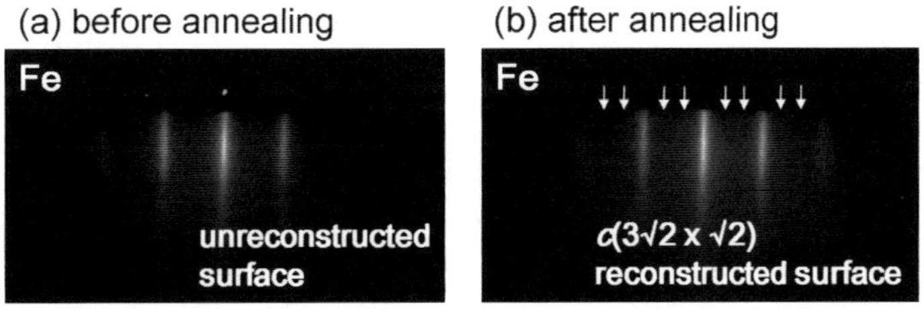

Figure 3. (Color online) RHEED patterns for the sample of $T_s = 400$ C at the surface of the rst Fe layer obtained in the Fe[110] direction. Pattern (a) shows the (1 1) unreconstructed surface and the arrows in (b) indicate the $c(3\sqrt{2} \quad \sqrt{2})$ reconstructed surface. By annealing at 500 C for 15 min, the RHEED pattern changes from (a) into (b). The Miller's indices of streak lines in (a) is the same as those in gure 2(a) excluding the reconstructed pattern (dotted line), and those in (b) is the same as in gure 2(b).

Table 1. Reconstructed structure observed at the surface of the rst Fe layer in each sample and relative order of the interfacial atness (larger number indicates relatively more disordered interface).

Sample	Reconstructed structure		Relative order of the interfacial atness (sample number)
	(before annealing)	(after annealing)	
$T_s = 300$ C (with annealing)	$c(2 \quad 2)$	$c(3\sqrt{2} \quad \sqrt{2})$	1
$T_s = 300$ C (without annealing)	$c(2 \quad 2)$	-	1'
$T_s = 400$ C (with annealing)	unreconstructed	$c(3\sqrt{2} \quad \sqrt{2})$	2
$T_s = 400$ C (without annealing)	unreconstructed	-	3

The reconstructed structures observed at the rst Fe layer in each sample are summarized in the table 1. For estimating the interfacial atness in each sample, we assumed that the reconstructed surface is more at than the unreconstructed surface. Based on the reconstruction, relative order of the interfacial atness is determined as shown in the rightmost column of the table 1. Here, the larger number indicates relatively more disordered interface. We note that the number 1 and 1' means nearly the same order of the interfacial atness as described above. We have also performed the RHEED observations at the other surfaces, but the order of the interfacial atness is not changed. Hence, each sample is labeled as the number.

3.2. Relaxation of the TRM

We measured magnetization curves for all samples from -50 kOe to +50 kOe, and observed magnetic hysteresis with coercive forces (about 1 Oe for the samples of $T_s = 300$ C and 30 100 Oe for $T_s = 400$ C) at 300 K for all samples, which indicates that our epitaxially grown Fe/Cr/Fe trilayers are ferromagnetic at least at 300 K. The result suggests that the internal disorder due to the interdi usion can be almost disregarded in our samples because the transition temperature becomes lower far below room temperature when such a bulk-like

Journal of Physics: Conference Series **320** (2011) 012042 doi:10.1088/1742-6596/320/1/012042

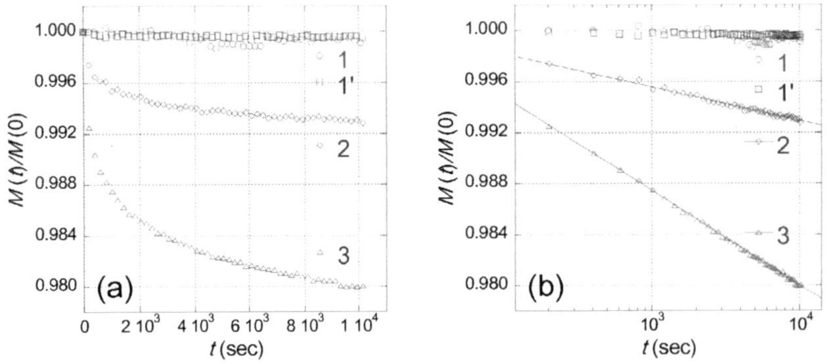

Figure 4. (Color online) Time dependence of the TRM in the epitaxial Fe/Cr/Fe(001) trilayers in the linear scale (a) and the logarithmic scale (b). Measurement conditions are as follows: H_{FC} = 100 Oe (Fe[100] direction), T_{m} = 250 K, and t_{w} = 0 min. The vertical axis is normalized by the rst measuring point $M(0)$. Sample number is marked in the graph (larger number indicates relatively more disordered interface).

frustration occurs [6, 7].

Figure 4 shows time dependence of the TRM in each sample in the linear scale (a) and the logarithmic scale (b). Measurement conditions were H_{FC} = 100 Oe, T_{m} = 250 K, and t_{w} = 0 min. We found that time decay of the TRM is dramatically changed depending on the interfacial atness as shown in the gure 4 (a). In the samples of number 2 and 3, which have relatively disordered interfaces, time decay of the TRM is well tted to the following logarithmic equation (1) as shown in the gure 4 (b),

$$M(t) = M(0) \quad S \ln t, \tag{1}$$

where $M(0)$ is the rst measuring point after removing the H_{FC}, t is the elapsed time after starting the measurement, and coe cient S is the magnetic viscosity which indicates the slope of the logarithmic decay. For the samples of number 1 and 1', which have relatively at interfaces, it is considered that the relaxation is very rapid, then, it has been already nished before measuring the rst point. Logarithmic decay observed in the samples 2 and 3 can be attributed to the e ect of the interfacial frustration. In addition, it is found that S/M_{s} increases with increasing the interfacial disorder, where S is divided by M_{s} (saturated magnetization obtained from the M-H curve at 300 K) to compare the relative value in each sample. The values of S/M_{s} for the samples 2 and 3 are 1.09 $10^{\ 3}$ and 3.39 $10^{\ 3}$, respectively. These results support consistently the schematic models demonstrated in the gure 1.

Figures 5 (a) and (b) show t_{w} dependences of the S/M_{s} in the samples 2 and 3, respectively. In the samples 2 and 3, particularly in the sample 3, S/M_{s} decreases substantially with increasing t_{w} as shown in the gures 5 (a) and (b). This t_{w} dependence has been generally observed in spin glasses [13]. These results suggest that dynamical behavior of the epitaxial Fe/Cr/Fe(001) trilayers with the interfacial magnetic frustration is close to that of the spin glasses. But, we emphasize that Curie temperature in the present Fe/Cr/Fe(001) system is above 300 K, that is, the epitaxally grown system is ferromagnetic re ecting dominantly to the interfacial disorder, therefore, to the interfacial magnetic frustration. Further experiments are needed to discuss the detailed relaxation mechanism.

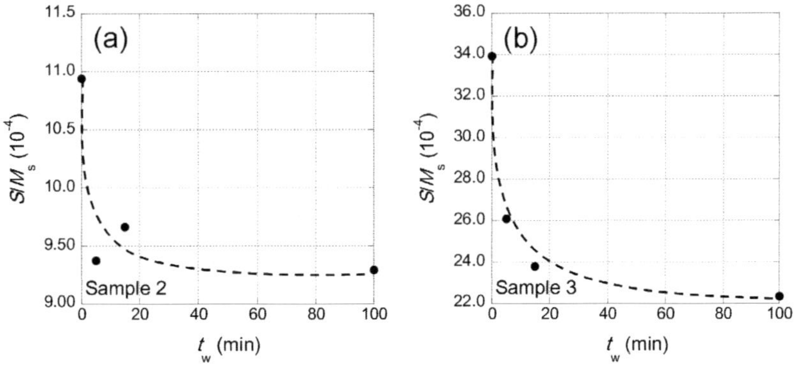

Figure 5. t_w dependence of the S/M_s in the sample 2 (a) and the sample 3 (b). S is divided by M_s (saturated magnetization obtained from the M-H curve at 300 K) to compare the relative value in each sample. Measurement conditions are as follows: $H_{FC} = 100$ Oe (Fe[100] direction), $T_m = 250$ K, and $t_w = 0$, 5, 15, and 100 min. The dashed line is a guide to the eyes.

4. Conclusion

We prepared epitaxial Fe/Cr/Fe(001) trilayers with di erent degrees of the interfacial atness and measured the relaxation of the TRM in each sample. The relative order of the interfacial atness was determined from the reconstructed structure observed in the RHEED. We found that in the samples with disordered interfaces TRM shows logarithmic time dependence and magnetic viscosity coe cient S increases with increasing the interfacial disorder. In addition, the most disordered sample shows clear t_w dependence as observed in spin glasses. These results de nitely indicate that the interfacial magnetic frustration increases with increasing the interfacial disorder in the epitaxial Fe/Cr(001) system. The e ect of ferromagnetic interlayer exchange coupling (IEC) of Fe layers on the interfacial frustration has not yet been clari ed. However, our preliminary investigation for a simple Fe/Cr bilayer system without IEC shows that the logarithmic TRM decay is also observed in the bilayer system as in the present Fe/Cr/Fe one, which indicates that the ferromagnetic IEC is not essential at least in the present sample structure. Recently, we have also found the logarithmic time dependence of the TRM in the epitaxial Fe/Cr/Fe(011) trilayers. But, in contrast to Fe/Cr/Fe(001) case, S decreases with increasing the interfacial disorder suggesting the decrease of the interfacial magnetic frustration. We will report elsewhere about the Fe/Cr/Fe(011) trilayers as second article.

References

[1] Baibichi M N, Broto J M, Fert A, Nguyen Van Dau F, Petro F, Eitenne P, Creuzet G, Friederich A and Chazelas J 1988 *Phys. Rev. Lett.* **61** 2472
[2] Wolf J A, Leng Q, Schreiber R, Grünberg P A and Zinn W 1993 *J. Magn. Magn. Mater.* **121** 253
[3] Berger A and Hopster H 1994 *Phys. Rev. Lett.* **73** 193
[4] Jiko N, Almokhtar M, Mibu K and Shinjo T 2005 *J. Phys.: Condens. Matter* **17** 2477
[5] Chambelin R V 1984 *Phys. Rev.* B **30** 5393
[6] Patel R S, Majumdar A K and Nigam A K 2007 *J. Magn. Magn. Mater.* **309** 256
[7] Drovosekov A B, Kreines N M, Kholin D I, Korolev A V, Milyaev M A, Romashev L N and Ustinov V V 2008 *JETP Letters* **88** 118
[8] Shikada K, Tabuchi K, Ohtake M, Kirino F and Futamoto M 2008 *J. Magn. Soc. Jpn.* **32** 296 (in Japanease)
[9] Jordan S M, Lawler J F, Schad R and van Kempen H 1998 *J. Appl. Phys.* **84** 1499
[10] Subagyo A, Oka H, Eilers G, Sueoka K and Mukasa K 2000 *Jpn. J. Appl. Phys.* **39** 3777
[11] Oka H, Subagyo A, Sawamura M, Sueoka K and Mukasa K 2001 *Jpn. J. Appl. Phys.* **40** 4334
[12] Oka H, Nakai A and Sueoka K 2007 *Jpn. J. Appl. Phys.* **46** 5602
[13] Refregier Ph, Vincent E, Hammann J and Ocio M 1987 *J. Physique* **48** 1533

Exotic ground state in triangle spin tubes

Hirotaka Manaka[1] **and Yoko Miura**[2]

[1]Graduate School of Science and Engineering, Kagoshima University, Korimoto, Kagoshima 890-0065, Japan
[2]Suzuka National College of Technology, Shiroko-cho, Suzuka, Mie 510-0294, Japan

E-mail: manaka@eee.kagoshima-u.ac.jp

Abstract. Triangle spin tubes have attracted much attention because they possess characteristics of one-dimensional Heisenberg antiferromagnets along the tubes and geometrical spin frustration in each triangle plane. We performed X-ray diffraction, magnetic susceptibility, and heat capacity experiments on triangle spin tubes composed of $CsCrF_4$ and α-$KCrF_4$. In $CsCrF_4$, no anomaly indicating an antiferromagnetic long-range order appeared, and a spin-liquid state was realized. In contrast, in α-$KCrF_4$, antiferromagnetic long-range order occurred at T_{N2} = 4.0(1) K and another phase transition occurred at T_{N1} = 2.5(1) K. Such successive phase transitions may suggest a simultaneous change of the spin and crystal structures owing to a magnetoelectric effect.

1. Introduction

One-dimensional Heisenberg antiferromagnets have attracted much interest for a long time. Recently, spin ladders have also been extensively studied because topological effects on one-dimensional antiferromagnets have attracted attention[1, 2, 3]. Odd-leg ladders have a gapless ground state and an algebraic decay of spin correlation functions, whereas even-leg ladders have a gapped ground state and an exponential decay of spin correlation functions[4, 5]. Furthermore, geometrically frustrated spin systems such as triangle lattice, kagomé lattice, and pyrochlore lattice have exotic and unconventional magnetic ground states. For the case of triangle lattice, a 120° spin structure is formed in the classical spin system, and Anderson proposed that a resonating-valence-bond (RVB) state is formed in the quantum spin system[6].

Triangle spin tubes belong to a new category of one-dimensional Heisenberg antiferromagnets that have geometrically frustrated spin systems. According to theoretical studies[7, 8, 9, 10], two kinds of topological effects will cause an exotic ground state: (1) a vector-chiral long-range ordered state, and (2) a gapless spin-liquid state (the Tomonaga-Luttinger-liquid state). However, only a few experimental candidate compounds for triangle spin tubes exist, such as $Na_2V_3O_7$[11, 12, 13] and $[(CuCl_2tachH)_3Cl]Cl_2$[14, 15]. Because such compounds consist of complex superexchange paths and have the same order for all paths, the magnetic properties of triangle spin tubes may still be obscure.

Recently, we discovered that chromium fluoride $CsCrF_4$ could form ideal equilateral triangle spin tubes having $S = 3/2$ spin, corresponding to the intermediate spin state between the quantum and classical models[16]. As schematically shown in Fig. 1, all the Cr ions are at equivalent sites, and each angle of Cr-Cr-Cr in the triangle plane is 60°. Because the geometrical frustration among the three $S = 3/2$ spins in the equilateral triangle plane is possibly robust,

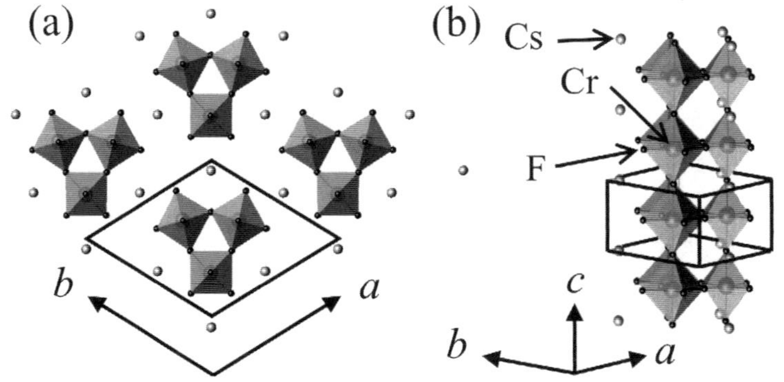

Figure 1. Schematic diagram of the crystal structure of $CsCrF_4$ (a) in the *ab*-plane and (b) along the tubes (*c*-axis). The identical Cr_3 triangles are stacked without rotation along the *c*-axis.

we propose an RVB-like spin-liquid state, i.e., resonating spin dimers in the equilateral triangle plane. The identical Cr_3 equilateral triangles are stacked without rotation along the *c*-axis, and Cs ions separate them. Thus, each spin tube is expected to show good one-dimensionality. On the basis of magnetic susceptibility and heat capacity experiments on $CsCrF_4$, we report that the magnetic behavior of α-$KCrF_4$ is very different from that of $CsCrF_4$. According to neutron experimental reports by Lacorre, α-$KCrF_4$ shows antiferromagnetic long-range order at 4 K and a distorted star arrangement of spin structure[17]. In addition, Nénert theoretically predicted a magnetoelectric (linear) effect in which a magnetic field in an antiferromagnetic ordered state induces electrical polarization[18]. Finally, we determined the difference not only in the magnetic ground states, but also in the crystal structures of these two compounds.

2. Sample preparation

We are interested in how the magnetic properties of $CsCrF_4$ change when its Cs ions are completely replaced with K ions. To this end, we prepared polycrystalline samples of α-$KCrF_4$ by the method similar to that employed for synthesis of $CsCrF_4$, i.e., by a conventional solid-state reaction method. We mixed the starting materials CsF and $CrF_3{\cdot}4H_2O$ for $CsCrF_4$, and KF and $CrF_3{\cdot}4H_2O$ for α-$KCrF_4$ and then heated them at 200 °C for more than 48 h under vacuum at $P < 1 \times 10^{-3}$ Pa to dehydrate the crystal water. Then, we kept them at 700 °C on $CsCrF_4$ and 745 °C on α-$KCrF_4$ for more than 50 h under a flow of N_2 gas (50 cc/min). The sample color was dark green or gray on $CsCrF_4$ and yellow green on α-$KCrF_4$. To further purify both samples, we annealed them at 550 °C under HF gas, and then etched them with a solution of HNO_3.

We performed powder X-ray diffraction experiments at room temperature to examine the crystal structure and phase purity of our samples. In addition, we performed Rietveld analyses for these data using RIETAN-2000[19]. As a result, we determined the space group and lattice parameters of both samples (see Table 1); our values were consistent with the previous reports[20, 21]. Because no additional peaks appeared and the values of the weighted reliability factor (R_{wp}) and the goodness-of-fit indicator ($S \equiv R_{\mathrm{wp}}/R_{\mathrm{e}}$) were very low, we determined that both samples were of high quality. Figure 2 shows a schematic diagram of the crystal structure of α-$KCrF_4$ and the atomic coordinates are show in elsewhere[22].

In the schematic diagram of $CsCrF_4$ shown in Fig. 1, the CrF_6 octahedra are corner-sharing, such as equilateral triangle tubes along the *c*-axis. We are interested in the strength of the superexchange interactions in the triangle plane (J_{rung}) and along the tubes (J_{chain}). The direct distance between Cr ions in the triangle plane is 3.702(4) Å, while that along the tubes is 3.847(3) Å. These distances are very similar. However, the angle of Cr-F-Cr in the triangle plane

Table 1. Results of our Rietveld analyses of $CsCrF_4$ and α-$KCrF_4$ at room temperature. R_{wp} and R_e are the weighted and expected reliability factors, respectively, and S is the goodness-of-fit indicator ($S \equiv R_{wp}/R_e$).

Compound	Space group	a [Å]	b [Å]	c [Å]	R_{wp}	R_e	S
$CsCrF_4$	$P\bar{6}2m$ (No. 189)	9.619(11)	9.619(11)	3.847(3)	4.04	2.98	1.35
α-$KCrF_4$	$Pnma$ (No. 62)	15.759(11)	7.447(5)	18.359(13)	4.13	3.03	1.36

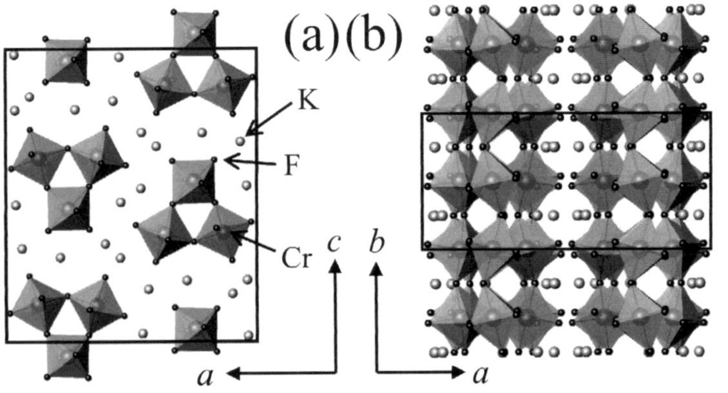

Figure 2. Schematic diagram of the crystal structure of α-$KCrF_4$ (a) in the ac-plane and (b) along the tubes (b-axis). Distorted CrF_6 octahedra are twisted and stacked along the b-axis.

is 147.8(2)°, while that along the tubes is 179.05(13)°. Thus, we believe that $|J_{chain}| > |J_{rung}|$, because $|J_{chain}|$ is almost along the 180° path.

In the schematic diagram of α-$KCrF_4$ shown in Fig. 2, distorted CrF_6 octahedra are twisted and stacked along the b-axis. The angles of Cr-Cr-Cr in the triangle plane are not 60°, they are 59.07(5)°, 59.40(6)°, and 61.53(6)°. Furthermore, the direct distances between Cr ions in the triangle plane are 3.664(3)$\sim$3.755(3) Å, while those along the spin tubes are 3.710(4)$\sim$3.738(4) Å. The angles of Cr-F-Cr in the triangle plane are 137.5(2)$\sim$149.5(2)°, while those along the spin tubes are 143.2(2)$\sim$158.6(2)°. Although there are three kinds of $|J_{rung}|$ and six kinds of $|J_{chain}|$, we assume that the effective $|J_{rung}|$ may be of the same order of strength as the effective $|J_{chain}|$, because the angles of Cr-F-Cr in the triangle plane and along the tubes are very similar.

3. Experimental results and discussion

Using a superconducting quantum interference device (SQUID) magnetometer, we measured the magnetization M induced by an external field H = 10 Oe between the temperature range 2 K and 350 K. In the present study, we define the magnetic susceptibility as $\chi \equiv M/H$. Figure 3 shows the temperature (T) dependence of χ for $CsCrF_4$. The field cooling (FC) and zero field cooling (ZFC) data were collected after applying H = 10 Oe at 350 K and 2 K, respectively. Because the $\chi(T)$ curves for the FC and ZFC data overlap each other above $\sim$4 K, the ZFC data were omitted in Figs. 3(a) and 3(c). The samples that went through the post chemical treatment clearly exhibited a broad $\chi(T)$ peak at $T_{max} \simeq 60$ K [Fig. 3(a)] that indicates one-dimensional antiferromagnets, because the magnetic impurity of these samples drastically decreases and is thus less than that of the as-grown sample[16]. We believe that the increase below 10 K as seen in Fig. 3(a) is probably due to paramagnetic impurities or some

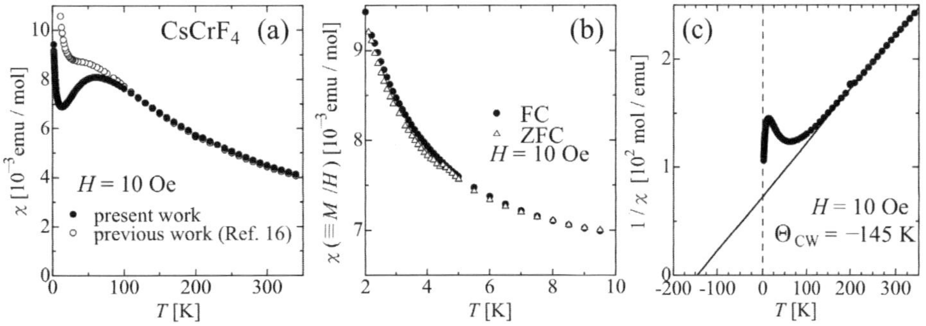

Figure 3. (a) Temperature dependence of magnetic susceptibility (χ) for $CsCrF_4$. (b) The lower-temperature part of (a). (c) Temperature dependence of $1/\chi(T)$. Solid line is the Curie-Weiss law; the resulting Curie-Weiss temperature is $\Theta_{CW} = -145$ K. Because the $\chi(T)$ curves for FC and ZFC data overlap each other above $\sim$4 K, the ZFC data are omitted in (a) and (c).

imperfect crystallization, although a weak ferromagnetic moment due to the Dzyaloshinskii-Moriya interaction may appear. Assuming the Curie law for the increasing part, we estimated that the ratio of $S = 3/2$ free spins to the total spins is 0.3$\sim$0.4 % in the present sample and 2 % in the as-grown sample. As shown in Fig. 3(b), a slight deviation between the FC and ZFC data appears below $\sim$4 K. We could not determine whether this small difference is intrinsic or extrinsic (an impurity effect). However, a gapless ground state is realized, and there is no anomaly indicating an antiferromagnetic long-range order above 2 K. Figure 3(c) shows the T dependence of $1/\chi$, whose curvature resembles that of Cr-Jarosites[23] in which two-dimensional Heisenberg antiferromagnets occur on a kagomé lattice. We fit the Curie-Weiss law ($\chi(T) = C/(T - \Theta_{CW})$) to the $1/\chi(T)$ data above 250 K; the resulting Weiss temperature (Θ_{CW}) is -145 ± 3 K, and the effective magnetic moment is $\mu_{eff} = 4.00\ \mu_B$. This value is almost the same as 3.87 μ_B for free Cr^{3+} ions. Because the experimental data only deviate from the solid line below 200 K, there are no effects on the spin frustration and short-range order above 200 K.

Figure 4 shows the T dependence of χ for α-$KCrF_4$. The FC and ZFC data were collected after applying $H = 10$ Oe at 350 K and 2 K, respectively. Because the $\chi(T)$ curves for the FC and ZFC data overlap each other above 4 K, the ZFC data were omitted in Figs. 4(a) and 4(c). As can be seen in Fig. 4(a), a broad $\chi(T)$ peak appears at $T_{max} \simeq 25$ K, which is lower than that for $CsCrF_4$ ($T_{max} \simeq 60$ K). We suppose that $|J_{chain}| \approx |J_{rung}|$ for α-$KCrF_4$ because the peak temperature is generally proportional to the strength of the exchange interaction in one-dimensional Heisenberg antiferromagnets[24]. Figure 4(c) shows the T dependence of $1/\chi$, and it shows that the antiferromagnetic long-range order occurs at $T_{N2} = 4.0(1)$ K, which agrees with the previous report[17]. Below 4 K, the $\chi(T)$ curves for the FC and ZFC data show a large deviation. This deviation is probably owing to the close correlation between the spin structure in the antiferromagnetic ordered state and the crystal structure, i.e., the magnetoelectric effect on the spin frustration in the distorted triangle plane[18]. We fit the Curie-Weiss law to the $1/\chi(T)$ data above 250 K as shown in Fig. 4(c); the resulting Θ_{CW} is -75 ± 3 K, and the effective magnetic moment is $\mu_{eff} = 3.84\ \mu_B$. This value agrees well with 3.87 μ_B for free Cr^{3+} ions. Because the experimental data only deviate from the solid line below 100 K, there are no effects on the spin frustration and short-range order above 100 K. The frustration factor $f(\Theta) = |\Theta_{CW}|/T_{N2}$ is estimated to be about 18.8. For Cr-Jarosites, $f(\Theta)$ is within 6.4$\sim$13.3[23]. We believe that good magnetic one-dimensionality and extreme spin frustration coexist in α-$KCrF_4$.

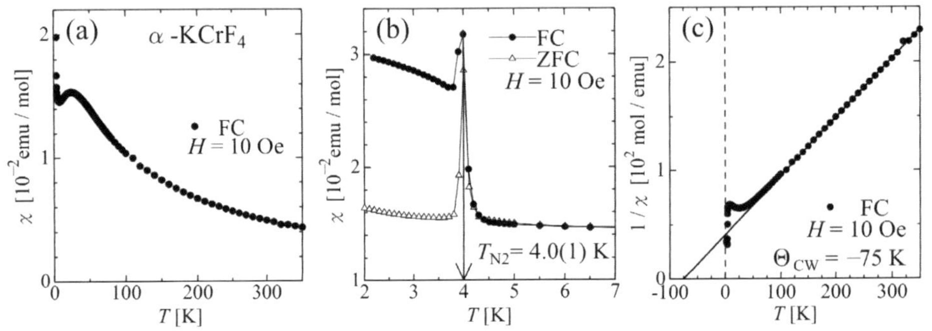

Figure 4. (a) Temperature dependence of magnetic susceptibility (χ) for α-$KCrF_4$. (b) The lower-temperature part of (a). Antiferromagnetic long-range order occurs at $T_{N2} = 4.0(1)$ K. (c) Temperature dependence of $1/\chi$. Solid line is the Curie-Weiss law; the resulting Curie-Weiss temperature is $\Theta_{CW} = -75$ K. Because the $\chi(T)$ curves for FC and ZFC data overlap each other above 4 K, the ZFC data are omitted in (a) and (c).

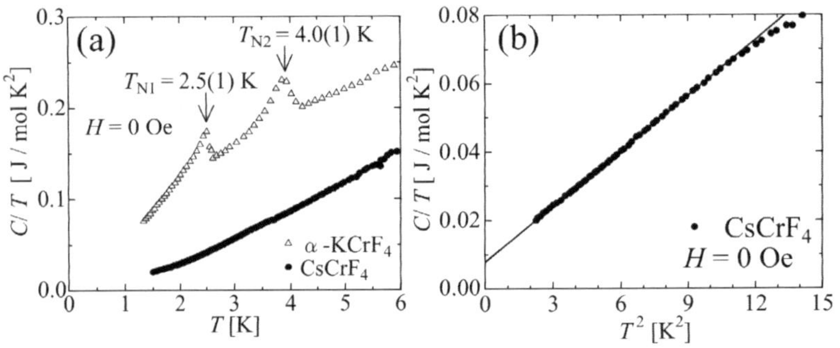

Figure 5. (a) Temperature dependence of the heat capacity (C/T) under zero magnetic field for $CsCrF_4$ and α-$KCrF_4$. (b) T^2 dependence of the heat capacity under zero magnetic field for $CsCrF_4$. Solid line is the best-fit line given by $C(T)/T = \gamma + AT^2$; thus, we found $\gamma > 0$ for $CsCrF_4$.

To confirm the phase transition including the magnetoelectric effect, we performed heat capacity [$C(T)$] measurements with the conventional adiabatic method under a zero magnetic field down to 1.5 K. The first-order phase transition (a structural phase transition) can be detected by the adiabatic method rather than by the relaxation method. Figure 5 shows the T dependence of the heat capacity curves. As can be seen in Fig. 5(a) for $CsCrF_4$, no anomaly that indicates a phase transition appears down to 1.5 K, i.e., the antiferromagnetic long-range order is absent. Because the heat capacity is generally given by $C(T) \equiv \gamma T + AT^3$, we found that a Fermi-liquid state is realized in the low T region. This is because a component linear in the temperature ($\gamma = 7.8$ mJ/mol K^2) exists [Fig. 5(b)] when the lattice part is determined to be $5.4 \times T^3$ mJ/ mol K^2. As a result, a gapless spin-liquid ground state consisting of resonating spin dimers in the equilateral triangle plane may be realized in $CsCrF_4$. On the other hand, the T dependence of C/T curve for α-$KCrF_4$ shows λ-type anomalies [Fig. 5(a)]: sharp peak at $T_{N1} = 2.5(1)$ K and broad peak at $T_{N2} = 4.0(1)$ K. These anomalies suggest that strong short-range order remains above T_{N2} owing to good magnetic one-dimensionality. The $C(T)$ anomaly at T_{N2} agrees well with the $\chi(T)$ anomaly [Fig. 4(b)]. However, the $C(T)$ anomaly at

T_{N1} is not clearly shown in the $\chi(T)$ curve. Because Lacorre determined the spin structure from the difference in the neutron diffraction pattern between 1.5 K and 5 K[17], we suspect that the spin structure for $T_{N1} < T < T_{N2}$ slightly varies from the distorted star arrangement of the spin structure below T_{N1}. If this scenario is true, then the sharp $C(T)$ peak at T_{N1} may suggest that both the spin and crystal structures slightly vary at the same time beyond the neutron experimental resolution limits owing to the magnetoelectric effect on the spin frustration in the distorted triangle plane[18].

In conclusion, we synthesized single-phase polycrystalline $S = 3/2$ spin compounds of equilateral and non-equilateral triangle spin tubes in $CsCrF_4$ and α-$KCrF_4$, respectively. From X-ray diffraction experiments, the angles of Cr-F-Cr in the triangle plane and along the tubes suggest $|J_{\rm chain}| > |J_{\rm rung}|$ for $CsCrF_4$ and $|J_{\rm chain}| \approx |J_{\rm rung}|$ for α-$KCrF_4$. In $CsCrF_4$, the $\chi(T)$ and $C(T)$ curves show that no anomaly indicating antiferromagnetic long-range order appeared, and also that the RVB-like spin-liquid state (the resonating spin dimers in the equilateral triangle plane) may be realized because the geometrical frustration is robust. In contrast, for α-$KCrF_4$, antiferromagnetic long-range order occurs at $T_{N2} = 4.0(1)$ K and another phase transition occurs at $T_{N1} = 2.5(1)$ K. Such successive phase transitions may suggest a simultaneous change of the spin and crystal structures owing to the magnetoelectric effect on the geometrical spin frustration in the distorted triangle plane.

Acknowledgments
We received valuable support from Frontier Science Research Center in Kagoshima University for powder X-ray diffraction experiments. This research was partially supported by Shimadzu Science Foundation, Kurata Memorial Hitachi Science and Technology Foundation, SEI Group CSR Foundation, and Grant-in-Aid for Scientific Research on Priority Areas "Novel States of Matter Induced by Frustration" (22014009).

References

[1] Azuma M, Hiroi Z, Takano M, Ishida K, and Kitaoka Y 1994 *Phys. Rev. Lett.* **73** 3463.
[2] Masuda T, Zheludev A, Manaka H, Regnault L -P, Chung J -H, and Qiu Y 2006 *Phys. Rev. Lett.* **96** 047210.
[3] Klanjšek M, *et al.* 2008 *Phys. Rev. Lett.* **101** 137207.
[4] Sierra G 1996 *J. Phys. A* **29** 3299.
[5] Dell'Aringa S, Ercolessi E, Morandi G, Pieri P, and Roncaglia M 1997 *Phys. Rev. Lett.* **78** 2457.
[6] Anderson P W 1973 *Mater. Res. Bull.* **8** 153.
[7] Kawano K and Takahashi M 1997 *J. Phys. Soc. Jpn.* **66** 4001.
[8] Sato M and Sakai T 2007 *Phys. Rev. B* **75** 014411.
[9] Sakai T, Sato M, Okunishi K, Otsuka K, Okamoto K, and Itoi C 2008 *Phys. Rev. B* **78** 184415.
[10] Nishimoto S and Arikawa M 2008 *Phys Rev. B* **78** 054421.
[11] Millet P, Henry J Y, Mila F, and Galy J 1999 *J. Solid State Chem.* **147** 676.
[12] Gavilano J L, Rau D, Mushkolaj S, Ott H R, Millet P, and Mila F 2003 *Phys. Rev. Lett.* **90** 167202.
[13] Gavilano J L, Felder E, Rau D, Ott H R, Millet P, Mila F, Cichorek T, and Mota A C 2005 *Phys. Rev. B* **72** 064431.
[14] Schnack J, Nojiri H, Kögerler P, Cooper G J T, and Cronin L 2004 *Phys. Rev. B* **70** 174420.
[15] Ivanov N B, Schnack J, Schnalle R, Richter J, Kögerler P, Newton G N, Cronin L, Oshima Y, and Nojiri H 2010 *Phys. Rev. Lett.* **105** 037206.
[16] Manaka H, Hirai Y, Hachigo Y, Mitsunaga M, Ito M, and Terada N 2009 *J. Phys. Soc. Jpn.* **78** 093701.
[17] Lacorre P, Leblanc M, Pannetier J, Ferey G 1987 *J. Mag. Mag. Mater.* **66** 219.
[18] Nénert G and Palstra T T M 2007 *J. Phys.: Condens. Matter* **19** 406213.
[19] Izumi F and Ikeda T 2000 *Mater. Sci. Forum* **321-324** 198.
[20] Babel V D and Knoke G 1978 *Z. anorg. allg. Chem.* **442** 151.
[21] Kissel D and Hoppe R 1987 *Z. Naturforsch.* **42b** 135.
[22] Manaka H, *et al.* in preparation.
[23] Morimoto T, Nishiyama M, Maegawa S, and Oka Y 2003 *J. Phys. Soc. Jpn.* **72** 2085.
[24] Bonner J C and Fisher M E 1964 *Phys. Rev.* **135** A640.

Journal of Physics: Conference Series **320** (2011) 012044 doi:10.1088/1742-6596/320/1/012044

Nonmagnetic impurity effect on equilateral triangle spin tubes $CsCrF_4$

Yoko Miura[1] **and Hirotaka Manaka**[2]

[1]Suzuka National College of Technology, Shiroko-cho, Suzuka, Mie 510-0294, Japan
[2]Graduate School of Science and Engineering, Kagoshima University, Korimoto, Kagoshima 890-0065, Japan

E-mail: miura@genl.suzuka-ct.ac.jp

Abstract. A chemical impurity substitution in gapped spin-liquid ground states often leads to antiferromagnetic long-range order. However, few experimental studies have been conducted on the substitution effect in gapless spin-liquid ground states. We performed X-ray diffraction and magnetic susceptibility experiments on equilateral triangle spin tubes composed of $CsCr_{1-x}Al_xF_4$ (x = 0∼0.06). Because the gapless spin-liquid ground state is realized in $CsCrF_4$ ($CsCr_{1-x}Al_xF_4$ with $x = 0$), for $x > 0$, we expected competition between the geometrical spin frustration in each triangular plane and the impurity-induced antiferromagnetic long-range order. We verified that for x = 0∼0.06 no anomaly that indicates an antiferromagnetic long-range order appeared and that the gapless spin-liquid ground state was robust with respect to the effect of nonmagnetic impurity; we suggested that this is because geometrical spin frustration exists in the equilateral triangular plane.

1. Introduction

One-dimensional Heisenberg antiferromagnets have attracted much attention because of Haldane gap systems that are characterized by the following properties: half-integer spin systems realize a gapless ground state, while integer spin systems realize a gapped ground state[1]. Recently, spin-gap systems in one-dimensional Heisenberg antiferromagnets with $S = 1/2$ such as two-leg ladders[2, 3, 4] have been extensively studied. Because an external magnetic field (H) drives the energy of low-lying magnons to zero by the Zeeman effect and induces their condensation at some critical field, field-induced antiferromagnetic long-range order that is regarded as a Bose-Einstein condensation of spin triplets occurs[5]. In contrast, a non-condensed ground state in three-leg ladders with $S = 1/2$ is realized in a gapless spin-liquid state, also known as a Tomonaga-Luttinger liquid (TLL) state[6]. TLLs are indentified by gapless low-energy properties, e.g., a k-linear energy dispersion with an incommensurate, and a spin correlation having an algebraic decay. TLLs are common in one-dimensional physics, such as in strongly correlated electron systems (solids[7], electron gases[8], and quantum wires[9]) and in trapped one-dimensional cold atoms[10]. If a periodic boundary along the rung direction joins the three-leg ladders, then triangle spin tubes form. Triangle spin tubes belong to a new category of one-dimensional Heisenberg antiferromagnets that have geometrically frustrated spin systems. According to theoretical studies[11, 12, 13, 14], two types of topological effects will cause an exotic ground state: (1) a vector-chiral long-range ordered state and (2) a gapless spin-liquid state corresponding to the TLL state.

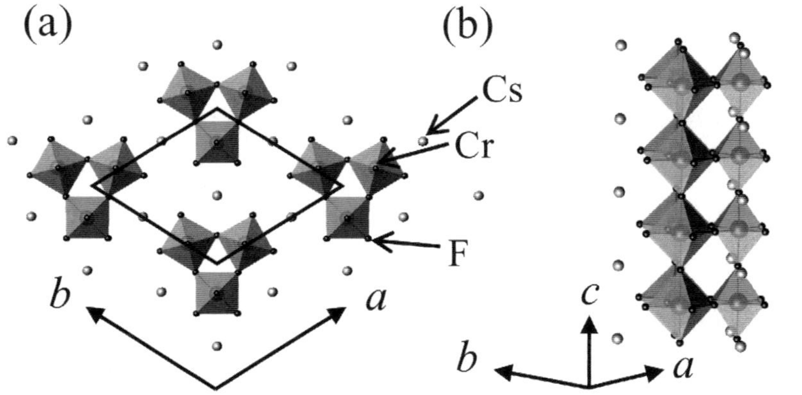

Figure 1. Schematic representation of the crystal structure of $CsCrF_4$ (a) in the *ab*-plane and (b) along the tubes (*c*-axis). Equilateral Cr_3 triangles are stacked without rotation along the *c*-axis.

A chemical impurity substitution turns a spin-gap ground state into an antiferromagnetic long-range ordered state. Several experimental studies have explored nonmagnetic-impurity-induced antiferromagnetic long-range order in two-leg ladders due to the edge spins of dangling bonds in various materials such as $Sr(Cu_{1-x}Zn_x)_3$[15] and IPA-$Cu(Cl_xBr_{1-x})_3$[16, 17]. However, few experimental studies have explored the effect of nonmagnetic impurity on the gapless spin-liquid state. In addition, nonmagnetic impurity effect weakens geometrical spin frustration. Therefore, we are interested in investigating the effect of nonmagnetic impurity on the triangle spin tubes because of the competition between the nonmagnetic-impurity-induced antiferromagnetic long-range order and the geometrical spin frustration in each triangular plane.

Recently, we discovered that chromium fluoride ($CsCrF_4$) could form ideal equilateral triangle spin tubes with $S = 3/2$[18]. Figure 1 shows a schematic representation of the crystal structure of $CsCrF_4$. All Cr ions reside at equivalent sites, and each angle of Cr-Cr-Cr in the triangular plane is 60°. The geometrical frustration among the three $S = 3/2$ spins in the equilateral triangular plane is robust with respect to nonmagnetic impurity effect. The identical Cr_3 equilateral triangles are stacked without rotation along the *c*-axis and Cs ions separate them. Thus, we expect each spin tube to show good one-dimensionality. The purpose of the present study is to determine the effect of nonmagnetic impurity on the gapless spin-liquid ground state in $CsCrF_4$, by substituting Al^{3+} ions ($S = 0$) for Cr^{3+} ions ($S = 3/2$).

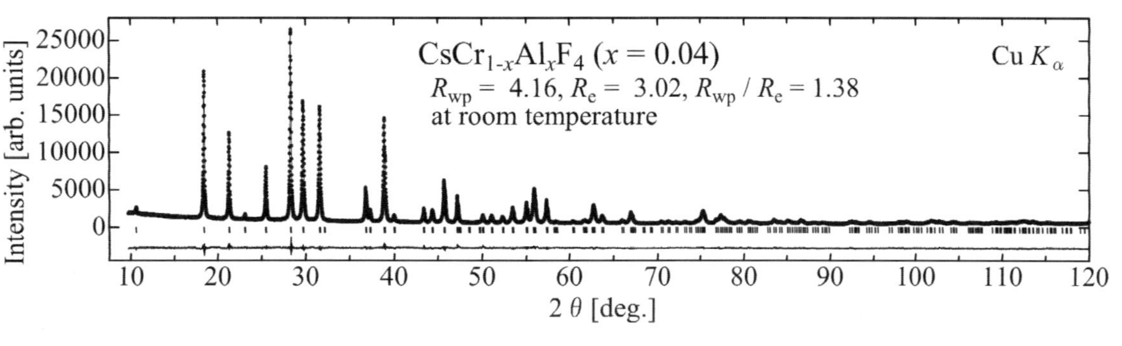

Figure 2. X-ray diffraction pattern and our Rietveld refinement results for $CsCr_xAl_{1-x}F_4$ ($x = 0.04$). Observed (circles), calculated (solid lines) profiles, and difference curve (at the bottom) along with the calculated Bragg peak positions (vertical lines at zero intensity).

2. Sample preparation

We prepared polycrystalline samples of $CsCr_{1-x}Al_xF_4$ by a conventional solid-state reaction method, which was also used to synthesize pure $CsCrF_4$. We mixed the starting materials (CsF, $CrF_3{\cdot}4H_2O$, and AlF_3) in stoichiometric proportions, and then to dehydrate the crystal water, heated them at 200 °C for more than 48 h under vacuum of $P < 1 \times 10^{-3}$ Pa. Then, we kept these samples at 550 °C for more than 50 h under a flow of N_2 gas (50 cc/min). The sample color was dark green or gray. To further purify the samples, we annealed them at 500 °C under an HF gas flow.

3. Experimental results and discussion

3.1. Crystal structures

We performed powder X-ray diffraction (XRD) experiments at room temperature to examine the crystal structure and phase purity of our samples. A Philips X'pert Pro MPD using Bragg-Brentano geometry with CuK_α radiation collected the XRD data. Figure 2 shows the XRD data of $CsCr_{1-x}Al_xF_4$ ($x = 0.04$). Furthermore, we performed Rietveld analyses of these data using RIETAN-2000 software[19]. We verified that the equilateral triangle tubes exist up to $x = 0.06$ because above $x = 0.06$, additional peaks appear and a few fundamental peaks split. Table 1 shows the lattice parameters and atomic coordinates of $CsCr_{1-x}Al_xF_4$ for $x = 0$ and 0.04 at room temperature. The $x = 0$ results (for $CsCrF_4$) are consistent with a previous report[20]. We determined that the samples up to $x = 0.06$ were of high quality, because the values of the weighted reliability factor ($R_{\rm wp}$) and the goodness-of-fit indicator ($R_{\rm wp}/R_{\rm e}$, where $R_{\rm e}$ is the expected reliability factor) were very low for $x \leq 0.06$. In addition, the values of the occupation factor of Al ions agree well with the values of x within our analytical errors.

From the results of our Rietveld analyses, as shown in Figs. 3(a) and 3(b), all lattice parameters [$a(= b)$, c, and V] decrease with increasing x, because in the case of six ligands, the radius of the Al^{3+} ion (0.68 Å) is smaller than that of the Cr^{3+} ion (0.76 Å). The direct distance between the Cr ions in the triangular plane is smaller than that along the tubes [Fig.

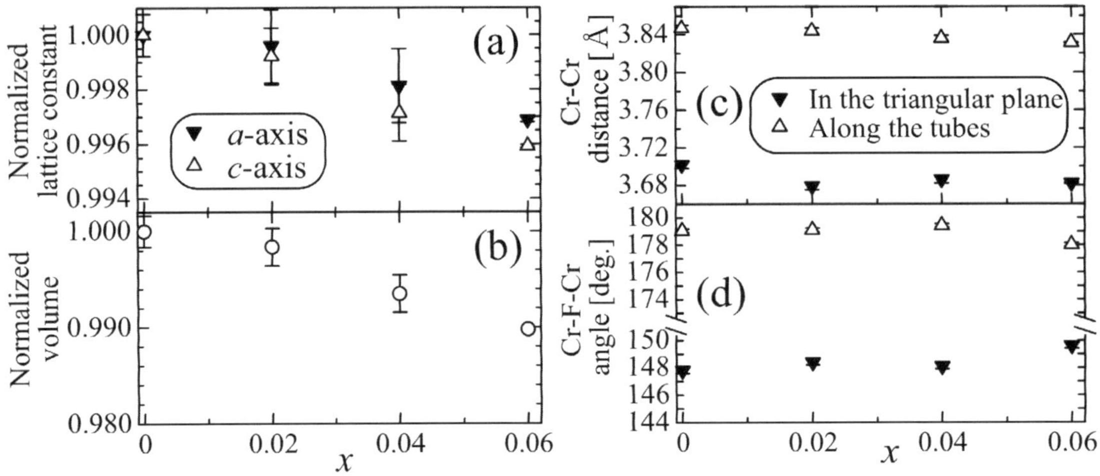

Figure 3. Results of our Rietveld analyses on $CsCr_xAl_{1-x}F_4$ for $x = 0{\sim}0.06$ at room temperature. Variation in the (a) lattice constants, (b) volume, (c) direct distance between Cr ions, and (d) Cr-F-Cr angle with x are shown. The ordinates in (a) and (b) are normalized by their respective values at $x = 0$: $a = 9.619(11)$ Å, $c = 3.847(3)$ Å, and $V = 308.2(6)$ Å^3.

Table 1. Atomic coordinates and displacement parameters of $CsCr_{1-x}Al_xF_4$ at room temperature for (a) $x = 0$ and (b) $x = 0.04$. The space group $P\bar{6}2m$ (No. 189) of $CsCr_{1-x}Al_xF_4$ remains up to $x = 0.06$. R_{wp} and R_e are the weighted and expected reliability factors, respectively, and R_{wp}/R_e is the goodness-of-fit indicator. We used the isotropic atomic displacement parameters $\exp[-8\pi^2 U_{iso}^2(\sin\theta/\lambda)^2]$ and the anisotropic atomic displacement parameters $\exp[-2\pi^2(U_{11}h^2a^{*2} + U_{22}k^2b^{*2} + U_{33}l^2c^{*2} + U_{12}hka^*b^* + U_{13}hla^*c^* + U_{23}klb^*c^*)]$ to calculate the Debye-Waller factor. For Cs ions, we applied the relations $U_{11} = U_{22}$ and $U_{13} = U_{23} = 0$ because of the space group $P\bar{6}2m$.
(a) For $CsCr_{1-x}Al_xF_4$ with $x = 0$ ($CsCrF_4$), the lattice parameters are $a = b = 9.619(11)$ Å, $c = 3.847(3)$ Å, and $V = 308.2(6)$ Å^3, resulting in R factors of $R_{wp} = 4.04$ and $R_e = 2.98$, and a goodness-of-fit indicator of $R_{wp}/R_e = 1.35$.

Atom	Wyckoff position	Occupation g	x	y	z	U_{iso}(Å^2)
Cs	$3g$	1(Fix)	0.57227(5)	0	1/2	-
Cr	$3f$	1(Fix)	0.22219(14)	0	0	0.0040(5)
F1	$3f$	1(Fix)	0.8339(3)	0	0	0.0131(7)
F2	$3g$	1(Fix)	0.2205(4)	0	1/2	$= U_{iso}$(F1)
F3	$6j$	1(Fix)	0.4311(3)	0.1538(3)	0	$= U_{iso}$(F1)

Atom	U_{11}(Å^2)	U_{22}(Å^2)	U_{33}(Å^2)	U_{12}(Å^2)	U_{13}(Å^2)	U_{23}(Å^2)
Cs	0.0400(7)	$=U_{11}$(Cs)	0.0124(6)	0.0295(6)	0	0

(b) For $CsCr_{1-x}Al_xF_4$ with $x = 0.04$, the lattice parameters are $a = b = 9.601(12)$ Å, $c = 3.836(4)$ Å, and $V = 306.2(6)$ Å^3, resulting in R factors of $R_{wp} = 4.16$ and $R_e = 3.02$, and a goodness-of-fit indicator of $R_{wp}/R_e = 1.38$.

Atom	Wyckoff position	Occupation g	x	y	z	U_{iso}(Å^2)
Cs	$3g$	1(Fix)	0.57209(5)	0	1/2	-
Cr	$3f$	0.964(5)	0.22165(14)	0	0	0.0053(6)
Al	$3f$	$= 1 - g$(Cr)	$= x$(Cr)	0	0	$= U_{iso}$(Cr)
F1	$3f$	1(Fix)	0.8343(3)	0	0	0.0145(8)
F2	$3g$	1(Fix)	0.2226(4)	0	1/2	$= U_{iso}$(F1)
F3	$6j$	1(Fix)	0.4301(3)	0.1537(3)	0	$= U_{iso}$(F1)

Atom	U_{11}(Å^2)	U_{22}(Å^2)	U_{33}(Å^2)	U_{12}(Å^2)	U_{13}(Å^2)	U_{23}(Å^2)
Cs	0.0444(8)	$= U_{11}$(Cs)	0.0145(7)	0.0329(6)	0	0

3(c)]. The Cr-F-Cr angle has a close relationship with the superexchange interaction in the triangular plane (J_{rung}) and with that along the tubes (J_{chain}). We consider $|J_{chain}| > |J_{rung}|$ because $|J_{chain}|$ is almost along the 180° path for $x = 0\sim0.06$ [Fig. 3(d)].

3.2. Magnetic properties
We measured the magnetization, M, induced by an external field ($H = 10$ Oe) between 2 K and 350 K using a superconducting quantum interference device (SQUID) magnetometer. In the present study, we define magnetic susceptibility as $\chi \equiv M/H$. We obtained the field cooling (FC) and zero-field cooling (ZFC) data by applying $H = 10$ Oe at 350 K and 2 K, respectively. Figure 4 shows the temperature (T) dependence of χ for $CsCr_{1-x}Al_xF_4$ ($x = 0$, 0.04, and 0.06).

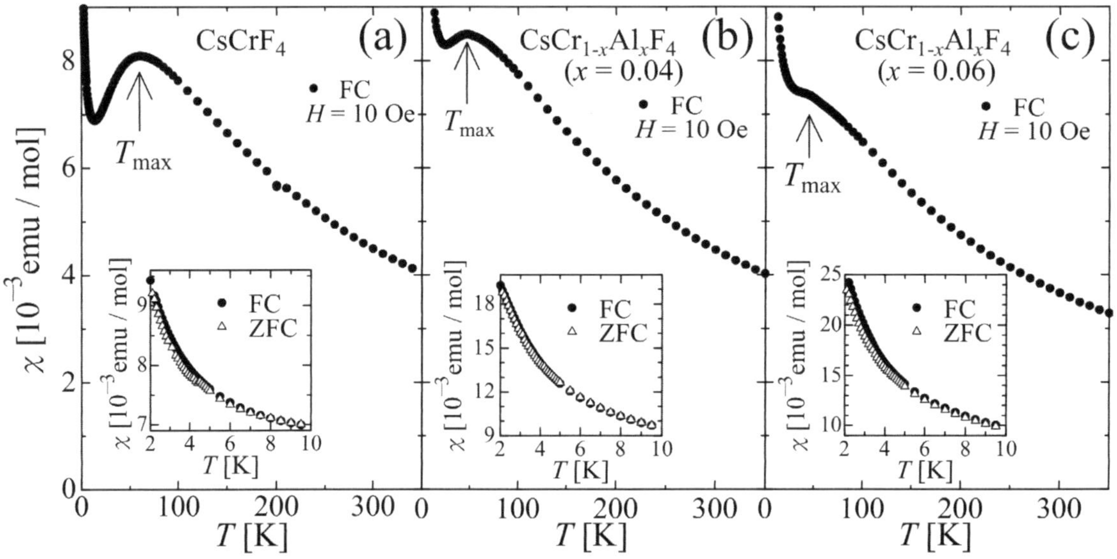

Figure 4. Temperature dependence of the magnetic susceptibility χ of $CsCr_{1-x}Al_xF_4$ for the following values of x: (a) 0, (b) 0.04, and (c) 0.06. Each broad $\chi(T)$ peak appears at T_{max}. Because the $\chi(T)$ curves for the FC and ZFC data overlap above ~4 K, the ZFC data are omitted. The insets show $\chi(T)$ in the lower-temperature region.

The $\chi(T)$ curves for the FC and ZFC data overlap above ~4 K (Fig. 4). A broad $\chi(T)$ peak that indicates one-dimensional antiferromagnets appears at T_{max}. As x increases, T_{max} decreases (Table 2), and the peak disappears due to the overlap with the Curie-like increasing $\chi(T)$. Below ~4 K (insets of Fig. 4), slight deviations between the FC and ZFC data appear, but they do not depend on the value of x; therefore, we believe that nonmagnetic impurity does not cause these slight deviations.

The $\chi(T)$ measurements for $x = 0 \sim 0.06$ (Fig. 4) indicate that gapless ground states are realized and that no anomalies indicating an antiferromagnetic long-range order above 2 K appear. The increase in $\chi(T)$ below 10 K (insets of Fig. 4) is probably due to the edge spins of dangling bonds. Assuming Curie law for the increase in $\chi(T)$, we estimated the ratio of the $S = 3/2$ free spins to the total spins (Table 2). Because the values of the ratio of the $S = 3/2$ free spins are almost equal to their respective values of x, nonmagnetic impurity effect is the major cause of the increasing $\chi(T)$. However, according to Ref. 21, a fractional spin with an $S = 1/2$ moment was observed by electron paramagnetic resonance in an ideal one-dimensional Heisenberg antiferromagnet with the $S = 1$ compound $Ni(C_2H_8N_2)_2NO_2(ClO_4)$ (NENP) doped with Cu^{2+} ions. Therefore, to verify the existence of edge spins of dangling bonds in $CsCr_{1-x}Al_xF_4$, we will perform electron paramagnetic resonance experiments in the future.

Applying Curie-Weiss law [$\chi(T) = C/(T - \Theta_{CW})$] to the $1/\chi(T)$ data above 250 K, we obtained the Weiss temperature (Θ_{CW}) and the effective magnetic moment (μ_{eff}) for $x = 0 \sim 0.06$ (Table 2). From molecular field theory, solving the Hamiltonian $\mathcal{H} = -2\Sigma_{\langle ij \rangle} J_{ij} \boldsymbol{S}_i \cdot \boldsymbol{S}_j$, the Weiss temperature is given by $\Theta_{CW} = 2zJS(S+1)/(3k_B)$, where z is the number of neighbors and J is the exchange interaction. Because this relation is valid for $T > |\Theta_{CW}|$, we carefully determined Θ_{CW} from the $1/\chi(T)$ data above 250 K. As x increases, the absolute value of $|\Theta_{CW}|$ decreases, because the number of dangling bonds increases, causing the effective value of z to decrease. Applying $\Theta_{CW} = -145(3)$ K and $z = 2$ considering $|J_{chain}| > |J_{rung}|$, we roughly estimated

Table 2. Variation in the magnetic parameters derived from the magnetic susceptibility data with x for $CsCr_{1-x}Al_xF_4$ ($x = 0 \sim 0.06$).

x	T_{max} [K]	Θ_{CW} [K]	μ_{eff} [μ_B]	Ratio of $S = 3/2$ free spins
0	$\simeq 60$	$-145(3)$	4.00(2)	0.003~0.004
0.02	$\simeq 55$	$-141(3)$	4.05(2)	0.02~0.03
0.04	$\simeq 47$	$-142(3)$	3.96(2)	0.03~0.04
0.06	~ 45	$-130(5)$	3.55(3)	0.04~0.05

$J_{chain}/k_B = -29.0(6)$ K in $CsCrF_4$. Similarly, the value of μ_{eff} decreases from 4.00(3) μ_B for $x = 0$ to 3.55(3) μ_B for $x = 0.06$; these values are consistent with the values of the free Cr^{3+} ions (3.87 μ_B).

In conclusion, we synthesized single-phase polycrystalline samples of equilateral triangle spin tubes composed of $CsCr_{1-x}Al_xF_4$, by substituting Al^{3+} ions ($S = 0$) for Cr^{3+} ions ($S = 3/2$) in $CsCrF_4$. As a result of refining the atomic coordinates by powder XRD profiles, we determined that the equilateral triangle tubes remained up to $x = 0.06$. $\chi(T)$ curves for $x = 0 \sim 0.06$ showed that no anomalies that indicated an antiferromagnetic long-range order appeared. In addition, geometrical spin frustration existed in the equilateral triangular plane, suggesting that the gapless spin-liquid ground state was robust with respect to the nonmagnetic impurity effect.

Acknowledgments

We received valuable support from the Frontier Science Research Center at Kagoshima University for performing the powder XRD experiments. This research was partially supported by the Shimadzu Science Foundation, the Kurata Memorial Hitachi Science and Technology Foundation, the SEI Group CSR Foundation, the Chubu Science and Technology Center Foundation, and Grant-in-Aid for Scientific Research on Priority Areas "Novel States of Matter Induced by Frustration" (22014009).

References

[1] Haldane F D M 1983 *Phys. Rev. Lett.* **50** 1153
[2] Masuda T, Zheludev A, Manaka H, Regnault L -P, Chung J -H, and Qiu Y 2006 *Phys. Rev. Lett.* **96** 047210
[3] Garlea V O, *et al.* 2007 *Phys. Rev. Lett.* **98** 167202
[4] Klanjšek M, *et al.* 2008 *Phys. Rev. Lett.* **101** 137207
[5] Giamarchi T, Rüegg C, and Tchernyshyov O 2008 *Nature Phys.* **4** 198
[6] Giamarchi T 2004 *Quantum Physics in One Dimension* (Oxford University Press, New York)
[7] Ogata M and Shiba H 1990 *Phys. Rev. B* **41** 2326
[8] Thomas K J, Nicholls J T, Simmons M Y, Pepper M, Mace D R, and Ritchie D A 1996 *Phys. Rev. Lett.* **77** 135
[9] Gold A and Ghazali A 2004 *Phys. Rev. B* **41** 7626
[10] Kinoshita T, Wenger T, and Weiss D S 2004 *Science* **305** 1125
[11] Kawano K and Takahashi M 1997 *J. Phys. Soc. Jpn.* **66** 4001
[12] Sato M and Sakai T 2007 *Phys. Rev. B* **75** 014411
[13] Sakai T, Sato M, Okunishi K, Otsuka K, Okamoto K, and Itoi C 2008 *Phys. Rev. B* **78** 184415
[14] Nishimoto S and Arikawa M 2008 *Phys Rev. B* **78** 054421
[15] Azuma M, Fujishiro Y, Takano M, Nohara M, and Takagi H 1997 *Phys. Rev. B* **55** R8658
[16] Manaka H, Yamada I, and Katori H A 2001 *Phys. Rev. B* **63** 104408
[17] Manaka H, Yamada I, Mitamura H, and Goto T 2002 *Phys. Rev. B* **66** 064402
[18] Manaka H, Hirai Y, Hachigo Y, Mitsunaga M, Ito M, and Terada N 2009 *J. Phys. Soc. Jpn.* **78** 093701
[19] Izumi F and Ikeda T 2000 *Mater. Sci. Forum* **321-324** 198
[20] Babel V D and Knoke G 1978 *Z. anorg. allg. Chem.* **442** 151
[21] Hagiwara M, Katsumata K, Affleck I, Halperin B I, and Renard J P 1990 *Phys. Rev. Lett.* **65** 3181

Journal of Physics: Conference Series **320** (2011) 012045 doi:10.1088/1742-6596/320/1/012045

Spin gapped behavior of a frustrated delta chain compound euchroite

Hikomitsu Kikuchi, Yutaka Fujii[1], Daisuke Takahashi, Masaki Azuma[2,5], Yuichi Shimakawa[2], Toshifumi Taniguchi[3], Akira Matsuo[4], Koichi Kindo[4]

Department of Applied Physics, Univerisity of Fukui, Fukui, Japan
[1]Research Center for Development of FIR, University of Fukui, Fukui, Japann
[2] Institute for Chemical Research, Kyoto University, Kyoto, Japan.
[3]Graduate School of Science, Osaka University, Toyonaka, Japan
[4]Institute for Solid State Physics, University of Tokyo, Kashiwa, Japan.

E-mail: kikuchi@apphy.u-fukui.ac.jp

Abstract. We nd $Cu_2(AsO_4)(OH)$ $3H_2O$ (euchroite) is a model compound for the frustrated delta chain which is composed of corner-shairng triangles. Magnetic susceptibility, speci c heat, high eld magnetization and ^{1}H-NMR measurements are carried out using natural mineral samples of euchroite to study magnetic properties. Large spin gap of about 100 K are shown to be presented in this compound.

1. Introduction

Quantum one-dimensional (1D) Heisenberg antiferromagnet (HAF) has attracted much interest because of its large quantum uctuations. New phenomena such as quantum phase transition are expected to occur if a geometrical frustration e ect is introduced into the quantum 1D HAF. Delta chain spin model is one of the frustrated 1D quantum spin systems, in which corner-sharing triangles run along 1D chain direction. Theoretical studies revealed that the delta chain system composed of regular triangle has a spin gap δE $0.22J$, where J is an exchange coupling constant[1]. Speci c heat is expected to have characteristic double peaks, and the peak at lower temperature is discussed in terms of kink-antikink soliton like excitations. Theoretical works on more general delta chain, which is composed of scalene triangles, were developed. Nakamura and his cowoker studied the delta chain with bond dimerization where one of the oblique bond is dimerized [2]. They show that the double peak structure of the speci c heat is taken over by single peak structure as increasing a degree of dimerization. A case where one of the exchange coupling is ferromagnetic is studied by Hida [3] and the ground state phase diagram is revealed to be divided into Haldane, ferromagnetic and ferrimagnetic phases. A few compounds which approximate the delta chain spin model have been found and their magnetic properties were investigated. The rst example of the delta chain is $YCuO_{2.5}$ in which large spin gap of about 650 K was found by NMR relaxation rate measurement [4]. However, some additional exchange interactions besides original delta chain interactions were found to need to explain experimental data of $YCuO_{2.5}$ [5]. $[Cu(bpy)(H_2O)][Cu(bpy)(mal)\text{-}(H_2O)](ClO_4)_2$ is a ferro-antiferromagnetic

[5] Present address : Materials and Structures Laboratory, Tokyo Institute of Technology, Tokyo, Japan.

Published under licence by IOP Publishing Ltd

delta chain whose ground state is ferrimagnetic without the spin gap [6]. These compounds thus are insu cient to be a representative compound for the delta chain. New material for the quantum delta chain is needed to investigate the properties of this spin model.

Recently, we found that Cu^{2+} ($S = 1/2$) ions in $Cu_2(AsO_4)(OH)\ 3H_2O$ (mineral name, euchroite) can be treated as the delta chain. The crystallographic structure of euchroite is orthorombic, space group $P2_12_12_1$, $a = 10.07$ A, $b = 10.52$ A, $c = 6.11$ A[7]. Its structure is shown in Fig. 1 (a). Magnetic Cu^{2+} ions are surrounded by six oxygen ions and exchange coupling between Cu^{2+} takes place through two oxygen ions. Cu^{2+} ions in euchroite form delta chain running parallel to crsytallographic c axis as schematically shown in Fig. 1 (b). Delta chains are well separated by AsO_4^3 ion groups. Bond lengths between Cu^{2+} in a unit triangle are di erent from each other (3.01, 3.06 and 3.17 A), but their di erence is within the range of several percent and not so large. Because no magnetic properties of euchroite have been reported so far, we carried out magnetic susceptibility, speci c heat, high eld magnetization and ^{1}H-NMR measurements using a natural mineral polycrystalline samples to study the magnetic properties, especially the spin gap behavior, of euchroite.

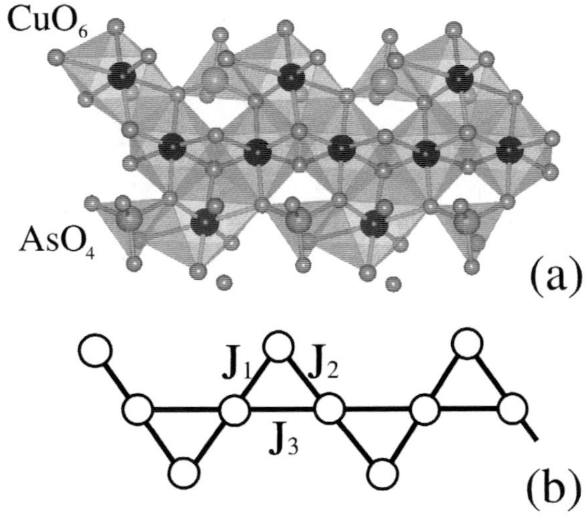

Figure 1. (a) Structure of euchroite, (b) schematic view of the delta chain.

2. Experiments

Polycrystalline natural mineral samples of euchroite were purchased at a mineral shop. Sample quality was checked using powder X-ray di raction. Impurity peaks were not observed. Magnetic susceptibility was measured using SQUID magnetometer (Quantum Design) from 2 to 350 K with an applied eld of 100 Oe. Speci c heat was measured using PPMS (Quantum Design) by relaxation method from 2 to 200 K . The high eld magnetization curve was measured up to about 55 T using a pulsed magnet at Ultra High Magnetic Field Laboratory, Institute for Solid State Physics, University of Tokyo. ^{1}H-NMR spectra and spin-lattice relaxation rate T_1 ^{1} were measured using a conventional pulsed NMR appratus down to 1.4 K.

3. Results and Discussion

Figure 2 shows the temperature dependence of the magnetic susceptibility $\chi(T)$ measured at applied eld of 100 Oe. $\chi(T)$ shows round peak at around 85 K, characteristic of low dimensional

Journal of Physics: Conference Series **320** (2011) 012045 doi:10.1088/1742-6596/320/1/012045

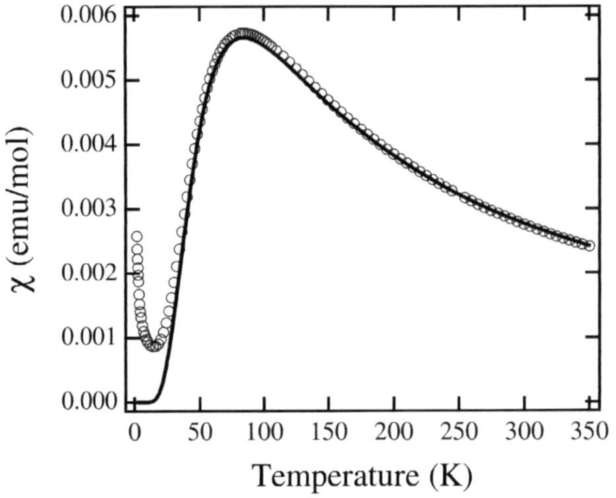

Figure 2. Temperature dependence of magnetic susceptibility of polycrystalline euchroite. The solid line is a tting curve of eq. (1) to the experimental data, with J/k_B=135 K and g =2.25.

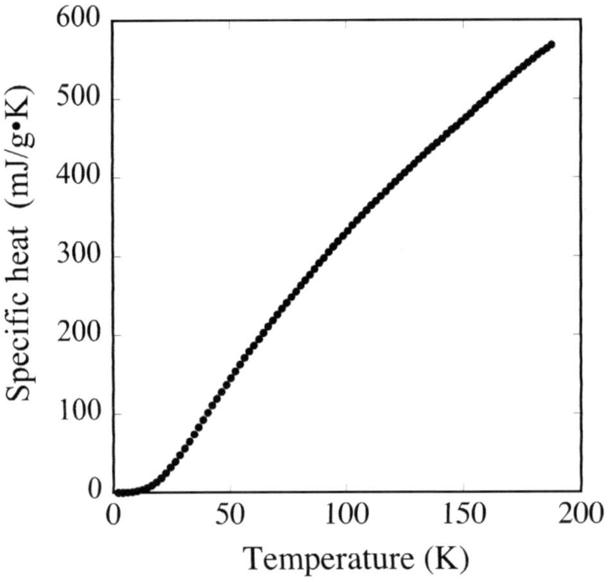

Figure 3. Temperature dependence of speci c heat of polycrystalline euchroite.

antiferromagnet, and decreases toward zero at low temperature, indicating the presence of a spin gap. Fitting to Curie-Weiss law ($\chi = C/(T$)) in higher temperature range ($>$ 150 K) gives 50 K and C = 0.51 emu K/mol. The negative Weiss temperature indicates dominant exchange coupling of euchroite is antiferromagnetic. Curie-like upturn observed in the lowest temperatures originates in very small amount (0.5 %) of magnetic impurities.

There is no theoretical calculation of magnetic susceptibility for the general delta chain whose exchange couplings are di erent from each other. At rst, we try to explain experimental data with a theoretical curve [4] for the delta chain whose all exchange couplings have same value but result was not good. This failure is expected because exchange constants of euchroite J_1, J_2 and J_3 must have di erent values each other. Next, we tentatively tried a simple equation (1) for a isolated $S = 1/2$ spin dimer system.

$$\chi_{dimer} = \frac{2Ng^2\mu_{\mathrm{B}}^2}{k_BT(3+\exp(J/k_{\mathrm{B}}T)}, \tag{1}$$

where N is Avogadro number, g is Lande's g-factor, μ_{B} is Bohr magneton, k_{B} is Boltzmann constant and J is exchange coupling between $S = 1/2$ spins. As shown in Fig. 2, eq. (1) with J/k_{B} =135 K and $g = 2.25$ reproduce fairly well the observed data. This result suggests that the ground state of euchroite is in a dimer phase [2]. Theoretical calculation of $\chi(T)$ of the general delta chain is desired to determine exchange constants quantitatively.

Figure 3 shows temperature dependence of the speicif heat $C(T)$ of euchroite. The lattice contribution is included in these data because nonmagnetic isostructural compound is not known. No sign of the magnetic long range ordering is observed. A broad peak is observed at around 80 K. Reminding that $\chi(T)$ data also show maximum at this temperature, it is natural to think that the broad peak in $C(T)$ originates in the magnetism.

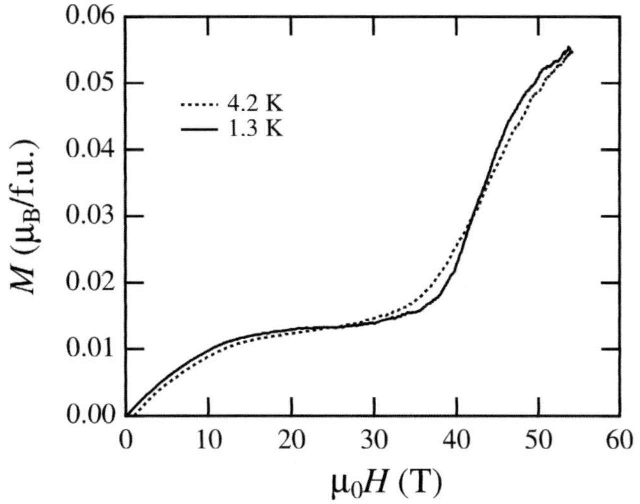

Figure 4. High eld magnetization curve of euchroite measured at 1.3 and 4.2 K up to about 55 T.

Figure 4 shows the high eld magnetization (M vs.H) curve of euchroite obtained at 1.3 and 4.2 K. Lower eld magnetization is due to the magnetic impurities and their concentration is consistent with that obtained from $\chi(T)$ data. An abrupt increase of the magnetization is observed above about 40 T. This behavior clearly indicates the presence of nite spin gap. As increasing magnetic eld, the rst excited magnetic energy crosses the nonmagnetic ground state at certain critical eld H_c and the magnetization appears. It is rather di cult experimentally to point out the accurate H_c. If we de ne H_c as the eld where the magnetization starts to increase, H_c is approximately determined to be 40 T. The energy level $_M$ of the excited state can be roughly estimated from H_c to be $_M = (g\mu_{\mathrm{B}}H_c)/k_{\mathrm{B}} = 56$ K. The magnetization tends to saturate at 50 T, suggesting that a magnetization plateau begins at about 50 T. Since full saturation magnetization M_{sat} per formula weight of euchroite is about 2 $\mu_{\rm B}$, the magnetization of this plateau will be about 1/40 of M_{sat}. To investigate detail of the new plateau, we need higher magnetic eld, which is beyond our accessible eld.

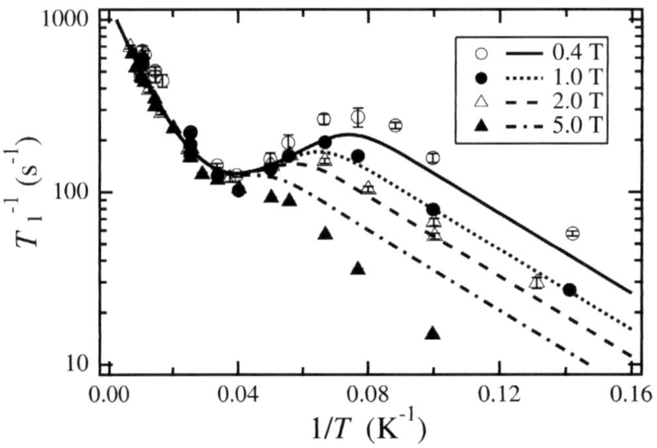

Figure 5. Temperature dependence of spin-lattice relaxation rate $T_1^{\ \ 1}$ of ^{1}H-NMR of euchroite measured at several magnetic elds. Fitting curves are calculated using eq. (3).

In order to investigate microscopic properties of euchroite, we measured [1]-H-NMR. Figure 5 shows the temperature dependence of $T_1^{\ \ 1}$ of euchroite. As decreasing temperature $T_1^{\ \ 1}$ decreases exponentially as expected for the spin gapped system. $T_1^{\ \ 1}$ above about 10 K is well tted with the Arrhenius- type equation $T_1^{\ \ 1}$ exp($/k_{\rm B}T$) with an activation energy 70 K. As further decreasing temperature, unexpected behavior is observed that $T_1^{\ \ 1}$ shows anomalous peak which shift toward lower temperatures as resonance frequency decreases. Similar behavior was reported for the S=1 1D anti eromagnet with the Haldane gap and was successfully explained by applying a general theory for the nuclear magnetic relaxation in nonmagnetic solid including paramagnetic impurities [8]. Taking into accounts of an e ect of spin di usion from the magnetic impurities to nuclear spin, following relation was derived as an impurity-induced part of the spin-lattice relaxation rate

$$T_1^{\ \ 1} \quad c_e(\frac{\tau_e}{1+\tau_e^2\omega_N^2})^{1/4}, \tag{2}$$

where c_e is impurity concentration, τ_e is the relaxation time of the impurity electron spin, ω_N is NMR angular frequency [8, 9].

If this relation is applicable to our system, temperature dependence of $T_1^{\ \ 1}$ in whole temperature region should be given as

$$T_1^{\ \ 1} = Ac_e(\frac{\tau_e}{1+\tau_e^2\omega_N^2})^{1/4} + B\exp(\quad {}_0/k_{\rm B}T), \tag{3}$$

where ${}_0$ is a modi ed activation energy and A and B are constants. The rst term of eq. (3) is impurity-indued relaxation and the second term is contribution from the bulk sample. Fitting curves with eq.(3) to the observed data measured at di erent applied magnetic eld (i. e., at di erent frequency) are shown in Fig. 5. Calculated curves agree well to the experimental data. Best t is obtained with ${}_0$ = 95±5 K. Notice that this value of ${}_0$ di ers slightly from which was obtained without considering the e ect of the impurity related relaxation. In uence of the impurity must be taken into accounts to estimate the precise value of the spin gap.

The spin gap $_0 = 95\pm5$ K obtained from NMR is larger than the gap energy $_M$ determined from the high eld magnetization. Considering that $_M$ is roughly half of $_0$, a new mid gap state can be present in this delta chain system.

4. Conclusion

We nd $Cu_2(AsO_4)(OH)\ 3H_2O$ (euchroite) is a model compound for the frustrated delta chain which is composed of corner-shairng triangles. Magnetic susceptibility, speci c heat, high eld magnetization and ^{1}H-NMR measurements are carried out using natural mineral samples of euchroite to study magnetic properties. Large spin gap of about $_0 = 90\pm5$ K are shown to be presented in this compound. The presence of a new mid-gap state is suggested.

5. Acknowledgement

This work was partly supported by a Grant-in-Aid for Scienti c Research on Priority Area (Grant No. 19052005) from the Ministry of Education, Culture, Sports, Science and Technology of Japan.

References

[1] Kubo K 1993 *Phys. Rev.* B **48** 10552
[2] Nakamura T and Kubo K 1996 *Phys. Rev.* B **53** 6393
[3] Hida K 2008 *J. Phys. Soc. Jpn.* **77** 044707
[4] Sen D, Shastry B S, Walstedt R E and Cava R 1996 *Phys. Rev.* B **53** 6401
[5] Le Bacq O, Pasturel A, Lacroix C and Nunez-Regueiro M D 2005 *Phys. Rev.* B **71** 014423
[6] Inagaki Y, Narumi Y, Kindo K, Kikuchi H, Kamikawa T, Kunimoto T, Okubo S, Ohta H, Saito T, Azuma M, Takano M, Nojiri H, Kaburagi M and Tonegawa T 2002 *J. Phys. Soc. Jpn.* **74** 2831
[7] Finney J J 1966 *Acta Cryst.* **21** 437
[8] Goto T, Satoh S, Matsumura Y and Hagiwara M 1997 *Phys. Rev.* B **55** 2709
[9] Abragam A 1961 *The Principles of Nuclear Magnetism* (Oxford: Clarendon Press) Chap. IX.

Journal of Physics: Conference Series **320** (2011) 012046 doi:10.1088/1742-6596/320/1/012046

Critical phenomena of canonical spin glass systems with large Dzyaloshinsky-Moriya anisotropy

T Taniguchi and K Makisaka

Graduate School of Science, Osaka University, Machikaneyama 1-1, Toyonaka 560-0043, Japan

ttani@ltfridge.ess.sci.osaka-u.ac.jp

Abstract. We have studied critical phenomena of spin glass transition for larger Dzyaloshinsky-Moriya anisotropy Heisenberg spin glass system, PtMn. The analysis of non-linear magnetizations in the framework of the existence of a phase transition at T_g yields the critical exponents $\delta = 2.5 \pm 0.3$, $\gamma = 1.5 \pm 0.3$ and $\beta = 1 \pm 0.3$. These results suggest that the universality class of spin glass transition on PtMn is the same as those of CuMn, AgMn and AuFe systems. Experimentally, there is an indication of a unique universality class for Heisenberg spin glass materials.

1. Introduction

In all real Heisenberg spin glass (SG) materials in addition to the pure spin–spin exchange interactions there are anisotropic interaction terms, principally of the Dzyaloshinsky–Moriya (D-M) type. When weak anisotropy terms are introduced into a numerical model, the spin sector becomes weakly re-coupled to the chiral sector and the onset of chiral ordering can be expected to be revealed through the spin behavior [1, 2, 3]. Careful and detailed experimental studies [4, 5, 6, 7, 8] established the existence of the critical divergence of the non-linear susceptibility thus demonstrating that the SG transition in laboratory samples is indeed a thermodynamic transition. The Heisenberg SG material which has been the most intensively studied is AgMn which, together with CuMn, is one of the canonical weak D-M anisotropy Heisenberg SGs [4, 5, 6]. These independent sets of careful measurements using either DC or AC techniques are entirely consistent with each other, whereas only a few experiments for the strong D-M anisotropy Heisenberg SG systems, AuMn and AuFe, have been done. It should be noted that these Au based SG systems show the same universality class of the SG transitions as that for the weak D-M anisotropy systems [7, 8]. It is very instructive to compare these experimental results with the most recent estimates from simulations for the chiral critical exponents at the pure chiral ordering transition in the 3d Heisenberg SG model. Thus the experimental exponent values for Heisenberg SG systems are very similar to the numerical values for the critical exponents at a pure chiral ordering transition.

It is well known that the D-M anisotropy of the canonical SG alloy depends on the atomic radius of the host metal [9]. The order of the magnitude of the D-M anisotropy is as follows; CuMn < AgMn < AuFe, AuMn < PtMn. Therefore the effect of the large D-M anisotropy, which might change the universality class, to the critical phenomena is the interesting problem. The purpose of this article is to obtain the critical exponents of the SG transition for PtMn, which has the largest D-M anisotropy in the canonical SG systems, and to compare the experimental results with those of the theoretical works based on the chirality scenario of SG transition [1, 2, 3].

2. Experiments and analysis

The samples used for the measurements are PtMn (8 at. % Mn) alloys. We prepared ingots of the sample alloys by melting constituent elements in an argon arc-furnace. A cylindrical sample of 6 mm in diameter and 10 mm length was cut out by a spark cutting machine from the ingot. The sample was sealed in a vacuous quartz ampoule and annealed at 1000 °C for 36 hours and quenched to iced water. The magnetization M as a function of the magnetic field H at the various temperatures around SG transition temperature T_g was measured with a commercial SQUID magnetometer MPMS-7XL. In order to accurately measure the small non-linear part of the magnetizations, we used a vibrating RSO mode in all the measurements and a magnet reset procedure for eliminating the remnant magnetic field of the superconducting magnet. The SG transition temperature, T_g=13.85 K, was determined from the cusp positions in the zero field cooled magnetization and in the AC magnetization measurements. Above high temperature region ($T > 200$ K) we obtained the effective moment, 6.0 μ_B, and the Weiss temperature, -42.5 K.

The magnetization M in the paramagnetic regime can be expanded as the odd powers of the applied field H as

$$M = \chi_0 H + \chi_2 H^3 + \chi_4 H^5 + \cdots \tag{1}$$

where χ_0 is a linear susceptibility and χ_2, χ_4 are non-linear susceptibilities respectively. The leading term of the non-linear susceptibilities χ_2 is obtained from the initial slope of M/H-χ_0 versus H^2 plot.

$$\frac{M}{H} - \chi_0 = \chi_2(H^2) + \chi_4(H^2)^2 + \cdots \approx \chi_2(H^2) = Q \tag{2}$$

It should be noted that the non-linear magnetization corresponds to the conjugate field H^2 induced SG order parameter in the small field limit. The critical exponents γ and δ of SG transition and the reduced temperature ε are defined as $\chi_2 \propto \varepsilon^{-\gamma}$ ($\varepsilon \equiv (T_g - T)/T_g$) and $Q \propto (H^2)^{1/\delta}$ at T_g.

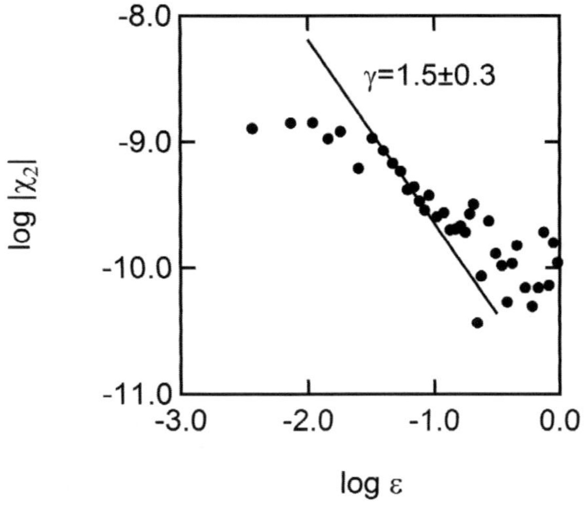

Figure 1. Reduced temperature ε dependence of the absolute value of the non-linear susceptibility χ_2 in a log-log scale. The critical exponent γ is determined as the slope of the plot shown in the figure.

Figure 1 shows the reduced temperature ε dependence of χ_2 derived from the procedure mentioned above in a log-log scale. We determined the value of $\gamma = 1.5$ from the slope of the plot. We note that the value $\gamma = 1.5$ is close to those reported for the weak anisotropy Heisenberg SG systems [2-4].

Figure 2 shows the squared field dependence of the non-linear magnetization divided by H in a log-log scale. Two asymptotic behaviors were observed and we find δ=2.5 in low fields and δ=4.9 in high fields, which was previously reported on CuMn + Au [4] system and was interpreted as a crossover from Ising to Heisenberg critical behavior of SG transition. However, it should be noted that 'fake' critical behavior is sometimes observed in the region far from the 'true' critical region. We can obtain the critical exponent β=1.0 from the scaling relation, $\beta\delta=\gamma+\beta$, of the second order phase transition.

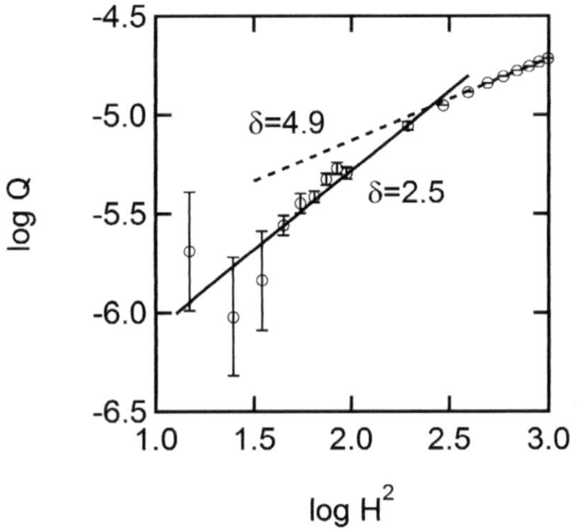

Figure 2. H^2 dependence of Q in a log-log scale. The critical exponent δ=2.5 is determined from the slope in low field region.

3. Discussion

The values of the critical exponents of SG transition was scattered in the early stage of the research [10]. The thoughtful analysis of the non-linear susceptibility in the restricted field and temperature range provides the same values of the critical exponents for the canonical SG transition at least [5, 7, 14]. Therefore, it has been thought that the SG transitions of the canonical SG systems belong to the same universality class and that they are the thermodynamic phase transition. However, a possibility that the universality class of AuFe system, which has strong D-M anisotropy, is different from the other canonical systems has recently been suggested [14].

The values of critical exponents of various canonical SG systems are summarized with this work in Table I. The exponents were obtained following the same experimental dc protocol as in the case of the AgMn, that is, measurements were all made in a restricted range of field, $\chi_2 < 0.1\ \chi_0$, and within a restricted range of temperatures, $\varepsilon < 0.1$. These conditions ensure that the estimates for the different materials are as reliable as the results for AgMn, and material-to-material comparisons are fully valid. It would be of great interest to re-measure Heisenberg SG exponents for a variety of different systems using AC and DC experimental techniques together with the improved form of data analysis, noting the 'true' critical region. In the region far from the 'true' critical region, the temperature and field dependences of non-linear magnetizations behave like the mean field one and provide the different critical exponents [5, 7, 14].

The values of critical exponents for all canonical SG alloys are almost the same, though the values of PtMn seems slightly small because of the strong antiferromagnetic interaction suggested from the large Weiss temperature in this system. The antiferromagnetic interaction surpresses the divergence of the non-linear susceptibility at T_g [15]. Therefore, we conclude that the values of the critical exponents are different from those of AuFe in [14] and that the universality class of PtMn belongs to the same universality class of the other canonical systems. These values well coincide with those of the chirality theory [13]. However, the direct observation of the critical phenomena in chiral susceptibilities, anomalous Hall effect [8, 16], is necessary for the final conclusion.

Table 1. Critical exponents of SG transition for various systems.

	β	γ	δ	Ref. No.
CuMn (+Au)	1.0	2.2	3.2	6
AgMn	0.9	2.1	3.3	4
AgMn	1.0	2.2	3.1	5
AuFe	1.0	2.0	3.0	7
AuMn	1.0	2.2	3.1	8
PtMn	1.0	1.5	2.5	This Work
(FeMn)TiO_3		4.0	8.4	11
Gaussian Ising		5.8	8.5	12
Chirality theory		2.0	2.75	13

In summary, we have studied critical phenomena of the SG transition for larger D-M anisotropy Heisenberg spin glass system, PtMn. The analysis of non-linear magnetizations in the framework of the existence of a phase transition at T_g yields the critical exponents $\delta = 2.5 \pm 0.3$, $\gamma = 1.5 \pm 0.3$ and $\beta = 1 \pm 0.3$. These results suggest that the universality class of SG transition on PtMn is the same as those of CuMn, AgMn and AuFe systems. Experimentally, there is an indication of a unique universality class for Heisenberg SG materials.

This work was supported by Grant-in-Aids for Scientific Research (No. 19540402) and on Priority Areas "Novel States of Matter Induced by Frustration" (No. 19052006).

References

[1] Kawamura H 1992 *Phys. Rev. Lett.* **68** 3785
[2] Imagawa D and Kawamura H 2004 *Phys. Rev. Lett.* **92** 077204
[3] Kawamura H 2010 *J. Phys. Soc. Jpn.* **79** 011007
[4] Levy L P and Ogielski T 1986 *Phys. Rev. Lett.* **57** 3288
[5] Bouchiat H 1986 *J. Phys. (Paris)* **47** 71
[6] de Courtenay N, Bouchiat H, Hurdequint H and Fert A 1986 *J. Phys. (Paris)* **47** 1507
[7] Taniguchi T and Miyako Y 1988 *J. Phys. Soc. Jpn.* **57** 3520
[8] Taniguchi T 2007 *J. Phys. Condens. Matter* **19** 145213
[9] Fert A, de Courtenay N and Bouchiat H 1988 *J. Phys. (Paris)* **49** 1173
[10] Binder K and Young A P 1988 *Rev Mod. Phys.* 58 801
[11] Gunnarsson K, Svedlindh P, Nordblad P, Lundgren L, Aruga H and Ito A 1991 *Phys. Rev.* B **43** 8199
[12] Katzgraber H G, Korner M and Young A P 2006 *Phys. Rev.* B **73** 224432
[13] Viet D X and Kawamura H 2010 *Phys. Rev.* B **80** 224432

[14] Campbell I A and Petit D C M C 2010 *J. Phys. Soc. Jpn.* **79** 011006
[15] Wada K and Takayama H 1980 *Prog. Theor. Phys.* **64** 327
[16] Taniguchi T *et al.* 2004 *Phys. Rev. Lett.* **93** 246605

Journal of Physics: Conference Series **320** (2011) 012047 doi:10.1088/1742-6596/320/1/012047

Magnetic properties of the triangular quantum spin tube $[(CuCl_2tachH)_3Cl]Cl_2$ studied by NMR and magnetization

Yuji Furukawa[1,2], Yuzuru Sumida[2], Ken-ichi Kumagai[2], Ferdinando Borsa[1,3], Hiroyuki Nojiri[4], Yusei Shimizu[2], Hiroshi Amitsuka[2], Ken-ichi Tenya[5], Paul Kögerler[1,6], Leroy Cronin[7]

[1]Ames Laboratory and Department of Physics and Astronomy, Iowa State University, Ames, Iowa 50011, USA

[2]Department of Physics, Graduate School of Science, Hokkaido University, Sapporo 060-0810, Japan

[3]Departmento Fisica "A. Volta", Universita di Pavis, Via Bassi 6, I27100 Pavia, Italy

[4]Institute for Materials Research, Tohoku University, Katahira 2-1-1, Sendai 980-8577, Japan

[5]Department of Physics, Faculty of Education, Shinshu University, Nagano 380-8544, Japan

[6]Institute of Inorganic Chemistry, RWTH Aachen University, 52074 Aachen, Germany

[7]Department of Chemistry, University of Glasgow, Glasgow, G12 8QQ, UK

E-mail: furukawa@ameslab.gov

Abstract. Static and dynamical properties of Cu^{2+} ($3d^9$; $S = 1/2$) spins in a twisted Heisenberg triangular spin tube, $[(CuCl_2tachH)_3Cl]Cl_2$, have been investigated by nuclear magnetic resonance (NMR) and magnetization measurements in the temperature range $T = 0.05$-300 K. Magnetization curves below 0.5 K show a clear change in slope around 5 T, attributed to a signature of 1/3 plateau in magnetization. From a systematic measurement of nuclear spin-lattice relaxation rates as a function of temperature and external field, the fluctuation frequency of Cu^{2+} spins is found to decrease with decreasing temperature, revealing an unusual slow spin dynamics at low temperatures.

1. Introduction

Magnetic properties of geometrically frustrated antiferromagnets have been attracted a great deal of interest, due to pronounced quantum phenomena, such as the suppression of the magnetic ordering and unconventional spin state like e.g. spin liquid [1]. Of particular interest is a new one-dimensional quantum spin system composed of one-dimensional arrays of Heisenberg triangular antiferromagnets,

Published under licence by IOP Publishing Ltd

called a triangular spin tube. Each Heisenberg antiferromagnetic triangle has a geometrical spin frustration effect so that the system has a new degree of freedom, spin chirality [2], in addition to the one-dimensional spin nature. Recent theoretical studies [3-6] indicate possible new quantum phenomena like a Tomonaga-Luttinger liquid state with spin chirality and spin gap formation, due to spin chirality in the quantum triangular spin tube.

$[(CuCl_2tachH)_3Cl]Cl_2$ (tach=cis,trans-1,3,5-triamino-cyclohexane), is a newly synthesized triangle spin tube [7]. With a trigonal crystal structure (space group $P6_3/m$, lattice parameters $a = b = 12.800$ Å, $c = 12.6287$ Å)[7], the triangles formed by three Cu^{2+} ($3d^9$, $S = 1/2$) ions are aligned to construct infinite stacks of one-dimensional antiprisms along the c direction. A schematic structure of the triangular copper chain is shown in Fig. 1. Each triangle rotates 180 degrees with respect to its neighboring triangles in the chain direction. From the temperature dependence of magnetic susceptibility measured between 2 and 300 K, antiferromagnetic interactions intra-triangle and inter-triangle are reported as $J_1/k_B \sim 0.9$ K and $J_2/k_B \sim 1.95$ K [8], respectively. Although initially the system was reported to have a spin singlet ground state with a spin gap $\Delta/k_B \sim 0.4$ K [8] from a magnetization curve at temperature $T = 0.5$ K, theoretical investigations indicate a gapless ground state for the twisted triangular spin tube [5-7]. Experimentally, a gapless ground state of the Cu spin tube was revealed by the observation of proton NMR line broadening at temperatures below 1 K [9]. Quite recently, specific heat measurements down to 100 mK also evidenced a gapless ground state of the Cu spin tube from the observed T linear dependence of the specific heat below 0.6 K, which is attributed to a Tomonaga-Luttinger liquid state [10].

In this paper, we report our results of the magnetization and proton NMR measurements at very low temperatures, down to 100 mK, using a ^3He-^4He dilution refrigerator. A change in slope of the magnetization curve around 5 T provides first evidence of 1/3 plateau in the magnetization. Nuclear spin lattice relaxation rate, $1/T_1$, reveals a peculiar slow spin dynamics of the Cu^{2+} spins in the system.

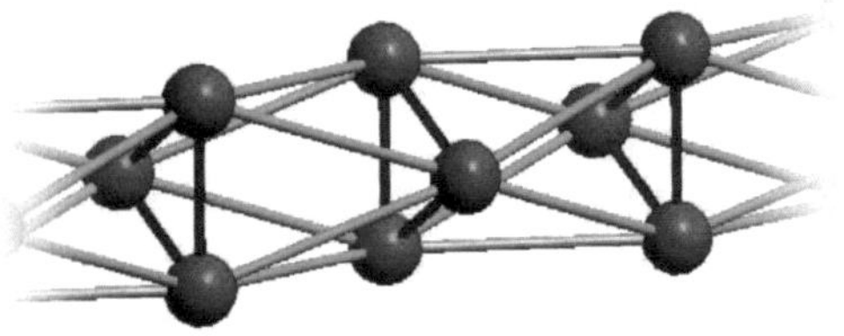

FIG 1. Schematic structure of the triangular spin tube, $[CuCl_2tachH]_3Cl]Cl_2$. The circles show Cu^{2+} ($S = 1/2$) ions. The blue and grey lines correspond to the intra-triangle exchange ($J_1/k_B \sim 0.9$ K) and inter-triangle exchange ($J_2/k_B \sim 1.95$ K) coupling paths.

2. Experimental

Polycrystalline samples of $[CuCl_2tachH]_3Cl]Cl_2$ were prepared as described in Ref. [7]. The temperature dependence of the magnetic susceptibility was measured in a temperature range of 1.8-300 K using a superconducting quantum interface device (SQUID) magnetometer (Quantum design MPMS-7T). DC magnetization measurements below 1 K were performed using a Faraday force capacitive magnetometer installed in the ^3He-^4He dilution refrigerator. The ^1H NMR measurements were achieved utilizing a phase coherent spin echo pulse spectrometer in magnetic fields of $H = 0.22$-3 T.

3. Magnetization Results and discussion

Open circles in Fig. 2 (a) show the temperature dependence of the magnetization, M, divided by the magnetic field, H, at $H = 1$ T measured by using a SQUID magnetometer down to 1.9 K. The M/H obeys the Curie-Weiss law with a Weiss temperature of ~ 1 K above 30 K and shows a Bonner-Fisher-like peak around 3 K. The observed behavior is in a good agreement with the one reported previously [7]. Temperature dependence of M/H at $H = 1$ T below 1.9 K is estimated from ^1H NMR line width [9] and from DC magnetization measurements shown by open triangles and closed circles, respectively. M/H shows a local minimum around 1 K and levels off at low temperatures, which is consistent with a gapless ground state of the system in agreement with specific heat measurements [10].

Fig. 2(b) shows the magnetization curve at T = 0.09, 0.3, 0.5 and 1.8 K. The magnetization has a small fraction which shows Brillouin function-like behavior in the low magnetic field region at low temperatures. This is attributed to spin contribution due to magnetic impurities or open end effects of spin triangle chains by defects. The fraction of the excess magnetization is estimated to be ~2 % of the total magnetization. The most important finding in our DC magnetization measurement is an obvious change in slope of the magnetization curves around 5 T below and above 0.5 K, which can be also seen in change of dM/dH as shown in the inset. This signature can be attributed to so-called 1/3 plateau in magnetization, since the magnetization around 5 T is ~ 1 μ_B/mol, which is one-third of an expected total magnetization of 3 μ_B/mol and 5 T is close to one-third of the saturation magnetic field of 14 T [8].

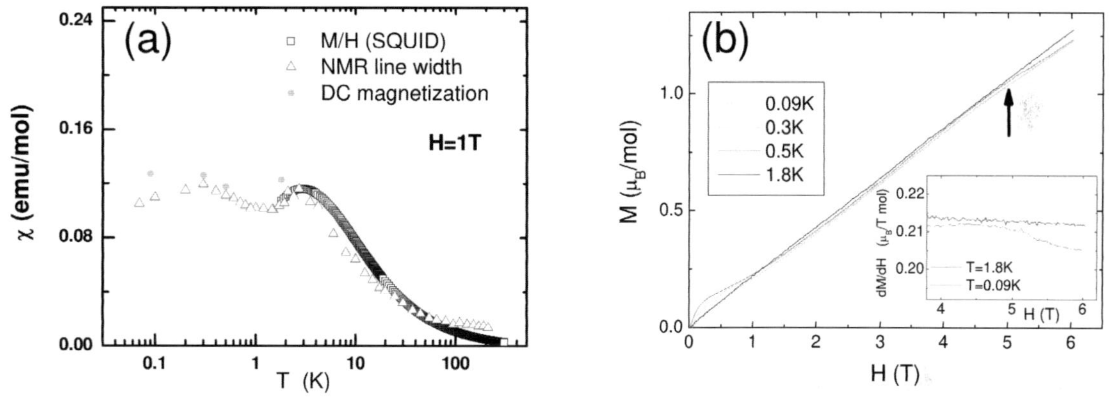

FIG 2. (a) Temperature-dependence of M/H at H = 1 T. Open squares from SQUID measurements for T = 1.8-300 K; open triangles from ^{1}H-NMR line width measurements for T = 0.1-250 K; closed circles from DC magnetization measurements for T = 0.09-1.8 K. (b) H-dependence of magnetization M at T = 0.09, 0.3, 0.5 and 1.8 K. The arrow shows the magnetic field, where the slope in magnetization changes below 0.5 K. The inset shows dM/dH for T = 1.8 K and 0.09 K around 5 T.

A phase diagram for a twisted spin tube as a function of J_2/J_1 and of magnetic field, H, has been proposed theoretically by Fouet et al.[4], using a density matrix renormalisation group calculation. When J_2/J_1 is smaller than 1.22, the system has a gapped ground state with a two-fold-degenerate state, due to spin chirality. On the other hand, if J_2/J_1 is large enough, the system is considered an effective S=3/2 chain with a gapless ground state. The phase transition between the gapped and gapless states takes place for a critical point at J_2/J_1 ~ 1.22. The 1/3 plateau in the magnetization curve is proposed below J_2/J_1 ~ 1.6. Although the ratio, J_2/J_1, is reported previously to be ~2.16 from the temperature-dependence of magnetic susceptibility, our finding of the 1/3 plateau with the gapless ground state indicates that the ratio should fall in the interval 1.22 < J_2/J_1 < 1.6, suggesting that the Cu-spin tube locates close to the critical point for the phase transition.

4. Proton spin lattice relaxation rate $1/T_1$

To investigate the dynamical properties of the Cu^{2+} spins, we completed proton T_1 measurements in a wide temperature range of T = 0.05-100 K. Figure 3(a) shows temperature-dependence of $1/T_1$ under various magnetic fields. With decreasing temperature, $1/T_1$ decreases gradually and starts to increase around 2 K, then shows a peak around 0.6 K. As the external magnetic field increases, the peak temperature of $1/T_1$ shifts to higher temperatures and at the same time the height of $1/T_1$ becomes smaller.

In general, $1/T_1$ is expressed by the Fourier transform of the time correlation function of the transverse fluctuating local field at nuclear sites as [11]

$$\frac{1}{T_1} = \frac{1}{2}\gamma_N^2 \int \langle h_\pm(t) h_\mp(0) \rangle e^{i\omega_L t} dt \quad , \tag{1}$$

with γ_N the nuclear gyromagnetic ratio and ω_L the Larmor frequency. When the time correlation function is assumed to decay as exp($-\Gamma t$), the $1/T_1$ can be written as [12]

$$\frac{1}{T_1} = A\chi T \frac{\Gamma}{\Gamma^2 + \omega_L^2} \quad , \tag{2}$$

where A is a parameter related to the hyperfine field and Γ corresponds to the inverse of the correlation time of the fluctuating hyperfine fields at proton sites due to the Cu^{2+} spins. If Γ is independent of temperature, the temperature-dependence of $1/T_1$ is simply expressed by χT. The open triangles in Fig. 3(a) show T-dependence of χT estimated from the T-dependence of NMR line width. As can be seen in Fig. 3(a), the T-dependence of $1/T_1$ scales with that of χT above 2 K. The magnetic field-dependence of $1/T_1$ in the region originates from the one-dimensional diffusive nature of spin correlation function, characterized by a $H^{1/2}$ dependence of $1/T_1$ [13,14]. Thus, nuclear spin relaxation above ~ 2 K is explained by the paramagnetic fluctuations of the Cu^{2+} spins and its fluctuation frequency Γ is independent of temperature.

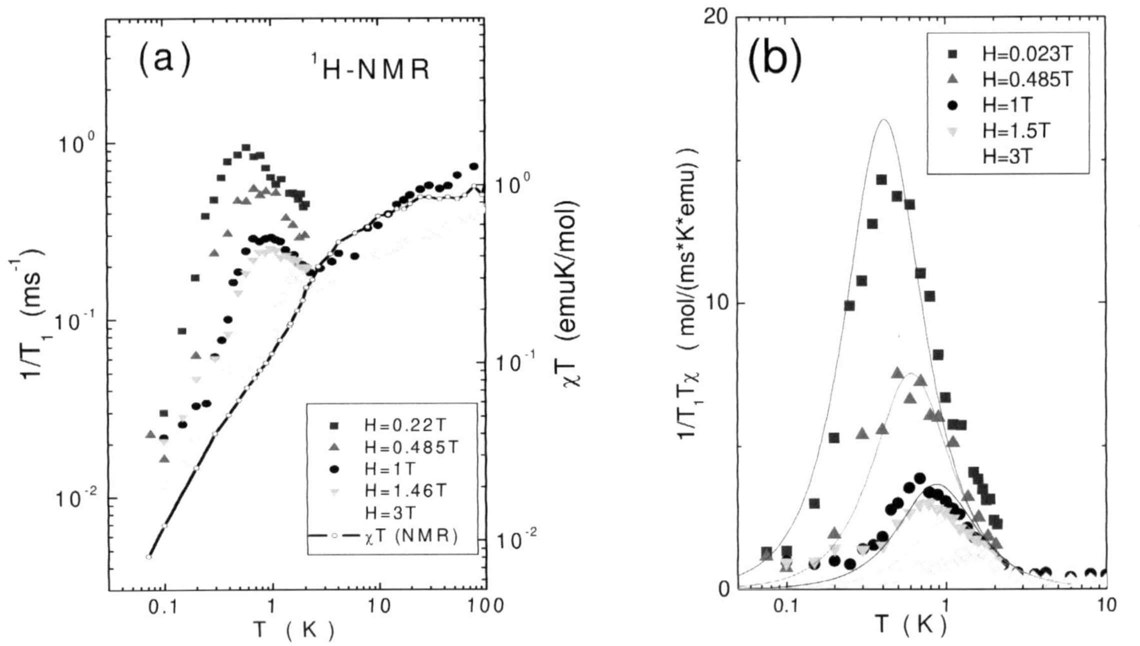

FIG 3. (a) Temperature-dependence of $1/T_1$ at various magnetic fields. Temperature-dependence of χT (at H = 1 T) estimated from NMR line width is plotted by open triangles. (b) Temperature-dependence of $1/T_1 T\chi$. Solid lines are calculated results of a model described in the text.

Below 2K, the simple paramagnetic fluctuations model cannot reproduce the experimental results. To analyze the T- and H-dependencies of $1/T_1$ at low temperatures by using Eq. (2), it is useful to re-plot the data by changing the vertical axis from $1/T_1$ to $1/T_1T\chi$ as shown in Fig. 3(b), where the χT values are from NMR line width data. According to Eq. (2), $1/T_1T\chi$ is proportional to $1/\Gamma$ when $\Gamma >> \omega_L$ (fast-motion regime), while $1/T_1T\chi$ is proportional to Γ/ω_L^2 in the case of $\Gamma << \omega_L$ (slow-motion regime). When $\Gamma = \omega_L$, $1/T_1T\chi$ shows a peak. Assuming $A = 1.96\times 10^{12}$ (rad^2·K·emu/mol/s^2) and $\Gamma = 3.48\times 10^8 T^2$ (rad/s), the experimental results are well reproduced by Eq. (2), as shown in Fig. 3(b) by solid lines for different magnetic fields. These results indicate that the peak observed in T-dependence of $1/T_1T\chi$ originates from a crossover between the fast-motion regime and the slow-motion regime, whereby the fluctuation frequency of Cu^{2+} spins below the peak temperature is less than the NMR frequency range which is of the order of MHz.

For a S = 1/2 one-dimensional quantum spin system with a gapless ground state, $1/T_1$ shows temperature-independent behavior at high temperatures and logarithmic increase at low temperatures without any crossover between fast-motion regime and slow-motion regime [15,16]. In the Tomonaga-Luttinger liquid state, $1/T_1$ is expected to be proportional to $1/T^{-0.5}$ [17]. These predictions are in sharp contrast with the present experimental results. On the other hand, similar slow spin dynamics are observed in some spin frustrated systems, such as the two-dimensional spin frustrated system $NiGa_2S_4$ [18] and in molecular magnets such as the spin ball Fe30 molecule [19]. Although the origin of slow spin dynamics in the current system is not clear at the present time, this unusual behavior could originate from spin chirality effects due to spin frustrations.

5. Conclusions

We have investigated the magnetic properties of $[CuCl_2tachH]_3Cl]Cl_2$, a Heisenberg triangular spin chain, by means of magnetic susceptibility (T = 1.8-300 K), DC magnetization (T = 0.09-1.8 K), and proton NMR (T = 0.05-300 K). The observation of 1/3 plateau in the magnetization curve below 0.5K suggests that the exchange coupling constant ratio J_2/J_1 should be in the range $1.22 < J_2/J_1 < 1.6$, and thus that the Cu-spin tube locates close to the critical point for the phase transition. T and H-dependences of ^{1}H-T_1 indicate that Cu spin fluctuations slow down at low temperature to a frequency range less than MHz.

Acknowledgments

The present work was in part supported by Grant-in-Aid for Scientific Researches (A) (No. 2024405) from the Ministry of Education, Culture, Sports, Science, and Technology of Japan. One of the authors (H.N.) has been supported by GCOE program: Material Integration. Ames Laboratory is operated for U.S. Department of Energy by Iowa State University under Contract No. DE-AC02-07CH11358. This work at Ames Laboratory was supported by the Office of Basic Energy Science. The work in Pavia was supported by NOE -MAGMAGNET.

References

[1] For example, *Frustrated Spin Systems*, edited by Diep H T, World Scientific, Singapore, 2005.
[2] Kawamura H 1998 *J. Phys.: Condensed Matter* 10 4707
[3] Sato M, and Sakai T 2007 *Phys. Rev B* **75** 014411
[4] Fouet J-B, Läuchli A, Pilgram S, Noack R M, and Mila F 2005 *Phys. Rev. B* **73** 014409
[5] Okunishi K, Yoshikawa S, Sakai T, and Miyashita S 2005 *Prog. Theo. Phys.* Supp. **159** 297
[6] Sakai T, Sato M, Okunishi K, Otsuka Y, Okamoto K, and Itoi C 2008 *Phys. Rev. B* **78** 184415
[7] Seeber G, Kögerler P, Kariuki B M, and Cronin L 2004 *Chem. Commun.* 1580
[8] Schnack J, Nojiri H, Kögerler P, Cooper G T J, and Cronin L, 2004 *Phys. Rev. B* **70** 174420
[9] Furukawa Y, Sumida Y, Kumagai K, Nojiri H, Kögerler P, and Cronin L 2009 *J. Phys. Conf. Ser.* **150** 042036
[10] Ivanov N B, Schnack J, Schnalle R, Richter J, Kögerler P, Newton G N, Cronin L, Oshima Y, and Nojiri H 2010 *Phys. Rev. Lett.* **105** 037206

Journal of Physics: Conference Series **320** (2011) 012047 doi:10.1088/1742-6596/320/1/012047

[11] Abragam A, *The Principle of Nuclear Magnetism* (Clarendon Press, Oxford, 1961)
[12] Baek S H, Luban M, Lascialfari A, Micotti E, Furukawa Y, Borsa F, Slageren J van, and Cornia A 2004 *Phys. Rev. B* **70** 134434
[13] Borsa F, and Mali M 1974 *Phys. Rev. B* **9** 2215
[14] Details of analysis of H dependence of $1/T_1$ will be published eleswhere.
[15] Sachdev S 1983 *Phys. Rev. B* **50** 13
[16] Takigawa M, Motoyama N, Eisaki H, and Uchida S 1996 *Phys. Rev. Lett.* **76** 4612
[17] Goto T, Ishikawa T, Shimaoka Y, and Fujii Y 2006 *Phys. Rev. B* **73** 214406 and references therein
[18] Takeya H, Ishida K, Kitagawa K, Ihara Y, Onuma K, Maeno Y, Nambu Y, Nakatsuji S, MacLaughlin D E, Koda A, and Kadono R 2008 *Phys. Rev. B* **77** 054429
[19] Schröder C, Fang X, Furukawa Y, Luban M, Prozorov R, Borsa F, and Kumagai K 2010 *J. Phys : Condensed Matter* **22** 216007

Low-temperature ^{31}P NMR study of the two-dimensional frustrated square lattice compound $BaCdVO(PO_4)_2$

Beas Roy[1], Yuji Furukawa[1], Ramesh Nath[1,2], David C. Johnston[1]

[1]Ames Laboratory and Department of Physics and Astronomy, Iowa State University, Ames, Iowa 50011, USA

[2]Indian Institute of Science Education and Research, Trivandrum-695016 Kerala, India

E-mail: broy@iastate.edu

Abstract. We report the ^{31}P NMR spectrum and spin-lattice relaxation rate $1/T_1$ as a function of temperature in the two-dimensional frustrated square lattice compound $BaCdVO(PO_4)_2$. The temperature dependence of $1/T_1$ indicates the existence of antiferromagnetic ordering at $T_N \sim 1.05$ K which is also evidenced by the broadening of the NMR spectrum below that temperature. The temperature dependences of $1/T_1$ and $1/T_1T\chi$ just above T_N demonstrate the importance of two-dimensional antiferromagnetic spin correlations for critical fluctuations.

1. Introduction

Geometrical frustration is the source of a variety of fascinating phenomena. Recently much attention has been given to the study of the frustrated $S = ½$ square lattice (FSL), also known as the J_1- J_2 model, where the frustration is caused by a next-nearest-neighbor exchange interaction J_2, along the diagonal of the square, competing with the nearest-neighbor exchange interaction J_1, along the side of the square. The spin Hamiltonian for this model is given by:

$$\mathrm{H} = (k_B J_1)\sum_{<ij>} \vec{S}_i \cdot \vec{S}_j + (k_B J_2)\sum_{<ik>} \vec{S}_i \cdot \vec{S}_k \, , \qquad (1)$$

where the first sum is over nearest-neighbor spin pairs and the second is over next-nearest-neighbor spin pairs. From the theoretical studies [1-3], an interesting phase diagram for this model has been proposed which includes three different magnetically ordered phases: ferromagnet [FM, wave vector $\mathbf{Q}_{FM} = (0, 0)$], Neel antiferromagnet [NAF, $\mathbf{Q}_{NAF} = (\pi, \pi)$ and columnar antiferromagnet [CAF, $\mathbf{Q}_{CAF} = (\pi, 0)$ or $(0, \pi)$]. In the classical point of view, first-order phase transition occurs at the phase boundaries. But quantum fluctuations destroy long range magnetic ordering and lead to a disordered ground state at the boundaries. A gapped spin singlet state is proposed for $J_2/J_1 \sim 0.5$, while a gapless nematic state is suggested in the region $-0.4 < J_2/J_1 < -0.7$.

The vanadium phosphate $BaCdVO(PO_4)_2$ is considered a prototype of the J_1-J_2 model. The compound has a layered crystal structure with square-like arrangements of V^{4+} ($3d^1$, $S = ½$) cations. The exchange interactions J_1 and J_2 between V^{4+} spins were reported to be $J_1 = -3.6$ K (ferromagnetic) and $J_2 = 3.2$ K (antiferromagnetic) from magnetic susceptibility and magnetization measurements [4]. A long-range antiferromagnetic (AFM) ordering at $T \sim 1$ K is reported from specific heat [4], magnetization [4,5] and muon spin resonance measurements [6]. Based on the ratio

$J_2/J_1 \sim -0.9$, the system is considered to be located close to the phase boundary between CAF and the gapless nematic phase. In this paper, we have carried out ^{31}P nuclear magnetic resonance (NMR) measurements at very low temperatures down to 100 mK in order to investigate magnetic properties of $BaCdVO(PO_4)_2$.

2. Experimental

The polycrystalline samples of $BaCdVO(PO_4)_2$ were synthesized following the same procedure as has been mentioned in Ref. [4]. The NMR measurements were conducted using pulsed NMR techniques on ^{31}P (nuclear spin $I = ½$ and gyromagnetic ratio $\gamma_N/2\pi = 17.237$ MHz/T) nuclei in the temperature range of $0.1 \text{ K} \leq T \leq 300$ K using a ^3He-^4He dilution refrigerator (Kelvinox MX100) installed at Ames Laboratory. There are two inequivalent phosphorus sites (P1 and P2) in the crystal structure. The $[VOPO_4]$ layers, which extend parallel to the *ab*-plane, are formed by the corner-sharing $V^{4+}O_5$ pyramids with $P1O_4$ tetrahedra while the other site forms isolated $P2O_4$ tetrahedra lying between $[VOPO_4]$ layers. Thus, the P1 site is expected to be strongly coupled and the P2 site weakly coupled to the V^{4+} spin. The nuclear spin-lattice relaxation rate $1/T_1$ was measured by the single saturation pulse method. The nuclear recovery slightly deviated from a single exponential function so that T_1 was determined by fitting the recovery curves versus time t with a stretched exponential function, $\exp[-(t/T_1)^{\alpha}]$. The magnetic susceptibility $\chi(T)$ was measured at two different magnetic fields (1 T and 5 T) using a commercial (Quantum Design) SQUID (Superconducting Quantum Interference Device) magnetometer.

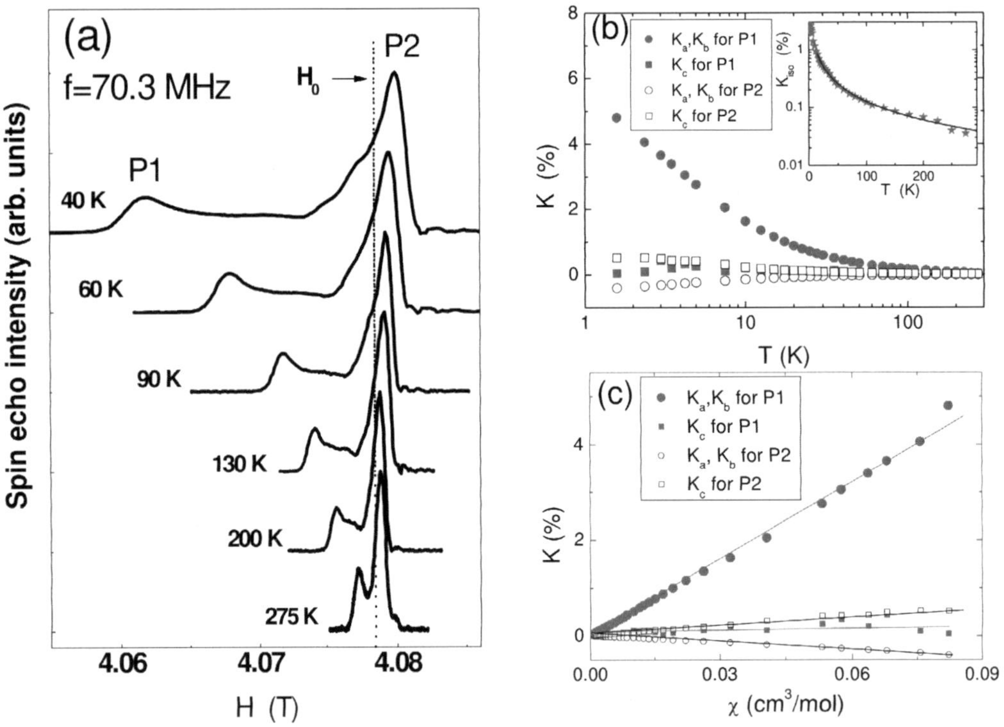

FIG. 1: (a) ^{31}P NMR spectra at different temperatures. The vertical line corresponds to the Larmor field H_0 determined by the reference sample H_3PO_4. (b) Temperature dependence of the NMR shift K (a, b and c axes) for each P site. The inset shows temperature dependence of K_{iso} for the P1 site. The solid line is the fit of K_{iso} by Eqs. (2) and (3). (c) K–χ plot for all three orientations for each P site.

3. Experimental results and discussion

Figure 1(a) shows the ^{31}P NMR powder spectrum at frequency $f = 70.3$ MHz at various temperatures. The two lines observed are due to the crystallographically inequivalent P sites, P1 and P2, and the spectrum is similar to the one observed in the isostructual $Pb_2VO(PO_4)_2$ [7]. The narrow line around the Larmor field H_0 can be assigned to the P2 site lying between vanadium layers and the other broad line is from the strongly-coupled P1 site. Each line has a characteristic shape of the anisotropic powder pattern. With decreasing temperature, both lines broaden and the broad line shifts to lower magnetic field, while the narrow line shifts to slightly higher magnetic field.

Figure 1(b) shows temperature T dependences of the NMR shifts K. For the P1 site, the shifts along the a and b axes are the same within our experimental uncertainty and strongly positive and the shift along the c axis is very small. For the P2 site, there is a relatively small negative shift along the a (and b) axis and a small positive shift along the c axis. The NMR shift varies linearly with the spin susceptibility χ, and $K(T)$ can be generally expressed in terms of $\chi(T)$ as

$$K(T) = K_0 + \frac{H_{\mathrm{hf}}}{\mu_{\mathrm{B}} N_{\mathrm{A}}} \chi(T), \qquad (2)$$

where K_0 is the temperature-independent chemical (orbital) shift, N_A is the Avogadro number, H_{hf} is the hyperfine field at the P site in Oe produced by the V^{4+} $S = ½$ spins and χ/N_A is expressed in units of μ_B/(Oe·spin). Shown in Fig. 1(c) is the K-χ plot with T as the implicit parameter. From the slope of the K-χ plot, the hyperfine fields at the P sites are estimated to be $H_a = H_b = (3045 \pm 32)$ Oe and $H_c = (149 \pm 21)$ Oe for P1. For P2, $H_a = H_b = -(264 \pm 2)$ and $H_c = (378 \pm 6)$ Oe. The isotropic hyperfine field H_{iso} [$= (H_a + H_b + H_c)/3$] is then (2080 ± 28) Oe for P1 and $-(50 \pm 3)$ Oe for P2. The isotropic component originates from the isotropic transferred hyperfine interaction (THI) from the V spin moments. The THI between the V and P1 ions is coming from the P(3s)-O(2p)-V(3d) covalent bond in the [$VOPO_4$] layers. The THI at the P2 site evidences that the V 3d spins polarize the 3s electron spins at the P2 site via covalent bonds between the O and Pb atoms, which will be responsible for three-dimensional magnetic interactions in the system. These hyperfine fields are compared with $H_a = (1882 \pm 40)$ Oe, $H_b = (1251 \pm 42)$ Oe and $H_c = -(1642 \pm 55)$ Oe for P1 site in the isostructural $Pb_2VO(PO_4)_2$ and its anisotropy in ab plane is due to the distortion in the square lattice with lower crystal structure [7]. On the other hand we did not detect any anisotropy in the ab plane in $BaCdVO(PO_4)_2$ within our experimental resolution. This indicates that the structural deviation from the square lattice is smaller in $BaCdVO(PO_4)_2$ than in $Pb_2VO(PO_4)_2$.

In order to estimate the exchange couplings J_1 and J_2, we fitted the temperature T dependence of the isotropic part of the NMR shift for P1 above 15 K by Eq. (2) where $\chi(T)$ is the high-temperature series expansion for the FSL model: [8]

$$\chi(T) = \frac{N_{\mathrm{A}} g^2 \mu_{\mathrm{B}}^2}{k_{\mathrm{B}} T} \sum_n \left(\frac{J_1}{k_{\mathrm{B}} T} \right)^n \sum_m c_{m,n} \left(\frac{J_2}{J_1} \right)^m, \qquad (3)$$

where $c_{m,n}$ are the coefficients listed in Table 1 of Ref. [8]. The temperature dependence of K_{iso} is well reproduced with the set of parameters $J_1 = -3.35$ K and $J_2 = 3.53$ K. These values are in good agreement with those reported previously [4].

Figure 2 shows the temperature dependence of the ^{31}P NMR spectrum below 4 K at $f = 13.8$ MHz. The P1 NMR line was found to broaden abruptly below $T_N \sim 1.05$ K and shoulder-like features are observed on either side of the P2 line. The broadening of the lines is due to the presence of the internal field at the P sites. The symmetric broadening indicates staggered moments in the antiferromagnetically ordered state below T_N which is consistent with the CAF ordered state. The inset of Fig. 2 shows the temperature dependence of the line-width (ΔH) for the P2 site, where a sudden increase of ΔH just below T_N can be observed. Since the ΔH is proportional to the V^{4+} sublattice magnetization, it is generally possible to estimate a critical exponent for the magnetic transition which depends on the spin and lattice dimensionalities. However we did not attempt to estimate it because of lack of a sufficient number of data points close to T_N. In order to obtain an accurate critical exponent, it is obviously needed to measure more data points close to T_N. This is currently in progress.

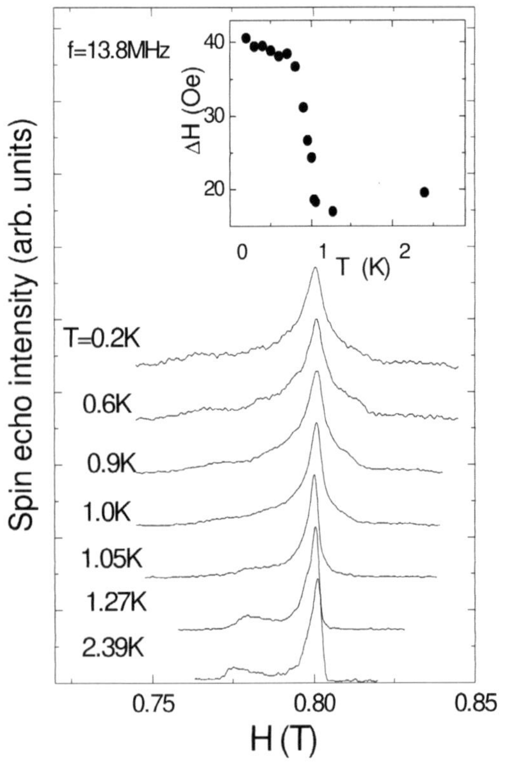

FIG. 2 Temperature dependence of the ^{31}P NMR spectrum at f = 13.8 MHz. The inset shows the temperature dependence of ΔH for the P2 site.

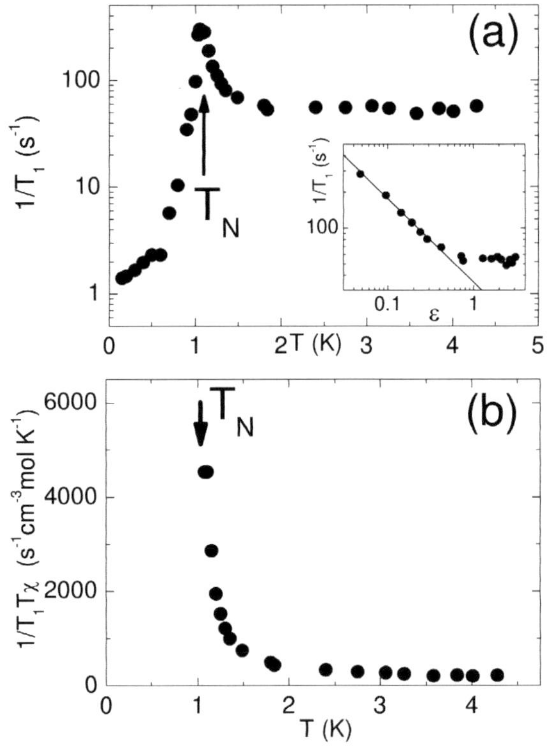

FIG. 3(a) Temperature dependence of $1/T_1$ for the P2 site at 13.8 MHz. The inset shows the plot of $1/T_1$ vs $\varepsilon = (T - T_N)/T_N$. The solid line is a fit as described in the text. (b) $1/T_1T\chi$ plotted as a function of temperature .

Figure 3(a) shows the temperature dependence of $1/T_1$ measured at the peak position of the narrow line (P2 site). Above 2 K, $1/T_1$ is temperature independent due to the exchange-narrowed hyperfine interactions between the V spin moments in the paramagnetic state. With lowering temperature close to T_N = 1.05 K, $1/T_1$ increases and shows a divergent behavior due to critical slowing down of spin fluctuations. As shown in the inset of Fig. 3(a) the divergent behavior of $1/T_1$ is approximately fitted by $1/T_1 \sim \varepsilon^{-\nu}$ with $\varepsilon = (T - T_N)/T_N$ ($T_N = 1.05$ K) and ν is estimated to be ~ 0.7 which is close to the value of 0.8 expected for a two-dimensional (2D) Heisenberg system [9].

In general, critical fluctuations at the antiferromagnetic wave vector dominate near the antiferromagnetic ordering temperature. To see these effects, it is useful to re-plot the data by changing the vertical axis from $1/T_1$ to $1/T_1T\chi$ as shown in Fig. 3(b). $1/T_1T$ can be expressed in terms of the dynamic susceptibility $\chi_M(\vec{q},\omega_0)$ per mole of electronic spins as [7, 10]

$$\frac{1}{T_1T} = \frac{2\gamma_N^2 k_B}{N_A^2} \sum_{\vec{q}} |A(\vec{q})|^2 \frac{\chi''_M(\vec{q},\omega_0)}{\omega_0}, \qquad (4)$$

where the sum is over the wave vectors $\vec{q}$ within the first Brillouin zone, $A(\vec{q})$ is the form factor of the hyperfine interactions and $\chi''_M(\vec{q},\omega_0)$ is the imaginary part of the dynamic susceptibility at the Larmor frequency ω_0. A plot of $1/T_1T\chi$ versus T shows the relative magnitude of $\sum_{\vec{q}} |A(\vec{q})|^2 \chi''_M(\vec{q},\omega_0)$ as compared to the uniform susceptibility $\chi(0,0)$. For high temperatures above 3 K, $1/T_1T\chi$ is a constant showing that the temperature dependence of $\sum_{\vec{q}} |A(\vec{q})|^2 \chi''_M(\vec{q},\omega_0)$ is equivalent to that of $\chi(0,0)$. With decrease in temperature, $1/T_1T\chi$ starts to increase. This implies dominance of $\sum_{\vec{q}} |A(\vec{q})|^2 \chi''_M(\vec{q},\omega_0)$ over $\chi(0,0)$, which is due to a growth of antiferromagnetic correlations. Thus we conclude that 2D antiferromagnetic spin fluctuations play an important role for driving the long range antiferromagnetic ordering in the system. Since the AFM ordered state must be columnar type, it is most likely that the spin correlations originate from the $\vec{q} = Q_{CAF}$ component.

Below T_N, $1/T_1$ decreases sharply and levels off at low temperatures below ~ 0.3 K. Similar behavior is observed in the temperature dependence of the muon relaxation rate λ in $BaCaVO(PO_4)_2$ [6]. The constant relaxation rates at low temperatures indicate temperature independent spin fluctuations, which is characteristic of a frustrated spin system with a degenerate ground state.

4. Conclusions

We have carried out ^{31}P NMR measurements in the 2D frustrated square lattice compound $BaCdVO(PO_4)_2$ down to 0.1 K using a ^3He-^4He dilution refrigerator. ^{31}P spin-lattice relaxation rates ($1/T_1$) measured at $H = 0.8$ T are almost independent of temperature above 2 K, show a peak at ~1.05 K and become constant below 0.4 K. The temperature dependence of $1/T_1$ indicates the existence of antiferromagnetic ordering at $T_N \sim 1.05$ K which is also evidenced by the broadening of the NMR spectrum below that temperature. The temperature dependences of $1/T_1$ and $1/T_1T\chi$, just above T_N, provide a prominent evidence of two-dimensional antiferromagnetic spin correlations for critical fluctuations.

5. Acknowledgments

Ames Laboratory is operated for U.S. Department of Energy by Iowa State University under Contract No. DE-AC02-07CH11358. This work at Ames Laboratory was supported by the Office of Basic Energy Science.

References

[1] Shannon N, Schmidt B, Penc K, and Thalmeier P 2004 *Eur. Phys. J. B* **38** 599
[2] Shannon N, Momoi T, and Sindzingre P 2006 *Phys. Rev. Lett.* **96** 027213
[3] Schmidt B, Shannon N, and Thalmeier P 2007 *J. Phys.:Condens. Matter* **19** 145211
[4] Nath R, Tsirlin A A, Rosner H, and Geibel C. 2008 *Phys. Rev. B* **78** 064422

[5] Tsirlin A A, Schmidt B, Skourski Y, Nath R, Geibel C, and Rosner H 2009 *Phys. Rev. B* **80** 132407
[6] Carretta P, Filibian M, Nath R, Geibel C, and King P J C, 2009 *Phys. Rev. B* **79** 224432
[7] Nath R, Furukawa Y, Borsa F, Kaul E E, Baenitz M, Geibel C, and Johnston D C 2009 *Phys. Rev. B* **80** 214430
[8] Rosner H, Singh R R P, Zheng W H, Oitmaa J, and Pickett W E, 2003 *Phys. Rev. B* **67** 014416
[9] Benner H and Boucher J P, *in Magnetic Properties of Layered Transition Metal Compounds*, edited by de Jongh L J (Kluwer Academic Publishers, Dordrecht, 1990)
[10] Moriya T 1963 *J. Phys. Soc. Jpn.* **18** 516

Journal of Physics: Conference Series **320** (2011) 012049 doi:10.1088/1742-6596/320/1/012049

Possibility of the field-induced spin-nematic phase in $LiCuVO_4$

M Hagiwara[1], L E Svistov[2], T Fujita[1], H Yamaguchi[1,5], S Kimura[1,6], K Omura[1], A Prokofiev[3], A I Smirnov[2], Z Honda[4]

[1] KYOKUGEN, Osaka Univ. 1-3 Machikaneyama, Toyonaka, Osaka 560-8531, Japan
[2] P. L. Kapitza Institute for Physical Problems RAS, 119334, Moscow, Russia
[3] Institut für Festkörperphysik Technische Universität Wien, A–1040 Wien, Austria
[4] Graduate School of Science and Engineering, Saitama University, 255 Shimo-okubo Saitama, Saitama 338-8570, Japan

E-mail: hagiwara@cqst.osaka-u.ac.jp

Abstract. We report on the magnetization of the frustrated S=1/2 chain compound $LiCuVO_4$. In addition to the transition from a planar spiral to a spin modulated structure observed recently by NMR, another transition was observed just below the saturation field. This magnetic phase could be a spin nematic, namely a condensation of two magnon bound states, phase which was predicted theoretically in the S=1/2 linear chain model with the nearest neighbor ferromagnetic and the next nearest neighbor antiferromagnetic exchange interactions. The slope of magnetization in this phase is in good agreement with a calculated one in a realistic quasi 2-dimensional model (M. E. Zhitomirsky and H. Tsunetsugu, Europhys. Lett. **92** 37001 (2010)). We compare the observed phase diagram with a numerically calculated one and discuss the possibility of the spin nematic phase.

1. Introduction

Unconventional magnetic orders and phases in frustrated quantum spin systems have attracted condensed matter scientists. A kind of frustration in quasi one-dimensional (1D) magnets is provided by competing interactions when the intra-chain nearest neighbor (NN) and the next-nearest neighbor (NNN) exchange interactions are ferromagnetic and antiferromagnetic, respectively. This type of exchange interactions is predicted to form bound magnon pairs in a high magnetic field. The condensation of bound magnon pairs could provide a nematic spin ordering near the saturation field which was indeed predicted by numerical and analytical calculations [1, 2, 3, 4].

$LiCuVO_4$ crystallizes in an inverse spinel structure AB_2O_4 with an orthorhombic distortion and copper ions are arranged in chains directed along the b-axis as shown in Fig. 1. These edge-sharing CuO chains are separated by the nonmagnetic ions of Li, V, O. The temperature dependence of magnetic susceptibility shows a round peak at about 28 K, which is typical of a low dimensional antiferromagnet, and the Néel ordering at T_N=2.3 K. The Curie-Weiss temperature is evaluated to be Θ_{CW}=−15 K, and $|\Theta_{CW}|/T_N \sim 6.5$ indicates a considerable effect of frustration

[5] Present address: Department of Physical Science, Osaka Prefecture Univ., 1-1 Gakuencho, Naka-ku, Sakai, Osaka 599-8531, Japan
[6] Present address:IMR, Tohoku Univ., 2-1-1 Katahira, Aoba-ku, Sendai, Miyagi 980-8577, Japan

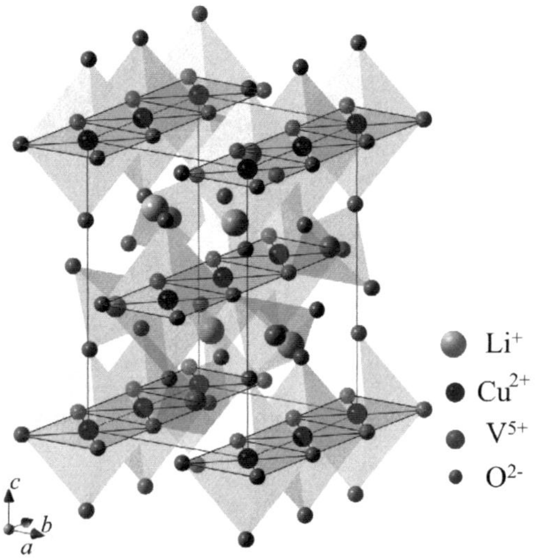

Figure 1. Crystal structure of $LiCuVO_4$. At the right bottom position, an edge-sharing copper-oxygen chain running along the b-axis is depicted.

and quantum fluctuation. It was reported from neutron diffraction experiments [5] that an incommensurate planar spiral spin structure occurs with a propagation vector k_{ic}=(0,0.532,0) directed along the b–axis at $T<T_N$=2.3 K and H=0. Inelastic neutron-scattering experiments on $LiCuVO_4$ confirmed that the NN (J_{NN}) and NNN (J_{NNN}) intrachain exchange interactions are ferromagnetic (J_{NN}/k_B=−18 K) and antiferromagnetic (J_{NNN}/k_B=49 K), respectively [6]. Accordingly, the ratio of J_{NN} to J_{NNN} defined as β is $\sim$−0.4, but a different value of β (−2) was reported from the analyses of magnetic susceptibility and the pitch angle of spiral spin structure [8].

2. Experimental

Single crystals of $LiCuVO_4$ of several cubic millimeters were prepared according to the method described in Ref. [7]. The composition of samples is characterized by the relations: Li/V=0.96±0.05, Li/Cu=0.95±0.04, Cu/V=0.99±0.01. Thus, the average composition is $Li_{0.97}CuVO_4$, which means a low concentration of vacancies in the Li-sublattice. The analysis of the quality of samples is discussed in Ref. [7]. High quality of our single crystal samples is obvious from the full width at the half of the peak height of field derivative of magnetization observed near the saturation field (about 1 T) compared to the previous work (about 5 T from Fig. 4 in Ref. [6]). Magnetization of $LiCuVO_4$ has been studied in pulsed magnetic fields of up to 55 T at KYOKUGEN in Osaka University. The magnetization was measured with an induction method using a pick-up coil and the calibration was done using the low field data (up to 7 T) taken with the use of a commercial SQUID magnetometer (Quantum design, MPMS-XL7).

3. Experimental results and discussion

Figure 2 shows the magnetization curve at 1.3 K for $H \| c$. As reported earlier in Ref. [9], besides the transition around 7 T (H_{c1}) which corresponds to the phase transition from a planar spiral to a spin modulated structure observed by NMR recently [10], an additional transition was observed at 41 T (H_{c2}) just below the saturation field at 44 T (H_{sat}). In the inset of Fig. 2, the expanded figure near the saturation field is depicted, and we found that the slope of magnetization ($\mathrm{d}M/\mathrm{d}H$) becomes gradual above 41 T.

This newly observed magnetic phase is considered as a spin nematic phase which was predicted theoretically. In Fig. 3, magnetization curve normalized by the saturation magnetization as

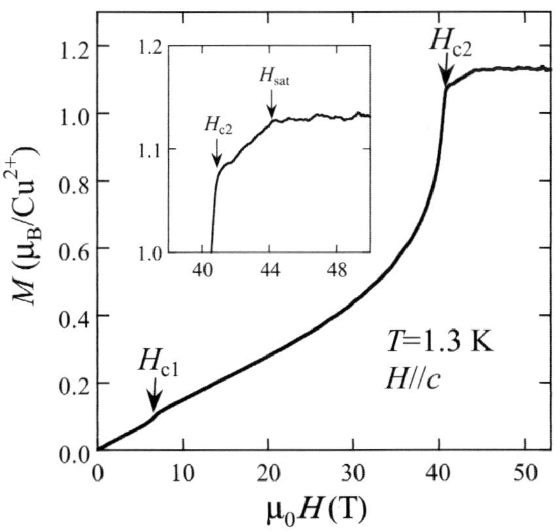

Figure 2. Magnetization curve of $LiCuVO_4$ along the c-axis at 1.3 K. In the inset, the expanded figure near the saturation field is depicted.

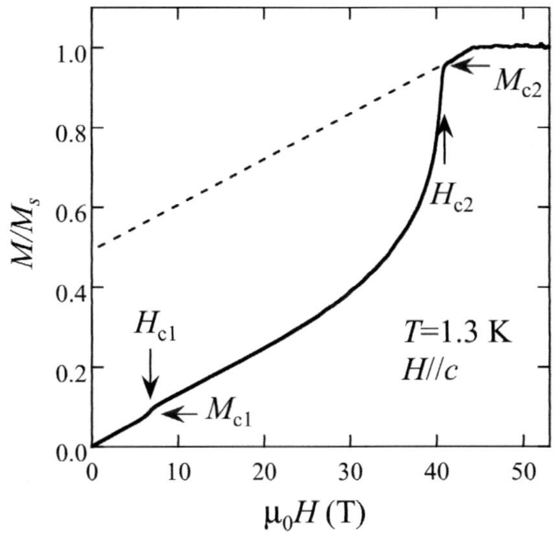

Figure 3. Magnetization curve of $LiCuVO_4$ normalized by the saturation magnetization along the c-axis at 1.3 K. The broken line is a linear extrapolation of magnetization between H_{c2} and H_{sat}. M_{1c} and M_{2c} correspond to the magnetization values at H_{c1} and H_{c2}, respectively.

a function of magnetic field is drawn. From the broken line in this figure which is a linear extrapolation of magnetization between H_{c2} and H_{sat}, $\mathrm{d}M/\mathrm{d}H$ in this phase is nearly a half of the slope of the classical magnetization (M_{sat}/H_{sat}) curve which is in good agreement with the calculation performed in a quasi 2D microscopic model [4]. Zhitomirsky and Tsunetsugu also calculated the transition fields using J_{NN}, J_{NNN} and the interchain exchange interaction constant along the a-direction evaluated from the inelastic neutron scattering experiments [6]. For the g value of 2.0, the transition field between a spin-cone (spiral) phase and a spin nematic one is 44.5 T, which is 2.6 T below the saturation field, calculated in the same model. The calculated transition field is close to the observed value of 41 T of this compound with g=2.3 for the c direction. Note that the calculated transition field should not be treated exactly, because the true spin structure at the magnetic field near 41 T is not yet known from the experiment. When extrapolated from lower fields, the spin structure should be a SDW-like spin modulated structure, and not a spiral one. Nevertheless, the critical field, calculated by a 2D model indicates that the nematic phase survives in a 2D case and that the transition from a spin ordered phase to the nematic one is expected near the field value about a few Tesla below the saturation field.

Next, we compare the magnetization values at the transition fields with calculated ones [1, 3] by 1D theory. If we take the β value of -0.4 evaluated from the inelastic neutron scattering experiments, magnetization at H_{c1} is nearly zero both in Fig. 10 (b) (Δ=1, β=-0.3 which is the nearest β value of -0.4) of Ref. [3] and Fig. 1 (b) (Phases are not shown above β=-1 and thus the boundary between the chiral vector phase and the SDW phase is smoothly extrapolated.) of Ref. [1]. On the other hand, if we take the β value of -2 evaluated from the analyses of magnetic susceptibility and the pitch angle of spiral spin structure [8], the magnetization at H_{c1} is about one tenth of the saturation magnetization which is close to our observed value. The numerical magnetization value at H_{c2}, however, is 0.7 of the saturation value, which is smaller than the observed one (0.95) (Fig. 3). In both cases, agreement between experiment and calculation is not satisfactory. This is probably caused by the interchain interaction which is not considered in 1D theory.

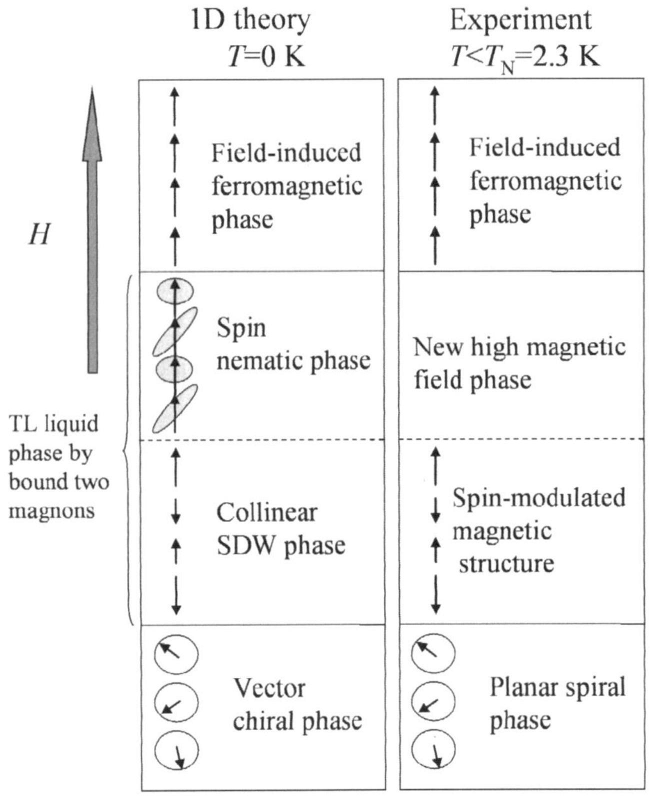

Figure 4. Schematic view of the field-induced phases calculated by 1D theory at 0 K (left) and observed in $LiCuVO_4$ below T_N=2.3 K (right).

Finally, we show the field-induced phases calculated by 1D theory at 0 K and observed in $LiCuVO_4$ below T_N in Fig. 4. In this figure, schematic view of each phase is depicted. According to the theory [1], both SDW and spin nematic states are formed by Tomonaga-Luttinger (TL) liquid of magnon bound pairs. In the left panel of Fig. 4, the vector chiral phase has a long range vector chiral correlations at T=0 K and the collinear SDW phase is a collinear phase with spin density wave correlations with power low decay. Then, the spin nematic phase, which is most interesting phase, is a planar not collinear phase with nematic correlations of spin components perpendicular to the field direction with power law decay. The highest phase is a usual field-induced ferromagnetic phase, namely the phase of the saturation magnetization. In the right panel of Fig. 4, two low phases are long-range ordered phases. In the lower phase, a planar spiral magnetic structure was observed and a spin-modulated magnetic structure was found in the higher one. Above these phases, we observed a new high magnetic phase and the field-induced ferromagnetic phase.

4. Conclusion

In conclusion, we performed high-field magnetization measurements on high-quality single crystals of $LiCuVO_4$. We observed two anomalies in the magnetization curve along the c-axis at 1.3 K. One at about 7 T corresponds to the transition from a planar spiral to a spin modulated magnetic states, and the other at about 41 T probably assign to the transition from the spin modulated state to a new phase. The low field magnetic phases of $LiCuVO_4$ are characterized by the same magnetic correlations as those in a frustrated $S=1/2$ chain in 1D theory. We suppose that the high field phase observed in $LiCuVO_4$ below the saturation field corresponds to a spin nematic phase with the long range nematic order of the perpendicular spin components with correlations: $< s_n^x s_{n+1}^y + s_n^y s_{n+1}^x >$. This proposal is supported by a reasonable correspondence of the field range of the new phase to a range just below the saturation field as well as by a good correspondence of the observed dM/dH value to the theoretical value calculated in a realistic 2D model [4].

Acknowledgments

We thank M. Zhitomirsky, T. Momoi and M. Sato for useful discussions. This work was carried out under the Foreign Visiting Researcher Program of KYOKUGEN and partly supported by Grants-in-Aid for Scientific Research (No.20340089), the Global COE Program (Core Research and Engineering of Advanced Materials-Interdisciplinary Education Center for Materials Science) (No. G10) from the MEXT, Japan, by the Grants 09-02-12341, 10-02-01105-a of the Russian Foundation for Basic Research, and Program of Russian Scientific Schools.

References

[1] Hikihara T, Kecke L, Momoi T and Furusaki A 2008 *Phys. Rev.* B **78** 144404
[2] Sudan J, Lüscher A and Läuchli A M 2009 *Phys. Rev.* B **80** 140402(R)
[3] Heidrich–Meisner F, McCulloch I P and Kolezhuk A K 2009 *Phys. Rev.* B **80** 144417
[4] Zhitomirsky M E and Tsunetsugu H 2010 *Europhys. Lett.* **92** 37001
[5] Gibson B J, Kremer R K, Prokofiev A V, Assmus W and McIntyre G J 2004 *Physica* B **350** e253
[6] Enderle M *et al.* 2005 *Europhys. Lett.* **70** 237
[7] Prokofiev A V, Vasilyeva I G, Assmus W 2005 *J. Crystal Growth* **275** e2009
[8] Sirker J 2010 *Phys. Rev.* B **81** 014419
[9] Svistov L E, Fujita T, Yamaguchi H, Kimura S, Omura K, Prokofiev A, Smirnov A I, Honda Z and Hagiwara M 2011 JETP Letters (Pis'ma v ŽÈTF) **93** 24
[10] Büttgen N *et al.* 2007 *Phys. Rev.* B **76** 014440

International Conference on Frustration in Condensed Matter (ICFCM) IOP Publishing
Journal of Physics: Conference Series **320** (2011) 012050 doi:10.1088/1742-6596/320/1/012050

Slow dynamics of Dy pyrochlore oxides $Dy_2Sn_2O_7$ and $Dy_2Ir_2O_7$

Kazuyuki Matsuhira[1], Makoto Wakeshima[2], Yukio Hinatsu[2], Chihiro Sekine[3], Carley Paulsen[4], Toshiro Sakakibara[5], Seishi Takagi[1]

[1] Kyushu Institute of Technology, Kitakyushu 804-8550, Japan
[2] Graduate School of Science, Hokkaido University, Sapporo 060-0810, Japan
[3] Muroran Institute of Technology, Muroran, Hokkaido 050-8585, Japan
[4] Institute Néel C.N.R.S - Universite Joseph Fourier, BP 166, 38042, Grenoble, France
[5] Institute for Solid State Physics, University of Tokyo, Kashiwa 277-8581, Japan

E-mail: `matuhira@elcs.kyutech.ac.jp`

Abstract. We report the magnetic properties of Dy pyrochlore oxides, $Dy_2Sn_2O_7$ and $Dy_2Ir_2O_7$. We show that the temperature dependence of the relaxation time $\tau(T)$ in the case of $Dy_2Sn_2O_7$ is very similar to that in the case of a typical spin ice, $Dy_2Ti_2O_7$. However, the time scale of $Dy_2Sn_2O_7$ is 10-100 times slower than that of $Dy_2Ti_2O_7$ over the entire temperature range. $Dy_2Ir_2O_7$ exhibits a metal-insulator transition at 134 K; it has a broad peak at 4.5 K in the DC magnetic susceptibility, indicating an antiferromagnetic correlation. However, the speci c heat has no sharp anomaly at 4.5 K. We show that the Dy moments in $Dy_2Ir_2O_7$ have no long-range ordering down to 100 mK. We elucidate the anomalous slow dynamics that $\tau(T)$ decreases and becomes inhomogeneous below 12 K.

1. Introduction

Pyrochlore magnets with a local ⟨111⟩ Ising magnetic anisotropy have been actively investigated owing to their strong geometrical frustration [1, 2, 3]. The ground state (GS) of Ising pyrochlore magnets is veri ed on the basis of theoretical studies [4]. The ferromagnetic coupling between the spins leads to stable spin con gurations, with two spins in the outward direction and two spins in the inward direction (the 2-in 2-out state) in a basic tetrahedron. There are six possible combinations in this spin con guration for every tetrahedron; hence, the GS of the entire spin system is highly degenerate. A static disordered (spin ice) state is attained at low temperatures, in spite of the structural order of the lattice [5, 6]. On the other hand, the antiferromagnetic (AFM) coupling leads to the GS of long-range ordering (LRO), consisting of alternate tetrahedra with 4 inward spins or 4 outward spins (the 4-in 4-out state). However, the e ects of electrical conductivity, spin-lattice coupling, *c*-*f* hybridization, etc. on the GS have not been veri ed. Further investigation of Ising pyrochlore magnets with a novel GS is required.

In this paper, we report the magnetic properties of Dy pyrochlore oxides, $Dy_2Sn_2O_7$ and $Dy_2Ir_2O_7$. In both compounds, the Dy^{3+} ion has a large magnetic moment of $10\mu_B$ with a local ⟨111⟩ Ising magnetic anisotropy. First, the insulating $Dy_2Sn_2O_7$ (Curie-Weiss temperature, $_{CW} = 1.69$ K) is considered as a spin ice compound because the anomaly in the AC magnetic properties is very similar to that of the typical spin ice compound, $Dy_2Ti_2O_7$, investigated in the previous study [7, 8]; the Sn^{4+} ion is non-magnetic. However, there have been no reports on the dynamics below 2 K thus far. In this study, we elucidate the slow spin dynamics below 2 K. Next, the semiconducting $Dy_2Ir_2O_7$ shows a metal-insulator transition (MIT) at $T_{MI} = 134$ K; we have adopted MIT for the sake of convenience. Recent studies have revealed that $Ln_2Ir_2O_7$ exhibits MITs, except for Ln = Pr [9]. In $Ln_2Ir_2O_7$, its electrical conductivity comes form the $5d$ electrons from Ir^{4+} as $4f$ electrons are generally well localized. The MITs in $Ln_2Ir_2O_7$ involve a magnetic ordering of *5d* electrons; however, the origin of the MITs has not been veri ed thus far. Consequently, the magnetic degrees of freedom of the Dy moments of $Dy_2Ir_2O_7$ persist below T_{MI}. We show that $Dy_2Ir_2O_7$ with an AFM spin correlation has no LRO down to 100 mK and that it exhibits a novel slow dynamics below 12 K which are quite di erent from those of a spin ice.

2. Experimental Procedure

Single-phase polycrystalline samples of $Dy_2Sn_2O_7$ and $Dy_2Ir_2O_7$ were prepared by a standard solid-state reaction method. For $Dy_2Sn_2O_7$, the sample was prepared in the same manner as the prevous report [7]. For $Dy_2Ir_2O_7$, the mixture was prepared using Dy_2O_3(99.99%) and IrO_2 (Tanaka Kikinzoku Kogyo). The molar ratio of Dy to Ir was 1:1.1. The mixture was pressed into pellets, which were then inserted into a Pt tube and heated at 1423-1523 K for around 10 days in a vacuum silica tube with several intermediate grindings. After adding 10 mol% of IrO_2, the process was continued for four days with several intermediate grindings. Then, the last step described above was repeated.

The reaction products were identi ed by powder X-ray di raction (XRD) measurements. The XRD pattern indicated a single phase with a cubic pyrochlore structure. DC magnetizations down to 2 K were measured using a commercial SQUID magnetometer (MPMS, Quantum Design (QD) Inc.). The AC magnetic susceptibilities, χ_{AC}, were measured using the ACMS option in PPMS (QD Inc.) at 1.8-300 K. χ_{AC} below 1.8 K was measured using a SQUID magnetometer developed at the Institute Neel, CNRS, Grenoble. Speci c heat measurements down to 0.35 K were performed by the thermal relaxation method (PPMS, QD Inc.).

3. Results and Discussion

3.1. Slow dynamics of spin ice type in $Dy_2Sn_2O_7$

Figures 1(a) and (b) show the temperature dependence of χ_{AC} for $Dy_2Sn_2O_7$ in an AC eld of 1.4 Oe. The real part, χ', and the imaginary part, χ'', are highly dependent on the frequency. χ' at 0.005 Hz shows a broad maximum at 0.75 K, and χ'' shows a peak below 0.65 K. As the frequency increases, the maximum of χ' and the peak of χ'' shift toward higher temperatures. The temperature dependence of the DC magnetic susceptibility, M/H, of $Dy_2Sn_2O_7$ at 100 Oe, measured under the zero- eld-cooled (ZFC) and eld-cooled (FC) conditions, is also shown in Fig. 1(a). The irreversibility is clearly visible below T_f 0.65 K. The ZFC curve has a sharp peak at T_f, and it steeply falls toward zero. On the other hand, the FC curve rises monotonically on cooling below T_f. This result is in agreement with the reported behavior of spin ice compounds [10, 11]. Thus, on cooling, $Dy_2Sn_2O_7$ exhibits slow spin dynamics without any LRO down to 100 mK.

It is important to study the slow dynamics in order to determine the temperature dependence of the relaxation time $\tau(T)$. We can estimate $\tau(T)$ from the frequency dependence of $\chi''(f)$. In the case of Debye relaxation with a single dispersion, $\chi''(f)$ becomes a symmetric function of log f, and it is maximum at the frequency de ned by $1/2\pi\tau$, where τ is a characteristic relaxation time. The frequency dependence of χ'' at various T is shown in Figs. 2(a) and 2(b). The experimental result is a slight asymmetric function of log f; hence, a narrow distribution is expected down to the lowest temperature. Therefore, the approximate $\tau(T)$ is obtained from the maximum frequency f_m in $\chi''(f)$, $\tau = 1/2\pi f_m$. Furthermore, from the frequency dependence of the peak temperature T_p of $\chi''(T)$ in Fig. 1(b), we can determine τ at T_p by using $\tau = 1/2\pi f$.

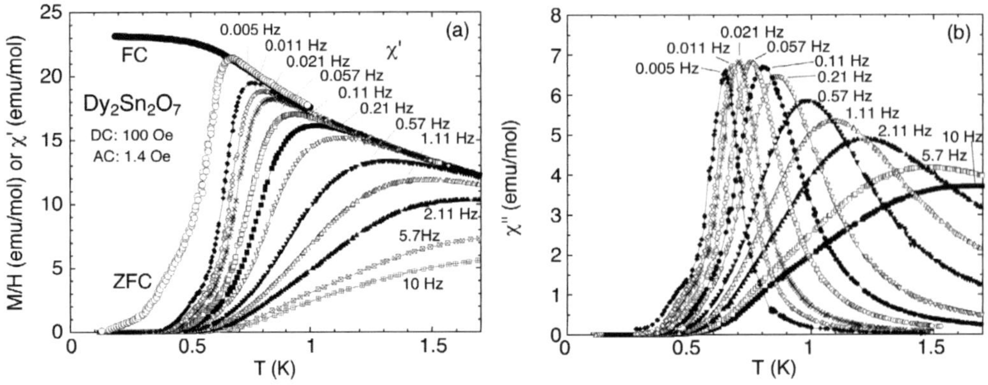

Figure 1. (a) Temperature dependence of M/H and χ' of $Dy_2Sn_2O_7$. ZFC and FC denote zero- eld cooling and eld cooling, respectively. (b) Temperature dependence of χ'' of $Dy_2Sn_2O_7$.

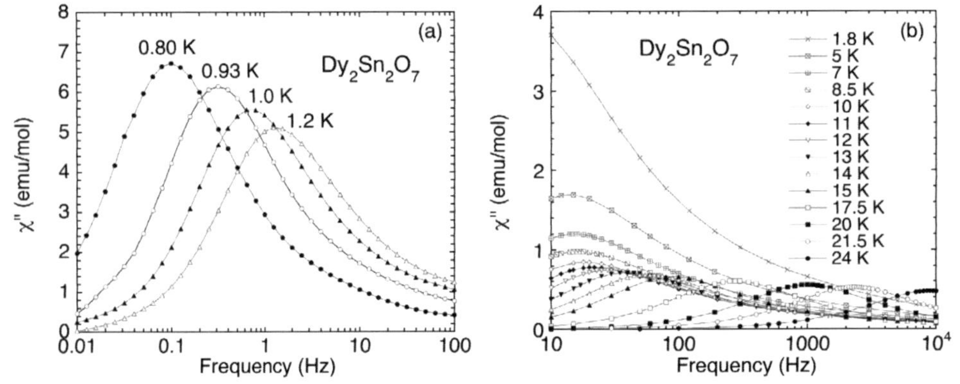

Figure 2. Frequency dependence of $\chi''(f)$ (a) below 1.2 K and (b) above 1.8 K.

Figure 3 shows $\tau(T)$ of $Dy_2Sn_2O_7$. The data show three distinct features. At temperatures above 10 K, $\tau(T)$ increases on cooling. In the temperature range of 2-10 K, $\tau(T)$ is virtually constant. Below 2 K, $\tau(T)$ increases again. For comparison, $\tau(T)$ of spin ice compounds are shown in Fig. 3 [10, 11, 12]. It should be noted that our de nition of τ estimated by χ_{AC} is di erent from the de nition ($\tau = 1/f_m$) provided by Snyder *et al.* [12]; in general, our de nition is used. The characteristics of $\tau(T)$ in the case of $Dy_2Sn_2O_7$ are very similar to those in the case of $Dy_2Ti_2O_7$. The present result strongly indicates that $Dy_2Sn_2O_7$ is a spin ice compound.

Next, we discuss the dynamics. Above 20 K, $\tau(T)$ is e ectively explained on the basis of the Arrhenius law, with the energy barrier $E_B = 246$ K. We can consider that E_B corresponds to the energy of the excited levels in the crystalline electric eld (CEF); this is related to Ising anisotropy [13]. The plateau regions of $\tau(T)$ in the temperature range of 2-10 K is explained by the quantum tunneling process through non-zero o -diagonal components of the dipolar interaction from neighboring moments [13]. The tunneling rate for $Dy_2Sn_2O_7$ is approximately 10 ^{2} s. This is speculatively attributed to a di erence between the dipolar interaction and the CEF level scheme. Below 2 K, the spin ice state is attained by the development of a short-ranged spin correlation. It should be noted that the time scale of $Dy_2Sn_2O_7$ is 10-100 times slower than that of $Dy_2Ti_2O_7$ over the entire temperature range. The dynamics of $Dy_2Sn_2O_7$ are much slower than those of $Dy_2Ti_2O_7$.

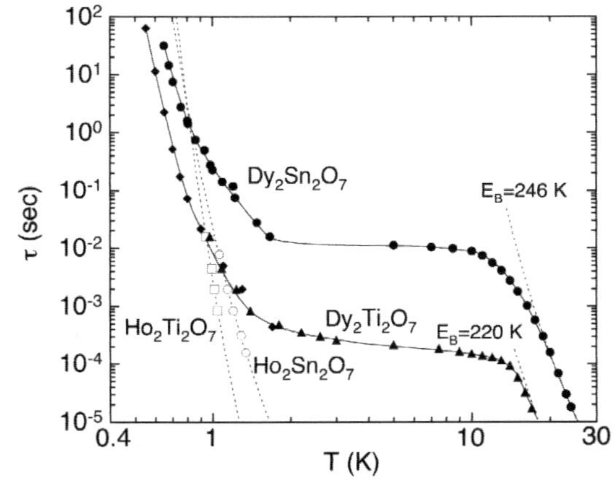

Figure 3. $\tau(T)$ of $Dy_2Sn_2O_7$. The data for $Ho_2Ti_2O_7$ and $Ho_2Sn_2O_7$ in the previous report are shown [10]. For $Dy_2Ti_2O_7$, the data above 1.8 K in the previous report are shown [11]. Data below 2 K for both powder and single crystal samples are shown [14].

3.2. Novel slow dynamics of $Dy_2Ir_2O_7$

The temperature dependence of the magnetization $M(T)$ for $Dy_2Ir_2O_7$, measured under ZFC and FC conditions, is shown in Fig. 4(a). A slight di erence between the FC and ZFC curves appears below $T_{\rm MI}$ = 134 K. This anomaly is caused by the appearance of a very weak ferromagnetic component (10 $^{3}\mu_{\rm B}$/f.u.) due to an unsolved magnetic ordering of Ir moments in MIT. The inset shows the reciprocal magnetic susceptibility (M/H) ^{1} under the ZFC condition below 60 K. (M/H) ^{1} exhibits Curie-Weiss behavior; hence, assuming the Curie-Weiss law at 20-60 K, the e ective moment and $_{\rm CW}$ are estimated to be 10.27 $\mu_{\rm B}$ and 3.4 K, respectively. The e ective moment is virtually in perfect agreement with the value of $Dy_2Sn_2O_7$; hence, the major contribution of DC magnetization below $T_{\rm MI}$ is attributed to the Dy moments. Furthermore, the degrees of freedom of the Dy moments persist much below $T_{\rm MI}$. Then, the negative value of $_{\rm CW}$ indicates an AFM coupling between the Dy moments. On cooling, $M(T)$ has a broad maximum at 4.5 K. In addition, a large di erence between the FC and ZFC curves appears below 0.8 K. Under the ZFC condition, $M(T)$ has a broad peak at 0.77 K, and it decreases toward a nite value. On the other hand, under the FC condition, $M(T)$ is virtually constant below 0.8 K. The results suggest a partial freezing of the Dy moments.

Figure 4(b) shows the magnetic contribution of the speci c heat $C_{\rm m}/T$ and the magnetic entropy $S_{\rm m}$ for $Dy_2Ir_2O_7$. In order to obtain the magnetic contribution $C_{\rm m}$, we subtracted the lattice contribution by using the lattice speci c heat of $Eu_2Ir_2O_7$. $C_{\rm m}/T$ shows a broad peak at 2.5 K. However, no sharp anomaly is observed at 4.5 K, where $M(T)$ shows a maximum. As $S_{\rm m}$ reaches 5.5 J/K Dy-mole ($0.95R\ln2$) at 20 K, the entropy due to the CEF GS doublet of Dy^{3+} is released below 20 K; there is no residual entropy, as in the case of a spin ice.

From the results presented above, we can conclude that the dynamics of $Dy_2Ir_2O_7$ are di erent from those of a spin ice. Figure 5(a) shows the temperature dependence of $\chi_{\rm AC}$ of $Dy_2Sn_2O_7$ in an AC eld of 5 Oe. The anomaly around 20 K is very similar to that of $Dy_2Sn_2O_7$ and $Dy_2Ti_2O_7$. As the frequency is lower, the maximum of $\chi'(T)$ and the peak of $\chi''(T)$ ($T_{\rm HT\,max}$) shift toward lower temperatures. On the other hand, $\chi'(T)$ at 10 Hz shows a broad maximum at 4.5 K; this is consistent with the anomaly in DC magnetization. However, as the frequency is higher, the maximum at 4.5 K at 10 Hz and the broad peak of $\chi''(T)$ ($T_{\rm LT\,max}$) shift to lower temperatures. Thus, the dynamics involved in this anomaly become fast on cooling. This feature is not responsible for a simple thermal process. The frequency dependence of χ'' at various T is shown in Fig. 5(b). On cooling, the maximum frequency shifts to lower frequencies down to

Journal of Physics: Conference Series **320** (2011) 012050 doi:10.1088/1742-6596/320/1/012050

12 K; below 12 K, it shifts to higher frequencies. To investigate the slow dynamics of $Dy_2Ir_2O_7$, we estimated $\tau(T)$ of $Dy_2Ir_2O_7$ from the data of χ_{AC}, as in the case of $Dy_2Sn_2O_7$.

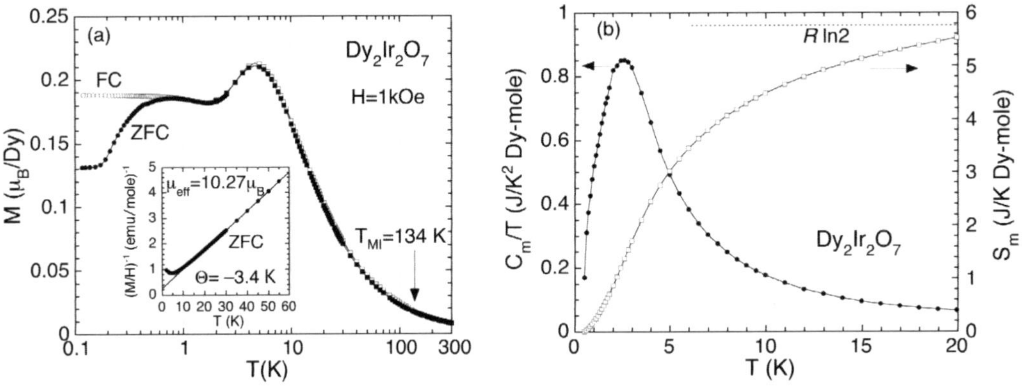

Figure 4. (a) $M(T)$ for $Dy_2Ir_2O_7$ measured under ZFC and FC conditions. The inset shows (M/H) ^{1} under the ZFC condition below 60 K. (b) C_m/T and S_m for $Dy_2Ir_2O_7$.

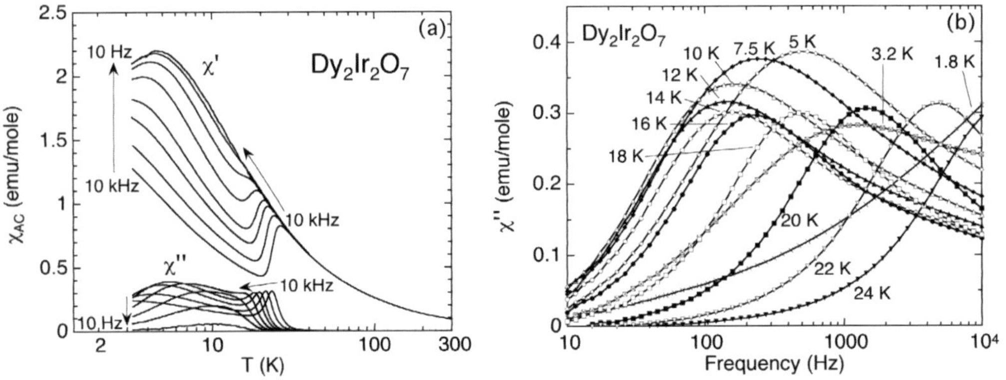

Figure 5. (a) Temperature dependence of χ_{AC} for $Dy_2Ir_2O_7$ at 10k, 3k, 1k, 500, 190, 80, 35, and 10 Hz. (b) Frequency dependence of χ'' for $Dy_2Ir_2O_7$ at various T.

Figure 6 shows $\tau(T)$ of $Dy_2Ir_2O_7$. Above 20 K, $\tau(T)$ is e ectively explained on the basis of the Arrhenius law, with $E_B = 260$ K. This value is virtually the same as that of $Dy_2Sn_2O_7$ and $Dy_2Ti_2O_7$. Therefore, we can assume that the anomaly is responsible for the Ising anisotropy of Dy moments. Above 16 K, $\tau(T)$ estimated by a maximum of $\chi''(f)$ corresponds to τ estimated by the peak of $\chi''(T)$. However, a discrepancy between their estimated τ appears below 16 K. On cooling, $\tau(T)$ shows a maximum at around 12 K, and it becomes short. This implies that τ is widely distributed below 12 K; the system becomes inhomogeneous below 12 K. Dy moments with a long component in τ cause the freezing in $M(T)$, as shown in Fig. 4(a). The other Dy moments with a short component in τ keep uctuating slowly down to 100 mK. To the best of our knowledge, the appearance of fast dynamics below 12 K in this study is a novel observation in the eld of magnetism.

Next, we discuss the origin of the anomalous dynamics. Although the local spin correlation between the Dy moments is AFM, there is no LRO down to 100 mK. Hence, we may speculate that a short-ranged ordering of the 4-in 4-out state is realized in the GS. At a nite temperature, the 1-in 3-out (or 3-in 1-out) and 2-in 2-out states are favored because they are highly degenerated states. From a point of view of the magnetic monopole picture, many pairs of magnetic monopoles are created on cooling, as opposed to the case of a spin ice [15]. It is speculated that this change in the spin con guration at a nite temperature is related to the anomalous dynamics. Furthermore, a slight structural change caused by MIT and an e ect of an internal eld induced by a magnetic ordering of *5d* electrons below T_{MI}, should also be considered. Further investigation is required to verify the origin.

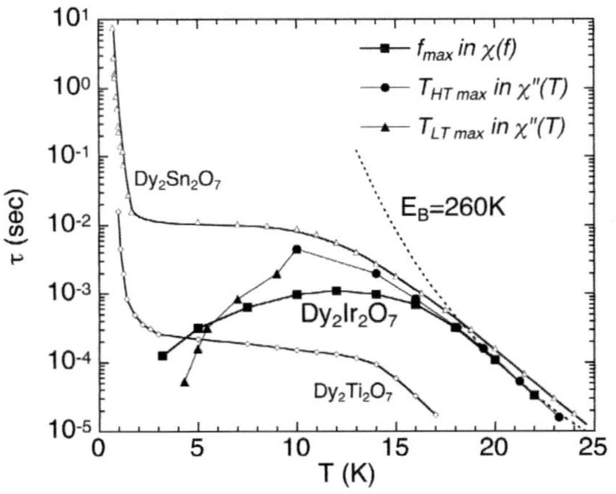

Figure 6. Temperature dependence of relaxation time $\tau(T)$ for $Dy_2Ir_2O_7$. $\tau(T)$ of $Dy_2Ti_2O_7$ and $Dy_2Sn_2O_7$ are shown for comparison. The data points are the same as those in Fig. 3.

Acknowledgment
We would like to thank M. Udagawa, K. Hattori, L. D. C. Jaubert, and P. C. W. Holdsworth for their insightful discussions. K.M. was supported by the Visiting Researcher's Program of the Institute for Solid State Physics, University of Tokyo. This work was supported by a Grant-in-Aid for Scienti c Research on Priority Areas, "Novel States of Matter Induced by Frustration," (No.19052005), a Grants-in-Aid for Scienti c Research on Innovation Areas "Heavy Electrons"(Nos. 21102518 and 20102007) from the MEXT, Japan and a Grant-in-Aid for Scienti c Research (Nos. 20244053 and 20540339) from the JSPS.

References

[1] Bramwell S T and Gingras M J P 2001 *Science* **294** 1495.
[2] Castelnovo C, Möessner R and S. L. Sondhi S L 2008 *Nature* **451** 42.
[3] Machida Y, Nakatsuji S, Onoda S, Tayama T and Sakakibara T 2010 *Nature* **463** 210.
[4] den Hertog B C and Gingras M J 2000 *Phys. Rev. Lett.* **84** 3430.
[5] Harris M J, Bramwell S T, McMorrow D F, Zeiske T and Godfre K W 1997 *Phys. Rev. Lett.* **79** 2554.
[6] Ramirez A P, Hayashi A, Cava R J, Siddharthan R and Shastry B S 1999 *Nature* **399** 333.
[7] Matsuhira K, Hinatsu Y, Tenya K, Amitsuka H and Sakakibara T 2002 *J. Phys. Soc. Jpn.* **71** 1576.
[8] Ke X, Ueland B G, West D V, Dahlberg M L, Cava R J and Schi er P, 2007 *Phys. Rev. B* **76** 214413.
[9] Matsuhira K, Wakeshima M, Nakanishi R, Yamada T, Nakamura A, Kawano W, Takagi S and Hinatsu Y 2007 *J. Phys. Soc. Jpn.* **76** 043706; Matsuhira K *et al.* submitted to *J. Phys. Soc. Jpn.*.
[10] Matsuhira K, Hinatsu Y, Tenya K and Sakakibara T 2000 *J. Phys.: Condens. Matter* **12** L649.
[11] Matsuhira K, Hinatsu Y and Sakakibara T 2001 *J. Phys.: Condens. Matter* **13** L737.
[12] Snyder J, Ueland B G, Slusky J S, Karunadasa H, Cava R J and Schi er P, 2004 *Phys. Rev. B* **69** 064414.
[13] Ehlers G, Cornelius A L, Orendac M, Kajnakova M, Fennell T, Bramwell S T and Gardner J S, 2003 *J. Phys.: Condens. Matter* **15** L9.
[14] Matsuhira K, Sekine C, and Paulsen C, unpublished. The detailed study on slow dynamics of $Dy_2Ti_2O_7$ is currently in progress.
[15] Jaubert L D C and Holdsworth P C W 2009 *Nature Physics* **5** 258.

Critical Phenomena in Long-Range RKKY Ising Spin Glasses

Yoshikazu Tabata, Satoshi Kanada, Teruo Yamazaki[A], Takeshi Waki, Hiroyuki Nakamura

Department of Materials Science and Engineering, Kyoto University, Kyoto, Japan
[A]The Institute for Soild State Physics, University of Tokyo, Chiba, Japan

E-mail: y.tabata@ht4.ecs.kyoto-u.ac.jp

Abstract.
We have investigated critical phenomena in spin glasses $R_xY_{1-x}Ru_2Si_2$ (R = Dy, Tb, Gd). These compounds, where the magnetic moments of rare-earth ions interact by the long-range Ruderman-Kittel-Kasuya-Yoshida (RKKY) interaction via conduction electrons, has uniaxial magnetic anisotropy. The separation of the zero- eld-cooled and eld-cooled magnetization was found only along the c-axis in all compounds, and hence, they are classi ed into the long-range Ising spin glass. The magnetic anisotropic energies in these compounds are di erent from each other in two orders of magnitude, from 330 K to 1.8 K, however, the critical exponents are similar. It clearly indicates a presence of the universality of the long-range RKKY Ising spin glasses.

1. Introduction

Spin glass (SG) state is an archetype of ordered states in random magnets where the ferromagnetic (positive) and antiferromagnetic (negative) spin-spin interactions are randomly distributed. The strong frustration due to the random distribution of the interaction produces a complicated magnetic ordered state, where the spins are frozen with random orientations, showing striking glassy behaviors such as the long-time relaxation, aging phenomenon, memory e ect, and chaos e ect [1; 2]. Although extensive investigations of the SG have been done since its discovery, many unresolved problems remains. One of them is whether the replica-symmetry-breaking (RSB) predicted by the mean- eld theory [3] occurs in real SG materials.

Two alternative pictures of the SG state are theoretically predicted; the mean- eld RSB picture and the droplet RS picture. According to theoretical studies of the mean- eld model, the SG phase transition is a 2nd order phase transition associated with the RSB [3; 4]. The RSB SG state has a complicated multi-valley structure of its free-energy [5] and is an essentially di erent state from RS states such as paramagnetic and (anti)ferromagnetic states, where simple one- and two-valley structures, respectively. Alternatively, the RS SG state is predicted by the droplet theory based on the short-range interaction model [6]. In the droplet picture, the SG state has a simple two-valley structure of its free energy as well as in an (anti)ferromagnetic state. The RSB and RS SG states can be distinguished by examination of stability of the SG phase under nite magnetic eld. The RSB SG state is stable in the presence of nite magnetic eld [7], whereas the droplet RS SG state is unstable in any magnetic eld [6].

Many numerical and experimental investigations were performed to examine the stability of the SG under magnetic eld [8–11]. The experiments of two di erent model SG materials, $Fe_xMn_{1\ x}TiO_3$ [9; 10] and $Dy_xY_{1\ x}Ru_2Si_2$ [11], are worthy of attention. The former $Fe_xMn_{1\ x}TiO_3$ and the latter $Dy_xY_{1\ x}Ru_2Si_2$ are model magnets of the Ising SGs with the short-range and long-range interactions, respectively. The short-range super-exchange interaction via oxigen-ion is dominant in $Fe_xMn_{1\ x}TiO_3$ and the long-range Ruderman-Kittel-Kasuya-Yoshida (RKKY) interaction via conduction electrons is dominant in $Dy_xY_{1\ x}Ru_2Si_2$. The experimental examinations on the stability of the SG under magnetic eld were performed in both materials to verify critical divergence of a characteristic relaxation time $\tau(T,H)$ with nite SG transition temperature $T_g(H)$. In $Fe_xMn_{1\ x}TiO_3$, $\tau(T,H)$ exhibits the critical divergence only at zero eld [9; 10]. On the other hand, the critical divergence of $\tau(T,H)$ in $Dy_xY_{1\ x}Ru_2Si_2$ was found in both zero and nite elds [11]. These results indicate that the RS and RSB SG states are realized in the short-range and long-range Ising SGs, respectively. Furthermore, the static and dynamic critical exponents of $Fe_xMn_{1\ x}TiO_3$ and $Dy_xY_{1\ x}Ru_2Si_2$ are quite di erent from each other, i.e., $= 4.0$, $\beta = 0.54$, $\delta = 8.4$, and $z\nu = 11$ in $Fe_xMn_{1\ x}TiO_3$ [9; 10; 12] and $= 1.1$, $\beta = 0.9$, $\delta = 1.7$, and $z\nu = 2.1$ in $Dy_xY_{1\ x}Ru_2Si_2$ [11], indicating that these SGs belong to di erent classes each other and a spacial-range of interaction is relevant to SG phase transitions. The critical exponents of $Dy_xY_{1\ x}Ru_2Si_2$ are similar to those of the mean- eld model, i.e., $_{MF} = 1$, $\beta_{MF} = 1$, $\delta_{MF} = 2$, and $(z\nu)_{MF} = 2$ [13].

In this article, we have reported on static critical phenomena in spin glasses $R_xY_{1\ x}Ru_2Si_2$ (R = Tb, Gd) in addition to $Dy_xY_{1\ x}Ru_2Si_2$. The parent compounds of these SGs, $DyRu_2Si_2$, $TbRu_2Si_2$, and $GdRu_2Si_2$, are isostructural intermetallic compounds, the tetragonal $ThCr_2Si_2$ structure, with paramagnetic-antiferromagnetic transitions of 29 K, 53 K, and 47 K, respectively [14]. In $TbRu_2Si_2$ and $GdRu_2Si_2$, the long-range RKKY interaction is dominant and spin glass (SG) phase also appears by substitution of non-magnetic Y for R, as well as in $DyRu_2Si_2$. Although magnetic anisotropic energies in these compounds are di erent in two orders of magnitude , from 330 K to 1.8 K, the magnetic anisotropy is the same, an Ising-like, and static critical exponents, and δ, are similar to each other. It clearly indicates a presence of the universality of the long-range RKKY Ising SGs.

2. Experimental details

The single crystalline samples of $R_xY_{1\ x}Ru_2Si_2$ (R = Dy, Tb, Gd) were grown by the Czhochralski method with a tetra-arc furnace. The concentrations of R were determined by comparing the saturated magnetizations of the diluted compounds along the magnetic easy c-axis with those of the pure compounds RRu_2Si_2. Ac and dc magnetizations were measured with the SQUID magnetometer MPMS (Quantum Design) equipped in the Research Center for Low Temperature and Material Sciences in Kyoto University. The dc magnetization was measured 10 minutes after stabilization at a certain temperature and a certain eld to avoid the e ect of very slow dynamics and to obtain the thermodynamic equilibrium magnetizations near the SG transition temperatures. The ac measurement of $Gd_xY_{1\ x}Ru_2Si_2$ was performed with an ac eld of 1 Oe and a frequency of 0.5 Hz. We use thin plate-shaped samples with weights of about 7.0 mg and sizes of 5 1 0.2 mm^3 for the measurements to inhibit the e ects of the diamagnetic eld and the eddy current.

To examine the static critical behaviors, we extracted the second nonlinear susceptibility $\chi_2(T)$ from the dc magnetic eld (H) dependences of the dc magnetization $M(H,T)$ and the real part of the ac susceptibility $\chi'(H,T)$. From a dc magnetization measurement, $\chi_2(T)$ can be obtained by tting $M(H,T)/H$ at each temperature as a function of H^2 in the form of

$$M(H,T)/H = \chi_0(T) + \chi_2(T)H^2 + \quad . \tag{1}$$

$\chi_2(T)$ can also be obtained from an ac susceptibility by tting $\chi'(H,T)$ as a function of H^2 in

Journal of Physics: Conference Series **320** (2011) 012051 doi:10.1088/1742-6596/320/1/012051

the form of

$$\chi'(H,T) = \chi'_0(T) + 3\chi'_2(T)H^2 + \quad , \tag{2}$$

because $\chi'(H,T)$ with small ac eld represents a di erential magnetization dM/dH. When an ac-frequency ω is so low that we can recognized the measurement to be in the dc-limit, $\chi'_2(T)$ obtained from the ac measurement is equivalent to $\chi_2(T)$ obtained from the dc measurement. In the present work, we obtained $\chi_2(T)$'s of $Dy_{0.103}Y_{0.897}Ru_2Si_2$ and $Tb_{0.065}Y_{0.935}Ru_2Si_2$ from the dc measurements because they show strong frequency dependence of the ac susceptibilities even very low frequency of 0.1 Hz. $Gd_{0.057}Y_{0.943}Ru_2Si_2$ hardly shows the frequency dependence and is within the dc limit below 1 Hz, and hence, we obtained its $\chi_2(T)$ from the ac measurement with $\omega = 0.5$ Hz. The eld-depedent nonlinear susceptibility $\chi_{\rm nl}(H,T) \equiv \chi_0(T) \quad M(H,T)/H$ was extracted by using the linear susceptibility $\chi_0(T)$ obtained from above-mentioned procedures.

3. Results and discussions

Figure 1 shows T dependences of dc magnetizations of (a) $Dy_{0.103}Y_{0.897}Ru_2Si_2$, (b) $Tb_{0.065}Y_{0.935}Ru_2Si_2$ and (c) $Gd_{0.057}Y_{0.943}Ru_2Si_2$ along the c- and a-axes with the applied magnetic eld of 2 Oe. All compounds exhibit distinct separations of the zero- eld-cooled (ZFC) and eld-cooled (FC) magnetizations along the c-axis around 1.9 K (Dy), 2.7 K (Tb), and 2.0 K (Gd). In contrast, no separations of the ZFC and FC magnetizations along the a-axis are found. These results imply that only the c-component of magnetic moments of the magnetic Dy^{3+}-, Tb^{3+}-, and Gd^{3+}-ions are randomly frozen and the a-component dynamically uctuates even at the lowest temperature of measurements. These component-separated freezing behaviors, which was theoretically predicted in the anisotropic SG based on the mean- eld model [15], indicate that these compounds should be categorized into the Ising SG with the long-range RKKY interaction.

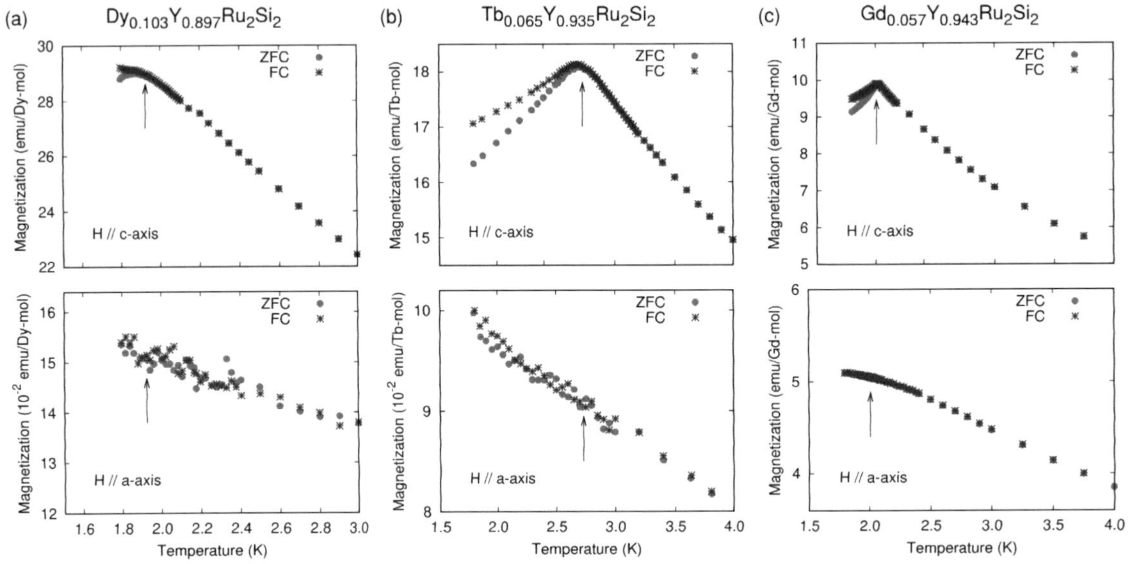

Figure 1. T dependences of dc magnetizations of (a) $Dy_{0.103}Y_{0.897}Ru_2Si_2$, (b) $Tb_{0.065}Y_{0.935}Ru_2Si_2$, and (c) $Gd_{0.057}Y_{0.943}Ru_2Si_2$. Upper and lower gures are the data with magnetic eld along the c-axis and a-axis, respectively. Arrows indicate the SG transition temperatures.

Figure 2 shows T dependences of second nonlinear susceptibilities $\chi_2(T)$'s, corresponding to the order-parameter susceptibility of the SG, (a) $Dy_{0.103}Y_{0.897}Ru_2Si_2$, (b) $Tb_{0.065}Y_{0.935}Ru_2Si_2$ and (c) $Gd_{0.057}Y_{0.943}Ru_2Si_2$ along the c-axis. In all compounds, clear divergent behavior of $\chi_2(T)$ are found. From the log-log plots of χ_2 vs ε ($\equiv T/T_g$ 1) shown in the lower gures of Fig. 2, we estimated the critical exponent of $\chi_2(T) \propto \varepsilon$ $^{\gamma}$. The best plots showing linear relations between log(χ_2) and log ε with T_g = 1.925 K (Dy), 2.740 K (Tb), and 2.008 K (Gd). The estimated 's are 1.08, 0.90, and 1.24 in $Dy_{0.103}Y_{0.897}Ru_2Si_2$, $Tb_{0.065}Y_{0.935}Ru_2Si_2$ and $Gd_{0.057}Y_{0.943}Ru_2Si_2$ respectively. These values of are similar to each other and are approximately the value of the mean- eld model $_{MF}$ = 1.

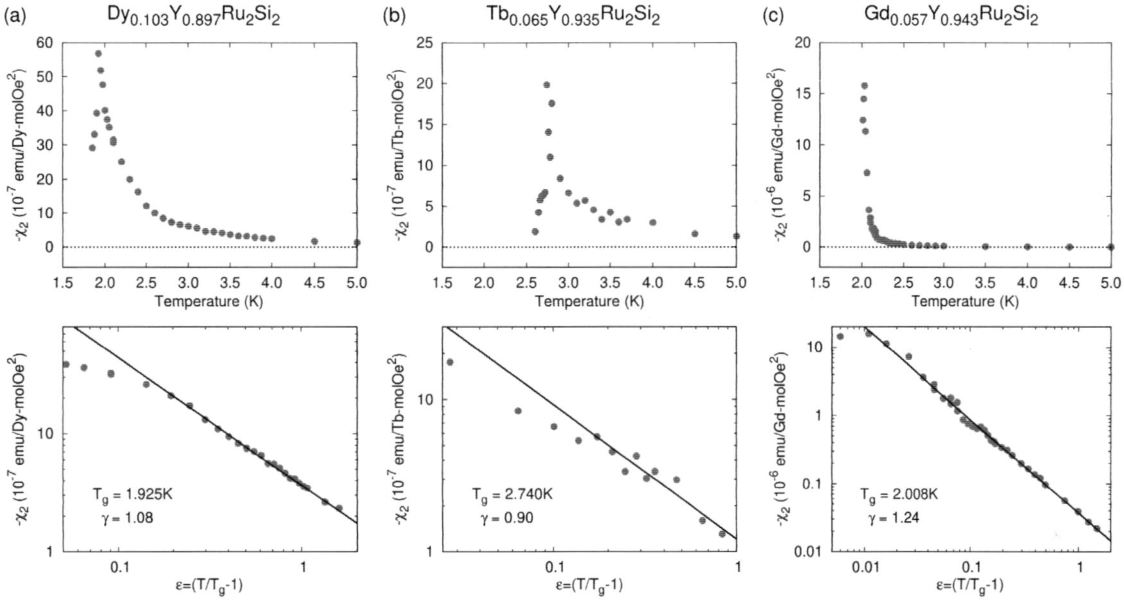

Figure 2. T dependences of second nonlinear susceptibilities of (a) $Dy_{0.103}Y_{0.897}Ru_2Si_2$, (b) $Tb_{0.065}Y_{0.935}Ru_2Si_2$, and (c) $Gd_{0.057}Y_{0.943}Ru_2Si_2$. Lower gures are the log-log plots of χ_2 vs ε ($\equiv T/T_g$ 1).

Next, we show H^2 dependences of eld-dependent nonlinear susceptibilities at T_g $\chi_{nl}(H, T_g)$, corresponding to the SG order-parameter, in Figs 3 (a) $Dy_{0.103}Y_{0.897}Ru_2Si_2$, (b) $Tb_{0.065}Y_{0.935}Ru_2Si_2$, and (c) $Gd_{0.057}Y_{0.943}Ru_2Si_2$ along the c-axis. In the gures, χ_{nl} vs. H^2 are plotted in double logarithmic scales. From these plots, the critical exponent δ of $\chi_{nl} \propto H^{2/\delta}$ is obtained as δ = 1.70 (Dy), 1.70 (Tb), and 2.16 (Gd). These values of δ are similar to each other and are approximately the value of the mean- eld model δ_{MF} = 2, as well as .

The similarity of the static critical exponents and δ indicates that the SG phase transitions of R_xY_1 $_xRu_2Si_2$ should belong to the same universality class, which is di erent from those of not only Fe_xMn_1 $_xTiO_3$ [12] but also so-called canonical SGs [16; 17]. The anisotropic behaviors shown in Fig. 1 clearly indicate the Ising nature of R_xY_1 $_xRu_2Si_2$. We roughly estimated magnetic anisotropic energies E_a's from the di erences between magnetization processes along the c- and a-axes; E_a = 250, 330, and 1.8 K/R^{3+}-ion in $Dy_{0.103}Y_{0.897}Ru_2Si_2$, $Tb_{0.065}Y_{0.935}Ru_2Si_2$ and $Gd_{0.057}Y_{0.943}Ru_2Si_2$ respectively. Moderate magnitude of E_a (T_g), which is enough large anisotropy to show the Ising nature of SG phase transition according to the theory [15], is found even in Gd-compound with zero orbital angular momentum (L = 0). It should be noted that

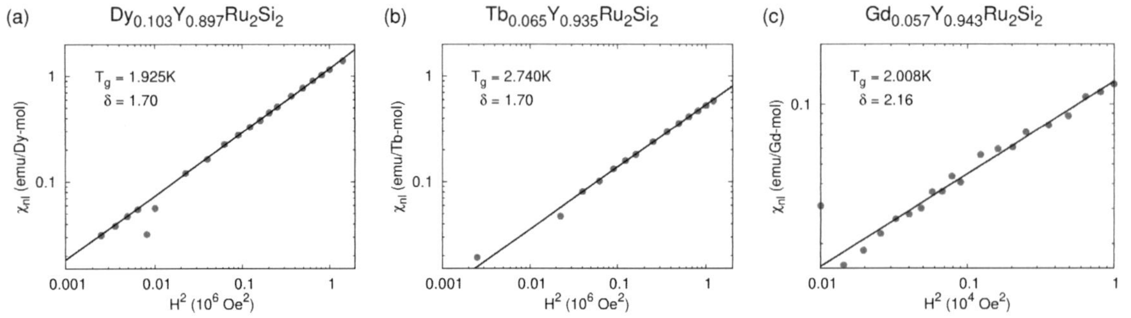

Figure 3. Log-log plots of H^2 dependences of eld-dependent nonlinear susceptibilities of (a) $\mathrm{Dy_{0.103}Y_{0.897}Ru_2Si_2}$, (b) $\mathrm{Tb_{0.065}Y_{0.935}Ru_2Si_2}$, and (c) $\mathrm{Gd_{0.057}Y_{0.943}Ru_2Si_2}$ at each T_g.

the critical exponents in $R_x\mathrm{Y}_{1\ x}\mathrm{Ru_2Si_2}$ are almost identical in spite of the large di erence of E_a, and are similar to those of the mean- eld model.

Here, we discuss on the validity of our static scaling analyses. The cusp-like anomalies of the ZFC magnetization of the $\mathrm{Dy_{0.103}Y_{0.897}Ru_2Si_2}$ and $\mathrm{Tb_{0.065}Y_{0.935}Ru_2Si_2}$ are rather broad, as shown in Figs 1 (a) and (b). It should be a dynamical e ect because very slow dynamics is found in both compounds [11]. The slow dynamics in the Dy- and Tb-compounds result from the strong Ising anisotropy and the consequent high energy barrier of the spin ipping. Due to this dynamical e ect, it is very di cult to observe the static critical properties very close to T_g in the Dy- and Tb-compounds. Thus, the regions of the measurements in these compounds shown in Figs. 2 and 3 are rather far from T_g, as comparing with that in the earlier work of the canonical SG AgMn [16]. The analysis of the data in rather high temperature and high eld may lead wrong critical exponents. On the other hand, the magnetic anisotropy of the $\mathrm{Gd_{0.057}Y_{0.943}Ru_2Si_2}$ is weak and the e ect of the slow dynamics is negligible above T_g, and hence, we can approach very close T_g, as shown in Figs. 2 (c) and 3 (c). Nevertheless, we obtained the critical exponents, $= 1.24$ and $\delta = 2.16$, which are roughly identical to the exponents of the Dy- and Tb-compounds and are much closer to the mean- eld exponents than to those of the short-range Ising SG or the canonical SG [18]. Consequently, we concluded that $R_x\mathrm{Y}_{1\ x}\mathrm{Ru_2Si_2}$, the long-range RKKY Ising SG, belongs to the mean- eld universality class.

4. Conclusion

We examined the static critical phenomena of the SG phase transition in $R_x\mathrm{Y}_{1\ x}\mathrm{Ru_2Si_2}$ (R = Dy, Tb, Gd), which are categorized as the long-range RKKY Ising SG, and found almost identical critical exponents 1 and δ 2. These values are similar to the mean- eld values $_{\mathrm{MF}} = 1$ and $\delta_{\mathrm{MF}} = 2$. It indicates a presence of the universality of the long-range RKKY Ising spin glasses, being the mean- eld universality class.

5. Acknowledgements

This study was supported by a Grand-in-Aid for Scienti c Research on Priority Areas "Novel States of Matter Induced by Frustration" (19052003), a Grand-in-Aid for the Global COE Program "International Center for Integrated Research and Advanced Education in Materials Science" from the Ministry of Education, Cluture, Sports, Science and Technology of Japan.

References

[1] Jonason K *et al.* 1998 *Phys. Rev. Lett.* **81** 3243

[2] Vincent E *et al.* 2009 *J Stat. Phys.* **135** 1105

[3] Parisi G 1980 *J. Phys. A* **13** 1101

[4] Sherrington D and Kirkpatrick S 1975 *Phys. Rev. Lett.* **35** 1792

[5] Mezard M *et al.* 1984 *Phys. Rev. Lett.* **52** 1156

[6] Fisher D S and Huse D A 1986 *Phys. Rev. Lett.* **56** 1601

[7] de Almeida J R L and Thouless D J 1978 *J. Phys. A: Math. Gen.* **11** 983

[8] Young A P and Katzgraber H G *Phys. Rev. Lett.* 2004 **93** 207203; Sasaki M *et al.* 2007 *Phys. Rev. Lett.* **99** 137202; Katzgraber H G *et al.* 2009 *Phys. Rev. Lett.* **102** 177205; Leuzzi L *et al.* 2009 *Phys. Rev. Lett.* **103** 267201

[9] Mattsson J *et al.* 1995 *Phys. Rev. Lett.* **74** 4305

[10] Jonsson P E *et al.* 2005 *Phys. Rev. B* **71** 180412(R)

[11] Tabata Y *et al.* 2010 *J Phys. Soc. Jpn.* **79** 123704

[12] Gunnarsson K *et al.* 1991 *Phys. Rev. B* **43** 8199

[13] Binder K and Young A P 1986 *Rev. Mod. Phys.* **58** 801

[14] Iwata N *et al.* 2003 *J. Magn. Magn. Mater.* **256** 189; Garnier A *et al.* 1995 *Physica B* **212** 343; Garnier A *et al.* 1996 *Physica B* **222** 80

[15] Roberts S A and Bray A J 1982 *J. Phys.* C **15** L527

[16] Levy L P and Ogielski A T 1986 *Phys. Rev. Lett.* **57** 3288

[17] Taniguchi T and Miyako Y 1988 *J Phys. Soc. Jpn.* **57** 3520

[18] According to the theoretical argument by Bray *et al.* (Bray A J *et al.*, 1986 *Phys. Rev. Lett.* **56** 2641), the RKKY Ising SG should belong to the same universality class of the short-range Ising SG. The discrepancy between the theoretical argument and our experimental result is an open question.

Journal of Physics: Conference Series **320** (2011) 012052 doi:10.1088/1742-6596/320/1/012052

Effect of DM interaction in a quantum antiferromagnet on a deformed kagome lattice, $Rb_2Cu_3SnF_{12}$

H Tashiro[1], M Nishiyama[1], A Oyamada[1], T Itou[1], S Maegawa[1], M Yano[2], T Ono[2] and H Tanaka[2]

[1] Graduate School of Human and Environmental Studies, Kyoto University, Kyoto 606-8501, Japan

[2] Department of Physics, Tokyo Institute of Technology, Oh-okayama, Meguro-ku, Tokyo 152-8551, Japan

E-mail: `h.tashiro@fs7.ecs.kyoto-u.ac.jp`

Abstract. We have performed ^{63,65}Cu-NQR measurements of a quantum antiferromagnet on a deformed kagome lattice, $Rb_2Cu_3SnF_{12}$. The spin-lattice relaxation rate T_1^{-1} of ^{63}Cu decreases steeply below 50 K. The rate shows no magnetic ordering down to 1.3 K and the existence of an energy gap. The temperature dependence of T_1^{-1} between 10 and 50 K is expressed by an Arrhenius-type function, with which the gap energy was estimated to be 50 K. This value is rather large compared to the value of 21 K obtained from the magnetic susceptibility and magnetization measurements. This difference can be ascribed to the dispersion of triplet excitations due to Dzyaloshinsky-Moriya (DM) interaction and four different antiferromagnetic (AF) interactions. A large state density in the triplet state with the dispersion gives a significant effect on the spin fluctuation. Eight peaks were observed between 45 and 55 MHz in ^{63,65}Cu-NQR spectra at 5 K. These peaks correspond to two isotopes and four crystallographic sites of Cu. This result implies that a crystal structure with two Cu sites at room temperature changes to a structure that has four Cu sites below the structural phase transition at 215 K.

1. Introduction

Antiferromagnets on the Kagome lattice have fascinating properties due to the geometrical frustration. A classical antiferromagnet on the Kagome lattice has continuous degeneracy, even at zero temperature. However, it is predicted that the co-planar 120° spin configuration with $\sqrt{3} \times \sqrt{3}$ structure is prefered due to thermal fluctuations: that is, order by disorder [1]. On the other hand, for quantum spins, it has been theoretically demonstrated that the ground state is singlet with a spin gap to the magnetic triplet excited state. This gap is continuously filled with nonmagnetic singlet states [2]. It has also been proposed that there is a nondispersive excited state in both classical and quantum spins [3, 4]. When the kagome lattice is deformed and has different exchange interactions, the excited states have a large dispersion [5].

Kagome lattice antiferromagnets inherently have an antisymmetric Dzyaloshinsky-Moriya (DM) interaction, since there is no inversion center in the middle of two neighboring spins. The

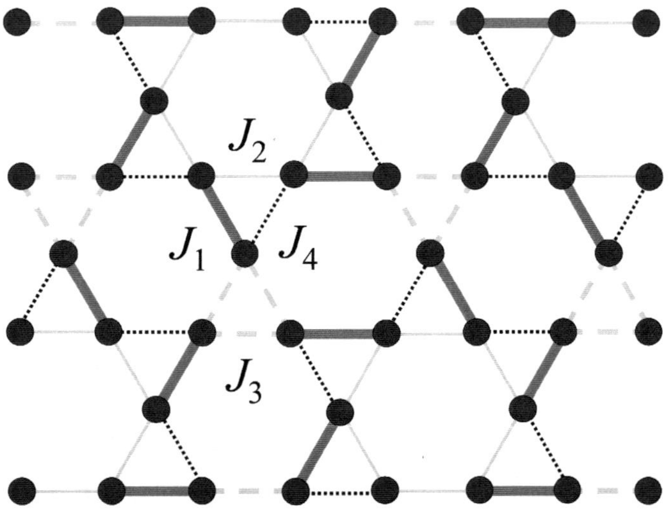

Figure 1. Exchange interaction network on a deformed kagome lattice, $Rb_2Cu_3SnF_{12}$. $J_1 = 234$ K, $J_2 = 211$ K, $J_3 = 187$ K and $J_4 = 108$ K [8].

DM interaction is described by [6, 7]

$$\mathcal{H}_{DM} = \sum_{\langle i,j \rangle} \boldsymbol{D}_{ij} \cdot (\boldsymbol{S}_i \times \boldsymbol{S}_j). \tag{1}$$

Here $\boldsymbol{D}_{ij}$ is the DM vector, obtained from a second-order purtubation of the spin-orbit interaction and antiferromagnetic (AF) interaction, and its direction is determined by the symmetry of the crystal structure. The DM interaction breaks the rotational symmetry of the Heisenberg spin system. The triplet excited state split into $S_z = 0$ and $S_z = \pm 1$ due to the DM interaction as $\langle S_z = 0 | \mathcal{H}_{DM} | S_z = 0 \rangle \neq 0$ and $\langle S_z = \pm 1 | \mathcal{H}_{DM} | S_z = \pm 1 \rangle = 0$. Since the DM interaction does not break time reversal symmetry, it does not lift the degeneracy of $S_z = \pm 1$. Here we present NQR studies of a quantum kagome antiferromagnet, $Rb_2Cu_3SnF_{12}$, to clarify the effects of DM interaction in a quantum antiferromagnet.

The crystal structure of $Rb_2Cu_3SnF_{12}$ is hexagonal with space group $R\bar{3}$ at room temperature [8]. CuF_6 octahedra are linked in the c-plane by sharing corners, and elongated along the principal axes, which are approximately parallel to the c-axis. Magnetic Cu^{2+} ions with $\boldsymbol{s} = 1/2$ are located at the center of the octahedra and form a kagome lattice in the c-plane. Since there are four different bond angles in the exchange pathway Cu^{2+}–F^-–Cu^{2+} in this compound, Cu^{2+} ions interact through four different antiferromagnetic interactions, $J_1 = 234$ K $> J_2 > J_3 > J_4 = 108$ K as shown in figure 1 [8]. Consequently, this system is regarded as a deformed kagome lattice antiferromagnet. The magnetic susceptibility decreases steeply at temperatures below 50 K, and magnetic ordering has not been observed down to 1.3 K [8]. Therefore, the ground state of this compound is considered to be a singlet with a spin gap. Because of the configuration of the strongest interaction J_1, this singlet state is considered to be a pinwheel valence-bond solid state [5]. A recent inelastic neutron scattering experiment has demonstrated that DM interaction plays an important role in the dispersion of triplet excitations [9]. In order to obtain microscopic informations regarding the effects of the DM interaction in the quantum antiferromagnet, we have performed ^{63,65}Cu-NQR on a single crystal of $Rb_2Cu_3SnF_{12}$.

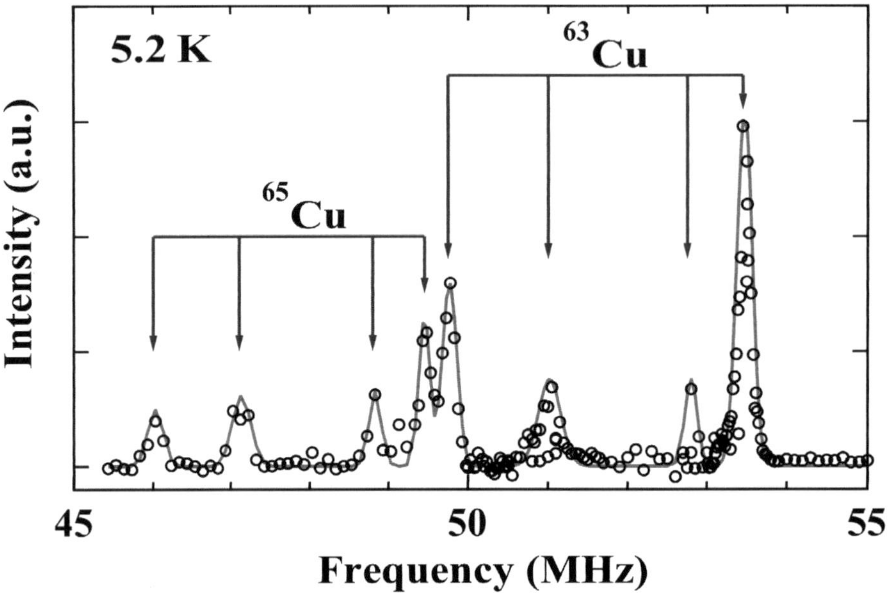

Figure 2. ^{63,65}Cu-NQR spectrum at 5.2 K.

2. Experimental Procedures

The detailed sample synthesis are shown in ref. 8. The ^{63,65}Cu NQR measurements of a single crystal were performed by a coherent pulsed spectrometer at zero field between 2 and 50 K. The oscillation field was applied perpendicular to the c-axis by a coil wound around the sample. The spin echo signals were measured at frequencies between 45 and 55 MHz to obtain the NQR spectrum. The spin-lattice relaxation rates T_1^{-1} were obtained from the recovery of the spin echo signals after the saturation pulses.

3. Results and Discussions

Figure 2 shows the ^{63,65}Cu-NQR spectrum at 5.2 K. Eight resonance peaks were observed. These peaks correspond to resonance with the nuclear quadrupole splitting, since there are neither an applied field nor an internal field. The nuclear quadrupole Hamiltonian is written as

$$\mathcal{H}_Q = \frac{eQV_{zz}}{4I(2I-1)}\left[(3I_z^2 - I^2) + \eta\frac{I_+^2 + I_-^2}{2}\right], \tag{2}$$

where I is the nuclear spin with $I = 3/2$ for both ^{63}Cu and ^{65}Cu, Q is the nuclear quadrupole moment, $^{63}Q = -0.211 \times 10^{-29}$ m^2 and $^{65}Q = -0.195 \times 10^{-29}$ m^2, respectively, $V_{\alpha\alpha}$ $(= \partial^2 V/\partial\alpha^2$, $\alpha = x, y, z)$ are the electric field gradients, where V is the electrostatic potential at the nuclear site, and η $(= (V_{xx} - V_{yy})/V_{zz})$ is the asymmetry parameter. The Hamiltonian $\mathcal{H}_Q$ can be diagonalized, and the NQR frequency ν_Q corresponding to the transition $I_z = \pm 1/2 \leftrightarrow \pm 3/2$ is obtained to be

$$\nu_Q = \sqrt{1 + \frac{\eta^2}{3}} \times \frac{3eQ|V_{zz}|}{2I(2I-1)h}. \tag{3}$$

The four peaks at ν_Q =49.8, 51.1, 52.8, and 53.5 MHz correspond to ^{63}Cu resonances, and the rest four peaks at ν_Q = 46.0, 47.1, 48.8, and 49.4 MHz correspond to ^{65}Cu resonances. Since ν_Q depends on Q and the electric field gradient parameters, $|V_{zz}|$ and η, the observed

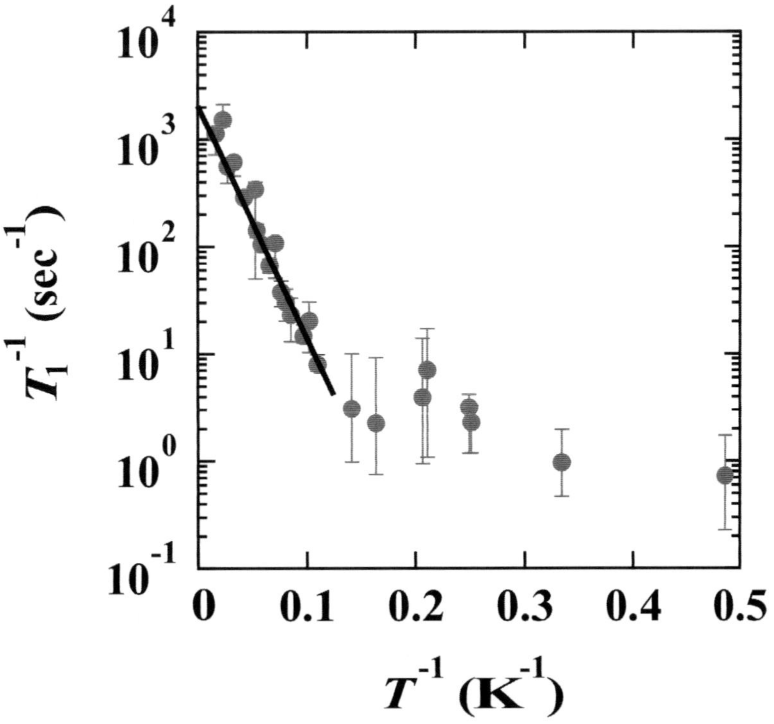

Figure 3. Temperature dependence of ^{63}Cu nuclear spin-lattice relaxation rate T_1^{-1} at 53.5 MHz. The solid line is an Arrhenius-type function, equation (5), with a gap energy of 50 K.

eight peaks could correspond to two isotopes with different quadrupole moments and four crystallographic Cu sites at which the electric field gradients are different. The crystal structure at room temperature has two kinds of Cu sites, and the structural phase transition at $T_S = 215$ K can be observed by X-ray diffraction [9, 10] and specific heat measurements [11]. However, the detailed crystal structure below T_S has not yet been determined so far. The NQR spectra indicate that the crystal structure below T_S should have four Cu sites.

Figure 3 shows the temperature dependence of the spin-lattice relaxation rate T_1^{-1} of ^{63}Cu on the site with the largest electric field gradient at 53.5 MHz. The recovery of nuclear magnetization above 10 K was expressed by a single exponential function. By contrast, the recovery of nuclear magnetization below 10 K was expressed by a stretched exponential function with $\beta < 1$, as

$$\frac{\mathrm{M}(\infty) - \mathrm{M}(t)}{\mathrm{M}(\infty)} = \exp\Big(- \Big(\frac{t}{T_1}\Big)^{\beta}\Big). \tag{4}$$

When $\beta = 1$, equation (4) is a single exponential function. Since $\beta < 1$, the recovery process below 10 K is considered to become inhomogeneous. The rate decreases steeply below 50 K ($T^{-1} = 0.02$ K^{-1}) and shows no critical divergence down to 2 K. This indicates that there is no magnetic ordering down to 2 K. When a gap energy Δ contributes to the relaxation process, the relaxation rate T_1^{-1} at $T \ll \Delta$ is proportional to an Arrhenius-type function [12],

$$\frac{1}{T_1} \propto \exp\Big(- \frac{\Delta}{T}\Big). \tag{5}$$

The temperature dependence of T_1^{-1} between 10 and 50 K indicates a gap energy of 50 K. Below

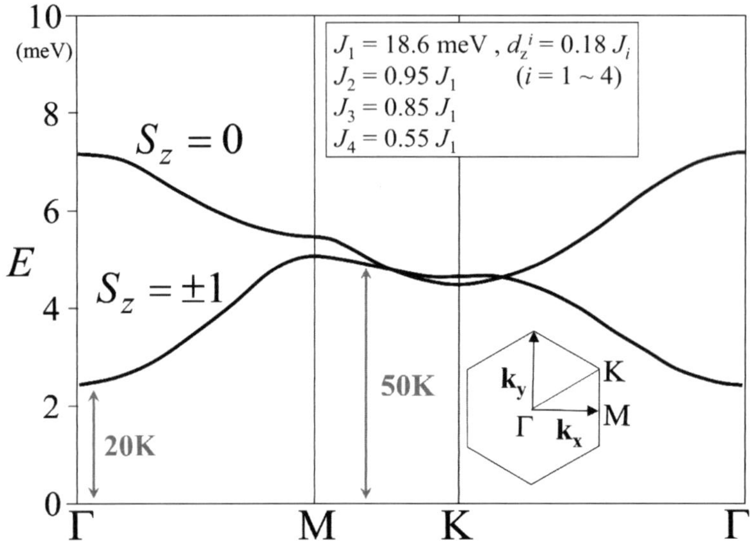

Figure 4. The dispersion of a triplet excitation obtained from the dimer series expansion of the pinwheel VBS state from ref. 9. The inset shows the Brillouin zone of the kagome lattice.

10 K, the temperature dependence of T_1^{-1} becomes weak. This behavior may be ascribed to the effect of relaxation due to impurity spins, since the recovery process becomes inhomogeneous below 10 K. Therefore the rate below 10 K may not reflect the intrinsic relaxation process.

The gap energy estimated from T_1^{-1} is rather large compared to the value of 21 K obtained from magnetic susceptibility and magnetization measurements [8]. This difference can be explained by the dispersion of the excited states, calculated theoretically by a dimer series expansion of the pinwheel VBS ground state using the Hamiltonian, which involves four different AF interactions and the DM interaction [9]. An illustration of the calculated dispersion is shown in figure 4. The splitting of the triplet excitation between $S_z = 0$ and $S_z = \pm 1$ is due to DM interaction and the large dispersions are derived from four AF interactions. As the relaxation rate T_1^{-1} depends on the state density of excitations, the NQR results between 10 and 50 K correspond to a magnetic excitation with a large state density with the gap energy 50 K. These excitations correspond to the excitations between the M-point and the K-point with flat dispersions in figure 4. On the other hand, the magnetic susceptibility at low temperatures is dominated by magnetic excitations with a minimum energy gap around the Γ-point.

In summary, a DM interaction and four different AF interactions produce a significant effect on the triplet excitation of a quantum antiferromagnet on a deformed Kagome lattice. Four AF interactions result in a large dispersion of the triplet excitation, and the DM interaction induces a splitting of the triplet excitation between $S_z = 0$ and $S_z = \pm 1$. The excitations with a large state density in the triplet state strongly affect the spin fluctuation, which arises a large relaxation of the nuclear spins.

Acknowledgments

This work was supported by Grant-in-Aid for Scientific Research on Priority Areas (19052005) from MEXT, Japan.

References

[1] Reimers J N and Berlinsky A J 1993 *Phys. Rev.* B **48** 9539
[2] Waldtmann Ch, Everts H -U, Bernu B, Lhuillier C, Sindzingre P, Lecheminant P and Pierre L 1998 *Eur. Phys. Jour.* **B2** 501
[3] Harris A B, Kallin C and Berlinsky A J 1992 *Phys. Rev.* B **45** 2899
[4] Yang B J, Kim Y B, Yu J and Park K 2008 *Phys. Rev.* B **77** 224424
[5] Yang B J and Kim Y B 2009 *Phys. Rev.* B **79** 224417
[6] Dzyaloshinsky I 1958 *J. Phys. Chem. Solids* **4** 241
[7] Moriya T 1960 *Phys. Rev.* **120** 91
[8] Morita K, Yano M, Ono T, Tanaka H, Fujii K, Uekusa H, Narumi Y and Kindo Y 2008 *J. Phys. Soc. Jpn.* **77** 043707
[9] Matan K, Ono T, Fukumoto Y, Sato J T, Yamaura J, Yano M, Morita K and Tanaka H 2010 *Nature Phys.* **6** 865-869
[10] Fukumoto Y 2009 private communication
[11] Yano M, Ono T and Tanaka H 2009 private communication
[12] Troyer M, Tsunetsugu H and Würtz D 1994 *Phys. Rev.* B **50** 13515

Magnetism of the Single Crystal Mg-V-O Spinel

Shigeo Hara[1], Hiroyuki Oshima[2], Takahiro Mastuda[3], Norio Umeyama[4], Shin-Ichi Ikeda[4], Miwako Takahashi[2], Tadataka Watanabe[3], and Keisuke Tomiyasu[5]

[1]Department of Physics, Chuo University, Kasuga, Bunkyo-ku, Tokyo 112-8551, Japan
[2]Institute of Materials Science, University of Tsukuba, Tennodai, Tsukuba, Ibaraki 305-8577, Japan
[3]Department of Physics, College of Science and Technology, Nihon University, Kanda Surugadai, Chiyoda-ku, Tokyo 101-8308, Japan
[4]Nanoelectronics Research Institute, National Institute of Advanced Industrial Science and Technology (AIST), Tsukuba, Ibaraki 305-8568, Japan
[5]Department of Physics, Tohoku University, Aoba-ku, Sendai, Miyagi 980-8578, Japan

E-mail: hrsgo@phys.chuo-u.ac.jp

Abstract. In MgV_2O_4 with normal spinel structure (AB_2O_4), there is magnetic frustration of V^{3+} ($3d^{2+}$) ions located on B-sites, if an antiferromagnetic interaction among B-site spins is essential for its magnetism. The excess Mg substituted materials such as $Mg_{1+x}V_{2-x}O_4$, may construct isolated magnetic triangular lattices, because some of B-sites forming magnetic tetrahedral lattice are substituted by non-magnetic Mg ions, while the normal spinel MgV_2O_4 contains tetrahedral magnetic ion networks. Here, we have succeeded in growing single crystals of Mg–V–O system spinel oxide by conventional floating-zone method. The valence numbers and abundance ratio of V ions in the single crystal were estimated by wavelength-dispersive X-ray fluorescence analysis and X-ray photoelectron spectroscopy analysis. Although the composition ratio was Mg:V = 1:2, a small amount of Mg and V ion partly replace each other between A- and B- sites in this crystal. The composition formula is described as $(Mg_{1-x}V_x)(V_{2-x}Mg_x)O_4$. The single crystal shows magnetic transition at around 12 K and large negative antiferromagnetic Weiss temperature about -840 K, and the frustration parameter $f = 70$ estimated by those temperatures. This result suggests the existence of strong magnetic frustration in our material.

1. Introduction

Geometrically frustrated spin systems have been merited attention to study those ground states and properties. The spinel compounds have one of the typical crystal structure for geometric frustrated systems, because magnetic ions on B-sites tetrahedral network (pyrochlore lattice) form the three-dimensional geometric frustration spin system, in the case that an antiferromagnetic interaction among B-site spins is essential for its magnetism.

There are reported about varieties of spinel oxide compounds such as MgV_2O_4, $Mg_{1.3}V_{1.7}O_4$, and Mg_2VO_4, in the Mg–V–O system [1,2]. Especially in MgV_2O_4 with normal spinel structure, it is known that there is a strong magnetic frustration at V^{3+} ($3d^{2+}$) ions located on B-sites. This compound shows large negative Weiss temperature around -750 K, and successive phase transitions in a low field. Although the one is a magnetic transition at 65 K accompanied with the cubic-tetragonal structural phase transition, the other is just a magnetic transition at 42 K without any structural modification [3-5]. Moreover, the effect of substitution on A or B sites such as $Li_xMg_{1-x}V_2O_4$ ($x = 0$–1) and $Mg(V_{1-x}Al_x)_2O_4$ ($x = 0$–0.25), had been studied. These materials also have normal spinel type structure. According to the previous reports [6], the effect of substitution of Mg^{2+} by Li^+ on the A-sites produces a change of the valence number on some of V ions from 3+ to 4+. As a result of variation in $Li_xMg_{1-x}V_2O_4$, while the successive phase transitions are observed on the magnetic susceptibilities for $0 \leq x < 0.07$, only a broad peak is observed for $0.07 \leq x \leq 0.7$. In $x = 0.8$ and 0.9, there is no sign of phase transition. In this case, the magnetic ions V^{3+} and V^{4+} are located on the whole B-sites.

On the other hand, when some of magnetic V^{3+} ions are substituted by non-magnetic Al^{3+} ions on B-sites [3,4], there is no change for the valence number. The substitution influences exchange interactions between magnetic ions located on the B-sites. In fact, on the magnetic susceptibility of $Mg(V_{1-x}Al_x)_2O_4$ in $x = 0.05$ and 0.15, spin glass behavior is observed instead of successive phase transitions. In addition, it seems that the magnetic frustration remains in these substituted compounds considering the relation between magnetic transition temperature and negative Weiss temperature.

Meanwhile, a Mg substituted Mg–V–O normal spinel oxide is also one of the examples with a different valence number and exchange interactions on the B-sites. Especially, $Mg_{1.3}V_{1.7}O_4$ with normal spinel type structure, its detail of crystal structure is already reported, contains not only magnetic tetrahedral but some of isolated triangular lattices [2], because some of B-sites forming magnetic tetrahedral lattice are substituted by non-magnetic Mg ions. In this material, V ions take valence number 3+ and 4+. Therefore, the different magnitude exchange interactions are produced between V ions. However, there is no report on magnetic properties for this Mg–V–O system normal spinel oxide. In order to understand the influence of compositive substitution, we have prepared high-quality single crystal grown by a floating zone (FZ) method. In this paper, we report the results of magnetic susceptibility measurements using the FZ single crystals of the frustrated magnet of Mg–V–O spinel.

2. Experiments

Polycrystalline Mg–V–O was synthesized by standard solid-state reaction techniques. The starting materials were MgO (99.9%) and V_2O_3 (99.9%) powders. We prepared several polycrystalline mixtures with molar ratio of MgO : $V_2O_3 = x : 1$ ($x = 1.0$–1.8). The mixed ingredients were formed into a rod with a diameter of 5 mm and pressed by a cold-isostatic press. Then, the rod was sintered at 1000°C for 6 h in a gas flow of Ar (99%) and H_2 (1%) at the rates 50 cc/min. The polycrystalline product was revealed to be black. For growing single crystals of Mg–V–O system spinel oxide, we employed a self-flux FZ method using image furnace with two-elliptical mirrors and two 1.5 kW halogen lamps (Model SC2-MDH-11020, NEC Machinery).

This furnace can form smaller molten zones. The length of molten zone was about 4 mm or less. The best condition was mixed Ar (99%) and H_2 (1%) 0.5 MPa, and scan speed 35 mm/h. We tried to grow the single crystal on several conditions, thereby we have succeeded in growing of single crystals. Typical dimensions of shaped crystals were $40 \times 3 \times 3$ mm^3. Single crystallizing was confirmed by the Laue method. Figure 1 shows a piece of single crystal MgV_2O_4 spinel, and the Laue photograph of a (*lll*) surface, as shown in Fig 2. The characterization of phases at room temperature was performed by powder X-ray diffraction (XRD) analysis diffractometer using Cu$K\alpha$ radiation (MiniFlex2, Rigaku Co.). In order to obtain the composition ratio of Mg and V element, we employed wavelength-dispersive X-ray fluorescence (XRF) analyses (SEA5120A, SII NanoTechnology Inc.). The valence number of V ion was confirmed by X-ray photoelectron spectroscopy (XPS) analysis (Quantum2000, ALVAC-PHI Inc.). Figure 3 shows a example of the diffraction pattern. The XRD pattern using crushed single crystals shows no impurity peaks. Furthermore, we investigated the structural phase transition between room temperature to 18 K (Ultrax18, Rigaku Co. and Model HC-2, APD CRYOGENICS Inc.). As a result, there is no

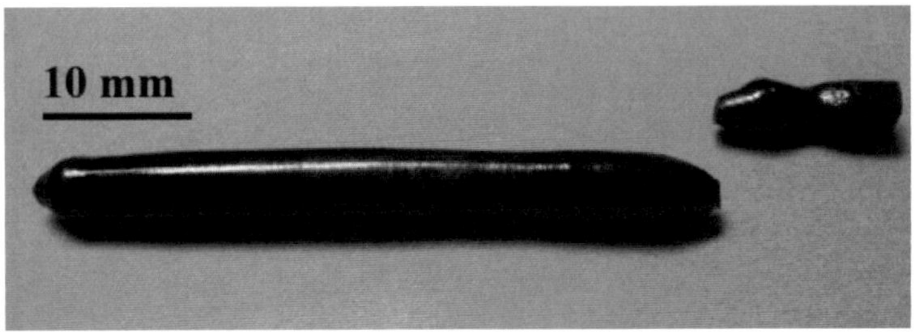

Figure 1. (Color online) The single crystal of MgV_2O_4 spinel.

Figure 2. The Laue photograph of (lll) surface for MgV_2O_4 spinel.

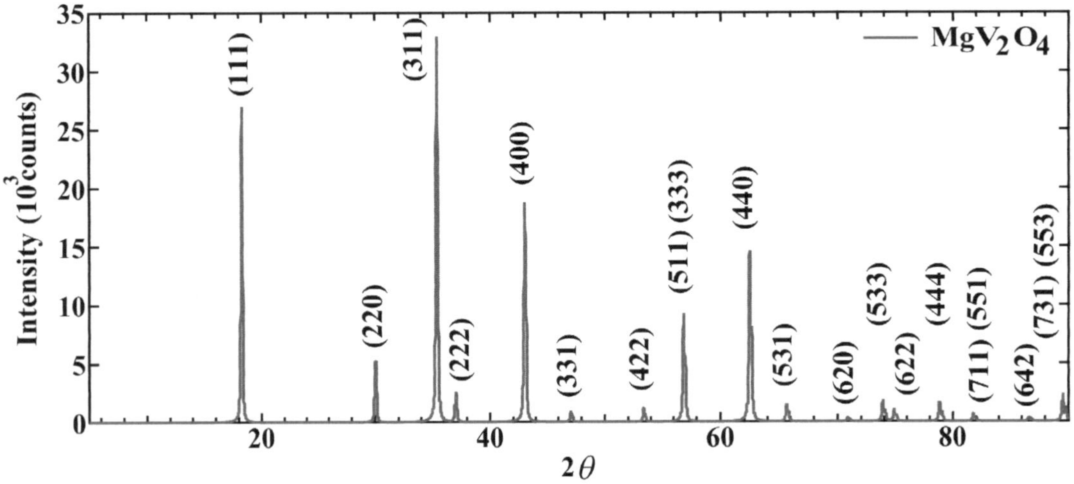

Figure 3. XRD patterns of MgV_2O_4 spinel at room temperatures.

change of the crystal structure down to 18 K.

Magnetic susceptibility, AC magnetic susceptibility and DC magnetization measurements were performed on the usual zero-field-cooled (ZFC) and field-cooled (FC) procedures with a superconducting quantum interference device (SQUID) magnetometer (Quantum Design MPMS-XL) in the temperature range between 2 and 300 K in a DC magnetic field from 0 to 50 kOe and an AC field amplitude 3.8 Oe, the frequency range from 1 to 1.5 kHz.

3. Results and discussion

From the results of spectral energy distribution in XRF analyses, we made sure the composition ratio of Mg:V = 1.0:2.0. The XPS analyses indicate that the valence number of V ion is 3+ [7]. Moreover, we tried to apply Rietveld analysis for the XRD pattern, and the XRD pattern showed that the crystal structure was the normal spinel structure. From XRD result, we confirmed that single-phase sample was obtained. These XRF, XPS, and XRD results proved the composition formula as $Mg^{2+}V^{3+}{}_2O_4$. However, the XRD pattern was not completely consistent with the MgV_2O_4, but almost similar to the pattern of $Mg_{1.3}V_{1.7}O_4$ [2]. In addition, the intensity ratio of this pattern has no accordance with the Mg_2VO_4 inverse-spinel type structures. So, it seems

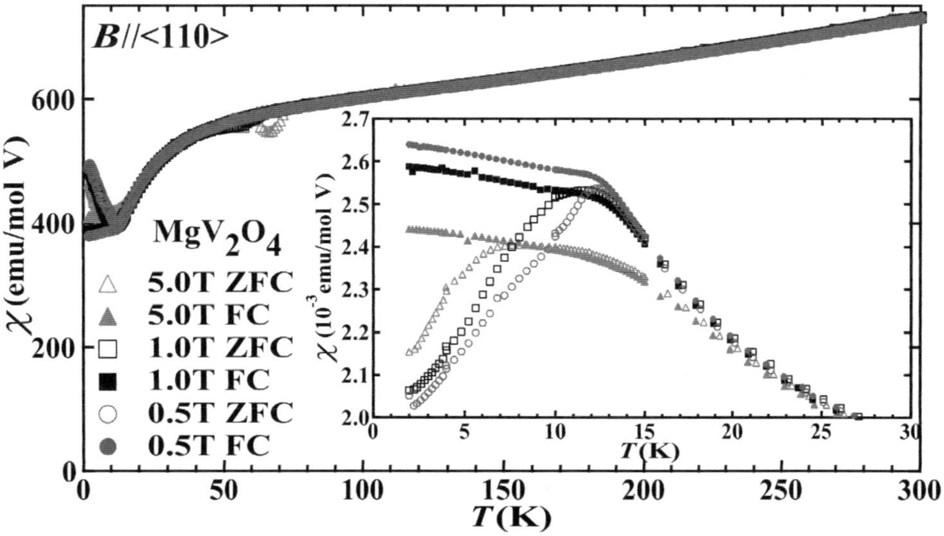

Figure 4. (Color online) Temperature dependence reciprocal susceptibility of single crystal $(Mg_{0.92}V_{0.08})(V_{1.92}Mg_{0.08})O_4$ spinel in whole temperature range measured under several magnetic fields. The inset shows the susceptibility in the low temperature region.

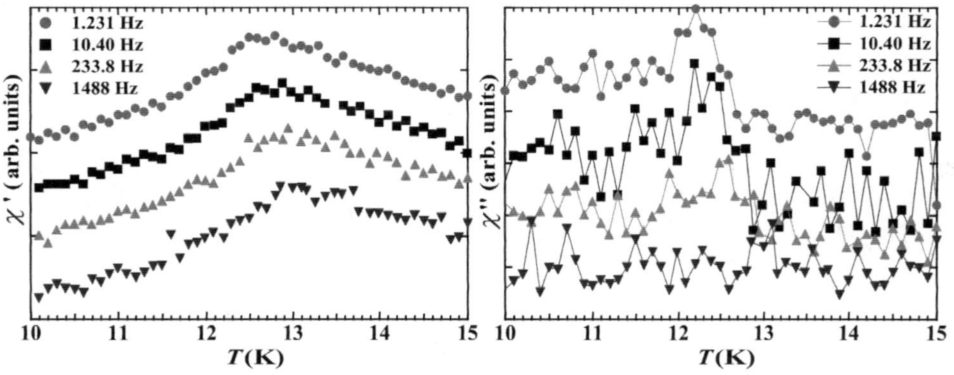

Figure 5. (Color online) Temperature dependence of AC susceptibility of $(Mg_{0.92}V_{0.08})(V_{1.92}Mg_{0.08})O_4$ spinel at around 13 K measured under various frequency, (left) real part and (right) imaginary part.

that a part of Mg and V ions replace each other in this material. Therefore, we presumed the improved composition formula as $(Mg_{1-x}V_x)(V_{2-x}Mg_x)O_4$. Rietveld analysis was applied for the XRD pattern in order to refine this composition formula. The best-fitted value of analysis result is $x = 0.08$, namely, $(Mg_{0.92}V_{0.08})(V_{1.92}Mg_{0.08})O_4$. There are sixteen equivalent B-sites per a unit-cell at the spinel structure. In the case of $x = 0.08$, it indicates that only one V ion should be substituted. In this partial inverse spinel case, almost all B-sites form magnetic tetrahedral lattices. Thus, the ratio also makes this material induce the magnetic frustration like MgV_2O_4. In this regard, however, there is a possibility to have effect upon the exchange interaction between A- and B-sites.

The temperature dependence DC magnetic susceptibilities under various fields show in Fig. 4.

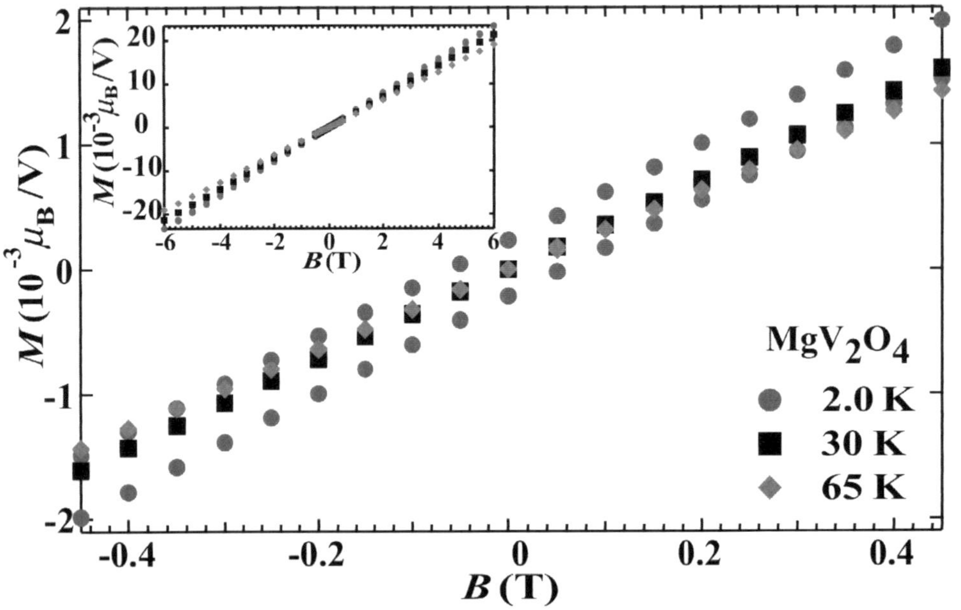

Figure 6. (Color online) Magnetic-field dependence of the magnetizations of $(Mg_{0.92}V_{0.08})(V_{1.92}Mg_{0.08})O_4$ spinel at several temperature. The inset shows the whole region of magnetizations.

The direction of the applied magnetic fields was ⟨110⟩ in this measurement. The spin-glass like behavior is observed at about $T_g = 12.5$ K. This behavior is similar to the $Mg(V_{0.85}Al_{0.15})_2O_4$. Figure 5 shows the AC magnetic susceptibility under various frequencies. The real part of AC magnetic susceptibilities show magnetic anomalies at around 12.5 K for all frequency. On the other hand, these anomalies at the imaginary part of one disappear above 233 Hz. However, because the range of error bars is large, we can not come to a decision whether this peak hides in the noise or vanishes at this stage. According to these results, these peaks on ZFC in χ_{DC} become broad in the higher field and those temperatures in χ_{AC} depend on the frequency, these anomalies ascribe spin glass transition.

Figure 6 shows the magnetic field dependence of the magnetizations for several temperatures. At 2.0 K, we observed a slight hysteresis in magnetization curve, although the loop did not appear above T_g. It seems that this remnant magnetization is caused by freezing spins in spin glass phase. The reason, why this partial inverse spinel shows spin glass behavior, may be explained by its characteristic alignment of V ions. In the case of our partial inverse spinel, some of Mg and V ions are exchanged each other, the normal spinel MgV_2O_4 reveals Mg(V) ions at A-(B-)sites. According to these results, the exchange interaction between A- and B-sites is produced and probably involves the ferromagnetic interaction. It seems that this complicated interaction brings on spin glass behavior.

Above T_g, the magnetic susceptibility seems to obey the Curie-Weiss law with a temperature-independent paramagnetic contribution. We fitted the observed susceptibility at temperatures between 100 and 300 K with the formula, $\chi(T) = \chi_0 + C / (T - \Theta_W)$, where χ_0, C, and Θ_W are the temperature-independent term, the Curie constant and the Weiss temperature, respectively. The best-fitted values of χ_0, C, and Θ_W are 1.8×10^{-5} emu/mol V, 0.63 emu·K/mol V, and 820 K, respectively. The frustration parameter $f = 70$ estimated by Weiss temperature suggests the existence of strong magnetic frustration. Assuming $g = 2.0$ and $L = 0$, the Curie constant reminds of the quantum spin number $S = 0.73$. Thus, this value reminds that the V ions take the valence number 3+ and 4+. However, as a result of XPS, V ions in this material take only 3+. The effective Bohr - magneton $p_{\text{eff}} = 2.25\ \mu_{\text{B}}$ is deduced from the Curie constant. This value is smaller than normal spinel MgV_2O_4 of its. This result suggests that the g-value is not 2. It is speculation that the spins on V ions are observed smaller than $S = 1$ by strong magnetic frustration. With increasing magnetic field, at around 60 K, a magnetic anomaly appears in susceptibility. This anomaly can not be observed in low field and the result of X-ray powder diffraction pattern establishes no structural phase transition at around 60 K. However, at this time, we do not have clear description of this transition.

4. Conclusions

Using FZ method, we have succeeded in growing single crystals of $(Mg_{0.92}V_{0.08})(V_{1.92}Mg_{0.08})O_4$ with normal spinel type structure oxide. The composition ratio was estimated by Rietveld analysis. We found the existence of strong magnetic frustration and observed spin glass behavior. There is no evidence for structural phase transition at T_g. To research this question is issue in the future.

5. Acknowledgments

This work was supported by the MEXT of Japan, Grants in Aid for Scientific Researches on Priority Areas (22014001). A part of this work was conducted at the Nano-Processing Facility, supported by IBEC Innovation Platform, AIST.

6. References

[1] W. Ruedorff and B. Reuter 1947 *Zeitschrift fuer Anorganische und Allgemeine Chemie* **153** 194
[2] H. Uchida, K. Righter, B. Lavina, M. M. Nowell, S. I. Wright, R. T. Downs AND H. Yang 2007 *American Mineralogist* **92** 1031
[3] H. Mamiya and M. Onoda 1995 *Solid State Commun.* **95** 217
[4] H. Mamiya, M. Onoda, T. Furubayashi, J. Tang and I. Nakatani 1997 *J. Appl. Phys.* **81** 5289
[5] E. M. Wheeler, B. Lake, A. T. M. Nazmul Islam, M. Reehuis, P. Steffens, T. Guidi and A. H. Hill 2010 *Phys. Rev.* B **60** 140406
[6] M. Onoda, H. Imai, Y. Amako and H. Nagasawa 1997 *Phys. Rev.* B **56** 3760
[7] unpublised

International Conference on Frustration in Condensed Matter (ICFCM) IOP Publishing
Journal of Physics: Conference Series **320** (2011) 012054 doi:10.1088/1742-6596/320/1/012054

Quantum Monte Carlo study of the transverse- eld Ising model on a frustrated checkerboard lattice

H. Ishizuka, Y. Motome, N. Furukawa[A,B]**, and S. Suzuki**[A]

Department of Applied Physics, University of Tokyo, Japan
[A]Department of Physics and Mathematics, Aoyama Gakuin University, Japan
[B]Multiferroics Project, ERATO, Japan Science and Technology Agency (JST), c/o Department of Applied Physics, University of Tokyo, Japan

E-mail: ishizuka@aion.t.u-tokyo.ac.jp

Abstract. We present the numerical results for low temperature behavior of the transverse-field Ising model on a frustrated checkerboard lattice, with focus on the effect of both quantum and thermal fluctuations. Applying the recently-developed continuous-time quantum Monte Carlo algorithm, we compute the magnetization and susceptibility down to extremely low temperatures while changing the magnitude of both transverse and longitudinal magnetic fields. Several characteristic behaviors are observed, which were not inferred from the previously-studied quantum order from disorder at zero temperature, such as a horizontal-type stripe ordering at a substantial longitudinal field and a persistent critical behavior down to low temperature in a weak longitudinal field region.

1. Introduction

Frustration due to the geometrical structure of systems has been one of the major topics in condensed matter physics. The interest covers a wide range of fields, such as frustrated magnetism [1], proton ordering in hydrogen-bonded systems [2, 3], and anomalous transport and multiferroics in transition metal oxides [4, 5, 6]. The geometrical frustration often prevents the system from selecting a unique ground state, instead gives rise to classically degenerate ground states; i.e., a macroscopic number of different configurations of the system variables lead to the same ground state energy at the classical level. As a result, the ground state remains to be disordered and bears finite residual entropy. Such macroscopic degeneracy provides a fertile ground for various peculiar phenomena as it makes the system to be extremely sensitive to perturbations, such as remnant interactions, external fields, and fluctuations.

Among such degeneracy-lifting mechanisms, quantum and thermal fluctuations have attracted particular interest. These fluctuations sometimes lift the ground-state degeneracy and induce some particular ordering, as known by order from disorder [7]. There, the ordering is selected from the manifold to maximize the associated entropy (zero-point entropy in the case of quantum fluctuation). This entropic effect plays a decisive role at low temperatures (T) and causes many fascinating phenomena in the frustrated systems.

One of the minimal models for studying the order-from-disorder phenomenon is the transverse-field Ising model (TIM). In the absence of geometrical frustration, in general, the model develops a long-range order at low T, while the ordering is suppressed and a quantum paramagnetic state is induced by the transverse field. When the classical ground state is macroscopically degenerate under strong frustration, quantum fluctuations by the transverse field and/or thermal fluctuations by temperature can induce a particular ordering via the order-from-disorder mechanism. Effect of thermal fluctuations has been studied mainly in the absence of the transverse field, i.e., for pure Ising models without the transverse field [1, 8]. Meanwhile, effect of quantum fluctuations has also been studied. For example, it was shown that TIM on a variety of frustrated lattices exhibit several nontrivial behaviors at $T = 0$, such as a bond ordering and Kosterlitz-Thouless transition [9, 10]. In general, the thermal and quantum fluctuations do not necessarily lead to the same effect, and the relation between them is of particular interest to explore yet another order-from-disorder phenomenon.

In this contribution, we present our numerical results for the order-from-disorder phenomena in TIM in a wide range of temperature and the transverse/longitudinal magnetic fields. We consider the TIM on a two-dimensional checkerboard lattice and study its low-T physics by a sophisticated quantum Monte Carlo (QMC) technique. As a result, in the intermediate longitudinal-field regime, we find instability toward a Neel order, in accordance with the previous report [10]. On the other hand, for both the weaker and stronger longitudinal fields, our results indicate different behaviors from the previous report. Under the weak field, featureless magnetic susceptibility is observed down to extremely low T, which implies a very weak proximity effect to the predicted Neel ground state or a possibility of another state in the $T = 0$ limit. On the contrary, in the strong field, the system indicates instability toward an unexpected horizontal-type stripe ordering.

2. Model and method

To investigate the effect of quantum and thermal fluctuations on the classically degenerate ground state in frustrated systems, here we focus on the frustrated checkerboard Ising model with transverse and longitudinal magnetic fields. The Hamiltonian is given as:

$$H = J\sum_{\langle i,j\rangle} s_i^z s_j^z + \quad \sum_i s_i^x + h\sum_i s_i^z, \tag{1}$$

where s_i^α is the Pauli spin operator at site i, and J is the Ising interaction between nearest-neighbor sites $\langle i,j\rangle$ on the checkerboard lattice (see Fig. 1); and h are the transverse and longitudinal fields, respectively. In the following, we consider the antiferromagnetic case, $J > 0$, and set the energy scale by $J = 1$.

When $= h = 0$, the model remains to be disordered down to zero T [8]. The lowest energy is achieved by enforcing a simple local constraint — a two-up two-down local spin configuration for all the plaquettes with crisscrossing interaction, similar to the so-called ice rule in water ice [2]. Consequently, a macroscopic number of different spin configurations give the same lowest energy: The ground state is macroscopically degenerate and the residual entropy is estimated to be $\sim \frac{N}{2}\log\frac{3}{2}$ (N is the number of spins). The spin correlation, however, is critical in the sense that it decays algebraically as a function of distance. The situation is unaltered as h increases up to $h = 2J$ at $= 0$. (For larger h, the ground state consists of "three-up one-down" configurations, which is also macroscopically degenerate.) When the quantum fluctuation sets in by switching on , it was predicted, from perturbative considerations in $/h$, that a Neel order is induced in the $0 < h < 2J$ region at $T = 0$ [10] [see Fig. 1(a)]. Hereafter, we focus on the $0 < h < 2J$ region.

To investigate the low-T behaviors of this model with taking account of both quantum and thermal fluctuations, we conducted numerical calculations by QMC method based on the path-integral approach [11, 12]. Here, we employed the recently-developed continuous-time algorithm [13] to approach extremely low T. At low T, the MC sampling suffers from the slow relaxation due to the severe frustration. To accelerate the relaxation process, we applied the

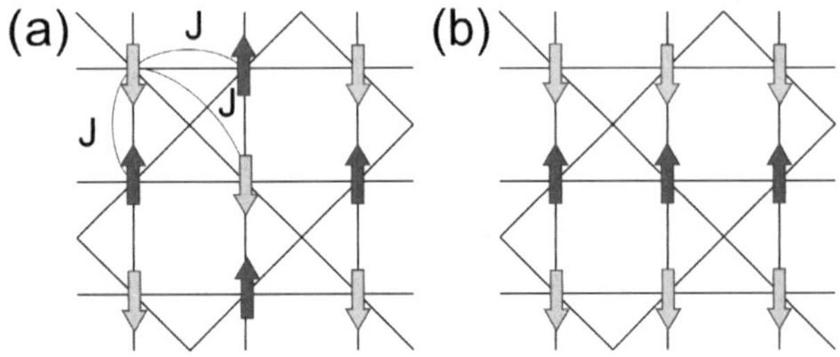

Figure 1. Schematic pictures of the transverse-field Ising model on the checkerboard lattice. (a) Neel type order and (b) horizontal stripe order.

loop-flip algorithm developed for frustrated Ising-type spin systems [14, 15]. In our calculations, the loops are formed in the real space at a particular imaginary-time slice, and flipped together with all the spin variables along the imaginary-time direction, i.e., on the torus defined by the loop in space and time. In addition, the replica exchange method was used to further suppress the slowing down [16]. For the present system, the replicas are exchanged along a constant- $/T$ line because the Boltzmann weight depends on the number of domain walls in the imaginary-time direction which is proportional to $/T$.

Calculations were done for $N = 4 \times L^2$ site systems with $L = 12$ under the periodic boundary conditions. Observables were computed typically for 120000 samplings after 30000 steps of thermalization. The results were divided into eight bins to evaluate the statistical errors.

3. Results and discussions

Figure 2 shows the QMC results for the susceptibility at a relatively weak longitudinal field $h = 0.4$ with varying . Figure 2(a) presents the susceptibility for the Neel-type order parameter [Fig. 1(a)],

$$_{\text{Neel}} = \frac{N}{T}(\langle m_{\text{Neel}}^2 \rangle - \langle |m_{\text{Neel}}| \rangle^2) \; ; \quad m_{\text{Neel}} = \frac{1}{N}\sum_i s_i^z (-1)^{i_x + i_y}, \tag{2}$$

while Fig. 2(b) is that for the horizontal-type stripe order parameter [Fig. 1(b)], which we consider as another candidate of order from disorder,

$$_{\text{stripe}} = \frac{N}{T}(\langle m_{\text{stripe}}^2 \rangle - \langle m_{\text{stripe}} \rangle^2) \; ; \quad m_{\text{stripe}}^2 = \Big(\frac{1}{N}\sum_i s_i^z (-1)^{i_x}\Big)^2 + \Big(\frac{1}{N}\sum_i s_i^z (-1)^{i_y}\Big)^2. \tag{3}$$

Note that this stripe state belongs to the two-up two-down manifold and is different from the diagonal one discussed in the three-up one-down manifold for $2J < h < 6J$ [10]. Figure 2(c) shows the local correlation parameter given by $= \frac{2}{N}\sum_p f(p)$, where the sum runs over all plaquettes. $f(p)$ is a function defined for each plaquette p as $f(p) = 1$ for $\sum_{i \in p} s_i^z = 0$ and otherwise $f(p) = -3/5$; $\to 1$ when all the plaquettes satisfy the two-up two-down spin configuration, and $\to 0$ when the spins are completely disordered, i.e., as $T \to \infty$ and/or $\to \infty$. As shown in Figs. 2(a) and 2(b), for weak $\lesssim 0.5$, both the susceptibilities increase with decreasing T in Curie-law like behavior, i.e., $\propto T^{-1}$. A shoulder-like feature observed at $T \sim 0.3$ in $_{\text{stripe}}$ corresponds to a formation of the ice-rule type local configurations, as indicated by the saturation of in Fig. 2(c). These behaviors suggest that the system is in the correlated regime with satisfying the local constraint, but remains critical, without choosing either Neel or stripe ordering, presumably because of the strong frustration. We also calculate the momentum dependence of the susceptibility, and confirm that it remains featureless without showing any peak in the momentum space. It is surprising that the critical behavior is robustly observed down to very low $T \sim 0.01$. This implies that the order-from-disorder mechanism is extremely weak or ineffective in this region. On the other hand, the results for $\gtrsim 0.6$ show

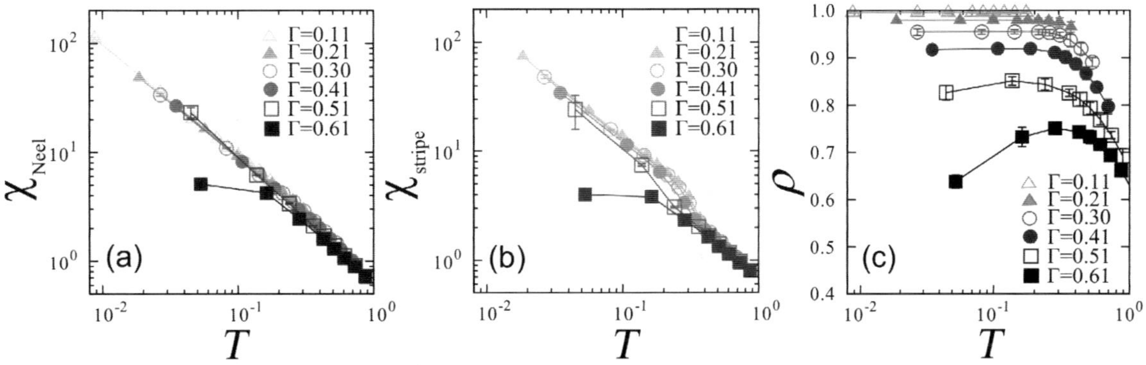

Figure 2. T dependence of (a) $_{\mathrm{Neel}}$, (b) $_{\mathrm{stripe}}$, and (c) local correlation parameter at $h = 0.4$ for the transverse field ranging from $= 0.11$ to 0.61.

saturating behavior at low T. In this region, is strong enough to disturb the ice-rule type configurations [Fig. 2(c)], and the ground state turns into a quantum paramagnetic state.

With increasing the longitudinal field, a contrasting behavior between the two susceptibilities shows up in the vicinity of phase transition to the quantum paramagnetic state. Figure 3 shows the results at $h = 0.9$. In the region $\lesssim 0.3$, both $_{\mathrm{Neel}}$ and $_{\mathrm{stripe}}$ diverge toward $T = 0$, in a similar manner to the results for $\lesssim 0.5$ in Fig. 2. On the other hand, before entering the quantum paramagnetic regime for $\gtrsim 0.5$ where both $_{\mathrm{Neel}}$ and $_{\mathrm{stripe}}$ saturate, a qualitatively different behavior is observed: At ~ 0.40, $_{\mathrm{stripe}}$ begins to deviate from the Curie-law like behavior and tends to saturate, while $_{\mathrm{Neel}}$ remains to diverge (indicated by the arrows in Fig. 3). This implies that the stripe-type fluctuation is suppressed and the Neel-type ordering is likely favored in the low-T limit. Similar behavior is observed in a finite region on the verge of the quantum paramagnetic phase for $0.5 < h < 1.0$ (see the phase diagram in Fig. 5 below).

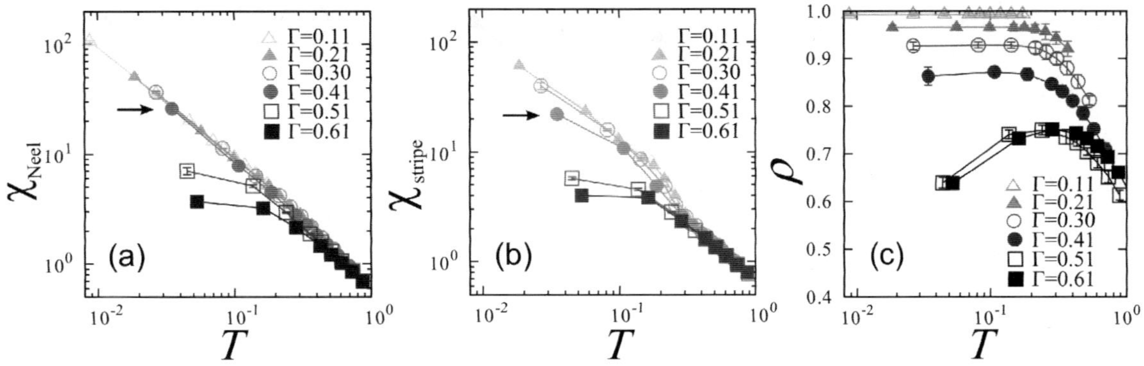

Figure 3. T dependence of (a) $_{\mathrm{Neel}}$, (b) $_{\mathrm{stripe}}$, and (c) local correlation parameter at $h = 0.9$ for the transverse field ranging from $= 0.11$ to 0.61. The arrows in (a) and (b) indicate a contrasting behavior between $_{\mathrm{Neel}}$ and $_{\mathrm{stripe}}$ in the intermediate region.

For larger longitudinal fields, yet another behavior appears. For $h \gtrsim 1.0$, the MC relaxation becomes very slow at low T even with the use of the loop-flip algorithm and the replica-exchange technique. The situation is demonstrated in Fig. 4. The results show the relaxation of the order parameters m_{Neel} and m_{stripe} at $h = 1.6$ along a constant $/T = \tan(85°)$ axis; the data are measured for 50000 MC steps after particular thermalization steps of N_{therm} starting from the initial configuration with corresponding perfect order: Figures 4(a) and 4(b) show the results starting from the perfectly-Neel-ordered state and the perfectly-horizontal-stripe-ordered state, respectively. As shown in Fig. 4, $m_{\rm Neel} \sim 0$ after $N_{\rm therm} = 10^5$ thermalization independent of the initial state. On the other hand, the $m_{\rm stripe}$ shows strong dependence on the initial state even after $N_{\rm therm} \sim 10^5$ steps. A possible origin of the freezing behavior is strong first order transition to the horizontal-stripe-ordered state. Although the extremely slow relaxation prevents us from obtaining numerically-converged results, the results suggest that the system has an instability toward the horizontal stripe ordering at low T in this high-h region.

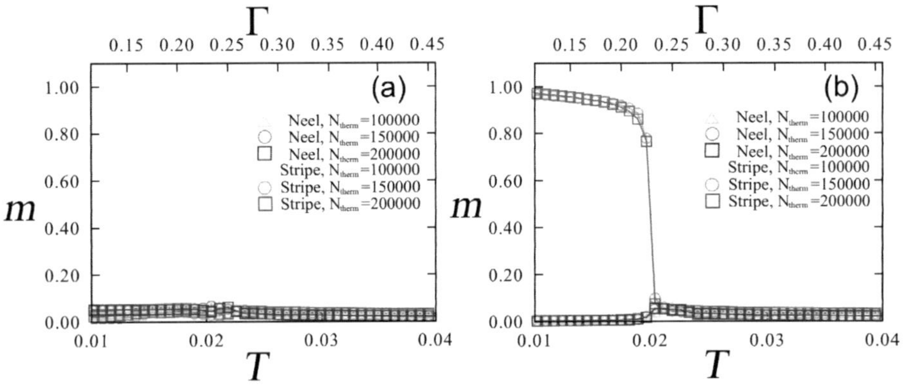

Figure 4. T dependence of $m_{\rm Neel}$ and $m_{\rm stripe}$ at $h = 1.6$ and $/T = \tan(85°)$ for varying thermalization steps $N_{\rm therm}$. (a) is the results starting from the perfectly-Neel-ordered configuration and (b) is from the perfectly-horizontal-stripe-ordered one.

Summarizing the results above, we deduce a 'phase diagram' in Fig. 5, which indicates the low-T instability anticipated from the MC data. In the strong longitudinal field region, $1.0 \lesssim h \lesssim 1.8$ and $\lesssim 0.2$, our MC results indicate an instability toward the horizontal-stripe-type ordering below some particular temperature, as exemplified in Fig. 4(b). Although we could not directly confirm the phase transition because of the extremely slow MC relaxation, we expect this phase to appear at a finite temperature. In the intermediate longitudinal field region, $0.5 \lesssim h \lesssim 1.0$ and $0.3 \lesssim \ \lesssim 0.5$, the results appear to favor the Neel-type instability. The Neel temperature appears to be extremely low, less than $T = 0.01$. In the remaining low-h region for $\lesssim 0.5$, we could not observe any sign of clear instability toward either Neel or stripe type, at least for $T > 0.01$; the system is strongly correlated but remains to be critical under the frustration.

Let us compare the result to the $T = 0$ argument by Moessner and Sondhi [10]. In the previous study, by a perturbation in $/h$, it was predicted that a Neel-type order occurs in the entire region of our consideration. Our result in Fig. 5 shows a similar tendency in the intermediate h, where the Neel-type order appears to be favored. However, we could not detect any sign of a particular ground state in the lower-h region, although the Neel-type order was expected to extend down to $h \to 0$ in the schematic $T = 0$ phase diagram in the previous study [10]. This apparent disagreement can be due to very weak degeneracy lifting by the order from disorder for the expected Neel ground state. Another possibility is that, for large $/h$, the perturbative argument in $/h$ no longer holds and a different phase emerges in the region. In order to identify the nature of the system in this region, further analysis down to lower T in larger system sizes is necessary. On the other hand, under relatively strong field $1.0 \lesssim h \lesssim 1.8$, we detected instability toward another phase from the slow relaxation behavior, namely, the horizontal-stripe phase. This might be related with the competition in the ground states between the two-up two-down Neel order and three-up one-down diagonal-stripe order at $h \simeq 2$. Our preliminary calculations show a finite-size effect in this regime; the horizontal-stripe type instability appears to extend to slightly lower-h region in larger size systems. More detailed results will be discussed elsewhere.

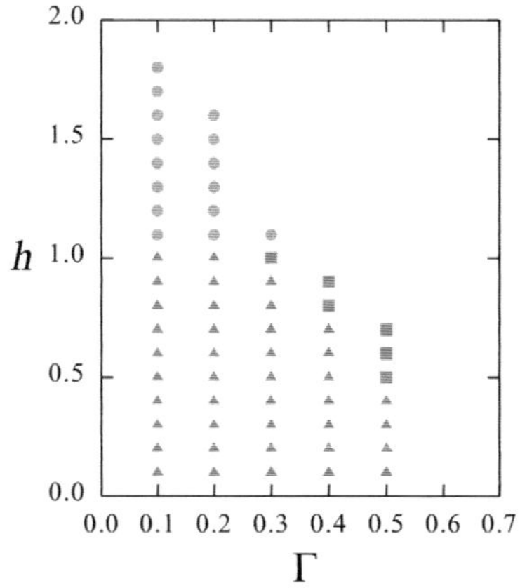

Figure 5. 'Phase diagram' for the model in Eq. (1) as functions of transverse field and longitudinal field h. The results indicate the low-T instabilities which are deduced from the QMC results down to $T \simeq 0.01$. The blue squares (orange circles) show the region in which the Neel(horizontal stripe)-type instability shows up. The green triangles show the region where both $_{\mathrm{Neel}}$ and $_{\mathrm{stripe}}$ remain to show the Curie-law like divergence down to the lowest T. The area without symbols for > 0 indicates the quantum paramagnetic region.

4. Summary

To summarize, we have investigated the effect of quantum and thermal fluctuations on the thermodynamics of the frustrated checkerboard Ising model with transverse and longitudinal magnetic fields. We have examined the magnetic instabilities induced by order-from-disorder mechanism by employing a continuous-time quantum Monte Carlo method with the loop-flip update and replica exchange algorithm. We identified characteristic behaviors, which were not inferred from the previous study for the quantum order from disorder in the ground state. One is the instability toward a horizontal-stripe-type ordering in a relatively high longitudinal-field regime. This is distinguished from the diagonal stripe predicted for a higher field. The other is no clear indication of dominant ordering or fluctuation down to $T \simeq 0.01J$ in a relatively low-field regime; the susceptibility shows a persistent Curie-law-like divergence at all the momenta. This could be due to the emergence of a novel state or simply by surprisingly weak order-from-disorder effect in the Neel ground state expected in the previous theory.

H.I. and Y.M. thank T. Misawa, Y. Motoyama, H. Shinaoka, and M. Udagawa for helpful comments. This research was supported by KAKENHI (No. 19052008 and No. 22540372), and Global COE Program"the Physical Sciences Frontier".

References

[1] For a recent review, Diep H T 2004 *Frustrated Spin Systems* (World Scientific Publishing)
[2] Pauling L 1935 *J. Am. Chem. Soc.* **57** 2680
[3] Slater J C 1941 *J. Chem. Phys.* **9** 16
[4] Anderson P W 1956 *Phys. Rev.* **102** 1008
[5] Taguchi Y, Oohara Y, Yoshizawa H, Nagaosa N and Tokura Y 2001 *Science* **291** 2573
[6] Kimura T, Goto T, Shintani H, Ishizaka K, Arima T, and Tokura Y 2003 *Nature* **426** 55
[7] Villain J, Bidaux R, Carton J P and Conte R J 1980 *J. Phys. (Paris)* **41** 1263
[8] Liebmann R 1986 *Statistical Mechanics of Periodic Frustrated Ising Systems* (Springer-Verlag, Berlin-Heidelberg-New York-Tokyo)
[9] Moessner R, Sondhi S L and Chandra P 2000 *Phys. Rev. Lett.* **84** 4457
[10] Moessner R and Sondhi S L 2001 *Phys. Rev. B* **63** 224401
[11] Suzuki M 1976 *Prog. Theor. Phys.* **56** 1454
[12] Suzuki M 1976 *Phys. Lett. A* **60** 478
[13] Nakamura T 2008 *Phys. Rev. Lett.* **101** 210602
[14] Rahman A and Stillinger F H 1972 *J. Chem. Phys.* **57** 4009
[15] Melko R G, den Hertog B C, and Gingras M J P 2001 *Phys. Rev. Lett.* **87** 067203
[16] Hukushima K and Nemoto K 1996 *J. Phys. Soc. Jpn.* **65** 1604

Novel electronic states close to Mott transition in low-dimensional and frustrated systems

D I Khomskii

II Physikalisches Institut, Universität zu Köln, Zülpicher Str. 77, 50937 Köln, Germany

E-mail: khomskii@ph2.uni-koeln.de

Abstract. Recent studies demonstrated that there may appear different novel states in correlated systems close to localized–itinerant crossover. Especially favourable conditions for that are met in low-dimensional and in frustrated systems. In this paper I discuss on concrete examples some of such novel states. In particular, for some spinels and triangular systems there appears a "partial Mott transition", in which first some finite clusters (dimers, trimes, tetramers, heptamers) go over to the itinerant regime, and the real bulk Mott transition occurs only later. Also some other specific possibilities in this crossover regime are shortly discussed, such as spin-Peierls–Peierls transition in TiOCl, spontaneous charge disproportionation in some cases, etc.

1. Introduction

The subdivision of electronic states into delocalized, itinerant, and localized, strongly correlated states is one of the most fundamental ones in the description of solids. Crudely, delocalized states exist when the intersite electron hopping is larger than the typical electron–electron interaction energy, and localized states exist in the opposite limit. In quantum chemistry these two limits correspond e.g. to the description of electrons in terms of molecular orbitals (Mulliken description, known in chemistry as molecular orbital–linear combination of atomic orbitals (MO–LCAO) method), and strongly correlated electrons are described by the Heitler–London approximation. For concentrated solids these two regimes give either standard band systems (metals or semiconductors), or Mott insulators.

The electron–electron repulsion, e.g. the Hubbard's U, is essentially an atomic property and does not depend on interatomic distance, whereas the electron hopping strongly depends on this distance. This leads to the concept, largely advocated e.g. by Goodenough [1], that there should exist a critical interatomic distance R_c (different for different transition metal (TM) ions) dividing the regions of localized and itinerant behaviour of electrons. When this distance changes, e.g. under pressure, there should occur an insulator–metal (Mott) transition.

Usually Mott transition in TM compounds such as TM oxides is supposed to occur homogeneously in the whole sample. But, as became clear recently, this is not the only possibility. There may appear novel, inhomogeneous states close to a Mott transition, so that we have, in a sense, a "partial" Mott transition: in some parts of the system, in some particular clusters, the interatomic distances may already be smaller than R_c, so that these clusters may be better described using molecular orbitals, whereas the distance *between* such clusters will still remain

Journal of Physics: Conference Series **320** (2011) 012055 doi:10.1088/1742-6596/320/1/012055

large, so that there will be no net metallic conductivity. The formation of such "metallic" clusters has much in common with the valence bond solids.

This phenomenon, i.e. such "fractional" Mott transition, is most often observed in low-dimensional and in frustrated systems. Indeed, in regular three-dimensional lattices such as perovskites we usually have homogeneous ordered states, e.g. 3d antiferromagnetic ordering. Molecular-type clusters such as valence bond states are much more plausible in low-dimensional and frustrated systems. Below I will describe this situation using several specific examples of systems with triangular and pyrochlore (spinel) lattices, and also on some quasi-one-dimensional systems. I will also shortly mention some other possible novel phenomena, which may appear in systems close to the localized–itinerant crossover.

2. Formation of dimers in spinels

Metal–insulator transitions with the formation of unusual structures are observed in many spinels. The recent examples are $MgTi_2O_4$ [2], $CuIr_2S_4$ [3] or AlV_2O_4 [4]. Structural studies have shown that in all these cases rather unusual structural modifications take place, which often can be described as the formation of molecular clusters. In $MgTi_2O_4$ there appears a "chiral" structure [2], in $CuIr_2S_4$ Ir octamers are formed [3], but both these phenomena can in fact be explained by the formation of Ti or Ir *singlet dimers*, which in a frustrated lattice of B-sites of a spinel finally give rise to these chiral or octamer structures [5]. And in AlV_2O_4 the molecular clusters formed consist of 7 vanadiums — real large "heptamer molecules".

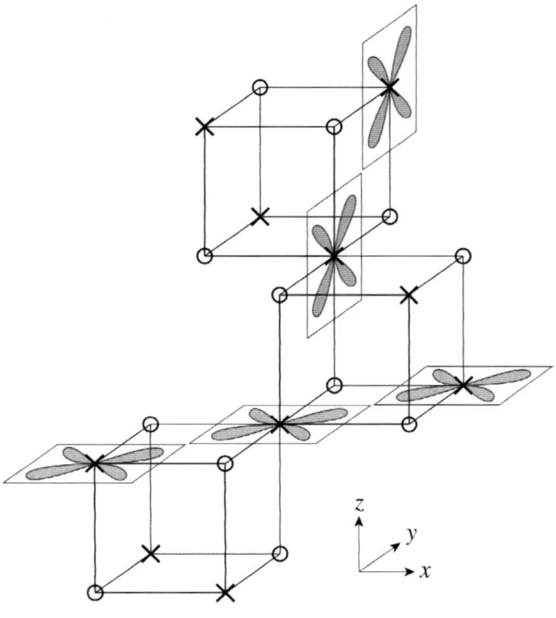

Figure 1. Formation of one-dimensional electronic spectrum for t_{2g} electrons in a lattice of B-sites of a spinel. The xy and yz orbitals are shown.

In all these cases the formation of these clusters is largely caused by specific features of the B-site spinel lattice, which may be represented as a collection of 1d chains running in xy (or $x\bar{y}$), xz and yz directions, see Fig. 1. The t_{2g}-orbitals of Ti or V ions on this lattice have a strong direct d–d overlap, such that e.g. the electrons from the xy-orbitals can hop to the same orbitals in neighbouring ions in the xy-direction, see Fig. 1, or similarly yz-electrons can hop in the yz-direction, etc. As a result the electronic structure in these, basically cubic crystals, has essentially one-dimensional character. And in the insulating phase one forms a Peierls state with singlet dimers in these 1d chains [5], which for a spinel lattice give the chiral structure in $MgTi_2O_4$, the heptamers in AlV_2O_4 and octamers (consisting actually of dimers) in $CuIr_2S_4$. Note that the interatomic distances in these clusters are rather short: they are shorter than the critical distance R_c for the localized–itinerant crossover [1], so that one can consider the electronic state in these clusters as forming molecular orbitals, i.e. in these clusters we are already on the "metallic" side of the Mott transition. But the distance *between* these clusters is larger than R_c and as a result the total compound is insulating.

The "metallic" clusters in the examples discussed above are singlet. But this is not the only possibility. Magnetic (e.g. triplet) clusters can also be formed in some cases. This seems to happen in ZnV_2O_4 [6]. V spinels such as MgV_2O_4, ZnV_2O_4, CdV_2O_4 show cubic–tetragonal structural transitions [7] with the formation of a very specific magnetic structure at lower temperatures, with $\uparrow\uparrow\downarrow\downarrow$ pattern in xz- and yz-chains (and with $\uparrow\downarrow\uparrow\downarrow$ ordering in the xy-direction) [8], see Fig. 2. Usually the properties of these systems are explained by some type of orbital ordering: the xy-orbitals are always occupied due to tetragonal distortion with $c/a < 1$, and the remaining second t_{2g} electron of the $V^{3+}(d^2)$ ions is supposed to form an additional orbital ordering of some type [9, 10, 5]. But our recent ab initio calculations [6] have shown that there should exist in ZnV_2O_4 a V dimerization in xz- and yz-chains, see Fig. 2, with, surprisingly, ferromagnetic bonds becoming shorter (thick bonds in Fig. 2)! One can explain this tendency by noticing that, in contrast to most examples of valence bond solids with singlet bonds, here we are dealing not with ions with one electron with $S = \frac{1}{2}$, but with d^2 ions with $S = 1$. One can argue that the extra delocalization of one of two electrons in a short bond enhances the ferromagnetic (actually double-exchange) interaction between these d^2, $S = 1$ ions. The ab initio calculations of another material of this family, CdV_2O_4, done by a different method, have confirmed the formation of these dimers, and demonstrated, in addition, that their formation can lead to the appearance of ferroelectricity, which was indeed observed in CdV_2O_4 [11].

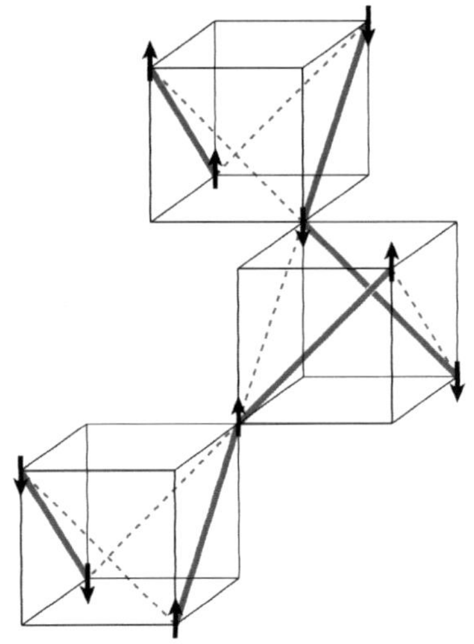

Figure 2. Formation of short (bold) and long (dashed) V–V bonds in ZnV_2O_4, following [6]; the same pattern also exists in CdV_2O_4 [11].

Interestingly enough, short V–V distances obtained theoretically for ZnV_2O_4 are only 2.92 Å — again shorter than the critical V–V distance 2.94 Å [1]. Thus again these V dimers may be considered as "metallic", though the material itself remains insulating, albeit with a rather small energy gap $\sim 0.2\,\mathrm{eV}$ [6].

3. "Metallic" clusters in layered materials

Some more examples of "metallic" clusters in an overall insulating matrix are found in layered materials, notably with triangular lattices, such as $LiVO_2$ [12], $LiVS_2$ [13], TiI_2 [14]. This lattice is usually considered as frustrated, meaning that it is not bipartite. But in $LiVO_2$, $LiVS_2$, TiI_2 we are dealing with ions Ti^{2+}, V^{3+} with two t_{2g} d-electrons, which, besides spin, have also triple orbital degeneracy. And, if one cannot subdivide a triangular lattice into two sublattices, one can naturally subdivide it into three! That is indeed what happens in these materials [12], see Fig. 3, in which the occupied orbitals are shown. After forming this orbital superstructure, we have strong antiferromagnetic exchange in the shaded triangles of Fig. 3, and in effect these bonds become shorter than the others, and there are singlet states formed *on such trimers* (three $S = 1$ ions with strong antiferromagnetic coupling have a singlet ground state with $S_{\mathrm{tot}} = 0$). Our recent detailed ab initio calculations [15] confirmed this picture and gave a strong antiferromagnetic exchange in a trimer, the exchange interaction between trimers being much weaker and ferromagnetic. Lattice optimization gave V–V distances in a trimer of 2.56 Å, in good agreement with the experimental value. Note that in this case the short distance in a cluster — here a trimer — is even shorter than the V–V distance in V metal, 2.62 Å, i.e. these clusters can indeed be considered as "metallic".

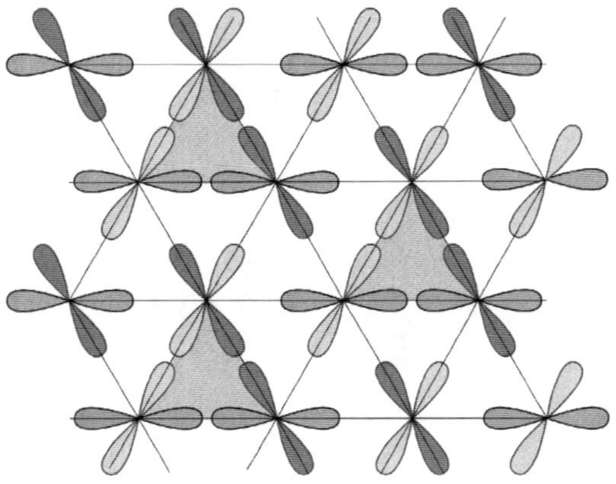

Figure 3. Orbital ordering in $LiVO_2$ with three orbital sublattices, leading to the formation of spin singlet states on tightly bound V triangles (shaded), following [12].

If however we increase the total lattice parameter, which we can do e.g. by going from $LiVO_2$ to $NaVO_2$, the situation changes drastically. As shown in [16], the orbital ordering in $NaVO_2$ is quite different from that in $LiVO_2$: only two types of orbitals are occupied, see Fig. 4. As a result, after such orbital ordering the system becomes topologically a square lattice, and as such it can develop the ordinary Néel magnetic ordering, which it actually does. Note also that, in contrast to $LiVO_2$, in $NaVO_2$ the bonds with strong orbital overlap become *not shorter, but longer*! One can explain this as a competition of two effects: stronger covalency tends to make such bonds shorter, but the bond–charge repulsion (Coulomb repulsion of electron clouds on orbitals directed towards each other) would tend to make such bonds longer. Apparently when the system has a relatively small lattice parameter, or small average TM–TM distance, i.e. when the system is close to the localized–itinerant crossover, the hybridization is rather strong, and the first effect, the tendency to increase it still further, prevails, so that such bonds become even shorter; this is what happens in $LiVO_2$ and $LiVS_2$. If, however, the system has localized d-electrons and is far from the Mott transition, the d–d overlap is exponentially small, and the second effect, the bond–charge repulsion, may become stronger, as a result of which it is favourable to make such bonds longer, to reduce this repulsion further. This is apparently what happens in $NaVO_2$.

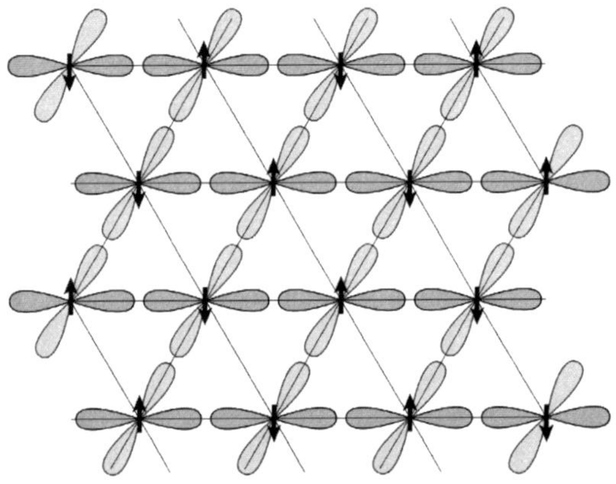

Figure 4. Orbital and magnetic ordering in $NaVO_2$ [16].

Until now we saw the formation of different types of TM clusters, mostly singlet ones: dimers ($MgTi_2O_4$, $CuIr_2S_4$), trimers ($LiVO_2$, $LiVS_2$, TiI_2), or larger clusters such as heptamers in AlV_2O_4. Also singlet tetramers can be formed, especially in layered materials. One such example is CaV_4O_9 [17], with the interesting "chiral" lattice, in which the singlets are formed by four V ions around an "empty" plaquette. In all these cases either low-dimensionality or the frustrated nature of the lattice apparently plays an important role. As mentioned above, valence bond solids are indeed especially easily formed in frustrated or quasi-one-dimensional systems — this is in fact the essence of the Peierls phenomena.

4. Spin-Peierls-to-Peierls transition in TiOCl

A very clear example of this phenomenon, in particular its features close to the localized–itinerant crossover, is observed in TiOCl. This quasi-one-dimensional material is insulating with localized d-electrons at ambient pressure, and it has a spin-Peierls transition at $\sim 60\,\mathrm{K}$ (with an incommensurate phase at $60\,\mathrm{K} < T \lesssim 90\,\mathrm{K}$). Under pressure its energy gap strongly decreases, and it approaches an insulator–metal transition at $\sim 10\,\mathrm{GPa}$ [18, 19].

However direct dc resistivity measurements have shown [20] that TiOCl actually remains insulating even for $P > 10\,\mathrm{GPa}$, at least up to $30\,\mathrm{GPa}$, albeit with a smaller gap. Surprisingly, ab initio calculations with lattice optimisation have shown [21] that in the high-pressure phase, in which indeed the energy gap strongly decreases, the Ti–Ti dimerization itself does not decrease but *increases* instead, see Fig. 5!

A qualitative explanation can be obtained, using again the idea of the proximity of this material to the localized–itinerant crossover. At $P = 0$ the d-electrons are localized, and we have a spin-Peierls transition, with T_c and with the degree of dimerization $\delta d/d$ proportional to

$$T_c \sim \frac{\delta d}{d} \sim J\,e^{-1/\lambda} \tag{1}$$

where J is the exchange interaction, $J \sim t^2/U$, and λ is the spin–phonon coupling constant. Note that the energy gap in this regime is still of the Mott–Hubbard type, i.e. it is large, $E_g \sim U$.

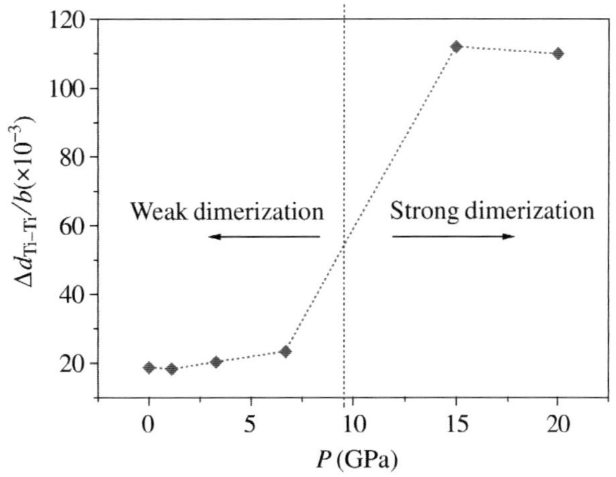

Figure 5. Change of dimerization in TiOCl with pressure [20].

Under pressure TiOCl approaches the insulator–metal transition, as observed in [18, 19]. But after crossing to the itinerant regime, the system still "knows" that it is one-dimensional, and as such it has a Peierls instability, with T_c ($\sim$ degree of dimerization $\sim$ energy gap)

$$T_c \sim E_g \sim \frac{\delta d}{d} \sim t\, e^{-1/\lambda'} , \tag{2}$$

with λ' being the electron–electron coupling constant in this regime. That is, the very nature of insulating state in this regime is *due to Peierls dimerization*, and is not of Mott insulator type. But the dimerization is in this regime proportional to the bandwidth, or the hopping matrix element t, and not to $J \sim t^2/U$. Thus apparently we have here, simultaneously with the localized–itinerant crossover, also a spin-Peierls–Peierls transition.

5. Other possible specific phenomena close to a Mott transition

As we have discussed above, a possible situation, especially in low-dimensional and frustrated systems, is that the Mott transition occurs not in the whole system simultaneously, but "piecewise", first in some finite clusters such as dimers, trimers, tetramers etc. But other nontrivial states may also occur in the vicinity of Mott transitions, different both from those typical for strong Mott insulators and for normal metals. These situations can occur also for simple lattices such as those of perovskites. In this chapter for completeness I will also shortly discuss some of these phenomena.

5.1. Spontaneous charge disproportionation vs orbital ordering

In some systems with strongly correlated electrons charge ordering occurs with decreasing temperature. Usually it appears in systems with noninteger electron concentration (fractional number of electrons per TM ion). The well-known examples are the Verwey transition in magnetite Fe_3O_4 or charge ordering in half-doped manganites such as $La_{0.5}Ca_{0.5}MnO_3$, see e.g. [22]. But there are also systems in which charge ordering, or rather spontaneous charge disproportionation, appears in case of, formally, integer electron occupation, or integer valence. Such is for example the situation in $CaFeO_3$, in which there occurs charge segregation

$$2Fe^{4+}(d^4) \Longrightarrow Fe^{3+}(d^5) + Fe^{5+}(d^3) . \tag{3}$$

A phase transition with the formation of a Fe^{3+}–Fe^{5+} superstructure occurs in $CaFeO_3$ at $T_c = 290\,K$ and is simultaneously a metal–insulator transition [23]. Note that a similar compound $SrFeO_3$ with somewhat broader bands remains metallic down to lowest temperatures. Apparently the phenomenon of charge disproportionation in this case occurs in the system close to the localized–itinerant crossover.

Similar spontaneous charge disproportionation occurs also in $La_{2/3}Sr_{1/3}FeO_3$ [23] ($3Fe^{4+} \Longrightarrow 2Fe^{3+} + Fe^{5+}$), but also e.g. in nickelates $RNiO_3$, especially those with small rare earths $R =$ Lu, Y, etc. [24, 25, 26]. In nickelates charge disproportionation may be described by the "reaction"

$$2Ni^{3+}(t_{2g}^6 e_g^1) \Longrightarrow Ni^{2+}(t_{2g}^6 e_g^2) + Ni^{4+}(t_{2g}^6) \quad (4)$$

(Ni^{3+} and Ni^{4+} here are in the low-spin state). In these systems charge disproportionation is also a metal–insulator transition.

Yet other systems showing the same phenomenon are materials formally containing $Au^{2+}(d^9)$, which at normal conditions always disproportionate into $Au^{1+}(d^{10}) + Au^{3+}(d^8)$. This occurs e.g. in $CsAuCl_3$ which at ambient pressure contains two inequivalent Au sites (its formula is usually even written as $Cs_2Au_2Cl_6$). Under pressure this system also becomes metallic with equivalent Au ions [27].

One can notice three specific features, which apparently are important for the very phenomenon of spontaneous charge disproportionation. First, in TM compounds such charge disproportionation occurs in systems in which the TM ion in the "original" state would have been a strong Jahn–Teller ion, with double orbital degeneracy ($Fe^{4+}(t_{2g}^3 e_g^1)$; low-spin $Ni^{3+}(t_{2g}^6 e_g^1)$; $Au^{2+}(t_{2g}^6 e_g^3)$). Thus it seems that this spontaneous charge disproportionation appears in Fe^{4+} and Ni^{3+} compounds *instead of* the Jahn–Teller or orbital ordering, when the systems are close to localized–itinerant crossover [28, 26]. But in "Au^{2+}" compounds such charge disproportionation is stabilized by the *enhanced* JT distortion instead [29]. Second, the TM ionic states involved typically correspond to states with unusually high oxidation states, or untypically high valence (Fe^{4+} and Fe^{5+}; Ni^{3+} and Ni^{4+}; Au^{3+}). As follows from the standard classification [30, 31], these states correspond to the situation with small or negative charge-transfer gaps and with a large contribution of ligand (e.g. oxygen) p-holes. In effect e.g. the reaction (3) should be rather written as

$$2Fe^{3+}\underline{L} \Longrightarrow Fe^{3+} + Fe^{3+}\underline{L}^2 \, , \quad (5)$$

where $\underline{L}$ denotes the ligand hole (e.g. $Fe^{3+}\underline{L}$ instead of Fe^{4+} denotes the state $Fe^{3+}(d^5)O^-(2p^5)$). Apparently the large contribution of ligand holes facilitates the process of charge disproportionation (the increase of Coulomb repulsion in the reaction (5) is definitely much smaller than if it would occur on the d-shells (3)).

These factors lead to the third conclusion: that the spontaneous charge disproportionation is typically observed in systems close to an insulator–metal transition. Experimentally this is indeed the case: in the examples mentioned above such disproportionation coincides with metal–insulator transition.

5.2. Phase separation

Without going into details, let us shortly mention yet another specific phenomenon, which is typically observed close to metal–insulator transitions. This is the phenomenon of phase separation, which often occurs especially in doped Mott insulators. It is found experimentally in many cases; in manganites [32, 33, 34]; in cobaltites [35]; and apparently also in cuprates [36]. The tendency towards phase separation is also seen theoretically in many models: in the Hubbard [37] and in the t–J models [38]; in the double-exchange models [39], etc. The resulting state may be visualised as a random system with metallic regions in an insulating matrix. The properties of such state can be described by percolation theory — a classical description of random systems, although quantum effects in this case can also be very important.

5.3. *Large spontaneous currents in frustrated systems*

Typical for many frustrated systems, which contain triangles or tetrahedra as building blocks, is an extra degeneracy, which can be characterized by spin chirality. Thus the ground state of an equilateral triangle of $S = \frac{1}{2}$ spins with an antiferromagnetic interaction is a quartet, with the total spin $S_{\rm tot} = \frac{1}{2}$ and with extra double degeneracy, which can be characterized by the scalar spin chirality

$$\chi_{123} \sim \boldsymbol{S}_1 \cdot [\boldsymbol{S}_2 \times \boldsymbol{S}_3] \; . \tag{6}$$

The physical meaning of the scalar chirality was clarified in [40] (see also [41]), where it was shown that the states of a triangle with nonzero chirality (6) carry a spontaneous circular electric current $j \sim \chi_{123}$. Correspondingly, such triangle would have an orbital magnetic moment $M \sim j \sim \chi_{123}$. In the nondegenerate Hubbard model with strong interaction $U \gg t$ the expression for this current, obtained in perturbation theory in $t/U \ll 1$, has the form

$$j = \frac{24et^3}{\hbar U^2}(\boldsymbol{S}_1 \cdot [\boldsymbol{S}_2 \times \boldsymbol{S}_3]) \; . \tag{7}$$

In this limit the current and the corresponding orbital moment are small. But one can show on general grounds, by using symmetry arguments, that such currents would be proportional to the chirality (6) in the general case, for $t \gtrsim U$, and also close to Mott transition. In such situations the corresponding current and orbital moments can be quite large. Indeed, estimates done in [42], using the exact solution for a triangle, show that for reasonable values of parameters the orbital magnetic moments can be as large as $\sim 0.7\mu_B$ — comparable with the spin moment. The authors of [42] suggested to use this degree of freedom for quantum computations. This is yet another nontrivial property of frustrated systems, which can lead to large and potentially useful effects close to Mott transitions.

6. Conclusion

In conclusion, I want to stress once again that the systems close to Mott transition, or to a localized–itinerant crossover, can show a variety of specific properties, different from those of states deep in the Mott insulator regime or those of conventional metals. Especially interesting is the situation of a "partial Mott transition", in which first certain finite clusters go over to the regime with uncorrelated ("itinerant") electrons, and only later there occurs a real Mott transition in the whole sample. As I tried to show on many examples, the most favourable conditions for this phenomenon are met in low-dimensional and in frustrated systems; they are less probable in simple lattices such as perovskites. But also other specific phenomena, less sensitive to the type of the lattice, such as spontaneous charge disproportionation, can occur close to insulator–metal transitions. Thus this crossover region between localized and itinerant electronic states presents a rich playground for interesting physics, with eventually some novel states appearing in this situation.

Acknowledgments

I am grateful to many people who contributed to works on which this paper is based, especially M. Abd-Elmeguid, C. Batista, L. Bulaevskii, Hua Wu, I. Mazin, M. Mostovoy, V. Pardo, R. Rivadulla, S. Streltsov and A. Ushakov. This work was supported by the German program SFB 608 and by the European project SOPRANO.

References

[1] Goodenough J B 1963 *Magnetism and the Chemical Bond* (New York: Wiley–Interscience)
[2] Schmidt M *et al* 2004 *Phys. Rev. Lett.* **92** 056402
[3] Radaelli P G *et al* 2002 *Nature* **416** 155
[4] Matsuno K *et al* 2001 *J. Phys. Soc. Japan* **70** 1456
[5] Khomskii D I and Mizokawa T 2005 *Phys. Rev. Lett.* **94** 156402
[6] Pardo V *et al* 2008 *Phys. Rev. Lett.* **101** 256403
[7] Furubayashi T *et al* 1994 *J. Phys. Soc. Japan* **66** 778
[8] Reehuis M *et al* 2003 *Eur. Phys. J.* B**35** 311
[9] Motome Y and Tsunetsugu H 2004 *Phys. Rev.* B **70** 184427
[10] Tchernyshev O 2004 *Phys. Rev. Lett.* **93** 157206
[11] Giovannetti G *et al* 2010 *Phys. Rev.* B – *Rapid Comm.* in press (*Preprint* arXiv:1009.4077)
[12] Pen H *et al* 1997 *Phys. Rev. Lett.* **78** 1323
[13] Katayama N *et al* 2009 *Phys. Rev. Lett.* **103** 146405
[14] Meyer G, Gloger T and Beekhuizen J 2009 *Z. Anorg. Allg. Chemie* **635** 1497
[15] Wu Hua and Khomskii D I, to be publ.
[16] McQueen T M *et al* 2008 *Phys. Rev. Lett.* **101** 166402
[17] Starykh O A *et al* 1996 *Phys. Rev. Lett.* **77** 2558
[18] Kuntscher C A *et al* 2006 *Phys. Rev.* B **74** 184402
[19] Kuntscher C A *et al* 2007 *Phys. Rev.* B **76** 241101
[20] Forthaus M K *et al* 2008 *Phys. Rev.* B **77** 165121
[21] Blanco-Canosa S *et al* 2009 *Phys. Rev. Lett.* **102** 056406
[22] Imada M, Fujimori A and Tokura Y 1998 *Rev. Mod. Phys.* **70** 1039
[23] Takano M *et al* 1981 *J. Solid State Chem.* **39** 75
[24] Alonso J A *et al* 1999 *Phys. Rev. Lett.* **82** 3871
[25] Mizokawa T, Khomskii D I and Sawatzky G A 2000 *Phys. Rev.* B **61** 11263
[26] Mazin I I *et al* 2008 *Phys. Rev. Lett.* **98** 176406
[27] Matsushita N *et al* 2007 *J. Solid State Chem.* **180** 1353
[28] Khomskii D I 2005 *Physica Scripta* **72** CC8–CC14
[29] Ushakov A V, Streltsov S V and Khomkii D I, to be publ.
[30] Bocquet A E *et al* 1992 *Phys. Rev.* B **45** 1561
[31] Bocquet A E *et al* 1992 *Phys. Rev.* B **46** 3771
[32] Moreo A, Yunoki S and Dagotto E 1999 *Science* **283** 2034
[33] Khomskii D I 2000 *Physica* B **280** 325
[34] Babushkina N A *et al* 1998 *J. Appl. Phys.* **83** 7389
[35] Kuhns P L *et al* 2003 *Phys. Rev. Lett.* **91** 127202
[36] Gor'kov L P and Sokol A 1987 *JETP Lett.* **46** 420
[37] Visscher P B 1974 *Phys. Rev.* B **10** 943
[38] Emery V J, Kivelson S A and Lin H Q 1990 *Phys. Rev. Lett.* **64** 475
[39] Kagan M Y, Khomskii D I and Mostovoy M V 1999 *Eur. Phys. J.* **12** 217
[40] Bulaevskii L N, Batista C D, Mostovoy M V and Khomskii D I 2008 *Phys. Rev.* B **78** 028402
[41] Khomskii D I 2010 *J. Phys. Cond. Matter* **22** 164209
[42] Georgeot B and Mila F 2010 *Phys. Rev. Lett* **104** 100502

Spontaneous Hall Effect in the Spin Liquid Phase of $Pr_2Ir_2O_7$

S. Nakatsuji[1], Y. Machida[1,4], J. J. Ishikawa[1], S. Onoda[2], Y. Karaki[1,3], T. Tayama[1,5], and T. Sakakibara[1]

[1] Institute for Solid State Physics, University of Tokyo, Kashiwa 277-8581, Japan.
[2] Condensed Matter Theory Laboratory, RIKEN, Wako 351-0198, Japan.
[3] Faculty of Education, University of the Ryukyus, 1 Senbaru, Nishihara, Okinawa 903-0213, Japan

E-mail: `satoru@issp.u-tokyo.ac.jp`

Abstract. An electric current flowing through a conductor in a magnetic field produces a transverse voltage drop known as the Hall effect. In the absence of the field, this effect also appears in ferromagnets in a plane normal to its spontaneous magnetization vector owing to the spin-orbit coupling. Generally, it may also detect a nontrivial order parameter breaking the time-reversal symmetry on a macroscopic scale, for example, scalar spin chirality. Here, we present our recent results in the study of the frustrated magnetism and Hall transport of the metallic magnet $Pr_2Ir_2O_7$. Strikingly, a spontaneous Hall effect is observed in the absence of both an external magnetic field and conventional magnetic long-range order. This strongly suggests the existence of a chiral spin liquid, a spin-liquid phase breaking the time-reversal symmetry. Both our measurements indicate that spin-ice correlations in the liquid phase lead to a non-coplanar spin texture forming a uniform but hidden order parameter: the spin chirality.

1. Introduction

Time-reversal symmetry (TRS) is one of the basic concepts in the broad fields of science. In condensed-matter physics, many prototypical TRS-broken states can be found in magnetic materials where electron spins $\boldsymbol{S}$ form various ordering patterns. Once the magnetic long-range order (LRO) sets in at low temperatures, the state loses the invariance under the time-reversal operation $\boldsymbol{S} \to -\boldsymbol{S}$. For magnetic materials, however, the source of the spontaneously broken TRS is not restricted to the LRO of spins. Since any product of odd number of spins is also odd under the time-reversal, it may in principle become a hidden but primary order parameter for the TRS breaking. The simplest one except the spin magnetic dipole moment is given by the scalar spin chirality [1], the solid angle subtended by three neighboring spins, $\kappa_{ijk} = \boldsymbol{S}_i \cdot \boldsymbol{S}_j \times \boldsymbol{S}_k$. Despite extensive researches over decades, however, no clear answer has been made experimentally on the intriguing possibility of the TRS breaking in the absence of conventional magnetic LRO until quite recently [2].

When the TRS is broken *macroscopically*, the anomalous Hall effect (AHE) can be a useful probe for the detection of such hidden TRS breaking in metallic magnets. The AHE

[4] Present address: Department of Physics, Tokyo Institute of Technology, Meguro 152-8551, Japan.
[5] Present address: Department of Physics, University of Toyama, Toyama 930-8555, Japan.

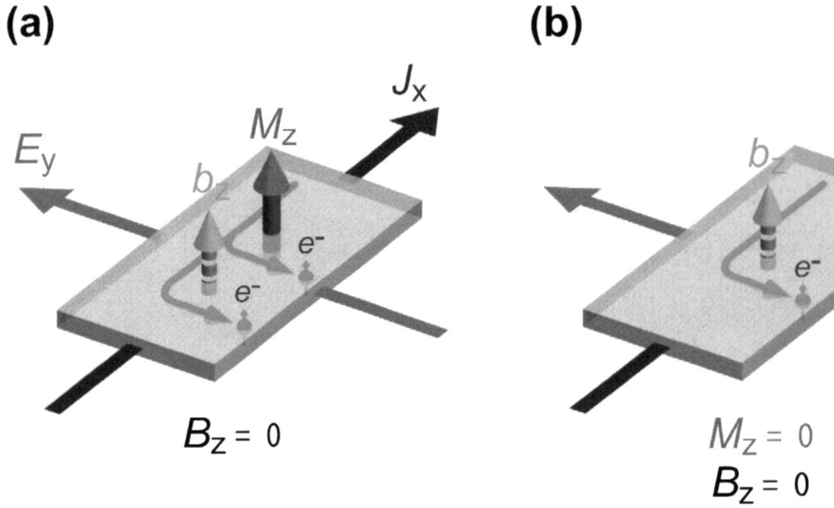

Figure 1. Two classes of the anomalous Hall effect in metallic magnets. (a) In a magnetic long range ordered state, the uniform spin magnetization M_z and the fictitious magnetic field b_z due to the spin-orbit interaction produce the anomalous Hall effect even at zero magnetic field $B_z = 0$. (b) In a spin-disordered state without uniform magnetization $M_z = 0$, the anomalous Hall effect may appear at $B_z = 0$ when the time-reversal symmetry is macroscopically broken with the finite fictitious magnetic field b_z produced by the uniform spin chirality.

is a fundamental transport phenomenon where the electric current generates the transverse voltage drop in the normal plane to the spontaneous magnetization $\boldsymbol{M}$ in ferromagnets [3] (Fig. 1a). This effect has attracted revived interest because of its topological and dissipationless character [4, 5, 6, 7, 8, 9, 10], and its potential application in spintronics [11]. In particular, it has been shown [8, 9, 10] that the intrinsic mechanism [4] of the AHE can capture the dominant part in moderately dirty metals. This intrinsic AHE can be understood in terms of the adiabatic motion of the Bloch electrons under the electric field $\boldsymbol{E}$ [6], which acquire a quantum geometrical Berry phase [12] in the wavevector ($\boldsymbol{k}$) space because of the spin-orbit interaction and the spin magnetization. This phase acts as a *fictitious magnetic field* $\mathbf{b}_{\boldsymbol{k}}$ in the k space and bends the orbital motion of electrons as in the case of the Lorentz force due to the real magnetic field $\boldsymbol{B}$. Thus, it causes the AHE characterized by the finite Hall conductivity σ_{H} at $B = 0$.

The condition for observing the AHE at $B = 0$ is the *macroscopically* broken TRS [13], which ensures a nonzero average $\boldsymbol{b}$ of the Berry phase over the occupied Bloch states. It does not necessarily require a finite ferromagnetic (FM) spin alignment (Fig. 1a), but a noncoplanar spatial distribution of spins. While unconventional scenarios directly relying on a noncoplanar spin texture with the uniform scalar spin chirality have been addressed experimentally so far for ferromagnets and for spin glasses [3, 14, 15, 16, 17], the macroscopically broken TRS and a resultant nonzero σ_{H} in the absence of a uniform spin magnetization have also been proposed theoretically, for example, in antiferromagnetic (AF) states and spin-liquid states with the scalar spin chirality [18] (Fig. 1b). In this exotic example of a chiral spin-liquid [1], the uniform scalar spin chirality shows the LRO, but the spin magnetic moment does not. Experimentally, however, the AHE at zero field has never been observed in the absence of the uniform spin magnetization associated with the ferromagnetism [3] or the spin freezing [19, 15].

Here, we discuss our recent results in the study of the frustrated magnetism and Hall transport of the metallic magnet $Pr_2Ir_2O_7$, which has revealed a macroscopically TRS-broken spin-liquid state [2]. The spontaneous Hall effect is found in the geometrically frustrated Kondo lattice $Pr_2Ir_2O_7$ [20] even above its spin freezing temperature $T_f \sim 0.3$ K. A clear hysteresis is observed in the Hall conductivity around zero field below the onset temperature $\theta \sim 1.5$ K, whereas that in the magnetization curve appears only below T_f within an experimental accuracy. Namely, a large anomalous Hall conductivity σ_H is found even at zero field where the magnetization practically vanishes, in sharp contrast to the conventional AHE in ferromagnets. This indicates an emergence of a hidden order at $T = \theta$ that macroscopically breaks the TRS without invoking a magnetic LRO. The phenomenon may be understood in terms of a formation of the uniform spin chirality out of "2-in, 2-out" configurations of localized magnetic moments of Pr^{3+} ions in an analogy to spin-ice systems [21].

2. Experimental

Single crystals of $Pr_2Ir_2O_7$ were grown by a flux method [20, 22]. Energy dispersive x-ray analysis found that the ratio between Pr and Ir is 1:1.03 and stoichiometric within experimental errors. The four-circle x-ray diffraction measurements confirmed the pyrochlore lattice structure of the crystals. The transport properties and magnetization were measured at low temperatures down to 0.03 K and under magnetic fields up to 7 T.

The field sweep measurements were performed with a small fixed rate of 1 Oe/s at each temperature $T \leq 0.5$ K, while at $T \geq 0.7$ K the field was fixed at each point for ~ 2 min. before the measurement. We also performed the measurements by using both procedures above at 0.5 K and confirmed that basically the same results can be obtained by the different field sweep methods. The field dependence was measured at each temperature after the same zero field cooling procedure. Namely, we first increased the temperature above 2 K before each measurement, and then cooled down the sample to a target temperature under zero filed.

The transport measurements were made by using a conventional four-probe method with a current path along the [110] direction. For the transverse Hall resistivity measurements, the longitudinal voltage drop was eliminated by reversing the field direction. For the field sweep measurement in Fig. 2c, the field was first increased from zero up to 7 T, and then a hysteresis loop was measured between 7 T and −7 T. Only the results in the positive field region are presented in Fig. 2c. The virgin curve between 0 T and 7 T is not shown because of the uncertainty in determining the longitudinal voltage drop.

The magnetization above and below 2 K was measured by the conventional SQUID magnetometer and by the Faraday method using a dilution refrigerator [23], respectively. The nonlinear susceptibility was obtained by fitting M vs. B curves to $M = \chi_1 B + \chi_3 B^3/3!$ at each temperature. For the analysis of the results at $T \leq T_f$, we added a constant term M_0 for the frozen component. The ac susceptibility measurements were performed by using the mutual inductance method with ac fields of $B_{ac} = 5\ \mu$T with frequency $f = 16$ Hz.

3. Experimental Results and Discussion

3.1. Spin Ice Correlation in the Spin Liquid Phase

The pyrochlore iridate $Pr_2Ir_2O_7$ is a cubic system consisting of Pr and Ir corner-sharing tetrahedra that are displaced by half the lattice constant. The Pr^{3+} ions provide localized $\langle 111 \rangle$ Ising-like $4f$ magnetic moments pointing inwards or outwards to the center of the Pr tetrahedron (Fig. 2a) with the effective size of $p_{\rm eff} \sim 3.0\ \mu_B$/Pr, while Ir $5d$ conduction electrons are weakly correlated and Pauli-paramagnetic with an appreciable spin-orbit coupling [20]. The localized moments interact with the conduction electrons through the AF Kondo coupling J_{fd}. Strong geometrical frustration [24] is apparent as the magnetic moments do not exhibit any LRO, but only a freezing behavior below $T_f \sim 0.3$ K, two orders of magnitude lower than the AF Weiss temperature $T^* \sim -20$ K [20].

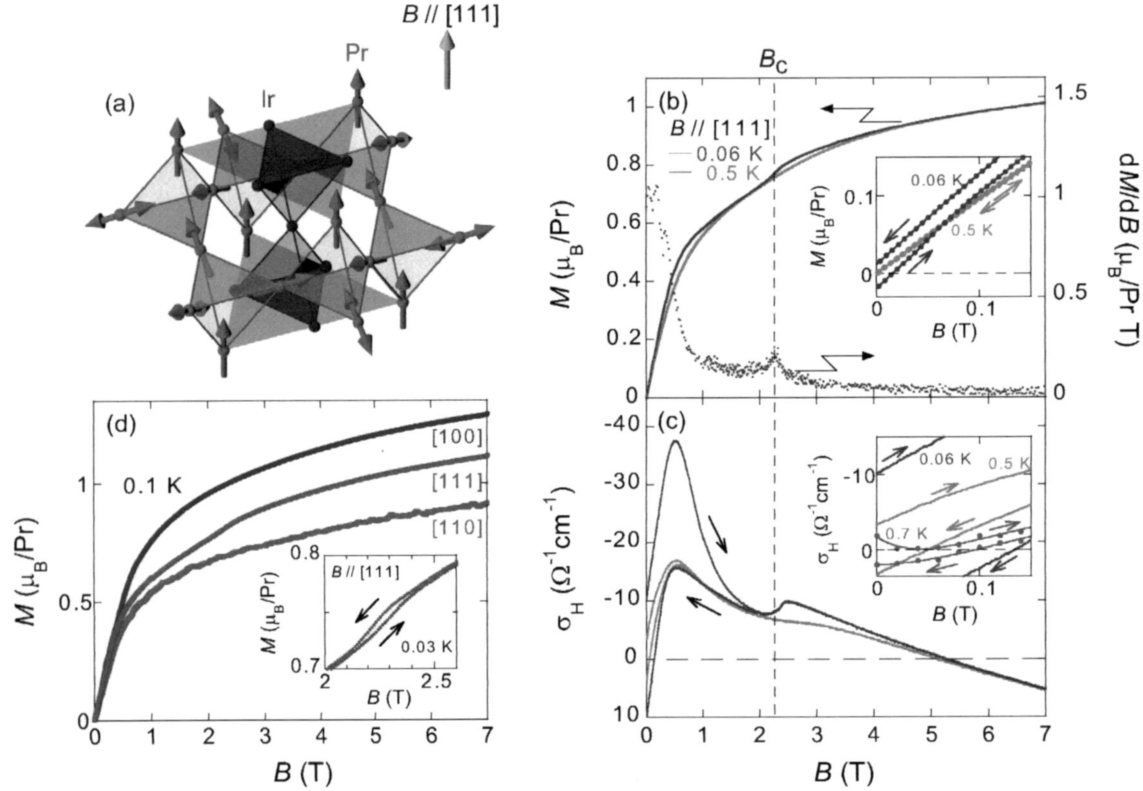

Figure 2. Magnetic and crystal structure, and field dependence of the magnetization and Hall conductivity of $Pr_2Ir_2O_7$ along the [111] field direction [2]. (a) The pyrochlore lattice is alternating stacking of kagome and triangular layers along the [111] direction. Under zero field, the Pr moments most likely form the "2-in, 2-out" configuration as denoted by three red arrows and one blue arrow in each tetrahedron. Application of the field B along the [111] direction flips the blue moments and stabilizes the "3-in, 1-out" (1-in, 3-out) configuration formed by four red arrows. (b) Field dependence of the magnetization M at 0.06 and 0.5 K (left axis) and its derivative $\mathrm{d}M/\mathrm{d}B$ at 0.06 K (right axis). (c) Field dependence of the Hall conductivity σ_{H} at 0.06 K. Dashed line represents the metamagnetic transition field $B_{\mathrm{c}} \sim 2.3$ T. Inset: Low field (b) M and (c) σ_{H} at fixed temperatures. During the measurements, the field was continuously swept with the rate of 1 Oe/s at temperatures $T \leq 0.5$ K, while it was fixed at each measurement field point at $T \geq 0.7$ K. The arrows indicate up and down field sweep sequences. (d) Field dependence of the magnetization $M(B)$ for fields along the [100], [110], and [111] directions at 0.1 K. Inset: Hysteresis in the magnetization $M(B)$ at the metamagnetic transition for fields along the [111] direction at 0.03 K.

Before discussing the results of the Hall transport measurements, we clarify low-temperature magnetic properties of $Pr_2Ir_2O_7$. First, we have performed the field-sweep measurements of the magnetization under the field along the [111] direction. Figure 2b represents the magnetization curve $M(B)$ obtained at 0.06 K and 0.5 K. In contrast with $M(B)$ at 0.5 K, a small but clear step-like feature is found at 0.06 K. This anomaly is also clearly observed as a kink in the field derivative (Fig. 2b). On further cooling down to 0.03 K, around the step-like feature there appears a hysteresis between the up and down sweep sequences, indicating a first-order metamagnetic transition (Fig. 2d inset). We also performed similar measurements under fields along [100] and [110] directions. However, we observed neither an anomaly nor a hysteresis, but a smooth increase of the magnetization (Fig. 2d).

The fact that the metamagnetic transition occurs only for $\boldsymbol{B} \parallel [111]$ provides an evidence for the formation of the "2-in, 2-out" configurations of $\langle 111 \rangle$ Ising moments on each tetrahedron. In general, four Ising moments on a tetrahedron form two distinct configurations, depending on the sign of the nearest-neighbor interaction: "all-in, all-out" and "2-in, 2-out" for AF and FM interactions, respectively. The "all-in, all-out" state has locally $M = 0$. Hence, to induce a finite M, a metamagnetic transition is expected for fields along all three directions. However, this is not the case in our experiment. In contrast, for "2-in, 2-out", the transition occurs only along the [111] direction as observed in our case as well as in the spin-ice compound $Dy_2Ti_2O_7$ [25], and stabilizes the "3-in, 1-out" state in high fields (Fig. 2a).

Further consistency with the "2-in 2-out" correlation can be found for the values of the magnetization M_c and the magnetic field B_c at the metamagnetic transition. In spin-ice systems, the magnetization just below the metamagnetic transition is known to reach $M_c = g_J J/3$ in the kagome-ice state, which can be viewed as the stacking sequence of the "2-in, 2-out" kagome-lattice layer and the fully polarized triangular-lattice layer along the [111] direction [21] (Fig. 2a). Using the Landé factor g_J for Pr^{3+} and J obtained from $p_{\rm eff} = g_J\sqrt{J(J+1)}$, M_c is estimated to be 0.9 μ_B/Pr, which is close to our observation of $M_c \sim 0.8$ μ_B/Pr. On the other hand, the effective nearest-neighbor FM coupling $J_{ff}^{\rm eff}$ in the spin-ice systems can be estimated from B_c as $J_{ff}^{\rm eff} \approx (1/3)g_J J\mu_B B_c/k_B$. In our case, the local "2-in 2-out" correlation inferred from the observed magnetization curves also indicates a FM coupling between nearest-neighbor $4f$ moment pairs. The observed $B_c = 2.3$ T indicates $J_{ff}^{\rm eff} \sim 1.4$ K which is close to the peak temperature (~ 2 K) of the magnetic specific heat C_m [20] as in spin-ice systems [21]. Furthermore, the nonlinear magnetic susceptibility χ_3 exhibits a steep negative increase below θ and saturate to a large negative value (not shown) [2], consistent with the FM correlation among $4f$ magnetic moments. Therefore, we conclude that at $T \leq J_{ff}^{\rm eff}$, the "2-in, 2-out" configuration appears with the highest probability. However, there is a sharp contrast to the spin-ice case where the spin freezing occurs at the same temperature scale, i.e., $T_f \sim J_{ff}^{\rm eff}$ [21], while the present material remains in a *spin-liquid* state down to $T_f \sim 0.3$ K even below the temperature $\theta \sim J_{ff}^{\rm eff}$.

Two possible sources of the FM interaction of the order of $J_{ff}^{\rm eff} \sim 1.4$ K can be considered. Here, the magnetic dipole interaction cannot be the origin because it is an order of magnitude smaller for Pr^{3+} (~ 0.1 K) than for Dy^{3+} (~ 1 K) of the dipolar spin-ice system $Dy_2Ti_2O_7$ [21]. Instead, one possibility is the superexchange coupling between Pr $4f$ moments through the superexchange path mediated by $2p$ orbitals at the oxygen O(1) site [26]. The significant deviation from 180° of the bond angle for the Pr-O-Pr path alters the sign of the superexchange and should make it ferromagnetic. Another possibility is the RKKY interaction mediated by conduction electrons via the Kondo coupling. Indeed, the RKKY interaction should be FM between the nearest-neighbor pairs, according to the analyses based on the simple Fermi gas model using the carrier concentration ~ 20 %/Pr found by the Hall effect measurements [27]. A first-principles calculation also predicts a single Fermi surface with a carrier concentration ~ 20 %/Pr [2]. A recent Monte-Carlo simulation [28] has indicated that this RKKY interaction with

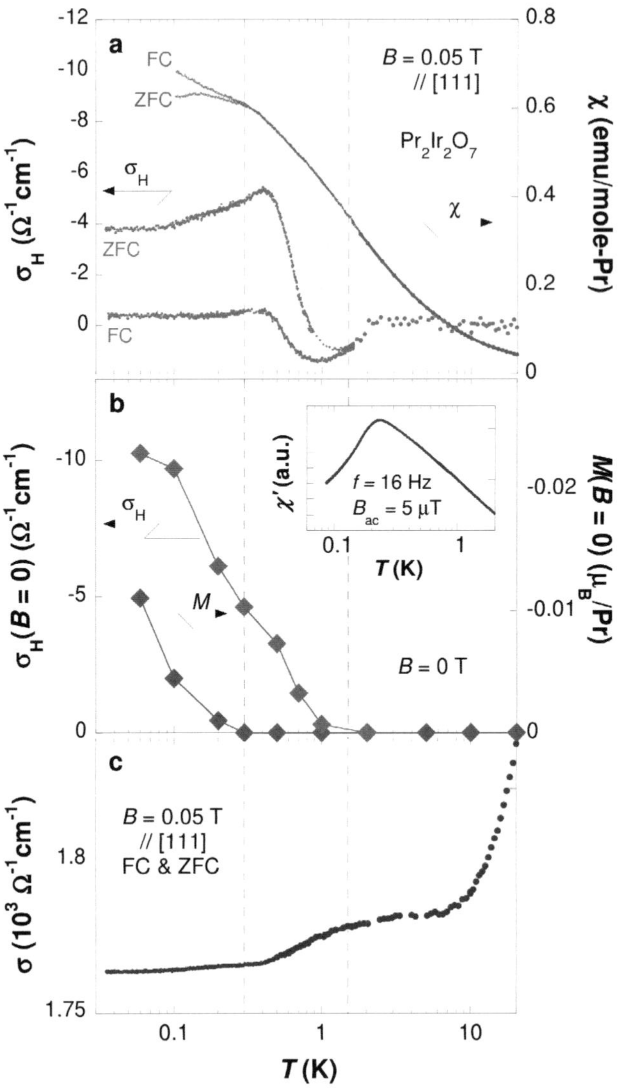

Figure 3. (a) Temperature dependence of the Hall conductivity $\sigma_{\rm H}$ (left axis) and the dc susceptibility χ (right axis) under a magnetic field of $B = 0.05$ T along the [111] direction [2]. Both the zero-field cooled (ZFC) and field-cooled (FC) results are plotted. Vertical dashed lines denote $\theta \sim 1.5$ K and $T_{\rm f} \sim 0.3$ K, respectively. (b) Temperature dependence of the remnant Hall conductivity $\sigma_{\rm H}(B = 0)$ (left axis) and remnant magnetization $M(B = 0)$ (right axis) at zero field, obtained after a field sweep down from 7 T in the hysteresis loop measurements [2]. Inset: Temperature dependence of the ac-susceptibility obtained under zero field with the ac-field of 5μT with frequency $f = 16$ Hz. The peak indicates (possibly partial) freezing of Pr moments below $T_{\rm f} \sim 0.3$ K [20]. (c) Temperature dependence of the longitudinal conductivity σ under $B = 0.05$ T along the [111] direction. No hysteresis is found between the results obtained in the ZFC and FC sequences.

the experimentally obtained k_F also explains the AF Weiss temperature in agreement with our experiment, while it predicts a magnetic dipole order for the ground state.

3.2. Spontaneous Hall Effect in the Spin Liquid Phase

Now, we present the experimental findings on the AHE and the associated macroscopic TRS breaking [2]. Figure 3a shows the temperature dependence of the Hall conductivity $\sigma_H(T) = -\rho_H/(\rho_H^2 + \rho^2)$ (left axis), and the dc susceptibility $\chi(T) = M/H$ (right axis) at $B = 0.05$ T along the [111] direction. Below 2 K, $\sigma_H(T)$ exhibits a strong temperature dependence and even bifurcation at $\theta \sim 1.5$ K between the zero-field cooled (ZFC) and field-cooled (FC) results, while the longitudinal conductivity $\sigma(T)$ shows no hysteresis (Fig. 3c). On the other hand, the onset temperature of the irreversibility in $\chi(T)$, $T_f \sim 0.3$ K (Fig. 3a, Fig. 3b and its inset), is significantly lower than that in $\sigma_H(T)$, $\theta \sim 1.5$ K. This hysteresis in $\sigma_H(T)$ suggests that the TRS is already broken macroscopically below θ in the spin-liquid state.

To confirm this possibility, we have performed the field-sweep measurements of $\sigma_H(B)$ with $\boldsymbol{B} \parallel [111]$ at various temperatures. First, let us discuss the results obtained in the spin-frozen state below T_f. Figure 2c presents the results at $T = 0.06$ K. Apparently, we do not find any field range where $\sigma_H(B)$ is proportional to $M(B)$. Instead, $\sigma_H(B)$ shows the following nonmonotonous field dependence: (i) a pronounced hysteresis loop around $B = 0$ between the up-sweep $\sigma_H^{up}(B)$ and the down-sweep $\sigma_H^{down}(B)$, (ii) a peak formation at $B_p \sim 0.5$ T, (iii) a kink around the metamagnetic transition at $B = B_c \sim 2.3$ T, (iv) a sign change around 5 T. The remnant Hall conductivity $\sigma_H(B = 0)$ amounts to 10 Ω^{-1} cm^{-1}, which is comparable to that for the ferromagnet $Nd_2Mo_2O_7$ [14, 29]. The size of the hysteresis, i.e., $\sigma_H^{up}(B) - \sigma_H^{down}(B)$, decreases with field and vanishes at $B \sim B_c$. In contrast with the clear hysteresis observed in $\sigma_H(B)$, the magnetization $M(B)$ shows only a small remnant component $\sim 0.01\mu_B$ due to the spin freezing (Fig. 2b, inset).

Notably, $\sigma_H(B = 0)$ does not vanish even above $T_f \sim 0.3$ K. Although it is weakened with temperature, we still observe a hysteresis in $\sigma_H(B)$ around $B = 0$ at 0.5 and 0.7 K (Fig. 2b, inset), while that in $M(B)$ is already absent within the experimental accuracy $\sim 10^{-3}\mu_B$ (Fig. 2b, inset). Thus, $\sigma_H(T, B = 0)$ decreases on heating, and it eventually vanishes at $\theta \sim 1.5$ K, which can be clearly distinguished from T_f where $M(T, B = 0)$ decreases to zero (Fig. 3b). The finite σ_H observed in the apparent absence of B and M indicates that the TRS is broken *spontaneously* and *macroscopically* in the spin-liquid state, pointing to an LRO or freezing of higher degrees of freedom than spin dipole moments, for instance, the net spin chirality. There exists a close link between the macroscopic TRS-breaking and the local "2-in, 2-out" spin correlation. The onset temperature $\theta \sim 1.5$ K almost coincides with the effective FM coupling $J_{ff}^{\text{eff}} \sim 1.4$ K estimated from the metamagnetic transition field B_c. Besides, the hysteresis observed in σ_H as a function of field disappears at B_c where a large portion of "2-in, 2-out" configurations are transformed into "3-in, 1-out" (Fig. 2b and c), as we have already explained. Therefore, the macroscopically TRS-broken spin-liquid state found in $T_f \leq T \leq \theta$ should comprise the "2-in, 2-out" configurations having a net spin chirality.

3.3. Possibility of Quantum Meting of Spin Ice

In $Pr_2Ir_2O_7$, most likely the spin-ice states with a uniform chirality component become dynamic at $T \geq T_f$, forming a chiral spin-liquid that does not show a freezing of dipole moments. With this regard, the recent theory by Onoda and Tanaka has concluded an interesting possibility of quantum melting of spin ice. They focus on the superexchange path made of Pr-O-Pr bond and found that it can be represented by an effective quantum pseudospin-1/2 model [26, 30]. They found that the combination of the small size of Pr $4f$ moments and the trigonal crystal electrical field at Pr site enhances the quantum terms to be as large as the classical Ising term in the Hamiltonian. Using the calculation for 16 sites, they further predict that each Pr tetrahedron may develop a spin quadrupole and spin chirality correlation at the ground state, as the results of the superposition of the "3-in, 1-out" configurations in the dominant "2-in, 2-out" states. The experimental determination of the profile of the magnetic correlations in the spin-liquid phase for $T_f \le T \le T_H$ is an important task for future studies.

4. Conclusion

The above results have demonstrated that a *macroscopic* breaking of the time reversal symmetry (TRS) can indeed occur even without a conventional magnetic order, and can be detected through a spontaneous Hall effect in the absence of external field and uniform magnetization, suggesting an unconventional anomalous Hall effect without the ferromagnetic or ferrimagnetic spin ordering. With this regard, the ac version of the anomalous Hall effect, i.e., a magneto-optical Kerr effect at zero field, may offer another useful probe of a macroscopically TRS-broken state, as actually reported in the pseudogap region of a high-temperature cuprate superconductor [31]. Future explorations searching for the spontaneous Hall effect might unveil a fertile field of exotic chiral states of matter in strongly correlated electrons.

5. Acknowledgments

We thank L. Balicas, H. Kawamura, Y. Maeno, Y. Matsumoto, N. Nagaosa, Y. Ohta and T. Taniguchi for useful support and discussions. This work is partially supported by Grant-in-Aid (No. 21684019) from the Japanese Society for the Promotion of Science, and also Grant-in-Aid for Scientific Research on Priority Area (No. 19052003) and on Innovative Areas (No. 20102007) from the Ministry of Education, Culture, Sports, Science and Technology, Japan.

6. References

[1] Wen X G, Wilczek F and Zee A 1989 *Phys. Rev. B* **39** 11413
[2] Machida Y, Nakatsuji S, Onoda S, Tayama T and Sakakibara T 2010 *Nature* **463** 210–213
[3] Nagaosa N, Sinova J, Onoda S, MacDonald A H and Ong N P 2010 *Rev. Mod. Phys.* **82** 1539–1592
[4] Karplus R and Luttinger J M 1954 *Phys. Rev.* **95** 1154
[5] Thouless D J, Kohmoto M, Nightingale M P and den Nijs M 1982 *Phys. Rev. Lett.* **49** 405
[6] Sundaram G and Niu Q 1999 *Phys. Rev. B* **59** 14915
[7] Jungwirth T, Niu Q and MacDonald A H 2002 *Phys. Rev. Lett.* **88** 207208
[8] Lee W L, Watauchi S, Miller V L, Cava R J and Ong N P 2004 *Science* **303** 1647
[9] Fang Z, Nagaosa N, Takahashi K S, Asamitsu A, Mathieu R, Ogasawara T, Yamada H, Kawasaki M, Tokura Y and Terakura K 2003 *Science* **302** 92
[10] Onoda S, Sugimoto N and Nagaosa N 2006 *Phys. Rev. Lett.* **97** 126602
[11] Ohno H 1998 *Science* **281** 951
[12] Berry M V 1984 *Proc. R. Soc. London Ser. A* **392** 45
[13] Landau L D and Lifshitz E M 1984 *Electrodynamics of Continuous Media* 2nd ed (Oxford: Elsevier) the macroscopically broken T means that the time-reversal operation, which inverts the spin and orbital angular momenta and the wavevector, $\boldsymbol{S} \to -\boldsymbol{S}$, $\boldsymbol{L} \to -\boldsymbol{L}$, and $\boldsymbol{k} \to -\boldsymbol{k}$, as well as the *fictitious magnetic field* $\boldsymbol{b}_{n\boldsymbol{k}} \to -\boldsymbol{b}_{n-\boldsymbol{k}}$, should not be compensated by any other symmetry operations of the crystal, e.g., translation, spatial inversion, reflection, rotation, and their combinations.
[14] Taguchi Y, Oohara Y, Yoshizawa H, Nagaosa N and Tokura Y 2001 *Science* **291** 2573
[15] Taniguchi T, Yamanaka K, Sumioka H, Yamazaki T, Tabata Y and Kawarazaki S 2004 *Phys. Rev. Lett.* **93** 246605
[16] Neubauer A, Pfleiderer C, Binz B, Rosch A, Ritz R, Niklowitz P G and Boni P 2009 *Phys. Rev. Lett.* **102** 186602
[17] Lee M, Kang W, Onose Y, Tokura Y and Ong N P 2009 *Phys. Rev. Lett.* **102** 186601
[18] Metalidis G and Bruno P 2006 *Phys. Rev. B* **74** 045327
[19] Tatara G and Kawamura H 2002 *J. Phys. Soc. Jpn.* **71** 2613
[20] Nakatsuji S, Machida Y, Maeno Y, Tayama T, Sakakibara T, van Duijn J, Balicas L, Millican] J N, Macaluso R T and Chan J Y 2006 *Phys. Rev. Lett.* **96** 087204
[21] Bramwell S T and Gingras M J P 2001 *Science* **294** 1495
[22] Millican J N, Macaluso R T, Nakatsuji S, Machida Y, Maeno Y and Chan J Y 2007 *Mater. Res. Bull.* **42** 928

[23] Sakakibara T, Mitamura H, Tayama T and Amitsuka H 1994 *Jpn. J. Appl. Phys.* **33** 5067
[24] Moessner R and Ramirez A 2006 *Physics Today* **59** 24
[25] Sakakibara T, Tayama T, Hiroi Z, Matsuhira K and Takagi S 2003 *Phys. Rev. Lett.* **90** 207205
[26] Onoda S and Tanaka Y 2010 *Phys. Rev. Lett.* **105** 047201
[27] Machida Y, Nakatsuji S, Maeno Y, Tayama T, Sakakibara T and Onoda S 2007 *Phys. Rev. Lett.* **98** 057203
[28] Ikeda A and Kawamura H 2008 *J. Phys. Soc. Jpn.* **77** 073707
[29] Yoshii S, Iikubo S, Kageyama T, Oda K, Kondo Y, Murata K and Sato M 2000 *J. Phys. Soc. Jpn.* **69** 3777
[30] Onoda S and Tanaka Y preprint availalble at arXiv:1011.4981
[31] Xia J, Schemm E, Deutscher G, Kivelson S A, Bonn D A, Hardy W N, Liang R, Siemons W, Koster G, Fejer M M and Kapitulnik A 2008 *Phys. Rev. Lett.* **100** 127002

International Conference on Frustration in Condensed Matter (ICFCM) IOP Publishing
Journal of Physics: Conference Series **320** (2011) 012057 doi:10.1088/1742-6596/320/1/012057

Dynamical Mean Field Study of the Kondo Lattice Model with Frustration

R Peters[1], N Kawakami[1], T Pruschke[2]

1 Department of Physics, Kyoto University, Kyoto 606-8502, Japan
2 Department of Physics, University of Goettingen, Goettingen 37077, Germany

E-mail: `peters@scphys.kyoto-u.ac.jp`

Abstract. We analyze the antiferromagnetic Kondo lattice model for the two-dimensional square lattice including frustration. The Kondo lattice model is especially interesting in the context of heavy fermion physics, as it possibly exhibits a quantum phase transition between an antiferromagnetically ordered state and a paramagnetic heavy Fermion state. The latter can, depending on the electronic properties be either a metallic state with dynamically generated flat bands at the Fermi energy, or an insulator, the so-called Kondo insulator. We analyze the effects of next nearest neighbor (NNN) hopping upon the properties of the Kondo lattice model, especially focusing on the paramagnetic and antiferromagnetic properties around half filling. To this end we use the dynamical mean field theory (DMFT) with the Numerical Renormalization Group (NRG) as an impurity solver. Whereas the paramagnetic results remain qualitatively unchanged when introducing the NNN hopping, the Néel state at half filling is replaced by a spin-density wave. Also the transition between the magnetic ordered state and the Kondo insulator depends on the NNN hopping. An interesting effect of the NNN hopping is the stabilization of Néel states away from half filling.

1. Introduction

Heavy fermion materials have been in the focus of intense research for over 30 years [1]. Their name originates from the fact that density of states (DOS) at the Fermi energy of approximately 100 or 1000 times higher than in ordinary metals has been measured. Besides this very large DOS, a common feature of these materials is a continuous transition between an antiferromagnetic and a paramagnetic phase at $T = 0$ driven by an external control parameter like pressure or magnetic field. The transition point at $T = 0$ is usually referred to as Quantum Critical Point (QCP) [2, 3], as quantum fluctuations dominate the behavior of the system at this point. Near this QCP new physics like unconventional superconductivity or non-Fermi-liquid behavior has been found.

Heavy fermion materials usually include rare earth ions, whose partially filled f-shells form a lattice of local moments. These local moments are coupled to the s- or p-bands of the material. A single localized moment coupled to a bath of non-interacting electrons is usually called an impurity model. These models have become famous for the Kondo effect [4, 5]. Thus, it is not surprising that Kondo physics plays an important role for describing heavy fermion compounds. The other major ingredient is the RKKY interaction [6, 7, 8] between the local moments favoring a magnetic phase. The quantum criticality comes as a consequence of the competition of these two mechanisms.

In the limit of strong local Coulomb interactions within the f-shells, this situation can be modeled by a lattice of spins $S_i = 1/2$, which are antiferromagnetically coupled to non-interacting electrons on a lattice. This Hamiltonian is the Kondo lattice Hamiltonian [9, 10, 11], reading

$$H = \sum_{i,j,\sigma} t_{i,j} c^\dagger_{i,\sigma} c_{j,\sigma} + J \sum_i \vec{S}_i \cdot \left(c^\dagger_{i,\sigma_1} \vec{\tau}_{\sigma_1,\sigma_2} c_{i,\sigma_2} \right),$$

for which $c^\dagger_{i,\sigma}$ creates an electron with spin σ on site i, and $t_{i,j}$ is the hopping amplitude for the electrons from site i to j. These electrons are coupled via a spin-spin interaction to the localized spins $\vec{S}_i$ of the rare earth ions with strength J, for which $\vec{\tau}_{\sigma_1,\sigma_2}$ denotes the Pauli spin matrices. In this article we assume an antiferromagnetic coupling being relevant for heavy fermion physics.

The phase diagram for the half filled Kondo lattice including only nearest neighbor (NN) hopping can be summarized as follows [10, 11, 12, 13, 14, 15]: For ferromagnetic or small antiferromagnetic coupling J, the ground state is an antiferromagnetic Néel state. For increasing antiferromagnetic coupling the ground state eventually changes into a paramagnetic insulating state, the Kondo insulator.

Recently, it was proposed that frustration will play an important role in explaining heavy fermion materials [16, 17]. It was asked how frustration influences the quantum phase transition. In this article we want to explore this question and analyze, if and how the Kondo insulator and the transition between the antiferromagnetic phase and the paramagnetic phase is changed, when a NNN hopping is included into the system. This additional hopping will impose frustration onto the antiferromagnetic state, but also influences the paramagnetic state. To determine the phase diagram we use the dynamical mean field theory (DMFT) [18, 19]. DMFT relates the lattice model to an impurity model, which has to be solved self-consistently. Thus DMFT neglects spatial fluctuations, which are of course important to the actual quantum critical point. Nevertheless, DMFT has proved to provide a thermodynamically consistent phase diagram being able to describe paramagnetic as well as antiferromagnetic phases and their properties. For solving the impurity model we employ the numerical renormalization group (NRG) [4, 20].

This article is organized as follows: In the next section we will introduce the lattice for which we have performed the calculations and show some results for non-interacting fermions. The next two sections deal with the paramagnetic and antiferromagnetic properties of the Kondo lattice model, which are followed by a conclusion to this article.

2. Two-dimensional frustrated square lattice

We have performed our calculations for a two-dimensional square lattice with NN and NNN hopping, see Fig. 1. Besides the usual NN hopping H_1, we introduced a special kind of NNN hopping H_2 consisting of two terms: a diagonal hopping t_2 and a NNN hopping along the axis with $t_2/2$. This allows us to perform the Hilbert transformation analytically, which increases the accuracy during the DMFT self-consistency. Using this kind of hopping, the NNN hopping H_2 can be related to the NN hopping H_1, as $H_2 = H_1^2 - 4$. Therefore, the Hilbert transformation and the local DOS of the whole hopping Hamiltonian can be deduced from the original NN hopping Hamiltonian. The dimensionality does not play a big role in our calculations as we use DMFT, i.e. only the local DOS enters the calculation. Nevertheless, from previous studies we have the experience that the detailed structure of the longer-range hopping does not play a significant role for the qualitative structures. The most important aspect is the generation of an asymmetry in the DOS [21]. Thus, our at first glance rather strange choice is mainly due to numerical reasons, but we do not expect strong qualitative changes when using a more physical DOS. Antiferromagnetic solutions within DMFT are obtained by breaking the spin-symmetry in the very first iteration and then using an AB-sublattice version of the Hilbert transformation, which for the frustrated lattice includes additional entries imposing frustration upon the antiferromagnetic Néel state.

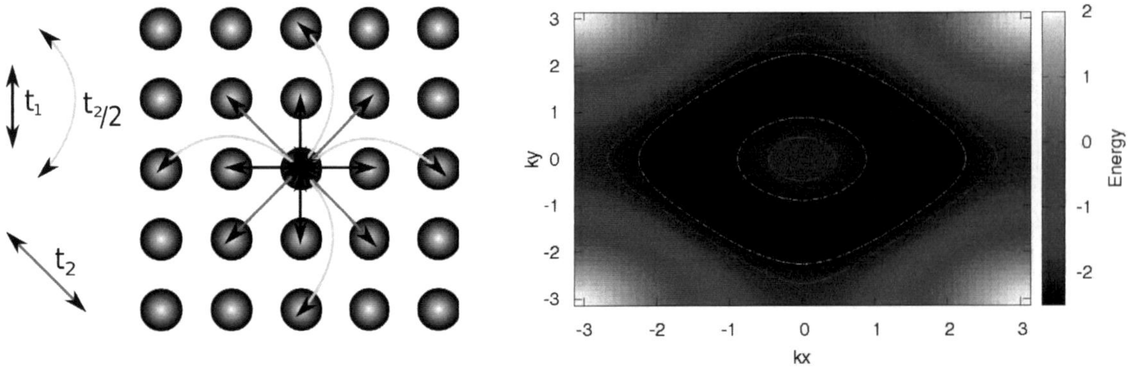

Figure 1. Left: Allowed hopping processes for the used two-dimensional lattice. Right: Energy landscape for different k-vectors for the frustrated case $t_1 = -1$ and $t_2 = 1$. Especially observe the local maximum at $(k_x, k_y) = (0, 0)$ which is completely absent in the unfrustrated case.

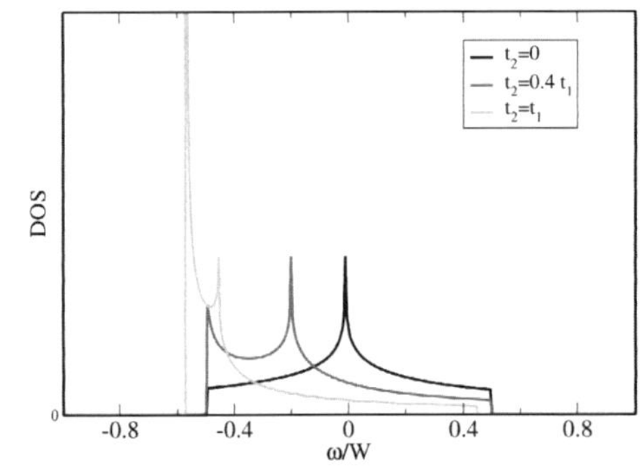

Figure 2. DOS for different strengths of NNN hopping t_2.

The right panel in Fig. 1 shows the energy landscape for non-interacting fermions for $t_1 = -1$ and $t_2 = 1$. Note that the sign of t_1 does not affect the DOS, thus has no effect on the DMFT results. Especially interesting is $(k_x, k_y) = (0, 0)$, where for $t_2 = 0$ a local minimum can be found. For increasing t_2 this completely changes and a local maximum is formed. This local maximum will especially change the physics for fillings $n < 1$. Finally, the local DOS for 3 different systems is shown in Fig. 2. The formation of the local maximum results in a divergence of the DOS at the lower band edge. The bandwidth W of the system is taken as energy unit.

3. Paramagnetic Phase

Let us briefly look at the paramagnetic results not allowing for antiferromagnetic ordering. Figure 3 summarizes the results for $J = 1/4W$. Qualitatively, the NNN hopping does not change the low temperature physics. Changing t_2, the k-dependent band-structure changes (see upper panels in Fig. 3), but the region around the Fermi energy shows the same effects: Exactly for half filling a gap at the Fermi energy is formed (bottom left panel), and there is a very narrow band near the Fermi energy for $(k_x, k_y) \approx (\pi, \pi)$ [23, 22, 24]. Finally this narrow band can be hole doped, thus crossing the Fermi-energy.

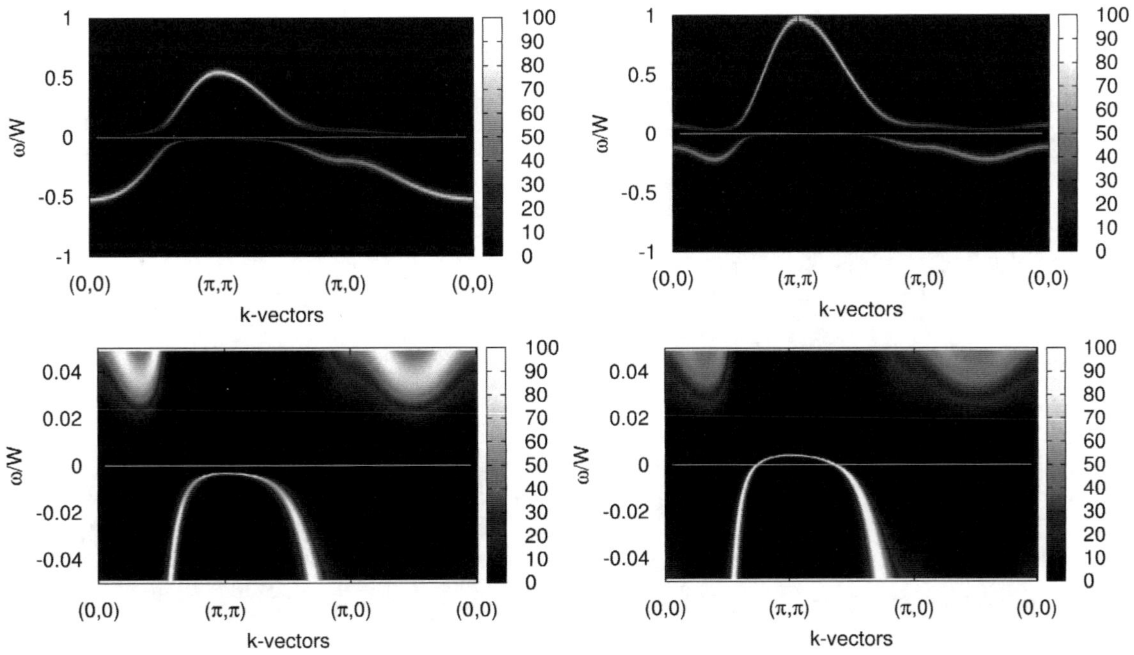

Figure 3. Frequency and k-dependent DOS: The upper panels show half filled solutions for $t_2/t_1 = 0.3$ (left panel) and $t_2/t_1 = 1$ (right panel) for $J = 1/4W$. The color encodes the weight. The lower panels show magnifications around the Fermi energy, $\omega = 0$, for $t_2/t_1 = 1$ and $J = 1/4W$. The bottom left panel shows the half filled solution, the bottom right panel shows a slightly hole doped solution (electron filling $n = 0.95$). The green line corresponds to $\omega = 0$.

This result is not very surprising. At half filling within the paramagnetic phase the screening of the local spins results in the mentioned gap at the Fermi energy and the narrow band near the Fermi energy. This screening will always take place as long as the non-interacting DOS has weight at the Fermi energy.

4. Antiferromagnetism

The major focus of this paper is however the antiferromagnetic solution, Fig. 4. The left panel shows a schematic phase diagram for different NNN hopping strengths and couplings J. For weak NNN hopping the phase diagram looks very similar to the unfrustrated case. For small J an antiferromagnetic Néel state is stabilized which finally vanishes via a second order phase transition and the Kondo insulator is formed. Increasing the NNN hopping, the Néel state is replaced by an incommensurate (IC) state for small J . This IC phase is characterized by oscillations during the DMFT iterations and therefore cannot be analyzed in more detail [25]. For strong J, near the QCP, the Néel state seems to remain stable for $t_2 < t_1$. However, the QCP moves to smaller values of J for increasing t_2. Therefore, one finds the expected line of second order phase transitions between the Néel state and the paramagnetic Kondo insulator. This behavior can be explained by the argument that for strong coupling J and $t_2 < t_1$, the system is finally dominated by the leading mechanism: the NN antiferromagnetic exchange, and for further increasing J the Kondo screening forming the Kondo insulator. Only for $t_2 = t_1$ the Néel state is completely destabilized towards the IC phase. One more thing should be mentioned: surprisingly, at large $t_2/t_1 = 0.8$ the Néel state can be stabilized for smaller values J than for $t_2/t_1 = 0.6$ or $t_2/t_1 = 0.4$. This behavior comes along with the formation of the local maximum

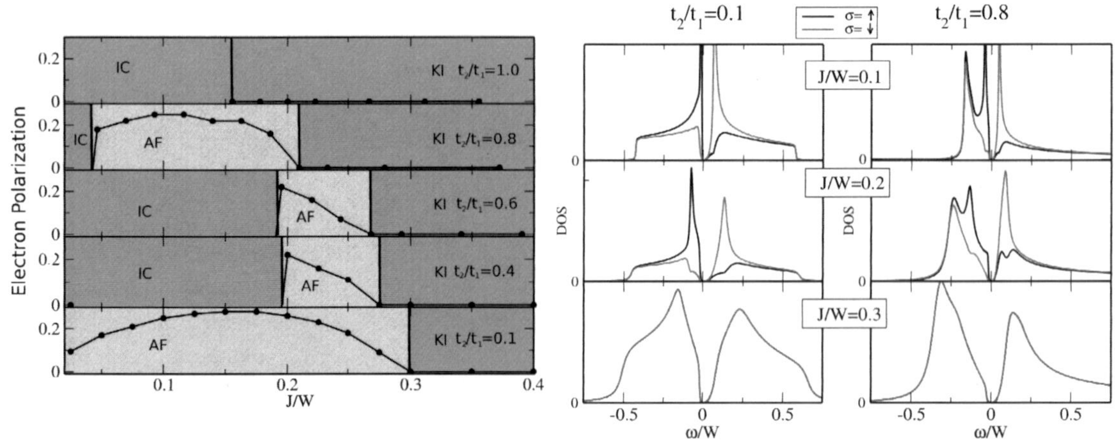

Figure 4. Antiferromagnetism at half filling. Left: The half filled DMFT solution for 5 different strengths of NNN hopping; IC = incommensurate, AF = antiferromagnetic Néel state, KI = Kondo insulator. The points in the antiferromagnetic region depict the electron polarization. Right: DOS for 2 different strengths of NNN hopping and 3 couplings J in the antiferromagnetic and Kondo insulating states.

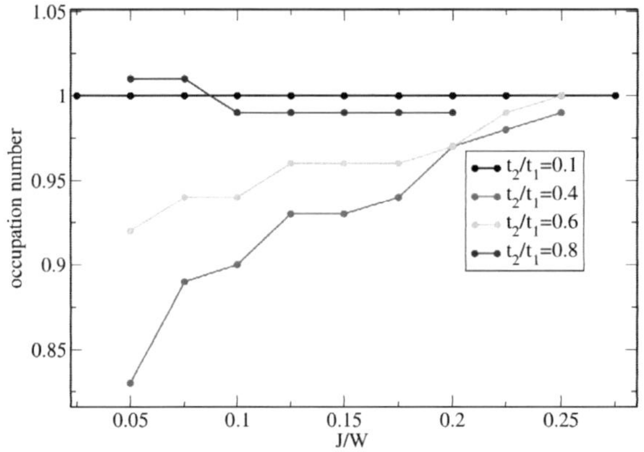

Figure 5. Antiferromagnetism away from half filling. Occupation number of antiferromagnetic regions (Néel states) for different NNN hopping strengths. The right end point of the lines mark the transition to the paramagnetic state. (The lines are meant as guide to the eye.)

in the energy landscape mentioned above. Thus, this stabilization of the Néel state quite likely is an artifact of the special kind of NNN hopping consisting of 2 parts used here, but the actual mechanism is still unclear.

The right panel in Fig. 4 shows the local DOS for two NNN hopping strengths and different couplings J. This figure depicts nicely how the non-interacting DOS, which still can be clearly seen for $J/W = 0.1$, is completely changed by increasing the coupling. For $J/W = 0.3$, in the paramagnetic Kondo insulating phase, the structures of the non-interacting DOS are not visible anymore.

Another interesting point, already mentioned in a former study [21], is that NNN hopping can stabilize antiferromagnetic Néel states away from half filling. Figure 5 shows for different strengths of NNN hopping and coupling J the occupation number around which a Néel state can be stabilized. These antiferromagnetic regions are rather small, in the meaning that doping $\Delta n \approx 0.02$ away from the shown occupation will destabilize the Néel state towards an IC phase. This figure shows again that the Néel states for $t_2/t_1 = 0.1$ as well as $t_2/t_1 = 0.8$ are located at half filling. However, as the coupling J has big influence on the occupation number, we cannot predict these values only from the non-interacting Fermi surface.

5. Conclusions

We have analyzed the paramagnetic and antiferromagnetic properties of the Kondo lattice model including NNN hopping. We find that the paramagnetic properties remain qualitatively unchanged when NNN hopping is included. While the k-dependent band structure is changed, the behavior around the Fermi energy qualitatively remains. Exactly at half filling we still find a gap with a narrow band very near the Fermi energy.

The situation is different when studying the antiferromagnetic phase. The NNN hopping destabilizes the Néel state at half filling towards an IC state for weak coupling J. However, for $t_2 < t_1$ and strong coupling J, we still find a second order phase transition between an antiferromagnetic Néel state and a paramagnetic state. The transition point now depends on the strength of the NNN hopping. Although NNN hopping destabilizes the Néel state at half filling, we show that it can stabilize antiferromagnetic Néel states away from half filling.

Acknowledgments

The calculations have been performed at the computer cluster "Kashiwa" of ISSP in Tokyo. RP thanks the Japan Society for the Promotion of Science (JSPS) together with the Alexander von Humboldt-Foundation for a postdoctoral fellowship. NK is supported by JSPS through its FIRST program and KAKENHI (21740232, 20104010). TP acknowledges support by the German Science Foundation (DFG) through SFB 602.

References

[1] Coleman P 2007 *Handbook of Magnetism and Advanced Magnetic Materials*, ed Helmut Kronmuller and Stuart Parkin (John Wiley and Sons), (arXiv:cond-mat/0612006v3).
[2] Coleman P and Schofield A 2005 *Nature* **443** 226.
[3] Gegenwart P, Si Q, and Steglich F 2008 *Nature Physics* **4** 186.
[4] Wilson K 1975 *Rev. Mod. Phys.* **47** 773.
[5] Hewson A 1997 *The Kondo Problem to Heavy Fermions* (Cambridge University Press).
[6] Ruderman M, and Kittel C 1954 *Phys. Rev.* **96** 99.
[7] Kasuya T 1956 *Progress Theoretial Physics* **16** 45
[8] Yosida K 1957 *Phys. Rev.* **106** 893.
[9] Doniach S 1977 *Physica* B **91** 231.
[10] Lacroix C and Cyrot M 1979 *Phys. Rev.* B **20** 1969.
[11] Fazekas P, and Müller-Hartmann E 1991 *Z. Phys. B: Condens. Matter* **85** 285.
[12] Costi T and Manini N 2002 *Journal of Low Temperature Physics* **126** 835
[13] Peters R and Pruschke T 2007 *Phys. Rev.* B **76** 245101.
[14] Otsuki J , Kusunose H, and Kuramoto Y 2009 *J. Phys. Soc. Jpn.* **78** 034719
[15] Hoshino S, Otsuki J, and Kuramoto Y 2010 *Phys. Rev.* B **81** 113108.
[16] Coleman P, and Nevidomskyy H 2010 *Journal of Low Temperature Physics* **161** 182.
[17] Motome Y, Nakamikawa K, Yamaji Y, and Udagawa M 2010 *Phys. Rev. Lett.* **105** 036403
[18] Metzner W, and Vollhardt D 1989 *Phys. Rev. Lett.* **62** 324.
[19] Georges A, Kotliar G, Krauth W, and Rozenberg M 1996 *Rev. Mod. Phys.* **68** 13.
[20] Bulla R, Costi T, and Pruschke T 2008 *Rev. Mod. Phys.* **80** 395.
[21] Peters R, and Pruschke T 2009 *New Journal Physics* **11** 083022.
[22] Martin L and Assaad F 2008 *Phys. Rev. Lett.* **101** 066404
[23] Lanatál N, Baronel P, and Fabrizio M 2008 *Phys. Rev.* B **78** 155127
[24] Otsuki J, Kusunose H, and Kuramoto Y 2009 *Phys. Rev. Lett.* **102** 017202
[25] Peters R and Pruschke T 2009 *Phys. Rev.* B **79** 045108

Journal of Physics: Conference Series **320** (2011) 012058 doi:10.1088/1742-6596/320/1/012058

A nonmagnetic effective Hamiltonian of a frustrated tetrahedron Kondo model

Kazumasa Hattori and Hirokazu Tsunetsugu

Institute for Solid State Physics, University of Tokyo, Kashiwanoha 5-1-5, Kashiwa, Chiba, Japan

E-mail: hattori@issp.u-tokyo.ac.jp

Abstract. We investigate a four-site impurity Kondo model with 1/2 spins locating on vertices of a tetrahedron, when the inter site superexchange interaction J is much larger than band width D of conduction electrons and magnetic Kondo coupling. In this case, four impurity spins form doubly-degenerate ground states and we derive their e ective Hamiltonian induced by coupling with conduction electrons. Quadrupole moments of the ground states couple to those of conduction electrons. For particle-hole asymmetric case, we nd that the model can be mapped to that with a pseudospin-1/2 impurity coupled with three-orbital conduction electrons with keeping U(1) charge conservation for each of the orbital.

1. Introduction

Frustration in condensed matter has been attracted much attention due to possible mechanism stabilizing spin liquid states without any symmetry breaking[1]. In the context of cluster impurity Kondo model, it has been shown that triangular cluster impurity con guration with antiferromagnetic interactions between the impurity spins leads to a stable non Fermi liquid (NFL) ground state on a basis of numerical renormalization group and boundary conformal eld theory[2]. This contrasts with an unstable NFL state in a two-impurity Kondo model studied more than two decades ago[3], where there is no magnetic frustration between the impurity spins. This suggests that, exotic NFL states emerge and are stabilized, when antiferromagnetic Kondo interactions are switched on for localized spins on geometrically frustrated clusters.

In this paper, we construct an e ective low-energy model in another fundamental cluster with frustrated geometry, a tetrahedron cluster, which has a possibility to realize NFL ground states. As in the case of triangular cluster, there remains nonmagnetic degeneracy in the ground state of tetrahedron cluster spins that antiferromagnetically interact with each other. We will concentrate on deriving an e ective Hamiltonian that describes low-energy dynamics of this degenerate ground state interacting with conduction electrons by using symmetry argument and scaling approach[4].

2. Model

2.1. Hamiltonian

In this paper, we investigate a tetrahedron-impurity Kondo model, in which four spin-1/2 impurities are located on the vertices of a tetrahedron. They interact antiferromagnetically to each other with Heisenberg coupling J and also interact with conduction electrons via Kondo exchange coupling J_K:

$$H = \int d\mathbf{k}\, \epsilon_{\mathbf{k}} c^{\dagger}_{\mathbf{k}\sigma} c_{\mathbf{k}\sigma} + J \sum_{i \neq j=1}^{4} \mathbf{S}_i \quad \mathbf{S}_j + H_K \tag{1}$$

$$H_{\mathrm{K}} = J_K \sum_{i=1}^{4} \sum_{\sigma\sigma'} \mathbf{S}_i \quad (\mathbf{s})_{\sigma\sigma'} \iint d\mathbf{k} d\mathbf{p}\; c^{\dagger}_{\mathbf{k}\sigma} c_{\mathbf{p}\sigma'} \exp[i(\mathbf{p} \quad \mathbf{k}) \quad \mathbf{x}_i] \tag{2}$$

where $\mathbf{S}_i (i = 1, 2, 3, \text{ or } 4)$ and $c^{\dagger}_{\mathbf{k}\sigma}$ represent the spin-1/2 operator at the site i and the conduction electron creation operator with wavevector $\mathbf{k}$ and spin σ, and $\mathbf{s} = \boldsymbol{\tau}/2$, where τ^x, τ^y, and τ^z are the Pauli matrices. The intersite antiferromagnetic superexchange interaction $J > 0$ is set to the largest energy scale in this paper. A schematic picture of this model is shown in Fig. 1(a).

For later purposes, let us de ne the following conduction electron operators[5], which directly couple to the impurity spins:

$$c_{ik\sigma} \equiv \int d\Omega_{\mathbf{k}}\; c_{\mathbf{k}\sigma} \exp(i\mathbf{k} \quad \mathbf{x}_i), \tag{3}$$

with $k = |\mathbf{k}|$ and $\int d\Omega_{\mathbf{k}}$ is the angular integration in the $\mathbf{k}$-space. $c_{ik\sigma}$'s, however, are not orthogonal with each other. An orthonormal set of conduction electron operators is obtained by using linear combinations of $c_{ik\sigma}$ as

$$c_{0k\sigma} = N^{\,1}_{kA_1}(c_{1k\sigma} + c_{2k\sigma} + c_{3k\sigma} + c_{4k\sigma})/2, \quad c_{ak\sigma} = N^{\,1}_{kT_2}(c_{1k\sigma} \quad c_{2k\sigma} + c_{3k\sigma} \quad c_{4k\sigma})/2, \tag{4}$$

$$c_{bk\sigma} = N^{\,1}_{kT_2}(c_{1k\sigma} \quad c_{2k\sigma} \quad c_{3k\sigma} + c_{4k\sigma})/2, \quad c_{ck\sigma} = N^{\,1}_{kT_2}(c_{1k\sigma} + c_{2k\sigma} \quad c_{3k\sigma} \quad c_{4k\sigma})/2, \tag{5}$$

where $N^2_{kA_1} = k^2[1 + 3\frac{\sin(ka_0)}{ka_0}]$ and $N^2_{kT_2} = k^2[1 \quad \frac{\sin(ka_0)}{ka_0}]$ with $a_0 = |\mathbf{x}_1 \quad \mathbf{x}_2|$. Note that $c_{0k\sigma}$ and $c_{lk\sigma} (l = a, b, \text{ or } c)$ are basis sets of A_1 and T_2 representations in T_d point group symmetry, respectively. For later purposes, we de ne the coe cients in Eqs. (4) and (5) as $c_{0k\sigma} = \sum_i f_{0i} c_{ik\sigma}/N_{kA_1}$, $c_{ak\sigma} = \sum_i f_{ai} c_{ik\sigma}/N_{kT_2}$ and similarly for b and c. Using these bases, the Hamiltonian (2) is rewritten as

$$\begin{aligned} H_{\mathrm{K}} = \sum_{\sigma\sigma'} \iint dk dp \Bigg[& J_0 \mathbf{S}_0 \quad \boldsymbol{s}_{\sigma\sigma'} c^{\dagger}_{0k\sigma} c_{0p\sigma'} + J_1 \mathbf{S}_0 \sum_{l=a,b,c} c^{\dagger}_{lk\sigma} c_{lp\sigma'} \boldsymbol{s}_{\sigma\sigma'} \\ + & J_2 \sum_{l=a,b,c} \left(\mathbf{S}_l \quad \boldsymbol{s}_{\sigma\sigma'} c^{\dagger}_{0k\sigma} c_{lp\sigma'} + \text{h.c.} \right) + \frac{J_3}{2} \sum_{lmn=a,b,c} |\varepsilon^{lmn}| \left(\mathbf{S}_l \quad \boldsymbol{s}_{\sigma\sigma'} c^{\dagger}_{mk\sigma} c_{np\sigma'} + \text{h.c.} \right) \Bigg], \end{aligned} \tag{6}$$

where $J_0 = J_K N^2_{A_1}/2$, $J_2 = J_K N_{T_2} N_{A_1}/2$, and $J_1 = J_3 = J_K N^2_{T_2}/2$, and ε^{lmn} is the antisymmetric tensor with $\varepsilon^{abc} = 1$. We have approximate $N_{kA_1,T_2} \quad N_{k_F A_1,T_2} \equiv N_{A_1,T_2}$ with k_F being the Fermi wave number. $\mathbf{S}_l (l = 0, a, b, \text{ and } c)$ are given as $\mathbf{S}_l = \sum_i f_{li} \mathbf{S}_i$. Here, $\mathbf{S}_0$ ($\mathbf{S}_{a,b,c}$) is A_1 (T_2) representation as in Eqs. (4) and (5). The conduction electron kinetic-energy part in the Hamiltonian (1) is approximated as

$$\int d\mathbf{k} \epsilon_{\mathbf{k}} c^{\dagger}_{\mathbf{k}\sigma} c_{\mathbf{k}\sigma} \simeq \int dk \sum_{l=0,a,b,c,d} \epsilon_{kl} c^{\dagger}_{kl\sigma} c_{kl\sigma} \equiv H_c, \tag{7}$$

since only these four components directly couple with impurity spins. We further approximate $\epsilon_{kl} \quad v_F(k \quad k_F)$, where v_F indicates the velocity at Fermi energy. The density of states of conduction electrons are set to constant ρ from $D \;=\; D$ to D as shown in Fig. 1(b). Note that the case $\;\neq 1$ represents a conduction electron band without particle-hole symmetry.

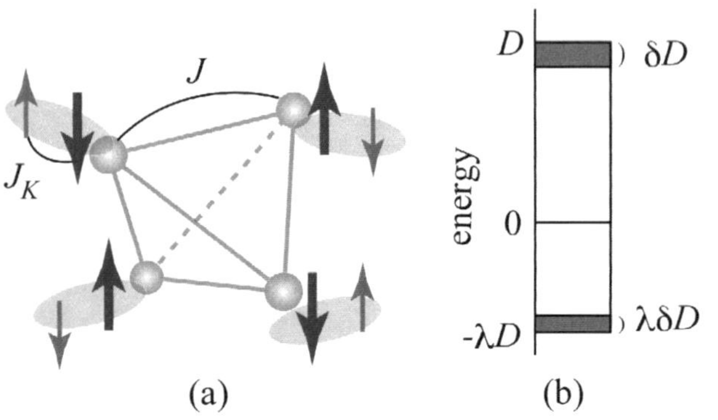

Figure 1. (a) Schematic picture of model (1). (b) Conduction electron density of states. Renormalization processes used in this work are also indicated.

2.2. Ground states and low-energy states of four local spins

As we mentioned before, the antiferromagnetic superexchange interaction J is the largest energy scale in the system. Let us rst examine the energy spectrum and the eigenstates for the isolated tetrahedron. Since the superexchange interaction term in the Hamiltonian (1) becomes $2J\mathbf{S}_0^2+$ constant, the ground state of a tetrahedron impurity is a nonmagnetic doublet, i.e., the eigenvalue of the total spin $\mathbf{S}(= 2\mathbf{S}_0)$ $S = 0$, which is E representation in T_d point group. The wavefunctions for these two states are given as

$$|e_1\rangle \;=\; \frac{1}{2}(1+\hat{I})\;|\uparrow\downarrow\uparrow\downarrow\rangle \quad |\uparrow\downarrow\downarrow\uparrow\rangle\;, \tag{8}$$

$$|e_2\rangle \;=\; \frac{1}{\sqrt{3}}(1+\hat{I})\Big\{|\uparrow\uparrow\downarrow\downarrow\rangle \quad \frac{1}{2}\;|\uparrow\downarrow\uparrow\downarrow\rangle + |\uparrow\downarrow\downarrow\uparrow\rangle\;\Big\}, \tag{9}$$

where $|\sigma_1\sigma_2\sigma_3\sigma_4\rangle$ represents a state with the z-component of spin $\sigma_i(=\uparrow$ or $\downarrow)$ at site i and $\hat{I}$ is the spin inversion operator.

The rst excited states are T_2 multiplet with the total spin $S = 1$, i.e., nine-fold degenerate states and the energy gap is J. The wavefunctions of T_2 states are given as

$$|0\;a\rangle = \frac{1}{\sqrt{2}}\;|\uparrow\downarrow\uparrow\downarrow\rangle \quad |\downarrow\uparrow\downarrow\uparrow\rangle\;, |0\;b\rangle = \frac{1}{\sqrt{2}}\;|\downarrow\uparrow\uparrow\downarrow\rangle \quad |\uparrow\downarrow\downarrow\uparrow\rangle\;, |0\;c\rangle = \frac{1}{\sqrt{2}}\;|\uparrow\uparrow\downarrow\downarrow\rangle \quad |\downarrow\downarrow\uparrow\uparrow\rangle\;, \tag{10}$$

$$|\Uparrow l\rangle = \sum_i f_{li}S_i\;|\uparrow\uparrow\uparrow\uparrow\rangle,\; |\Downarrow l\rangle = \quad \sum_i f_{li}S_i^+|\downarrow\downarrow\downarrow\downarrow\rangle, \tag{11}$$

where $|S_z l\rangle$ is a T_2 state with the z-component of spin S_z and orbital index $l = a, b,$ or c. There is also an $S = 2$ excited state with the excitation energy $3J$, but it is not relevant to our discussion in the following, because of vanishing matrix element of S_0 and $S_l(l = a, b,$ and $c)$ between the ground states and the second excited states. The matrix elements of S_l between the ground and the rst excited states are nite, while those of S_0 vanishes since the ground states are spin singlet. We show the nonvanishing matrix elements in Table 1.

3. Effective Hamiltonian

3.1. Symmetry

Now, we derive an e ective Hamiltonian describing interactions between nonmagnetic E states and conduction electrons. Since the ground state is doublet, we can express any operators projected to these two states in terms of the Pauli matrices. Relevant operators in this paper are $\tau^x = |e_1\rangle\langle e_2| + |e_2\rangle\langle e_1|$ and $\tau^z = |e_1\rangle\langle e_1| \quad |e_2\rangle\langle e_2|$, which form two-dimensional E representation and represent quadrupole moment of the ground states, or alternatively scalar chirality of the ground state[6], when we employ the complex representation for the E state: $[|e_1\rangle \pm i|e_2\rangle]/\sqrt{2}$.

Table 1. List of nonvanishing matrix elements between the ground states and rst excited states $\langle S_z\ l|S_{l'}|e_i\rangle$ $(i = 1,$ or $2; l, l' = 0, a, b,$ or $c)$. $S_l^{\pm}$ is de ned as in a usual manner $S_l^{\pm} = S_l^x \pm iS_l^y$. Elements not shown are all zero.

$\langle S_z l\vert S_{l'}^{z,\pm}\vert e_1\rangle$		$\langle S_z l\vert S_{l'}^{z,\pm}\vert e_2\rangle$	
$\langle 0\ a\vert S_a^z\vert e_1\rangle = \langle 0\ b\vert S_b^z\vert e_1\rangle$	$\frac{1}{\sqrt{2}}$	$\langle 0\ a\vert S_a^z\vert e_2\rangle = \langle 0\ b\vert S_b^z\vert e_2\rangle$	$\frac{1}{\sqrt{6}}$
		$\langle 0\ c\vert S_c^z\vert e_2\rangle$	$\sqrt{\frac{2}{3}}$
$\langle \Uparrow a\vert S_a^+\vert e_1\rangle = \langle \Uparrow b\vert S_b^+\vert e_1\rangle$	1	$\langle \Uparrow a\vert S_a^+\vert e_2\rangle = \langle \Uparrow b\vert S_b^+\vert e_2\rangle$	$\frac{1}{\sqrt{3}}$
		$\langle \Uparrow c\vert S_c^+\vert e_2\rangle$	$\frac{2}{\sqrt{3}}$
$\langle \Downarrow a\vert S_a\ \vert e_1\rangle = \langle \Downarrow b\vert S_b\ \vert e_1\rangle$	1	$\langle \Downarrow a\vert S_a\ \vert e_2\rangle = \langle \Downarrow b\vert S_b\ \vert e_2\rangle$	$\frac{1}{\sqrt{3}}$
		$\langle \Downarrow c\vert S_c\ \vert e_2\rangle$	$\frac{2}{\sqrt{3}}$

As for the conduction electron part, operators consisting of two fermion operators with E representation are the most relevant and can couple with the impurity. It turns out that there is only one such E representation for the conduction electron part, which is given as

$$\int dkdp \sum_{\sigma} \frac{1}{\sqrt{3}}[c_{ka\sigma}^{\dagger}c_{pa\sigma} + c_{kb\sigma}^{\dagger}c_{pb\sigma} \quad 2c_{kc\sigma}^{\dagger}c_{pc\sigma}],\ \int dkdp \sum_{\sigma}[c_{kb\sigma}^{\dagger}c_{pb\sigma} \quad c_{ka\sigma}^{\dagger}c_{pa\sigma}]. \tag{12}$$

Thus, taking into account trivial potential scattering terms in A_1 and T_2 sectors, the e ective interactions in low-energy sector should be

$$\begin{aligned} H_K^{\text{eff}} &= g\int dkdp \sum_{\sigma}\left[\tau^z \frac{1}{\sqrt{3}}\ c_{ka\sigma}^{\dagger}c_{pa\sigma} + c_{kb\sigma}^{\dagger}c_{pb\sigma} \quad 2c_{kc\sigma}^{\dagger}c_{pc\sigma}\ + \tau^x\ c_{kb\sigma}^{\dagger}c_{pb\sigma} \quad c_{ka\sigma}^{\dagger}c_{pa\sigma}\right] \\ &\quad + V_0\int dkdp\sum_{\sigma} c_{k0\sigma}^{\dagger}c_{p0\sigma} + V_1\int dkdp \sum_{\sigma}\sum_{l=a,b,c} c_{kl\sigma}^{\dagger}c_{pl\sigma}. \end{aligned} \tag{13}$$

Note that operator τ^y does not appear, since τ^y belongs to A_2 representation and there is no such representation in conduction electron part as far as we consider quadratic operators of fermions. The e ective interaction (13) is purely nonmagnetic and all the terms break particle-hole symmetry ($c_{kl\sigma} \leftrightarrow \sigma c_{kl\bar{\sigma}}^{\dagger}$), which is not broken in the magnetic Kondo couplings in H_K. Thus, the particle-hole asymmetry in the conduction electron band $\neq 1$ is essential to realize the model (13)[7]. Since there is no magnetic terms in the e ective Hamiltonian (13), which is present in the original interactions (2), the magnetic interactions play a role of "source" of the nonmagnetic interactions. It is noted that emergent U(1)⊗U(1)⊗U(1) symmetry is present in the e ective model (13), which corresponds to the charge conservation of each conduction electron orbital. Such an emergent charge conservation was also found in the NFL state in triangular Kondo model[2].

3.2. Renormalization group equations

Let us demonstrate how the model (13) is generated in the low-energy sector of the tetrahedron Kondo model (1). Here, we employ a renormalization group method[4] rather than perturbation theory in terms of $1/J$. Following the discussion in Ref. [4], we can derive one-loop renormalization group equations for this system. We take into account the virtual scattering processes in the second order in $H_K + H_K^{\text{eff}}$, in which conduction electrons near the Fermi surface are scattered into the band edge within the width δD and δD as shown in Fig. 1(b). Initially, the parameters g, V_0 and V_1 are all zero. They are generated by the second order processes of the magnetic interactions J's in Eq. (6). Since we project out all the excited states, the magnetic couplings are not renormalized. Owing to the selection rule, the possible virtual states are T_2 states and the matrix elements are given in Table 1. The renormalization group equations are, indeed, trivial because of no renormalization in magnetic exchange interactions and commutativity in the conduction electron part of the e ective Hamiltonian (13):

$$\frac{\partial g}{\partial D} = \frac{\sqrt{3}f(D, \ , J)}{8} \ \rho {J_2}^2 \quad \rho J_3^2 \ , \tag{14}$$

$$\frac{\partial V_0}{\partial D} = \frac{3f(D, \ , J)}{4} \rho {J_2}^2, \tag{15}$$

$$\frac{\partial V_1}{\partial D} = f(D, \ , J) \ \frac{\rho {J_2}^2}{4} + \frac{\rho {J_3}^2}{2} \ , \tag{16}$$

$$\frac{\partial J_0}{\partial D} = \frac{\partial J_1}{\partial D} = \frac{\partial J_2}{\partial D} = \frac{\partial J_3}{\partial D} = 0, \tag{17}$$

where $f(D, \ , J) = \frac{(1 \ \ \lambda)J}{(D+J)(\lambda D+J)} \quad (1 \quad)/J$ for $J \gg D$. The scaling stops when the coupling constant $|g(D^*)| \quad D^*$ and D^* is given as $D^* \simeq D \ /(1 + \ ) \simeq \ D$ with $\ = \sqrt{3}\rho|(1 \quad)(J_2^2 \quad J_3^2)|/(8J)$ for $J \gg D \gg \rho|J_2^2 \quad J_3^2|$. Since A_1 component of conduction electrons is decoupled in the e ective Hamiltonian (13), we can concentrate on the T_2 components hereafter.

Now, we have obtained the e ective coupling constant g and V_1 in the e ective Hamiltonian (13). Next, in order to capture physical picture of the e ective model, we consider strong coupling limit, where the conduction electron kinetic part H_c can be neglected. Diagonalizing the e ective Hamiltonian (13), we obtain various ground states as shown in Fig. 2. Each state is characterized by the occupation numbers in T_2 conduction electrons $n_l \equiv \int dk \sum_\sigma n_{kl\sigma}$: (n_a, n_b, n_c) and the ground state energy is given as $\ |g|\sqrt{\frac{2}{3}}\sqrt{(n_a \quad n_b)^2 + (n_b \quad n_c)^2 + (n_c \quad n_a)^2} + V_1(n_a + n_b + n_c) \ $. Except for the lines $|g| = \pm\sqrt{3}V_1/2$, the ground states are nonmagnetic. The states (0,0,0) and (2,2,2) are both doublet, where orbital degeneracy of E states remains, i.e., $(0,0,0)|e_1\rangle$ and $(0,0,0)|e_2\rangle$, while (2,2,0) and (2,0,0) are both orbital triplet. Here, for example, (2,0,0) means the three states: $(2,0,0)$, $(0,2,0)$ and $(0,0,2)$. In the cases of $(2,2,0)$ and $(2,0,0)$ ground states, further step renormalization generates e ective interactions, in which residual degrees of freedom couple to the pair hopping processes between di erent orbitals of conduction electrons. Further analysis of this e ective model is now in progress by using boundary conformal eld theory[8]. The high symmetry present in the e ective model also enables us to carry out an e cient numerical renormalization group calculations in the model with three orbital degrees of freedom. The results support the present one and will be published elsewhere[8].

4. Summary and perspective

In summary, we have discussed an e ective Hamiltonian of a tetrahedron Kondo model for $J \gg J_K$. The point group symmetry restricts the type of Kondo coupling between conduction electrons and degenerate impurity E states. Particle-hole asymmetry also plays an important role to realize the Kondo-like e ective model. In the e ective model, conduction electrons in each orbital are not scattered into di erent orbitals by scatterings due to the impurity. This means U(1) charge for each of the conduction electron orbital conserves during scattering processes. Detailed analyses of the xed point and possible xed points for $J_K \gg J$ are now in progress. We also study a tetrahedron Anderson impurity model by using continuous time quantum Monte Carlo method and results will be published elsewhere[9].

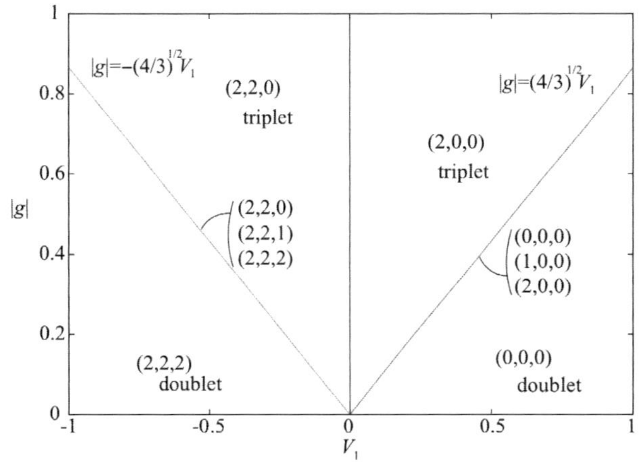

Figure 2. Ground state phase diagram of $H_K^{\rm eff}$ at the scale D^*.

Acknowledges
This work is supported by a Grant-in-Aid for Scienti c Research on Priority Areas "Novel State of Matter Induced by Frustration" (19052003).

References
[1] Diep H T 2004 *Frustrated Spin Systems* (World Scienti c Publishing)
[2] Ingersent K, Ludwig A W W, and Affleck I 2005 *Phys. Rev. Lett.* **95** 257204
[3] Jones B A, and Varma C M 1987 *Phys. Rev. Lett.* **58** 843
[4] Anderson P W 1970 *J. Phys. C* **3** 2436
[5] Cox D L and Zawadowski A 1997 *Adv. in Phys.* **47** 599
[6] Tsunetsugu H 2000 *J. Phys. Soc. Jpn.* **70** 640; 2001 *Phys. Rev. B* **65** 024415
[7] For particle-hole symmetric case, it is important to take into account two-particle scattering at the impurity site made by four fermion operators, which also couple to the impurity.
[8] Hattori K and Tsunetsugu H, to be published
[9] Hattori K and Tsunetsugu H, to be published

Noncoplanar spin canting in lightly-doped ferromagnetic Kondo lattice model on a triangular lattice

Yutaka Akagi and Yukitoshi Motome

Department of Applied Physics, University of Tokyo, Hongo 7-3-1, Bunkyo-ku, Tokyo 113-8656

E-mail: akagi@aion.t.u-tokyo.ac.jp

Abstract. E ect of the coupling to mobile carriers on the 120° antiferromagnetic state is investigated in a ferromagnetic Kondo lattice model on a frustrated triangular lattice. Using a variational calculation for various spin orderings up to a four-site unit cell, we identify the ground-state phase diagram with focusing on the lightly-doped region. We nd that an electron doping from the band bottom immediately destabilizes a 120° coplanar antiferromagnetic order and induces a noncoplanar three-sublattice ordering accompanied by an intervening phase separation. This noncoplanar phase has an umbrella-type spin con guration with a net magnetic moment and a nite spin scalar chirality. This spin-canting state emerges in competition between the antiferromagnetic superexchange interaction and the ferromagnetic double-exchange interaction under geometrical frustration. In contrast, a hole doping from the band top retains the 120°-ordered state up to a nite doping concentration and does not lead to a noncolpanar ordering.

1. Introduction

Coupling between charge and spin degrees of freedom is a fertile ground for various fascinating phenomena in strongly correlated electron systems. One of the fundamental models for describing the interplay is the ferromagnetic Kondo lattice model, or equivalently, the double-exchange (DE) model [1]. The model includes the ferromagnetic Hund's-rule coupling between itinerant electrons and localized spins as well as the antiferromagnetic (AF) superexchange (SE) interaction between localized spins. In addition to the early researches [1, 2, 3], the rediscovery of the colossal magnetoresistance (CMR) e ect has stimulated intensive studies and many interesting properties have been revealed [4, 5, 6]. Particularly, in the lightly-doped regions, intricate behaviors show up because of the keen competition between the AF SE interaction and the ferromagnetic DE interaction induced by mobile carriers. In the early stage of the study, de Gennes predicted a spin canting state with a spin- op type ordering, which smoothly connects collinear AF ordering in the undoped insulator and ferromagnetism in a doped metal [3]. Later, the scenario has been revisited; it was argued that a phase separation (PS) takes place to hinder the canting state and play an important role in the CMR e ect [6, 7].

Recently, increasing interest has been devoted to the e ect of geometrical frustration of the lattice structure on the charge-spin coupled systems. This has been inspired by the discovery of peculiar transport properties, such as an unconventional anomalous Hall e ect, in some pyrochlore oxides [8, 9, 10]. Theoretically, it was shown that, in the ferromagnetic Kondo lattice model on a two-dimensional kagome lattice, the coupling to a noncoplanar spin ordering leads to the anomalous Hall e ect via the so-called Berry phase from a nite spin scalar chirality [11]. The idea was extended to other frustrated lattices, such as a face-centered-cubic lattice [12], a triangular lattice [13, 14], and a pyrochlore lattice [15, 16]. Among them, in the triangular-lattice systems that we focus on in the present study, it was pointed out that a perfect nesting of the Fermi surface at 3/4 electron lling might lead to a noncoplanar four-sublattice ordering and the anomalous Hall e ect [13]. The authors performed careful energy comparison among possible spin orderings, and found that a noncoplanar four-sublattice spin ordering with a nite spin scalar chirality emerges near 1/4 lling in a wider parameter range than in the 3/4 lling case [14]. The 1/4 lling phase was later con rmed by Monte Carlo simulation [17, 18]. Such triangular-lattice charge-spin coupled systems have also drawn attention experimentally, e.g., in some delafossite oxides [19, 20, 21, 22].

In this contribution, we extend the analysis of the ferromagnetic Kondo lattice model on the triangular lattice with focusing on the lightly-doped region. As mentioned above, the geometrical frustration brings about new aspects in the charge-spin coupled systems, but the keen competition between the AF SE and the ferromagnetic DE interactions in the lightly-doped region has not been studied in detail. On the triangular lattice, the AF SE interaction stabilizes a three-sublattice 120 AF order when the electron band is empty or ful lled. We clarify the e ect of both electron and hole doping to the 120 AF states by a variational calculation of the ground-state energy for various spin states up to four-sublattice ordering. In the electron doped case, we nd that the 120 order is immediately collapsed by doping and a three-sublattice spin canting phase appears with accompanying PS. In contrast to the unfrustrated cases, this canting state has a noncoplanar umbrella-type spin con guration, which shows a nite scalar spin chirality together with a net magnetic moment. On the other hand, in the hole doped case, the 120 AF state persists up to a nite doping rate and PS takes place to a ferromagnetic metal; a noncoplanar canting phase is not observed. The contrastive behavior between the electron and hole doping is discussed from the change of the density of states for the spin ordering patterns.

2. Model and method

We consider the ferromagnetic Kondo lattice model on the triangular lattice. The Hamiltonian is given by

$$\mathcal{H} = \quad t \sum_{\langle i,j \rangle, } (c^\dagger_{i, } c_{j, } + \text{h.c.}) \quad J_\text{H} \sum_{i, ,\beta} c^\dagger_{i, } \vec{\sigma} \ _\beta c_{i,\beta} \ \vec{S}_i + J_\text{K} \sum_{\langle i,j \rangle} \vec{S}_i \ \vec{S}_j, \tag{1}$$

where $c^\dagger_{i, }$ ($c_{i, }$) is a creation (annihilation) operator for a conduction electron with spin at site i, $\vec{\sigma} \ _\beta = (\sigma^x \ _\beta, \sigma^y \ _\beta, \sigma^z \ _\beta)$ is a vector of Pauli matrices, and $\vec{S}_i$ is a localized spin on site i. Here, t is the transfer integral between the nearest-neighbor sites on the triangular lattice, J_H is the Hund's-rule coupling between conduction electrons and localized spins, and J_K is the antiferromagnetic (AF) superexchange (SE) interaction between localized spins. We take $t = 1$ as an energy unit and consider classical spins for $\vec{S}_i$ with $|\vec{S}_i| = 1$.

We investigate the ground state of the model given by Eq. (1) following the variational method taken in Ref. [14]. Namely, we compare the ground-state energies for di erent ordered states of the localized spins and determine the most stable ordering, while varying the electron density $n = \frac{1}{N} \sum_i \langle c^\dagger_{i, } c_{i, } \rangle$ (N is the total number of sites), J_H, and J_K. In the calculation, we consider 13 di erent types of ordered states, up to a four-sublattice unit cell, as shown in Fig. 1 in Ref. [14]. For the states with spin canting, we optimize the canting angle .

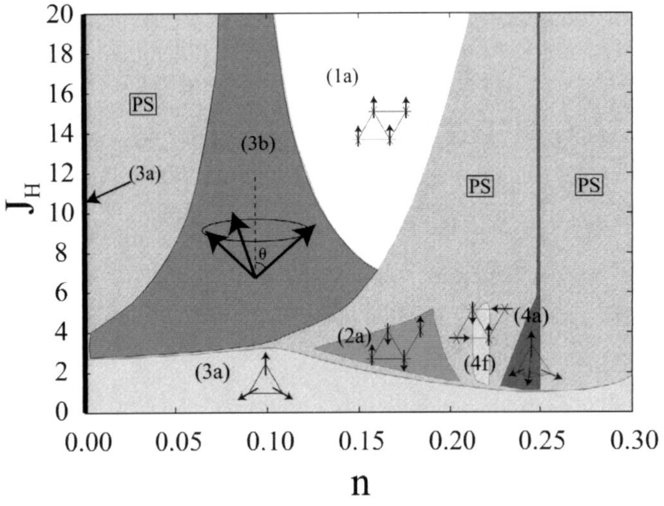

Figure 1. Ground-state phase diagram as functions of n and $J_{\rm H}$ at $J_{\rm K} = 0.03$. Labels for the phases are common to those in Ref. [14]. (3b) is the umbrella-type noncoplanar phase appearing in between the ferromagnetic metallic state (1a) and the phase separation (PS). The thick black line at $n = 0$ indicates that (3a) 120 coplanar ordered phase is stabilized at zero doping.

3. Result and Discussion

Figure 1 shows the result of the ground-state phase diagram at $J_{\rm K} = 0.03$ in the lightly-doped region near $n = 0$. For comparison, the results at smaller $J_{\rm K}$ are found in Fig. 3 in Ref. [14]. A new feature emerges in the low-density region in Fig. 1, which was not seen in the previous results for smaller $J_{\rm K}$. That is, a spin-canting phase (3b) appears on the lower-n side of the ferromagnetic metallic phase (1a). This state has a three-sublattice ordering with an umbrella-type noncoplanar spin con guration, as schematically shown in the gure [see also the inset of Fig. 2(a)]. Spin canting itself is expected in general in lightly-doped region because of the competition between the AF SE interaction $J_{\rm K}$ and the ferromagnetic DE interaction [3]. For example, in the cubic lattice case, a plausible canting state, which naturally connects two-sublattice collinear AF state and doped ferromagnetic state, is a simple coplanar spin- op state. However, it is not trivial what type of spin-canting order is realized in geometrically-frustrated systems. In the present triangular-lattice case, there are several possibilities of di erent spin canting orders, as listed in Fig. 1 in Ref. [14]; even for the three-sublattice orders, there are three di erent types of canting states, (3b) umbrella, (3c) coplanar cant, and (3d) 2:1 cant (see Fig. 1 in Ref. [14]). Among the possibilities, the umbrella-type one is energetically favored in the present model. Note that the umbrella state has a nite scalar spin chirality in each triangle unit, but this spatial con guration does not lead to the anomalous Hall e ect.

The umbrella phase is developed by increasing $J_{\rm K}$ in the region near the border of the ferromagnetic phase to PS as $J_{\rm K}$ increases. PS occurs to the $n = 0$ state, in which itinerant electrons are absent and the localized spins are 120 ordered by $J_{\rm K}$. It is noteworthy that the umbrella phase is always accompanied by PS. The situation is demonstrated in Fig. 2. Figure 2(a) shows the energy per site at a xed canting angle for various values of . Here, $= 0$ and $= 90$ correspond to the ferromagnetic phase (1a) and the three-sublattice 120 -ordered phase (3a), respectively. As decreases from 90 , the energy at $n = 0$ increases because of the loss of the SE energy. At the same time, the initial slope $dE/dn|_{n=0}$ becomes steeper because the kinetic energy of electrons becomes larger for smaller via the DE mechanism. Hence, the envelope of the energy curves $E(n, \ )$ becomes convex upward in the low-density limit. The convex behavior is a sign of PS. The phase-separated region is determined by drawing a tangent line from the energy at $n = 0$, $E(n = 0, \ = 90\ )$, to the envelope of the energy curves $E(n > 0, \ < 90\ )$. In the present case at $J_{\rm K} = 0.03$ and $J_{\rm H} = 10$, the tangent line touches the envelope at $n \simeq 0.06$ with the energy curve for $\simeq 50$; i.e., the system exhibits PS between the 120 -ordered state at $(n, \ ) = (0, 90\ )$ and the umbrella-type canting state at $(n, \ ) \simeq (0.06, 50\ )$.

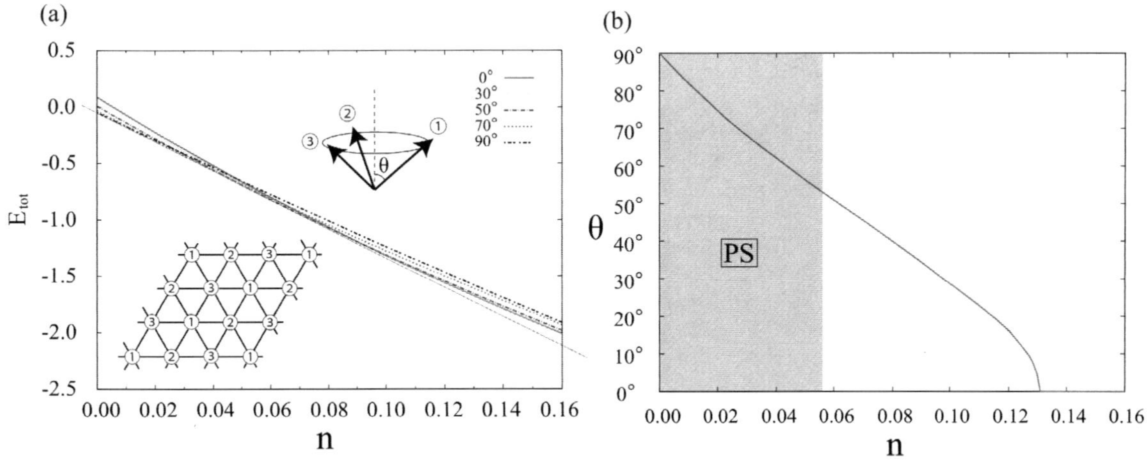

Figure 2. (a) Energy per site of the umbrella-type noncoplanar ordered state. Di erent curves represent the energies for a xed di erent canting angle . The data are calculated at $J_K = 0.03$ and $J_H = 10$. Note that $= 0$ and $= 90$ are equivalent to (1a) ferromagnetic phase and (3a) three-sublattice 120 -ordered phase, respectively. The straight black line shows a tangent line to de ne the phase-separated region. The insets show schematic pictures of the spin con guration of the umbrella state and the triangular lattice. (b) n dependence of which gives the lowest energy. The gray region denotes the phase-separated region. See the text for details.

Similar behavior is observed for all the regions of the umbrella phase, which accounts for the reason why the umbrella phase always appears with accompanied by PS.

The canting angle which gives the lowest-energy envelope changes continuously in the umbrella phase, as seen in Fig. 2(a). Figure 2(b) gives a plot of the optimized as a function of n. The result shows that, as n increases, starts from about 50 at the border of PS at $n \simeq 0.06$, and gradually decreases to zero when approaching the boundary to the ferromagnetic phase at $n \simeq 0.13$.

The value of the optimized depends not only on n but also on J_H (and J_K). The behavior is summarized in Fig. 3 at $J_K = 0.03$. As shown in the gure, decreases as n and J_H increase. This is reasonable since larger n and J_H enhance the DE ferromagnetic interaction which favors smaller canting angle (larger net magnetic moment). This change of clearly demonstrates that the noncoplanar umbrella phase is stabilized in the competition between the AF SE interaction J_K, which is dominant at less carrier doping and small J_H, and the DE ferromagnetic interaction enhanced by doping and J_H.

In contrast to the above results for the electron doped case, a hole doping from the band top $n = 1$ does not lead to a noncolpanar canting state. In the hole-doped case, 120 coplanar ordering persists up to a nite doping concentration, and PS takes place between the doped 120 AF metal and a ferromagnetic metal when J_H is su ciently large. The situation is unchanged against the value of J_K; a typical phase diagram for n 1 is found in the result at $J_K = 0.01$ in Fig. 3 in Ref. [14].

The reason why the 120 AF order is stable against the hole doping is understood from the density of states (DOS). Figure 4 shows DOS for the 120 coplanar AF phase, in comparison with typical DOS for the umbrella canting states including the ferromagnetic state. The data are calculated at $J_H = 10$ and $J_K = 0$. The result shows that the band top energy $\varepsilon = 13t$ is unchanged for . This is con rmed by analytical calculations. In particular, a consideration in the limit of $J_H \to \infty$ provides simple understanding of the form of DOS in terms of the renormalization of the transfer integrals and the ux introduced by the Berry phase [2, 23].

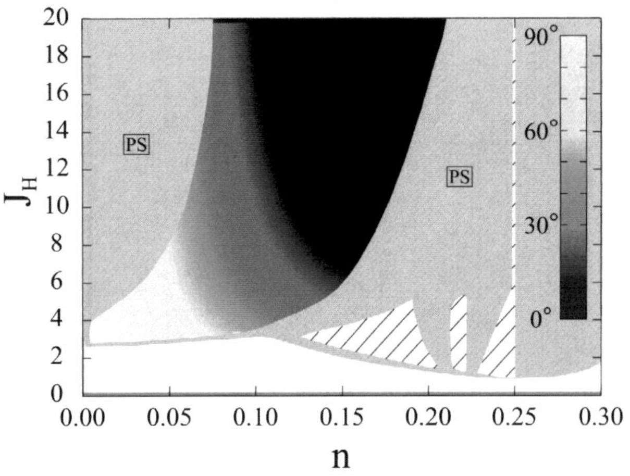

Figure 3. Canting angle in the umbrella phase as functions of n and J_{H} at $J_{\mathrm{K}} = 0.03$. The ferromagnetic phase ($= 0$) and the 120 -ordered phase ($= 90$) are also shown by black and white, respectively, whereas other phases are omitted (the hatched areas). The gray areas are the phase separation.

While the band top energy is unchanged, the value of DOS at the band top is smaller for > 0 compared to the $= 0$ ferromagnetic state. For these situations, the kinetic energy by hole doping becomes lower for > 0 states. Among the > 0 states, the 120 AF state has the lowest AF SE energy, and therefore, the 120 AF ordering remains stable against the hole doping for $J_{\mathrm{K}} > 0$.

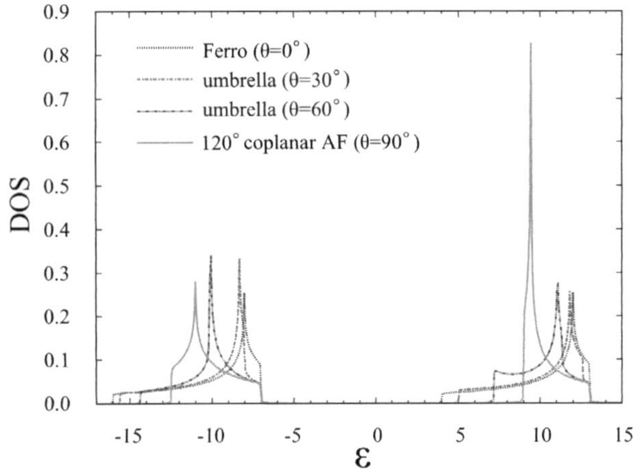

Figure 4. The density of states (DOS) for the umbrella-type canted states including the 120 coplanar AF state and the ferromagnetic state. The data are calculated at $J_{\mathrm{H}} = 10$ and $J_{\mathrm{K}} = 0$.

DOS in Fig. 4 also accounts for the reason why electron doping from $n = 0$ immediately collapses the 120 AF state and causes PS to a canted state. As shown in the gure, the band-bottom energy is lower for smaller , and hence, the electron doping to the 120 AF state costs the energy than the canted states. At $n = 0$, the 120 AF ordering becomes stable by a small energy gain from the AF SE coupling, $3J_{\mathrm{K}}$ per site. The energy gain is compensated by in nitesimally small doping to a canted state with smaller because of the lower band bottom. Therefore the electron doping leads to a sudden PS to a canted state, as demonstrated above.

The contrastive behavior between electron and hole doping is expected to be observed around an insulating state at half lling $n = 0.5$. At $n = 0.5$, the 120 coplanar AF ordering is stable and the system becomes insulating for $J_{\mathrm{H}} > 2$ (see Fig. 3 in Ref. [14]). In the insulating state for larger J_{H}, DOS just below and above the gap behave similar to the band top and bottom, respectively, as shown in Fig. 4. In fact, as a preliminary result, we nd a similar contrastive behavior around $n = 0.5$; a noncoplanar canting state appears against electron doping with

Journal of Physics: Conference Series **320** (2011) 012059 doi:10.1088/1742-6596/320/1/012059

PS, while the 120 AF order retains up to a nite hole doping. The details will be reported elsewhere.

Finally, let us discuss brie y the e ect of the band structure on these properties by including the next-nearest-neighbor hopping t' in the model (1). We nd that, when t' has an opposite sign to t, dependence of DOS at the band edges is qualitatively unchanged. For t' with the same sign as t, the band top becomes higher for smaller , while the band bottom remains lower for smaller , at least, for small t'. Hence, we expect the similar results for the models including t' irrespective of its sign. Further analyses including the detailed energy comparison among di erent spin states will be reported elsewhere.

4. Summary

We have investigated the lightly-doped regions of the ferromagnetic Kondo lattice model on the triangular lattice by variational calculations. We have found that a three-sublattice noncoplanar phase is stabilized by electron doping from the band bottom. The noncoplanar spin canting phase has an umbrella-type spin con guration with a nite spin scalar chirality in each triangular plaquette. This phase always appears accompanied by a phase separation to the 120 coplanar antiferromagnetic insulating state at zero doping. On the contrary, hole doping from the band top does not lead to such a noncoplanar spin canting state. These peculiar properties are the consequence of geometrical frustration in the charge-spin coupled system in which the antiferromagnetic superexchange interaction and the ferromagnetic double-exchange interaction compete with each other.

We acknowledge helpful discussions with Takahiro Misawa, Masafumi Udagawa, Youhei Yamaji, and Junki Yoshitake. This work was supported by Grants-in-Aid for Scienti c Research (No. 19052008 and 21340090), Global COE Program "the Physical Sciences Frontier", the Next Generation Super Computing Project, and Nanoscience Program, from MEXT, Japan.

References

[1] Zener C 1951 *Phys. Rev.* **82** 403
[2] Anderson P W and Hasegawa H 1955 *Phys. Rev.* **100** 675
[3] de Gennes P G 1960 *Phys. Rev.* **118** 141
[4] Kaplan T A and Mahanti S D 1999 *Physics of Manganites* (Kluwer Academic/Plenum Pub.)
[5] Tokura Y 1999 *Colossal Magnetoresistive Oxides* (Gordon & Breach Science Pub.)
[6] Dagotto E, Hotta T, Moreo A 2001 *Phys. Rep.* **344** 1
[7] Yunoki S, Hu J, Malvezzi A L, Moreo A, Furukawa N, and Dagotto N 1998 *Phys. Rev. Lett.* **80** 845
[8] Taguchi Y, Oohara Y, Yoshizawa H, Nagaosa N, and Tokura Y 2001 *Science* **291** 2573
[9] Nakatsuji S, Machida Y, Maeno Y, Tayama T, Sakakibara T, van Duijn J, Balicas L, Millican J N, Macaluso R T and Chan J Y 2006 *Phys. Rev. Lett.* **96** 087204
[10] Machida Y, Nakatsuji S, Maeno Y, Tayama T, Sakakibara T and Onoda S 2009 *Phys. Rev. Lett.* **98** 057203
[11] Ohgushi K, Murakami S, and Nagaosa N 2000 *Phys. Rev.* B **62** R6065
[12] Shindou R and Nagaosa N 2001 *Phys. Rev. Lett.* **87** 116801
[13] Martin I and Batista C D 2008 *Phys. Rev. Lett.* **101** 156402
[14] Akagi Y and Motome Y 2010 *J. Phys. Soc. Jpn.* **79** 083711
[15] Ikoma D, Tsuchiura H and Inoue J 2003 *Phys. Rev.* B **68** 014420
[16] Chern G W 2010 *Phys. Rev. Lett.* **105** 226403
[17] Sanjeev Kumar and Jeroen van den Brink 2010 *Phys. Rev. Lett.* **105** 216405
[18] Kato Y, Martin I, and Batista C D preprint (arXiv:1009.3059).
[19] Okuda T, Jufuku N, Hidaka S, and Terada N 2005 *Phys. Rev.* B **72** 144403
[20] Okuda T, Beppu Y, Fujii Y, Onoe T, Terada N, and Miyasaka S 2008 *Phys. Rev.* B **77** 134423
[21] Takatsu H, Yoshizawa H, Yonezawa S and Maeno Y 2009 *Phys. Rev.* B **79** 104424
[22] Takatsu H, Yonezawa S, Fujimoto S and Maeno Y 2010 *Phys. Rev. Lett.* **105** 137201
[23] Wang Y F, Gong C D and Zhu S Y 2005 *Europhys. Lett.* **69** 404

International Conference on Frustration in Condensed Matter (ICFCM) IOP Publishing
Journal of Physics: Conference Series **320** (2011) 012060 doi:10.1088/1742-6596/320/1/012060

Orbital degeneracy and Mott transition in Mo pyrochlore oxides

Yukitoshi Motome and Nobuo Furukawa[A,B]

Department of Applied Physics, University of Tokyo, Japan
[A]Department of Physics and Mathematics, Aoyama Gakuin University, Japan
[B]Multiferroics Project, ERATO, Japan Science and Technology Agency (JST), c/o Department of Applied Physics, University of Tokyo, Japan

E-mail: motome@ap.t.u-tokyo.ac.jp

Abstract. We present our theoretical results on an effective two-band double-exchange model on a pyrochlore lattice for understanding intricate phase competition in Mo pyrochlore oxides. The model includes the twofold degeneracy of e_g' orbitals under trigonal field splitting, the interorbital Coulomb repulsion, the Hund's-rule coupling between itinerant e_g' electrons and localized a_{1g} spins, and the superexchange antiferromagnetic interaction between the a_{1g} spins. By Monte Carlo simulation with treating the Coulomb repulsion at an unrestricted-type mean-field level, we obtain the low-temperature phase diagram as functions of the Coulomb repulsion and the superexchange interaction. The results include four dominant phases with characteristic spin and orbital orders and the metal-insulator transitions among them. The insulating region is characterized by a 'ferro'-type orbital ordering of the e_g' orbitals along the local $\langle 111 \rangle$ axis, irrespective of the spin ordering.

1. Introduction

Mo pyrochlore oxides, whose chemical formula is generally given by $R_2Mo_2O_7$ (R is rare earth), are known to exhibit interesting phase competition and related phenomena. By chemical substitution of R cations, the system changes from a ferromagnetic metal (FM) to a spin-glass (SG) insulator systematically depending on the ionic radius of R cation [1]. On the other hand, in applied external pressure, FM collapses and turns into a peculiar paramagnetic metal (PM) with showing an intermediate SG metallic state, whereas the SG insulator remains rather robust against pressure [2]. This variety of phases and competition among them offer an opportunity to systematically study many fascinating phenomena in strongly correlated electron systems, such as metal-insulator transition and anomalous transport properties [3].

The basic lattice structure of $R_2Mo_2O_7$ is composed of two intervening pyrochlore lattices: One is formed by Mo cations and the other is by R. The pyrochlore lattice is a three-dimensional network of corner-sharing tetrahedra, which is known to possess strong geometrical frustration. Mo cation is nominally tetravalent and has two electrons on average. The crystal field from trigonal distortion of MoO_6 octahedra splits three-fold t_{2g} levels of Mo $4d$ electrons into a_{1g} singlet and e_g' doublet. The trigonal axis is along the direction toward the center of each Mo tetrahedron, which is called the local $\langle 111 \rangle$ axis. With aligning their spins parallel according to the Hund's rule, one of two electrons occupies the lower a_{1g} level, and the other comes in the doubly-degenerate e_g' levels. (See, for example, Fig. 1 in Ref. [2].) Hence spin, charge, and orbital degrees of freedom are all active in these compounds, and the frustration of the lattice structure promotes the competition among them by preventing a simple ordering. In addition, the coupling between Mo $4d$ electrons and R-site rare-earth moments brings about further complications. In the following, however, we will focus on the phenomena commonly observed regardless of magnetic or nonmagnetic R cations, such as the phase competition mentioned above, and neglect the R sites: We will concentrate on the Mo pyrochlore network alone and the intrinsic physics from Mo $4d$ electrons in this paper.

Several theoretical studies have been done for understanding the phase competition in the Mo pyrochlore oxides. A pioneering work was done for the chemical substitution at ambient pressure by Solovyev by using the LDA+U calculation [4]. The calculated electronic structure indicates that the lower a_{1g} level has rather localized nature with a narrow band, while the e'_g levels form a relatively wide band and bears itinerant nature. The observation suggests that the system can be viewed as a double-exchange (DE) system, in which localized a_{1g} spins are coupled with itinerant e'_g electrons via the ferromagnetic Hund's-rule coupling. This explains well the emergence of FM in the compounds with large ionic radius of R, such as R=Nd and Sm. The calculation while changing the value of U shows that the localization from FM to SG insulator by the substitution is basically understood by the correlation effect (or the bandwidth control). These calculations, however, were performed with assuming a simple collinear ferro or antiferromagnetic (AF) ordering; hence, the detailed nature of the frustrated insulating phase, such as the SG behavior and the orbital state, was not clarified.

Recently, the authors have studied the phase competition under applied pressure [5, 6, 7]. The results show that the change from FM, SG metal, to PM is well reproduced by Monte Carlo (MC) calculations for an extended DE model including the superexchange (SE) AF coupling between localized spins. In particular, the peculiar diffusive nature of PM is explained by the strongly incoherent state originating from competition between the DE ferromagnetic interaction and the SE AF interaction [5]. The SG metallic state is also accounted by an instability toward phase separation related to the competition [7]. The agreement implies that the physical pressure causes a different effect from the chemical pressure (chemical substitution), as implied in the experimental report [2]: The former tends to increase the SE AF coupling between the a_{1g} spins dominantly, while the latter mainly modifies the bandwidth of itinerant e'_g electrons. MC calculations so far were performed for a single-band model with neglecting the Coulomb repulsion between itinerant electrons; for comprehensive understanding of the phase competition in the Mo pyrochlore oxides, it is necessary to extend the analysis to incorporate the electron correlation as well as the e'_g orbital degree of freedom.

In this contribution, we investigate the effect of electron correlation and orbital degeneracy on the phase competition in the extended DE model. The model explicitly includes the twofold orbital degeneracy and the Coulomb repulsion between itinerant electrons. Thermodynamic properties of the model are studied by MC calculations, in which the electron correlation is handled by an unrestricted-type Hartree approximation. The results indicate that, in addition to the phase transition from FM to PM previously obtained in the single-band model, a metal-insulator transition is caused by the electronic correlation accompanied by a 'ferro'-type orbital ordering. We discuss the phase diagram in relation to the distinct behaviors in experiments for the chemical substitution and the applied pressure in the Mo pyrochlore oxides.

2. Model and method

As an extension of the previous studies [5, 6, 7], we here consider the DE model with twofold orbital degeneracy. The Hamiltonian is given by

$$\mathcal{H} = \sum_{\langle ij \rangle} \sum_{\alpha} \sum t_{\alpha} \, (c^{\dagger}_{i\alpha} \, c_{j} \quad + \mathrm{H.c.}) + \frac{1}{2} \sum_{i} \sum_{\alpha \, \alpha' \, ' \, ,} \sum_{'} U_{\alpha \, \alpha' \, '} c^{\dagger}_{i\alpha} \, c^{\dagger}_{i} \, {}_{'} c_{i} \, {}_{' \, '} c_{i\alpha'}$$

$$-J_{\mathrm{H}}\sum_{i}\sum_{\alpha} c^{\dagger}_{i\alpha\ } \vec{\ }\ {}_{\ '} c_{i\alpha\ '} \cdot \vec{S}_i + J_{\mathrm{AF}}\sum_{\langle ij\rangle}\vec{S}_i\cdot\vec{S}_j, \tag{1}$$

where $c^{\dagger}_{i\alpha\ }$ ($c_{i\alpha\ }$) creates (annihilates) an electron with orbital α and spin   at site i, $\vec{\ }$ denotes the Pauli matrix, and $\vec{S}_i$ represents the localized spin at site i. The first term describes the electron hopping between nearest-neighbor sites $\langle ij\rangle$ on the pyrochlore lattice; the transfer integrals depend on the orbital indices α and   which represent two e'_g orbitals. The second term denotes the onsite Coulomb interactions between the e'_g electrons, including the intra and interorbital repulsions, the Hund's-rule coupling, and the pair hopping. The third term denotes the Hund's-rule coupling between itinerant electrons and localized spins $\vec{S}_i$. The last term is the SE AF coupling between the neighboring localized spins. A similar model has been studied on the square or cubic lattices in the context of colossal magnetiresistive (CMR) manganites [8].

Our purpose is to clarify thermodynamics of the model given by Eq. (1) and to elucidate the competing phases including correlated disordered states, such as the diffusive PM or the Mott insulating state. For this purpose, some sophisticated method beyond a simple mean-field approximation is necessary. The model (1), however, is very complicated and difficult to handle beyond a simple mean-field approximation. Hence, we introduce several simplifications as follows. First we take the limit of large J_{H}. In this limit, as often considered in the study of CMR, spins of itinerant electrons are aligned parallel to the localized spin at each site, and the transfer integrals are modulated by the relative angle between the neighboring localized spins. At the same time, among may contributions in the second term in Eq. (1), only the interorbital interaction between the same spin electrons remains relevant. Furthermore, we consider the localized spins $\vec{S}_i$ as classical vectors with the renormalized length $|\vec{S}_i| = 1$. The simplifications were shown to give reasonable results in the study of a single-band model for the phase competition under pressure in the wide-bandwidth compounds [5, 6, 7]. Consequently, by taking the local spin axis along the direction of localized spin at each site, the model can be described by spinless fermions in the form

$$\mathcal{H} = \sum_{\langle ij\rangle}\sum_{\alpha\ } \tilde{t}_{i\alpha j\ }\,(\tilde{c}^{\dagger}_{i\alpha}\tilde{c}_{j\ } + \mathrm{H.c.}) + \frac{\tilde{U}}{2}\sum_{i}\sum_{\alpha\neq\ }\tilde{n}_{i\alpha}\tilde{n}_{i\ } + J_{\mathrm{AF}}\sum_{\langle ij\rangle}\vec{S}_i\cdot\vec{S}_j, \tag{2}$$

where $\tilde{c}_{i\alpha}$ is the spinless fermion operator and $\tilde{n}_{i\alpha} = \tilde{c}^{\dagger}_{i\alpha}\tilde{c}_{i\alpha}$. Here, $\tilde{U}$ is considered as an effective interorbital interaction, which controls the double occupancy in the e'_g orbital with the same spin electrons under the strong Hund's-rule coupling; its origin is traced back to the e'_g interorbital Coulomb repulsion and the Hund's-rule couplings between e'_g-e'_g and e'_g-a_{1g} orbitals in Eq. (1). Hereafter, we consider the filling with one electron per site on average.

In Eq. (2), the effective transfer integral is given by

$$\tilde{t}_{i\alpha j\ } = t_{\alpha\ }\left[\cos\frac{\theta_i}{2}\cos\frac{\theta_j}{2} + \sin\frac{\theta_i}{2}\sin\frac{\theta_j}{2}\exp\{-i(\phi_i-\phi_j)\}\right], \tag{3}$$

where (θ_i, ϕ_i) denotes the angle of $\vec{S}_i$. To estimate the value of $t_{\alpha\ }$, it is necessary to take account of the anisotropy of the e'_g orbitals and the relative angle of Mo-O-Mo bond. Following the calculation in Ref. [9], we estimate the orbital diagonal and off-diagonal elements: it turns out that the ratio between them depends solely on the Mo-O-Mo bond angle   as

$$\frac{t_{\alpha\neq}}{t_{\alpha=}} = \frac{3-\cos\ }{3+\cos\ }. \tag{4}$$

Thus, the orbital off-diagonal element is larger than the diagonal one for $> 90°$. In Mo pyrochlore oxides,   is estimated around 130° at ambient pressure [10]. In the following calculations, for simplicity, we consider only the dominant off-diagonal element and fix $t_{\alpha\neq} = -1$ and $t_{\alpha=} = 0$, although the transfer integrals might change depending on both chemical and physical pressure.

We study the model (2) by MC simulation in a similar way as in the previous studies for the single-band model [5, 6, 7]. To facilitate the MC calculations without negative sign problem, we introduce an unresticted-type mean-field decoupling to the Coulomb interaction in the second term in Eq. (2). Here, as the simplest approximation, we employ the Hartree approximation; $\tilde{n}_{i\alpha}\tilde{n}_{i} \rightarrow \langle\tilde{n}_{i\alpha}\rangle\tilde{n}_{i} + \tilde{n}_{i\alpha}\langle\tilde{n}_{i}\rangle - \langle\tilde{n}_{i\alpha}\rangle\langle\tilde{n}_{i}\rangle$. The MC sampling is performed to update the configurations of localized spins $\{\vec{S}_i\}$. In each MC step, the mean fields $\{\langle n_{i\alpha}\rangle\}$ are recalculated by using the eigenvalues and eigenvectors obtained from the exact diagonalization of the Hamiltonian (2). We confirmed the convergence of the mean fields within the statistical errorbars after thermalization. We also employ the averaging over the twisted boundary conditions to suppress the finite size effect [6]. In the following, we show the result for 16 site system with averaging over 8^3 different twisted boundary conditions. Typically, we perform 2000 MC steps for measurement after 1000 MC steps for thermalization.

3. Result

Figure 1 shows the MC results for the model (2) at a low temperature $T = 0.04$ while changing the Coulomb repulsion $\tilde{U}$ and the SE AF coupling J_{AF}. We compute the square of magnetization m^2, the charge compressibility ${}_{\mathrm{c}}$, and the z component of orbital polarization ^{z}, by

$$m^2 = \Big\langle\Big(\frac{1}{N_{\mathrm{s}}}\sum_i \vec{S}_i\Big)^2\Big\rangle, \tag{5}$$

$${}_{\mathrm{c}} = \frac{\partial\langle\tilde{n}\rangle}{\partial} = \frac{\partial}{\partial}\Big\langle\frac{1}{N_{\mathrm{s}}}\sum_{i\alpha}\tilde{n}_{i\alpha}\Big\rangle, \tag{6}$$

$${}^{z} = \Big\langle\Big|\frac{1}{N_{\mathrm{s}}}\sum_i(\tilde{n}_{i\alpha} - \tilde{n}_{i})\Big|\Big\rangle, \tag{7}$$

respectively. Here the bracket $\langle\cdots\rangle$ denotes the thermal average, N_{s} is the number of sites, is the chemical potential, and $\tilde{n}$ is the electron density.

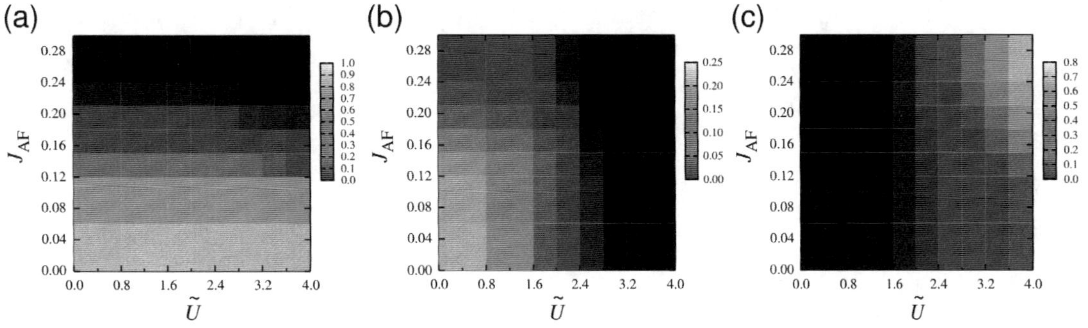

Figure 1. MC results for the model given by Eq. (2) at $\langle\tilde{n}\rangle = 1$ and $T = 0.04$: (a) square of the magnetization m^2 [Eq. (5)], (b) charge compressibility ${}_{\mathrm{c}}$ [Eq. (6)], and (c) z component of orbital polarization ^{z} [Eq. (7)]. The results are obtained for 16 site cluster by taking average over twisted boundary conditions with 8^3 different fluxes.

As shown in Fig. 1(a), in the small J_{AF} region, the magnetization is large, indicating that the system is ferromagnetic. The ferromagnetism collapses as J_{AF} increases. On the other hand, the magnetism is relatively insensitive to $\tilde{U}$. For small $\tilde{U}$, the charge compressibility $_{\mathrm{c}}$ is finite as shown in Fig. 1(b), indicating that the system is metallic. Hence, the system is FM for small J_{AF}, while it turns into PM as J_{AF} increases in the small $\tilde{U}$ region. This behavior is qualitatively the same as observed in the single-band model [5]. In the single-band case, phase separation appears between the two phases [7]; we expect a similar behavior in the present two-band case, but it is not clearly seen in the present calculation presumably because of the small size cluster.

As $\tilde{U}$ increases, $_{\mathrm{c}}$ decreases rapidly to zero, as shown in Fig. 1(b). This signals the metal-to-insulator transition. The insulating state is characterized by an orbital ordering. Figure 1(c) shows that the orbital polarization becomes nonzero almost simultaneously with the disappearance of the charge compressibility. This means that the insulating state shows a 'ferro'-type orbital order by choosing one of two e_g' orbitals. The reason for the 'ferro'-type orbital order is understood from the strong coupling picture as follows. In the large $\tilde{U}$ limit, the second order perturbation in the transfer integrals leads to the so-called Kugel-Khomskii-type spin-orbital exchange couplings [11]. In the present model, as seen in Fig. 1(a), the spin configuration is governed by the competition between DE and SE interactions, not by $\tilde{U}$. Then, the 'ferro'-type orbital configuration is favored via the spin-orbital exchange couplings, irrespective of spin configurations, because the dominant hopping processes in the perturbation are the orbital off-diagonal ones by $t_{\alpha\neq}$ in the present model. The situation is contrastive to the 'myth' for the Kugel-Khomskii-type interactions, in which the system usually favors complementary spin-orbital configurations, such as spin-AF orbital-ferro or spin-ferro orbital-AF.

It is worthy to note that the 'ferro'-type orbital order is not a uniform orbital order. The trigonal distortion is along the local $\langle 111 \rangle$ axis at each site, and hence, the e_g' orbital polarization takes place along the site-dependent local axis. In other words, the 'ferro'-type orbital order is a $\vec{q}=0$ four-sublattice order. Thus it does not break the cubic symmetry of the pyrochlore lattice structure.

Summarizing the results, we obtain four dominant phases at low temperatures in the plane of $\tilde{U}$ and J_{AF}, as schematically shown in Fig. 2. The spin-ferro orbital-para metal in the small $\tilde{U}$ and small J_{AF} region is stabilized by the DE mechanism. The spin-para orbital-para metal in the small $\tilde{U}$ and large J_{AF} region is induced by the competition between the DE ferromagnetic interaction and the SE AF coupling under the geometrical frustration. In these two phases, the orbital degree of freedom is less important, and hence, the physics is common to the single-band case [5, 6, 7]. On the other hand, the spin-ferro and orbital-'ferro' insulator in the large $\tilde{U}$ and small J_{AF} region is a correlation-driven insulator with development of the orbital ordering. The spin-para orbital-'ferro' insulator is driven by the increase of J_{AF} and the spin frustration.

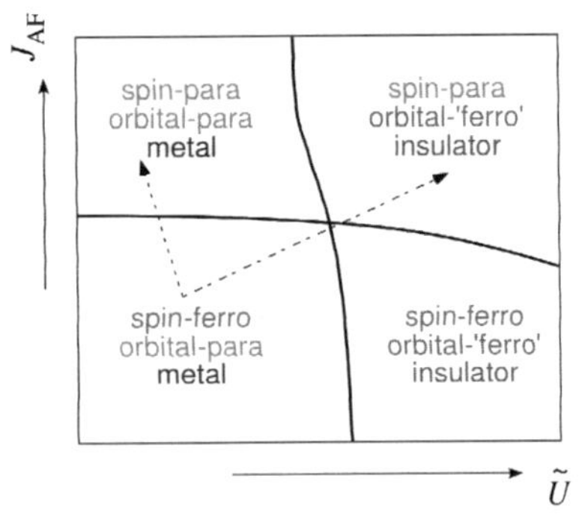

Figure 2. Schematic phase diagram at low temperatures suggested from the results in Fig. 1. The dotted (dot-dashed) arrow in the figure indicates a possible parameter change while changing the physical (chemical) pressure in the Mo pyrochlore oxides. See the text for details.

4. Discussion and concluding remark

In Mo pyrochlore oxides, as mentioned in Sec. 1, the chemical substitution leads to the transition from FM to SG insulator, while the pressure induces the transition from FM to PM. The latter is qualitatively explained by the change of $J_{\rm AF}$ in the single-band model [5], and the present calculation for the extended two-band model also gives similar results in the small $\tilde{U}$ regime; the parameter change is, e.g., schematically shown by the dotted arrow in Fig. 2. On the other hand, the metal-insulator transition by the chemical substitution cannot be understood by the single-band physics. From our results, we deduce that the transition from FM to SG insulator is described by the transition from the spin-ferro orbital-para metal to the spin-para orbital-'ferro' insulator, as indicated by the dot-dashed arrow in Fig. 2. In the spin-para orbital-'ferro' insulating state, the spin degree of freedom remains paramagnetic because of the strong frustration, but short-range AF correlations are expected by the dominant $J_{\rm AF}$. The remaining frustration might be released by remnant perturbations neglected in the present model. In fact, one of the author and his collaborators recently demonstrated that the frustrated spins have instability toward a SG state through the coupling to local lattice distortions in the presence of bond disorder, reasonably in accord with the experiments in Mo pyrochlore oxides [12]. Thus, our results provide a hint for understanding the nature of the frustrated Mott insulating state in the Mo pyrochlore oxides. From these observations, we consider that, despite several simplifications introduced, our effective two-band model given by Eq. (2) can offer a simple starting point for comprehensively understanding the intricate physics $R_2Mo_2O_7$. Further analysis in larger system sizes is desired to elucidate the thermodynamics as well as transport properties.

Why and how the two effects, the chemical substitution and the pressure, lead to such different behaviors remain unclear. In the present scenario, these effects are attributed to the different control of the Coulomb repulsion (or the bandwidth) and the SE AF coupling between a_{1g} spins. The bandwidth is basically determined by the transfer integrals between the itinerant e_g' orbitals, i.e., t_α in our model, on the other hand, the SE AF coupling $J_{\rm AF}$ is approximately set by $t_{a_{1g}}^2/U_{a_{1g}}$, where $t_{a_{1g}}$ and $U_{a_{1g}}$ are the transfer integral between the a_{1g} orbitals and the intraorbital Coulomb repulsion in the a_{1g} orbital, respectively. Both t_α and $t_{a_{1g}}$ are modified by the chemical or physical pressure via the change of the angle and/or bond length of Mo-O-Mo bonds. In order to clarify how they are modified and to distinguish the two effects, it is highly desired to study the microscopic change of the lattice parameters by, e.g., the detailed x-ray scattering experiment or the first-principles calculations with structural optimization.

The authors thank Y. Tokura, S. Iguchi, and H. Shinaoka for fruitful discussions. This work was supported by Grant-in-Aids (Nos. 19052008 and 21340090), Global COE Program "the Physical Sciences Frontier", and the Next Generation Super Computing Project, Nanoscience Program, MEXT, Japan.

References

[1] Greedan J E, Sato M, Ali N and Datars W R 1987 *J. Solid State Chem.* **68** 300
[2] Iguchi S, Hanasaki N, Kinuhara M, Takeshita N, Terakura C, Taguchi Y, Takagi H and Tokura Y 2009 *Phys. Rev. Lett.* **102** 136407
[3] For a recent review, see Gardner J S, Gingras M J P and Greedan J E 2010 *Rev. Mod. Phys.* **82** 53
[4] Solovyev I V 2003 *Phys. Rev.* B **67** 174406
[5] Motome Y and Furukawa N 2010 *Phys. Rev. Lett.* **104** 106407
[6] Motome Y and Furukawa N 2010 *J. Phys.: Conf. Ser.* **200** 012131
[7] Motome Y and Furukawa N 2010 *Phys. Rev.* B **82** 060407(R)
[8] For a review, see Dagotto E, Hotta T and Moreo A 2001 *Phys. Rep.* **344** 1
[9] Ichikawa H, Kano L, Saitoh M, Miyahara S, Furukawa N, Akimitsu J, Yokoo T, Matsumura T, Takeda M and Hirota K 2005 *J. Phys. Soc. Jpn.* **74** 1020
[10] Moritomo Y, Xu Sh, Machida A, Katsufuji T, Nishibori E, Takata M, Sakata M, and Cheong S-W 2001 *Phys. Rev.* B **63** 144425
[11] Kugel K I and Khomskii D I 1973 *Zh. Eksp. Teor. Fiz.* **64** 1429 [*Sov. Phys. JETP* **37** 725]
[12] Shinaoka H, Tomita Y and Motome Y 2010 Spin-glass transition in bond-disordered Heisenberg antiferromagnets coupled with local lattice distortions on a pyrochlore lattice *Preprint* arXiv:1010.5625

International Conference on Frustration in Condensed Matter (ICFCM) IOP Publishing
Journal of Physics: Conference Series **320** (2011) 012061 doi:10.1088/1742-6596/320/1/012061

Resonating Hartree-Fock studies on magnetic states in the Hubbard model on the uniform triangular lattice

Takaaki Maruyama, Fumikazu Sato, and Norikazu Tomita
Department of Physics, Yamagata University, Japan
E-mail: s11e107d@st.yamagata-u.ac.jp

Abstract. We investigate the quantum fluctuations in magnetic states of the half-filled Hubbard model on the uniform triangular lattice, by means of the resonating Hartree-Fock approximation where a many-body wavefunction is constructed by non-orthogonal multi-Slater determinants. It is shown that the quantum fluctuations are described by large deviations from the 120 ° Néel order even at U/t=10, which results in significant reduction of the spin correlations.

1. Introduction

Spin structures on the triangular lattice have been attracting much interest since discovery of the non-magnetic insulating state in κ-(BEDT-TTF)$_2$Cu$_2$(CN)$_3$.[1] So far, the path integral renormalization group (PIRG) theory has pointed out such non-magnetic insulating phase.[2, 3] Now, what we need would be much more detailed information on quantum fluctuations. Conventional numerical methods give physical information mainly through the correlation functions, and we cannot see how and why such electronic structures are realized. Such information would be important also for materials design using the complicated strongly correlated systems. In this research, we describe the spin structures on the uniform triangular lattice by means of a resonating Hartree-Fock (Res-HF) method. We will give physics to large quantum fluctuations in this system.

This paper is organized as follows. A brief review on the Res-HF method is given in §2. Our results on the quantum fluctuations and spin correlation function are shown in §3. A short summary is given in §4.

2. Res-HF method

How to describe strongly correlated electron systems has been one of the central subjects in solid state physics. Especially, it has been quite difficult to give reasonable physics to such large quantum fluctuations. To overcome this difficulty, we introduce the Res-HF method, where a many electron wavefunction is constructed by superposition of non-orthogonal Slater determinants, such as

$$|\Psi >= \sum_{f=1}^{N_S} C_f \sum_G P^G |f > . \tag{1}$$

All the orbitals of the Slater determinants $|f>$, as well as superposition coefficients C_f, are variationally determined. Details of this orbital optimization are given in ref.[4]. In this research, to describe the coplanar spin state, the Slater determinants have general spin orbits (GSO), which are characterized by the unitary transformation,

$$(c_1, ..., c_n, c_{n+1}, ..., c_{2n}) = (a_{1\uparrow}, ..., a_{n\uparrow}, a_{1\downarrow}, ..., a_{n\downarrow}) \begin{pmatrix} U^{\uparrow\uparrow} & U^{\uparrow\downarrow} \\ U^{\downarrow\uparrow} & U^{\downarrow\downarrow} \end{pmatrix}. \quad (2)$$

Restricted HF orbitals correspond to $U^{\uparrow\downarrow} = U^{\downarrow\uparrow} = 0$ and $U^{\uparrow\uparrow} = U^{\downarrow\downarrow}$, while different orbitals for different spins (DODS) correspond to $U^{\uparrow\downarrow} = U^{\downarrow\uparrow} = 0$ and $U^{\uparrow\uparrow} \neq U^{\downarrow\downarrow}$. GSO allow non-zero off-diagonal terms, that is, $(U^{\uparrow\downarrow})^{\dagger} = U^{\downarrow\uparrow} \neq 0$. As a result, dimension of the matrices become double. In addition, GSO-type Slater determinants break the original symmetries of the system, such as a translation and rotation. As shown below, this is very important to visualize the quantum fluctuations, but we have to recover the original symmetries of the system for the wavefunction. For this purpose, we adopt symmetry projections for each constituting S-det, which are symbolically denoted by P^G in eq. (1). The symmetry projection corresponds to superposition of the Goldstone set for each symmetry broken state.

The most important feature of this method is that we can visualize the quantum fluctuations by analyzing the structures of the optimized Slater determinants.

We apply this method to the Hubbard model on the uniform triangular lattice, whose Hamiltonian is given by

$$H = -t \sum_{<i,j>\sigma=\uparrow\downarrow} (a^{\dagger}_{i\sigma} a_{j\sigma} + a^{\dagger}_{j\sigma} a_{i\sigma}) + U \sum_i n_{i\uparrow} n_{i\downarrow}, \quad (3)$$

where t and U represent the transfer integral between the nearest neighbor sites and the on-site Coulomb interaction, respectively. In the following calculations, the system size is 6×6, and a periodic boundary condition is imposed. This system has a D_6 rotational symmetry as well as a translational symmetry.

3. Results and Discussion

The phse diagram of the triangular Hubbard model has intensively been investigated by PIRG. The purpose of the present work is not a follow-up of PIRG results, but to explicitly show the quantum fluctuations in this system. Figure 1 shows the spin correlation function at $U/t = 10$, where the PIRG concluded that the ground state is the 120° Néel state. We can see a significant peak at $(2\pi/3, 2\pi/3)$, which suggests a triple periodicity of the spin structure. This is consistent with the PIRG result. However, all the correlation function say is that the system has a triple

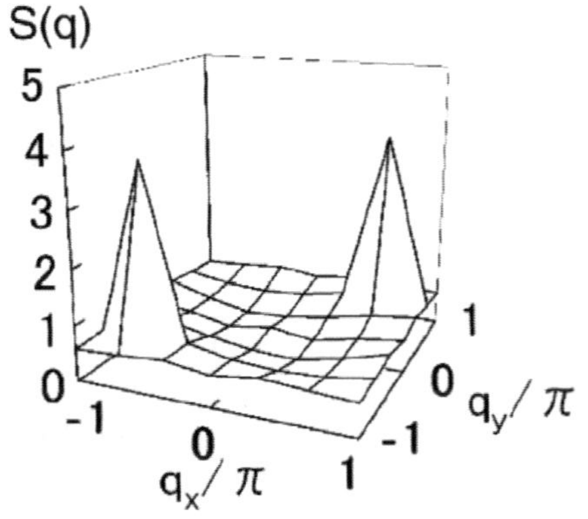

Figure 1. Spin correlation function at $U/t = 10$. There exits a significant peak at $(2\pi/3, 2\pi/3)$, which corresponds a triple periodicity of the spin structure.

Journal of Physics: Conference Series **320** (2011) 012061 doi:10.1088/1742-6596/320/1/012061

periodicity and we cannot conclude that such spin structure is certainly a 120° Néel state. Furthermore, the peak at $(2\pi/3, 2\pi/3)$ is significantly reduced, compared to the uniform 120° Néel state. The origin of this reduction is also unclear.

Before showing the Res-HF results, we briefly discuss the unrestricted HF (UHF) ground and low energy excited states. As shown in Fig.2(a), the UHF ground state at $U/t = 10$ is the 120° Néel state. On the other hand, the excited states have one (b) or two (c) flipped spins marked by circles, and the 120° Néel structure is modified around the flipped spins.

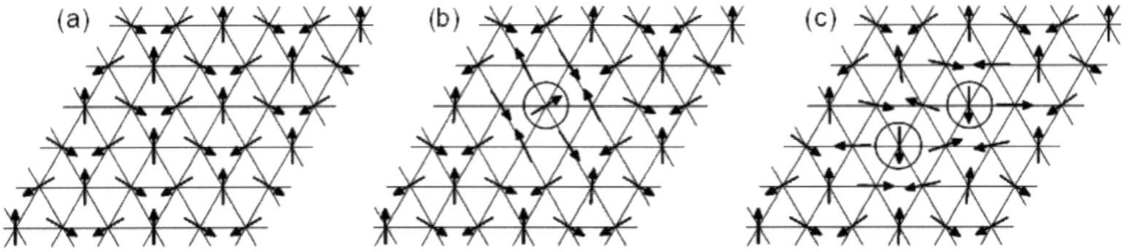

Figure 2. Spin structures of UHF ground (a), and excited states(b,c) at $U/t = 10$.

Now, we show in Fig.3 the structures of two typical Slater determinants generating the Res-HF wavefunction. We can see that 120° Néel state is largely modified, especially around spins marked by circles. This modulation is much large and wide ranging compared to the UHF excited states. Thus, we can conclude that the modulation from the 120° Néel spin structure makes dominant quantum fluctuations in the triangular Hubbard model and these quantum fluctuations reduce the uniform 120° Néel spin structure. At the same time, we can say that the triple periodicity in the spin correlation function actually comes from the 120° Néel structure. On the other hand, we should be careful to conclude that this spin correlation is really long-ranged, since this correlation is largely reduced, compared to the UHF result. We need larger systems to be investigated. In addition, the spin structure in the non-magnetic phase is another interesting subject. These challenging researches will be done in the near future.

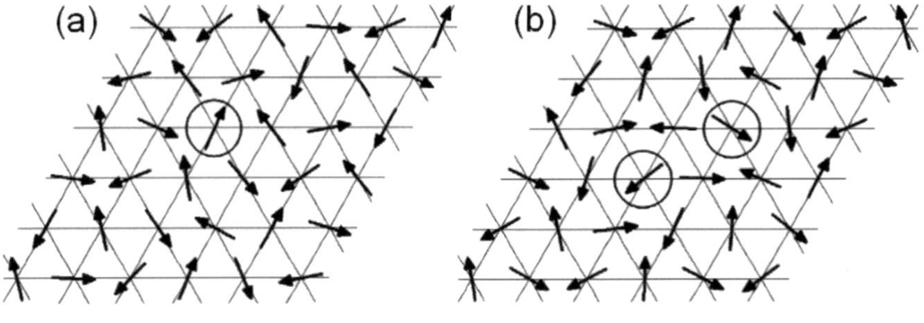

Figure 3. Spin structures of two typical Slater determinants generating the Res-HF wavefunction at $U/t = 10$.

Finally, it might be meaningful to see how superposition of such Slater determinants lowers the energy compared to the uniform UHF ground state. For convenience, we denote the UHF ground state by $|UHF>$, while Slater determinants having defects or deviations from the uniform UHF spin structure, as in Fig.3, are denoted by $|S1>$, $|S2>$,... . The UHF ground state corresponds to the state having the lowest diagonal component $<UHF|H|UHF>$. However, it does not lower the energy by the off-diagonal components, such as $<UHF|HP^G|UHF>$,

Table 1. N_S-dependence of the ground state energies by the Res-HF method at $U/t = 10$. UHF ground state energy is -15.226.

N_S	Energy
1	-16.224
3	-16.468
5	-16.574
7	-16.636
9	-16.671

or $< UHF|H|S1 >$, because the UHF ground state has a uniform spin structure. On the other hand, the Slater determinants, shown in Fig.3, have large off-diagonal components, such as $< S1|HP^G|S1 >$, or $< S2|H|S1 >$. Our results indicate that such off-diagonal components are important to lower the many-body ground state energy. We show how the Res-HF energy depends on the number of generating Slater determinants (N_S) in Table I. The Res-HF energy is slightly lower than the PIRG result[3] when N_S is larger than 7.

4. Summary

The Res-HF method has been applied to the Hubbard model on the uniform triangular lattice. The spin correlation function shows the triple periodicity of the spin structure, and we have shown that such periodicity really comes from the 120° Néel structure. Furthermore, we have explicitly shown the deviations from the 120° Néel structure make quantum fluctuations in the triangular Hubbard model. The 120° Néel spin correlation is significantly reduced by these quantum fluctuations.

This work was supported by the Next Super Computing Project, Nanoscience Program, MEXT, Japan.

References

[1] Y. Shimizu, K. Miyagawa, K. Kanoda, M. Maesato, and G. Saito, Phys. Rev. Lett. **91**, 107001(2003).
[2] H. Morita, S. Watanabe, and M. Imada, J. Phys. Soc. Jpn. **71**, 2109(2002).
[3] T. Yoshioka, A. Koga, and N. Kawakami, Phys. Rev. Lett. **103**, 036401(2009).
[4] A. Ikawa, S. Yamamoto, and H. Fukutome, J. Phys. Soc. Jpn. **62**, 1653(1993).

Journal of Physics: Conference Series **320** (2011) 012062 doi:10.1088/1742-6596/320/1/012062

Orbital Compass Model in a Checkerboard Lattice

Joji Nasu and Sumio Ishihara

Department of Physics, Tohoku University, Sendai 980-8578 Japan

E-mail: `ishihara@cmpt.phys.tohoku.ac.jp`

Abstract.
The orbital compass model in a checkerboard lattice is studied. There is a competition between the interaction for the z component of the orbital pseudo-spin along the vertical/horizontal direction and the interaction for the x component along the diagonal direction in a checkerboard lattice. In a frustration point where the two interactions compete with each other, a macroscopic number of the orbital pseudo-spin configurations are degenerated in the classical grounds state. This degeneracy is lifted by the thermal and quantum fluctuations and the z component long range order is realized. The tricritical point appears due to the coexistence of the orbital frustration and the geometrical frustration.

1. Introduction

Orbital degree of freedom in the transition-metal compounds and its correlation with spin, charge and lattice degrees of freedom are one of the central issues in the recent correlated electron systems[1]. Essence of the orbital degree of freedom in a solid is its directional and frustration character. Since the orbital degree of freedom represents the anisotropy of the charge distribution, the inter-site interaction between the orbitals explicitly depends on the direction of the bonds. When one considers an orbital configuration which minimizes the bond energy in a direction, this configuration does not minimize the bond energy in other directions. Therefore, all bond energies are not able to be minimized simultaneously. This is a kind of the frustration even without geometrically frustrated lattice[3].

The inter-site interaction between the orbitals is mainly due to the superexchange-type and the cooperative Jahn-Teller type[2]. These originate from the virtual exchange of an electron under the Coulomb interaction and that of a Jahn-Teller phonon. The superexchange type of the interaction is usually represented by the coupled form of the spin and orbital degrees of freedom, although the cooperative Jahn-Teller type is written by the orbital degree of freedom alone. One of the well known orbital model is the e_g orbital model in a cubic lattice given by

$$\mathcal{H}_{e_g} = J \sum_{\langle ij \rangle} \tau_i^l \tau_j^l. \tag{1}$$

We define $\tau_i^l = \cos(2\pi n_l/3)T_i^z - \sin(2\pi n_l/3)T_i^x$ where $\mathbf{T}_i$ is the pseudo-spin operator for the doubly degenerate e_g orbital, and the numerical factor $n_l = (1, 2, 3)$ where $l = (x, y, z)$ represents the bond direction in a cubic lattice. It is known that the classical ground states in this model are macroscopically degenerated, and the long-range orbital order is realized by the thermal and quantum fluctuations[4, 5]. Another well known orbital model is the orbital compass model in a square lattice defined by

$$\mathcal{H}_{compass} = J\sum_{\langle ij\rangle_x} T_i^x T_j^x + J\sum_{\langle ij\rangle_z} T_i^z T_j^z, \tag{2}$$

where $\langle ij\rangle_x$ and $\langle ij\rangle_z$ represent the nearest neighboring (NN) ij pair along the x and z directions in a two dimensional square lattice[2]. Only the x (z) component of the pseudo-spin is concerned in the interaction along the x (z) direction. It is known that, instead of the conventional long-range order, the directional order characterized by $\sum_i \langle (T_i^z)^2 - (T_i^x)^2\rangle$ appears below a certain temperature[6].

In this paper, we study the competition and cooperation of the intrinsic orbital frustration and the geometrical frustration, by analyzing the orbital compass model in a checkerboard lattice. This model is derived from the Hubbard-type model where the doubly degenerate d_{xz} and d_{yz} orbitals are located at each site in a checkerboard lattice in the strong coupling limit. As seen in the compass model in a square lattice, there is a competition between the z component order along the horizontal/vertical directions and the x component order along the diagonal direction. The classical and quantum versions of this model are analyzed by the mean-field methods, the Monte-Carlo simulation, and the mixed method of the mean-field and the Jordan-Wigner transformation. We calculate the finite temperature phase diagram and show that, at the frustration point, the long range order is realized due to the thermal and quantum fluctuations. A competition between the different kinds of the interactions provides the tricritical point.

2. Model Hamiltonian

We consider the doubly degenerate d_{yz} and d_{zx} orbitals in a two dimensional checkerboard lattice. The two orbitals are located at each site in the lattice[7]. These orbitals are denoted by $\sigma = \uparrow$ for d_{zx} and $\sigma = \downarrow$ for d_{yz}. We start from the Hubbard-type Hamiltonian given by

$$\mathcal{H}_0 = \mathcal{H}_t + \mathcal{H}_U, \tag{3}$$

where the electron transfer term and the on-site Coulomb interaction term are defined by

$$\mathcal{H}_t = -\sum_{\langle ij\rangle_l} t_0^{(l)\sigma\sigma'} c_{i\sigma}^\dagger c_{j\sigma'} - \sum_{(ij)_m} t_1^{(m)\sigma\sigma'} c_{i\sigma}^\dagger c_{j\sigma'} + H.c., \tag{4}$$

and

$$\mathcal{H}_U = U\sum_i n_{i\uparrow} n_{j\downarrow}, \tag{5}$$

respectively. Here, c_i and $n_i (\equiv c_i^\dagger c_i)$ represent the annihilation operator and the number operator of a spin-less fermion at site i, respectively. In the electron transfer term $\mathcal{H}_t$, the first term represents the electron hopping along the [10] and [01] directions in the checkerboard lattice and $t_0^{(l)\sigma\sigma'}$ is the intensity between the NN sites i and j along the direction $l (= x, y)$. The second term in $\mathcal{H}_t$ represents the electron transfer along the [11] and [1$\bar{1}$] directions and $t_1^{(m)\sigma\sigma'}$ is its intensity along $m (= xy, x\bar{y})$. The symbols $\sum_{\langle ij\rangle_l}$ and $\sum_{(ij)_m}$ represent the summation for the NN ij pairs along the direction l, and that for the next NN ij pair along m, respectively. Since the d_{zx} and d_{yz} orbitals have the transfer integrals along the x and y directions, the matrix elements in the basis of (d_{zx}, d_{yz}) are given by $t_0^{(x)} = -t_0(1+\sigma^z)/2$ and $t_0^{(y)} = -t_0(1-\sigma^z)/2$ where σ are the Pauli matrices. On the other hand, the electron transfer along the [11] ([1$\bar{1}$]) direction is finite between the $d_{yz} + d_{zx}$ ($d_{zx} - d_{yz}$) orbitals. Therefore, the matrix elements of the transfer integrals in these directions are given by $t_1^{(xy)} = -t_1(1+\sigma^x)/2$ and $t_0^{(x\bar{y})} = -t_1(1-\sigma^x)/2$.

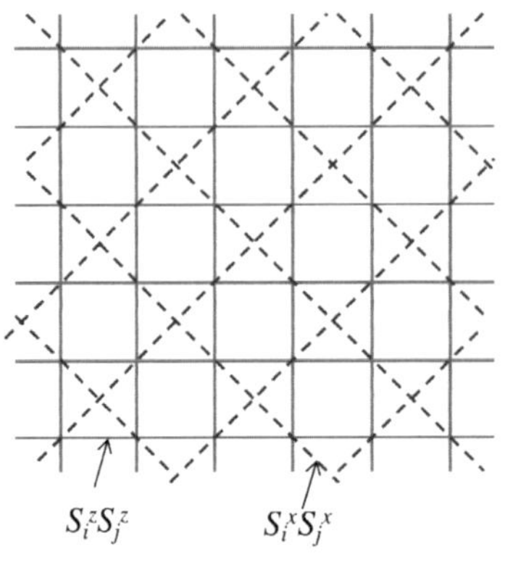

Figure 1. (Color online) A schematic view of the orbital compass model in a checkerboard lattice.

In the strong coupling limit of $U \gg t_0, t_1$ at half filling, the effective Hamiltonian for the Hubbard-type model is obtained by the perturbational calculation. The result is given by

$$\mathcal{H} = -J_z \sum_{\langle ij \rangle} S_i^z S_j^z - J_x \sum_{(ij)} S_i^x S_j^x \tag{6}$$

where we introduce the pseudo-spin operator for the orbital degree of freedom defined by $\mathbf{S}_i = (1/2) \sum_{\sigma\sigma'} c_{i\sigma}^\dagger \sigma_{\sigma\sigma'} c_{i\sigma'}$ and the symbols $\sum_{\langle ij \rangle}$ and $\sum_{(ij)}$ represent the summation for the NN ij pairs, and those for the next NN pairs, respectively. The exchange constants are given by $J_{z(x)} = -2t_{0(1)}^2/U$ which are negative. However, these signs are not essential, since the signs are reversed by rotation of the pseudo-spins in one of the sublattices in the checkerboard lattice with respect to the S^y axis. We define a ratio of the two interaction as $J \equiv J_x/(2J_z)$. In this Hamiltonian, the z-component of the pseudo-spin operators is concerned in the interaction along the [10] and [01] directions. On the other hand, the x component appears in the interaction along [11] and $[1\bar{1}]$ directions. Therefore, this is recognized as a orbital compass model defined in the checkerboard lattice.

We analyze the classical and quantum versions of the orbital compass model in a checkerboard lattice. In the classical and quantum models, the pseudo-spins are treated as the classical vectors in the $S^x - S^z$ plane, and the quantum spin operators of an amplitude of 1/2, respectively.

3. Results of Calculation

First we show the numerical results of the classical version of the compass model. It is obvious that in the limits of $J = 0$ and $J \to \infty$, the system corresponds to the Ising model in a square lattice, and an independent set of the Ising models in a one-dimensional chain, respectively. In the mean-field approximation at zero temperature, there are two phases as a function of the interaction ratio J. In the region of $J < 1$, the ferromagnetic long-range order of the z component is realized. On the other hand, in the region of $J > 1$, the ferromagnetic order of the x component is stabilized. It is worth noting that in this region, since each one-dimensional chains are independent, reverse of all spins in any chains along the [11] and $[1\bar{1}]$ directions does not change the mean-field energy. Thus, a macroscopic number of the mean-field ground state exists in the region of $J > 1$. At the frustration point $J = 1$, the two interactions compete with each other. The mean field solutions have a rotational symmetry in the two dimensional $S^x - S^z$ plane. Since this is not seen in the Hamiltonian, this high symmetry in the mean-field solution is accidental. From now on, the point $J = 1$ is termed a frustration point.

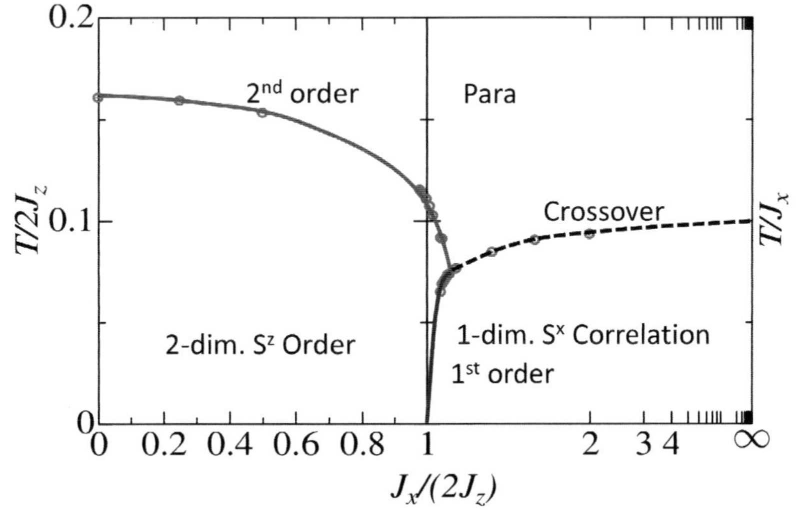

Figure 2. (Color online) Finite temperature phase diagram in the classical version of the orbital compass model in a checkerboard lattice obtained by the classical Monte-Carlo method. The broken line shows a cross over line where the one-dimensional pseudo-spin correlation develops.

In Fig. 2, the finite temperature phase diagram obtained by the classical Monte-Carlo simulation is presented. A two-dimensional $20^2 - 40^2$ site cluster with the periodic boundary condition is used. We adopt the Wang-Landau algorithm[8] where 2×10^8 MC steps is used for standard measurements. The phase diagram is obtained by the correlation functions and the specific heat. As expected, in the region of $J < 1$, the S^z Ising order is realized. The transition temperature gradually decreases with increasing J. At the point of $J = 1$, where the classical ground state is not determined uniquely, the two dimensional order of the S^z component is stabilized at finite temperature. This is a kind of the order by thermal fluctuation mechanism. As results the reentrant feature of the two-dimensional order appears around the frustration point.

The observed S_z order at finite temperature in $J > 1$ is explained by the density of excited state in the ordered state as follows. The low-energy excitations in the S_z ordered state are mainly governed by the one-dimensional like S_x fluctuation. This is the larger density of excited state than that in the S_x mean-field ordered state where the two-dimensional S_z fluctuation dominates the low energy excitation. As a result, the large entropy stabilizes the S_z order at finite temperature. This is not contradict to the fact mentioned previously that there is a macroscopic number of the mean-field ground-state for the S_x ordered state. This degeneracy originates from the inter-chain configuration of the one-dimensional S_x order and does not contribute to the entropy in the thermodynamic limit.

With increasing of J furthermore, temperature dependence of the specific heat shows a broad peak which corresponds to the cross over where the spin correlation develops in the one-dimensional chain as well known in the one-dimensional Ising model. This is plotted by the broken line in Fig. 2. One noticeable point in the phase diagram is that the order of the phase transition of the S^z component is changed from the first order to the second one with decreasing temperature. Thus, the tricritical point exists in the connecting point of the first and second order phase transition lines.

In particular, the first order phase transition line is identified by the Wang-Landau method in the MC simulation. In this method, the free energy is obtained as a function of energy. Near

Journal of Physics: Conference Series **320** (2011) 012062 doi:10.1088/1742-6596/320/1/012062

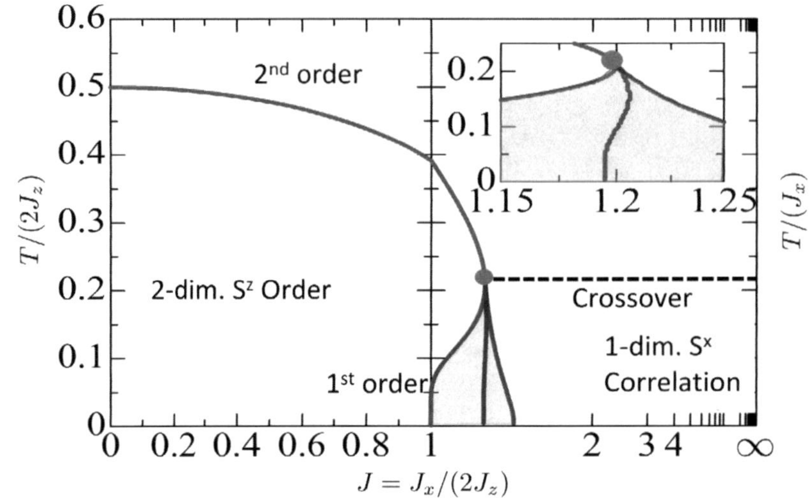

Figure 3. (Color online) Finite temperature phase diagram in the quantum version of the orbital compass model in a checkerboard lattice obtained by the mixed method of the mean-field approximation and the Jordan-Wigner transformation. The broken line shows a cross over line where the one-dimensional pseudo-spin correlation develops. A circle represents the tricritical point. Inset shows a detailed phase transition lines around the tricritical point.

the phase transition line, there are double minima in the free energy function. By changing the temperature, the free energy at one of the minima becomes lower than that at another minimum. This is a direct evidence of the first order phase transition at low temperature around $J = 1$.

Next we show the results in the quantum version of the orbital compass model in the checkerboard lattice. We analyze this model by utilizing the spin wave approximation, the exact diagonalization method based on the Lanczos procedure, and the mixed method of the Jordan-Wigner transformation and the mean-field approximation. Here we introduce the last method and its results. In this method, the interaction between the one-dimensional chain is treated by the mean-field approximation. In a chain, the Ising interaction between the S^x component under the mean-field is treated exactly. The effective Hamiltonian is given by

$$H/(2J_z) = -J \sum_{(ij)} S_i^x S_j^x - h \sum_i S_i^z + L \langle S^z \rangle^2, \tag{7}$$

where the effective field is defined by $h = 2\langle S^z \rangle$, and L is a size of the chain. This is the one-dimensional Ising model under the transverse magnetic field, and is solved by using the Jordan-Wigner transformation.

The phase diagram in the quantum version of the orbital compass model is shown in Fig 3. This phase diagram is obtained by the free energy calculation in the above mixed method of the mean-field approximation and the Jordan-Wigner transformation. We have checked that the qualitative results are reproduced by the quantum Monte-Carlo simulation [9]. The cross over behavior of the one-dimensional S_x correlation is also consistent with the prediction form the generalized Elitzur's theorem [10].

Even at $T = 0$ and $J = 1$, the S^z long range order is realized. This is the order by quantum fluctuation mechanism. The reentrant feature of the phase boundary around $J = 1$ is survived in the quantum calculation. The tricritical point represented by the circle in Fig. 3 is also confirmed by a direct calculation of the free energy. Origin of this tricritical point is interpreted by the following facts: 1) the phase transition for the two-dimensional square lattice expected in $J \ll 1$ is the second order, and 2) the quantum phase transition between the S^z Ising order and the S^x one at $T = 0$ is the first order. Therefore, there is a tricritical point at the connecting point of the first and second order phase transition lines. This is a consequence of the orbital frustration and the geometrically frustration effect.

4. Summary

In summary, the orbital compass model in a checkerboard lattice is studied. The intrinsic orbital frustration and the geometrical frustration coexist. There is a competition between the interactions between S^z along the vertical/horizontal direction and S^x along the diagonal direction. The two interactions compete with each other at $J = 1$ where a macroscopic number of the classical pseudo-spin configurations are degenerated. This degeneracy is lifted by the thermal and quantum fluctuations and the S^z Ising order is realized. The tricritical point appears due to the coexistence of the orbital frustration and the geometrical frustration.

5. Acknowledgments

The authors would like to thank M. Matsumoto and S. Todo for valuable discussions. This work was supported by JSPS KAKENHI, TOKUTEI from MEXT, CREST, and Grand Challenges in Next-Generation Integrated Nanoscience. JN is supported by the global COE program of MEXT Japan.

6. References

[1] Tokura Y and Nagaosa N 2000 *Science* **288** 462
[2] Kugel K I and Khomskii D I 1982 *Sov. Phys. Usp.* **25** 231
[3] Khomskii D I and Mostovoy M V 2003 *J. Phys.* A: Math. Gen. **36** 9197
[4] Kubo K 2002 *J. Phys. Soc. Jpn.* **71** 1308
[5] Nussinov Z and Biskup M L. Chayes and J. van den Brink 2004 *Europhys. Lett.* **67** 990
[6] Mishra A, Ma M, Zhang F -C, Guertler S, Tang L -H and Wan S 2004 *Phys. Rev. Lett.* **93** 207201
[7] Clay R T, Li H, Sarkar S, Mazumdar S and Saha-Dasgupta T 2010 *Phys. Rev.* B **82** 035108
[8] Wang F G and Landau D P 2001 *Phys. Rev. Lett.* **86** 2050
[9] Nasu J, Todo S, and Ishihara S (in preparation).
[10] Batista C D and Nussinov Z 2005 *Phys. Rev. B* **72** 045137

Functional renormalization group approach for extended Hubbard model in a triangular lattice

Hirokazu Takashima, and Sumio Ishihara

Department of Physics, Tohoku University, Sendai 980-8578 Japan

E-mail: takashima@cmpt.phys.tohoku.ac.jp

Abstract.
The extended Hubbard model in an anisotropic triangular lattice is analyzed at quarter filling by the functional renormalization group (fRG) approach. We obtain the finite temperature phase diagram in the plane of nearest inter-site Coulomb interaction V_p and next-nearest Coulomb interaction V_c. When $V_p \approx V_c$, charge order does not occur down to the low temperature of $0.1t_p$, where the t_p is the electron transfer between the nearest neighbor sites. We interpret the result using the random-phase approximation(RPA).

1. Introduction

The charge ordered and disordered system with strong electron correlation is one of the attractive issues in the recent transition metal oxides as well as the low dimensional organic salts. Above all, the charge order associated with the macroscopic electric polarization is an interesting topic. Since the charge order without the inversion symmetry is responsible for the macroscopic electric polarization, this kind of material is known as the electronic ferroelectricity [1] . Layered iron oxide $LuFe_2O_4$ is a candidate for the electronic antiferroelectricity where the charge order of Fe^{2+} and Fe^{3+} in the paired triangular lattice produces the electric dipole moment [2]. Another example is κ-(BEDT-TTF)$_2$Cu$_2$(CN)$_3$ [3, 4], in which the anomaly in the dielectric constant is observed in the experiment. This material is a 1/4 filled system, since the number of the hole carrier is 0.25 per BEDT-TTF molecule. It has a frustrated lattice structure, since two BEDT-TTF molecules form a dimer to consitute a triangular lattice. In the strong dimer coupling limit, one electron per dimer occupies the bonding orbital originated from the two molecule orbitals. From this point of view, studying the 1/4 charge order models in a triangular lattice is an interesting theoretical topic. In fact, seveal authors have studied the so-called 1/4 charge ordered models in a triangular lattice [5, 6, 7, 8, 9]. They showed the existence of competition of several charge ordered phases in an anisotropic triangular lattice with two kinds of the electron transfer and the inter-site Coulomb interactions. In the parameter region where there is strong competition, they expect some non-trivial phenomena such as the metallic phase or superconductivity.

In this paper, we present a theoretical study of the extended Hubbard model in an anisotropic triangular lattice at quarter-filling by weak coupling approach. We analyze the model by using the functional renormalization group (fRG)method. It is known that this method reproduces several static quantities in the Hubbard model calculated by the quantum Monte-Carlo simulation up to the intermediate coupling regime. We obtain the finite-temperature phase diagram in the plane of the two inter-site Coulomb interactions. When there is strong competition among the two Coulomb interactions, the charge order does not occur down to the low temperature of $0.1t_p$ where the t_p is the electron transfer between the nearest neighbor sites. We interpret the result using the random-phase approximation(RPA).

2. Model and method

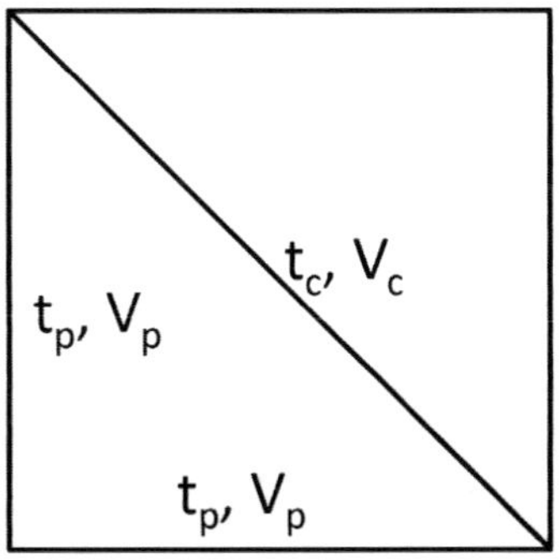

Figure 1. The electron transfers and the inter-site Coulomb interactions in an anisotropic triangular lattice

We consider the extended-Hubbard model in a triangular lattice at quarter-filling. The Hamiltonian is given by

$$H = \sum_{\langle ij \rangle \sigma} t_{ij} c^{\dagger}_{i\sigma} c_{j\sigma} + U \sum_{i} n_{i\uparrow} n_{i\downarrow} + \sum_{\langle ij \rangle} V_{ij} n_i n_j, \tag{1}$$

where $c^{\dagger}_{i\sigma}$ is the electron creation operator with spin σ at site i, and $n_i (\equiv \sum_\sigma n_{i\sigma} = \sum_\sigma c^{\dagger}_{i\sigma} c_{i\sigma})$ is the number operator. We introduce the electron transfer t_{ij}, the on-site Coulomb interaction U, and the inter-site Coulomb interaction V_{ij}. We consider the square lattice with diagonal bonds along $[1\bar{1}]$ direction as shown in Fig. 1, which is equivalent to the triangular lattice. The transfer integral and the inter-site Coulomb interaction between the nearest neighbor sites along [10] and [01] directions are denoted by t_p and V_p, respectively, and those along the $[1\bar{1}]$ direction are denoted by t_c and V_c, respectively. It is obvious that this system is an isotropic triangular lattice when $t_p = t_c$ and $V_p = V_c$. We mainly report the result at $t_c = 0$.

We used the functional renormalization group (fRG) method [10, 11, 12, 13, 14]. The basic procedure of the fRG calculation is as follows: The susceptibility is expressed as a functional derivative of a generating functional of the green function. Let us consider an infinitesimal change in a certain parameter in a system. This parameter, termed a flow-parameter, determines the quadratic part of the action and remains independent in the remainder part of the action. By differentiating the generating functional of the green function with respect to the flow parameter, we can obtain the renormalization group equation, where the derivative of the generating functional is written as a functional of the generating functional. It is straightforward to rewrite this equation for the susceptibility by taking functional derivatives. Integrating the renormalization group equation from the bare-coupling regime, we can obtain the susceptibilities at strong correlated regime.

For numerical calculation, there are two major approximations; 1) two-loop truncation which neglects the two-loop contribution and 2) the static approximation, which neglects the Matsubara frequency dependence of the four-point vertex. These approximations are partly justified because the neglected diagrams often have small contribution in phase volume even at intermediate coupling regime. The fRG method has the advantage of taking account of all Feynman diagrams at one-loop level and including full mode-couplings. In the present calculation, we adopt the temperature flow fRG [15], which is convenient for the finite temperature calculation.

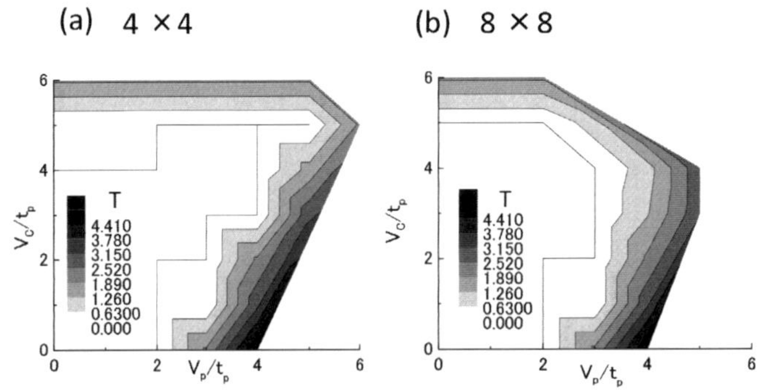

Figure 2. Finite temperature phase diagram The number of discretization in the momentum space of the four point vertex function are taken to be 4×4 in (a) and 8×8 in (b). The charge ordering temperature is identified by divergence of the charge susceptibiliyu. The metallic phase impies that there is no divergence in susceptibility nor vertex function down to T=0.1. The parameter values are taken to be $t_c = 0$, $U = 10$ and $t_p = 1$.

3. Numerical results

In Fig. 2, we show the finite temperature phase diagram in the plane of V_p and V_c. Other parameters are chosen to be $U = 10$ and $t_c = 0$ in a unit of t_p. The electron numbers per site are fixed to be 1.5. In this figure, we plot the temperature where the charge susceptibility diverges. The parameter region with $T = 0$ means that there is no divergence in the charge susceptibility nor the vertex function down to $T/t_p = 0.1$. In the regions of large V_p limit or large V_c limit, the charge ordered phase appears respectively, with the characteristic momentum of each charge ordered phase being mentioned later. The metallic phase is stable until large inter-site Coulomb interactions down to $T = 0.1t_p$ when $V_p \approx V_c$. The metallic state in the vicinity of the $V_p = V_c$ line becomes less stabilized as we increase the discretization number of four-point vertex, The global feature in the present results is consistent with the previous phase diagram at $T = 0$ obtained by the exact diagonalization method [16].

Next, we show the momentum dependence of the charge susceptibility in Fig. 3 for several parameter sets of (V_c, V_p). The parameter values are chosen to be $(V_c = 5, V_p = 0)$ and $T/t_p = 0.017$ in (a), $(V_c = 6, V_p = 2)$ and $T/t_p = 2.03$ in (b), and $(V_c = 3, V_p = 4)$ and $T/t_p = 1.64$ in (c). In Fig. 3(a), the peak structures are seen in $(\pi, 0)$ and $(0, \pi)$ which corresponds to the so-called diagonal charge order, and the plateau appears along the line of $q_x + q_y = \pi$. In Fig. 3(b), in addition to the peaks in $(\pi, 0)$ and $(0, \pi)$, a new peak appears at about $(\pi/2, \pi/2)$ which corresponds to the horizontal charge order. In the case of $Vc > V_p$ as shown in Fig. 3(c), the peak position shifts to $(2\pi/3, 2\pi/3)$.

Now we interpret the result by the random-phase approximation (RPA). The charge

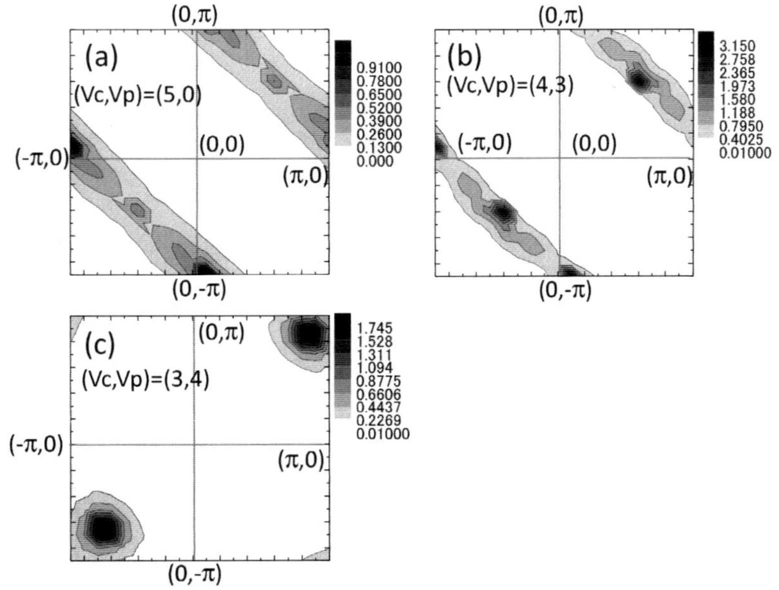

Figure 3. Momentum dependence of the charge susceptibility at $(V_c, V_p) = (5, 0)$ in (a), $(V_c, V_p) = (6, 2)$ in (b), and $(V_c, V_p) = (3, 4)$ in (c). Temperatures are chosen to be $T = 0.017$ in (a), 2.03 in (b) and 1.64 in (c). Other paramters are chosen to be $U = 10$, $t_c = 0$, and $t_p = 1$.

susceptibility $\chi(q)$ in RPA is given by

$$\chi_c(q) = \frac{\chi_0(q)}{1 + [U + 2V(q)]\chi_0(q)}. \tag{2}$$

We introduce the bare charge susceptibility as

$$\chi_0(q) = -\sum_p \frac{n_f(\epsilon_p) - n_f(\epsilon_{p+q})}{\epsilon_p - \epsilon_{p+q}}, \tag{3}$$

where $\epsilon_p = 2t_p[\cos(p_x) + \cos(p_y)] + 2t_c \cos(p_x + p_y)$ is the bare band dispersion, $n_f(q) = 1/[1 + \exp\{(\epsilon_p - \mu)/T\}]$ is the Fermi distribution function, and $V(q) = 2V_p[\cos(q_x) + \cos(q_y)] + 2V_c \cos(q_x + q_y)$ is the Fourier transform of the inter-site Coulomb interaction. The charge susceptibility is enhanced at the momentum where the denominator of Eq. (2) takes its minimum. We have confirmed that, around $T/t_p = 1$, the momentum dependence of $\chi_0(q)$ is about 10% of its average value. Since this momentum dependence is weaker than that in the inter-site interaction $V(q)$, the wavenumbers where the denominator of Eq. (2) takes its minimum are governed by $V(q)$. Let us consider the following limit in the expression in Eq. (2). i) in $V_c = 0$, $V(q)$ takes its minimum at $q = (\pi, \pi)$. With increasing V_c from $V_c = 0$, the minimum point of $V(q)$ remains at $q = (\pi, \pi)$ until a certain value of V_c which is termed V_c^*. In the region of $V_c > V_c^*$, the minimum point shifts to an incommensurate wave number. ii) in the case of $V_p = V_c (\equiv V)$, $V(q)$ takes its minimum at $q = (\frac{2}{3}\pi, \frac{2}{3}\pi)$, regardless of the value of V. iii) in $V_p = 0$, the momentum where $V(q)$ takes its minimum satisfies the condition $q_x + q_y = \pi$.

From the above considerations based on the momentum dependence of $V(q)$, the numerical results by fRG methods shown in Fig. 3(a) and (c) are consistent with the expectations from the RPA expression. Large peak structures around $(\pi, 0)$ and $(\pi/2, \pi/2)$ in Fig. 3(b) may be due to the effects of $\chi_0(q)$ and/or the renormalization.

4. Conclusion

To summarize, we have analyzed the extended Hubbard model in an anisotropic triangular lattice by using the functional renormalization group method. The finite-temperature phase diagram is obtained in the plane of the two inter-site Coulomb interactions, V_p and V_c. In the limit of large V_p, and V_c, the charge order with the momentum of $q = (\pi, \pi)$ and $q = (\pi, 0)$ are realized respectively. When $V_p \approx V_c$, the charge order does not occur down to the low temperature of $0.1t_p$. The obtained numerical results are interpreted by the random-phase approximation.

One possible extension of the present calculation is applying this method to the triangular lattice with the internal degree of freedom, such as the dimer degree of freedom at each site. Since some electronic ferroelectric materials such as κ-(BEDT-TTF)$_2$Cu$_2$(CN)$_3$ are located near the boundary of the metal-insulator transition, this theoretical approach may provide supplemental information to that of the strong coupling approach in resolving the mechanism of ferroelectricity

Acknowledgments

The authors would like to thank M. Naka for valuable discussions. This work was supported by JSPS KAKENHI, TOKUTEI from MEXT, CREST, and Grand Challenges in Next-Generation Integrated Nanoscience.

5. References

[1] Ishihara S 2010, *J. Phys. Soc. Jpn.* **79**, 011010.
[2] Ikeda N, Ohsumi H, Ohwada K, Ishii K, Inami T, Kakurai K, Murakami Y, Yoshii K, Mori S, Horibe Y and Kito H, 2005 *Nature* **436** 1136.
[3] Abdel-Jawad M, Terasaki I, Sasaki T, Yoneyama N, Kobayashi N, Useu Y and Hotta C, 2010 *Phys. Rev. B* **82** 125119.
[4] Naka M, and Ishihara S, 2010 *J. Phys. Soc. Jpn.* **79**, 063707.
[5] Seo H, 2000 *J. Phys. Soc. Jpn.*, **69** 805.
[6] Kaneko M, Ogta M, 2006 *J.Phys. Soc. Jpn.* ,**75**, 014710.
[7] Kuroki K, 2006 *J. Phys. Soc. Jpn.*, **75**,114716.
[8] Watanabe H, and Ogata M, 2006 *J.Phys. Soc. Jpn.* ,**75** 063702.
[9] Nishimoto S, Shingai M and Ohta Y 2008 *Phys. Rev. B* **78** 035113.
[10] Salmhofer M and Honerkamp C, 2001 *Prog.Thoer. Phys.* **105**, 1.
[11] Zaqnchi D and Schulz D J , 2000 *Phys. Rev. B*, **61**,13609.
[12] Wieczerkowski C, 1998 *Commun. Math. Phys.* **120**, 149.
[13] Polchinski J, 1983 *Nucl.Phys.B*, **231**,269.
[14] Halboth C and Meizner W, 2000 *Phys. Rev. B*, **61**, 7364.
[15] Honerkamp C and Salmhofer M 2001 *Phys. Rev. B* **64** 184516.
[16] Merino J, Seo H, and Ogata M, 2005 *Phys. Rev. B* **71**, 125111.

Optical conductivity of geometrical frustrated electronic systems

Toshihiro Sato, Kazumasa Hattori and Hirokazu Tsunetsugu
Institute for Solid State Physics, University of Tokyo, 5-1-5 Kashiwanoha, Kashiwa, Chiba 277-8581, Japan
E-mail: t.sato@issp.u-tokyo.ac.jp

Abstract. We study optical conductivity of the one-band Hubbard model on a two-dimensional isotropic triangular lattice at half filling by using the cellular dynamical mean field theory and particularly investigate its behavior near the critical end point of the first order Mott transition. In the metallic phase, the enhancement of effective mass and the significant frequency dependence of scattering rate is identified near the Mott transition. From the results in the insulating state near the Mott transition, we find that the frequency dependence of optical conductivity decays quite smoothly toward zero frequency and exhibits a power-law-like behavior in an intermediate frequency region. We also study optical conductivity of the Hubbard model on a kagomé lattice near the critical end point of the Mott transition. It shows a larger incoherent peak both in the metallic and insulating phases and a stronger frequency dependence of effective mass and scattering rate.

1. Introduction

Transport properties are one of the central topics in strong correlated electronic systems. As a typical example, the measurement of optical conductivity provides us useful information on charge dynamics, in particular, effective mass, scattering process and electric structure. As for geometrically frustrated systems, it is interesting to investigate how spin frustration influences charge dynamics. This is the issue of this paper and we study optical conductivity of geometrical frustrated electronic systems. The organic compound κ-$(ET)_2Cu_2(CN)_3$ has a structure of the two-dimensional (2D) isotropic triangular lattice of dimer pairs with a half-filed conduction band. At ambient pressure, the compound is a Mott insulator and a transition to metal has been observed under pressure [1]. In the insulating phase, the system exhibits a spin liquid behavior, which shows no magnetic long-range order [2, 3, 4]. In a recent work, interesting behaviors of optical conductivity have been observed in this insulating state, such as the smooth decay of the spectra toward zero frequency at low temperatures [5]. In contrast, the anisotropic triangular lattice organic compound κ-$(ET)_2Cu[N(CN)_2]Cl$ exhibits an antiferromagnetic order and a clear gap in the optical conductivity [6]. Another example is the spinel compound LiV_2O_4. It exhibits heavy fermion behaviors, showing large Sommerfeld coefficient of specific heat and temperature square coefficient of resistivity. Optical conductivity shows a Drude peak at low temperatures, indicating the formation of coherent quasiparticles [7]. The spectra gradually decays towards high frequencies unlike a simple Drude formula.

In this paper, we study transport properties in strongly correlated electronic systems on the isotropic triangular lattice. We focus on optical conductivity near the critical end point of the

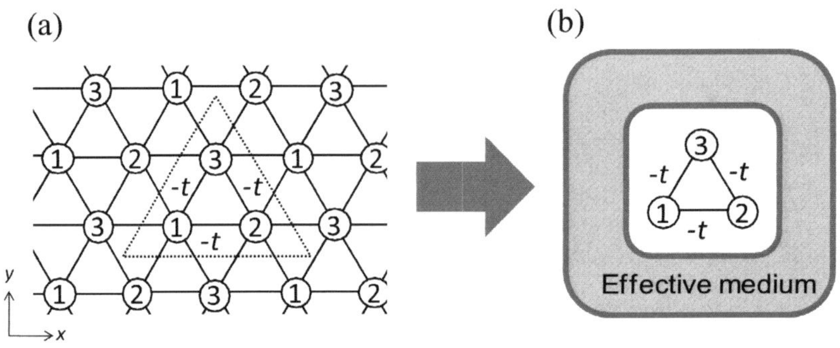

Figure 1. (a) Isotropic triangular lattice and (b) effective cluster model employing in the three-site cluster CDMFT.

first order Mott transition and discuss its behavior.

2. Model and Method

We consider the one-band Hubbard model on a two-dimensional isotropic triangular lattice at half filling

$$H = -t \sum_{\langle i,j \rangle, \sigma} c_{i\sigma}^{\dagger} c_{j\sigma} + U \sum_{i} n_{i\uparrow} n_{i\downarrow} - \mu \sum_{i\sigma} c_{i\sigma}^{\dagger} c_{i\sigma}, \quad (1)$$

where t is nearest-neighbor hopping amplitude, U is on-site Coulomb repulsion and μ is chemical potential. $c_{i\sigma}^{\dagger}(c_{i\sigma})$ is an electron creation (annihilation) operator at site i with spin σ and $n_{i\sigma} = c_{i\sigma}^{\dagger} c_{i\sigma}$. To take into account strong electronic correlations and geometrical frustration, we use the cellular dynamical mean field theory (CDMFT) [10], employing a triangular three-site cluster to calculate various dynamical correlation functions at finite temperatures as shown in Fig. 1. We numerically obtain the electron Green's function by using the continuous-time Monte Carlo method based on the strong coupling expansion [11]. We first confirm the finite temperature Mott transition with varying U and then calculate optical conductivity.

Optical conductivity $\sigma(\omega)$ is defined by the Kubo formula and is described as

$$\sigma(\omega) = -ie^2 \lim_{\mathbf{q} \to 0} \frac{\chi(\mathbf{q}, \omega) - \chi(\mathbf{q}, 0)}{\omega}, \quad (2)$$

where $\chi(\mathbf{q}, \omega)$ denotes the current-current correlation functions and e is the elementary charge. In our calculation, we do not consider vertex corrections in $\chi(\mathbf{q}, \omega)$ and calculate

$$\chi(\mathbf{q} = \mathbf{0}, i\omega_n) = -\frac{2T}{N} \sum_{\mathbf{k}} \sum_{i\nu_l} v(\mathbf{k})^2 G(\mathbf{k}, i\nu_l) G(\mathbf{k}, i\nu_l + i\omega_n), \quad (3)$$

where $v(\mathbf{k}) = 2t\{\sin(k_x) + \sin(k_x/2)\cos(\sqrt{3}k_y/2)\}$ is the x-component of velocity in the case of the isotropic triangular lattice. $G(\mathbf{k}, i\omega_n)$ is the electron Green's function with the Matsubara frequencies $i\omega_n$ and wave vector $\mathbf{k}$, T is a temperature and N is the total number of sites. From $\chi(\mathbf{0}, i\omega_n)$, we calculate the real part of $\sigma(\omega)$ by analytic continuation $i\omega_n \to \omega + i\delta$ by using the maximum entropy algorithm [12].

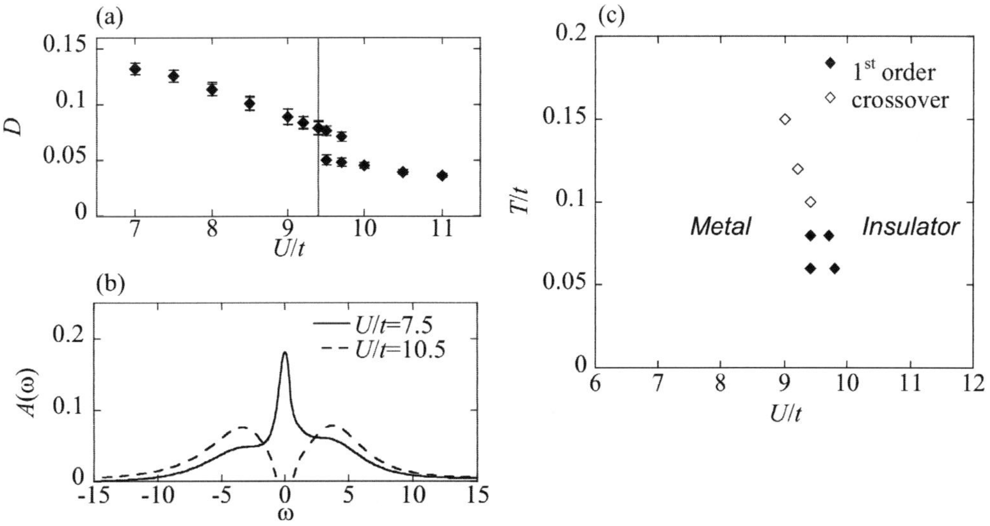

Figure 2. (a) Coulomb repulsion U dependence of double occupancy D at $T/t = 0.08$. Solid line represents $U_c/t = 9.4$, where the first order Mott transition takes place with decreasing U. (b) Density of state $A(\omega)$ at $T/t = 0.08$. (c) Phase diagram in the parameter space of T and U.

3. Results

We first identify the location where the first order Mott transition takes place in the parameter space of temperature T and Coulomb repulsion U. Figures 2 (a) and (b) present U-dependence of double occupancy $D = \langle n_{i\uparrow} n_{i\downarrow} \rangle$ and density of state $A(\omega) = -(1/\pi N)\sum_{\mathbf{k}} \mathrm{Im}\, G(\mathbf{k}, \omega + \mathrm{i}0)$ at $T/t = 0.08$. At $U/t = 7.5$, $A(\omega)$ shows the formation of coherent quasiparticles around $\omega = 0$ and thus, the system is metallic. At $U/t = 10.5$, these quasiparticles disappear and there emerges a gap around $\omega = 0$ and the system is in the insulating phase. Near $U_c/t = 9.4$, D jumps and the first order Mott transition takes place. Repeating calculations for various T, we obtain the U-T phase diagram shown in Fig. 2 (c).

Let us start with investigation of optical conductivity $\sigma(\omega)$. We concentrate on $\sigma(\omega)$ near $T_c/t \sim 0.09$, which corresponds to the critical end point of the line of the first order Mott transition. For simple metals at low temperatures, it is well known that the frequency dependence of $\sigma(\omega)$ is described by a simple Drude formula $\sigma(\omega) = \omega_p^2/[4\pi(-i\omega + 1/\tau)]$, where ω_p is plasma frequency and $1/\tau$ is scattering rate. On the other hands, the simple Drude formula does not work in Mott insulators, since a Drude peak around $\omega = 0$ disappears and an incoherent broad peak appears near $\omega \sim U$. Figure 3 (a) shows the real part of $\sigma(\omega)$ at $T/t = 0.08$ with varying U. The first order Mott transition takes place at $U_c/t = 9.4$. For $U \leq U_c$, a Drude peak is identified, indicating the formation of coherent quasiparticles. As U increases, the spectral weight at low frequencies is depressed at low frequencies and an incoherent peak around $\omega = 5$ evolves significantly. Moreover, ω-dependence of $\sigma(\omega)$ is not described by a simple Drude formula. From these results, it is expected that these characteristics are attributed to a renormalization of the scattering rate $1/\tau(\omega)$ and the effective mass $m^*(\omega)$ due to strong electronic correlations and the effects of the frustration. In order to clarify the charge dynamics in the metallic state, we analyze the frequency dependence of $\sigma(\omega)$ by using an extended Drude formula:

$$\sigma(\omega) = \frac{\omega_p^2}{4\pi} \frac{1}{-i\omega \frac{m^*(\omega)}{m_b} + \frac{1}{\tau(\omega)}}, \tag{4}$$

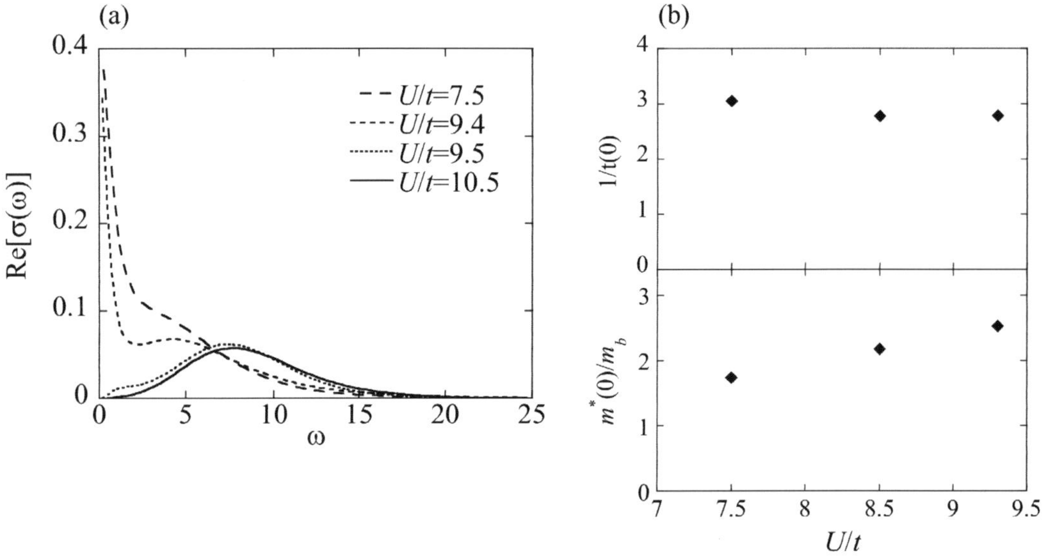

Figure 3. (a) Real part of the optical conductivity $\sigma(\omega)$ for various U/t at $T/t = 0.08$. The first order Mott transition takes place at $U_c/t = 9.4$. (b) Coulomb repulsion U dependence of scattering rate $1/\tau(0)$, effective mass $m^*(0)$ estimated by an extended Drude formula Eq. (4).

where m_b is the unrenormalized band mass. ω_p^2 is estimated from the integration of Re$[\sigma(\omega)]$. From the results of $1/\tau(\omega)$, rather unexpectedly, we find a suppression in $1/\tau(\omega)$ with increasing U. On the other hands, $m^*(\omega)/m_b$ exhibits an enhancement with increasing U. From these results, we also estimated the mean free path $l(\omega) = p_F\tau(\omega)/m^*(\omega)$, where p_F is Fermi momentum. We have found that $l(\omega)$ decreases at low frequencies with increasing U as expected. To discuss the low-frequency behavior, we analyze the most characteristic value, namely, the static limit $\omega = 0$, of $1/\tau(\omega)$ and $m^*(\omega)$. Near the Mott transition, it is expected that $1/\tau(0)$ or $m^*(0)$ increase with increasing U. Their U-dependence is plotted in Fig. 3 (b). While $1/\tau(0)$ shows little change with U, $m^*(0)$ shows a noticeable increase with increasing U. From these results, we find that the increase of mass is more related to the Mott transition rather than scattering rate.

For $U > U_c$, the system is insulating and, correspondingly, $\sigma(\omega)$ exhibits two features. First, as shown in Fig. 3 (a), a Drude peak disappears and $\sigma(\omega)$ shows a large incoherent peak around $\omega \sim U - 2.5t$. This value $2.5t$ is the renormalization of the energy difference between the lower and upper Hubbard bands due to charge fluctuation. This comes from the kinetic energy of a pair of doublon and holon. We numerically estimated this renormalization from the peak position of the incoherent part of $\sigma(\omega)$. Secondly, it is also interesting that the ω-dependence of $\sigma(\omega)$ decays quite slowly toward $\omega = 0$. For this behavior, we have found that $\sigma(\omega)$ follows a power-law-like behavior in an intermediate frequency region near the Mott transition.

Finally, we also investigate $\sigma(\omega)$ of another frustrated system, the Hubbard model on a kagomé lattice, and compare $\sigma(\omega)$ of the two models near the critical end point of the first order Mott transition. To make fair comparison between the two models, we normalized ω by the effective bandwidth W of the dimension of energy. We use $W = \langle(\omega - \langle\omega\rangle)^2\rangle^{1/2}$ calculated from density of state $A(\omega)$. For the isotropic triangular lattice, the critical end point is near $T/t = 0.08$ ($U_c/t \sim 9.4$) and $W/t = 7.6034$ at $U/t = 9.4$, $W/t = 5.6249$ at $U/t = 9.5$. For the kagomé lattice, the critical end point is near $T/t = 0.07$ ($U_c/t \sim 8$) and $W/t = 7.6438$ at $U/t = 8$, $W/t = 5.9929$ at $U/t = 8.1$. As for $\sigma(\omega)$, the Kubo formula shows that it has the

Journal of Physics: Conference Series **320** (2011) 012064 doi:10.1088/1742-6596/320/1/012064

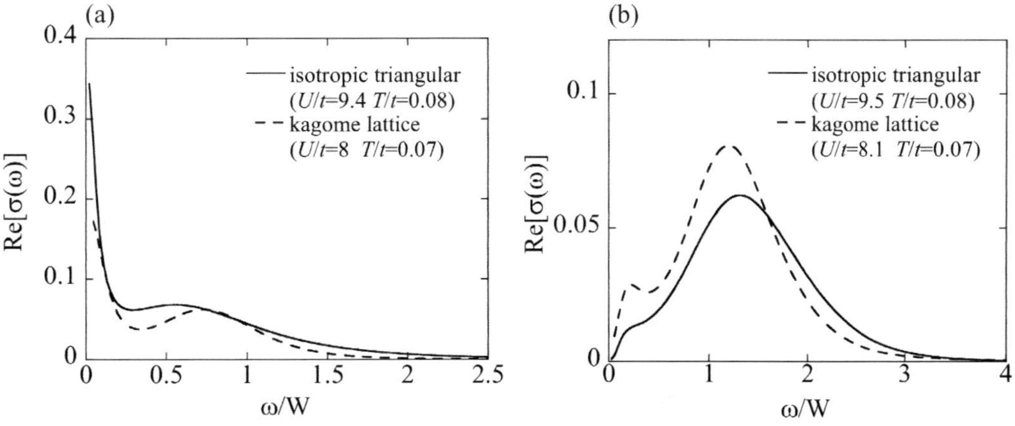

Figure 4. Real part of optical conductivity $\sigma(\omega)$ in the metallic phase (a) and in the insulating phase (b) near the Mott transition for the isotropic triangular lattice ($U_c/t \sim 9.4$ at $T/t = 0.08$) and kagomé lattice ($U_c/t \sim 8$ at $T/t = 0.07$). W is the effective band width $W = \langle(\omega - \langle\omega\rangle)^2\rangle^{1/2}$ calculated from density of state $A(\omega)$.

dimension zero in our units, and therefore we do not need to renormalize $\sigma(\omega)$ for comparison. Figure 4 shows $\sigma(\omega)$ as a function of dimensionless frequency ω/W near the critical end point of the first order Mott transition of the two models. As shown in Figs. 4 (a) and (b), an incoherent peak of $\sigma(\omega)$ emerges clearly both in the metallic and insulating phases. From the results of an extend Drude analysis in the metallic phase, we find that $1/\tau(\omega)$ and $m^*(\omega)$ in the kagomé case show the larger ω-dependence than in the isotropic triangular one corresponding to sharper incoherent peak. In the insulating phase, we find similar power-law-like behaviors of $\sigma(\omega)$ of the kagomé lattice in an intermediate frequency region. The detailed analysis of $\sigma(\omega)$ is in progress and will be published elsewhere.

4. Summary

In this paper, we have studied optical conductivity of the one-band Hubbard model on a two-dimensional isotropic triangular lattice at half filling by using the cellular dynamical mean field theory. We calculate optical conductivity near the critical end point of the first order Mott transition. In the metallic phase, an extend Drude analysis indicates the enhancement of the mass and the significant frequency dependence of the scattering rate with increasing Coulomb repulsion. In the insulating phase near the Mott transition, we have found that the frequency dependence of optical conductivity decays quite smoothly toward zero frequency and exhibits a power-law-like behavior in an intermediate frequency region. Optical conductivity of the Hubbard model on a kagomé lattice near the critical end point of the Mott transition also shows similar behaviors. We have also found the larger incoherent peak both in the metallic and insulating phases and stronger frequency dependence of effective mass and scattering rate.

In future publication, we will take into account vertex corrections in current-current correlation function and discuss their effects as well as detailed analysis of these frustrated lattice systems.

Acknowledgments

The authors are grateful to T. Ohashi and H. Kusunose for helpful discussions. The present work is financially supported by MEXT Grant-in-Aid for Scientific Research on Priority Areas "Novel States of Matter Induced by Frustration" (19052003), and by Next Generation Supercomputing

Project, Nanoscience Program, MEXT, Japan. The computation in the present work was performed on computers at the Supercomputer Center, Institute for Solid State Physics, University of Tokyo.

Reference

[1] Kanoda K 2006 *J. Phys. Soc. Jpn.* **75** 051007
[2] Shimizu Y, Miyagawa K, Kanoda K, Maesato M and Saito G 2003 *Phys. Rev. Lett.* **91** 107001
[3] Yoshioka T, Koga A and Kawakami N, 2009 *Phys. Rev. Lett.* **103** 036401
[4] Morita H, Watanabe S and Imada M, 2002 *J. Phys. Soc. Jpn.* **71** 2109
[5] Kézsmárki I, Shimizu Y, Mihály G, Tokura Y, Kanoda K and Saito G 2006 *Phys. Rev. B* **74**201101(R)
[6] Kornelsen K, Eldridge J E, Wang Hau H, Charlier Henry A and Williamst Jack M 1992 *Solid State Commun.* **81** 343
[7] Jönsson P E, Takenaka K, Niitaka S, Sasagawa T, Sugai S and Takagi H 2007 *Phys. Rev. Lett.* **99** 167402
[8] Dordevic S V, Basov D N, Dilley N R, Bauer E D and Maple M B 2001 *Phys. Rev. Lett.* **86** 684
[9] Degiorgi L, Anders F B B, and Grüner G 2001 *Eur. Phys. J. B* **19** 167
[10] Kotliar G, Savrasov S Y, Palsson G and Biroli G 2001 *Phys. Rev. Lett.* **87** 186401
[11] Werner P, Comanac A, Medici L, Troyer M and Millis A J 2006 *Phys. Rev. Lett.* **97** 076405
[12] Jarrel M and Gubernatis J E 1996 *Phys. Rep.* **269** 133

Effective quantum pseudospin-1/2 model for Yb pyrochlore oxides

Shigeki Onoda

Condensed Matter Theory Laboratory, RIKEN, 2-1, Hirosawa, Wako 351-0198, Saitama, JAPAN

E-mail: s.onoda@riken.jp

Abstract. An effective quantum pseudospin-1/2 Hamiltonian for $Yb_2TM_2O_7$ (TM = Ti and Sn) is obtained in terms of the atomic Kramers ground doublet of the LS coupling and the crystalline electric field, which is almost described with $J^z = \pm 1/2$. It is calculated microscopically as the sum of Anderson's superexchange interaction and the magnetic dipole interaction. It is found that it shows a strong exchange anisotropy.

1. Introduction

Geometrical frustration and quantum fluctuation provide a promising route to unconventional states in magnetism, including various types of spin liquids characterized by the absence or emergence of unconventional orders of chirality or multipoles in three spatial dimensions [1, 2, 3]. The pyrochlore lattice structure offers a prototypical stage where the geometrical frustration prevents the magnetic dipole long-range order (LRO) [1, 4, 5]. One class of magnetic pyrochlore oxides [6] which have been most intensively studied is a dipolar spin ice, e.g., $Dy_2Ti_2O_7$ and $Ho_2Ti_2O_7$ [6, 7, 8, 9]. Recently, quantum variants of the spin ice have attracted great interest [10, 11, 12], which include $Tb_2Ti_2O_7$ [13, 14, 15], $Tb_2Sn_2O_7$ [15], and $Pr_2Sn_2O_7$ [16, 17], $Pr_2Ir_2O_7$ [18, 19], $Yb_2Ti_2O_7$ [20], and $Er_2Ti_2O_7$ [20].

In the dipolar spin ice, large $\langle 111 \rangle$ Ising magnetic moments ($\sim 10\mu_B$) of Dy^{3+} and Ho^{3+} ions, each of which points either inwards ("in") to or outwards ("out") from the center of the tetrahedron, interact with each other mainly through the magnetic dipole interaction [21]. This gives a ferromagnetic nearest-neighbor coupling, forming the "2-in, 2-out" ice rule [22] and leaving macroscopically degenerate ground states because of geometrical frustration. The spin dynamics is quite slow and the quantum effects are almost negligible.

Recently, quantum effects have been observed with inelastic neutron-scattering experiments on the spin-ice related compounds, $Tb_2Ti_2O_7$ [13, 14, 15], $Tb_2Sn_2O_7$ [15], and $Pr_2Sn_2O_7$ [17]. $Pr_2TM_2O_7$ (TM = Zr, Sn, and Ir) shows no magnetic dipole LRO but a (partial) spin freezing at $T_f \sim$ 0.1-0.3 K [16, 17, 18, 19, 23]. $Pr_2Ir_2O_7$ shows a metamagnetic transition at low temperatures only when the magnetic field is applied in the [111] direction [19], as in the spin ice. A chiral spin liquid [3] has been detected through the anomalous Hall effect [24] at zero magnetic field without magnetic dipole LRO in $Pr_2Ir_2O_7$ [19]. Unlike the dipolar spin ice, the Curie-Weiss temperature T_{CW} is antiferromagnetic for the zirconate [23] and iridate [18]. The stannate shows a significant level of low-energy short-range spin dynamics in the energy range up to a few Kelvin [17], which is absent in the classical spin ice. Vital roles of the planar components have also been experimentally observed in $Yb_2Ti_2O_7$ and $Er_2Ti_2O_7$ [20]. Obviously, quantum fluctuations enrich the otherwise classical properties of the spin ice.

The mechanism in which the quantum fluctuations indicated by these observations appear depend on the rare-earth ions, including the difference in Kramers and non-Kramers ions. A relatively small localized magnetic moment reduces the magnetic dipole interaction, allowing for an important role of the superexchange interaction, as in the case of Pr^{3+} ions [11, 12] and also expected for Nd^{3+}, Sm^{3+}, and Yb^{3+} ions because of their small moment amplitudes, $3.3\mu_B$, $0.7\mu_B$, and $4\mu_B$, respectively, for isolated cases. In the case of Tb^{3+} ions, the presence of low-energy crystalline electric field levels also plays a role [10]. Remarkably, the large LS coupling and the crystalline electric field for localized $4f$ electrons makes the superexchange interaction highly anisotropic [11, 12]. In fact, the strength of the anisotropy depends on the properties of local magnetic doublets of rare-earth ions, producing the variety of experimentally observed low-temperature magnetic properties which are mentioned above. Theoretically, it is expected that this anisotropic superexchange interaction drives *quantum phase transitions among the spin ice, quadrupolar states having nontrivial chirality correlations, and the quantum spin ice* for non-Kramers magentic doublets.

As for Kramers doublets, on the other hand, understandings of $Yb_2Ti_2O_7$ are rather controversial. Yb localized $4f$ magnetic moments are believed to almost lie perpendicular to the $\langle 111 \rangle$ directions, unlike Dy, Ho, and Pr cases. It has a ferromagnetic Curie-Weiss temperature ~ 0.65 K and a first-order transition has been observed at 0.24 K with the specific heat measurement [25]. It has been reported that a neutron-scattering experiment on the single crystal shows a LRO in the low-temperature phase [26], while a combined study using neutron diffraction and Mössbauer and μSR spectroscopies [27] and the polarized newtron-scattering [28] on the polycrystals have indicated that the magnetic dipole correlation remains dynamic. Recent neutron-scattering experiments have clarified a [111] rod scattering intensity in the paramagnetic phase [29], which has been phenomenologically analyzed by taking into account the anisotropic superexchange interaction [30].

In this paper, we present a microscopic derivation of an effective quantum pseudospin-1/2 model, which takes the sum of the magnetic dipole interaction and the superexchange interaction, in terms of atomic Kramers doublets of Yb^{3+} ions for the frustrated magnet $Yb_2TM_2O_7$ (TM = Ti and Sn) on the pyrochlore lattice.

2. Local Kramers doublets of Yb^{3+} ions

The ground state of an isolted Yb^{3+} ion is characterized by the eightfold degenerate manifold $^2F_{\frac{7}{2}}$. In the pyrochlore oxide $Yb_2Ti_2O_7$, the D_{3d} crystalline electric field partially lift the degeneracy, yielding the Kramers ground doublet

$$|\sigma\rangle_D = -\alpha\sigma|J_z = \frac{7}{2}\sigma\rangle + \beta|J_z = \frac{1}{2}\sigma\rangle + \gamma\sigma|J_z = -\frac{5}{2}\sigma\rangle, \tag{1}$$

with $\sigma = \pm$, real coefficients $\alpha \approx 0.388$, $\beta \approx 0.889$, and $\gamma \approx 0.242$ [31, 32], and the eigenstate $|J_z = M\rangle$ of the total angular momentum J^z, where the local coordinates $(\boldsymbol{x}_i, \boldsymbol{y}_i, \boldsymbol{z}_i)$ $(i = 0, \cdots, 3)$ are site-dependent, as explicitly given in Appendix A. In particular, the quantization axis $\boldsymbol{z}_i$ is always taken to be parallel to the $\langle 111 \rangle$ direction that points from the Yb site to the center of one of the two tetrahedrons to which the Yb^{3+} ion belongs. This ground doublet is energetically well separated from the first excitated doublet by 620 K [31].

The ground doublet has the following nonvanishing Ising and transverse components of the total angular momentum $\boldsymbol{J}$

$$_D\langle\sigma|\hat{J}^z|\sigma'\rangle_D = \frac{1}{2}J_{\parallel}\sigma^z_{\sigma,\sigma'}, \tag{2}$$

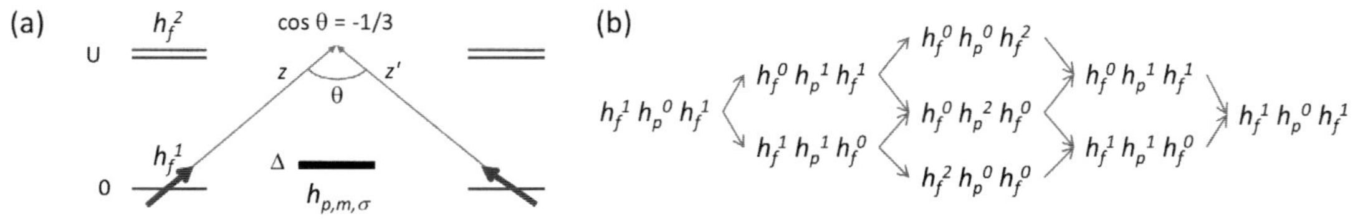

Figure 1. (a) Local level scheme of O $2p$ holes and Yb $4f$ holes. (b) Virtual hopping processes of holes contributing to the superexchange interaction. n (n') and ℓ in the state $h_f^n h_p^\ell h_f^{n'}$ represent the number of f holes at the Yb site $\boldsymbol{r}$ ($\boldsymbol{r}'$) and that of p holes at the O site.

$$ {}_D\langle\sigma|\hat{J}^{x,y}|\sigma'\rangle_D \;=\; \frac{1}{2} J_\perp \sigma^{x,y}_{\sigma,\sigma'}, \tag{3}$$

where $J_\parallel = 7\alpha^2 + \beta^2 - 5\gamma^2$ and $J_\perp = 2\sqrt{7}\alpha\gamma + 4\beta^2$. On the other hand, it does not contribute to the atomic quadrupole moment, as is clear from

$$ {}_D\langle\sigma|\left\{\hat{J}^\pm, \hat{J}^z\right\}|\sigma'\rangle_D = {}_D\langle\sigma|\hat{J}^+\hat{J}^+|\sigma'\rangle_D = {}_D\langle\sigma|\hat{J}^-\hat{J}^-|\sigma'\rangle_D = 0. \tag{4}$$

Let us asign the Pauli matrices $\hat{\boldsymbol{\sigma}}_{\boldsymbol{r}}$ to a psedospin-$\frac{1}{2}$ that operates on the subspace for the the atomic doublet at the site $\boldsymbol{r}$. Then, the localized magnetic dipole moment is given by

$$\hat{\boldsymbol{m}}_{\boldsymbol{r}} = g_J \mu_B \hat{\boldsymbol{J}}_{\boldsymbol{r}} = \frac{1}{2}\mu_B \left[g_\perp \left(\hat{\sigma}^x_{\boldsymbol{r}} \boldsymbol{x}_{\boldsymbol{r}} + \hat{\sigma}^y_{\boldsymbol{r}} \boldsymbol{y}_{\boldsymbol{r}} \right) + g_\parallel \hat{\sigma}^z_{\boldsymbol{r}} \boldsymbol{z}_{\boldsymbol{r}} \right], \tag{5}$$

where $g_\perp = g_J J_\perp \approx 4.18$ and $g_\parallel = g_J J_\parallel \approx 1.77$ are reasonably close to experimental values 4.27 and 1.79, respectively, for $Yb_2Ti_2O_7$ [31], with $g_J = 8/7$ being the Landé factor. This sharply contrasts to the non-Kramers magnetic doublet of Pr^{3+} ions, where $\hat{\sigma}^z_{\boldsymbol{r}}$ only contributes to the magnetic dipole moment and $\hat{\sigma}^\pm_{\boldsymbol{r}}$ to the magnetic quadrupole moment [11, 12].

3. Magnetic dipole interaction

The magnetic dipole interaction takes the form

$$\hat{H}_{\mathrm{D}} = \frac{\mu_0}{4\pi} \sum_{\langle \boldsymbol{r},\boldsymbol{r}'\rangle} \left[\frac{\hat{\boldsymbol{m}}_{\boldsymbol{r}} \cdot \hat{\boldsymbol{m}}_{\boldsymbol{r}'}}{(\Delta r)^3} - 3 \frac{(\hat{\boldsymbol{m}}_{\boldsymbol{r}} \cdot \Delta \boldsymbol{r})(\Delta \boldsymbol{r} \cdot \hat{\boldsymbol{m}}_{\boldsymbol{r}'})}{(\Delta r)^5} \right], \tag{6}$$

with $\Delta\boldsymbol{r} = \boldsymbol{r} - \boldsymbol{r}'$ and the summation $\sum_{\langle \boldsymbol{r},\boldsymbol{r}'\rangle}$ over all the pairs of atomic sites, where the localized magnetic moment is given by Eq. (5).

4. Superexchange interaction

To derive an anisotropic superexchange interaction between Yb magnetic moments, it is convenient to take the hole picture, where the local ground state is described by a single $4f$ hole at each Yb site and no hole at O stites. The local Coulomb repulsion among Yb $4f$ holes and the energy level Δ of the O $2p$ holes compared with that of a Yb $4f$ hole provide the largest energy scales of the problem, as schematically shown in Fig. 1 (a). In fact, the local ground state is modified by the nonlocal electron transfer. The amplitude is described by two Slater-Koster parameters $V_{pf\pi}$ and $V_{pf\sigma}$, which represent the transfer integrals between p_x/p_y and $f_{x(5z^2-r^2)}/f_{y(5z^2-r^2)}$ orbitals and between p_z and $f_{(5z^2-3r^2)z}$ orbitals, respectively [33]. Then, we perform the fourth-order strong-coupling perturbation expansion of the f-p electron transfer, by fully taking into account the virtual electron transfer processes among O $2p$ orbitals located at

Journal of Physics: Conference Series **320** (2011) 012065 doi:10.1088/1742-6596/320/1/012065

the center of the tetrahedron and $4f$ orbitals located at adjacent Yb sites, which are shown in Fig. 1 (b), in the local coordinate frames. The superexchange Hamiltonian is then obtained as

$$\hat{H}_{ff} = \frac{2}{\Delta^2} \sum_{\langle \boldsymbol{r},\boldsymbol{r}'\rangle}^{\text{n.n.}} \sum_{m_1,m_2} \sum_{m_1',m_2'} \sum_{\sigma_1,\sigma_2} \sum_{\sigma_1',\sigma_2'} V_{m_1} V_{m_1'} V_{m_2} V_{m_2'} \hat{f}^{\dagger}{}_{\boldsymbol{r},m_1,\sigma_1} \hat{f}_{\boldsymbol{r},m_2,\sigma_2} \hat{f}^{\dagger}{}_{\boldsymbol{r}',m_1',\sigma_1'} \hat{f}_{\boldsymbol{r}',m_2',\sigma_2'}$$
$$\times (\frac{1}{\Delta} + \frac{1}{U}) \left(R^{\dagger}_{\boldsymbol{r}} R_{\boldsymbol{r}'}\right)_{m_1,m_2';\sigma_1,\sigma_2'} \left(R^{\dagger}_{\boldsymbol{r}'} R_{\boldsymbol{r}}\right)_{m_1',m_2;\sigma_1',\sigma_2} . \quad (7)$$

with $V_{\pm 1} = V_{pf\pi}$ and $V_0 = V_{pf\sigma}$ as well as the rotation operator $R_{\boldsymbol{r}}$ of the angular momentum from the global coordinates to the local ones at the Yb site $\boldsymbol{r}$, which is given in Appendix A. Here, $\hat{f}^{\dagger}{}_{\boldsymbol{r},m,\sigma}$ and $\hat{f}_{\boldsymbol{r},m,\sigma}$ represent the creation (annihilation) and annihilation (creation) operators of $4f$ electron (hole) with the orbital m and the spin σ at the Yb site $\boldsymbol{r}$. $R_{\boldsymbol{r}}$ is the rotation matrix of the total angular momentum from the glocal coordinates to the local, as given in Appendix A. Projecting Eq. (7) onto the space of the doublets, we obtain

$$\begin{aligned}\hat{H}_{ex} &= -J_{n.n.} \sum_{\langle \boldsymbol{r},\boldsymbol{r}'\rangle}^{n.n.} \Big[g^{\parallel} \hat{\sigma}^z_{\boldsymbol{r}} \hat{\sigma}^z_{\boldsymbol{r}'} + g^{\perp} \left(\hat{\sigma}^x_{\boldsymbol{r}} \hat{\sigma}^x_{\boldsymbol{r}'} + \hat{\sigma}^y_{\boldsymbol{r}} \hat{\sigma}^y_{\boldsymbol{r}'} \right) \\ &\quad + g^q \left(\left(\hat{\vec{\sigma}}_{\boldsymbol{r}} \cdot \vec{n}_{\boldsymbol{r},\boldsymbol{r}'} \right) \left(\hat{\vec{\sigma}}_{\boldsymbol{r}'} \cdot \vec{n}_{\boldsymbol{r},\boldsymbol{r}'} \right) - \left(\hat{\vec{\sigma}}_{\boldsymbol{r}} \cdot \vec{n}'_{\boldsymbol{r},\boldsymbol{r}'} \right) \left(\hat{\vec{\sigma}}_{\boldsymbol{r}'} \cdot \vec{n}'_{\boldsymbol{r},\boldsymbol{r}'} \right) \right) \\ &\quad + g^K \left(\hat{\sigma}^z_{\boldsymbol{r}} \left(\hat{\vec{\sigma}}_{\boldsymbol{r}'} \cdot \vec{n}_{\boldsymbol{r},\boldsymbol{r}'} \right) + \left(\hat{\vec{\sigma}}_{\boldsymbol{r}} \cdot \vec{n}_{\boldsymbol{r},\boldsymbol{r}'} \right) \hat{\sigma}^z_{\boldsymbol{r}'} \right) \Big] , \end{aligned} \quad (8)$$

with $J_{n.n.} = \frac{(2\beta V_{pf\sigma})^4}{3^3 (7\Delta)^2} \left(\frac{1}{U} + \frac{1}{\Delta} \right)$ and the dimensionless coupling constants as functions of $x = V_{pf\pi}/V_{pf\sigma}$,

$$g^{\parallel} = 1 - 8\sqrt{6}x - \frac{9}{2}x^2 - 3\sqrt{6}x^3 + \frac{63}{16}x^4, \quad (9)$$

$$g^{\perp} = 1 + 4\sqrt{6}x + \frac{45}{2}x^2 - 3\sqrt{6}x^3 + \frac{9}{16}x^4, \quad (10)$$

$$g^q = -2\left(1 - 2\sqrt{6}x + 9x^2 - 3\sqrt{6}x^3 + \frac{9}{4}x^4\right), \quad (11)$$

$$g^K = 2\sqrt{2}\left(1 + \sqrt{6}x - \frac{45}{4}x^2 + \frac{15}{4}\sqrt{6}x^3 - \frac{9}{8}x^4\right). \quad (12)$$

Here, we have introduced $\hat{\vec{\sigma}}_{\boldsymbol{r}} = (\hat{\sigma}^x_{\boldsymbol{r}}, \hat{\sigma}^y_{\boldsymbol{r}})$ as well as two unit vectors $\vec{n}_{\boldsymbol{r},\boldsymbol{r}'} = (\cos\phi_{\boldsymbol{r},\boldsymbol{r}'}, -\sin\phi_{\boldsymbol{r},\boldsymbol{r}'})$ and $\vec{n}'_{\boldsymbol{r},\boldsymbol{r}'} = (\sin\phi_{\boldsymbol{r},\boldsymbol{r}'}, \cos\phi_{\boldsymbol{r},\boldsymbol{r}'})$. In the choice of the local coordinate frames given in Appendix A, the phase takes $\phi_{\boldsymbol{r},\boldsymbol{r}'} = -2\pi/3$, $2\pi/3$, and 0, when the pair of Yb sites $\boldsymbol{r}$ and $\boldsymbol{r}'$ corresponds to that of $\boldsymbol{R} + \boldsymbol{a}_i$ and $\boldsymbol{R} + \boldsymbol{a}_j$ with $(i,j) = \{(0,1),(2,3)\}$, $\{(0,2),(1,3)\}$, and $\{(0,3),(1,2)\}$, respectively, where $\boldsymbol{a}_0 = -\frac{a}{8}(1,1,1)$, $\boldsymbol{a}_1 = \frac{a}{8}(-1,1,1)$, $\boldsymbol{a}_2 = \frac{a}{8}(1,-1,1)$, and $\boldsymbol{a}_3 = \frac{a}{8}(1,1,-1)$ with the lattice constant a and an fcc lattice vector $\boldsymbol{R}$.

Previously, this form of the nearest-neighbor Hamiltonian with and without the g^K term is found from the symmetry analysis for the Kramers doublets [12] and microscopically for the non-Kramers doublets [11, 12], respectively. The present microscopic calculation of the exchange coupling constants for the Kramers doublets of Yb^{3+} ions remarkably reveals that the symmetric exchange part is ferroic in terms of pseudospins, i.e., "in" or "out", and antiferromagnetic in terms of real magnetic moments, unlike the case of Pr moments [11, 12], and that the four coupling constants are of the same order in the magnitude with their ratios controlled by a single parameter $V_{pf\pi}/V_{pf\sigma}$.

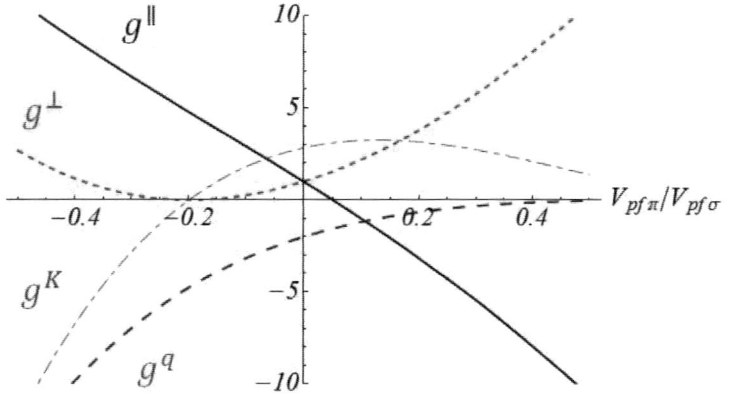

Figure 2. Dimensionless coupling constants as functions of $V_{pf\pi}/V_{pf\sigma}$.

5. Summary

We have obtained a highly anisotropic effective pseudospin-1/2 Hamiltonian for interacting Yb Kramers doublets of $Yb_2TM_2O_7$, which includes four adjustable parameters; α and γ describing crystal-field doublets, $J_{n.n.}$ for the magnitude of whole superexchange terms, and $V_{pf\pi}/V_{pf\sigma}$ specifying the ratios of aniotropic exchange couplings. Theoretical analyses of the model in comparison with experiments will be published elsewhere.

Acknowledgments

The author thanks M. J. P. Gingras for a critical reading of the manuscript. The work was partly supported by Grants-in-aid for Scientific Research under Grant No. 19052006 from the Ministry of Education, Culture, Sports, Science, and Technology of Japan and No. 21740275 from Japan Society for the Promotion of Science.

Appendix A. Local coordinate frames

The local coordinate frames are chosen as

$$\boldsymbol{x}_0 = -\frac{1}{\sqrt{6}}(1,1,-2), \quad \boldsymbol{y}_0 = -\frac{1}{\sqrt{2}}(-1,1,0), \quad \boldsymbol{z}_0 = \frac{1}{\sqrt{3}}(1,1,1), \tag{A.1}$$

$$\boldsymbol{x}_1 = -\frac{1}{\sqrt{6}}(1,-1,2), \quad \boldsymbol{y}_1 = -\frac{1}{\sqrt{2}}(-1,-1,0), \quad \boldsymbol{z}_1 = \frac{1}{\sqrt{3}}(1,-1,-1), \tag{A.2}$$

$$\boldsymbol{x}_2 = -\frac{1}{\sqrt{6}}(-1,1,2), \quad \boldsymbol{y}_2 = -\frac{1}{\sqrt{2}}(1,1,0), \quad \boldsymbol{z}_2 = \frac{1}{\sqrt{3}}(-1,1,-1), \tag{A.3}$$

$$\boldsymbol{x}_3 = -\frac{1}{\sqrt{6}}(-1,-1,-2), \quad \boldsymbol{y}_3 = -\frac{1}{\sqrt{2}}(1,-1,0), \quad \boldsymbol{z}_3 = \frac{1}{\sqrt{3}}(-1,-1,1), \tag{A.4}$$

for the Yb sites at $\boldsymbol{R} + \boldsymbol{a}_i$ ($i = 0, \cdots, 3$). They are invariant under the twofold rotations about x, y, and z axes that include the center of the tetrahedron. In particular, all the local z axes given above point outwards from the center of the tetrahedron. The coordinate frame for the spins is always attached to that for the orbital space in each case. The rotation matrix of the total angular momentum $\boldsymbol{j} = \boldsymbol{l} + \boldsymbol{s}$ with the orbital $\boldsymbol{l}$ and the spin $\boldsymbol{s}$ of a single electron takes the form

$$\hat{R}_{\boldsymbol{r}} = \exp\left[-i\varphi_i \hat{j}^z\right] \exp\left[-i\vartheta_i \hat{j}^y\right] \exp\left[-i\pi \hat{j}^z\right]. \tag{A.5}$$

with $\varphi_0 = \pi/4$, $\vartheta_0 = \arccos\left(1/\sqrt{3}\right)$, $\varphi_1 = 3\pi/4$, $\vartheta_1 = -\pi + \arccos\left(1/\sqrt{3}\right)$, $\varphi_2 = -\pi/4$, $\vartheta_2 = -\pi + \arccos\left(1/\sqrt{3}\right)$, $\varphi_3 = -3\pi/4$, and $\vartheta_3 = \arccos\left(1/\sqrt{3}\right)$.

Appendix B. Single-hole f-electron states

The Kramers doublet given in Eq. (1) can be expressed in terms of single f-hole states $|l^z, s^z\rangle_h$ with the orbital l^z and the spin s^z using

$$\begin{aligned}
|J^z = \frac{7}{2}\sigma\rangle &= |-3\sigma, -\frac{\sigma}{2}\rangle_h, \\
|J^z = \frac{5}{2}\sigma\rangle &= \frac{1}{\sqrt{7}}|-3\sigma, \frac{\sigma}{2}\rangle_h + \sqrt{\frac{6}{7}}|-2\sigma, -\frac{\sigma}{2}\rangle_h, \\
|J^z = \frac{\sigma}{2}\rangle &= \sqrt{\frac{3}{7}}|-\sigma, \frac{\sigma}{2}\rangle_h + \frac{2}{\sqrt{7}}|0, -\frac{\sigma}{2}\rangle_h.
\end{aligned} \quad \text{(B.1)}$$

References

[1] Anderson P W 1956 *Phys. Rev.* **102** 1008
[2] Anderson P W 1973 *Mater. Res. Bull.* **8** 153
[3] Wen X G, Wilczek F and Zee A 1989 *Phys. Rev.* B **39** 11413
[4] R. Moessner and J. T. Chalker, *Phys. Rev. Lett.* **80**, 2929 (1998).
[5] Isakov S V, Gregor K, Moessner R and Sondhi S L 2004 *Phys. Rev. Lett.* **93** 167204
[6] Gardner J S, Gingras M J P and Greedan J E, *Rev. Mod. Phys.* **82** 53
[7] Harris M J, Bramwell S T, McMorrow D F, Zeiske T and Godfrey K W 1997 *Phys. Rev. Lett.* **79** 2554
[8] Ramirez A P *et al.* 1999 *Nature (London)* **399** 333
[9] Bramwell S T and Gingras M J P 2001 *Science* **294** 1495
[10] Molavian H R, Gingras M J P and Canals B 2007 *Phys. Rev. Lett.* **98** 157204
[11] Onoda S and Tanaka Y 2010 *Phys. Rev. Lett.* **105** 047201
[12] Onoda S and Tanaka Y 2010 *Preprint* arXiv:1011.4981
[13] Gardner J S *et al.* 1999 *Phys. Rev. Lett.* **82** 1012
[14] Gardner J S, Gaulin B D, Berlinsky A J, Waldron P, Dunsiger S R, Raju N P and Greedan J E 2001 *Phys. Rev.* B **64** 224416
[15] Mirebeau I, Bonville P and Hennion M 2007 *Phys. Rev.* B **76** 184436
[16] Matsuhira K *et al.* 2002 *J. Phys. Soc. Jpn.* **71** 1576
[17] Zhou H D, Wiebe C R, Janik J A, Balicas L, Yo Y J, Qiu Y, Copley J R D and Gardner J S 2008 *Phys. Rev. Lett.* **101** 227204
[18] Nakatsuji S *et al.* 2006 *Phys. Rev. Lett.* **96** 087204
[19] Machida Y, Nakatsuji S, Onoda S, Tayama T and Sakakibara T 2010 *Nature (London)* **463** 210
[20] Cao H, Gukasov A, Mirebeau I, Bonville P, Decorse C and Dhalenne G 2009 *Phys. Rev. Lett.* **103** 056402
[21] den Hertog B C and Gingras M J P 2000 *Phys. Rev. Lett.* **84** 3430
[22] Pauling L 1938 *The Nature of the Chemical Bonds* (Ithaca: Cornel University Press)
[23] Matsuhira K *et al.* 2009 *J. Phys.: Conf. Series* **145** 012031
[24] Nagaosa N, Sinova J, Onoda S, MacDonald A H and Ong N P 2010 *Rev. Mod. Phys.* **82** 1539
[25] Blöte W J and Wielinga R F and Huiskamp W J 1969 *Physica (Amsterdam)* **43** 549
[26] Yasui Y *et al.* 2003 *J. Phys. Soc. Jpn.* **72** 3014
[27] Hodges J A *et al.* 2002 *Phys. Rev. Lett.* **88** 077204
[28] Gardner J, Ehlers G, Rosov N, Erwin R W and Pertovic C 2004 *Phys. Rev.* B **70** 180404(R)
[29] Ross K A, Ruff J P C, Adams C P, Gardner J S, Dabkowska H A, Qiu Y, Copley J R D and Gaulin B D 2009 *Phys. Rev. Lett.* **103** 227202.
[30] Thompson J D, McClarty P A, Ronnow H M, Regnault L P, Sorge A and M J P Gingras 2010 *Preprint* arXiv:1010.5476.
[31] Hodges J A, Bonville P, Forget A, Rams M, Królas K and Dhalenne G 2001 *J. Phys.: Condens. Matter* **13** 9301
[32] Similar CEF parameters have been obtained in Ref. [20].
[33] Sharma R R, Phys. Rev. B **19**, 2813 (1979).

Bond Orders in a Frustrated Itinerant System on a Kagome Lattice

Tsutomu Hasegawa, Nobuo Furukawa

Department of Physics and Mathematics, Aoyama Gakuin University, 5-10-1 Fuchinobe, Chuo-ku, Sagamihara-shi, Kanagawa 252-5258, Japan

E-mail: `furukawa@phys.aoyama.ac.jp`

Abstract. Extended double-exchange model on a Kagomé lattice is investigated numerically. We introduce a new Monte Carlo algorithm to perform finite-temperature calculation in the strong coupling limit. Carrier-induced hexamer formation, where electron hoppings are made along hexagonal bond clusters, is found at certain ranges of carrier densities. As a mechanism, we discuss the order-by-disorder phenomena that the macroscopic degeneracy of the geometrically frustrated spin system is lifted by the kinetic energy of itinerant electrons.

1. Introduction

Systems with macroscopic degeneracies are expected to exhibit various thermodynamical anomalies through residual perturbative interactions, which may lead to exotic phenomena. One of such example is the extended double-exchange model defined on a pyrochlore lattice [1, 2, 3]. The model has been introduced to investigate ferromagnetic to spin-glass transitions and metal-insulator transitions in Mo pyrochlore oxides $R_2Mo_2O_7$ [4, 5], where such phase transitions are driven by interplay between magnetism and transports in these strongly correlated electron systems.

More precisely, the model is given by

$$\mathcal{H} = \mathcal{H}_{\rm AF} + \mathcal{H}_{\rm DE}, \tag{1}$$

where the first term gives the antiferromagnetic interaction between localized spins, while the second term gives the ferromagnetic double-exchange interaction [6]. In $\mathcal{H}_{\rm DE}$, itinerant electrons are strongly coupled to localized spins. Thus the system is essentially in the strong coupling regime, and the magnetism and transports colossally affects each other.

Although many interesting properties are observed through numerical calculations, the system size is limited by numerical complexities of the system. In this paper, we introduce a two-dimensional analogue of the model to simplify the situation. Furthermore, when we take both strong coupling and Ising limits, it is shown that the the numerical complexities is much reduced by introducing a new algorithm for Monte Carlo calculations. This enables us to perform finite temperature calculations for larger-sized systems in detail. Through such calculations, we show some thermodynamical properties of the model, including the carrier-induced bond orders.

2. Model

We introduce the extended double-exchange model on a Kagomé lattice in Eq. (1), where frustrated spin system are interacting with itinerant electrons, in detail. In the present study, the first term represents an anteferromagnetic Ising model defined on a Kagomé lattice,

$$\mathcal{H}_{\rm AF} = J \sum_{<i,j>} S_i S_j, \tag{2}$$

where $J > 0$ and $S_i = \pm 1$. Summation is taken over nearest neighbor bonds. This model is one of the most canonical geometrically frustrated systems, and its ground state has macroscopic degeneracies. The ground state has no magnetic orders, but satisfies the so-called ice-rule. The ice-rule on a Kagomé lattice is that in all unit cell triangles, three spins take the configuration of either "2 up 1 down" or "1 up 2 down".

The second term of the Hamiltonian (1) gives a double-exchange model [6] where conduction electrons interact with localized spin through Hund's rule coupling as

$$\mathcal{H}_{\rm DE} = -t \sum_{<i,j>,\sigma} c^\dagger_{i,\sigma} c_{j,\sigma} + h.c. \; - J_H \sum_i \sigma^z_i S_i - \mu \sum_{i,\sigma} c^\dagger_{i,\sigma} c_{i,\sigma}. \tag{3}$$

Here $c^\dagger_{i,\sigma}$ ($c_{j,\sigma}$) represents creation (annihilation) operator for conduction electrons, while $\sigma^z_i = c^\dagger_{i\uparrow} c_{j\uparrow} - c^\dagger_{i\downarrow} c_{j\downarrow}$ is the z-component of the spin of the conduction electron at i-th site. Electron hopping integrals are expressed by t, while J_H is the Hund's coupling between local spins and conduction electrons. Chemical potential is given by μ. Note that, since the localized spins are Ising spins, we have only z-component of the Hund's rule coupling.

We now restrict ourselves to the case of strong Hund's rule coupling limit $J_H/t \to \infty$. Then, we may energetically project out the states where directions of S_i and σ^z_i are antiparallel to each other [7]. Therefore, in the present case of Ising spins, electron hoppings are allowed only within ferromagnetic clusters, *i.e.*, sites connected by bonds with ferromagnetic spin configurations. This is sometimes referred to as the "go/no go" rule. In order to gain of kinetic energies, ferromagnetic configuration of local spins are favored.

The effective Hamiltonian can be described as

$$\mathcal{H}'_{\rm DE} = \sum_{\rm Cluster} \left\{ -t \sum_{<i_n,j_n>} \tilde{c}^\dagger_{i_n} \tilde{c}_{j_n} + h.c. \right\} - \mu \sum_i \tilde{c}^\dagger_i \tilde{c}_i. \tag{4}$$

Here, the first summation is taken with respect to ferromagnetic clusters. The second summation is taken over neighboring sites on n-th ferromagnetic clusters. Here, $\tilde{c}^\dagger_i$ ($\tilde{c}_j$) represents creation (annihilation) operator for electrons with spins parallel to local S_i. Although the Hamiltonian (4) may effectively be regarded as a spinless model, local spin quantization axes depend on S_i and are different from site to site.

In this model (1), the double-exchange ferromagnetic interactions compete with the antiferromagnetic exchange interactions J. As J is increased, ferromagnetic metal phase become unstable. However, due to frustrated behavior of the system, the transition toward antiferromagnetism is also suppressed. Then, it is expected that various instabilities in spin as well as charge sectors may occur, as in the case of the pyrochlore systems where magnetic phase competitions accompany metal-insulator as well as metal-metal transitions [3].

Let us also note that the range of ferromagnetic interactions are not limited to nearest neighbors, due to the extended nature of the wave functions for itinerant electrons which can gain kinetic energy within the ferromagnetic clusters. Then, when the long-range part of the double-exchange interactions are introduce perturbatively to the frustrated antiferromagnetic systems with nearest neighbor interactions, they may act as residual interactions to lift macroscopic degeneracies and stabilize exotic phases.

3. New Monte Carlo Algorithm

In the Monte Carlo calculation of the model, the Boltzmann weight of each spin configuration is calculated by a sum of the kinetic energy part and the Ising spin exchange part. While the latter is easily obtained, the calculation of the former part needs quantum mechanical evaluation of the electronic Hamiltonian. In most cases, matrix diagonalization is performed for each Monte Carlo updates. The numerical complexity increases as $O(N^4)$, where N is the system size. Moreover, in frustrated systems in general, typical temperature scale to observe interesting phenomena is much smaller and the typical relaxation time scale for thermal equilibrium is longer, compared to systems without frustrations. These give us further difficulties for detailed investigation of the system.

Here, in some limited parameter spaces, we show that the numerical complexity is much reduced by introducing a new algorithm. We restrict ourselves to the strong exchange coupling region $J \gg t$. At low temperatures, we may only consider the macroscopically degenerate ground state of $\mathcal{H}_{AF}$, which have spin configurations obeying the ice-rule. In this case, double-exchange hopping is allowed only on one out of three bonds on a unit triangle which has a ferromagnetic spin configuration, since "2 up 1 down" or "1 up 2 down" rule is satisfied in the ground state (See Fig. 1). Between neighboring unit triangles, such bonds may be connected to form longer chain of bonds. However, they do not branch off into multiple chains, since multiple ferromagnetic bonds on a unit triangle are prohibited by the ice-rule. Then, such ferromagnetic bonds form simple one-dimensional clusters of spins; clusters are classified as "open chains" when they have open ends, while the closed ones are referred to as "rings". Electronic Hamiltonian can analytically be solved for each finite-sized chain/ring; electronic eigenstates depends only on the size and the boundary of the cluster but not on its shape. Total electronic energies can be calculated by counting the length of each open chains/rings, without matrix diagonalization. Labeling of the connected clusters are made using the Hoshen-Kopelman algorithm for percolation systems [8].

As a conclusion, Boltzmann weights for a given spin configuration can easily be calculated by this method, if the system is in the strongly correlated limit $(J, J_H \gg t, T)$ where antiparallel-spin electrons are completely projected out and the configurations of local spins are within the manifold of macroscopically degenerate ground state. The numerical complexity scales as $O(N)$ to $O(N\sqrt{N})$, depending on the average length of the one-dimensional clusters. This allows us to perform calculations on larger-sized systems. In typical cases, maximum of the system size using the matrix diagonalization method is limited to $N \sim 10^2$ sites, while the new algorithm easily allows us to calculate systems with $N \sim 10^3$ sites.

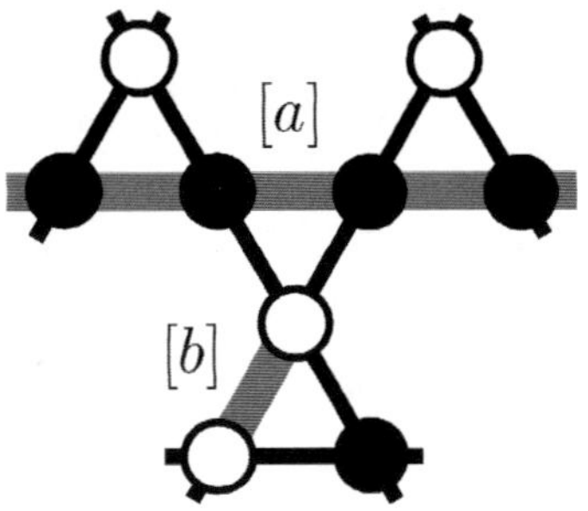

Figure 1. Ferromagnetic bonds at the ground state of the antiferromagnetic Ising model on a Kagomé lattice. Filled (open) circles represent sites with up (down) spins. Thick Ired) lines express bonds with ferromagnetic spin configurations. For the case of the bond [a], both ends of are connected to other bonds, while for bond [b] one of the end is terminated. Note that electrons are allowed to hop only along the thick bonds.

Journal of Physics: Conference Series **320** (2011) 012066 doi:10.1088/1742-6596/320/1/012066

4. Results

Here we show our results. We have performed Monte Carlo calculation using the replica exchange method [9], typically in the range $T/t = 0.010$, 0.015, 0.020, 0.025 and 0.030. We have run 100,000 steps of Monte Carlo measurements after 100,000 steps for thermalizations. In this paper, we focus on the low carrier concentration region $\mu/t = -1.85$. In this region, the carrier concentration is stabilized at $\rho \sim 0.1$. In Fig. 2, we show the spin correlation function at $T/t = 0.01$. We see typical peak features in our model. Since the largest part of the Hamiltonian, $\mathcal{H}_{\rm AF}$, should show spin liquid state without sharp peaks in spin correlations, these features are due to the gain of kinetic energies for itinerant electrons, creating possible order-by-disorder phenomena.

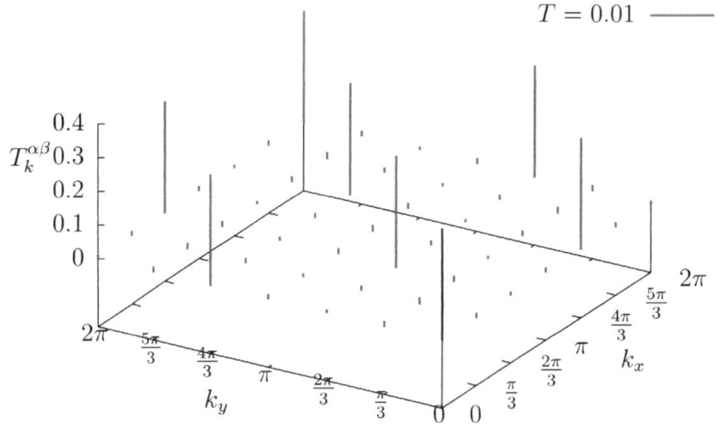

Figure 2. Spin correlation function at $T/t = 0.01$ and $\mu/t = -1.85$.

Let us discuss what types of ordering may stabilize the system. Since open chains have larger kinetic energies than rings of identical sizes, we should consider some ways to tile the lattice with rings of ferromagnetic clusters in periodic manners under the constraint of the ice-rule. Simplest candidates are (a) stripe order and (b) hexamer order, as depicted in Fig. 3. In the former case, one-dimensional stripes extend toward the boundaries of the system and form a semi-macroscopic sized rings. Contrarily, the latter case has rings with the minimum size allowed in this system (length=6). (Note that trimers on unit trialngles are prohibited by the ice-rule.) Through comparison with the numerical data, the position and the structure factor of the spin correlation function is consistent with the assumption of the hexamer formation.

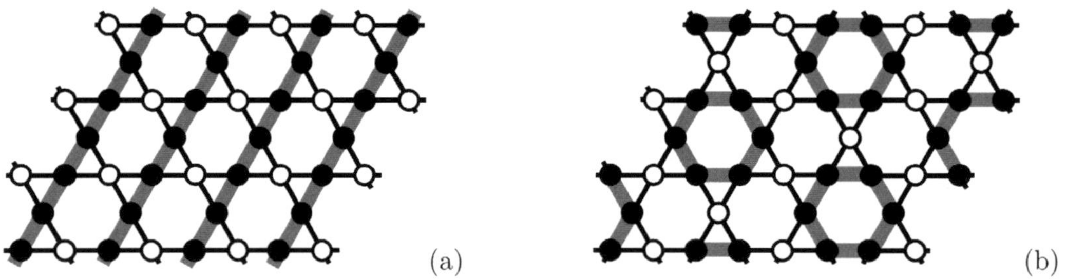

Figure 3. Possible bond orders without open boundary clusters under the restriction of ice-rule. These patterns are repeated periodically throughout the lattice. Thick bonds represent ferromagnetic bonds which are electron hopping paths. (a) Stripe order. (b) Hexamer order.

Journal of Physics: Conference Series **320** (2011) 012066 doi:10.1088/1742-6596/320/1/012066

The mechanism for hexamer order can be understood as follows. Let us first consider the density of states (DOS) of the stripe-ordered state in the thermodynamic limit. The order is constructed by one-dimensional hopping paths with parallel spins, and isolated sites with antiparallel spins. Then, they make a one-dimensional electronic DOS at $|\varepsilon| \leq 2t$ by stripes, plus a delta-functional DOS at the band center due to localized states, as shown in Fig. 4. In the low carrier concentration region, electrons fill the bottom of the DOS which has a square-root singularity $D(\varepsilon) \sim 1/\sqrt{4t^2 - \varepsilon^2}$.

On the other hand, in a thermodynamic limit of a ring-multimer order at the ground state, each ring consists of a finite-sized one-dimensional clusters with a periodic boundary condition so that DOS is discretized. Each multimer has the minimum energy eigenvalue of $\varepsilon = -2t$ irrespective of its size. Then, in the low carrier concentration region at $\mu \sim -2t$, the bottom part of the DOS, $D(\varepsilon) \sim \delta(\varepsilon + 2t)$, is occupied. Therefore, at low but finite carrier densities, electronic energies are pinned at $\varepsilon = -2t$ until the lowest level is completely occupied, whereas in the stripe ordered state the energies gradually increase as carrier is doped. Thus, multimer orders are more stabilized due to the gain of electronic energies in these regions, compared to the stripe orders.

Among possible ring-multimer orders within the ice-rule, hexamers are the smallest in the size so that they can tile the entire Kagome-lattice with a maximum number of rings. Then the number of the bottom energy levels with $\varepsilon = -2t$ is maximum. (Equivalently, the prefactor of the delta-functional DOS is maximum). Thus hexamers are mostly favored in the low carrier concentration regions, due to the energetics of the system. The DOS is also show in Fig. 4.

In the hexamer ordered state, prefactors of the delta-functional DOS are given by $N/9$ and $2N/9$ for $\varepsilon = \pm 2t$ and $\pm t$, respectively, since the number of hexamers N_{h} is given by $N_{\mathrm{h}} = N/9$. The delta-functional DOS at $\varepsilon = 0$ has the prefactor $N/3$ which the number of isolating sites. Note that a complete occupation of the lowest energy levels of hexamers occurs at $\rho = 1/9$. Our numerical data for carrier concentration are also consistent with the hexamer formation.

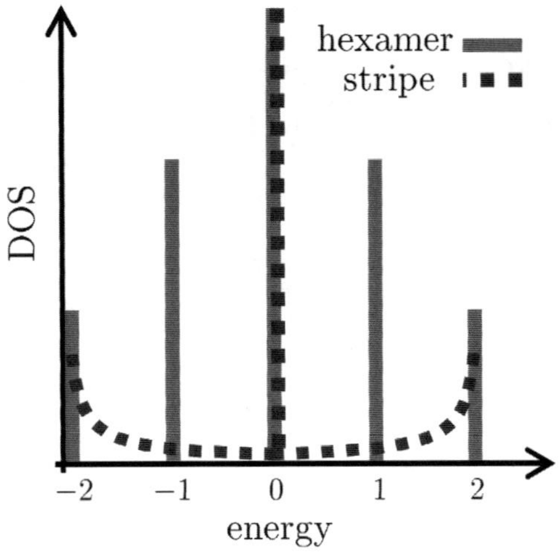

Figure 4. Electronic density of states for the stripe order and the hexamer order.

5. Summary

We have performed Monte-Carlo simulation for extended double-exchange model on a frustrated Kagomé lattice. A new algorithm has been introduced. In the strong coupling region, we have found the carrier induced bond orders. Thus, formation of the one-dimensional clusters not merely provides us a method for calculation but is an essential nature for this spin-charge coupled model. Such formation of bond orders may be regarded as the order by disorder process to lift the macroscopic degeneracies. In the present investigations, we have showed the presence of characteristic hexamer ordered state in the strong coupling and low carrier concentration region at low temperatures, but alternative bond ordering may appear in different parameter regions. Details will be reported elsewhere.

It is worthwhile to mention that the shape and the size of the rings in this bond ordering are determined by the geometry of the lattice. This is in sharp contrast with the weak-coupling electron systems where spin/bond orderes are determined by the Fermiology, *i.e.*, size and shape of the Fermi surfaces. This may be considered as a sign of the strongly-correlatedness of the system.

Let us finally note that such cluster ordering is not a unique feature in this model but can be seen ubiquitously as one of the mechanisms to release the residual entropies in low temperatures. Experimentally, various transition metal oxides with frustrated interaction exhibit spontaneous cluster formations, including trimers in $LiVO_2$ [10], hexamers in $ZnCr_2O_4$ [11], heptamers in AlV_2O_4 [12, 13], and octamers in $CuIr_2S_4$ [14, 15]. Cluster formations are also in geometrically frustrated double exchange systems defined on a Kagomé lattice with uniaxial anisotropies [16]. In this case, formation of 12-site clusters or dodecamers has been found at low temperatures. The mechanism of such dodecamer formation is similar to the present case. Namely, the macroscopic degeneracy of the frustrated system is lifted by residual double-exchange interactions. However, the way the residual entropy is released is different. We expect that there should be many other candidates which exhibit more exotic orderings and transport anomalies accompanied by them.

We acknowledge fruitful discussions with S. Aoyama, S. Miyahara and M. Sato. This work has been partially supported by Grants-in-Aid for Scientific Research (KAKENHI) Scientific Research (C) and a High-Tech Research Center project for private universities from the MEXT of Japan.

References

[1] Motome Y and Furukawa N 2010 *Phys. Rev. Lett.* **104** 106407
[2] Motome Y and Furukawa N 2010 *J. Phys.: Conf. Ser.* **200** 012131
[3] Motome Y and Furukawa N 2010 *Phys. Rev. B* **82** 060407(R)
[4] Greedan J E, Sato M, Ali N and Datars W R 1987 *J. Solid State Chem.* **68** 300
[5] Iguchi S, Hanasaki N, Kinuhara M, Takeshita N, Terakura C, Taguchi Y, Takagi H and Tokura Y 2009 *Phys. Rev. Lett.* **102** 136407
[6] Zener C 1951 *Phys. Rev.* **82** 403
[7] Anderson P W and Hasegawa H 1955 *Phys. Rev.* **100** 675
[8] Hoshen J and Kopelman R 1976 *Phys. Rev.* **14** 3438
[9] Hukushima K and Nemoto K 1996 *J. Phys. Soc. Jpn.* **65** 1604
[10] Pen H F, van den Brink J, Khomskii D I and Sawatzky G A 1997 *Phys. Rev. Lett.* **78** 1323
[11] Lee S H, Broholm C, Ratcliff W, Gasparovic G, Kim T H, Huang Q and Cheong S W 2002 *Nature* **418** 856
[12] Horibe Y, Shingu M, Kurushima K, Ishibashi H, Ikeda N, Kato K, Motome Y, Furukawa N, Mori S and Katsufuji T 2006 *Phys. Rev. Lett.* **96** 086406
[13] Matsuda K Furukawa N and Motome Y 2006 *J. Phys. Soc. Jpn* **75** 124716
[14] Radaelli P G, Horibe Y, Gutmann M J, Ishibashi H, Chen C H, Ibberson R M, Koyama Y, Hor Y S, Kiryukhin V and Cheong S W 2002 *Nature* **416** 155
[15] Khomskii D I and Mizokawa T 2005 *Phys. Rev. Lett.* **94** 156402
[16] Shimomura Y Miyahara S and Furukawa N 2004 *J. Phys. Soc. Jpn.* **73** 1623

Journal of Physics: Conference Series **320** (2011) 012067 doi:10.1088/1742-6596/320/1/012067

Critical behavior in a Kondo-screening partially-ordered antiferromagnet CePdAl

Akira Oyamada[1], Tomomi Kaibuchi[1], Masahide Nishiyama[1], Tetsuaki Itou[1], Satoru Maegawa[1], Yosikazu Isikawa[2], Andreas Dönni[3], and Hideaki Kitazawa[3]

[1]Graduate School of Human and Environmental Studies, Kyoto University, Kyoto 606-8501, JAPAN

[2]Graduate School of Science and Engineering, University of Toyama, Toyama 930-8555, JAPAN

[3]National Institute for Materials Science, Tsukuba 305-0047, JAPAN

E-mail: oyamada.akira.2s@kyoto-u.ac.jp

Abstract. A rare-earth compound CePdAl is a good example of the Kondo-screening partially-ordered antiferromagnet below $T_N = 2.7$ K. In this compound, the formation of the Kondo-singlet state is realized as a new route of the relief of the frustration. The spin dynamics of CePdAl are, however, still controversial. In order to further investigate the spin dynamics of CePdAl, we have performed ^{27}Al-NMR and speci c heat single crystal measurements and reanalysed published powder neutron di raction data. NMR intensities decrease with decreasing temperature in the ordered state suggesting the slow spin uctuations below $T_N = 2.7$ K, while the intensity of magnetic Bragg peak measured by neutron di raction increases monotonously in the same temperature range. The critical exponents derived from the neutron di raction and the speci c heat measurements are not coincident with those of known universality classes. An unconventional magnetic order parameter or a charge degree of freedom originated from the coexistence of frustration and the Kondo e ect may in uence the critical behavior.

1. Introduction

The interplay between the Kondo e ect and the geometrical frustration is one of the most attractive issues [1]. The competition between RKKY interactions and the Kondo e ect are often observed in f-electron compounds with the magnetic ions of Ce, Sm, Eu, Yb and U. The RKKY interactions are the origin of magnetic orderings, while the Kondo e ect tends to quench the magnetic moments and competes with the magnetic orderings. This competition results in a quantum critical point (QCP) at the boundary between a non-magnetic heavy-fermion state and a magnetically ordered state. Anomalous phenomena like non-Fermi-liquid state around QCP attract much attention. On the other hand, the geometrical frustration prevents the usual magnetic orderings and often cause a disordered state or a peculiar magnetic ordered state. Furthermore new universality class may be originated by the frustration, for example in XY stacked triangular antiferromagnet [2]. If the RKKY interactions, the Kondo e ect and the geometrical frustration coexist, a new quantum state may be originated.

A Kondo-screening partially-ordered state is one example of an attractive quantum state. A part of the magnetic moments are ordered below the transition temperature and the others

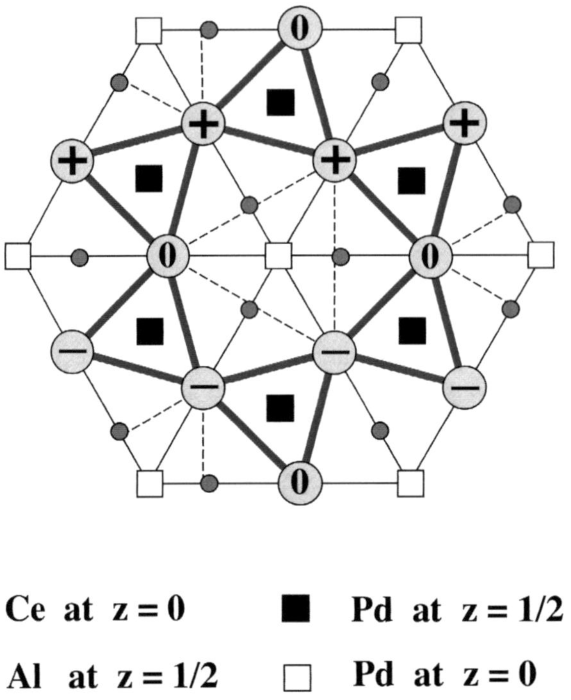

Figure 1. Projection onto the c-plane of the crystal and magnetic structure of CePdAl. The symbols +, and 0 show up spin, down spin and disordered spin, respectively.

are disordered. Although the partially-ordered state has been observed in localized spin system with the frustration and the Ising-type anisotropy [3], the partially-ordered state in Kondo lattice is a di erent quantum state. The disordered moments remained in the ordered state are not paramagnetic but in the Kondo singlet state.

A rare-earth compound CePdAl is a Kondo-screening partially-ordered antiferromagnet as shown in gure 1. In CePdAl, two thirds of magnetic moments are in the ordered state with sine wave structure along the c-axis below T_N = 2.7 K and a third of moments are disordered [4]. NMR measurements have shown that the disordered moments are in the Kondo singlet state [5]. This situation brings a new route of relief from the frustration to light. The frustration is relieved by the disappearance of one third of moments due to the formation of the Kondo singlet state and the partially-ordered state is stabilized down to 0 K.

The crystal structure of CePdAl is hexagonal ZrNiAl-type as shown in gure 1 and its space group is $P\bar{6}2m$ [6]. The magnetic Ce^{3+} ions are on the network of the regular triangles in the c-plane. The network is similar to the Kagome lattice, however the hexagons are distorted. The distance between the nearest neighbor Ce atoms are considerably shorter in the c-plane than in the c-axis, suggesting the strong two-dimensional magnetic correlations. The degeneracy of the total spin $J = 5/2$ of Ce^{3+} lifts to three doublets in the crystal eld. The ground state doublet is regarded as the e ective spin 1/2 with a strong Ising-type anisotropy [7].

Neutron di raction measurements of powder samples have revealed that the Ce moments order antiferromagnetically below 2.7 K (T_N) with a propagating vector k = [1/2, 0, τ], τ 0.35, as shown in Fig. 1 [4]. The magnetic structure is partially ordered in the c-

plane and incommensurate sine waves along the c-axis. The magnetic structure of this partially ordered state in CePdAl was explained by a two-dimensional mean eld approximation including the Kondo e ect [8], however, the spin dynamics of CePdAl are still controversial. The most remarkable feature of the spin dynamics is an anomalous temperature dependence of the NMR relaxation rate ($T_1^{\;1}$). The rate $T_1^{\;1}$ is proportional to $\exp(\delta/T)$ with $\delta = 2.7K$ between $T_{\rm N} = 2.7$ K and 50 K, which is similar to that of two dimensional Heisenberg system although CePdAl has strong Ising-type anisotropy [5].

In order to further investigate the spin dynamics of CePdAl, we have performed NMR and speci c heat single crystal measurements and reanalysed published powder neutron di raction data [4].

2. Experimental Procedures

Single crystals of CePdAl were prepared by Czochralski method. We performed ^{27}Al-NMR measurements of a single crystal using a coherent spectrometer. The magnetic eld is applied along the c-axis, that is one of the principal axes of the electric eld gradient tensor. The operating frequency was 18.78 MHz. Spin echo spectra were measured by a eld sweep.
Speci c heat measurements of a single crystal were performed using a thermal relaxation method between 2 and 10 K with Quantum Design PPMS.

3. Results and Discussions

Figure 2 shows the temperature variation of spin echo spectra of a single crystal above and below $T_{\rm N}$. The ve peaks of the spectra come from the rst order quadrupolar e ect for $I = 5/2$ of ^{27}Al nuclei. The intensity of the spectra starts to decrease above $T_{\rm N}$ signi cantly and disappears below 2.8 K without remarkable broadening.

Figure 3 shows the temperature dependence of the integrated spin echo intensity of a single crystal. The integrated intensity of powder samples are also plotted for comparison [14]. The e ects of the temperature and spin-spin relaxation time (T_2) have been corrected. The temperature dependence of T_2 is rather weak and has only a small increase around $T_{\rm N}$. The spectra of the powder samples at lowest temperature consists of seven peaks and is in the eld range of Fig.5 [5]. Thus the intensity is proportional to the number of detected nuclei. The integrated intensity considerably decreases below 3.5 K although no temperature dependence is expected. However, the intensity of powder samples recovers below 2 K while the intensity of a single crystal is completely lost below 2.5 K. Here we have to note that the compound is a metallic system. In general, the intensity decreases with decreasing of the resistivity due to the skin e ect. However, the resistivity decreases below $T_{\rm N}$ while the integrated intensity decreases below 3.5 K, that is signi cantly higher than $T_{\rm N}$ = 2.7 K. Furthermore since the change of the resistivity is small [15, 16], the skin e ect has only slight e ect on the intensity. Thus the anomalous decrease and the di erence between a single crystal and powder samples are intrinsic. This decrease of the intensity means the decrease of number of detected nuclei and can be explained as follows. The relaxation time T_2 in a part of the sample becomes much shorter than 10 μs and NMR can not detect the part of the sample. While T_2 is larger in the other part of the sample, NMR can detect the signal of the other part. This situation suggests that the slow spin uctuation with the order of MHz frequency remains below $T_{\rm N}$ inhomogeneously. There is no de nitive explanation of the slow spin uctuation for the moment. We will discuss the slow spin uctuation later.

On the other hand, the neutron di raction measurements showed that the magnetic Bragg peak intensity increases with decreasing temperature monotonously in the temperature range where the NMR intensity decreases signi cantly [4]. In order to investigate whether the temperature dependence of the magnetic Bragg peak intensity is conventional or not, we have reanalyzed the neutron di raction data [4] and estimated the critical exponent β. A sublattice

Journal of Physics: Conference Series **320** (2011) 012067 doi:10.1088/1742-6596/320/1/012067

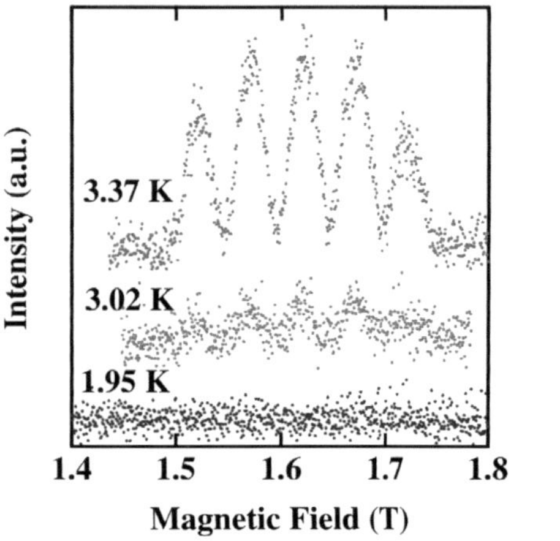

Figure 2. Temperature variation of spin echo spectra of a single crystal above and below T_N.

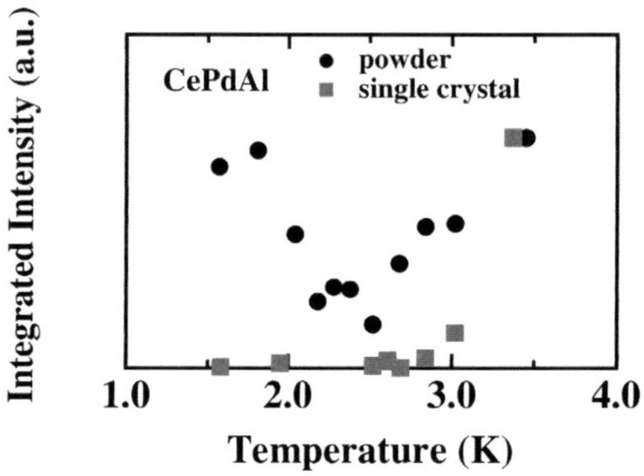

Figure 3. Temperature dependence of the integrated intensity of spin echo spectra. The temperature and T_2 effects are corrected.

magnetization or the Fourier component of the magnetic structure obeys the relation, $M \propto \epsilon^\beta$ below and close to T_N, where $\epsilon = (T_N - T)/T_N$. The Fourier component is proportional to the root of Bragg peak intensity. Figure 4 shows the Fourier component versus ϵ plot. The solid line shows the best fit of the experimental data to a power law with $T_N = 2.77$ K. The value of β is estimated to be 0.22 ± 0.02. There are two other reports on neutron diffraction experiments. One is using powder samples [9] and the other is using a single crystal [10]. Within the experimental error, our result is compatible with the β values estimated from these two experimental data. This value is considerably smaller than the usual values of three-dimensional(3D) Ising (0.33), 3D XY (0.35) and 3D Heisenberg model (0.37). Now we have to keep one point in mind. In CePdAl, the long range RKKY or dipolar interactions are important. Thus β may become close to the value derived from the mean field approximation or infinite range model, that is 0.5 [11, 12]. Therefore the smaller value obtained for β is quite anomalous.

Figure 5 shows the temperature dependence of the specific heat between 2 K and 10 K. A sharp peak was observed around 2.7 K that confirms the existence of the magnetic phase transition. The tail above 2.7 K shows that the development of the short range correlations is extending over wide temperature range. In order to estimate the critical exponent α and α', a single logarithmic plot of the specific heat against the $|\epsilon|$ is shown in figure 6. In this plot, convex and concave curves mean minus and positive signs of α, respectively. This plot shows that the critical exponent α above T_N and α' below T_N seem to be different although α should be equal to α' because of the scaling law. The solid lines show the best fits of the data with $T_N = 2.73$ K, $\alpha = 0.1 \pm 0.1$ and $\alpha' = -0.3 \pm 0.1$.

The estimated value of the β is close to the value of $Z_2 \times SO(2)$ order parameter (0.25) which is proposed in XY stacked triangular antiferromagnet [2], however α and α' are different from the theoretical value of 0.44. We compare our results with the metallic triangular-lattice Heisenberg antiferromagnet $PdCrO_2$ [13]. It was reported that in $PdCrO_2$ $\beta = 0.18$, $\alpha = 0.13$ and $\alpha' = 0$. The reason why $\alpha \neq \alpha'$ was suggested as follows. In the frustrated systems the critical behavior is extending to a rather large temperature range above T_N and it becomes difficult to distinguish between critical behavior and short range fluctuations. This meas that α above T_N was not estimated correctly. On the other hand, in the case of CePdAl the exchange interactions are thought to be dominated by long-range RKKY interactions. Therefore the true

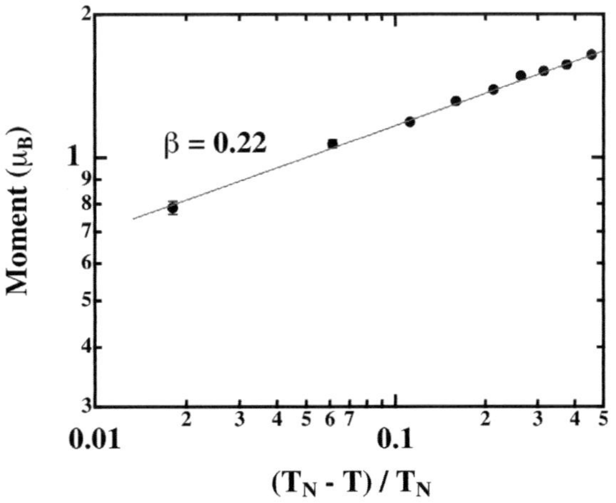

Figure 4. Fourier component of the magnetic ordered structure versus the reduced temperature $(T_{\mathbf{N}}\ \ T)/T_{\mathbf{N}}$. The solid line shows the best t of the experimental data to a power law with $T_{\mathrm{N}} = 2.77$ K and $\beta = 0.22$.

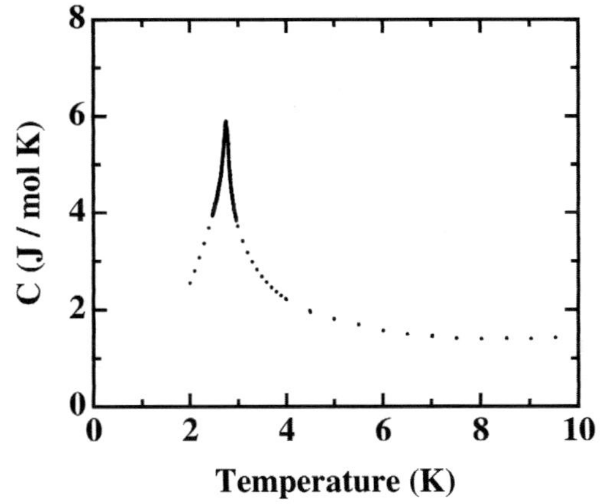

Figure 5. Temperature dependence of the speci c heat between 2 K and 10 K.

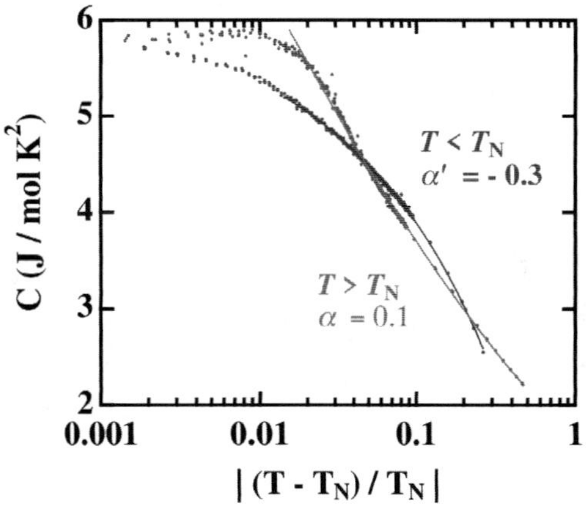

Figure 6. Speci c heat versus the reduced temperature plot.

critical region may be smaller than the measured ϵ region and the estimation of the critical exponents may be di cult although the similarity to $PdCrO_2$ is remarkable.

As previously mentioned, there is no de nitive explanation of the slow spin uctuation below T_{N} and the unconventional critical exponents. Here we note that the magnetic structure in CePdAl is the incommensurate sine wave along the c-axis stabilized by the Kondo e ect. If only the phase degree of freedom in the sine wave uctuates, the local eld which NMR observes uctuates but the amplitude of the Fourier component orders. This explains the decrease of NMR intensity and the monotonous development of the Bragg peak observed by neutron di raction measurements. Since the phase uctuations are accompanied by the exchange of the spin-up sites and the spin-down sites in the c-plane, the phase uctuations may be related with the frustration in the c-plane. If the defects in the crystal lock the phase uctuations, the di erence between powder samples and a single crystal may be explained as we expect the larger amount of defects in the powder samples. This scenario also seems to relate with the unconventional β and the temperature dependence of the k-vector. Similar decrease of the NMR intensity below T_N was observed in the two-dimensional Heisenberg triangular antiferromagnets $NaCrO_2$, $LiCrO_2$ and $NiGa_2S_4$ [17, 18, 19]. In those compounds, the possibility of the spin uctuations due to the excitations of Z_2 vortices was proposed. Since CePdAl has a strong Ising-type anisotropy and the Kondo e ect, the origin of the slow spin uctuation must be di erent from those materials. The interesting point is that the similar properties to those two-dimensional Heisenberg systems emerges in CePdAl due to the coexistence of the frustration and the Kondo e ect.

Recently, variational Monte-Carlo simulations of the Kondo lattice model on the triangular lattice with Ising-type anisotropy have been reported [20]. In this theory, the charge disproportionation occurs in the partially ordered state. This result pose an interesting question that how the charge degree of freedom is involved in the magnetic phase transition. In CePdAl, both spin and charge degrees of freedom may also play important roles in the anomalous critical behavior.

In summary, the critical behavior of CePdAl was investigated by ^{27}Al-NMR, neutron di raction and speci c heat measurements. The anomalous decrease of the NMR intensity was observed suggesting the slow spin uctuation below T_N. While the Bragg peak observed by neutron di raction measurements develops monotonously with decreasing temperature but with an unconventional β value. The slow spin uctuation and the unconventional critical exponents seems to be closely related with the coexistence of frustration and the Kondo e ect.

Acknowledgments
This study was supported by a Grant-in-Aid for Scienti c Research on Priority Areas "Novel States of Matter Induced by Frustration" (19052005).

References

[1] Ballou R, Lacroix C and NunezRegueiro M D 1991 *Phys. Rev. Lett.* **66** 1910
[2] Kawamura H 1986 *J. Phys. Soc. Jpn.* **55** 2095
[3] T. Takagi and M. Mekata 1993 *J. Phys. Soc. Jpn.* **62** 3943
[4] Dönni A, Ehlers G, Maletta H, Fischer P, Kitazawa H and Zolliker M 1996 *J. Phys.: Condens. Matter* **8** 11213
[5] Oyamada A, Nishiyama M, Maegawa S, Kitazawa H and Isikawa Y 2008 *Phys. Rev.* B **77** 064432
[6] Hulliger F 1995 *J Alloys Compounds* B **218** 44
[7] Isikawa Y, Mizushima T, Fukushima N, Kuwai T, Sakurai J and Kitazawa H 1996 *J. Phys. Soc. Jpn.* **65** Suppl.B 117
[8] Dolores M, Regueiro N, Lacroix C, Canals B 1997 *Physica* C **282-287** 1885
[9] Keller L, Dönni A, Kitazawa H and van den Brandt B 2002 *Appl. Phys.* A **74** S686
[10] Prokes K, Manuel P, Adroja D T, Kitazawa H, Goto T and Isikawa Y 2006 *Physica* B **385-386** 359
[11] Kotzler J 1976 *J. Phys.* C: Solid State Phys. **9** 1291
[12] Larkin A I and Khmelnitskii D E 1969 *Sov. Phys. -JETP* **29** 1123
[13] Takatsu H, Yoshizawa H, Yonezawa S and Maeno Y 2009 *Phys. Rev.* B **79** 104424
[14] Oyamada A, Kamioka K, Hashi K, Maegawa S, Goto T and Kitazawa H 1996 *J. Phys. Soc. Jpn.* **65** Suppl. B 123
[15] Kitazawa H, Matsushita A, Matsumoto T and Suzuki T 1994 *Physica* B **199-200** 28
[16] Goto T, Hane S, Umeo K, Takabatake T and Isikawa Y 2002 *J. Phys. Chem. Solids* **63** 1159
[17] Olariu A, Mendels P, Bert F, Ueland B G, Schi er P, Berger R F and Cava R J 2006 *Phys. Rev. Lett.* **97** 167203
[18] Alexander L K, Buttgen N, Nath R, Mahajan A V and Loidl A 2007 *Phys. Rev.* B **76** 064429
[19] Takeya T, Ishida K, Kitagawa K, Ihara Y, Onuma K, Maeno Y, Nambu Y, Nakatsuji S, MacLaughlin D E, Koda A and Kadono R 2008 *Phys. Rev.* B **77** 054429
[20] Motome Y, Nakamikawa K, Yamaji Y and Udagawa M 2010 *Phys. Rev. Lett.* **105** 036403

Journal of Physics: Conference Series **320** (2011) 012068 doi:10.1088/1742-6596/320/1/012068

Anomalous slow dynamics in the metallic helimagnet $Gd_{1\ x}Y_x$

T Yamazaki[1], Y Tabata[2], T Waki[2], H Nakamura[2]

[1]Institute for Solid State Physics, University of Tokyo, Chiba 277-8581, Japan
[2]Dept. of Materials Science and Engineering, Kyoto University, Kyoto 606-8501, Japan

E-mail: `t.yamazaki@issp.u-tokyo.ac.jp`

Abstract. AC-suseptibility measurements were performed in the metallic helimagnet $Gd_{1-x}Y_x$ alloy. A remarkable increase of the imaginary part of the ac-suseptibility was observed in the temperature range of the helimagnetic phase. Moreover, a strong nonlinearity of the magnetization was observed at paramagnetic-helimagnetic transition temperature T_N. On the other hand, These anomalous behavior were not observed in similar rare-earth helimagnets Ho and $Ho_{1-x}Y_x$. It strongly suggests that the weak magnetic anisotropy of the Gd-moments is responsible for the anomalous slow dynamics in the helimagnetic phase and for the strong nonlinearity of the magnetization at T_N. The slow dynamics may result from a chiral-domain motion, or a long-time variation of the period of the helimagnetic structure.

1. Introduction

The rare-earth alloy $Gd_{1\ x}Y_x$ exhibits interesting magnetic properties, such as the Lifshits transition between the long-period helical magnetic state and the ferromagnetic state in zero magnetic eld, and has been investigated extensively for decades [1, 2, 3, 4, 5]. The crystal structure of the $Gd_{1\ x}Y_x$ alloys is hexagonal-closed-pack (hcp) one. Many heavy rare-earth hcp metals and their yttrium alloys $R_{1\ x}Y_x$ exhibit long-period modulated structures,which can be explained by the geometry of their Fermi surfaces [6, 7]. A phase transition from a modulated magnetic structure to a collinear ferromagnetic one is generally found in the heavy rare-earth metals. Most of these transitions are actually rst order phase transitions accompanied by lattice distortions because of the strong spin-orbit interaction [8], which cause the strong magnetic anisotropy. Among of them, $Gd_{1\ x}Y_x$ alloys are important exceptions, where the spin-orbit interaction does not act because Gd^{3+}-ion has a zero orbital-angular moment ($L = 0$) [11]. And hence, a continuous transition from the modulated structure to the collinear ferromagnetic one can be realized in $Gd_{1\ x}Y_x$. An interesting second-and-a-half order transition, which may be a Lifshitz transition, was reported actually in $Gd_{1\ x}Y_x$ [5]. In the Y-concentration range of 0.32 $< x <$ 0.40, $Gd_{1\ x}Y_x$ exhibits a magnetic transition at T_N from the paramagnetic state to the proper-type helimagnetic state and a successive transition at T_C to the collinear ferromagnetic state. The propagation vector $\boldsymbol{q}$ parallel to the c*-direction continuously decreases to zero with decreasing temperature from T_N to T_C [3].

In this paper, we report an anomalous slow dynamics and a strong nonlinearity of magnetization newly discovered in $Gd_{1\ x}Y_x$. These anomalous phenomena most likely result from the very weak magnetic anisotropy of Gd-moments because they were not observed in the similar rare-earth helimagnets Ho and $Ho_{1\ x}Y_x$. We speculate two possible origins of the anomalous slow dynamics in the $Gd_{1\ x}Y_x$. The rst one is a collective spin dynamics with keeping its helical spin arrangements, which is a motion of the chiral-domains themselves. The second is a long-time variation of the period of the helimagnetic structure.

2. Experimental

Single crystalline samples of $Gd_{0.62}Y_{0.38}$ were grown by the Czochralski pulling method with tetra-arc furnace. The obtained crystals wrapped in the Ta lms and sealed in the evacuated quartz tubes and annealed for 1 week at 973 K, and then, quenched in cold water. Polycrystalline samples of Ho and $Ho_{0.60}Y_{0.40}$ prepared by arc melting with the mono-arc furnace were also used for the comparison study. The size of $Gd_{0.62}Y_{0.38}$ samples for the ac-susceptibility measurements along the a*- and c*-axes are 0.9 0.3 1.9 mm^3 and 0.4 0.2 1.9 mm^3 respectively, and the size of Ho and $Ho_{0.60}Y_{0.40}$ samples are 0.5 0.5 9.0 mm^3.

The ac-susceptibility measurements were performed by using the SQUID-magnetometer (Quantum Design, MPMS) equipped in the Research Center for Low Temperature and Material Sciences in Kyoto University. In the measurements, dc-magnetic- eld was also applied parallel to the ac-magnetic- eld.

3. Results and Discussion

Figures 1 (a) and (b) show temperature (T) dependences of the real and imaginary parts of the ac susceptibility, χ' and χ'', in $Gd_{0.62}Y_{0.38}$, respectively. In the gures, we show the data with applying the ac-magnetic eld along the a*- and c*-directions, being in and perpendicular to the helical plane. In these measurements, the magnitude and frequency of the ac-magnetic- eld h_{ac} and f were 3 Oe and 10 Hz, respectively. In the T-dependences of χ', a sharp cusp-type anomaly is found at 198 K, corresponding to the para-helimagnetic phase transition at T_N. A signature of the heli-ferromagnteic transition is found in χ' along the a*-direction, a rapid increase of χ', around 155 K. On the other hand, the increase of χ' along the c*-direction is more gentle. These are consistent with the fact previously reported, the ferromagnetic ordered moment lies around the c*-plane. The most striking feature we found in Fig. 1 is a remarkable increase of χ'' in the whole T-range of the helimagnetic phase. It indicates a presence of a slow dynamics in the helimagnetic phase. It should be noted that the increase of χ'' in the helical plane is much stronger than that perpendicular to the helical plane.

To obtain further information of the anomalous slow dynamics, we performed ac-susecptibility measurements along the a*-direction with various frequency in the range of 0.01 Hz $\leq f \leq$ 10

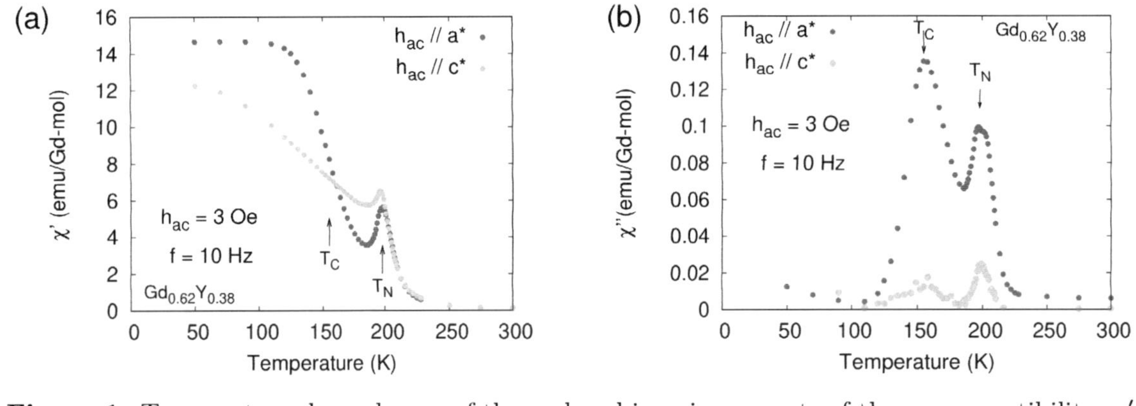

Figure 1. Temperature dependences of the real and imaginary parts of the ac-susceptibility, χ' and χ'', in $Gd_{0.62}Y_{0.38}$ alloy with applying the ac-magnetic- eld along the a*- and c*-directions.

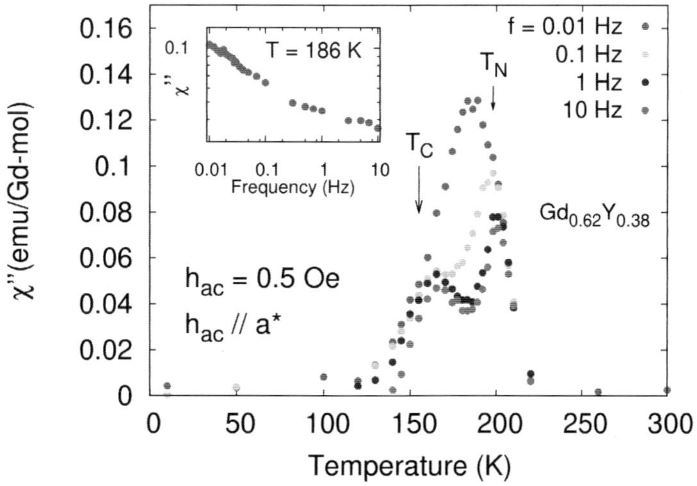

Figure 2. Temperature dependences of χ' of $Gd_{0.62}Y_{0.38}$ along the a*-direction, being in-plane direction, with various frequency. Inset: Frequency dependence of χ' at 186 K with a double logarithmic scale.

Hz, shown in Fig. 2. In this measurements, the smaller ac-magnetic- eld of 0.5 Oe was applied to avoid the nonlinear e ect appeared in the helimagnetic phase, which will be described later. The increase of χ'' becomes more remarkable with decreasing the frequency from 10 Hz to 0.01 Hz. At 0.01 Hz, a large broad peak of χ'' is found around 186 K. The results clearly indicate that the dynamics in the helimagnetic phase is extremely slow. The log-log plot of the frequency dependence of χ'' at 186 K is shown in the inset of Fig. 2. χ'' rapidly increases with decreasing the frequency down to 0.01 Hz, indicating that the characteristic frequency of the helimagnetic state is smaller than 0.01 Hz.

Figure 3 (a)-(d) show temperature dependences of χ' and χ'' of $Gd_{0.62}Y_{0.38}$ along the a* and c*-directions, respectively, with applying the dc-magnetic- eld of $0 \text{ Oe} \leq H \leq 100$ Oe. The magnitude and frequency of the ac-magnetic- eld were 3 Oe and 10 Hz, respectively. Along the a*-direction, χ' and χ'' are strongly suppressed by the dc-magnetic- eld in the helimagnetic state (see Fig.3 (a), (b)). The suppression of χ' along the c*-direction is much weaker than that along the a*-direction, while χ'' is suppressed strongly as well as that along the a*-direction is (see Figs. 3 (c) and (d)). The suppression of χ' implies a strong nonlinearity of the magnetization and the eld dependence of χ'' indicates that the anomalous slow dynamics in the helimagnetic state can be easily suppressed by magnetic eld. For the quantitation of the nonlinearity of the magnetization, we estimated the nonlinear susceptibility χ_2. If h_{ac} is much smaller than H, χ' can be treated as the di erential magnetization given by

$$\chi' \simeq \frac{dM}{dH}\bigg|_{H=H_{DC}} = \chi_0 + 3\chi_2 H^2 + 5\chi_4 H^4 + \quad . \tag{1}$$

Then, the nonlinear susceptibility χ_2 was obtained from the slant of the χ' vs H^2 plots in the limit of $H^2 \to 0$. In this analysis, demagnetization correction was operated carefully. The obtained χ_2 was shown in Fig.3 (e). We can nd a very strong negative peak of χ_2 at T_N along the a*-direction.

We performed the same ac-susceptibility measurements in the similar rare-earth helimagnets Ho and $Ho_{0.60}Y_{0.40}$ for comparison. These materials also exhibit para-helimagnetic transition

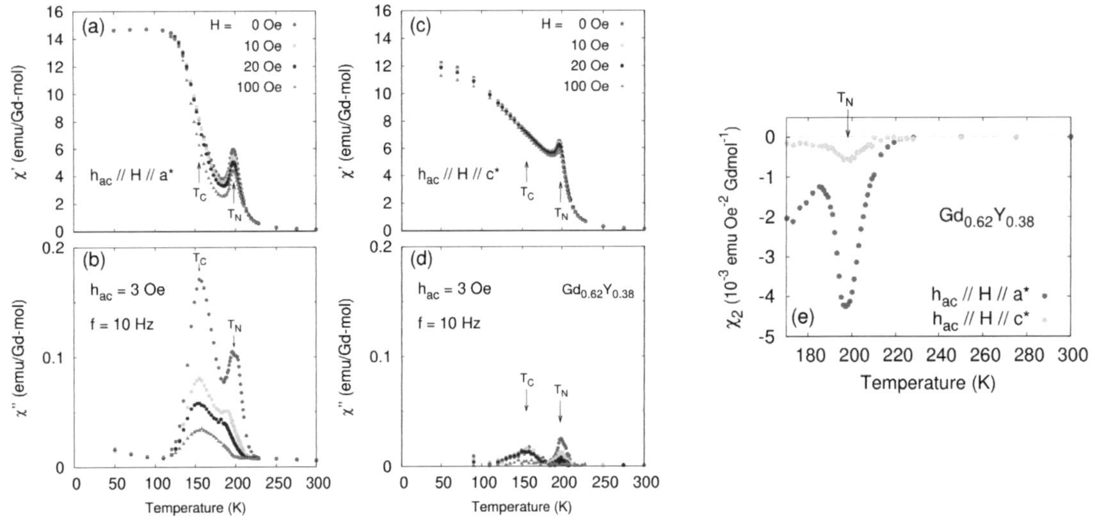

Figure 3. (a)-(d) Temperature dependences of χ' and χ'' of $Gd_{0.62}Y_{0.38}$ with applying the various dc-magnetic- eld along the a*-direction and c*-direction, respectively. (e) The temperature dependences of the nonlinear susceptibility χ_2 along the a*- and c*-directions.

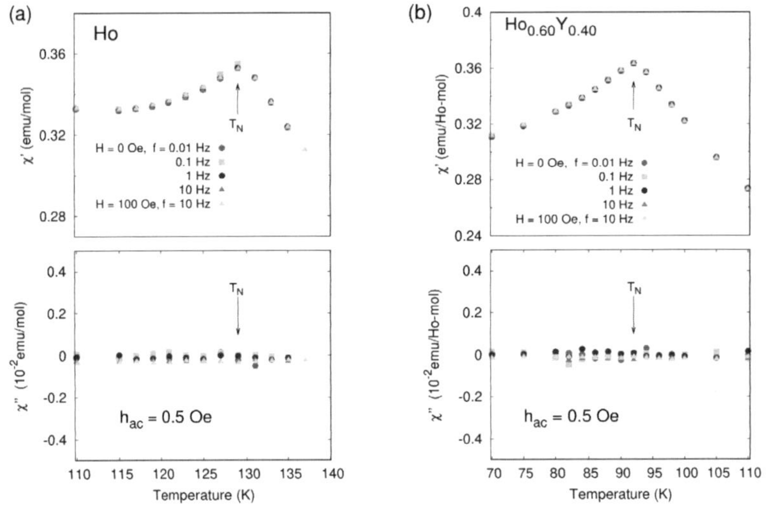

Figure 4. Temperature dependences of the real and imaginary parts of the ac-susceptibility, χ' and χ'', in (a) Ho and (b) $Ho_{0.60}Y_{0.40}$.

at $T_N = 130$ K and 92 K. Figure 4 shows temperature dependences of χ' and χ'' around T_N in polycrystalline Ho and $Ho_{0.60}Y_{0.40}$ with the frequency of 0.01 Hz $\leq f \leq$ 10 Hz. The magnitude of the ac-magnetic- eld was 0.5 Oe. To con rm the nonlinearity of the magnetization, the measurement with applying the dc-magnetic- eld of 100 Oe was also performed. No frequency dependence of χ' was observed in both materials. Correspondingly, an increase of χ'' was not observed at all in this frequency range. No dc-magnetic- eld dependences of χ' and χ'' were observed in these helimagnets.

Here, we discuss on the peculiar properties newly observed in $Gd_{0.62}Y_{0.38}$, the anomalous slow dynamics and the strong nonlinearity of the magnetization in the helimagnetic state. In the diluted system such as $Gd_{0.62}Y_{0.38}$, randomness can be an origin of the slow dynamics, as the case of spin glasses. Nevertheless, the increase of χ'' and nonlinearity of the magnetization were not observed in the similar dilute rare-earth helimagnet $Ho_{1\ x}Y_x$ (see Fig. 4 (b)). Therefore, we conclude that the randomness due to the dilution is not the origin of the slow dynamics and nonlinearity of the magnetization in $Gd_{1\ x}Y_x$. It should be noted that similar nonlinearity has been reported in $Gd_{1\ x}La_x$ in a previous work [10]. The previous and our experimental results suggest that these anomalous behaviors result from the weak magnetic anisotropy of Gd-moment. In particular, the in-plane anisotropy is vanishingly weak in $Gd_{1\ x}Y_x$ [11, 12]. And hence, the facts found in this work, the increase of χ'' and nonlinearity of the magnetization observed remarkably only in-plane direction, are consistent with this suggestion.

Now, we propose two di erent speculations on the origin of the anomalous slow dynamics in $Gd_{1\ x}Y_x$. The rst one is a possible collective motion of a large number of spins in the helimagnetic structure. A excitation mode that spins rotate dynamically in the helical plane with keeping its helical spin arrangement, which can be excited by the ac-magnetic eld along the in-plane direction, has a very low excitation energy because of very weak magnetic in-plane anisotropy in $Gd_{1\ x}Y_x$. Such a low-energy collective spin excitation can be responsible for the slow dynamics. In this excitation mode, the spins themselves can be rotated by the ac-magnetic-eld, however, the vector chirality is preserved and a motion of the chiral-domains emerges. It suggests that a chirality is decoupled from spins in the helimagnetic phase of $Gd_{1\ x}Y_x$. In $Gd_{1\ x}Y_x$, the inversion symmetry is preserved globally, and hence, the left-handed and right-handed chiral domain should coexist in the helimagnetic state. The domain-wall of these two chiral domain can be also moved by ac-magnetic- eld very slowly. By the dc-magnetic eld or the magnetic anisotropy, spins are pinned along the magnetic eld or the magnetic easy direction, and the chiral-domain and chiral-domain-wall motions can be xed. The suppression of χ'' by applying the magnetic eld and the lack of χ'' in Ho and $Ho_{1\ x}Y_x$ are manifestations of these pinning e ects. The second is a long-time variation of the period of the magnetic structure, which was recently observed in $Ca_3Co_2O_6$ [13]. It is known that the period of the helimagnetic structure gradually changes with changing temperature in zero magnetic eld in $Gd_{1\ x}Y_x$. It indicates that energy levels of states with a little di erent wave vector $\boldsymbol{q}$ are close and degenerate mostly each other, which may result the slow dynamics. The observation of the time-variation of the magnetic wave vector by the neutron scattering experiments with epithermal neutron or X-ray magnetic scattering is exciting future issue.

4. Conclusion

An anomalous slow dynamics and a strong nonlinearity of magnetization were observed in the metallic helimagnet $Gd_{1\ x}Y_x$. These peculiar phenomena should result from the weak magnetic anisotropy of Gd-moments. The anomalous slow dynamics may be chiral-domain and chiral-domain-wall motions or a long-time variation of the period of the magnetic structure.

Acknowledgments

This work was partly supported by a Grant-in-Aid for Scienti c Research, the Ministry of Education, Culture, Sports, Science and Technology, Japan (the Global COE Program "Integrated Materials Science" on Kyoto University and Priority Areas "Novel States of Matter Induced by Frustration").

References

[1] Legvold S, Burgardt P, and Beaudry B J 1980 *Phys. Rev. B* **22** 2573
[2] Legvold S, Ito T and Beaudry B J 1980 *Phys. Rev. Lett.* **45** 1275
[3] Bates S, Palmer S B, Sousa J B, McIntyre G J, Fort D, Legvold S, Beaudry B J and Koehler W C 1985 *Phys. Rev. Lett.* **55** 2968

[4] Melville R J, and Eccleston R S, McIntyre G J and Palmer S B 1992 *J. Phys.: Condens Matter* **4** 10045
[5] Andrianov A V and Savelieva O A 2008 *EPL* **82** 47012
[6] Fretwell H M, Dugdale S B, Alam M A, Hedley D C R, Rodriguez-Gonzalez A and Palmer S B, 1999 *Phys. Rev. Lett.* **82** 3867
[7] Hughes I D, Däne M, Ernst A, Hergert W, Lüders M, Poulter J, Staunton J B, Svane A, Szotek Z, Temmerman W M 2007 *Nature* **446** 650
[8] Evenson W E and Liu S H, 1969 *Phys. Rev.* **178** 783
[9] Jensen J *et al* 1991 *Rare Earth Magnetism (Clarendon Press, Oxford)*
[10] Larica C, Nunes E and Freitas J C C 1999 *J. Phys.: Condens Matter* **11** 821
[11] Jensen J and Macintosh A R, 1991 *Rare Earth Magnetism (Clarendon Press, Oxford)*
[12] Brooks M S S and Goodings D A, 1968 *J. Phys. C: Solid State Phys.* **1** 1279
[13] Moyoshi T *et al* 2010 Int. *Conf. on Strongly Correlated Electron Systems* (Santa Fe: New Mexico Consortium)

Journal of Physics: Conference Series **320** (2011) 012069 doi:10.1088/1742-6596/320/1/012069

Metamagnetism of η-carbide-type transition-metal carbides and nitrides

T Waki, S Terazawa, Y Umemoto, Y Tabata, K Sato[†1], A Kondo[†], K Kindo[†], H Nakamura

Department of Materials Science and Engineering, Kyoto University, Kyoto 606-8501, Japan
[†] The Institute for Solid State Physics, The University of Tokyo, Kashiwa 277-8581, Japan

E-mail: `nakamura.hiroyuki.2w@kyoto-u.ac.jp`

Abstract. η-carbide-type transition-metal compounds include the frustrated *stella quadrangula* lattice. Due to characteristics of the lattice, we expect subtle transitions between frustrated and non-frustrated states. Here, we report metamagnetic transitions newly found in η-carbide-type compounds Fe_3W_3C, Fe_6W_6C and Co_6W_6C.

1. Introduction

Transition-metal nitrides and carbides with the cubic -carbide-type structure (prototype Fe_3W_3C) are expected to show new and various electronic phenomena, because they are typical intermetallic compounds with narrow d-electron bands sitting at around the Fermi level [1]. However, despite that a huge number of the compounds are known to exist, their electronic properties have been less studied in contrast to extensive investigations in material science, for example, as hard and refractory materials, and as potential catalysts [2]. In the -carbide-type structure, one can nd a geometrically frustrated sublattice called the *stella quadrangula* lattice (16d and 32e sites in the space group $Fd\bar{3}m$; gure 1, center) [3]. This lattice is obtained by inserting a small regular tetrahedron into each tetrahedron in the pyrochlore lattice (gure 1, left). However, the nature of the frustration is essentially di erent from the pyrochlore lattice. The element of this lattice, which is referred to as a *stella quadrangula* (stellate tetrahedron; gure 1, right), is two nested regular tetrahedrons having the same center of gravity. There are two types of near-neighbor bonds, 16d–32e and 32e–32e. When considering only J_1 and J_2 corresponding to these bonds, respectively, it is found that (i) the frustration presents only when J_2 is negative and dominant, (ii) J_1 works to release the frustration, and (iii) these features do not depend on the sign of J_1. As a result, the frustrated and non-frustrated states are switched depending on the ratio of $J_2/|J_1|$ [4]. Thus we expect rich variety of magnetic states, and various types of transitions between frustrated and non-frustrated states associated with the variation in the $J_2/|J_1|$ ratio.

Recently, we have studied systematically electronic properties of the -carbide-type compounds, and already reported several examples of interesting discoveries [4, 5, 6, 7]. In this paper, we focus on the metamagnetism. As summarized in gure 2, we have already

[1] Present address: Department of Natural Science, Ibaraki National College of Technology, Hitachinaka 312-8508, Japan

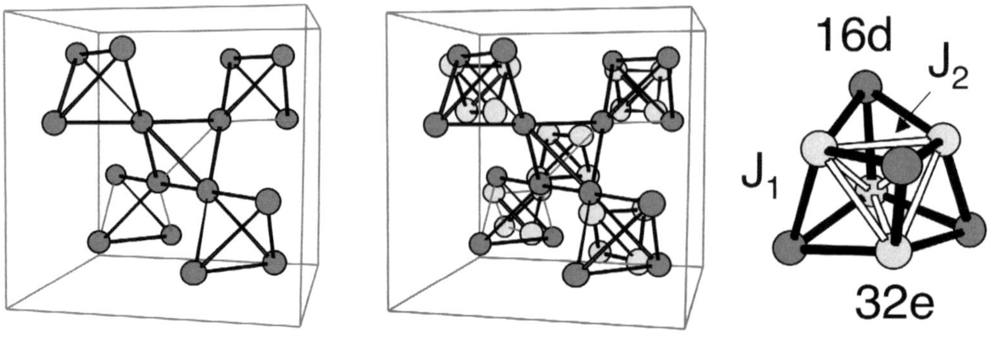

Figure 1. The *stella quadrangula* lattice (center) (Wycko positions 16*d* and 32*e* in the space group $Fd\bar{3}m$) can be obtained by inserting a small regular tetrahedron into each tetrahedron in the pyrochlore lattice (left). Right: The *stella quadrangula* as an element of the *stella quadrangula* lattice. J_1 (solid bonds) and J_2 (open bonds) correspond to NN (16*d*–32*e*) and NNN (32*e*–32*e*) interactions, respectively.

reported a quite sharp metamagnetic transition at 14 T [4] from a non-Fermi liquid ground state in Fe_3Mo_3N [5], and a rather gradual itinerant-electron metamagnetic transition in Co_3Mo_3C at 37 T [6]. In the gure, data of related materials Fe_3Mo_3C and Co_3Mo_3N, which show no metamagnetic transition, are also included. The magnetism of Fe_3Mo_3C is less enhanced, resulting in the small susceptibility and no metamagnetic transition. On the other hand, in the Co systems, although the low- eld susceptibilities of Co_3Mo_3C and Co_3Mo_3N are comparable, only Co_3Mo_3C shows a metamagnetic transition. We ascribed this di erence to two di erent type of criticality, only one of which leads to the susceptibility peak and the itinerant metamagnetism [6]. These behaviors of magnetization curves are quite contrastive depending on not only the transition-metal element but also the interstitial atoms (C and N), suggesting the presence of competing interactions in the frustrated *stella quadrangula* lattice. In this paper, we show metamagnetic transitions newly found in other -carbide-type compounds Fe_3W_3C, Co_6W_6C and Fe_6W_6C.

2. Experimental procedures

Polycrystalline samples of Fe_3W_3C, Fe_6W_6C and Co_6W_6C were synthesized by standard solid-state reactions from powdered pure elements in evacuated silica tubes [8]. The magnetization at low elds was measured by using a SQUID magnetometer, MPMS (Quantum Design) equipped in the LTM centre, Kyoto University. The high- eld magnetization up to 45 T was measured by using a pulse magnet equipped in ISSP, the University of Tokyo at 4.2 K.

3. Results

3.1. Fe_3W_3C

Fe_3W_3C is known as the prototype of the -carbide, which was originally found as an impurity phase in steel [9]. Magnetic properties of Fe_3W_3C have not been reported so far, although electronic structure calculations have been performed recently [10, 11, 12]. We have measured the temperature dependence of the susceptibility, which obeys the Curie-Weiss law at high temperatures (the e ective moment $\mu_{\rm eff} = 0.85\ \mu_{\rm B}$/Fe and the Weiss temperature = 1 K) and exhibits a broad peak at 75 K [7]. This behavior looks like an aspect of the enhanced Pauli paramagnet, although the magnetic ground state has not been con rmed microscopically, The magnetization curve of Fe_3W_3C at 4.2 K is shown in gure 3(a). A slight ferromagnetic component coming from ferromagnetic impurities is seen as non-zero magnetization at zero eld. A small jump of the magnetization was observed at 27 T, with a small eld hysteresis of 1 T. Since this feature is similar to that of Co_3Mo_3C, we speculate that this transition is a kind of itinerant-electron metamagnetism (polarization of $3d$ bands) from an enhanced Pauli paramagnetic state to a eld-induced ferromagnetic state. However, the transition is not as simple as the typical itinerant-electron metamagnetism as in YCo_2 [13]; the magnetization jumps only slightly and does not tend to saturate even at the eld of 45 T. These anomalies would be ascribed to the non-uniform magnetization in the Fe sublattice with two di erent chemical sites, namely bands of only one of the sites may be polarized at the transition eld. It is notable that locally di erent Fe moments, 0.75 and 0.11 μ_B at the $16d$ and $32e$ sites, respectively, have been estimated from band calculations [11].

3.2. Fe_6W_6C

Fe_6W_6C forms the Ni_6Mo_6C-type structure [14], which is another -carbide-type structure. For both Fe_3W_3C- and Ni_6Mo_6C-type structures, the space group is identical of cubic $Fd\bar{3}m$ (227). The di erence is the site occupied by C; $16c$ in Fe_3W_3C and $8a$ in Ni_6Mo_6C. The temperature dependence of the susceptibility of Fe_6W_6C shows a Curie-Weiss behavior at high temperatures ($\mu_{\mathrm{eff}} = 1.50\ \mu_B$/Fe and $= 28$ K) and a sharp cusp at 20 K [7], which looks like of a typical antiferromagnetic transition. The positive Weiss temperature suggests the coexistence of ferromagnetic interaction. The magnetization curve of Fe_6W_6C at 2 K, shown in gure 3(b), exhibits a relatively sharp jump at a low eld of 0.57 T, and a saturating tendency at higher elds. It is reasonable to ascribe this transition to a spin ip from an antiferromagnetic ground state to a eld-induced ferromagnetic state, because the transition eld decreases with increasing temperature in contrast to an opposite tendency generally expected for the itinerant-electron metamagnetism; such a tendency was indeed observed in the cases of Fe_3Mo_3N [4] and Co_3Mo_3C [6]. From band calculations, Fe moments, 0.58 and 0.25 μ_B, have been estimated at the $16d$ and $32e$ sites, respectively [11]. It is notable that the sum is roughly equivalent to the saturation magnetization. Naively we expect similar electronic structures for Fe_3W_3C and Fe_6W_6C because the metallic elements form the same sublattices and C-$2p$ bands are much apart from the Fermi

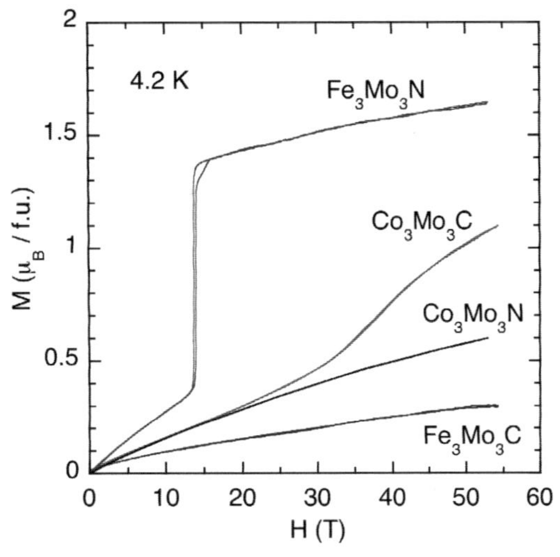

Figure 2. Magnetization curves of Fe_3Mo_3N (cited from [4]), Fe_3Mo_3C, Co_3Mo_3N (cited from [6]) and Co_3Mo_3C (cited from [6]) at 4.2 K.

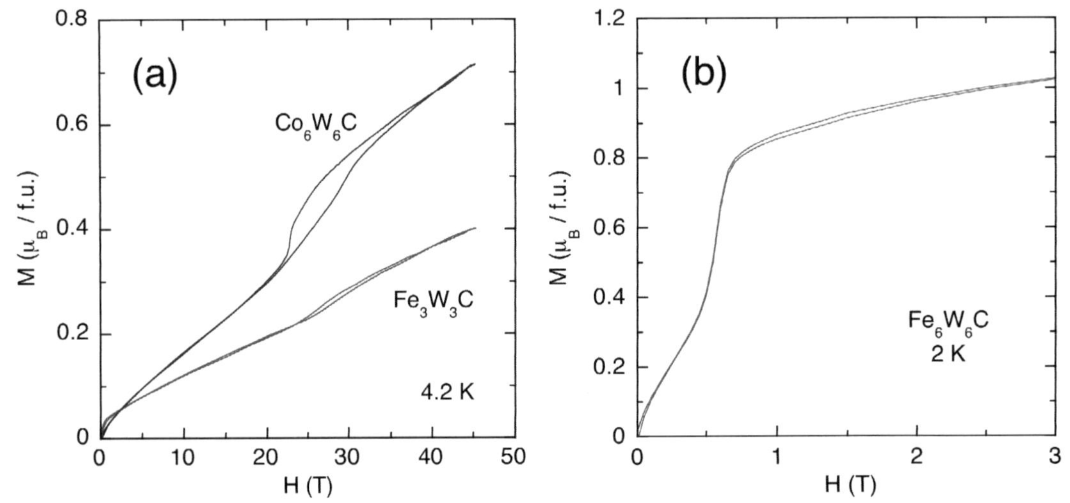

Figure 3. Magnetization curves of Fe_3W_3C and Co_6W_6C at 4.2 K (a), and of Fe_6W_6C at 2 K (b).

level. By contrast, however, the experimental observations are qualitatively di erent, suggesting the delicate and competitive interactions in the Fe sublattice.

3.3. Co_6W_6C

Co_6W_6C forms the Ni_6Mo_6C-type structure [15, 16] as Fe_6W_6C, and is known as a material with extremely high bulk modulus [17]. The temperature dependence of the susceptibility is similar to that of Fe_3W_3C; the susceptibility follows the Curie-Weiss law at high temperatures ($\mu_{\rm eff} = 1.13\ \mu_B$/Co and $= 93K$) and shows a broad peak at 80 K. As expected from the broad peak in the susceptibility, we observed a metamagnetic transition as in gure 3(a) with a relatively large eld hysteresis. The transition elds are 22.8 and 29.6 T at eld-increasing and decreasing processes, respectively. Considering the similar behaviors of the susceptibility and the magnetization, this transition is probably classi ed as the itinerant-electron metamagnetism as those observed in Co_3Mo_3C and Fe_3W_3C.

4. Summary

We have shown metamagnetic transitions found in Fe_3W_3C, Fe_6W_6C and Co_6W_6C. Together with previous data, we summarize as a number of -carbide-type compounds exhibit metamagnetic transitions, indicating that the magnetism is, in general, highly enhanced in these compounds. The transitions of Fe_6W_6C and Co_6W_6C are possibly the itinerant-electron metamagnetism as in Co_3Mo_3C. Compared with these moderate transitions, the quite sharp discontinuity of Fe_3Mo_3N from the nonmagnetic non-Fermi liquid state [4] appears to be contrastive. The origin of this transition remains for further detailed studies. The ground state of Fe_6W_6C is likely to be antiferromagnetic, which is relatively unstable against external eld. The rich variety of the magnetism is probably related with competing interactions in the frustrated *stella quadrangula* sublattice in the -carbide-type structures.

Acknowledgments

This study was supported by a Grant-in-Aid for Scienti c Research on Priority Areas "Novel States of Matter Induced by Frustration" (19052003), a Grant-in-Aid for the Global COE

Program, "International Center for Integrated Research and Advanced Education in Materials Science", and a Grant- in-Aid for Young Scientists (B) 21760531 from the Ministry of Education, Culture, Sports, Science and Technology of Japan.

References

[1] for example,
Suetin D V, Shein I R and Ivanovskii A L 2010 *Rus. Chem. Rev.* **79** 611
[2] for example,
Toth L E 1971 *Transition Metal Carbides and Nitrides* (New York: Academic Press)
Jacobsen C J H 2000 *Chem. Commun.* 1057
Errandonea D *et al* 2010 *Phys. Rev.* B **82** 174105
[3] Nyman H, Andersson S and O'Kee e M 1978 *J. Solid State Chem.* **26** 123
[4] Waki T *et al* 2011 *EPL* **94** 37004
[5] Waki T *et al* 2010 *J. Phys. Soc. Jpn.* **79** 043701
[6] Waki T *et al* 2010 *J. Phys. Soc. Jpn.* **79** 093703
[7] Waki T *et al* 2010 *Proc. Int. and Interdisciplinary Workshop on Novel Phenomena in Integrated Complex Sciences: from Non-living to Living Systems (Kyoto)* to appear in *J. Phys.: Conf. Ser.*
[8] Ettmayer P and Suchentrunk R 1970 *Monatsh. Chem.* **101** 1098
[9] Westgren A 1933 *Jernkontorets Ann.* **7** 1
[10] Ilyasov A V, Ryzhkin A A, Ilyasov V V 2008 *J. Struct. Chem.* **49** 795
[11] Suetin D V *et al* 2009 *Physica* B **404** 3544
[12] Li Y *et al* 2010 *Physica* B **405** 1011
[13] for example,
Goto T, Fukamichi K and Yamada H 2001 *Physica* B **300** 167
[14] Leciejewicz J 1964 *J. Less-Common Metals* **7** 318
[15] Pollock C B and Stadelmaier H H 1970 *Metall. Trans.* **1** 767
[16] Ramnath V and Jayaraman N 1987 *J. Mater. Sci. Lett.* **6** (1987) 1414
[17] Dubrovinskaia N A *et al* 1999 *J. Alloys Compd.* **285** 242

Journal of Physics: Conference Series **320** (2011) 012070 doi:10.1088/1742-6596/320/1/012070

Magnetic interactions and orbital state in double chain systems investigated by NMR measurements

Masayuki Itoh[1], Hikaru Takeda[1], Yasuhiro Shimizu[1,2], Hiroya Sakurai[3], Masahiko Isobe[4] and Yutaka Ueda[4]

[1]Department of Physics, Graduate School of Science, Nagoya University, Furo-cho, Chikusa-ku, Nagoya 464-8602, Japan

[2]Institute for Advanced Research, Nagoya University, Furo-cho, Chikusa-ku, Nagoya 464-8601, Japan

[3]National Institute for Materials Science, 1-1 Namiki, Tsukuba 305-0044, Japan

[4] Institute for Solid State Physics, University of Tokyo, 5-1-5 Kashiwanoha, Kashiwa 277-8581, Japan

E-mail: mitoh@nucc.cc.nagoya-u.ac.jp

Abstract. We have made NMR measurements to clarify the relationship between the orbital state and the magnetic properties of the double chain compounds NaV_2O_4 and $K_2Cr_8O_{16}$. In both the oxides, the t_{2g} orbital is split into a non-degenerated d_{xy} and a degenerated $d_{yz,zx}$ orbital due to the lattice symmetry. Based on the anisotropy of the hyperfine coupling tensor determined by the present NMR measurements, we conclude the predominant d_{xy} orbital and the partially occupied $d_{yz,zx}$ orbital, naturally leading to the double exchange mechanism of the ferromagnetic interaction. In addition to the ferromagnetic interaction, there is an antiferromagnetic interaction via the direct d_{xy}-d_{xy} path along the double chain direction. Thus, the ferromagnetic transition takes place in $K_2Cr_8O_{16}$, whereas the helical spin order appears in NaV_2O_4 due to the competition between the ferromagnetic and antiferromagnetic interactions.

1. Introduction

Itinerant electron systems on a geometrically frustrated lattice with a triangular, pyrochlore or spinel struture show a variety of fascinating phenomena near the metal-insulator (MI) transition such as charge order with a spin-single ground state, heavy-mass Fermi liquid and superconductivity [1]. Recently, quasi-one-dimensional (quasi-1D) oxides with the calcium ferrite AM_2O_4 (A : alkali metal; M : transition metal) and the hollandite $A_2M_8O_{16}$ structures as shown in Figs. 1 (a) and (b), respectively, have been intensively investigated as a new frustrated quasi-1D system with a double chain shown in Fig. 1 (c) [2, 3, 4, 5, 6]. Among them, NaV_2O_4 with the mixed valence of V^{3+}:V^{4+}=1:1 (the average electron number $3d^{1.5}$) was reported to be an antiferromagnetic (AFM) metal with T_N=140 K in spit of the strong ferromagnetic (FM) correlation which appears in magnetic susceptibility [4, 5] and the Knight shift [7]. Also an incommensurate helical spin structure was proposed in the AFM phase [8, 9]. A novel MI transition from the FM metallic to the FM insulating phases was reported to take place at T_{MI}=95 K in the FM state below T_C=180 K in $K_2Cr_8O_{16}$ with a rare mixed valence of Cr^{3+}:Cr^{4+}=2:6 ($3d^{2.25}$) [6]. In particular, the mechanism of the MI transition has attracted much attention from the experimental and theoretical point of views [6, 10]. In both oxides,

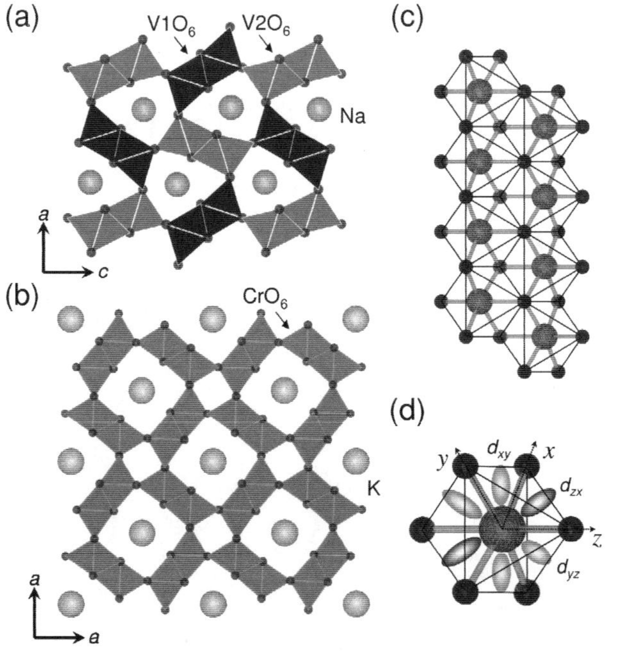

Figure 1. (Color online) Crystal structures of (a) NaV_2O_4 and (b) $K_2Cr_8O_{16}$ viewed along the b axis. (c) The double chain composed of edge-shared MO_6 (M= V, Cr) octahedra. (d) The d_{xy}, d_{yz}, and d_{zx} orbitals in the MO_6 octahedron with the local xyz axes.

the orbital degree of freedom plays an important role for the magnetic and electric properties. As shown in Fig. 1 (d), the t_{2g} orbital is considered to be split into a non-degenerated d_{xy} and a degenerated $d_{yz,zx}$ (d_{yz}, d_{zx}) orbital due to the local symmetry. However, the orbital state and the frustration effect on the magnetic properties have not been understood clearly in the frustrated double chain systems.

In this study, we have performed ^{51}V and ^{53}Cr NMR measurements to study the orbital state and the magnetic interactions in NaV_2O_4 and $K_2Cr_8O_{16}$. We determine the hyperfine coupling constant A and show that the ratio of the axially anisotropic to isotropic components, $A_{\rm ax}/A_{\rm iso}$, is a useful measure of the $3d$ electron occupation number. The magnetic interactions in the double chain are discussed on the basis of the t_{2g} orbital, and the competition between the FM and AFM interactions is concluded to play an important role for the helical spin order in NaV_2O_4 and the ferromagnetic order in $K_2Cr_8O_{16}$.

2. Experiments

A single crystal of NaV_2O_4 and a polycrystalline sample of $K_2Cr_8O_{16}$ were prepared for the present NMR measurements as reported in Refs. [5], and [6], respectively. ^{51}V and ^{53}Cr Fourier-transformed (FT) NMR spectra were measured in a magnetic field of H_0= 5.8701 or 0.5 T by using a pulsed spectrometer. The ^{51}V Knight shift was determined as ^{51}K=($\nu_{\rm res}$-ν_0)/ν_0 where $\nu_{\rm res}$ is the ^{51}V resonance frequency in the NaV_2O_4 sample and ν_0 (=65.704 MHz) is that of the aqueous $NaVO_3$ solution measured at 5.8701T. Frequency-swept ^{53}Cr NMR spectra in $K_2Cr_8O_{16}$ were taken point by point at 4.2 K and 0.5 T.

3. Results and discussion

In order to determine the angular dependence of the ^{51}V Knight shift ^{51}K in NaV_2O_4, we made ^{51}V NMR measurements with the single crystal rotated in H_0 = 5.8701 T. We observed two NMR spectra with no electric quadrupole splitting from the V1 and V2 sites, which are crystallographically inequivalent V sites in the orthorhombic lattice (space group $Pnma$), and each spectrum with H_0 applied in the ca plain was furthermore split into two spectra due to the

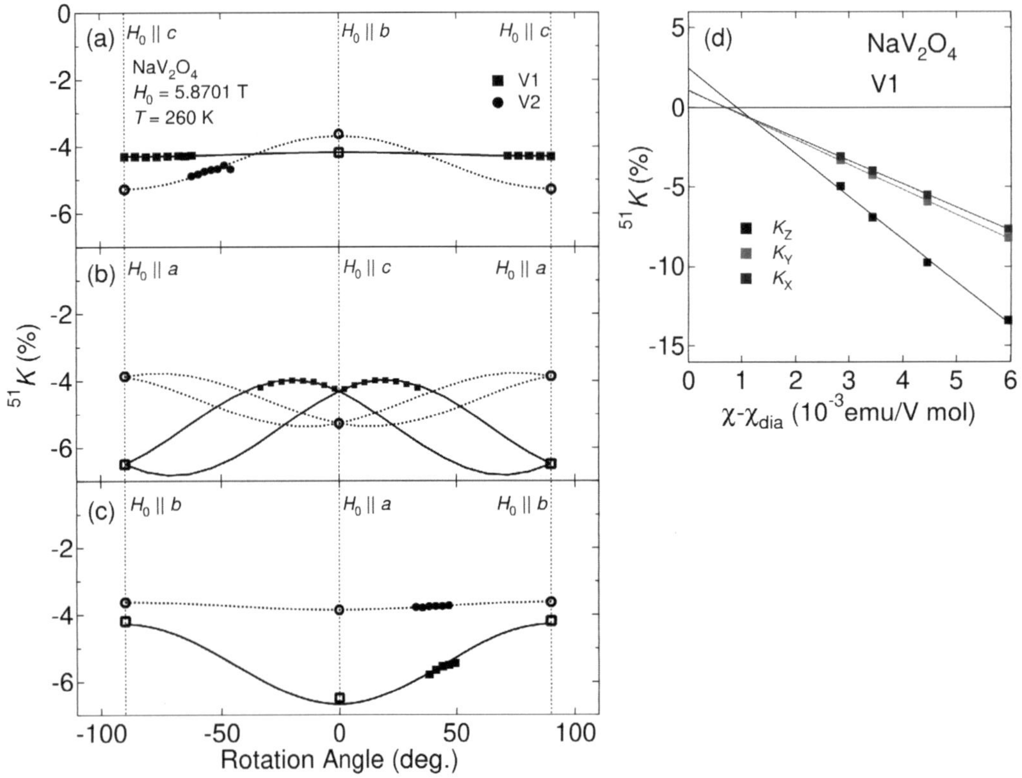

Figure 2. (Color online) Angular dependence of the ^{51}V Knight shift at 260 K with H_0 (= 5.8701 T) rotated in the (a) *bc*, (b) *ca*, and (c) *ab* planes of NaV_2O_4. The solid and dotted curves represent the fitted results to the equation in the text. (d) The ^{51}V Knight shift versus magnetic susceptibility plots for the V1 site. K_Z, K_Y, and K_X are principal components of the ^{51}V Knight shift tensor.

local symmetry axes different from each other at the V sites on the mirror plane perpendicular to the *b* axis [11]. Also the angle range where the spectrum can be observed was limited due to anisotropy of the nuclear spin-spin relaxation time T_2. Figures 2 (a)-(c) show the angular dependence of ^{51}K at 260 K in the *bc*, *ca*, and *ab* planes, respectively. The angular dependence in the *bc* plain is well fitted to the equation $^{51}K = k_1 + k_2 \cos 2\theta_a + k_3 \sin 2\theta_a$ where k_1, k_2, and k_3 are the fitting parameters including Knight shift components expressed in the *abc* coordinate system and θ_a is the angle between the *b* axis and H_0 [12], as shown by the solid and dotted curves in Fig. 2 (a). The equations for the H_0 rotation in the *ca* and *ab* planes are obtained by cyclic permutation. After diagonalizing the Kinigh shift tensor in the *abc* coordinate system, we obtain the X, Y, and Z principle components of the shift tensor as ($^{51}K_X$, $^{51}K_Y$, $^{51}K_Z$)=(−4.0, −4.3, −6.9%) at 260 K for the V1 site and (−3.7, −3.8, −5.4%) for the V2 site.

In 3*d* transition metal oxides, the Knight shift is generally composed of the orbital and spin parts as $K = K_{\text{spin}} + K_{\text{orb}}$, and magnetic susceptibility χ is similarly composed of the orbital, spin, and diamagnetic ones as $\chi = \chi_{\text{spin}} + \chi_{\text{orb}} + \chi_{\text{dia}}$ where $\chi_{\text{dia}} = -3.5\times10^{-5}$ emu/V mol [13] for NaV_2O_4. Also K_{spin} is expressed as $K_{\text{spin}} = \frac{A}{N_A \mu_B}\chi_{\text{spin}}$ where A is the hyperfine coupling constant, N_A is the Avogadro's number and μ_B is the Bohr magneton. From the standard analysis of the ^{51}K versus χ plots in NaV_2O_4 as shown in Fig. 2(d), we obtain the principal values of the hyperfine coupling tensor as ($^{51}A_X$, $^{51}A_Y$, $^{51}A_Z$) = (−81, −87, −150) for the V1 site and (−66, −64, −100) in the units of kOe/μ_B for V2. The result of $^{51}A_X \sim {}^{51}A_Y$ supports the presence of the degenerate $d_{yz,zx}$ orbital.

The Knight shift and hyperfine coupling tensors governed by the magnetic hyperfine interaction $\mathcal{H}_{\rm mag}$ are useful to study the electron orbital in $3d$ transition metal oxides. In general, $\mathcal{H}_{\rm mag}$ is expressed as

$$\mathcal{H}_{\rm mag} = -2\mu_B \gamma_n \hbar \left\langle r^{-3} \right\rangle_{\rm mag} \left[\kappa \mathbf{S} \cdot \mathbf{I} - \mathbf{L} \cdot \mathbf{I} + \xi \mathbf{S} \cdot \mathbf{q} \cdot \mathbf{I} \right], \tag{1}$$

where γ_n is the nuclear gyromagnetic ratio, $\hbar$ the Planck's constant, $\left\langle r^{-3} \right\rangle_{\rm mag}$ the expectation value of r^{-3} for the $3d$ orbital, κ a parameter for the Fermi contact interaction due to the core-polarization effect, $\mathbf{S}$ the total electron spin, $\mathbf{L}$ the total orbital momentum, and $\mathbf{I}$ the nuclear spin operator. Also $\xi = \frac{2l+1-4S}{S(2l-1)(2l+3)(2L-1)}$ (l=2, S=1/2, and L=2 for $3d^1$) and $\mathbf{q}$ is the electron quadrupole moment tensor defined as $q_{\alpha\beta} = \frac{3}{2}(L_\alpha L_\beta + L_\beta L_\alpha) - \delta_{\alpha\beta} L^2$ (α, β = x, y, z) [14, 15]. The first term of eq.(1) is the Fermi contact interaction, the second is the orbital one, and the third is the dipole one. The anisotropy of the Knight shift tensor is mainly governed by the dipole interaction. For example, for d^1_{xy}, the expectation value of $\mathbf{q}$, which is proportional to the anisotropic part of the Knight shift tensor, is expressed as (q_{xx}, q_{yy}, q_{zz})=−3(1, 1, −2) where the z axis of $\mathbf{q}$ is parallel to the symmetric axis of the d_{xy} orbital, closely related to the principal axis of the Knight shift tensor. In NaV_2O_4, the Z axis of the ^{51}V Knight shift tensor is parallel to the direction to the apical oxygen of VO_6 octahedron as seen in Fig.1 (d). This is consistent with the local symmetry leading to the splitting of the $3d$ orbital to the non-degenerate d_{xy} and the degenerated $d_{yz,zx}$ orbital.

We discuss the $3d$ electron occupation number from a ratio $A_{\rm ax}/A_{\rm iso}$ where $A_{\rm iso}$ and $A_{\rm ax}$ are isotropic and axial hyperfine coupling constants defined as $A_{\rm iso}=(A_X + A_Y + A_Z)/3$ and $A_{\rm ax}=(2A_Z - A_Y - A_Z)/6$, respectively. Theoretically $A_{\rm ax}/A_{\rm iso}=\frac{2}{7\kappa}$=0.57 for d^1_{xy} and $-\frac{1}{7\kappa}$=−0.29 for $(d^1_{yz}+d^1_{zx})/2$ for a standard κ=0.5. Comparing these values with the experimental value of $^{51}A_{\rm ax}/^{51}A_{\rm iso}\sim$0.2, the ratio of the electron number n_i (i= xy, yz, and zx) on the d_i orbital is estimated as $n_{xy} : n_{yz} : n_{zx}$ = 0.85 : 0.32 : 0.32, showing the predominant occupied d_{xy} orbital. Recently the band calculation in the generalized gradient approximation was made for the hollandite oxide $K_2V_8O_{16}$ with the similar crystal structure to that of NaV_2O_4 [16]. Based on the band calculation, they proposed a picture of the relatively localized d_{xy} and itinerant d_{yz}/d_{zx} electrons. This picture is applicable to NaV_2O_4 as discussed above on the occupation number, reasonably resulting in the double exchange mechanism as the origin of the ferromagnetic interaction which was observed by the magnetic susceptibility [4] and the NMR [7] experiments of NaV_2O_4.

In the paramagnetic state of $K_2Cr_8O_{16}$, the poor signal to noise ratio due to the small natural abundance of 9.5% prevented us from observing a ^{53}Cr NMR spectrum. However, we could observe the ^{53}Cr NMR spectrum in the ferromagnetic insulating phase below 95 K for the powdered sample. In a ferromagnetically ordered state, one usually observes NMR signals from domain and domain wall in H_0=0 T, whereas only the domain spectrum is observed above a magnetic field, ∼0.5 T at 5 K in $K_2Cr_8O_{16}$ [6], where magnetization is saturated and becomes parallel to H_0. Figure 3 shows the ^{53}Cr domain spectrum at 4.2 K and 0.5 T in $K_2Cr_8O_{16}$. It has no electric quadrupole splitting which is consistent with the small electric quadrupole frequency ν_Q<1 MHz reported in Cr compounds [17]. Thus the spectrum is governed by the internal field $\mathbf{H}_{\rm n}=\mathbf{^{53}AM}$ due to the spontaneous magnetization $\mathbf{M}$. In the ferromagnetic phase of $K_2Cr_8O_{16}$, the ^{53}Cr NMR frequency ν is expressed as $\nu={}^{53}\gamma|\mathbf{^{53}AM} + \mathbf{H}_0|$ where $^{53}\gamma=2\pi\times2.406$ MHz/T and $\mathbf{M}||\mathbf{H}_0$ for H_0>0.5 T. The observed spectrum is well reproduced by assuming two powdered spectra with ($^{53}A_{\rm iso}$, $^{53}A_{\rm ax}$, $^{53}A_{\rm aniso}$)= (−92, −8.2, −6.1) for the Cr1 spectrum, where $A_{\rm aniso}$ is defined as $A_{\rm aniso}=(A_Y - A_X)/2$, and (−82, −9.2, −8.2) in the units of kOe/μ_B for Cr2,

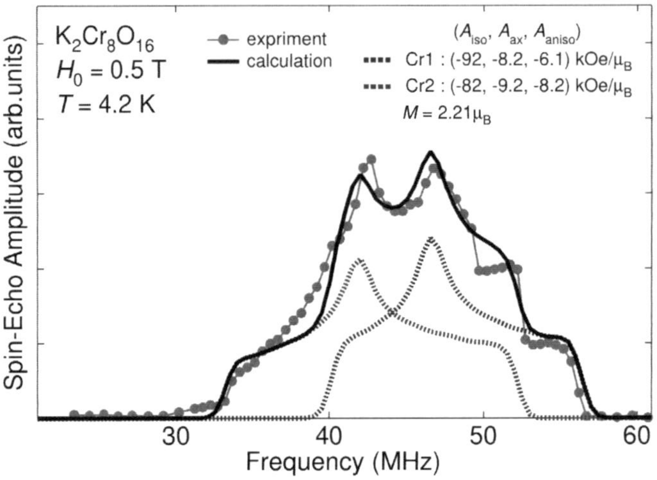

Figure 3. (Color online) ^{53}Cr NMR spectra at 4.2 K and 0.5 T in $K_2Cr_8O_{16}$. The solid curve represents the sum of the dotted curves which are powder spectra calculated by assuming the hyperfine coupling constants given in the figure and spontaneous magnetization $2.21\mu_B$.

and the saturation moment $M = 2.21\mu_B$ [6] as shown in Fig. 3. The observation of the two spectra and $^{53}A_{\rm aniso}$ indicates that the lattice symmetry in the ferromagnetic insulating phase lowers from the tetragonal one (space group $I4/m$) in the paramagnetic metallic phase. Also the ratio $^{53}A_{\rm ax}/^{53}A_{\rm iso}\sim0.1$ shows the $d_{yz,zx}$ orbital more occupied than that in NaV_2O_4. This is consistent with the result of the first principle electronic calculation by Sakamai *et al.* [10]. They also found the rather dispersionless narrow band of predominantly d_{xy} and the dispersive t_{2g} bands of predominantly d_{yz} and d_{zx} character with strong admixture of the $2p_z$ of the oxygen and proposed the double exchange FM mechanism in $K_2Cr_8O_{16}$ [10].

Finally we discuss the magnetic properties of NaV_2O_4 and $K_2Cr_8O_{16}$. In these double chain systems, the FM interaction is ascribed to the double exchange mechanism via the d_{yz}/d_{zx} orbital along the rung directions in the double chain as discussed above. In addition to the double exchange, an AFM interaction is expected to exist due to the direct d_{xy}-d_{xy} path along the double-chain direction. Consequently there should appear the competition between the FM and AFM interactions. Indeed, NaV_2O_4 shows the helical spin order [8, 9] which comes from this competition as reported in double chain spin systems with the competing nearest and second nearest exchange couplings, J_1 and J_2, respectively, where a helical spin order appears for the ratio $|J_2/J_1|>1/4$ [18]. On the other hand, in $K_2Cr_8O_{16}$, the FM interaction is considered to overcome the AFM one, resulting in the FM spin order. The dominant FM double exchange interaction may be ascribed to the strong hybridization between the oxygen and chromium ions which makes the degenerate $d_{yz,zx}$ band width broader, as discussed in a negative charge transferred oxide CrO_2 [19].

4. Conclusion

We have performed ^{51}V and ^{53}Cr NMR measurements to investigate the local magnetic properties and the orbital state in the double chain systems NaV_2O_4 and $K_2Cr_8O_{16}$. From the analysis of the hyperfine coupling constant, we concluded that the $3d$ electron occupation of the d_{xy} orbital is predominant compared with that of the degenerated orbital $d_{yz,zx}$ orbital in both oxides, reasonably leading to the double exchange ferromagnetic interaction. The competition between the ferromagnetic interaction along the rung directions and the antiferromagnetic one along the leg of the double chain direction plays an important role for the helical spin order in NaV_2O_4, whereas the predominant double exchange interaction results in the ferromagnetic order in $K_2Cr_8O_{16}$.

Acknowledgement
We would like to acknowledge Y. Kobayashi and M. Sato for fruitful discussion, and also S. Inoue for technical assistance. This study was supported by the Grant-in-Aid for Scientific Research on Priority Area" Novel State of Matter Induce by Frustration" (Grant No. 22014006) from the Ministry of Education, Culture, Sports, Science, and Technology of Japan.

References

[1] Gardner J S, Gingras M J and Greedan J E 2010 *Rev. Mod. Phys.* **82** 53.
[2] Isobe M, Koishi S, Kouno N, Yamaura J, Yamauchi T, Ueda H, Goutou H, Yagi T and Ueda Y 2006 *J. Phys. Soc. Jpn.* **75** 073801.
[3] Horiuchi S, Shirakawa T and Ohta Y 2008 *Phys. Rev. B* **77** 155120.
[4] Yamaura K, Arai M, Sato A, Karki A B, Young D P, Movshovich R, Okamoto S, Mandrus D and Takayama-Muromachi E 2007 *Phys. Rev. Lett.* **99** 196601.
[5] Sakurai H 2008 *Phys. Rev.* B **78** 094410.
[6] Hasegawa K, Isobe M, Yamauchi. T, Ueda H, Yamaura J, Gotou H, Yagi T, Sato H and Ueda Y 2009 *Phys. Rev. Lett.* **103** 146403.
[7] Takeda H, Itoh M and Sakurai H 2010 *J. Phys.: Conf. Ser.* **200** 012200.
[8] Ofer O. Ikeda Y, Goko T, Månsson M, Sugiyama J, Ansaldo E J, Brewer J H , Chow K H and Sakurai H 2010 *Phys. Rev.* B **82** 094410.
[9] Takeda H, Itoh M and Sakurai H (unpublished).
[10] Sakamaki M, Konishi T and Ohta Y 2009 *Phys. Rev. B* **80** 024416.
[11] Hahn Th (ed) 2002 *International Table for Crystallography* (Dordrecht : Kluwer Academic) vol A.
[12] Volkoff G M, Petch H E and Smellie D W 1952 *Can. J. Phys.* **30** 270.
[13] Hellwege K H and Hellwege A M (ed) 1986 *Atomic and Molecular Physics* (Berlin: Springer) Landolt-Börnstein New Series Group II vol 16.
[14] Abragam A and Bleaney B 1970 *Electron Paramagnetic Resonance in Transition Ions* (Oxford : Clarendon Press).
[15] Kiyama T and Itoh M 2003 *Phys. Rev. Lett.* **91** 167202.
[16] Sakamaki M, Horiuchi S, Konishi T and Ohta Y *arXiv:0811.4338.*
[17] Narath A 1965 *Phys. Rev.* **140** A854.
[18] Bursill R, Gehring G A, Farnell D J J, Parkinson J B, Xiang T and Zeng C 1995 *J. Phys.: Condens. Matter* **7** 8605.
[19] Korotin M A, Anisimov V I, Khomskii D I and Sawatzky G A 1998 *Phys. Rev. Lett.* **80** 4305.

International Conference on Frustration in Condensed Matter (ICFCM) IOP Publishing
Journal of Physics: Conference Series **320** (2011) 012071 doi:10.1088/1742-6596/320/1/012071

Local spin susceptibility in the orbital-frustrated vanadium oxide $BaV_{10}O_{15}$

Yasuhiro Shimizu[1,2], Ken-ichiro Matsudaira[2], Masayuki Itoh[2], Tomomasa Kajita[3], Takuro Katsufuji[3]

[1] Institute for Advance Research, Nagoya University, Furo-cho, Chikusa-ku, Nagoya 464-8601, Japan
[2] Department of Physics, Nagoya University, Furo-cho, Chikusa-ku, Nagoya 464-8602, Japan
[3] Department of Physics, Waseda University, Tokyo 169-8555, Japan

E-mail: yasuhiro@iar.nagoya-u.ac.jp

Abstract. $BaV_{10}O_{15}$ having a bilayer triangular lattice exhibits a semiconductor-insulator transition accompanied by the structural phase transition at T_c ~130 K. We show the microscopic properties of the phase transition investigated by ^{51}V NMR measurements on a single crystal. We have observed ^{51}V NMR spectra from two of the three inequivalent V sites at high temperatures. The Knight shift tensors are determined at 300 K for the two V sites. The local spin susceptibility strongly depends on the V sites and temperature despite the nearly independent magnetic susceptibility. Each spectrum is split into two below 200 K far above T_c, suggesting a charge/order ordering transition. The local spin susceptibilities of the three sites gradually vanish with decreasing temperature, consistent with the V trimer formation. The microscopic evidence for the spin singlet state is clearly observed below T_c.

1. Introduction

Geometrical frustration effects in charge and orbital degrees of freedom are less understood than in spin one. Among the strongly correlated materials, vanadium oxides are conventional examples of the metal-insulator transition that involves changes in the orbital state, as recently highlighted in V_2O_3 and VO_2 [1]. A series of the layered V oxides with triangular lattice, such as $BaV_{10}O_{15}$, $BaV_{13}O_{18}$, and $Ba_2V_{13}O_{22}$, offers the interesting stage to investigate the frustration effects on the metal-insulator transition with charge and orbital degrees of freedom [2, 3, 4, 5]. Among them, $BaV_{10}O_{15}$ exhibits a semiconductor-insulator transition at T_c ~130 K [2, 6, 7, 8, 9]. The crystal structure consists of a strongly-coupled bilayer of triangular lattice with a hole of V triangle, as shown in Fig. 1. The average vanadium valence is 2.8+, leading to 2.2 electrons per one V site. There are three inequivalent V sites, V1, V2, and V3, in the original lattice of orthorhombic *Cmce* at 300 K [2]. A structural transition at T_c makes two inequivalent V2 and V3 sites, called as V2, V2B, V3, and V3B. From the atomic displacements across T_c, Kajita *et al.* proposed formation of the V trimer of the V2B-V3-V3B triangle [2], instead of the V3-V3B dimer model [7]. Thus it can be a model material of the orbitally-induced spin singlet proposed theoretically on triangular lattice and spinel [10, 11]. However, the magnetic susceptibility rather increases at T_c and shows an antiferromagnetic transition at 43 K. Hence, the direct observation of the spin singlet has been desired by using microscopic experimental probes.

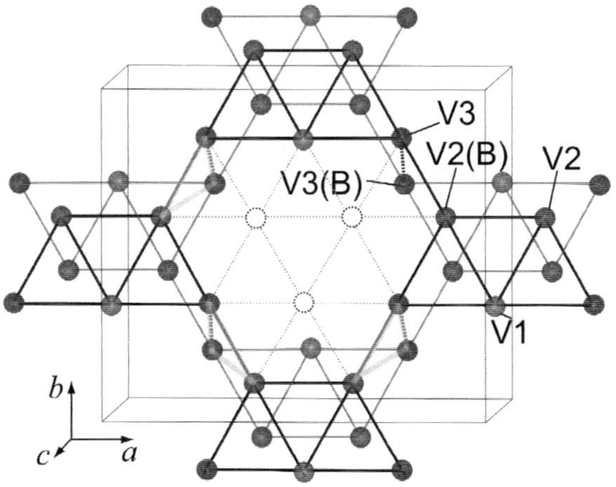

Figure 1. (Color online) The bilayer structure of the V lattice in $BaV_{10}O_{15}$.

NMR is one of the most powerful probes to detect the local spin susceptibility at the atomic site. Nonmagnetic ion sites, if any, can be distinguished from the magnetic ion sites, as observed in $Ba_2V_{13}O_{22}$ and AlV_2O_4 [3, 12]. In this paper, we report ^{51}V NMR measurements on a single crystal of $BaV_{10}O_{15}$. Based on the temperature dependence of the ^{51}V NMR spectra, we show a precursor of the spin-singlet trimer formation above T_c.

2. Experimental

The single crystal sample was grown by the floating-zone technique [2]. The obtained crystal was cut into a cubic with a side of 2 mm for NMR experiments. The frequency-swept ^{51}V NMR spectra were obtained from the Fourier transformation of the spin echo signals at a constant magnetic field $H_0 = 9.40$ T. The magnetic field was aligned to the a axis, based on the angle dependence of the Knight shift K. K was defined as a relative shift from the resonance frequency of an aqueous solution of $NaVO_3$. The typical width of the rf pulse was 2 μs with the time spacing between the first and second pulses, 6 μs. The nuclear spin-lattice relaxation rate $1/T_1$ below T_c was measured by the saturation recovery method and evaluated by the single exponential fit to the recovery curve in the magnetic field parallel to the c axis close to the magic angle where the nuclear quadrupole splitting diminishes.

3. Experimental results

3.1. Knight shift tensor

We determine the ^{51}V Knight shift tensor and assign the V sites from the angle dependence of the ^{51}V NMR spectra of the $BaV_{10}O_{15}$ single crystal. The representative data of the angle dependence of ^{51}V NMR spectra are shown in Fig. 2(a), where the magnetic field is rotated in the c plane at 300 K. We observed two spectra at the magnetic field direction parallel to the a and b axes. The spectrum at a high-frequency side in $H_0||a$ consists of sharp equally-spaced seven lines arising from the electric quadrupole splitting. On the other hand, the low-frequency spectrum is broader and has no quadrupole splitting. Rotating the crystal from the a to b axis, the low-frequency spectrum disappears in the intermediate angles but shows a negative shift, whereas the high-frequency spectrum shows a much smaller positive shift, as shown in Figs. 2(b) and (c). Since the spin hyperfine coupling constant would be negative in the V oxides due to the predominant core polarization, the high-frequency spectrum arises from the V site with

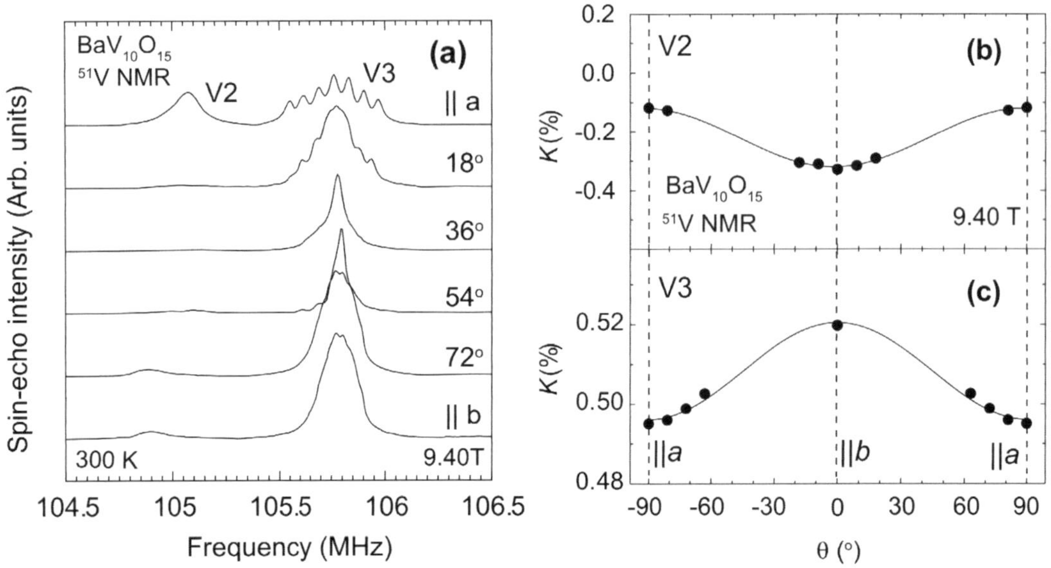

Figure 2. Angle dependences of (a) ^{51}V NMR spectra and the Knight shifts for the (b) V2 and (c) V3 sites of the $BaV_{10}O_{15}$ single crystal rotated in the c plane. The solid curves in (b) and (c) are the fitting results to the general formula of Eq. (1).

the smaller spin susceptibility than the low-frequency one. The wiping out of the low-frequency spectrum indicates the fast spin-echo decay time T_2.

Referring to the crystal structure, the vanishing local spin susceptibility is expected in the V3 site with the strong dimerization even at 300 K. The positive Knight shift of the high-frequency spectrum indicates the predominant orbital shift due to the Van-Vleck susceptibility. Among the three V sites, V1 is most isolated with the longer distances to the neighboring V ions, which is expected to have the largest spin susceptibility. It is natural that the spectrum from the V1 site is wiped out due to the fast T_2. Therefore, the low-frequency spectrum originates from the V2 site.

To determine the Knight shift tensor K, the angle dependence of K around the c axis is fitted by the general formula given by

$$K = k_{1,c} + k_{2,c}\cos 2\theta + k_{3,c}\sin 2\theta, \quad (1)$$

where $k_{1,c} = (K_{aa} + K_{bb})/2, k_{2,c} = (K_{aa} - K_{bb})/2$, and $k_{3,c} = K_{ab}$ with $K_{\alpha\beta}$ $(\alpha, \beta = a, b,$ and $c)$ is the components of K [13]. Diagonalizing the obtained K, the principal components of K are given as $(0.13\%, -0.42\%, -0.73\%)$ for the V2 site and $(0.44\%, 0.53\%, 0.63\%)$ for the V3 site.

3.2. Temperature dependence of local spin susceptibility

Figure 3 shows the temperature dependence of ^{51}V NMR spectra of $BaV_{10}O_{15}$. As temperature decreases, the NMR spectra from the V2 and V3 sites shift to higher frequencies, indicating a decrease in the local spin susceptibility. Since the magnetic susceptibility $\chi_{\rm bulk}$ rather increases [2], the local spin susceptibility of the magnetic V1 site is expected to increase. Below 200 K, each spectrum is gradually split into two, indicating appearance of the inequivalent V sites. It is noted that $\chi_{\rm bulk}$ becomes slightly anisotropic around 200 K [2]. Although no structural change is reported, the charge gap tends to open with decreasing temperature in the semiconducting

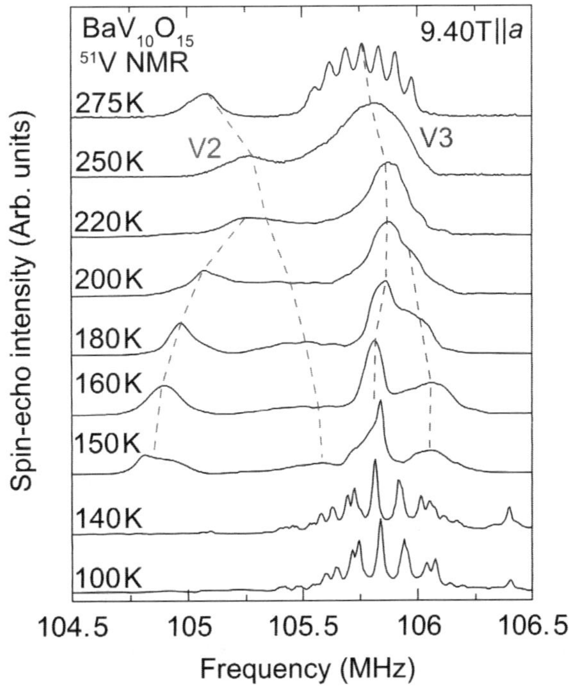

Figure 3. (Color online) Temperature dependence of ^{51}V NMR spectra of the single crystal in $BaV_{10}O_{15}$ at the magnetic field parallel to the a axis.

state [2]. These results indicate an electronic phase transition above T_c, possibly involved in the charge and orbital orders.

The shift in one of the V2 sites is positive, whereas that of another one is negative. Namely, the local spin susceptibility at one V2 (V2B) site decreases as temperature decreases. Since the spin parts of the Knight shifts at the two V3 sites already diminish, the three V sites come to spin singlet. This can be regarded as a precursor of the trimer formation. Below 140 K, the spectral profile abruptly changes and a number of sharp lines from the nonmagnetic sites appear. The spectra from the magnetic V ions are wiped out. It manifests the first order transition into the spin singlet state. The emergence of the large quadrupole splitting indicates the orbital order and/or the distortion of the VO_5 octahedra.

3.3. Nuclear spin-lattice relaxation rate

The temperature dependence of $1/T_1$ in the insulating phase is shown in Fig. 4(a). Below T_c, $1/T_1$ is nearly independent of temperature, as expected in the localized spin system. It shows a steep increase below 70 K. A divergent increase indicates the critical slowing down due to the antiferromagnetic order. $1/T_1$ shows a peak at $T_N = 46.3$ K and steeply decreases with decreasing temperature.

In order to evaluate the critical exponent of the antiferromagnetic transition, we show in Fig. 4(b) the logarithmic plot of $1/T_1$ against the reduced temperature $1-T/T_N$. Above T_N, it follows a power law, $1/T_1 \propto (1 - T/T_N)^n$, with $n = 0.80$. This is in good agreement with the critical exponent of the two-dimensional antiferromagnetic Heisenberg model [14]. Here the magnetic V site is expected to be V1 and V2, which forms a square lattice separated by triangular spin singlet lattice. Hence the two-dimensional antiferromagnetic interactions can work between the

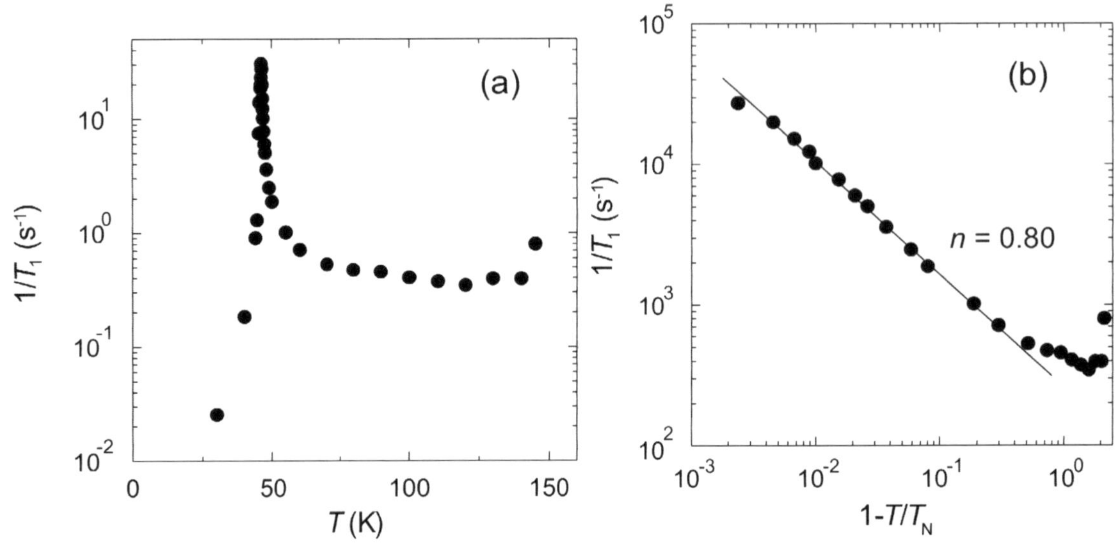

Figure 4. (a) Temperature dependence of the ^{51}V nuclear spin-lattice relaxation rate $1/T_1$ in $BaV_{10}O_{15}$ under the magnetic field of 9.40 T parallel to the c axis close to the magic angle. (b) Log plot of $1/T_1$ vs reduced temperature $1 - T/T_N$. The solid line is a fitting result of $1/T_1 = a(1 - T/T_N)^n (n = 0.80)$.

square lattice by the superexchange via the spin singlet. Actually, we confirmed the presence of the spin singlet sites even below T_N in the NMR spectrum.

4. Conclusion

We investigated the local spin susceptibility in the geometrically frustrated compound $BaV_{10}O_{15}$ by the ^{51}V NMR measurements on the single crystal. We observed the ^{51}V NMR spectra from two of the three inequivalent V sites, having the smaller local spin susceptibility. The presence of the spin-singlet nonmagnetic V sites was clearly observed in the drastic change in the NMR spectrum across the phase transition temperature T_c. Even above T_c, the two spectra were split into four. The temperature dependence showed the local spin susceptibilities at three of the four V sites already diminish above T_c. The result suggests that the charge/orbital order occurs below 200 K and gradually forms the spin-singlet trimer. We observed the critical slowing down toward the antiferromagnetic long-range order in $1/T_1$. The critical exponent behaves as the two-dimensional antiferromagnetic Heisenberg model.

Acknowledgments

This work was financially supported by the Grant-in-Aid for Scientific Research on Priority Areas "Novel State of Matter Induced by Frustration" (Grant No. 22014006) and Special Coordination Funds for Promoting Science and Technology from the MEXT, and also by Grants-in-Aid for Scientific Research (Grant No. 19740224) from the JSPS.

References

[1] Imada M, Fujimori A and Tokura Y 1998 *Rev. Mod. Phys.* **70** 1039
[2] Kajita T, Kanzaki T, Suzuki T, Kim J E, Kato K, Takata M and Katsufuji T 2010 *Phys. Rev. B* **81** 060405(R)
[3] Miyazaki J, Matsudaira K, Shimizu Y, Itoh M, Nagamine Y, Mori S, Kim J E, Kato K, Takata M and Katsufuji T 2010 *Phys. Rev. Lett.* **104** 207201

[4] Ikeda M, Nagamine Y, Mori S, Kim J E, Kato K, Takata M and Katsufuji T 2010 *Phys. Rev. B* **82** 104415
[5] Miyazaki J, Sonehara T, Akahoshi D, Kuwahara H, Kim J E, Kato K, Takata M and Katsufuji T 2009 *Phys. Rev. B* **79** 180410(R)
[6] Liu G and Greedan J E 1996 *J. Solid State Chem.* **122** 416
[7] Bridges C A and Greedan J E 2004 *J. Solid State Chem.* **177** 1098
[8] Bridges C A, Greedan J E and Kleinke H 2004 *J. Solid State Chem.* **177** 4516
[9] Bridges C A, Hansen T, Wills A S, Luke G M and Greedan J E 2006 *Phys. Rev. B* **74** 024426
[10] Pen H F, van den Brink J, Khomskii D I and Sawatzky G A 1997 *Phys. Rev. Lett.* **78** 1323
[11] Khomskii D I and Mizokawa T 2005 *Phys. Rev. Lett.* **94** 156402
[12] Shimizu Y, Tanaka M, Itoh M and Katsufuji T 2008 *Phys. Rev. B* **78** 144423
[13] Volkoff G M 1953 *Can. J. Phys.* **31** 820
[14] Benner H and Boucher J P 1990 *Magnetic Properties of Layered Transition Metal Compounds*, ed de Jonh L J (Dordrecht/Boston/London: Kluwer Academic Publishers)

International Conference on Frustration in Condensed Matter (ICFCM) IOP Publishing
Journal of Physics: Conference Series **320** (2011) 012072 doi:10.1088/1742-6596/320/1/012072

Successive Transition in Rare-earth Intermetallic Compound $GdNiC_2$

N. Hanasaki[1], K. Mikami[1], S. Torigoe[1], Y. Nogami[1], S. Shimomura[2], M. Kosaka[3], and H. Onodera[4]

[1]Department of Physics, Okayama University, Tsushimanaka 3-1-1, Okayama 700-8530, Japan
[2]Department of Physics, Kyoto Sangyo University, Motoyama, Kamigamo, Kita, Kyoto 603-8555, Japan
[3]Graduate School of Science and Engineering, Saitama University, Shimo-Okubo, Saitama 338-8570, Japan
[4]Institute for Materials Research, Tohoku University, Katahira 2-1-1, Aoba, Miyagi 980-8577, Japan

E-mail: hanasaki@psun.phys.okayama-u.ac.jp

Abstract. In $GdNiC_2$, the magnetic state is investigated under the high magnetic fields and the low temperatures. We find the intermediate phases, in which the magnetization value is close to a half of the full moment, and the successive transition in the phase boundary between the antiferromagnetic, the metamagenetic, and the paramagnetic phases. The mechanism of the intermediate phase is discussed.

1. Introduction

The correlation between the conduction electrons and the local moments provides interesting phenomena. In case of the low-dimensional conduction, the Peierls instability produces the ground state correlating with the local moments. Since the external magnetic field can control the ground state through the interaction between the conduction electrons and the local moments, one can expect the gigantic response and the nontrivial electronic state due to the competitive interactions. In $RNiC_2$, as shown in Fig.1(a), the rare-earth R and the Ni atoms form the chains along the c axis, which make the one-dimensional conduction coming from the d orbital of the rare-earth R and Ni. The R and the Ni chains form the triangular lattice. The warped sheet-like Fermi surface is predicted in different R compounds by the band calculations [1]. Various ground states such as the superconducting state, the antiferromagnetic (AF) state, the ferromagnetic (FM) state and the charge-density-wave (CDW) state are reported so far in $RNiC_2$, which are suggestive of the competition between these electronic states [2-5]. Recently, the X-ray diffraction in $SmNiC_2$ reveals the lattice modulation with the wave number $\boldsymbol{q} = (0.5, \eta, 0)$ ($\eta \sim 0.5$) [5]. The band calculation also suggests that there is a Peierls instability for the modulation of this wave number $\boldsymbol{q}$ [1,5]. In this paper, we report the electronic phase diagram at the low temperatures and under the high magnetic fields in $GdNiC_2$.

2. Experimental

The sample crystal is synthesized by a conventional argon arc technique. The characterization of the sample is performed by the X-ray diffraction. The magnetization is measured by the Quantum Design PPMS.

3. Results and Discussion

Figure 1(b) displays the temperature dependence of the magnetization under the magnetic field applied along the *c* axis. Above 24K, the magnetization shows the paramagnetic behavior. The anomaly of the magnetization is observed around 20K. As illustrated in Fig.2, we defined the phase boundary between AF1 phase and PM1 (paramagnetic 1) phase by this magnetic anomaly. As shown in the inset of Fig.1(b), the Weiss temperature is estimated to be ~ –25K, implying the AF interaction [2]. In $TbNiC_2$, the AF state also exists at the low temperature. The neutron diffraction reveals the spin

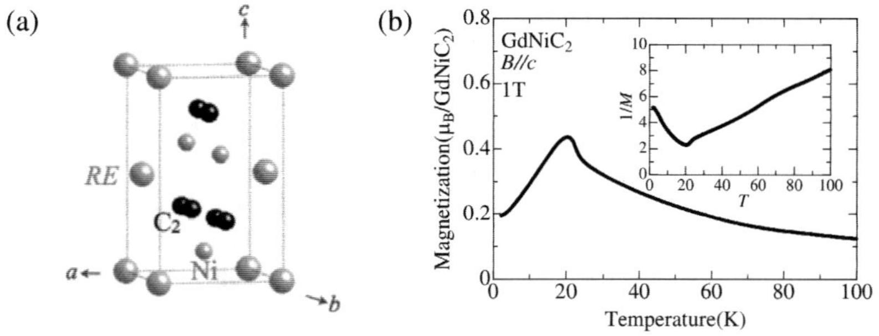

Figure 1. (a) Crystal structure of $RNiC_2$. The grey, the white, and the black circles indicate the rare-earth *R*, the Ni, C atom, respectively. (b) Temperature dependence of the magnetization under the magnetic field parallel to the *c* axis at the higher temperatures. Inset : Reciprocal of the magnetization.

superstructure having the non-collinear spin configuration with the wave number $\boldsymbol{q}$ = (0.5, ~0.5, 0) in the AF state [6]. If the sheet-like Fermi surface in $TbNiC_2$ is similar to those in $SmNiC_2$ and $GdNiC_2$, this wave number $\boldsymbol{q}$ is close to the nesting vector of the Fermi surface, and the AF state is anticipated to be ascribed to the spin-density-wave (SDW) state coming from the Peierls instability. Thus, as a possible origin of the AF1 in $GdNiC_2$, we speculate the spin-density-wave state, in which the spin of the conduction electrons is antiferromagnetically coupled with the 4f moment of *R*. To reveal the mechanism of the AF1 phase in $GdNiC_2$, it is necessary to reveal the magnetic periodicity of the spins for the future study.

Figure 3(a) displays the magnetic-field dependence of the magnetization in $GdNiC_2$. At 2K, the magnetization increases steeply in 6T, as indicated by the arrow (A) in the main panel [3]. The clear hysteresis of the magnetization curve is also seen, suggesting the character of the first-order transition. Since the full moment of Gd is $7\mu_B$ (M_s), the magnetization value just below the transition in 6T corresponds to one third of the full moment ($M_s/3$). As iliustrated in Fig.2, this transition corresponds to the phase boundary between the AF2 phase and the MM (metamagnetic) phase. The MM transition field increases with the temperature.

Moreover, at the higher temperatures, other anomalies emerge in 4.5T and 6T, as indicated by the arrows (E) and (D) in the inset of Fig.3(a), respectively. The hysteresis of the magnetization is seen in 4.5T and 6T, implying the first-order transition. In the phase diagram of Fig.2, the anomaly in 4.5T corresponds to the transition between the AF2 phase and the IM2 (intermediate 2) state, while the anomaly in 6T to the transition between the IM2 phase and the IM1 (intermediate 1) phase. In the transition between the IM2 and the IM1 phases, the magnetization has a value of ~2.6 μ_B, which is close to one third of the full moment (M_s/3). In the transition between the IM2 and the AF2 phases, the magnetization value is ~1.7 μ_B, which corresponds to a quarter of the full moment (M_s/4). As the temperature increases, the transition field (D) approaches the transition (E), indicating the disappearance of the IM2 phase at the higher temperature. One can see the spin flop around 1.2T at 2K, as shown by the arrow (B) of Fig.3(a). The spin-flop field increases with the temperature. As a result, the spin-flop field (B) verges on the transition field (E). The AF2 phase is suppressed at the higher temperature, as well.

Figure 3(b) displays the temperature dependence of the magnetization. In 6.5T, one can see the hump of the magnetization in the temperature region between 18K and 21K. This hump corresponds to the IM1 phase. In 6T, there is the dip of the magnetization in the temperature region between 12K and 18K. This magnetization dip is ascribed to the IM2 phase.

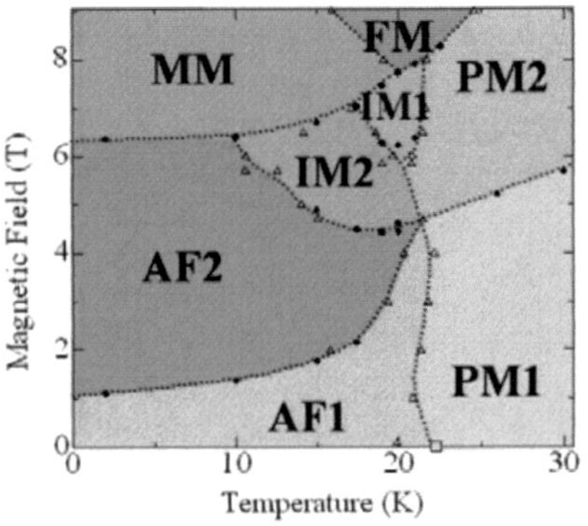

Figure 2. Temperature dependence of the magnetization of $GdNiC_2$ under the magnetic field along the *c* axis. Here, AF, FM, IM, PM, MM denote the antiferromagnetic, the ferromagnetic, the intermediate, the paramagnetic, and the metamagnetic states, respectively.

As shown by the arrow (C) in the inset of Fig.3(a), the enhancement of the magnetization near 7T corresponds to the MM transition mentioned above. The magnetization just below the MM transition has the value of ~ 3.3 μ_B, which is close to a half value of the full moment (M_s/2). If there are four independent spin sites, it is reasonable to consider that three spins of the four sites are parallel to the

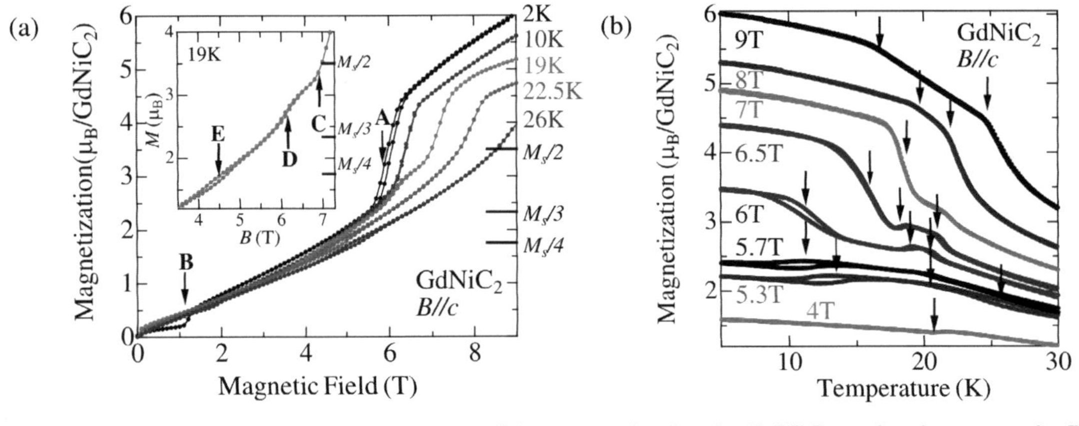

Figure 3. (a) Magnetic-field dependence of the magnetization in $GdNiC_2$ under the magnetic field along the *c* axis. Inset : Magnetization at 19K. (b) Temperature dependence of the magnetization. The arrows indicate the anomaly of the magnetization.

magnetic field, and one spin is antiparallel in the IM1 phase. Here, for simplicity, one assumes the wave number $\boldsymbol{q}$ = (0.5, 0.5, 0) of the modulation, which gives independent four spin sites. Since the amplitude in the SDW decreases by the applied magnetic field, the SDW gap is anticipated to be reduced. To clarify the correlation between the CDW and the SDW in the IM phases is left for the future study.

In summary, we investigate the electronic state in $GdNiC_2$ at the low temperatures and under the high magnetic fields, and find the successive transition and its associated the IM1, the IM2, and the PM2 phases. The magnetization value at the transitions in the IM1 phase is close to a half of the full moment of Gd. The SDW gap is anticipated to be reduced in the subphases under the magnetic field.

This work was supported by Grant-in-Aid for Scientific Research on Priority Areas "Novel States of Matter Induced by Frustration (19052003)".

References

[1] J.Laverock, T.D.Haynes, C.Utfeld, and S.B.Dugdale, Phys.Rev.B **80**, 125111 (2009). I.Hase and T.Yanagisawa, J.Phys.Soc.Jpn. **78**, 084724 (2009).
[2] H.Onodera, Y.Koshikawa, M.Kosaka, M.Ohashi, H.Yamaguchi, and Y.Yamaguchi, J.Mag.Mag.Mater. **182**, 161 (1998).
[3] S.Matsuo, H.Onodera, M.Kosaka, H.Kobayashi, M.Ohashi, H.Yamaguchi, and Y.Yamaguchi, J.Mag.Mag.Mater. **161**, 255 (1996).
[4] A.D.Hillier, J.Quintanilla, and R.Cywinski, Phys.Rev.Lett. **102**, 117007 (2009).
[5] S.Shimomura, C.Hayashi, G.Asaka, N.Wakabayashi, M.Mizumaki, and H.Onodera, Phys.Rev.Lett. **102**, 076404 (2009).
[6] N.Uchida, H.Onodera, M.Ohashi, Y.Yamaguchi, N.Sato, and S.Funahashi, J.Mag.Mag.Mater. **145**, L16 (1995). J.K.Yakinthos, P.A.Kotsanidis, W.Schafer, and G.Will, J.Mag.Mag.Mater. **81**, 163 (1989).

Hydrothermal Synthesis of Hexagonal Vanadium Oxide

T.Tokuyama[1], T.Okusako[1], K.Mikami[1], N.Hanasaki[2], and Y.Nogami[1]

[1]Department of Physics, Okayama University, Tsushimanaka 3-1-1, Okayama 700-8530, Japan

[2]Department of Physics, Osaka University, Machikaneyama 1-1, Toyonaka 560-0043, Japan

hanasaki@phys.sci.osaka-u.ac.jp

Abstract. We synthesize the hexagonal vanadium oxide containing the alkaline-metal ion (Li, Na, and K). The crystal belongs to the magnetoplumbite-type structure. The title compound, in which the vanadium valance is not an integer, shows the metallic and the paramagnetic behavior. The electronic specific heat coefficient γ is 34.7 mJ/K^{2}mol, which is perhaps due to the spin fluctuation.

1. Introduction

In the strongly correlated electron system, the degree of freedom of the electron provides rich physics, such as the charge, spin, and orbital orders and the gigantic response to the external stimuli. This is also effective for the enhancement of the thermoelectric effect, as reported in $NaCo_2O_4$ [1]. The t_{2g} orbital systems are good candidate for the thermoelectric material, since one can expect the orbital degeneracy under the sixfold oxygen coordination. Since the transition-metal atom of the IV and V groups tends to have the highest oxidization state, it is necessary to synthesize under the highly reducing atmosphere. To introduce the alkaline-metal ion in the crystal is advantageous to the control of the band filling. Taking account of these synthesis conditions, we try the hydrothermal synthesis under the high temperature and the high pressure and under the condition of the acid or the base. Recently, we obtain the new compound of the hexagonal vanadium oxide, and report its physical property.

2. Experimental

The single crystal of the hexagonal vanadium oxide is obtained by the hydrothermal synthesis method. The LiOH, NaOH, KOH, V_2O_3, and VO_2 are mixed in the glove box. Then, these starting materials are put in the autoclave with the purified water. The typical synthesis condition is 700℃ and 135MPa (24 hours). The resistivity, the heat capacity, and the Seebeck effect are measured under the magnetic field by use of the Quantum Design PPMS. The magnetization is acquired by the Quantum Design MPMS2. The molar unit in the magnetization corresponds to one vanadium. The characterization of the sample is performed by the scanning electron microscopy (SEM) and the energy dispersive X-ray analysis (EDX), and the X-ray diffraction (XRD).

3. Results and discussion

By the hydrothermal synthesis, we obtain the black hexagonal crystal, as shown in the SEM image in Fig.1 (c). Figures 1(a) and (b) show the crystal structure determined by the X-ray diffraction of the single crystal at room temperature. The refined atomic coordinates and the site of occupancy

Table 1 Atomic coordinates and SOF in $Li_xNa_{0.6}K_{0.5}V_{10.5}O_{19}$ (Z= 2)

Atom	Wyckoff position	x/a	y/b	z/c	SOF
Alkaline metal (1) site	2d	2/3	1/3	1/4	1.00
$V(1)^{oct}$	12k	0.3378(2)	0.1689(1)	0.10687(6)	1.00
$V(2)^{oct}$	4f	1/3	2/3	0.1922(1)	1.00
$V(3)^{bip}$	2b	0	0	1/4	1.00
$V(4)^{tet}$	4f	1/3	2/3	0.0265(7)	0.25
$V(5)^{oct}$	2a	0	0	0	1.00
O(1)	4e	0	0	0.1578(4)	1.00
O(2)	4f	2/3	1/3	0.0594(4)	1.00
O(3)	6h	0.642(1)	0.8210(4)	1/4	1.00
O(4)	12k	0.1483(4)	0.8516(6)	0.0549(2)	1.00
O(5)	12k	0.4997(6)	0.998(9)	0.1505(2)	1.00

The superscripts on V(1)–(5) represent oxygen coordinations; oct (octahedral), bip (bipyramidal), tet(tetrahedral).

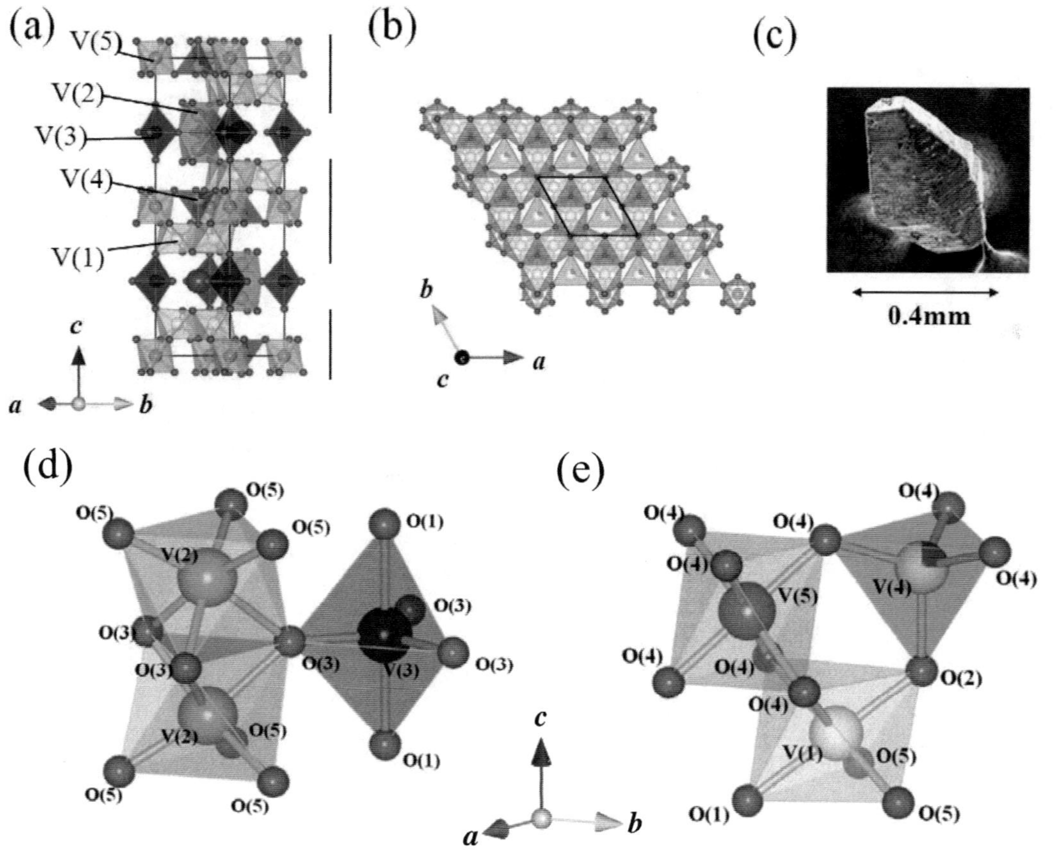

Figure 1. (color online) (a) Crystal structure of the title compound viewed from the direction perpendicular to the c axis. The alkaline-metal ion and oxygen correspond to the purple and red ball, respectively. The vertical thick bars indicate the position of the layer containing the Kagomé lattice. (b) Crystal structure of the layer containing the Kagomé lattice viewed from the c axis. The black parallelogram indicates the unit cell. (c) SEM image of the sample. (d) Views of V-O polyhedral located around z = 1/4, and (e) z = 0.

factor (SOF) are given in Table 1. The space group is the $P6_3/mmc$. The lattice parameters are a = 5.7905(1)Å, c = 22.663(1)Å. This crystal, whose atomic composition estimated by the EDX is $Li_xNa_{0.6}K_{0.5}V_{10.5}O_{19}$, belongs to the magnetoplumbite-type structure with the alkaline-earth metal [2]. Figures 1(d) and (e) show the local structure of the oxygen and the vanadium ions. The crystal analysis suggests the alkaline-metal ions exist in one kind of site, that is, the alkaline metal (1) site listed in Table 1. In this crystal analysis, this ion is assumed to have the intermediate number of electrons between the Na and the K ions. In addition, the diffuse scattering indicating the short-range order of the alkaline-metal ions is not observed. Thus, the Li, whose amount is anticipated to be much smaller than other alkaline-metal ions, Na, and K ions exist randomly in the alkaline-metal (1) site. It is difficult to determine the amount of the Li ion by the measurement of the EDX. The title compound can be obtained, when the LiOH is included in the starting material. Therefore, the title compound seems to contain a slight amount of the Li ion, which is expected to be less than ~0.2 by comparing other magnetoplumbite oxides [3]. We performed the refinement of the crystal analysis by adjusting the SOF in the V(4). As a result, we found that the SOF of V(4) is about 1/4. In this magnetoplumbite-type structure, the V(1) forms the Kagomé lattice. The layer containing this Kagomé lattice is stacked along the c axis and is separated by the layer containing the alkaline-metal ion. The average of the V(1), V(2), V(3), and V(5) valence estimated by the atomic composition of $Li_xNa_{0.6}K_{0.5}V_{10.5}O_{19}$ is about 3.4. Here, the valence of the V(4) is assumed to be 5.

Figure 2 shows the temperature dependence of the resistivity. The resistivity gradually decreases on lowering the temperature. This increases below 100K and then reaches maxima around 25K. Below 20K, the resistivity is reduced by the application of the magnetic field. This negative magnetoresistance is perhaps due to the reduction of the spin scattering.

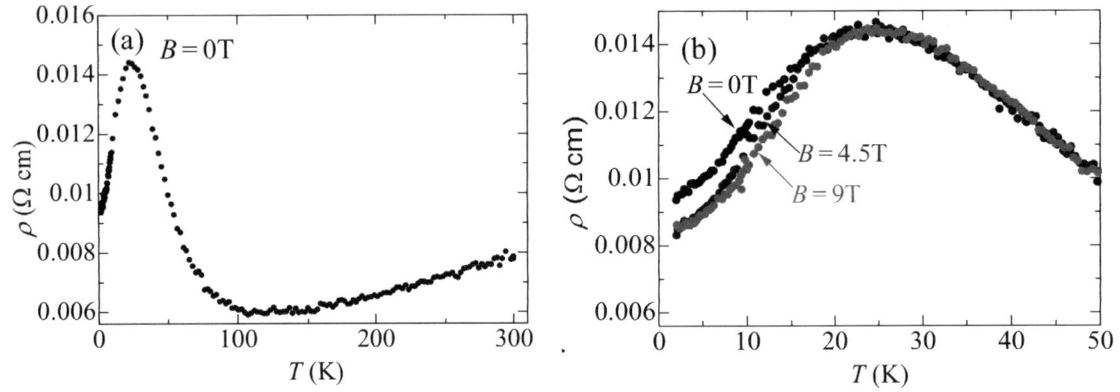

Figure 2. (color online) Temperature dependence of the resistivity ρ in the title compound in the temperature region between 300K and 2K (a), and in the low temperature region (from top to bottom, 0T(black), 4.5T(blue), and 9T(green)) (b).

Figure 3 displays the temperature dependence of the magnetic susceptibility χ measured in 1T. The title compound shows the Pauli-like paramagnetism at the higher temperatures. On lowering the temperature, the magnetic susceptibility increases below 50K. The increment of the magnetization is reduced below 8K. The magnetization is enhanced sublinearly at 2K, as the magnetic-field strength increases.

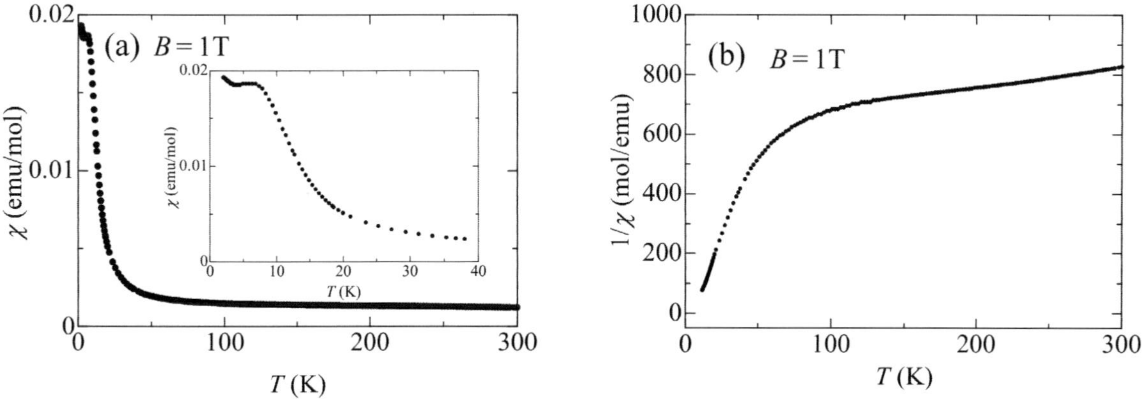

Figure 3. (a) Temperature dependence of the magnetic susceptibility χ in the title compound in the temperature region between 300K and 2K. Inset: χ in the low temperature region. (b) Reciprocal of χ as a function of the temperature.

As shown in Fig.4, the heat capacity divided by the temperature C/T exhibits a broad peak around 8K. This peak temperature coincides with the temperature of the shoulder-type anomaly of the magnetic susceptibility, suggesting that the peak of the C/T is due to the magnetic anomaly. This peak of the C/T is not sharp, indicating the character of the short-range magnetic order. In the absence of a magnetic field, the electronic specific heat coefficient γ is estimated to be 34.7 mJ/K^{2}mol by the exptrapolation toward zero temperature. The peak of the heat capacity and the electronic heat capacity coefficient are reduced by applying the magnetic field, since the spin fluctuation is suppressed by the magnetic field.

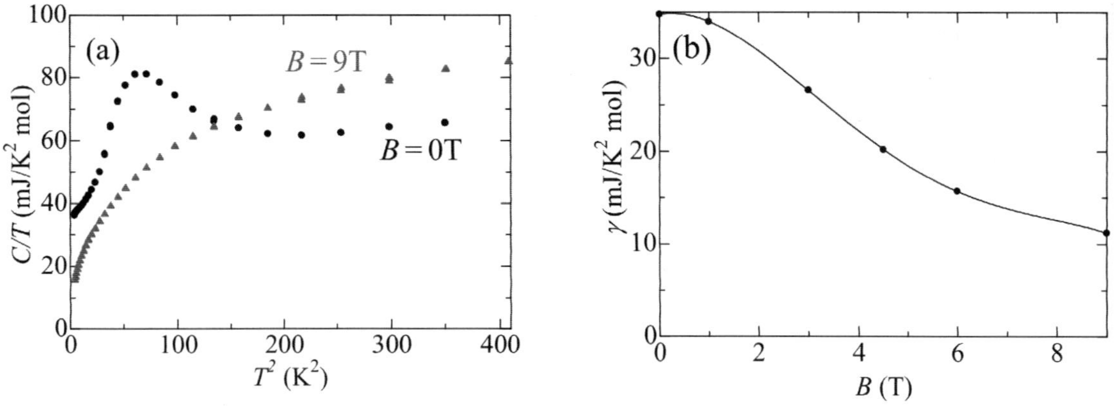

Figure 4. (color online) (a) Heat capacity divided by the temperature C/T as a function of T^2 in the title compound (circle: 0T, triangle: 9T), (b) Magnetic-field dependence of γ. The curve is a guide to the eyes.

Figure 5 displays the temperature dependence of the Seebeck coefficient S. The Seebeck effect increases on lowering the temperature, and then reaches the maxima, whose value is 35μV/K around

26K. The peak of S is associated with that of the resistivity peak. The magnetic-field dependence of the Seebeck effect is not observed.

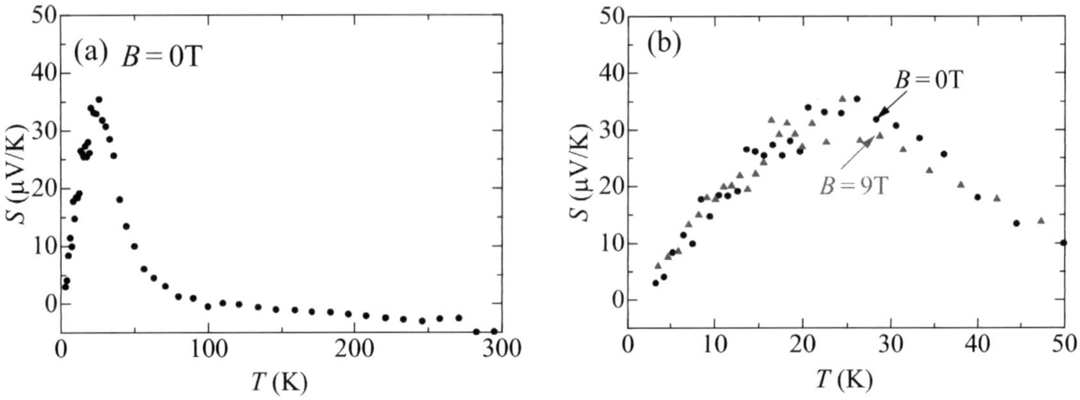

Figure 5. (color online) Temperature dependence of the Seebeck coefficient S in the title compound in the temperature region between 300K and 2K (a), and in the low temperature region (circle: 0T, triangle: 9T) (b).

In summary, we synthesize the new compound of the hexagonal vanadium oxide. The title compound shows the metallic and the paramagnetic behavior. The heat capacity is perhaps enhanced by the spin fluctuation. The electronic specific heat coefficient is 34.7mJ/K^2mol.

This work was supported by Grand-in-Aid for Scientific Research on Priority Areas "Novel States of Matter Induced by Frustration" (19052003).

Reference

[1] I. Terasaki, Y. Sasago, K. Uchinokura, Phys. Rev. B 56, R12685 (1997).
[2] S. Ishiwata, I. Terasaki, M. Azuma, M. Takano, J. Solid State Chem. 181, 1273 (2008).
[3] S. Ito, H. Kurosawa, K. Akashi, Y. Michiue, M. Watanabe, J. Solid State Ionics. **86**, 745 (1996).

Growth Temperature Dependence of the $LaAlO_3/SrTiO_3$ Interfacial Structure

Yusuke Wakabayashi[1], Yoshihiro Yamasaki[1], Christopher Bell[2], Yasuyuki Hikita[2], Harold Y. Hwang[2,3], and Tsuyoshi Kimura[1]

[1]Division of Materials Physics, Graduate School of Engineering Science, Osaka University, Toyonaka 560-8531, Japan

[2]Department of Advanced Materials Science, University of Tokyo, Kashiwa, Chiba 277-8561, Japan

[3] Japan Science and Technology Agency, Kawaguchi, Saitama 332-0012, Japan

E-mail: wakabayashi@mp.es.osaka-u.ac.jp

Abstract. The growth temperature dependence of the interfacial structure of $LaAlO_3$ thin films on $SrTiO_3$ substrates is clearly observed by means of surface x-ray diffraction, combined with a holographic analysis. Although the interfacial structure that is tuned by the fabrication condition is known as a key to the control of various film properties, such a minute change in structure has not been studied in detail. Our observations show that a microscopic evaluation of the interfacial structure, as a function of the growth conditions, is achievable.

Transition metal oxide thin films are widely studied for their possible use in applications. In some cases, the interface between the substrate and the film possesses electronic or magnetic properties that are very different from that of either the film or the substrate. One of the most famous examples is the interface between two band insulators with the perovskite ABO_3 structure, $LaAlO_3$ and $SrTiO_3$.[1, 2] With certain growth conditions, this interface exhibits metallic conduction to low temperatures, showing a surprisingly large carrier mobility[1, 3]. Various groups have reported wide range of properties of the interface, and the differences are attributed to the film thickness[4] and the growth conditions such as the oxygen pressure and the temperature for the pre-anneal, growth, and post-anneal processes for pulsed laser deposition (PLD).[2] In other words, films fabricated under different condition provide different conductivity. This implies that the seeds of these differences must be in the film structure. However, the structural features as a function of growth conditions have not been studied in detail.

Structural characterization for transition metal oxide films has been mostly made by reflection high-energy electron diffraction (RHEED) during the film growth process and atomic force microscopy (AFM) measurements, post-fabrication, to observe the surface topography. X-ray diffraction is also often used to obtain the in-plane and out-of-plane lattice parameters and the homogeneity of the thickness of the film. Less is known about the detailed atomic distributions, and chemical intermixing at the interface. To clarify the interfacial structure, only a limited numbers of studies have been made by scanning transmission electron microscope (STEM) or similar electron microscopy on thinned section.[5] Recent developments in the analysis techniques of x-ray diffraction make it possible to observe the detailed atomic arrangements at the interface

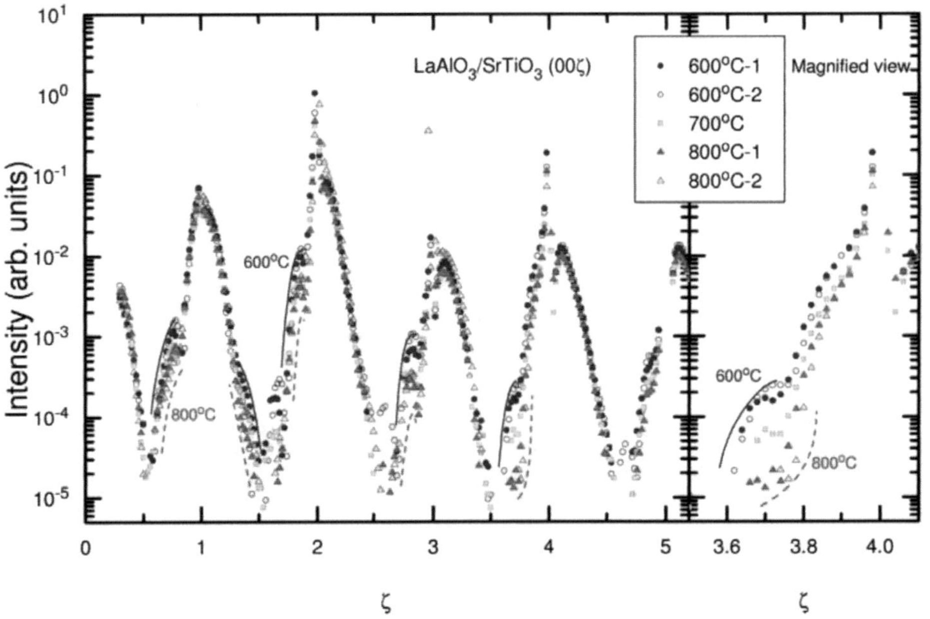

Figure 1. X-ray scattering intensity profiles along (00ζ) for two 600°C-samples, one 700°C-sample and two 800°C-samples. Right hand panel shows a magnified view around the (004) Bragg reflection of the substrate.

of ultrathin films[6, 7, 8]. In this paper, we demonstrate that it is possible to observe minute structural differences caused by variations in the growth temperature of $LaAlO_3$ film by means of the surface x-ray diffraction, which is a nondestructive technique.

The $LaAlO_3$ ultrathin films were fabricated by means of PLD, using 7.5 mm by 7.5 mm TiO_2 terminated (001) $SrTiO_3$ substrates. Following a pre-anneal at 950°C in 5×10^{-6} torr of oxygen for 30 minutes, the $LaAlO_3$ deposition was carried out at a growth pressure of 10^{-5} torr of oxygen with a total laser energy of 40 mJ, (spot size at the target = 2.5 mm^2), using a laser repetition rate of 2 Hz. The thickness of the films were monitored by RHEED oscillations and controlled to 5 unit cells. The samples were cooled to room temperature at the deposition pressure. During the deposition, the temperature of the substrate was set to 600, 700 or 800 °C. In the following, we refer to the samples using this temperature label. Five films were prepared: two 600°C-samples, one 700°C-sample and two 800°C-samples. X-ray diffraction measurements were performed at BL-3A of the Photon Factory, KEK, Japan. A standard four-circle diffractometer equipped with a scintillation counter is installed in this beamline. The x-ray energy for the measurements was 14 keV.

The structure of a ultrathin film made on a (001) substrate is reflected in the x-ray scattering intensity profile along the c^*-direction.[9] One of the simplest examples is the scattering intensity profile along (00ζ), which includes the reflectivity profile. This profile is given by the absolute-square of the Fourier transformation of the electron density along the depth direction. In normal reflectivity measurements, only a limited part of q-space is analyzed, resulting in a poor spatial resolution. In order to achieve atomic resolution, we measured a larger range of q-space including several Bragg reflections in this study. This method is known as crystal truncation rod scattering.

X-ray scattering intensity profiles along (00ζ) up to ζ=5.2 were obtained by performing rocking curve measurements at various ζ positions. Throughout this paper, we use the reciprocal

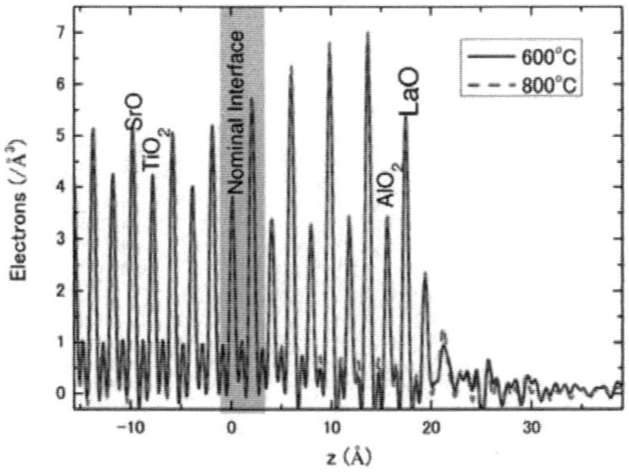

Figure 2. Electron density distribution of a 600°C-sample and a 800°C-sample around the interface.

lattice defined by the substrate crystal lattice. Figure 1 shows the scattered intensity per unit illuminated area for all of the five samples. The results clearly show that the intensity profiles of the two 600°C-samples are similar to each other, and the two 800°C-samples are also similar, while the 600°C-samples data can be clearly distinguished from the 800°C-samples. Moreover, the intensity profile for the 700°C-sample is intermediate between the higher and lower growth temperature data. Since the intensity profile is directly connected to the structure, these results clearly indicate that the film structure changes with growth temperature.

An important question is what structural feature is controlled by the growth temperature? To answer this question, we utilized a recently developed analyzing method for surface x-ray diffraction data: coherent Bragg rod analysis[6, 7]. This method is a kind of holography, and has been utilized to analyze not only oxide films[8] but also an extremely complicated organic semiconductor structure[10]. In this method, the electron density of the unknown structure around the surface is obtained as the difference from that of a reference structural model. Therefore, this method is ideal to find minute changes between films having very similar structures. The data for samples 600°C-1 and 800°C-1 in Fig. 1 are processed with this method. Since an electron density analysis requires quasi-continuous intensity data, we interpolated the experimental results. For the regions with no intensity data available because of low intensity or too close proximity either to the direct beam or substrate Bragg reflections, we used the calculated intensity profile for the reference structure model that reproduces the experimental result rather well. This analysis procedure follows that of our previous report on an organic semiconductor[10].

The resultant depth (z) profiles of the electron densities for the two samples are shown in Fig. 2. The alternating peaks in the electron density, corresponding to the AO and BO_2 planes, are clearly visible. The flat and very low electron density in the region $z > 23$Å, corresponding to the vacuum beyond the sample, ensures that the electron density given by this analysis is correct. The peaks close to the surface, i.e., those between 17Å and 23Å, show significant suppression, meaning that there is some surface roughness as well as the miscut of the substrate. Although the profiles shown in Fig.2 look very similar to each other at the first sight, there *are* some differences in interfacial atomic mixing. In order to present the differences more clearly, we plot the ratio of the electron number in the AO planes and the averaged electron number in the neighboring BO_2 planes in Fig.3. This ratio should be 1.2 in $SrTiO_3$ (electron number of $[Sr^{2+}+O^{2-}]/[Ti^{4+}+2O^{2-}] = [36+10]/[18+20]$) and 2.1 in $LaAlO_3$ ($[La^{3+}+O^{2-}]/[Al^{3+}+2O^{2-}] = [54+10]/[10+20]$)). For both the 600°C-1 and 800°C-1 samples, these ratios are 1.2 inside the substrate, and more than 2 in the film, indicating good stoichiometry of the films. The depth profile of the ratio shows the interfacial structure including the atomic mixing. According to Ref.[8], the interface between $LaAlO_3$ and TiO_2-terminated $SrTiO_3$ has one layer of $LaTiO_3$; this structure is schematically shown in the inset of Fig. 3. The open triangles in the main

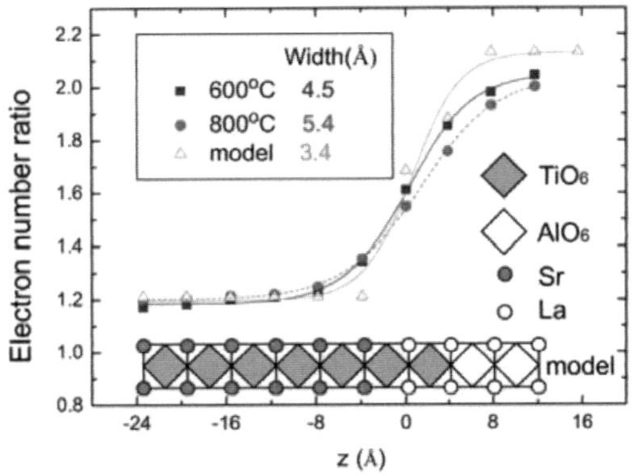

Figure 3. Depth dependence of the ratio of the electron number in the AO planes and the averaged electron number in the neighboring BO_2 planes. Closed squares and circles show the results for the 600°C-1 and 800°C-1 samples, and the open triangles show the values for the model interface that has one layer of $LaTiO_3$ at the interface. Inset:(inset) Schematic view of the model structure, which is made bya simplification simplified version of the result of Ref.[8].

panel show the calculated electron number ratio for an interface having such a structure, i.e., the ($SrTiO_3$/)SrO/TiO_2/LaO/TiO_2/LaO/AlO_2(/$LaAlO_3$) interface. Note that our surface x-ray scattering result supports the structural model. The slope of the ratio at the interface for the model structure is steeper than the experimental values, meaning that the experimental results show some interfacial roughness. In order to quantify this interfacial roughness, these profiles were fitted using the error function. As a result, we obtained interfacial r.m.s. roughness of 4.5Å, 5.4Å, and 3.4Å for the 600°C-1 sample, the 800°C-1 sample and the model structure, respectively, which are very close to the interfacial roughness reported by STEM measurements[5]. Our results allows us to find minute changes in structure caused by the deposition condition, and this increase in interfacial roughness for the higher growth temperatures may be related to the adatom mobility and thermodynamic driving force towards intermixing.

To summarize, surface x-ray diffraction experiments combined with the coherent Bragg rod analysis make it possible to clarify minute changes in the interfacial structure caused by changes in the film growth conditions. This method provides an opportunity to study the growth condition dependence of the film properties in detail.

Acknowledgement
This work was supported by the TORAY science foundation and the support center for advanced telecommunications technology research. This work was also supported by KAKENHI (21740274,19052002) and Global COE Program (G10).

References
[1] A. Ohtomo and H.Y. Hwang, Nature **427**, (2004) 423.
[2] M. Huijben, A. Brinkman, G. Koster, G. Rijnders, H. Hilgenkamp and H. A. Blank, Adv. Mater. **21**, (2009) 1665.
[3] S. Thiel G. Hammerl, A. Schmehl, C.W. Schneider, J. Mannhart, Science, **313**, (2006) 1942.
[4] C. Bell, S. Harashima, Y. Hikita, and H. Y. Hwang, Appl. Phys. Lett. **94**, (2009) 222111.
[5] N. Nakagawa, H.Y. Hwang and D.A. Muller, Nature Materials, **5**, (2006) 204.
[6] Y. Yacoby, M. Sowwan, E. Stern, J. O. Cross, D. Brewe, R. Pindak, J. Pitney, E. M. Dufresne and R. Clarke, Nature Materials **1**, (2002) 99.
[7] M. Sowwan, Y. Yacoby, J. Pitney, R. MacHarrie, M. Hong, J. Cross, D. A. Walko, R. Clarke, R. Pindak, and E. A. Stern, Phys. Rev. B **66** (2002) 205311.
[8] P. R. Willmott, S. A. Pauli, R. Herger, C. M. Schlepütz, D. Martoccia, B. D. Patterson, B. Delley, R. Clarke, D. Kumah, C. Cionca, and Y. Yacoby, Phys. Rev. Lett. **99**, (2007) 155502.
[9] V. Vonk, M. Huijben, K. J. I. Driessen, P. Tinnemans, A. Brinkman, S. Harkema, and H. Graafsma, Phys. Rev. B **75**, (2007) 235417.
[10] Y. Wakabayashi, J. Takeya, and T. Kimura, Phys. Rev. Lett. **104**, (2010) 066103.

Anomalous behavior of linear resistivity in the vanishing process along the intergrain ordering of weak-sintered ceramic superconductor $YBa_2Cu_4O_8$

M Hagiwara[1], A Fujii[1], T Shima[2] H Deguchi[3], T Shoho[3] and K Koyama[4]

[1] Faculty of Engineering Design, Kyoto Institute of Technology, Kyoto, 606-8585, Japan

[2] Advanced Technology Center, Kyoto Institute of Technology, Kyoto, 606-8585, Japan

[3] Faculty of Engineering, Kyushu Institute of Technology, Kitakyushu 804-8550, Japan

[4] Institute of Socio-Arts and Sciences, The University of Tokushima, Tokushima 770-8502, Japan

E-mail: hag@kit.ac.jp

Abstract. Intergrain ordering nature of weak-sintered ceramic of $YBa_2Cu_4O_8$ (Y124) is studied experimentally by ultra precise electric resistivity observation together with magnetic measurements for linear and nonlinear susceptibilities. For this purpose, a new method to detect pure linear resistivity down to the order of 1 nΩm has been developed with the help of combination technique of pulse delta method and FFT treatment. Making use of this experimental system for several weak-sintered Y124 samples, it has been revealed that linear resistivity first drops toward zero steeply with decreasing temperature, and then turns abruptly upward to forms faint maximum at lower temperature. Such peaking of electric resistivity has been found to occur below the intergrain ordering temperature that was estimated by the magnetic behavior. This newfound phenomenon, to say 'revival of resistivity', is being regarded as general nature for the weak-sintered ceramics, and considered to be an intrinsic frustration effect for actual d-wave type ceramics of irregular network structure.

1. Introduction

In sintered materials as aggregates of fine size grains of cuprate high-T_c superconductor, of which the pairing symmetry is d-wave kind, frustration effect against the intergrain ordering is expected, because π-junctions as well as ordinary ones are necessarily included in the intergrain Josephson-coupling network.[1-5] This effect may restrain the grain system form becoming simply unified superconductive state, and thus some novel characteristics is expected in the ordering and the ordered state. On the theoretical viewpoint, a deeply studied ordering mechanism is 'chiral glass transition' that is random freezing of spontaneous loop-supercurrents at local intergrain circuits.[4,5]

Published under licence by IOP Publishing Ltd

As experimental approaches, ceramic material of stoichiometric cuprate superconductor $YBa_2Cu_4O_8$ (Y124) has been studied continuously by the authors, and the following facts were reported. Through the magnetic measurements, two-step-wise dc magnetization behavior against temperature clearly reflects the transition process from intragrain to intergrain orderings,[6] and a singular peak of nonlinear magnetic susceptibility is formed at the intergrain ordering temperature T_{c2}.[7] Besides, through the measurements of the electric ac resistivity, tailing of linear resistivity was detected around temperature region below T_{c2}, and was discussed to be a possible frustration effect of the intergrain ordering.[8] Afterward recently, nonlinear magnetic susceptibility of the Y124 samples prepared by various sintering conditions were studied, and it has been found that another anomaly may appear apart from the chief negative singular peak at T_{c2} when the sintering duration is not so long.[9] This result has suggested that the intergrain ordering process advances with some inhomogeneous nature,[10] and that successive change in the ordering structure might be caused. If that is the case, the successive behavior is thought to be reflected naturally by electric transport phenomena for the samples of the frustrated ceramic superconductor.

Thus the present subject is directed to clarify detailed behavior of linear resistivity in the process of intergrain ordering for the weak-sintered ceramic Y124. Here we should note that possible resistivity brought by the fluctuation in the ordered state holding frustration must be reflected by linear resistivity, yet the signal revel is estimated to be order of nΩm or less.[5] In order to examine such intrinsically small resistivity, therefore, a new experimental method of enough performance in both sensitivity and offset elimination is now required. Accordingly, the authors designed and constructed a new measurement system for linear and nonlinear resistivity observation, using combination techniques of 'delta method' [11] by pulse applied current and Fourier analysis method for the periodic measurements by sinusoidally sequenced excitation current pulses. With this system, intrinsic temperature dependent behavior of very weak but physically valid signal of linear resistivity is searched near and below T_{c2} for several samples of the weak-sintered ceramic Y124.

2. Experimental

First of all, fine-powder form Y124 was obtained as the calcined product of the developed synthetic method.[12,13] Then, the calcined powder was sieved by mesh size down to 10 μm, and then was press-molded into pellets using pressure in range of 9.3-18.5 MPa. These obtained several pellets were sintered at 780°C for 25-50 h, to be the weak-sintered ceramic series with varied forming and sintering conditions.

Our new measurement system for very weak electric resistance is outlined as follows. Elemental working is 4-wire voltage-drop detection by 'pulse-delta-method' that can compensate time-dependent offset signal using three excitation current pulses alternating quickly between a set value and zero. As the further advanced method (named SSPD), the current pulse height I is sequentially swept along sinusoidal waves as $I(n) = I_0\sin\{2\pi n/512\}$ (with $n = 0\text{-}511$), and the stationary response of the voltage drop $V(n)$ for each step is measured and stored. The unit periodic measurement of 512 points is repeated optionally settled times, and the accumulated multi periods data are averaged into one periodic data in order to increase the S/N ratio or the final sensitivity.

Such obtained periodic voltage data are analyzed by FFT to be decomposed into fundamental and harmonic components with a computer. The derived in-phase n-th harmonics is represented by $V'_{n\omega}$, and the quantity $V'_{1\omega}/I_0$ correspond to first term approximation for linear resistance as the first expansion coefficient of $V(I)$. It is noted that the analyzed data cannot inform us of any physically dynamic characteristics, because each unit pulse measurement is executed at stationary state.

Using this SSPD method with current $I_0 = 8$ mA, precise resistivity behaviors for the several prepared ceramic samples are traced against temperature. At each controlled temperature, the periodic sweep was repeated 4 - 16 times, which was changed with the signal level. The resistivity results in this report are chiefly shown in quantity $\rho_{1\omega}$ as the first-term-approximated linear resistivity that was reduced by taking into account the sample size factor.

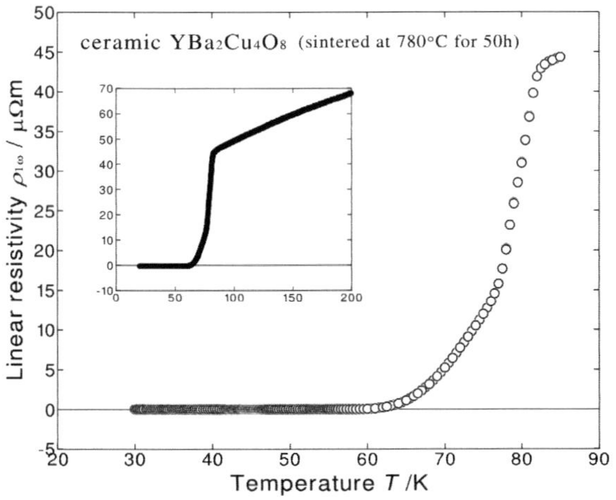

Figure 1. Full range temperature variation of linear resistivity for weak-sintered ceramic of $YBa_2Cu_4O_8$. The main view results were measured by SSPD (see the text) method by excitation pulse height of 8 mA, and the values were shown by fundamental resistivity $\rho_{1\omega}$. The results of the insert were measured by ordinary delta-method of ecitation current 4 mA.

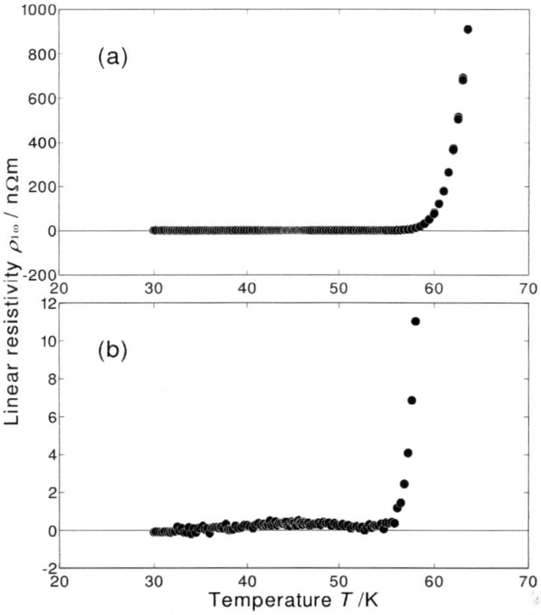

Figure 2. Perpendicularly enlarged scopes of the linear resistivity, by re-plotting the same data in the main view of Fifure 1. Applied magnification is 50 for (a) and 4200 for (b).

For one of the typical samples, magnetic behaviors were also observed by a SQUID system. AC susceptibilities χ' and χ'', and harmonic magnetizations $M'_{n\omega}$ were measured with weak excitation magnetic field of frequency 1.0 Hz and amplitude of 0.1 Oe, and then the temperature dependences were examined in comparison with the electric resistivity data.

3. Results and discussion

In the present report, we primarily focus on a sample, that was obtained by 50 h sintering, among the several prepared weak-sintered ceramic samples. The experimental result will show a new novel behavior of linear resistivity just below the intergrain ordering temperature T_{c2}. To begin with, the full range behavior of $\rho_{1\omega}(T)$ is shown in Figure 1. The characteristic dropping curve of resistivity certainly reflects typical temperature dependence of ceramic Y124 showing the intergrain ordering.[8] To describe from higher temperature side, $\rho_{1\omega}$ begins steep dropping around 81 K (i.e. T_c of crystalline Y124), then gradually lowers during intergrain incoherently disordered state, and finally tends to (almost) zero toward intergrain ordering state. Possible minute resistivity near and below this ordering temperature is a matter of the present concern.

Now, the data around the critical region in Figure 1 are re-plotted in two-stepwise expanded views of Figure 2 (a) and (b), whose respective longitudinal scopes are 1/50 and ~1/4200 of the full area. Seeing in comparison the temperature regions where $\rho_{1\omega}$ appears to reach zero in Figures 1 and 2, we notice that the $\rho_{1\omega}(T)$ curve remains non-zero tailing in the more expanded views. When the scope becomes of Figure 2 (b), instead of such apparent lowering of the vanishing temperature, an unexpected anomalous curve appears at the lower temperatures. To show precisely, further enlarged view is drawn in Figure 3 with the help of moving-average-filtering. Seen from high temperature side in this figure, while $\rho_{1\omega}$ first approaches zero monotonously, the dropping abruptly changes to rising at ~52.6 K, forming a minimum just the above zero level. Below this temperature, $\rho_{1\omega}(T)$ shapes a

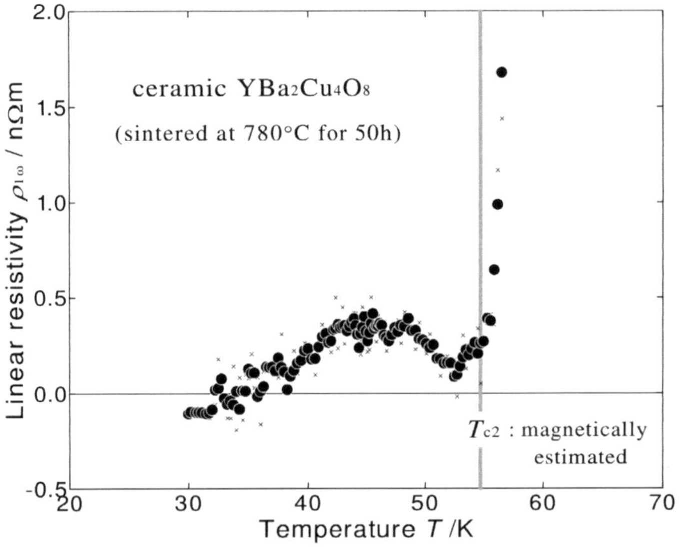

Figure 3. Perpendicularly 2500 times enlarged view of the linear resistivity, which are treated from the same data as in Figure 1. The solid circles are three points moving-average-treated data, and the small crosses are the original data.

somewhat broad peak around ~46 K, and then it decreases again with the lowering temperature. Such found behavior of $\rho_{1\omega}$, 'revival of resistivity' as it were, is a very novel phenomenon as against ordinary resistivity change for superconductive transitions.

Here we take the magnetic behaviors into consideration. First, in-phase linear component of ac magnetic susceptibility, χ', and out-of-phase one, χ'', are shown together in Figure 4. The curve of $\chi'(T)$ qualitatively traces typical temperature variation of zero-field-cooled dc magnetization whose absolute value shows drastic rising from intragrain Meissner signal to full shielding signal owing to the intergrain-ordering phenomenon.[6] As for the behavior of $\chi''(T)$, the positive peak indicates initial-phase-lead effects around the ordering process. These behaviors confirm that this ceramic is a typical sample of the present subject of the intergrain ordering. Next, nonlinear magnetic susceptibility χ_2 is examined in order to clarify the critical temperature of this ordering. Using the previously reported method by the authors',[14] temperature variation of χ_2 is estimated from series summation of the measured harmonic magnetizations of $M'_{n\omega}$ with n=3, 5, 7, and the result is revealed as shown in Figure 5. A conspicuous negative peak with singular feature is formed at around 55 K. This behavior is a reflection of cooperative phase transition, and the peak position informs precisely that the critical point T_{c2} is ~54.5 K. Now this temperature position is marked with the vertical line on the already presented Figure 3. It has been then clarified that the found 'revival phenomenon' of linear electric resistivity is certainly an occurrence below T_{c2} that is decided as the singular point of magnetic nonlinear susceptibility.

The experimentally observed signal of the 'revival phenomenon' seen in Figure 3 is no larger than the order of 1 nV level that may be easily affected by various drift-error origins, so that we should confirm reproducibility and generality for the phenomenon. For this purpose, it is noted first that the $\rho_{1\omega}(T)$ feature was well reproduced by several repeated temperature sweeps near the revival peak region, and actually, several plotted points around 45 K in Figure 3 were such repeat-checked data. Next, the generality can be speculated as follows. In the early stage of this study, it was varied with the samples that the 'revival' behavior was observed or not. However, it is being known that more sensitive measurement, that is attained by increasing of the accumulation times of the voltage data, often reveals the weak 'revival'-like phenomenon, though the S/N condition is severe. One of such cases has resulted for the ceramic sample prepared by 25 h sintering, and the revealed behavior of $\rho_{1\omega}(T)$ is illustrated in Figure 6, in which the extremely enlarged view was obtained through both 16-periods accumulation for each measurement point and post filtering by moving-average. Though the data points are fluctuating, we can recognize the 'revival phenomenon' in this figure. To observe the

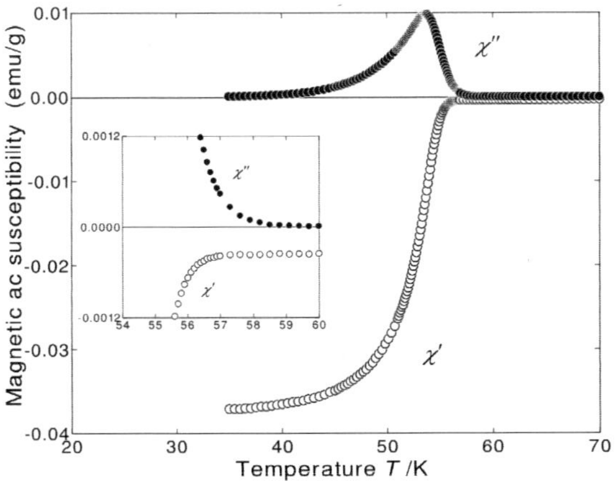

Figure 4. Temperature dependences of linear magnetic susceptibilities χ' and χ'' measured by excitation field of frequency of 1.0 Hz and amplitude of h=0.1 Oe, for the same ceramic sample of $YBa_2Cu_4O_8$ as shown in Figures 1-3. The insert shows enlarged view around the rising region of χ' and $-\chi''$.

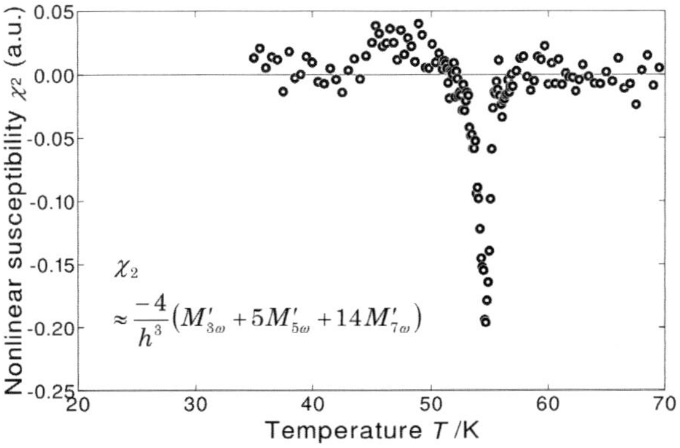

Figure 5. Temperature dependence of nonlinear magnetic susceptibility χ_2 estimated from series summation of harmonic magnetizations $M'_{n\omega}$ (n=3, 5, 7) obtained together with the data in Figure 4. The applied series formula is shown in the figure.

behavior from higher temperature side, $\rho_{1\omega}$ first drops monotonously with decreasing temperature down to ~55 K, and once attains almost zero around 40-45 K, and then looks to turn upward and to form a maximum at around 30 K. The restored resistivity from zero is a purer 'revival' as it were. It has been also noticed for the other weak-sintered samples, such a revival-like phenomenon can be often observed, though each signal level and peak position are diverse for the samples. Accordingly, it is rather natural to regard the revival of resistivity as a quantitatively general phenomenon inherent in our weak-sintered ceramic samples. In this understanding, impurity effects or other extrinsic sample problems may be eliminated, because temperature position of the 'revival' peak changes correlating with T_{c2} for each sample.

Our interest is now taken in possible origins of the revival phenomenon. Whereas decisive discussion cannot be done at present, a clue to the understanding is noticed to consist in structural inhomogeneity of the ordering process. For a theoretical model of frustrated gain network on regular lattice system, non-zero linear resistivity has been suggested to remain at the ordered state.[6] Differently, the ceramic systems in the present study are formed as assembly of actual grain particles and include various irregularity or inhomogeneity. In such irregular systems, the intergrain short-range ordering might progress in mesoscopically inhomogeneous manner. As a possibility, it is considered that intergrain (ordinary-type) superconductive ordering of dispersed non-frustrated regions is advanced ahead the whole ordering including frustration. If such like mixed ordering structure is realized, it seems rather natural that the resistivity first drops toward zero steeply, because lower resistivity region must rule whole resistivity through shunting effects of many conductive passes, in the structurally mixed systems. However, on the other hand, frustration origin certainly exist in the ceramic will exert final ordered state, because cooperative ordering as a whole system must require reconstruction of the incompletely ordered state. We consider that the observed 'revival phenomenon' of resistivity is an essential and detectable reflection by the frustration in such irregular ceramic systems.

A problem to be studied is in the relation between the resistivity phenomenon and the magnetic behavior. The speculated reconstruction of the ordered structure must be nothing but a kind of phase transition. So, it is naturally considered that the magnetic responses, including linear and nonlinear

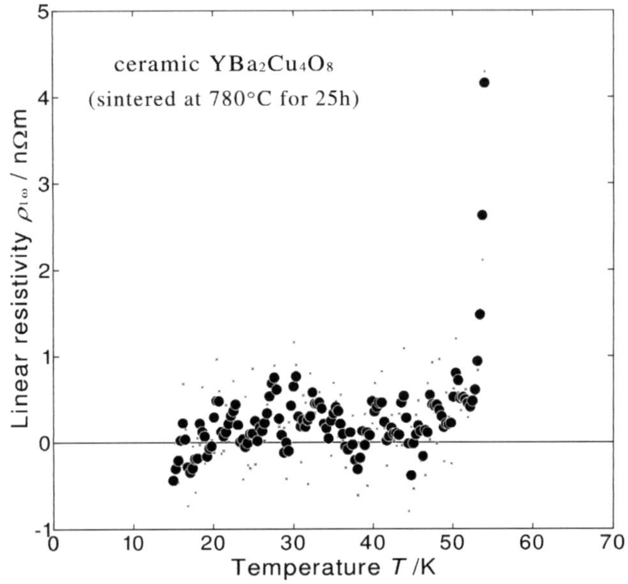

Figure 6. Exremely enlarged view of the linear resistivity vs. temperature for another sample (different from Figure 1-5) of the weak-sintered ceramic $YBa_2Cu_4O_8$. The solid circles are three points moving-average-treated data, and the small crosses are the original data.

terms, may form some anomaly at around 45 K in Figures 1-5 for the main sample in this report. While such an indication may be found in $\chi_2(T)$ of Figure 5, the linear susceptibilities do not show any apparent anomaly in Figure 4. However, besides, the authors has preliminarily observed that some samples show detectable anomalies of χ' and χ'' as well as the nonlinear magnetic responses. Comprehensive experiments for both magnetic and resistive observations using the authors' latest method are progressing, in order to discuss more decisively and reveal the ordering nature affected by frustration in actual ceramic superconductor systems.

Acknowledgment
The authors express finally that this work was supported by both Grant-in-Aid for Exploratory Research (40655001) and Grant-in-Aid for Scientific Research on Priority Areas (19052006).

References
[1] Kusmartsev F.V *Phys. Rev. Lett.* **74** 2268
[2] Sigrist M, Rice T.M *Rev. Mod. Phys.* **67** 503
[3] Choi M.Y and Stroud D, *Phys. Rev.* B **35** 7109, and the references therein
[4] Kawamura H *J. Phys. Soc. Jpn.* **64** 711, Kawamura H and Suan Li M *Phys. Rev.* B **54** 619, Kawamura H and Suan Li M *J. Phys. Soc. Jpn.* **66** 2110
[5] Kawamura H *J. Phys. Soc. Jpn.* **69** 281
[6] Kawachi M, Hagiwara M, Koyama K and Matsuura M *J. Phys. Soc. Jpn.* **63** 3405
[7] Matsuura M, Kawachi M, Miyoshi K, Hagiwara M and Koyama K *J. Phys. Soc. Jpn.* **64** 4540
[8] Yamao T, Hagiwara M, Koyama K and Matsuura M *J. Phys. Soc. Jpn.* **68** 871
[9] Hagiwara M, Yamao T, Shima T, Deguchi H and Matsuura M *Physica* C **412-414** 94
[10] Hagiwara M, Shima T, Kitada R, Deguchi H and Koyama K *Physica* C **470** 1052
[11] Daire A *Online article titled as 'Low-Voltage Measurement Techniques'*
http://www.evaluationengineering.com/ archives/ ArchiveWrapper.aspx?file=0605/0605low_voltage.asp
[12] Koyama K, Junod A, Graf T, Triscone and G and Muller J *Physica* C **185-189** 461
[13] Hagiwara M, Yamao T and Matsuura M *Physica* C **392-396** 66
[14] Hagiwara M, Shima T, Yamao T, Deguchi H and Matsuura M *Physica* E **29** 534

Magnetic Memory Effects in a Chiral-Glass Phase of a Superconductive Ceramic $YBa_2Cu_4O_8$

H Deguchi[1], T Ashida[1], T Shoho[1], Y Kato[1], M Mito[1], S Takagi[1], M Hagiwara[2], and K Koyama[3]

[1]Faculty of Engineering, Kyushu Institute of Technology, Kitakyushu 804-8550, Japan

[2]Faculty of Engineering and Design, Kyoto Institute of Technology, Kyoto 606-8585, Japan

[3]Faculty of Integrated Arts and Science, The University of Tokushima, Tokushima 770-8502, Japan

E-mail: deguchi@tobata.isc.kyutech.ac.jp

Abstract. The ceramic $YBa_2Cu_4O_8$ composed of sub-micron size grains is considered as random Josephson-coupled networks of 0 and π junctions and shows successive phase transitions. The upper transition occurs inside each grain at T_{c1} and the lower transition occurs among the grains at T_{c2} ($< T_{c1}$), where a negative divergence of nonlinear susceptibility is found. This critical phenomenon at T_{c2} suggests the onset of the chiral-glass phase predicted by Kawamura and Li. We investigated the magnetic memory phenomena in the temperature dependence of thermoremanent magnetization (TRM). The memory effects are observed in TRM measured on heating after the cooling process with a halt at the temperature T_s below T_{c2}. The results indicate the existence of a glassy phase such as the chiral-glass one.

1. Introduction

Ceramic high-T_c superconductors , which are composed of homogeneous sub-micron-size grains, may be regarded as random Josephson-coupled networks that contain so-called π junctions. The circulation of a local loop-supercurrent is generated spontaneously in a zero field when there is an odd number of π junctions in a closed loop that consists of some Josephson junctions. Theoretically it is suggested that these spontaneous current loops are affected by an external magnetic field and a net positive magnetization is produced. [1] Furthermore, the frustration effect due to the random distribution of π junctions should lead to the chiral-glass state predicted by Kawamura and Li.[2] They investigated the chiral-glass phase of ceramic high-T_c superconductors using Monte Carlo simulations. In experimental studies, ceramic high-T_c superconductors such as $YBa_2Cu_4O_8$[3,4], $YBa_2Cu_3O_{7-\delta}$[5] and $Bi_2Sr_2CaCu_2O_8$[6] samples has been investigated by several magnetic and transport measurements. The critical phenomenon at the transition temperature and the collective behavior in the glass state of these real samples suggest the existence of the chiral-glass phase.

In the ceramic samples, the magnetic memory behavior was investigated only in $Bi_2Sr_2CaCu_2O_8$ one.[7,8] The memory effects in $Bi_2Sr_2CaCu_2O_8$ samples were observed in the zero-field-cooled (ZFC) and TRM magnetization *vs.* temperature curves measured on heating after certain cooling

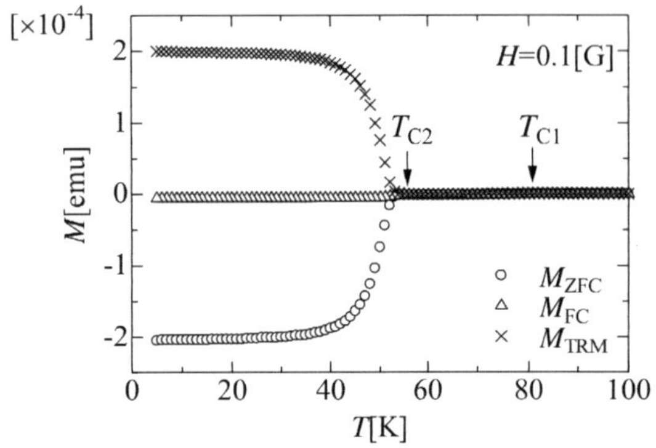

Figure 1. Temperature dependence of M_{ZFC}, M_{FC} and M_{TRM}.

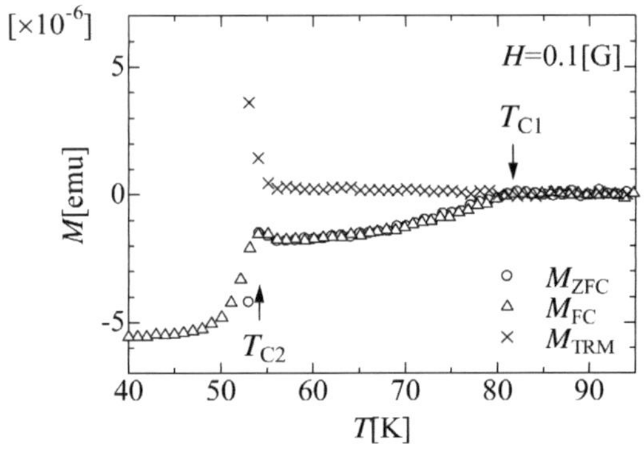

Figure 2. The longitudinal enlargement of the region close to T_{c1} and T_{c2} in Figure 1.

protocols. However, the sample is a strong Josephson-coupled system, and the transition temperature of inter-grain superconductivity is close to that of intra-grain glass phase. Therefore, it is difficult to distinguish between the intra-grain and inter-grain ordering properties. The ceramic $YBa_2Cu_4O_8$, and $YBa_2Cu_3O_{7-\delta}$ have shown successive superconducting transitions under zero field. With decreasing temperature, a transition occurs at first inside each grain at T_{c1} and the second transition occurs among the grains at T_{c2}. The discrepancy between the field–cooled (FC) and ZFC magnetization levels appears at below T_{c2}, and negative divergence of nonlinear susceptibility is observed at T_{c2}, which may reflect a chiral-glass transition.[9] In such extremely weak-Josephson-coupled system, the transition in the intra-grain and inter-grain regions are well separated in temperature. In the present work, we investigate the magnetic memory effects for the typical weak-link system of the ceramic $YBa_2Cu_4O_8$ in order to clarify the properties of the glass phase.

2. Sample and experiments

The sample of ceramic $YBa_2Cu_4O_8$ is synthesized by the procedure reported in Ref.[10]. The sub-micron powder was prepared using the citerate pyrolysis method. The precursor was calcined for 120 h at 778 ℃ to yield the pure $YBa_2Cu_4O_8$ phase, which was sieved and pressed, and then sintered for 50 h at 780 ℃. The dc magnetization and the ac susceptibility were measured with a SQUID magnetometer (Quantum Design MPMS-5) using the ultra-low-field option and the temperature sweep operation. The sample space in the magnetometer was shielded with μ -metal. As a result, the residual field was reduced to less than 10 mG. The nonlinear susceptibility was derived from the harmonics in-phase Fourier component for the ac field response.

3. Results

We measured the temperature dependence of the dc magnetization and ac susceptibility, to determine the transition temperatures T_{c1} and T_{c2}. Figures 1 and 2 show the temperature dependence of the ZFC, FC and TR magnetizations at H = 0.1G. The upper transition at T_{c1}=81 K was identified as the inter-grain superconducting ordering, in which the small diamagnetism due to the Meissner effect appears in ZFC and FC magnetizations. The TRM and the discrepancy between the FC and ZFC magnetizations appeared at around 54 K. Below 54K, the equality M_{ZFC} -M_{FC} = M_{TRM} is closely satisfied for all temperatures.

The linear susceptibilities, in-phase susceptibility χ' and out-of-phase one χ'', are measured in the frequency of 1.0 Hz with an ac-field amplitude of 0.1 G under the zero external field. As shown in Figure 3, χ' changes abruptly at around 54 K and the behavior is similar to that of M_{ZFC}. χ'' has a maximum at around 53 K and exhibits strong dissipation even in low-frequency region. The

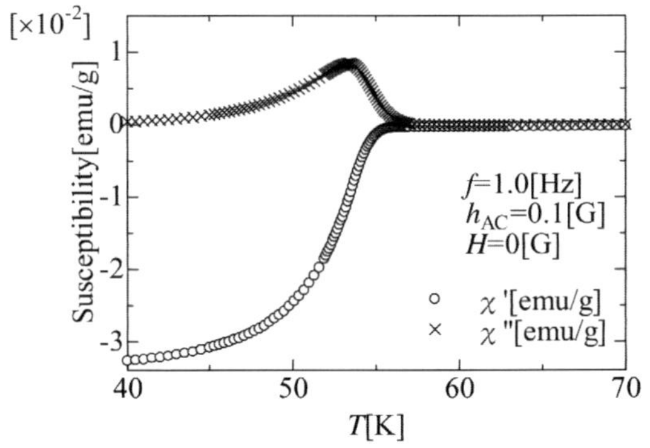

Figure 3. Temperature dependence of linear susceptibilities χ' and χ''.

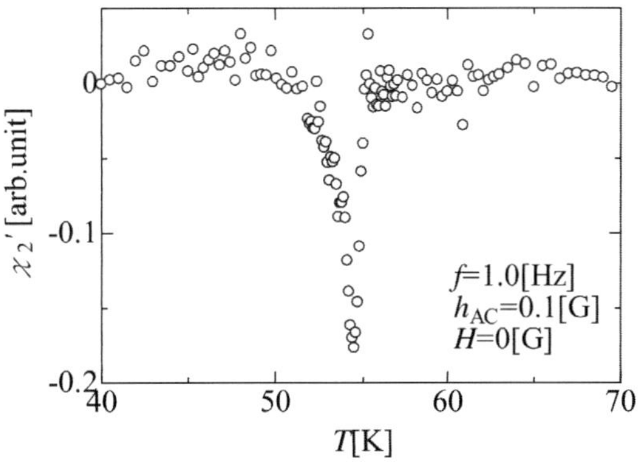

Figure 4. Temperature dependence of the nonlinear susceptibility χ_2.

nonlinear susceptibility χ_2 estimated from the first three terms of the series of in-phase odd-harmonic responses is shown in Figure 4. Negative divergence of χ_2 was observed at T_{c2}=54.5 K, at which temperature the ceramic $YBa_2Cu_4O_8$ underwent a chiral-glass transition.

The magnetic memory experiments are performed using the temperature dependence of TRM. The sample was rapidly cooled (cooling rate of 3.5 K/min) in H = 0.1 G from 90 K above T_{c1} down to a temperature T_s below T_{c2}, at which a halt was made for t_w=10000 s. The cooling was then resumed down to 30 K with cooling rate of 3.5 K/min, and the magnetic field is turned off. The TRM was then recorded on slow heating process (heating rate of 0.14 K/min). We have measured TRM experiments with the fixed wait time t_w=10000 s for three stop temperatures T_s= 52 K, 47 K and 42 K. Figure 5 shows the influence of a temperature stop of t_w=10000 s at T_s = 52 K close to T_{c2} on the TRM. The TRM plots corresponding to the stop experiment around T_s lies slightly above the reference plots measured with the same cooling and heating rates. Figure 6 shows the difference ΔM_{TRM}= M_{TRM}(reference) − M_{TRM}(T_s=52 K) in Figure 5. The influence of the halt is confined to a narrow temperature region near the halt temperature T_s=52K. The maximum of ΔM_{TRM} is small in magnitude (about 3 % of M_{TRM}), but well defined. Figures 7 and 8 show the difference ΔM_{TRM} with T_s = 47 K and 42 K, respectively. The temperature of the peak of ΔM_{TRM} is almost same to each T_s in the three cases. However, the peak of ΔM_{TRM} becomes smaller and broader with decreasing T_s.

4. Discussion and conclusions

As seen from ΔM_{TRM} in Figures 6,7 and 8 there is a memory of the halt at T_s imprinted in the system on the re-heating the sample. The memory of the halt is observed in a temperature region surrounding T_s. When cooling from above T_{c2}, the halt at T_s below T_{c2} is made during t_w, allowing the system to relax towards its equilibrium state at T_s. The state becomes frozen in for further cooling down to the lowest temperature, and is retrieved on reheating. Similar memory effects were observed in the ceramic $Bi_2Sr_2CaCu_2O_8$ samples [7,8] and also in disordered magnetic systems such as spin glasses.[11,12] The $Bi_2Sr_2CaCu_2O_8$ sample is a strong Josephson-coupled system, so the transition temperature of inter-grain superconductivity is close to that of intra-grain one. Therefore, it is difficult to distinguish the memory effects due to the intra-grain or those due to inter-grain superconducting transition. Also, the nonlinear susceptibility does not diverge at the transition temperature.

The ceramic $YBa_2Cu_4O_8$ is considered as a typical weak-link system. The inter-grain transition at T_{c2}=54.5 K separate well from the intra-grain transition at T_{c1} = 81 K in temperatures. Also the nonlinear susceptibility shows the negative divergence at T_{c2}. The memory phenomena of TRM are observed below T_{c2}. The peak of ΔM_{TRM} is large in the case for T_s = 52 K near the temperature at the

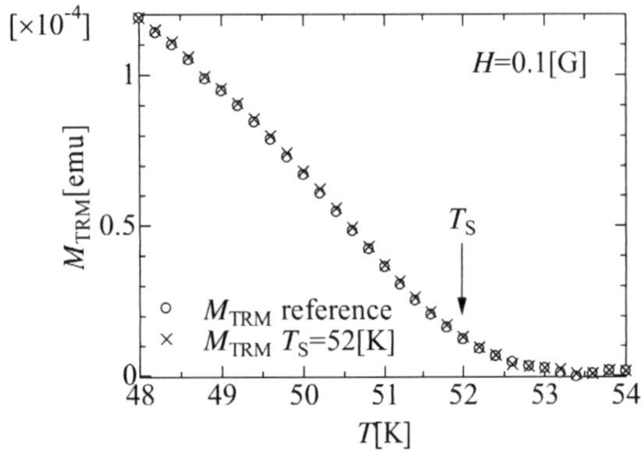

Figure 5. The TRM *vs.* Temperature. The plots (◦) are measured without a halt (reference) and the others (×) are measured with a halt of 10^4 s at 52 K.

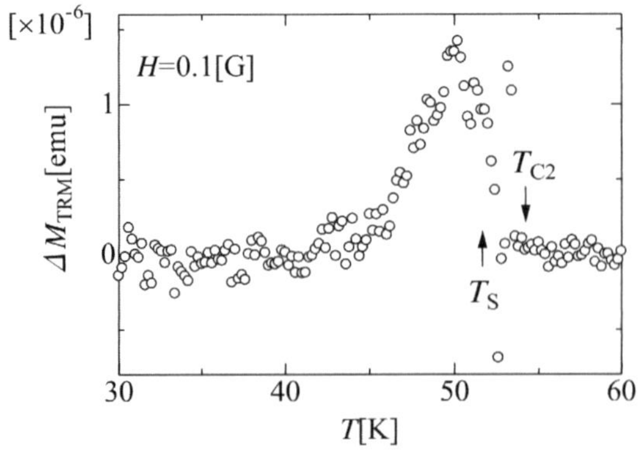

Figure 6. The difference ΔM_{TRM} of two pairs of the plots in Figure 5.

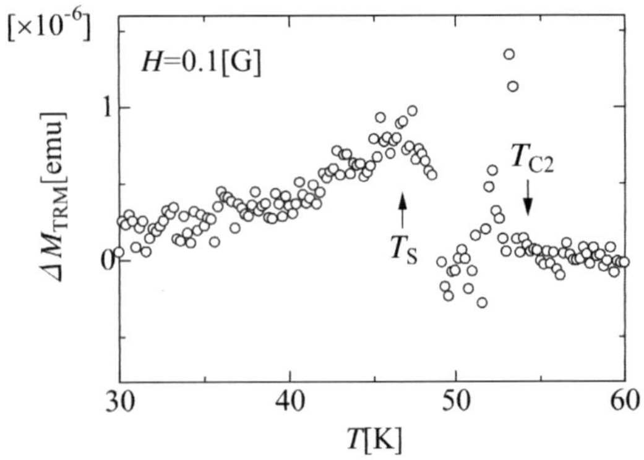

Figure 7. The difference ΔM_{TRM} between the reference TRM and the TRM with a halt of 10^4 s at 47 K.

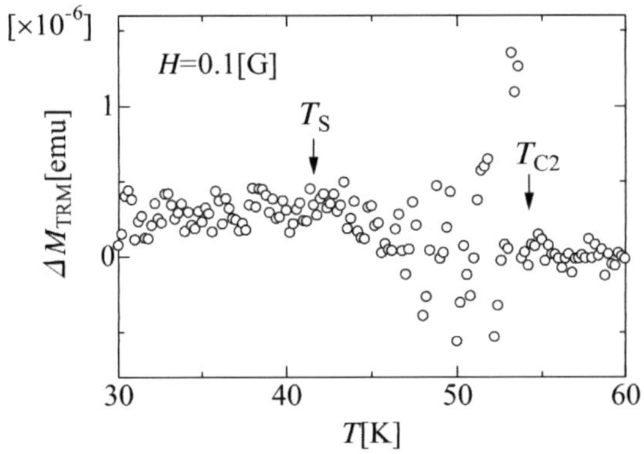

Figure 8. The difference ΔM_{TRM} between the reference TRM and the TRM with a halt of 10^4 s at 42 K.

maximum of χ'' and becomes small in the case for T_s = 42 K where χ'' is close to zero. The observation of χ'' at low frequency shows the existence of a slow dynamics near the glass-transition. The origin of the memory effects may relate to the slow dynamics in chiral-glass phase. The results suggest that chiral-glass ordering occurs at T_{c2} in the ceramic $YBa_2Cu_4O_8$.

Acknowledgments
The authors express their sincere thanks to Prof. H. Kawamura and Prof. M. Matsuura for their continual interest and helpful discussions. The SQUID measurement was carried out at the Center for Instrumental Analysis, Kyushu Institute of Technology. This work was supported by Grant-in-Aid for Scientific Research on Priority Areas (19052006).

References
[1] Sigrist M and Rice T M 1995 *Rev. Mod. Phys.* **67** 503
[2] Kawamura H and Li M S 1997 *Phys. Rev. Lett.* **78** 1556
[3] Kawachi M, Hagiwara M, Koyama K and Matsuura M 1994 *J. Phys. Soc. Jpn.* **63** 3405

[4] Yamao T, Hagiwara M, Koyama K and Matsuura M 1999 *J. Phys. Soc. Jpn.* **68** 871
[5] Koyama K, Deguchi H, Takagi S and Mizuno K 2003 *J. Low Temp. Phys.* **131** 595
[6] Papadopoulou E L, Nordblad P and Svedlindh P 2004 *Phys. Rev. B* **70** 064501
[7] Papadopoulou E L and Nordblad P 2001 *Eur. Phys. J. B* **22** 187
[8] Gardchareon A, Mathieu R, Jönsson P E and Nordblad P 2003 *Phys. Rev. B* **67** 052505
[9] Matsuura M, Kawachi M, Miyoshi K, Hagiwara M, Koyama K 1995 *J. Phys. Soc. Jpn.* **64** 4540
[10] Yamao T, Hagiwara M, Shima T and Matsuura M 2004 *Physica C* **412-414** 98
[11] Jonason K, Vincent E, Hammann J, Bouchaud J P, Nordblad P 1998 *Phys. Rev. Lett.* **81** 3243
[12] Mathieu R, Jönsson P, Nam D N H, and Nordblad P 2001 *Phys. Rev. B* **63** 092401

Journal of Physics: Conference Series **320** (2011) 012077 doi:10.1088/1742-6596/320/1/012077

Superconducting behaviours in reduction-treated mixtures of fine Pr124 and Pr123 ceramics

K Koyama[1], R Nakashima[1], K Magishi[1], T Saito[1], T Shima[2] and M Hagiwara[2]

[1]Institute of Socio-Arts and Sciences, The University of Tokushima, Tokushima 770-8502, Japan

[2]Faculty of Engineering and Design, Kyoto Institute of Technology, Kyoto606-8585, Japan

E-mail: koyama@ias.tokushima-u.ac.jp

Abstract. Superconducting behaviours in reduction-treated mixtures of fine Pr124 and Pr123 ceramics synthesized using citrate pyrolysis precursor method are examined by electrical resistivity and magnetic measurements. With increasing the ratio of Pr123, the temperature dependence of $\rho(T)$ changes to be semiconductor-like behaviour hugely, and there exist no broad maximum for as-sintered Pr124/Pr123 (1:1.5) and (1:2) samples contrary with the case of (1:1). By reduction treatment, the $\rho(T)$ curve shows the maximum according to the increase of carrier concentration in the CuO double chain in Pr 124 grains. By subtracting the Pr^{3+} paramagnetic contribution from the temperature dependence of magnetization, the remnant magnetization shows weak diamagnetism below ~17 K, indicating small Meissner effect. The diamagnetic susceptibility is independent of magnetic field in the measured field range below 20 Oe.

1. Introduction

For Y-base cuprate superconductors, Pr substitution for Y sites restrains the CuO_2 plane conductivity. So, possible superconductive nature at CuO double chains can be searched experimentally in such Pr-substituted materials. As known, the compounds include several related phases distinguished by layered pattern of crystalline blocks of CuO double chain (-D-) and/or single chain (-S-), alongside of CuO_2 plane. In these, $Pr_2Ba_4Cu_7O_{15-\delta}$ (Pr247) has been found to show electron-doped superconductivity at –D block when the ceramic was reduction treated [1,2]. Then Pr247 has being regarded as an essential studying material bringing new viewpoints of cuprate oxide superconductors.

Citrate pyrolysis (CP) is a method to produce reactive precursor for syntheses of oxide compounds [3]. From the precursor, many kinds of oxide superconductors can be synthesized by each single calcination process at ambient pressure condition. Rather variable sintering condition in this method may have been utilized not only to our experimental study of inter-grain ordering in ceramic superconductors [4] but also to prepare $Pr_2Ba_4Cu_7O_{15-\delta}$ (Pr247) ceramic superconductors [5], and furthermore, may be applicable to prepare multi-phase mixed fine ceramic systems.

Recently, superconductive characteristics of multi-phased ceramic samples consist simply of $PrBa_2Cu_4O_8$ (Pr124) and $PrBa_2Cu_3O_{7-\delta}$ (Pr123) with different ratio prepared by citrate pyrolysis (CP) method [6] are examined experimentally. After reduction treatment by vacuum-heating at 450°C, the

Published under licence by IOP Publishing Ltd

multi-phased samples of Pr124+Pr123 (1:1) and (2:1) show onset of abrupt electrical resistivity dropping and also weak Meissner magnetization below ~20 K. The results suggest that superconductivity at CuO double chains in Pr124 grains is caused by charge (electron) transfer from neighboring Pr123 grains. [7,8]

The increase of Pr 123 grain (source of electron supply) ratio is expected to enhance the inter-grain charge transfer effect from Pr123 to Pr124 submicron grains of mixed ceramics by reduction treatment. In the present work, superconductive signs in electrical resistivity and magnetic behaviours of mixed ceramic samples with the ratio of Pr124 : Pr123 = 1:1.5 and 1:2 are searched and discussed, to advance preceding discussion about the superconductive region induced by inter-grain charge (electron) transfer effect.

2. Samples and experiments

An intentionally multi-phased ceramic sample composed purely of Pr124 and Pr123 was synthesized by the mix-sintering method. For this method, various mass ratios of pure calcined powders of Pr124 and Pr123 were prepared independently by the citrate pyrolysis (CP) precursor method at the respective conditions. Then various mass ratios of these (Pr124:Pr123 = 1:1, 1:1.5 and 1:2) were mix-grinded and pressed, and was sintered at 870°C (below the formation temperatures of both Pr124, Pr247 and Pr123). Judged from the XRD observation, the obtained samples consist simply of Pr124 and Pr123, except for a little impurity of $BaCuO_2$ and $PrBaO_3$ formed at the calcinations. Seen from TEM experiments, no Pr247 structure is formed any where in the samples within sampling tries.

Pieces of these ceramic samples were vacuum-heated for 24 h at 300-550°C to be various reduction grade samples. In the present work, the reduction treatments were done at 350°C and 400°C for the mix-sintered samples of Pr124:Pr123 = 1:1.5 and 1:2, respectively. Attained oxygen reduction degree was estimated by measuring weight loss % ($\delta_{wt\%}$) of the pieces.

Electrical resistivities were measured in temperature range 4-300 K. AC 4-wire method by lock-in amplifying was applied using ac current of 23.0 Hz, 1-10 mA. Temperature dependence of DC magnetization was measured by Quantum Design's MPMS with SQUID magnetometer for both reduced and as-sintered (O_2 annealed) samples.

3. Results and discussion

3.1. Electrical resistivity measurements

At first, temperature dependences of electrical resistivity for the as-sintered multi-phased Pr124+Pr123 (1:1, 1:1.5 and 1:2) samples are compared as shown in Figure 1. With increasing the ratio of Pr123,

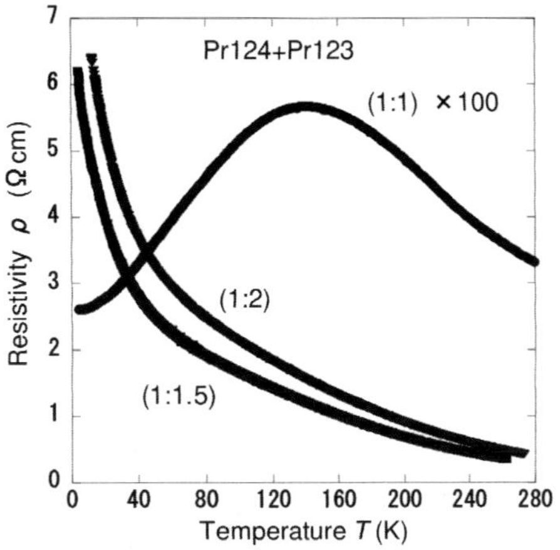

Figure 1. Resistivities vs. temperature for as-sintered mixed ceramic samples of Pr124+Pr123 with the ratio of (1:1), (1:1.5) and (1:2).

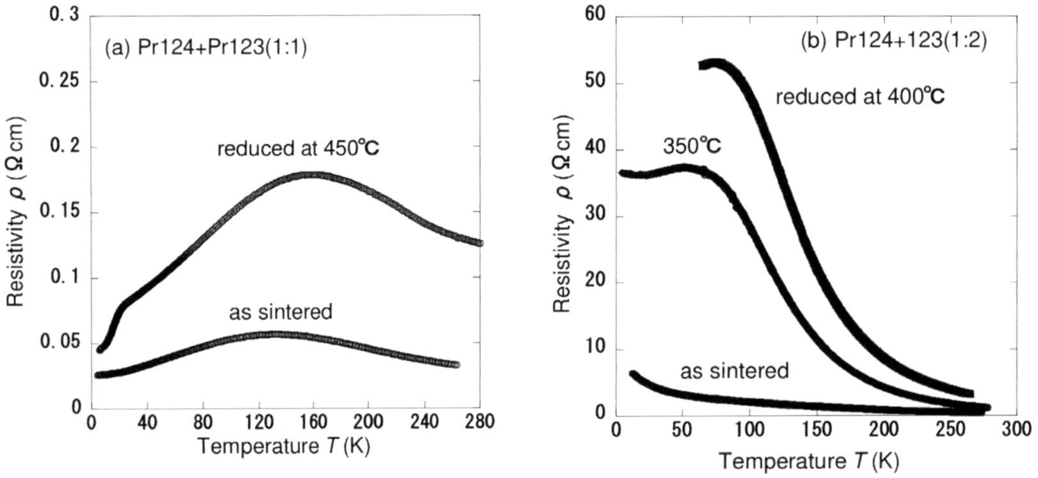

Figure 2. Resistivities vs. temperature for as-sintered and reduced mixed ceramic samples of Pr124+Pr123 with the ratio of (a) (1:1) and (b) (1:2).

the temperature dependence of $\rho(T)$ changes to be semiconductor-like behaviour hugely contrary with the metallic behaviour in the Pr124+Pr123 (1:1) sample below 150K. There exist no broad maximum for Pr124+Pr123 (1:1.5) and (1:2) samples as observed in the case of (1:1).[7,8] This is considered to be the semicondutive nature of Pr123 phase is dominant comparing with the metallic behaviour of CuO double chain in Pr124 phase also in low temperature region. It is quite different with the previous result of Pr124+Pr123 (1:1).[7,8]

Reduction effects on $\rho(T)$ curve for the Pr124+Pr123 (1:1) sample is shown in Figure 2(a).[7] The curve of as-sintered sample shapes broad maximum at about 150K similar to the simple Pr247. As for the reduced sample of $\delta_{wt\%}$=0.835 (vacuum-heated for 24h at 450°C), the $\rho(T)$ curve shows delicate but meaningful characteristics; the maximum shifts a little toward higher temperature, and rather steep drop is revealed around about 20K. Reduction effects on $\rho(T)$ curve for the Pr124+Pr123 (1:2) sample is shown as comparison with the as-sintered one in Figure 2(b). The curve of as-sintered sample is semiconductor-like behaviour in all temperature range. As for the reduced sample of $\delta_{wt\%}$=0.996 (reduced temperature: 350°C), the $\rho(T)$ curve shows the maximum at about 50K. With increasing the reduced treatment temperature, the maximum temperature of $\rho(T)$ curve increases slightly to 80K for the 400°C reduced sample in spite of almost the same $\delta_{wt\%}$ value with that for 350°C reduced one. This is according to the increase of carrier concentration in the CuO double chain in Pr 124 grains, and shows the weak metallic behaviour below the $\rho(T)$ maximum temperature.

The result for the reduction effects for the sample of Pr124+Pr123 (1:1.5) is similar to that for the Pr124+Pr123 (1:2). However, any kinds of resistivity dropping just like the case of the reduced Pr124+Pr123 (1:1) sample at 450°C [7,8] are not observed in both Pr124+Pr123 (1:2) and (1:1.5) samples reduced below 400°C. The carrier transfer effect from Pr123 grain to the CuO double chain in Pr 124 grain may not be enough in these reduction treatment conditions.

3.2. Magnetic measurements

Peculiar magnetic behaviour reflected by the superconductive(-like) contribution is then discussed. DC field-cooling magnetizations for the reduced Pr124+Pr123 (1:2) sample (reduced temperature: 350°C) under the magnetic field of H=0.4 Oe and 16 Oe are shown against temperature in Figure 3(a) and (b), respectively. Main part of magnetization is considered to be from Pr^{3+} paramagnetic contribution, which is estimated as Curie-Weiss law (fitting by using the experimental data above 20K), as shown by dotted line in the figures. Besides, magnetization increases abruptly below 12 K, indicating the ferromagnetic spontaneous component caused by the impurity $PrBaO_3$ phase. [9]

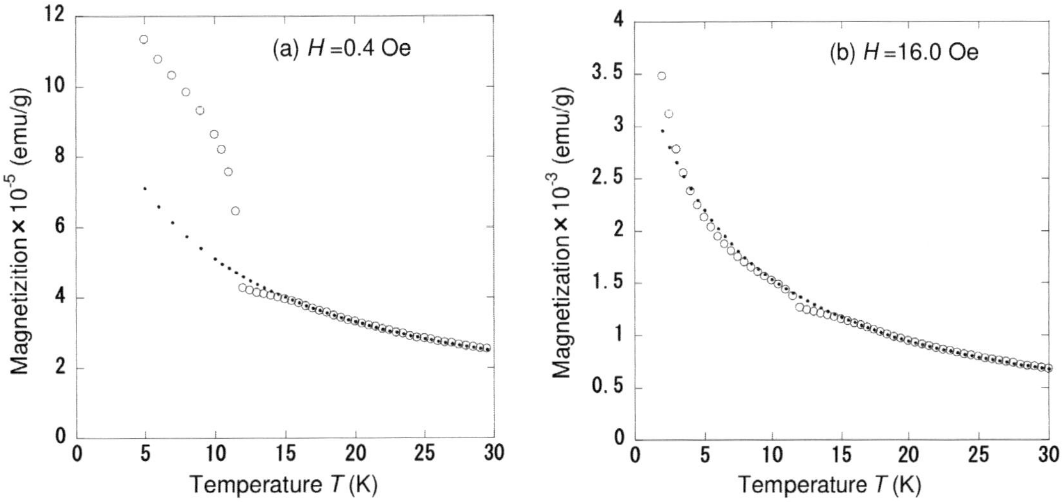

Figure 3. Magnetization vs. temperature for 350°C-reduced mixed sample of Pr124+Pr123 (1:2) under the field of (a) 0.4 Oe and (b)16 Oe.

The amount of increase by ferromagnetic moment are almost independent of magnetic field. The weak diamagnetic component is distinctly observed below ~17 K as the discrepancy from the Curie-Weiss fitting curve of paramagnetic Pr^{3+}.

The Curie-Weiss parameters are estimated in Figure 4. The Curie constant C and residual susceptibility χ_0 are estimated as gradient of Figures 4(a) and 4(b), respectively. From the value of Curie constant, the effective number of Bohr magneton is estimated as about 3.2, comparing with the theoretical value of 3.58 for Pr^{3+} ion. The Curie Weiss temperature θ is about 7K, and independent of magnetic field. As the result, the paramagnetic susceptibility of Pr^{3+} ions in both Pr124 and Pr123 grains are considered as ordinary behavior.

By subtracting the Pr^{3+} paramagnetic contribution, the difference between experimental magnetic moment and Pr^{3+} paramagnetic moment ΔM (= M_{exp}-M_{para}) is plotted against temperature as shown in Figure 5(a) under H=0.4 Oe and 5(b) H=16 Oe, respectively. ΔM shows weak diamagnetism below

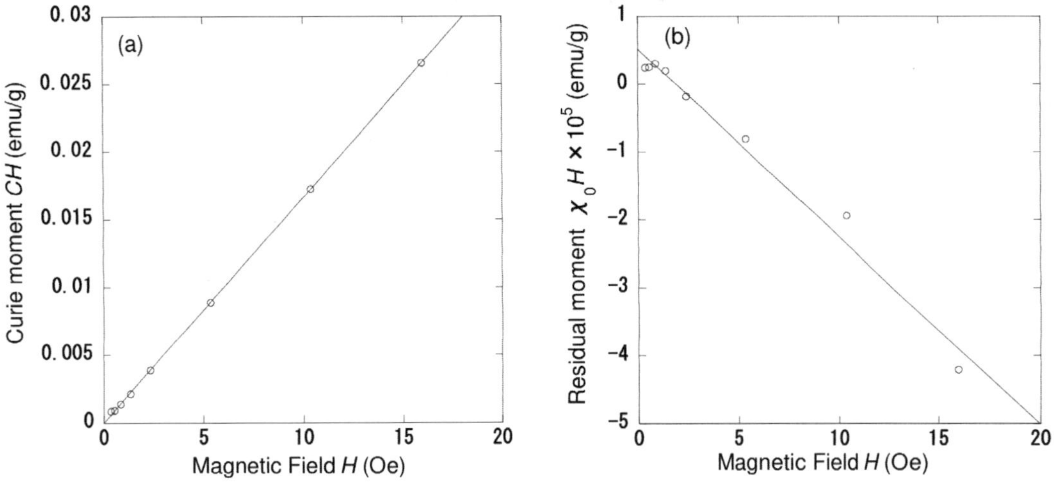

Figure 4. Curie-Weiss parameters: (a) Curie moment (*CH*) and (b) Residual moment ($\chi_0 H$) against magnetic field *H* for the 350°C-reduced mixed sample of Pr124+Pr123 (1:2).

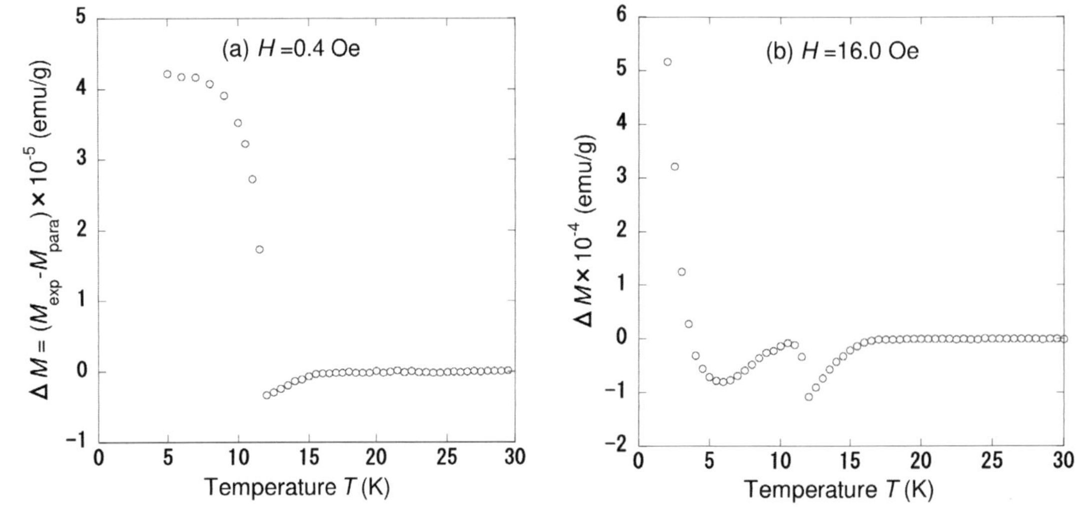

Figure 5. ΔM after subtraction of Pr^{3+} paramagnetic contribution vs. temperature for 350°C-reduced mixed ceramic sample of Pr124+Pr123 (1:2) under the field of (a) 0.4 Oe and (b) 16 Oe, respectively.

~17 K, indicating small Meissner effect, but changes abruptly toward positive value below 12 K, implying spontaneous magnetic moment. A small impurity $PrBaO_3$ is considered as the ferromagnetic origin.[9] Ferromagnetic spontaneous component is almost independent on magnetic field. The other hand, the diamagnetic component increases with magnetic field strength. Moreover, the effect of another paramagnetic impurity ($BaCuO_2$ or other impurities) becomes remarkable in the low temperature and strong magnetic field region.

In order to neglect the effect of ferromagnetic ordering of impurity $PrBaO_3$ below 12K, the amounts of ΔM against temperature above 12K are shown as in Figure 6 under various magnetic field

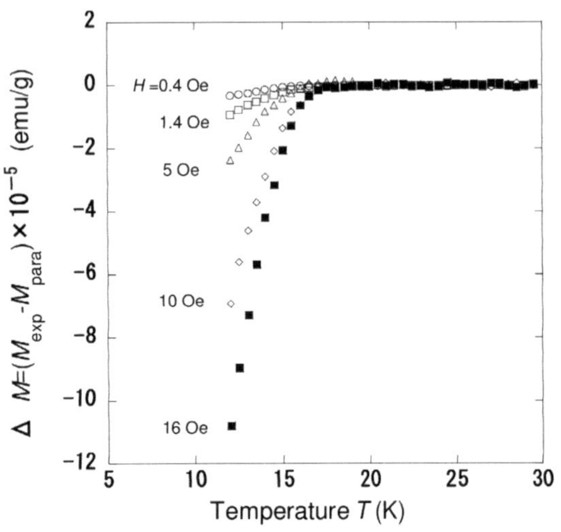

Figure 6. ΔM against temperature above 12K under various magnetic field for the 350°C -reduced Pr124/Pr123 (1:2) sample.

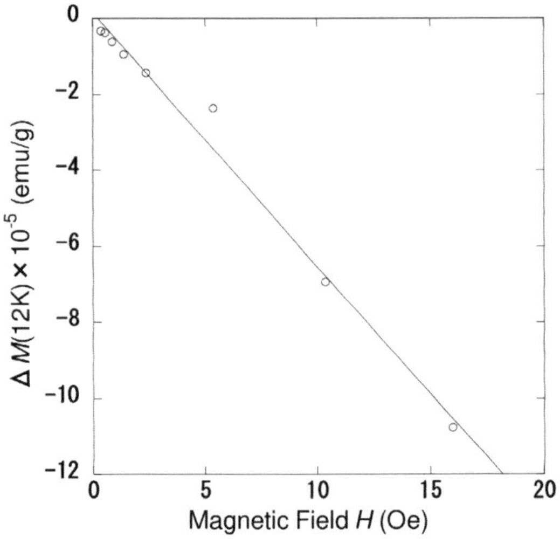

Figure 7. Magnetic field dependence of diamagnetism ΔM at 12K for the 350°C -reduced Pr124/Pr123 (1:2) sample.

for the 350°C-reduced Pr124/Pr123 (1:2) sample. The onset temperature of small diamagnetism caused by Meissner effect is about 17K. The diamagnetic moment increases with increasing magnetic field in the measured field region. The magnetic field dependence of diamagnetism ΔM at 12K (without the influence of ferromagnetic impurity $PrBaO_3$) is shown as in Figure 7. The diamagnetic susceptibility is constant below 16Oe. However, estimated Meissner volume fraction is considered as quite small (about 0.1%). This is because the inter-grain charge transfer effect from Pr123 to Pr124 grains is considered to be occurred in the quite small region in the sample. The other possibility is that the appearance of charge in Pr123 grain is not enough by the reduction treatment below 400°C in the present study.

4. Conclusion

Reduction treatments in vacuum for 24 h below 400°C for the mixed ceramic samples with the ratio of Pr124 : Pr 123 = 1:2 and 1:1.5 bring the appearance of broad maximum in $\rho(T)$ curve. This is considered to be caused by the inter-grain charge transfer effect from CuO single chain site of Pr123 grains to CuO double chain site of Pr124 grains. By subtracting paramagnetic component of Pr^{3+} ions, ferromagnetic trace below 12K of impurity phase $PrBaO_3$ and diamagnetic Meissner signal is observed in the same sample below 17 K, just like the previous result for the Pr124 : Pr 123 = 1:1 sample reduced at 450°C. The inter-grain charge transfer effects not only in the mixed samples with different Pr124:Pr123 ratio but also reduced at higher temperature are now on progress.

References

[1] Matsukawa M, Yamada Y, Chiba M, Ogasawara H, Shibata T, Matsushita A and Takano Y 2004 *Physica* C **411** 101
[2] Yamada Y and Matsushita A 2005 *Physica* C **426-431** 213, and the references therein
[3] Koyama K, Junod A, Graf T, Triscone and G, Muller J 1991 *Physica* C **185-189** 461
[4] Yamao T, Hagiwara M, Koyama K and Matsuura M 1999 *J. Phys. Soc. Jpn.* **68** 871
[5] Hagiwara M, Shima T, Tanaka S, Nishio K, Isshiki T, Saito T and Koyama K 2007 *Physica* C **463-465** 161
[6] Hagiwara M, Shima T, Shugano T, Koyama K and Matsuura M 2006 *Physica* C **445-448** 111
[7] Hagiwara M, Kitada R, Shima T, Nishio K, Kanda S, Deguchi H, Koyama K and Matsuura M 2009 *J. Phys.: Conf. Series* **150** 052065
[8] Hagiwara M, Shima Tanaka S and Koyama K 2010 *Physica* **470** S65
[9] Felner I et al. *Phys. Rev.* 1992 B **46** 9132

X-ray Absorption Spectra in Pyrochlore Niobates

S. Torigoe[1], Y. Ishimoto[1], N. Hanasaki[2], Y. Nogami[1], D. Matsumura[3], K. Yoshii[3], Y. Yoneda[3] and Y. Nishihata[3]

[1]Department of Physics, Okayama University, Tsushimanaka 3-1-1, Okayama 700-8530, Japan
[2]Department of Physics, Osaka University, Machikaneyama 1-1, Toyonaka 560-0043, Japan
[3]Synchrotron Radiation Research Center, Japan Atomic Energy Agency, Hyogo 679-5148, Japan

E-mail: sc19222@s.okayama-u.ac.jp

Abstract. The X-ray absorption spectra reveal that the valence of the Nb ion, and the displacement of the Nb ions from the center of the oxygen octahedra in the pyrochlore niobates $YCaNb_2O_7$ and $NdCaNb_2O_7$, which is consistent with the previous report of the neutron diffraction.

1. Introduction

The spin frustration in the pyrochlore oxides, which have the chemical formula of $A_2B_2O_7$, provides interesting phenomena such as the spin ice state, where the two-in-two-out spin configuration is produced in the tetrahedra of the *A* site. If it is possible that the occupation of the electron (hole) in the *B* site in the pyrochlore structure corresponds to the out (in) state of the spin, one can expect the "charge ice" state as the analogy of the spin ice. If this local charge order is also coupled with the lattice distortion, the local dipole moment is anticipated to be formed in the Nb tetrahedra. For the realization of the charge ice state, the average valence of the transition-metal ion has to be a half integer in the pyrochlore oxides. Istomin *et al.* synthesized the pyrochlore niobates $YCaNb_2O_7$ and $NdCaNb_2O_7$ [1], in which the Nb valence is +4.5. Recently, Cava *et al.* also reported the neutron diffraction in $YCaNb_2O_7$ and performed its structure analysis [2]. In this paper, we report the X-ray absorption spectra, in order to obtain the information of the Nb valence and its position.

2. Experimental

The pyrochlore niobates $YCaNb_2O_7$ and $NdCaNb_2O_7$ are synthesized under the Ar and H_2 gas by use of the floating zone furnace. The characterization of the samples is performed by the X-ray diffraction, the energy dispersive X-ray spectrometry (EDX), and the thermobalance. The X-ray absorption spectra near the Nb *K* edge of the pyrochore niobates are measured at SPring-8 BL14B1 with the energy range from 18.2 to 20.3 keV. The sample disks are prepared by pressing the mixture of the BN and the pyrochlore niobates. The X-ray absorption coefficient of the samples is obtained by measuring the intensity of the incident and the transmitted X-ray beam.

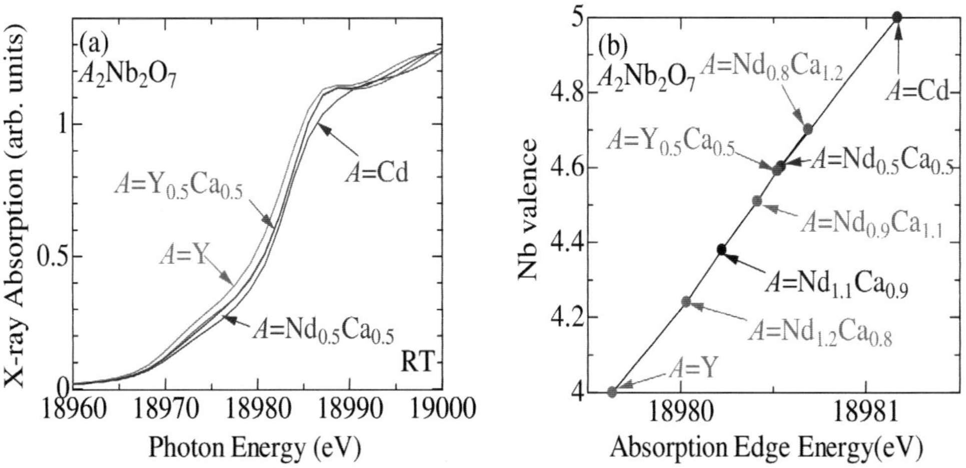

Figure 1. (color online) (a) X-ray absorption spectra near the Nb *K* edge in $A_2Nb_2O_7$ (A = Cd, $Nd_{0.5}Ca_{0.5}$, $Y_{0.5}Ca_{0.5}$, and Y) at room temperature. (b) Relation between the Nb valence and the energy of the Nb *K* edge.

3. Results and discussion

The energy dependence of the X-ray absorption in $A_2Nb_2O_7$ is shown in Fig.1(a). In order to investigate the Nb valence, we also measure the isostructural compounds $Y_2Nb_2O_7$ and $Cd_2Nb_2O_7$, whose valences are also checked by the thermogravimetric analysis. Here, we average the energy position at the X-ray absorption of 0.2, 0.4, 0.6, 0.8 and 1 in Fig.1(a), and determine the energy at the absorption edge. The Nb valence is estimated at about +4.6 in $YCaNb_2O_7$ and $NdCaNb_2O_7$, as shown in Fig.1(b). As the ratio of Nd in the *A* site increases, the Nb valence is reduced. Indeed, the absorption edge position shifts to the low energy side, as shown in Fig.2(b). This result indicates that we can control the Nb valence by changing the Nd(Ca) ratio in the *A* site. Figure 2 displays the result of the

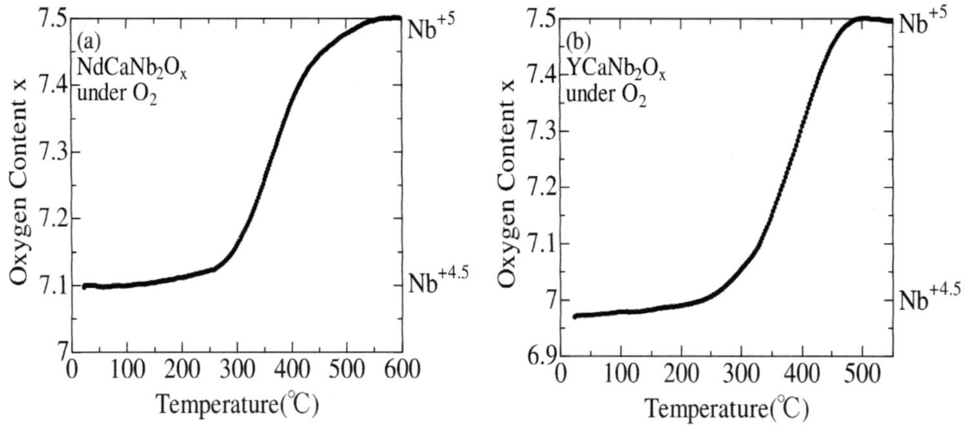

Figure 2. Temperature dependence of the oxygen content determined by the thermobalance in the Nd compound (a) and the Y compound (b).

thermogravimetric analysis in $NdCaNb_2O_7$ and $YCaNb_2O_7$. Here, the samples are heated under the oxygen gas. If one assumes that the Nb valence becomes +5 at the higher temperatures, the Nb valence is +4.59 in the Nd compound, and +4.49 in the Y compound.

Figure 3 (a) shows the X-ray absorption fine structure (XAFS) in the wide energy region at room temperature. The small oscillation of the extended X-ray absorption fine structure (EXAFS) is observed in the energy range between 19keV and 20keV. The energy measured from the absorption edge is transformed into the wave number k. Figure 3 (b) is obtained by the Fourier transformation of Fig.3(a). The peak around r =1.5Å corresponds to the nearest oxygens from the Nb ion, while the peak around r =3.3Å to the nearest neighbor Nb ion and the atoms in the A site. In $Cd_2Nb_2O_7$, the Nb ion locates at the center of the oxygen octahedra at room temperature. The distance between the Nb ion and the nearest oxygen is the same as other nearest oxygens. Thus, the peak ascribed to the nearest oxygens becomes sharp. On the other hand, as shown in Fig.3(b), the peak height around r =1.5Å is small in $YCaNb_2O_7$, $Y_2Nb_2O_7$ and $NdCaNb_2O_7$, as compared to $Cd_2Nb_2O_7$. The suppression of the peak around r =1.5Å indicates that the Nb-O distance is not the same as other oxygens. It is because the wave amplitude of the electron reflected by the oxygens is reduced by the interference.

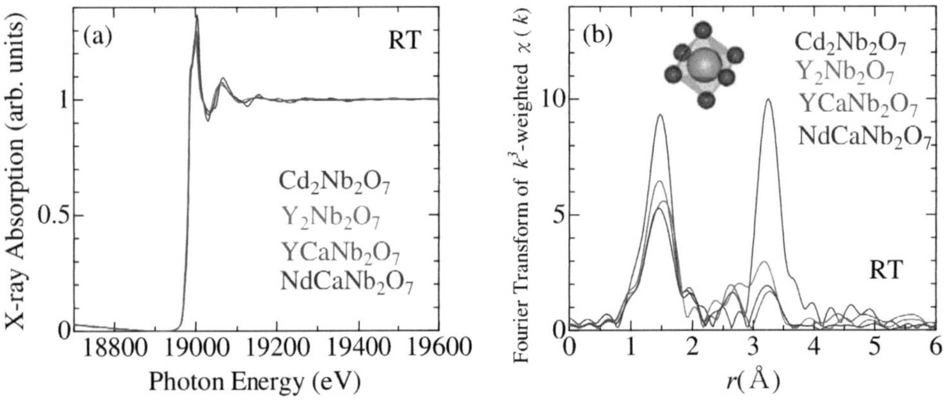

Figure 3. (color online) (a) X-ray absorption spectra in $A_2Nb_2O_7$ (A = Cd, $Nb_{0.5}Ca_{0.5}$, $Y_{0.5}Ca_{0.5}$, and Y) at room temperature. (b) Fourier transform of k^3-weighted EXAFS function $\chi(k)$ in $A_2Nb_2O_7$ (A = Cd, $Nb_{0.5}Ca_{0.5}$, $Y_{0.5}Ca_{0.5}$, and Y). Inset : the Nb ion and the oxygen octahedra.

This experimental result is consistent with the displacement of the Nb ion from the center of the oxygen octahedra, which is reported in $YCaNb_2O_7$ by Cava *et al.* [2]. In $YCaNb_2O_7$ and $NdCaNb_2O_7$, we find the Debye relaxation of the dielectric constants suggesting the ferroelectric state. As a possible origin of this ferroelectricity, we speculate the local charge order such as two-electron-two-hole configuration in the Nb tetrahedra, and its resultant displacement of the Nb ion. To clarify the local charge density at each Nb sites and the correlation between the displacements of the Nb ions is left for the future study.

This work was supported by Grant-in-Aid for Scientific Research on Priority Areas "Novel States of Matter induced by Frustration" (19052003). The synchrotron radiation experiments were performed at the BL14B1 of SPring-8 under the Common-Use Facility Program of JAEA and with the approval of the Japan Synchrotron Radiation Research Institute (JASRI) (Proposal No.2010A3614).

References

[1] S. Ya. Istomin, O. G. D'yachenko, E. V. Antipov and G. Svensson, Mater. Res. Bull. **32**, 421 (1997).

[2] T. M. McQueen, D. V. West, B. Muegge, Q. Huang, K. Noble, H. W. Zandbergen and R. J. Cava, J. Phys: Condens. Matter. **20**, 235210 (2008).
[3] H. Fukazawa and Y. Maeno, Phys. Rev. B. **67**, 054410 (2003).

Low temperature magnetism of the metallic pyrochlore oxide $Pr_{2+x}Ir_{2-x}O_{7+\delta}$

Kenta Kimura, Yasuo Ohta, and Satoru Nakatsuji

Institute for Solid State Physics, University of Tokyo, Kashiwa, Chiba, 277-8581, Japan

k_kimura@issp.u-tokyo.ac.jp and y-ohta@issp.u-tokyo.ac.jp

Abstract. $Pr_2Ir_2O_7$ is a metallic pyrochlore compound which has been revealed to form a spin liquid phase by recent studies. Here we report results of magnetic and electrical transport properties of a newly synthesized polycrystalline sample of $Pr_{2+x}Ir_{2-x}O_{7+\delta}$ with excess Pr. A clear anomaly in the temperature dependence of the electrical resistivity is observed at 0.8 K, suggesting a phase transition into some sort of ordered state.

1. Introduction

It has been widely recognized that geometrical frustration due to certain lattice symmetries based on triangle motifs can be a source stabilizing novel states of matter [1, 2]. Particularly, physics of spins residing on a three dimensional pyrochlore lattice consisting of a corner sharing network of tetrahedra has been extensively studied experimentally and theoretically [3, 4]. A well-known example is a pyrochlore compound with the general formula $A_2B_2O_7$ (*Fd*-3*m* space group), in which both *A* and *B* sites individually form a pyrochlore lattice [5]. A wide variety of exotic states have been discovered in insulating pyrochlore compounds, such as spin ice in $Ho_2Ti_2O_7$ [6] and $Dy_2Ti_2O_7$ [7] and spin liquid in $Tb_2Ti_2O_7$ [8].

Much attention has been currently devoted to pyrochlore $Pr_2Ir_2O_7$, a rare example of a *metallic* frustrated spin system [9-12]. In this compound, Pr^{3+} and Ir^{4+} ions carry the localized frustrated magnetism and conduction electrons, respectively, providing a unique opportunity to investigate the correlation between frustrated magnetism and mobile electrons. Indeed, a single crystal study revealed that despite relatively high Weiss temperature of $\approx$ 20 K, the system exhibits no long-range order and instead forms spin liquid state at least down to T_f = 110 mK, where a partial spin freezing was found. Besides, the Kondo effect is observed, indicating that the frustrated magnetism correlates with conduction electrons through the Kondo coupling [10]. Furthermore, a spontaneous Hall effect in the absence of uniform magnetization has been discovered in the spin-liquid phase [12]. This observation suggests that $Pr_2Ir_2O_7$ is the first example of a chiral spin liquid where time reversal symmetry is macroscopically broken without magnetic dipole order.

In frustrated spin systems, an external perturbation such as magnetic field and chemical disorder often destabilizes a spin liquid state, and, as a result, a novel ordered phase might emerge. Here, we have investigated magnetic and electrical transport properties of a newly synthesized polycrystalline sample of $Pr_{2+x}Ir_{2-x}O_{7+\delta}$ with an off-stoichiometric composition, x = 0.4(3), and found a possible phase transition into some sort of ordered phase. An origin of the phase transition will be discussed.

2. Experimental

The polycrystalline sample of $Pr_{2+x}Ir_{2-x}O_{7+\delta}$ was prepared according to Ref. [13]. The appropriate amounts of Pr_6O_{11} (99.9%), and IrO_2 (> 99.9%) were well mixed together and pressed into a pellet. The pellet was wrapped with a Pt-foil, placed in a silica tube, sealed under vacuum, and then fired at 1423 K for about 5 days with several intermediate gridings. Powder x-ray diffraction confirmed the single pyrochlore phase (*Fd*-3*m*) of the sample. Scanning electron microscopy coupled with energy dispersive x-ray was used to analyse the composition and yields x = 0.4(3), off-stoichiometry with excess Pr [14]. Magnetization measurements were performed using Quantum Design magnetic property measurement system (MPMS). The electrical resistivity was measured by means of a standard four-probe technique in Quantum Design physical property measurement system (PPMS).

3. Results and discussion

We show in Figure 1 the magnetic susceptibility $\chi \equiv M/H$ as a function of temperature (2-300 K). No magnetic phase transition is observed in this temperature range. As seen in the inverse susceptibility ($1/\chi$), the data above 100 K can be well fitted by the Curie-Weiss law with an additive constant term, $\chi = \chi_0 + C/(T - \theta_{CW})$, and the fit yields $\chi_0 = 1.3\times10^{-3}$ emu/mol-Pr, effective moment $p_{eff} = 3.0\ \mu_B$, and Weiss temperature $\theta_{CW} = -13$ K (antiferromagnetic). These values are consistent with the previous studies based on polycrystalline samples [9] and stoichiometric single crystals of $Pr_2Ir_2O_7$ [10]. Note that, in the previous polycrystalline study [9], no information on the chemical composition (i.e. Pr/Ir ratio) of the sample was reported.

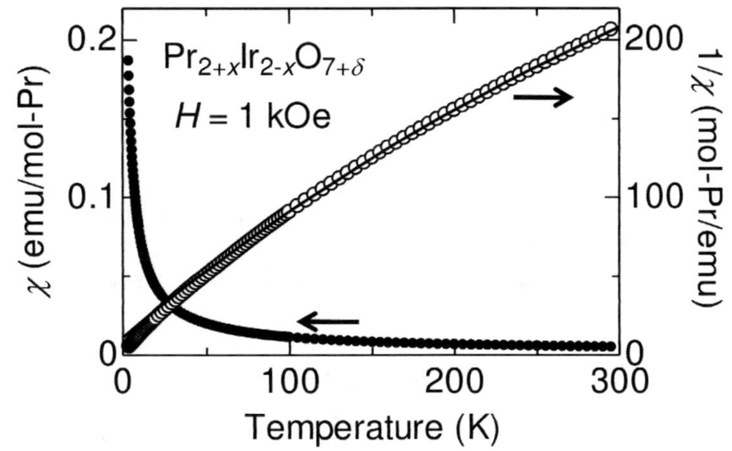

Figure 1: Temperature dependence of the magnetic susceptibility χ (filled circles, left axis) and inverse χ (open circles, right axis). Solid line denotes a fitted curve to $\chi_0 + C/(T - \theta_{CW})$ above 100 K. The data was measured under a field H = 1 kOe. .

Figure 2 provides the temperature dependence of the electrical resistivity $\rho(T)$ down to 0.4 K at 0 T. At a high temperature region, $\rho(T)$ gradually decreases as T decreases, confirming the metallic property of the present sample. In addition, $\rho(T)$ shows a minimum at around 20 K, similarly to the case for single crystals. What is remarkable here is a clear drop in $\rho(T)$ at $T \approx 0.8$ K. No corresponding anomaly was reported in previous studies in this temperature range. The most reasonable explanation for this phenomenon is the onset of phase transition into some sort of ordered state which reduces the spin scattering. The phase transition is confirmed by our specific heat and magnetic susceptibility measurements, which will be published elsewhere [15].

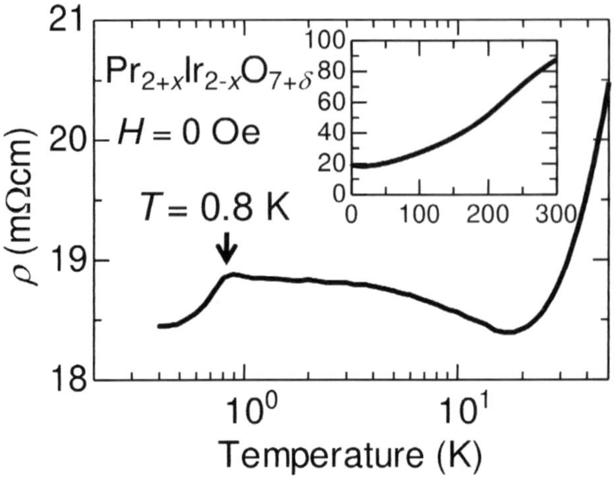

Figure 2: Temperature dependence of the electrical resistivity, $\rho(T)$, at 0 Oe below 50 K (main panel) and below 300 K (inset).

We now discuss the possible origin of the difference in magnetism between present and previous samples. A likely candidate is the sample off-stoichiometry. There are two types of off-stoichiometry: the deviation of the ratio Pr/Ir from 1 and oxygen deficiency. The former modifies magnetic interactions and introduces randomness, which obviously affects the magnetism. The latter is also important because this should affect anisotropy of Pr^{3+} magnetic moments through the variation of crystalline electric field made by O^{2-} ions around Pr^{3+}. Furthermore, a recent theory for Pr-based pyrochlores has revealed the significant importance of Pr^{3+}-Pr^{3+} superexchange coupling mediated by an intervening O^{2-} ion, being also affected by the oxygen deficiency [16]. Finally, it can also alter the carrier concentration and as a result the Kondo effects and RKKY interaction. Another possibility is the difference in structural parameters among samples, that is, lattice constant and oxygen position at 48*f* site being only structural parameters in pyrochlore compounds. Indeed, this may be responsible for the mysterious observation in the insulating Tb-based pyrochlores: $Tb_2Ti_2O_7$ shows spin liquid state [8] while $Tb_2Sn_2O_7$ shows an ordered state [17], despite the fact that only main differences between them are lattice constant and oxygen position. It cannot be ruled out that this is also the case for the $Pr_2Ir_2O_7$ system. At present, we cannot identify the origin for the phase transition in the present sample. Systematic studies to check the above possibilities are on progress.

It is not clear whether the possible phase transition is accompanied by a conventional magnetic dipole order. The above-mentioned recent theory has proposed that, based on an effective quantum pseudospin-1/2 model for Pr-based pyrochlores, spin quadrupolar order formed by four spins on a tetrahedron can be stabilized by the Pr^{3+}-Pr^{3+} superexchange interaction [16]. Further investigations on the ordered state by means of neutron and μSR experiments are highly required to explore such possibility. Furthermore, the understanding of how the ordered phase is connected to the chiral spin liquid is an interesting future subject.

4. Conclusion

We have investigated the temperature dependence of the magnetic susceptibility and electrical resistivity of a newly synthesized polycrystalline sample of the metallic pyrochlore $Pr_{2+x}Ir_{2-x}O_{7+\delta}$ with excess Pr. Although the magnetic susceptibility of the present sample is similar to the results of previous studies, a clear drop in resistivity is found at 0.8 K, which has been never observed. This strongly suggests that the system experiences a phase transition into some sort of ordered state, in sharp contrast to the previously reported spin liquid behavior. Although the origin of the phase transition is not clear at present, our results spark further interests in $Pr_2Ir_2O_7$.

Acknowledgements
We thank Y. Machida, E. Nishibori, H. Sawa, C. Stock, C. Broholm, and Y. Nambu for discussions. This work is partially supported by Grant-in-Aid (No. 21684019) from the Japanese Society for the Promotion of Science, and also Grant-in-Aid for Scientific Research on Priority Area (No. 19052003) from the Ministry of Education, Culture, Sports, Science and Technology, Japan.

References
[1] *Frustrated Spin Systems*, edited by H. T. Diep (World Scientific, Singapore, 2004).
[2] L. Balents, Nature **464** (2010) 199.
[3] S. T. Bramwell and M. J. P. Gingras, Science **294** (2001) 1495.
[4] J. S. Gardner, M. J. P. Gingras, and J. E. Greedan, Rev. Mod. Phys. **82** (2010) 53.
[5] M. A. Subramanian, G. Aravamudan, and G. V. Subba Rao, Prog. Solid State Chem. **15** (1983) 55
[6] M. J. Harris, S. T. Bramwell, D. F. McMorrow, T. Zeiske, and K. W. Godfrey, Phys. Rev. Lett. **79** (1997) 2554.
[7] A. P. Ramirez, A. Hayashi, R. J. Cava, R. Siddharthan and B. S. Shastry, Nature **399** (1999) 333.
[8] J. S. Gardner, S. R. Dunsiger, B. D. Gaulin, M. J. P. Gingras, J. E. Greedan, R. F. Kiefl, M. D. Lumsden, W. A. MacFarlane, N. P. Raju, J. E. Sonier, I. Swainson, and Z. Tun, Phys. Rev. Lett. **82** (1999) 1012.
[9] D. Yanagishima and Y. Maeno, J. Phys. Soc. Jpn. **70** (2001) 2880.
[10] S. Nakatsuji, Y. Machida, Y. Maeno, T. Tayama, T. Sakakibara, J. van Duijn, L. Balicas, J. N. Millican, R. T. Macaluso, and Julia Y. Chan, Phys. Rev. Lett. **96** (2006) 087204.
[11] Y. Machida, S. Nakatsuji, Y. Maeno, T. Tayama, T. Sakakibara, and S. Onoda, Phys. Rev. Lett. **98** (2007) 057203.
[12] Y. Machida, S. Nakatsuji, S. Onoda, T. Tayama, and T. Sakakibara, Nature **463** (2010) 210.
[13] J. N. Millican, R. T. Macaluso, S. Nakatsuji, Y. Machida, Y. Maeno, and J. Y. Chan, Mater. Res. Bull. **42** (2007) 928.
[14] Because of the polycrystalline sample the data were much scattered, resulting in a large error of x.
[15] Y. Ohta *et al.*, private communication.
[16] S. Onoda and Y. Tanaka, Phys. Rev. Lett. **105** (2010) 047201.
[17] I. Mirebeau, A. Apetrei, J. Rodriguez-Carvajal, P. Bonville, A. Forget, D. Colson, V. Glazkov, J. P. Sanchez, O. Isnard, and E. Suard, Phys. Rev. Lett. **94** (2005) 246402.

Crystalline electric field study in the pyrochlore $Nd_2Ir_2O_7$ with metal-insulator transition

M. Watahiki[1], **K. Tomiyasu**[1,*], **K. Matsuhira**[2], **K. Iwasa**[1], **M. Yokoyama**[3], **S. Takagi**[2], **M. Wakeshima**[4] **and Y. Hinatsu**[4]

[1] Department of Physics, Tohoku University, Sendai 980-8578, Japan
[2] Faculty of Engineering, Kyusyu Institute of Technology, Kitakyusyu 804-8550, Japan
[3] Faculty of Science, Ibaraki University, Mito 310-8512, Japan
[4] Division of Chemistry, Graduate School of Science, Hokkiado University, Sapporo 060-0810, Japan
* E-mail: tomiyasu@m.tohoku.ac.jp

Abstract. We carried out inelastic neutron scattering measurements to identify crystalline electric field (CEF) states of Nd^{3+} (a total angular momentum J = 9/2) in $Nd_2Ir_2O_7$, which exhibits a metal-insulator transition at T_{MI} = 36 K. Excitation peaks observed around 26 and 41 meV are interpreted as CEF excitations within five Kramers doublets. It is notable that splitting of the ground doublet by 1.3 meV was microscopically observed at 3 K for the first time, indicating emergence of magnetic long-range order despite geometrical frustration. The splitting width decreases with increasing the temperature towards T_{MI} like an order parameter of the transition. We discuss the results in light of Nd-Ir interactions.

1. Introduction

Geometrical frustration has attracted much attention. A pyrochlore $A_2B_2O_7$ is the representative system, in which both A and B sites form pyrochlore sublattices. Recently, the interesting transport properties were discovered, e.g. metal-insulator (MI) transition in $Hg_2Ru_2O_7$ and $Cd_2Os_2O_7$, superconductivity in $Cd_2Re_2O_7$ and anomalous Hall effect in $Nd_2Mo_2O_7$ [1; 4].

$Nd_2Ir_2O_7$ with magnetic ions Nd^{3+} and Ir^{4+} exhibits an MI transition at T_{MI} = 36 K [5]. In magnetic susceptibility measurements, zero-field-cooling and field-cooling curves do not trace each other below T_{MI}, indicating magnetic ordering/freezing at T_{MI} [6]. In specific heat measurements, magnetic entropy of Nd^{3+} was estimated to be R ln 2 at T_{MI} (R: the gas constant), meaning that the crystalline-electric-field (CEF) ground state is a doublet [6]. Schottky anomaly was also observed at around 5 K, suggesting that the doublet splits by 1.2 meV below T_{MI} [6]. This splitting probably comes from a static internal magnetic field at Nd sites, induced by the magnetic anomaly.

In this paper, we study the magnetic state with MI transition in light of CEF by inelastic neutron scattering. The ground doublet splitting is microscopically verified for the first time, indicating the emergence of magnetic long-range order despite the expected geometrical frustration. A model of CEF level scheme is also proposed.

2. Experiments

Powder samples of $Nd_2Ir_2O_7$ and $Y_2Ir_2O_7$ were synthesized by K. Matsuhira, as described in Ref. [5]. The 4.5 g $Nd_2Ir_2O_7$ sample and the 4.2 g $Y_2Ir_2O_7$ one were enclosed in thin aluminium foils, and were shaped to hollow cylinders with 0.7 mm thickness in order to diminish an effect of the strong neutron absorption of Ir nuclei as low as possible. Then the cylinder was set in an aluminium container, which was placed under a cold head in a ^{4}He closed-cycle refrigerator.

Inelastic neutron scattering experiments were performed using the triple axis spectrometers HER installed at the cold-neutron guide of JRR-3, JAEA, Tokai, Japan and TOPAN installed at the thermal beam hole of the same reactor. On HER, the final energy of the neutrons was fixed at E_f = 3.6 meV. Higher-order contamination was eliminated by a pyrolytic graphite (PG) filter or a cooled Be filter before a sample and another Be filter after it. The horizontal collimation sequence was guide-blank-radial collimator (with 2 blades)-blank. The scattered neutrons were merged by a horizontal focusing analyzer in 5 degrees scattering angle. The energy resolution (full width at half maximum FWHM) was 0.13 meV at the elastic condition. On TOPAN, E_f = 13.5 meV was selected. A sapphire filter and a PG filter cut the fast and higher-order neutrons. The collimation was blank-100'-100'-blank. The energy resolution was 1.5 meV.

3. Results

Figures 1(a) and 1(b) show the constant-Q scan data on $Nd_2Ir_2O_7$ and $Y_2Ir_2O_7$, respectively. $Nd_2Ir_2O_7$ shows a clear peak at around 1.3 meV at 3 K, as shown in Fig. 1(a), which coincides with the splitting width estimated from the Schottky anomaly [6]. As the temperature increases, both the excitation energy and the intensity decrease towards T like order parameters of the transition. In contrast, $Y_2Ir_2O_7$ shows no distinct excitation down to 5 K, as shown in Fig. 1(b). Therefore, the former excitations most likely correspond to be the ground doublet splitting of Nd^{3+}. Below 21 K, a weak broad signal is also observed in the lower energy side of the main peak even at 3 K (Fig. 1(a)).

In order to shed further light on the magnetic state, we performed measurements in higher energy transfer where the CEF levels can be detected. Figure 2(a) shows the constant-Q scan data measured at 3.1, 3.5 and 3.9 Å^{-1} at 4 K, and Fig. 2(b) shows that measured at 3.5 Å^{-1} at 40 K. The $Y_2Ir_2O_7$ data, measured with the same experimental conditions, was subtracted from the $Nd_2Ir_2O_7$ data to eliminate phonon contribution and background. Around 26 and 41 meV, appreciable peaks are observed for all the Q's below and above T_{MI}, which are identified as higher-energy CEF excitations of Nd^{3+}.

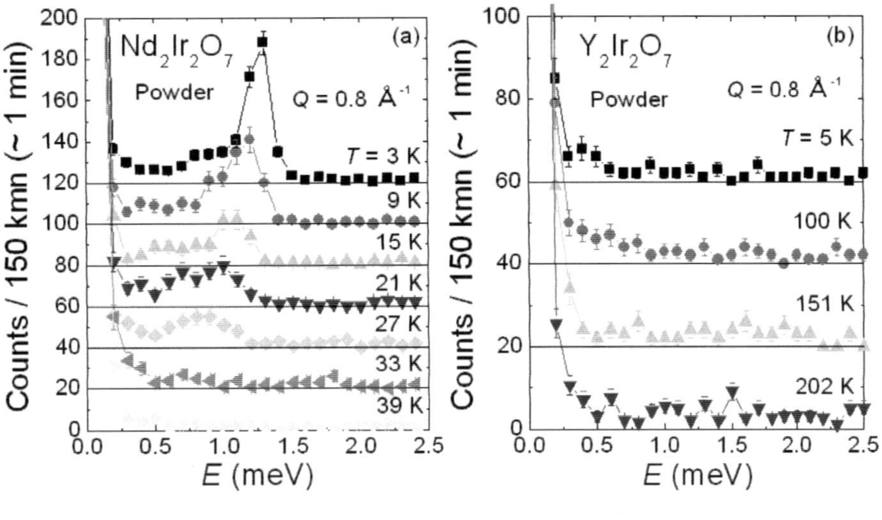

Figure 1. Constant-Q scan data of $Nd_2Ir_2O_7$ (a) and $Y_2Ir_2O_7$ (b). Q is the magnitude of scattering vector and E is the energy transfer.

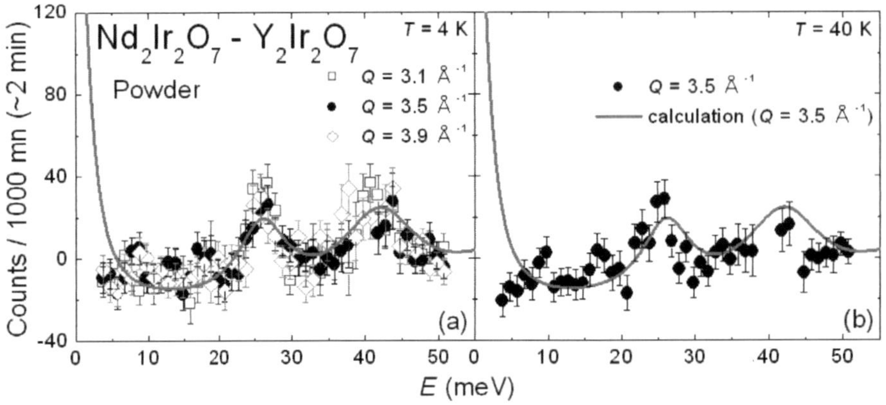

Figure 2. Constant-Q scan data in a high-energy region measured at 3.1, 3.5 and 3.9 Å^{-1} at 4 K (a) and that measured at 3.5 Å^{-1} at 40 K (b). The solid curves show calculated results (See the text).

4. Analyses

Within a CEF level scheme of Nd^{3+} (a total angular momentum $J = 9/2$), we simultaneously analyzed the spectra with the 26-meV and 41-meV excitations and the magnetic susceptibility data [6]. The CEF Hamiltonian $\hat{H}_{CEF}$ with trigonal symmetry ($\boldsymbol{D_{3d}}$) is described by

$$\hat{H}_{CEF} = \sum_{n,m} B_n{}^m \hat{O}_n{}^m = B_2{}^0\hat{O}_2{}^0 + B_4{}^0\hat{O}_4{}^0 + B_4{}^3\hat{O}_4{}^3 + B_6{}^0\hat{O}_6{}^0 + B_6{}^3\hat{O}_6{}^3 + B_6{}^6\hat{O}_6{}^6, \quad (1)$$

where the quantisation z axis is along the [111] three-fold axis, $B_n{}^m$ are coefficients to determine scale of the CEF eigenenergy E_i and order of the eigenstate $|\Gamma_i\rangle$, and $\hat{O}_n{}^m$ are Stevens' operators, e.g. $\hat{O}_2{}^0 = 3\hat{J}_z{}^2 - J(J+1)$ [7; 8]. The internal magnetic field generating splitting of ground doublet (~ 1 meV) is ignored, since it is too small to be resolved from the present thermal neutron scattering data. By using the E_i and $|\Gamma_i\rangle$ obtained by numerically diagonalising Eq. (1), inelastic neutron scattering cross section for a transition from $|\Gamma_i\rangle$ to $|\Gamma_j\rangle$ in a non-interacting ion can be calculated by

$$\frac{d^2\sigma}{d\Omega dE} = C_0 \frac{k_f}{k_i} \frac{\exp[-E_i/k_BT]}{Z} \left|\langle\Gamma_i|\hat{J}_\perp|\Gamma_j\rangle\right|^2 R\left(E_i - E_j + E\right), \quad (2)$$

where C_0 is a constant, k_i and k_f are wave numbers of incident and final neutrons, respectively, T is temperature, and Z is the CEF partition function [9]. The matrix element $\left|\langle\Gamma_i|\hat{J}_\perp|\Gamma_j\rangle\right|^2$ is represented by $(2/3)\sum_{\alpha=1}^{3}\left|\langle\Gamma_i|\hat{J}_\alpha|\Gamma_j\rangle\right|^2$ for a powder sample, where $\hat{J}_\alpha$ is a total angular momentum operator and α means the components of the Cartesian coordinate [10]. The function R is an instrumental energy resolution, for which we assumed a Lorentzian with FWHM estimated by the Cooper-Nathans formula [11]. Then, based on the above $\hat{H}_{CEF}$, the powder-averaged magnetic susceptibility is evaluated by

$$\tilde{\chi}_{CEF} = \frac{Ng_J\mu_B}{H} \sum_{\alpha=1}^{3} \frac{\exp\left[-E_i^{(\alpha)}/k_BT\right]}{Z^{(\alpha)}} \frac{\left|\langle\Gamma_i^{(\alpha)}|\hat{J}_\alpha|\Gamma_i^{(\alpha)}\rangle\right|}{3}, \quad (3)$$

where N is number of Nd^{3+} in a sample, $|\Gamma_i^{(\alpha)}\rangle$ and $E_i^{(\alpha)}$ are eigenstate and eigenenergy of another Hamiltonian $\hat{H}'_{CEF} = \hat{H}_{CEF} - g_J\mu_B\hat{J}_\alpha H$, respectively, $Z^{(\alpha)}$ is the partition function, g_J is the Landé's g-factor, and H is magnitude of an applied magnetic field [12].

We carried out a fitting procedure of the CEF parameters B_n^m for the experimental data using Eqs. (1) to (3). Figures 2(a) and 2(b) show the calculated curves of scattering intensity for a found model, which are in good agreement with the experimental data. Figure 3 shows the calculated lines of $\tilde{\chi}_{CEF}$ and $\tilde{\chi}_{CEF}^{-1}$ for H = 0.1 T, which are also in good agreement with the experimental data [6]. Table 1 summarizes the obtained CEF parameters, eigenenergies and eigenstates. The J = 9/2 decatet splits into five Kramers doublets. The ground doublet exhibits a highly anisotropic magnetic dipole moment $g_J\langle\vec{J}\rangle = (0, 0, \pm 2.37)\ \mu_B$ per ion.

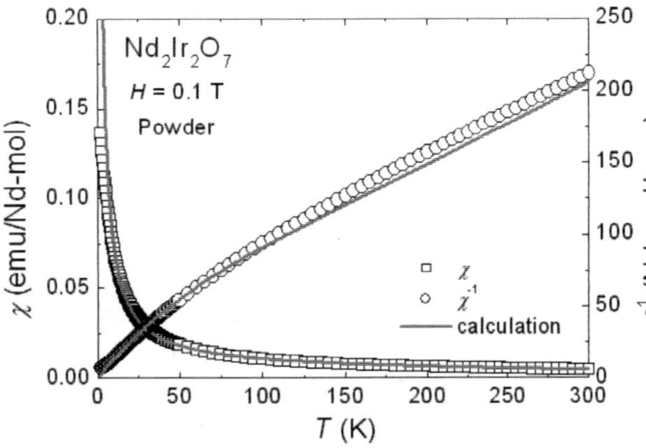

Figure 3. Magnetic susceptibility and the inverse one. Open squares and circles indicate experimental data [6], and solid curves show calculated results.

Table 1. Calculated CEF parameters, eigenenergies and eigenstates for Nd^{3+} (J = 9/2). Ten coefficients with respect to $|J_z\rangle$ are given for each eigenstates.

CEF parameters [meV]										
B_2^0		B_4^0		B_4^3	B_6^0		B_6^3		B_6^6	
-0.28(0)		-0.95(9)×10^{-2}		-0.33(9)	-0.44(3)×10^{-3}		0.17(7)×10^{-2}		-0.45(7)×10^{-2}	
eigenenergies [meV] and eigenstates										
E_i	|-9/2>	|-7/2>	|-5/2>	|-3/2>	|-1/2>	|1/2>	|3/2>	|5/2>	|7/2>	|9/2>
123	0	0	0.375	0	0	0.533	0	0	-0.758	0
123	0	-0.758	0	0	-0.533	0	0	0.375	0	0
57	0.135	0	0	-0.289	0	0	-0.772	0	0	0.550
57	0.550	0	0	0.772	0	0	-0.289	0	0	-0.135
42	0	0	-0.439	0	0	-0.617	0	0	-0.653	0
42	0	0.653	0	0	-0.617	0	0	0.439	0	0
26	0	0	0.816	0	0	-0.578	0	0	-0.003	0
26	0	0.003	0	0	-0.578	0	0	-0.816	0	0
0	-0.057	0	0	-0.286	0	0	-0.489	0	0	-0.822
0	-0.822	0	0	0.489	0	0	-0.286	0	0	0.057

5. Discussion

We succeeded in microscopically verifying the splitting of ground doublet below T_{MI} (Fig. 1(a)), indicating an appearance of static internal magnetic field at Nd sites. The energy spectrum shows a single sharp peak at 1.3 meV at 3 K, suggesting that the internal field is almost equivalent at all Nd sites. Therefore magnetic long-range order seems to uniformly occur despite geometrical frustration.

On the other hand, a weak broad signal is also observed in the spectrum, suggesting the coexistence of an additional random-field component like short-range order. Thus it was revealed that frustration is predominantly relieved below T_{MI}.

The coexistence of long-range order and short-range order is observed in frustrated spinel insulators ACr_2O_4 (A = Mn, Co) with magnetic A^{2+} and Cr^{3+}, in which A sites form a non-frustrated diamond lattice and Cr sites form a frustrated pyrochlore lattice [13; 14]. In lowest temperature phases the chromates exhibit a conical spin structure, resolved into an Ising-type ferrimagnetic long-range component and a spiral short-range one. This partially frustrated state is interpreted as a result of the Cr-site frustration suppressed by A-Cr exchange interactions [13; 15]. In analogy with the chromates, the coexistence of two components in $Nd_2Ir_2O_7$ would be also due to Ir-site frustration suppressed by Nd-Ir exchange interactions.

The internal field at Nd sites might also come from these interactions. In fact, the value 1.3 meV (~ 15 K) is too large for dipole-dipole interaction energy with 2.37 μ_B, and interactions between well-localized 4f moments of Nd^{3+} can be normally neglected in an insulator without RKKY interactions.

6. Conclusions

We studied CEF states of Nd^{3+} in $Nd_2Ir_2O_7$ by powder inelastic neutron scattering. A CEF model with five Kramers doublets was proposed. The ground-doublet splitting below T_{MI} was microscopically verified for the first time, indicating emergence of magnetic long-range order with static internal field at Nd sites despite geometrical frustration. The internal field can be interpreted as Nd-Ir exchange field, which probably relieves frustration and causes the magnetic ordering. Further experimental and theoretical works will be needed to elucidate the magnetic state and MI transition.

Acknowledgments

We thank Mr. T. Asami, Mr M. Ohkawara and Mr. K. Nemoto for their supports in JAEA and Mr. M. Onodera for his supports in Tohoku University. The neutron experiments were performed under User Programs conducted by ISSP, University of Tokyo. This work was supported by the MEXT of Japan, Grants in Aid for Young Scientists (B) (22740209), Priority Areas (22014001), Scientific Researches (S) (No. 21224008), (A) (22244039), Scientific Researches on Priority Areas ''Novel States of Matter Induced by Frustration'' (19052005) and Innovation Areas ''Heavy Electrons'' (21102518).

References

[1] Klein W, Kremer R K and Jansen M 2007 *J. Mater. Chem.* **17** 1356
[2] Sleight A W, Gillson J L, Weiher J F and Bindloss W 1974 *Solid State Commun.* **14** 357
[3] Hanawa M, Muraoka Y, Tayama T, Sakakibara T, Yamaura J and Hiroi Z 2001 *Phys. Rev. Lett.* **87** 187001
[4] Taguchi Y, Oohara Y, Yoshizawa H, Nagaosa N and Tokura Y 2001 *Science* **291** 2573
[5] Matsuhira K, Wakeshima M, Nakanishi R, Yamada T, Nakamura A, Kawano W, Takagi S and Hinatsu Y 2007 *J. Phys. Soc. Jpn.* **76** 043706
[6] Matsuhira K *et al.*, (submitted to *J. Phys. Soc. Jpn.*)
[7] Walter U 1984 *J. Phys. Chem. Solids* **45** 401
[8] Hutchings M T 1964 *Solid State Physics* **16** 227
[9] Trammell G T 1953 *Phys. Rev.* **92** 1387
[10] Alekseev P A, Goremychkin E A, Lippold B, Mühle E and Sadikov I P 1983 *Phys. Stat. Sol. (b)* **119** 651
[11] Cooper M J and Nathans R 1967 *Acta Cryst.* **23** 357
[12] Dunlap B D 1983 *J. Magn. Magn. Mater.* **37** 211
[13] Tomiyasu K, Fukunaga J, and Suzuki H 2004 *Phys. Rev. B* **70** 214434
[14] Tomiyasu K, Horigane K, Yokobori T, Kousaka Y, Takahashi M, Hiraka H, Akimitsu J, and Yamada K 2009 *J. Phys. Soc. Jpn.* **78** 084704
[15] Kaplan T A and Menyuk N 2007 *Phil. Mag.* **87** 3711

International Conference on Frustration in Condensed Matter (ICFCM) IOP Publishing
Journal of Physics: Conference Series **320** (2011) 012081 doi:10.1088/1742-6596/320/1/012081

Probing local polar structures in PZN-xPT and PMN-xPT relaxor ferroelectrics with neutron and x-ray scattering

Guangyong Xu
Condensed Matter Physics and Materials Science Department, Brookhaven National Laboratory, Upton, New York 11973, USA
E-mail: gxu@bnl.gov

Abstract.
"Polar nano-regions" (PNR) are nanometer sized local polar structures found in relaxor ferroelectric materials. They play key roles in understanding many relaxor properties. Here I will discuss some of our recent work using neutron scattering and x-ray diffraction to probe the properties of the PNR in two prototypical lead based perovskite relaxor systems $Pb(Zn_{1/3}Nb_{2/3})O_3$ (PZN), $Pb(Mg_{1/3}Nb_{2/3})O_3$ (PMN), and their solid solutions with $PbTiO_3$ (PT). The local atomic structures in the PNR are rather complicated, consisting of polarizations/atomic shifts along both ⟨110⟩ and ⟨100⟩ directions. These different (components of) local polarizations respond differently to external electric fields along [001] or [111] directions; and are associated with phonons of different polarizations (T1 and T2 phonon modes). Our work also indicates that the PNR are dynamic at high temperatures well above the Curie temperature T_C, and gradually freeze with cooling. A significant portion of the PNR has already become static before a long-range polar order can be established in the bulk. The short-range polar order in the PNR is quite robust, and persists when the system is cooled (with or without external electric field) into the ferroelectric phase. Because of the frustration between charge neutrality and lattice strain in relaxor systems, the low temperature phase of these relaxor materials usually does not have perfect long-range order, but rather a phase where both short-range and long-range polar orders coexist and compete.

1. Introduction

Relaxor ferroelectrics differ from conventional ferroelectrics by having highly frequency dependent dielectric response (ϵ) with a broad maximum in temperature. In recent years, lead based perovskite relaxors taking the form of Pb(B'B")O_3 have attracted a lot of attention because of their unusual dielectric and piezoelectric response and great potential for industrial applications [1, 2, 3]. Although the origin of many relaxor properties are still not entirely understood, it is widely believed that the composition disorder on the B-site plays key role in determining the bulk response [4, 5]. Here the B-site is normally occupied by two cations with different valencies. For example, in the two prototypical lead based perovskite relaxors $Pb(Zn_{1/3}Nb_{2/3})O_3$ (PZN) and $Pb(Mg_{1/3}Nb_{2/3})O_3$ (PMN), Zn^{2+}/Mg^{2+} and Nb^{5+} take a 1 : 2 ratio on B-site to achieve an average valence of 4+ for charge conservation. Charge neutrality would naturally lead to chemical order on the B-site, while this 1 : 2 type order is not favored by lattice strain. The frustration between the two makes these systems unique in ferroelectric matierials [6], and the resulting local charge imbalance due to the randomness of B-site occupancy creates local random fields, which is the key in understanding many relaxor properties [7, 8, 9, 10, 11].

In addition to the chemical disorder/short-range order, short-range polar order is also present in relaxor systems. The concept of "polar nano-regions" (PNR) was initially proposed by Burns and Dacol [12], to explain the deviation from linear behavior in their optical measurements from a series of relaxor systems including PZN and PMN. It was suggested that while the majority of the lattice still remain unpolarized, local polar clusters start to form at a temperature T_d, called the "Burns temperature", which is typically a few hundred degrees above the Curie temperature T_C. The PNR are believed to be several nanometers in size, grow with cooling, and contribute to many important relaxor properties. For example, the large frequency dependency of the dielectric response ϵ can be naturally explained with the relaxation process of the PNR [4].

Diffuse scattering measurements probes how local atomic positions differ from the average crystal structures, and are therefore very sensitive to local inhomogeneities and short-range orders in solid materials. The PNR in lead based relaxors have been extensively studied by both x-ray [13, 14, 15] and neutron [16, 17, 18, 19, 20, 21, 22, 23, 24] diffuse scattering measurements, as well as other bulk probes including Raman and dielectric measurements [25, 26], and high resolution piezoelectric force microscopy [27, 28]. It has been shown that the PNR not only compete with, but also can coexist with long-range polar orders in relaxor compounds [29, 30]. They do respond to external electric fields [31, 32, 29, 33], but cannot be fully suppressed. The low temperature phase of these relaxor ferroelectric materials is actually a very special phase where long-range (the bulk) and short-range (the PNR) polar orders coexist [30], and accompanied by large lattice strains induced by the PNR [18, 34, 30].

In this paper, I will review our work on the polar nanoregions in PMN, PZN, and their solid solutions with $PbTiO_3$ (PT) with neutron and x-ray scattering over the last decade. The local atomic structures in the polar nano-regions are complicated, consisting of polarizations both along the $\langle 110\rangle$ and $\langle 100\rangle$ directions [35]. They respond differently to external electric fields along different directions. The PNR also start with being partially dynamic at high temperature and gradually freeze with cooling - the process can be probed with neutron spin-echo measurements which has a much better energy resolution than conventional neutron scattering techniques [36, 35]. Finally, I will discuss how the PNR interact with the bulk through their coupling to phonon modes, and its implications on various bulk properties.

2. Structures, polarizations of the PNR and their electric field response

2.1. $\langle 110\rangle$ polarizations and the "pancake model"

With small PT doping, the solid solutions of PMN-xPT and PZN-xPT still exhibit relaxor properties [37, 38], and diffuse scattering from these systems are very similar in general [15, 39]. The diffuse scattering intensities appear above T_C, and increase monotonically with cooling. In order to obtain information on the structure, shape, and polarizations of the PNR, one needs to measure the distribution of diffuse scattering intensities in the reciprocal space. An example of these measurements is shown in Fig. 1. Here a mesh scan from neutron diffuse scattering measurements taken at 200 K from a single crystal PMN, around the (100) Bragg peak in the (H0L) plane is plotted. The intensity is highly anisotropic in the reciprocal space, taking a "butterfly" shape in the (H0L) plane. A more detailed mapping of the three-dimensional distribution of diffuse scattering intensities from single crystals of PZN-xPT for x=0, 4.5% and 8% has been performed with high energy x-ray diffraction [15]. It was found that the diffuse scattering is dominated by rod type intensities along various $\langle 110\rangle$ directions. Although there are totally six different $\langle 110\rangle$ rods, they do not always show up with the same intensity across different Bragg peaks. A sketch of the intensity distribution in the 3-D reciprocal space is plotted in Fig. 2.

Figure 1. (Color online)A smoothed logarithmic plot of the neutron elastic diffuse scattering intensity measured at 200 K from a single crystal PMN near the (100) Bragg peak in the (H0L) scattering plane (see Ref. [24]).

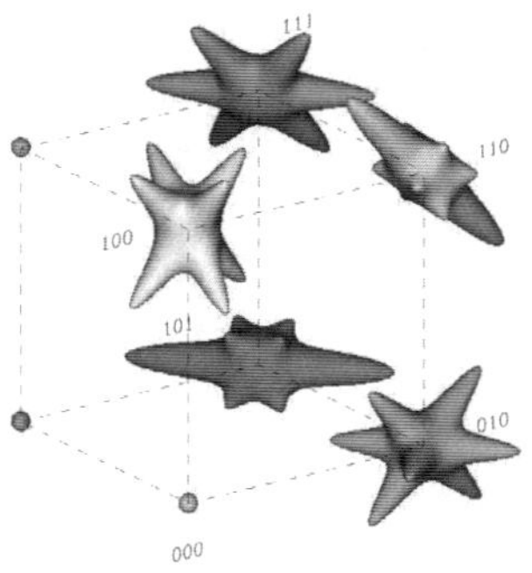

Figure 2. (Color online) Sketch of the diffuse scattering distribution in the 3-D reciprocal space around (100), (110), (111), (010), and (011) reciprocal lattice points from PZN-xPT single crystals for x=0, 4.5% and 8% (see Ref. [15]).

Since rod type intensities in reciprocal space correspond to planar type correlations/structures in real space, a model (the "pancake" model) is proposed that PNR must take a planar shape in real space [15]. In this model, there are six possible orientations/polarizations of PNR, with $\langle 1\bar{1}0\rangle$ type polarizations, correlated in $\{110\}$ planes, that give rise to $\langle 110\rangle$ diffuse rods (see Fig. 3). The in-plane and out-of-plane correlation lengths (or, the diameter and thickness of the "pancake" PNR, respectively) can be estimated from the line widths of diffuse scattering perpendicular and along the intensity rod directions. A rough estimate will give a in-plane correlation length of 10 to 20 nm (20 to 40 lattice spacings) while the out-of-plane correlation length is about four times smaller [24, 15].

More x-ray and neutron diffuse scattering measurements have confirmed that the "butterfly"-shaped diffuse scattering is dominant in most PMN-xPT and PZN-xPT samples [13, 15, 39, 40] for small x. The "butter-fly" diffuse are results of (short-range) ordered $\langle 110\rangle$ type polarizations, extending in the perpendicular $\langle 110\rangle$ directions in reciprocal space - similar to T2 phonon modes in these perovskite systems, which are the transverse phonon modes with $\langle 110\rangle$ polarizations propagating along the perpendicular $\langle 110\rangle$ directions. We therefore also refer to the "butter-fly" diffuse as the "T2-diffuse". In fact, later in the paper, I will show that the "butter-fly" diffuse scattering can interact strongly with the T2 phonon modes in relaxors.

When more $PbTiO_3$ is added into the solid solution, the relaxor properties gradually disappear and the system becomes more of a conventional ferroelectric. Diffuse scattering studies for systems on the ferroelectric side of the PT-doping phase diagram have shown that the "butter-fly" diffuse completely disappears [41, 39], indicating that no PNR exists in the ferroelectric side of the phase diagram.

There are, however, arguments suggesting that the "butter-fly" diffuse scatterings from relaxors are in fact Huang-scattering due to defects and strains in the system. If the "butter-fly" diffuse scattering intensities were only strain related, their intensities should scale with $\mathbf{Q}^2|F(\mathbf{G})|^2$ where $\mathbf{Q}$ is the measurement wave-vector and $|F(\mathbf{G})|^2$ is the lattice structure factor that gives the Bragg peak intensity at the reflection $\mathbf{G}$ near $\mathbf{Q}$. This is clearly not the

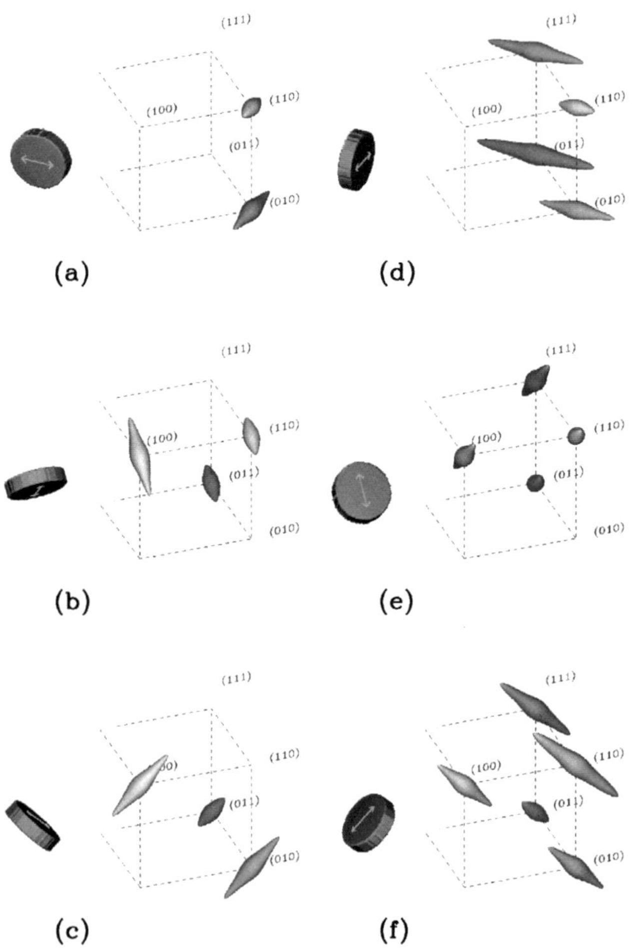

Figure 3. (Color online) PNR in the real space and their contributions to the diffuse scattering in the reciprocal space. A "pancake" shaped PNR in real space corresponds to rod type diffuse scattering in reciprocal space. From (a) to (f), we show PNR with in-plane polarizations along the $[01\bar{1}]$, $[10\bar{1}]$, $[1\bar{1}0]$, [011], [101], and [110] directions, correlated in the (011), (101), (110), $(01\bar{1})$, $(10\bar{1})$, and $(1\bar{1}0)$ planes, and contributing to the diffuse rods along [011], [101], [110], $[01\bar{1}]$, $[10\bar{1}]$, and $[1\bar{1}0]$ directions, respectively (see Ref.[15]).

case, shown by measurements on the relative diffuse scattering intensities across various Bragg peaks [18, 20, 30]. In fact, the atomic shifts in the PNR that contribute to the "butter-fly" diffuse scattering intensities can always be decomposed into two components, one is a relative shift of various atoms in the unit cell without movement of the center-of-mass – this is optic type component and contribute to local polar moments; another is a "uniform phase shift" which is the movement of all atoms in the unit cell together, corresponding to the acoutic type component that contribute to strain [18, 30].

The "butter-fly" diffuse scattering from the PNR also respond to external electric fields in a very unusual manner. An external electric field along the [001] direction applied to single crystal PMN-xPT [42] and PZN-xPT [35] samples has been shown to have virtually no effect on the "butter-fly" diffuse scattering pattern. On the other hand, an external field along the [111] direction can induce dramatic changes on the diffuse pattern. Naively one would expect an external field along [111] direction to stablize the long-range polar order in the low temeprature rhombohedral (R) phase, and suppress the diffuse scattering intensity from the PNR. Our neutron [29] and x-ray [33] diffuse scattering measurements indicate otherwise. In PZN-xPT single crystal samples, the "butter-fly" diffuse scattering does not simply disappear with field cooling (E along [111]). Instead, a "redistribution" of diffuse scattering intensities across different $\langle 110\rangle$ rods has been observed [29, 33] (see Fig. 4). The portion of diffuse scattering coming from

PNR with polarizations perpendicular to the external [111] field has been enhanced, while the part from PNR with polarizations not perpendicular to the external field has been partially suppressed. The overall diffuse scattering intensity appears to be conserved (compared to that measured under the zero-field-cooled condition).

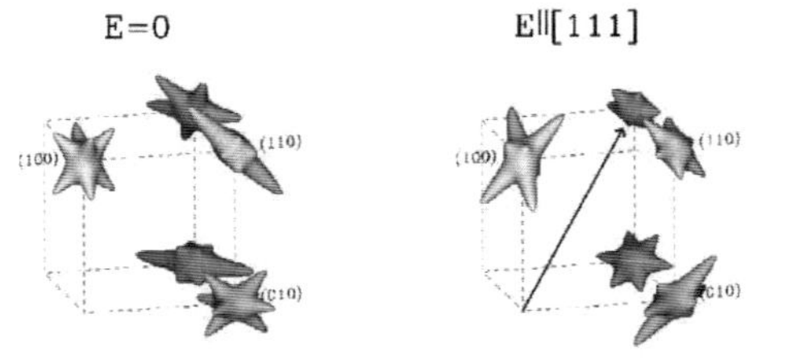

Figure 4. (Color online) Sketch of the three-dimensional diffuse scattering distribution from single crystal PZN. They are plotted in the 3-D reciprocal space around (100), (110), (111), (010), and (011) reciprocal lattice points for (a) E=0, and (b) E along [111]. In (b), the diffuse rods along the [110], [101], and [011] directions are enhanced, while the diffuse rods along the [$1\bar{1}0$], [$10\bar{1}$], and [$01\bar{1}$] are suppressed (see Ref. [33]).

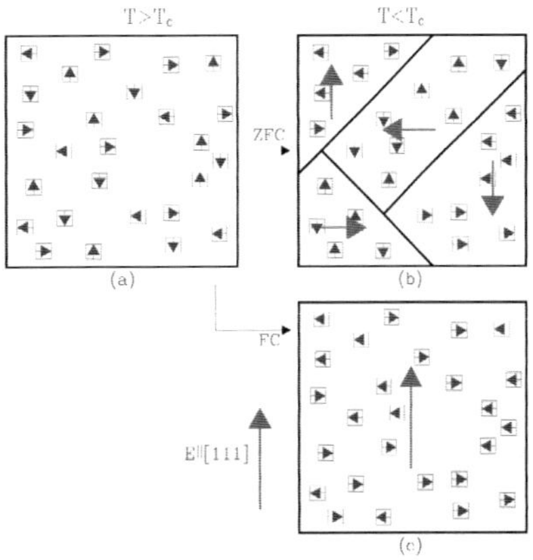

Figure 5. (Color online) A schematic showing the PNR configurations in a relaxor system in (a) the paraelectric phase, (b) ZFC into the ferroelectric phase, and (c) FC into the ferroelectric phase. The large arrows indicate the polarization of the ferroelectric domains separated by domain walls (solid lines). The small squares represent the PNR (see Ref. [30]).

These results can be understood in a simple picture. In the paraelectric high temperature phase, there is no long-range ferroelectric domains, and PNR with different ⟨110⟩ polarizations are equivalent under the cubic symmetry. When the system enters the low temperature phase, long-range polar order is established and ferroelectric domains with ⟨111⟩ polarizations are formed. Within these domains, the symmetry is lowered to R, and the different ⟨110⟩ directions are not equivalent any more. If within a certain ferroelectric domain, the PNR would prefer to exist in a configuration where their polarizations are perpendicular to that of the domain, all our previous results can be explained easily. As shown in Fig. 5 (b), under ZFC, after multi-domain averaging, in the crystal there is no macroscopic preferred ⟨110⟩ polarization of the PNR, resulting in the symmetric "butterfly" diffuse scattering shown in Figs. 1, 2. However, with FC, the volume of the ferroelectric domain polarized along the field [111] direction is greatly enhanced [see Fig. 5 (c)], and our measurements thus provide direct information on how the PNR reside in a [111] polarized ferroelectric lattice - they tend to have polarizations perpendicular to that of their surrounding lattice.

The case for pure PMN is an exception. Once an external electric field along [111] direction is applied, in addition to the redistribution of diffuse scattering intensities between different ⟨110⟩ directions, there is also an overall suppression of the diffuse scattering by the field [43]. In the mean time, Bragg peak intensities increase, indicating an enhancement of long-range order in the system. Note that in pure PMN, no long-range polar order is established without external field and therefore no macroscopic ferroelectric domains exist at low temperature. However, there could be polar-orders developing in the system at low temperature in the mesoscopic range (e.g. ~ 1 μm) that provides local $\langle 111 \rangle$ polarized lattice environment for the PNR. With an external electric field, the re-arrangements of these mesoscopic polar lattice can give rise to the redistribution of PNR and therefore diffuse scattering. On the other hand, not having a long-range ferroelectric order seems to also make the short-range polar order in the PNR less stable and more sensitive to external fields.

This situation where the two competing orders can co-exist, and having the long-range order helps stabilizing the short-range polar order, is quite unusual. In addition, as the temperature decreases, both orders will develop (with the exception of pure PMN) - the long-range polar order develops as evidenced by the increase of rhombohedral distortion with cooling; and the short-range order develops shown by the increase of diffuse scattering intensities. The robustness of these local polar orders within the long-range polarized lattice is yet another indication of charge-lattice frustration in relaxor systems.

2.2. $\langle 001 \rangle$ polarizations and the "T1" diffuse

As mentioned previously in the article, the "butter-fly" diffuse scattering is barely affected by an external field along the [001] direction. However, there is evidence suggesting possible diffuse scattering components away from the $\langle 110 \rangle$ rods in the "butter-fly" diffuse. For instance, neutron diffuse scattering experiments [32] on a single crystal of PZN-8%PT show that the diffuse scattering measured along (H03) is strongly suppressed by an external [001] field, whereas that measured along (30L) remains unaffected. This effect has been confirmed later in our measurements on PZN-4.5%PT [35] [see panels (c) and (d) of Fig. 6] and cannot be explained by a simple redistribution of the T2-diffuse scattering.

From Fig. 6, we notice that this extra component of diffuse scattering that can respond to an external field along [001] direction is measured at q along [001] or [100] away from the Bragg peak. Detailed analysis [35] indicates that the local atomic shifts contributing to this extra component are also along the $\langle 001 \rangle$ directions perpendicular to q. Considering its similarities to the T1 phonon modes, where transverse phonons with $\langle 001 \rangle$ type polarizations propagate along the perpendicular $\langle 001 \rangle$ directions, we therefore refer to this component of diffuse scattering as the "T1" diffuse. The "T1" component of the diffuse scattering follows similar temperature trend as the "butterfly" diffuse, but is considerably weaker in intensity, and can only be clearly identified by examining the change of diffuse scattering intensity when an external field along [001] direction is applied. The structure factors for the T1-diffuse also differ from those of the "butterfly" diffuse, yet one can still decompose the atomic shifts (along $\langle 001 \rangle$ directions) that contribute to the T1-diffuse into an acoustic strain component plus an optic polar component [35].

Neutron pair-density-function (PDF) measurements on PMN [45] suggest that the average local Pb displacements are along the $\langle 111 \rangle$ directions. There is no inconsistency between the PDF measurements and the diffuse scattering measurements. The PDF measurements probes the average atomic positions, while the diffuse scattering measurements probe the (short-range) order of various atomic displacements. An local environment with Pb shifted along $\langle 111 \rangle$ directions but only the $\langle 001 \rangle$ and $\langle 110 \rangle$ components of the shifts are short-range ordered (with different correlations), will be able to explain the experimental data from different techniques.

Overall, the local structure in these relaxor compounds are complicated. Our diffuse scattering measurements have found clear contributions from at least two type of (short-range ordered) local atomic displacements. They have similar temperature dependence, and both have strain/polar components; while they respond quite differently to external electric fields.

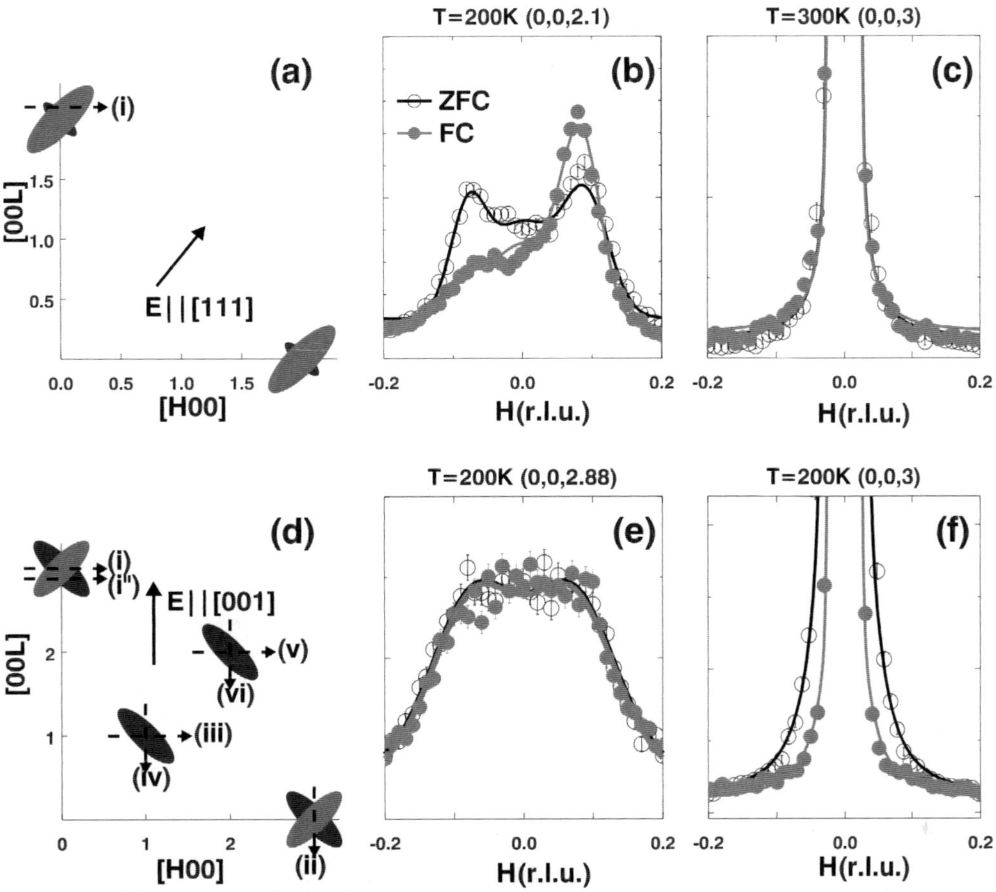

Figure 6. (Color online) (a) Schematic diagram of the neutron scattering measurements made in the (H0L) scattering plane with an electric field applied along [111]. The large (red) and small (blue) ellipses reflect how the "butter-fly" diffuse scattering intensity is redistributed after the sample is field cooled (FC). (b) Measurements of the diffuse scattering intensity near (002) along the dashed arrow shown in panel (a) at 200 K (from Ref. [44]). Scans were made along (H,0,2.1) because they intercept both wings of the butterfly sufficiently far from (002) that significant contamination from unwanted Bragg peak scattering can be avoided, thus making it possible to monitor how intensities of the two wings change with field and temperature. Open circles represent zero-field cooled (ZFC) data and closed circles represent FC data . (c) Same schematic diagram as in panel (a) except with the electric field applied along [001]. Dashed lines denote linear q-scans performed across Bragg peak positions. (d) Linear q-scans at 200 K along [100] measured across (003) [see scan (i) in panel (c)]. Error bars in (b) and (d) represent the square root of the number of counts. The figure is taken from Ref. [35]

3. Dynamic PNR and interaction with phonons

3.1. Dynamic PNR

One important question on the PNR is whether they have a static or dynamic origin. Most of our previous work has only been focused on the static structures of the PNR using elastic diffuse scattering, or, at least, "quasi-elastic". Nevertheless, our neutron diffuse scattering measurements are limited by the energy resolution of the instruments, which could range between

0.1 meV to $\sim$ 1 meV depending on the instrument set-up and the type of instrument used. The x-ray diffuse scattering measurements, on the other hand, cannot really distinguish between the static and dynamic nature of the diffuse scattering because of its very coarse energy resolution.

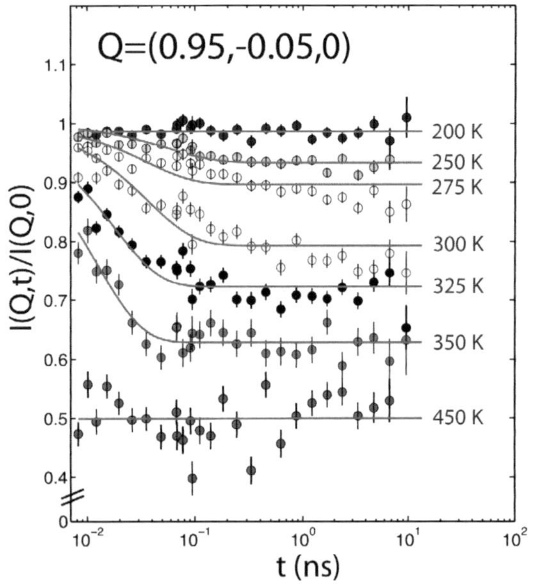

Figure 7. (Color online)Neutron spin-echo data from PMN for different temperatures (see Ref. [36]).

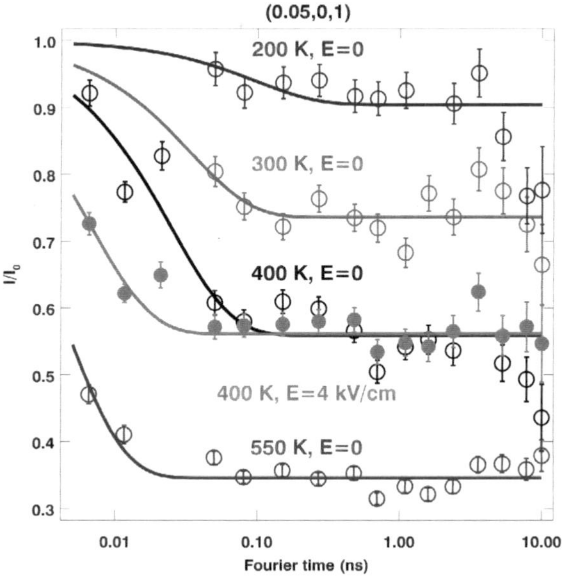

Figure 8. (Color online) Neutron spin-echo data from PZN-4.5%PT for different temperatures and electric field effect (see Ref. [35]).

The neutron spin-echo (NSE) spectroscopy is ideal for this purpose [46]. In principle, NSE measurements probe how a certain mode decays with time. In our case, we can use NSE to probe how long-lived the diffuse scattering from the PNR is. Figs. 7 and 8 show NSE measurements on single crystals of PMN and PZN-4.5%PT at $\mathbf{Q} = (0.95, -0.05, 0)$ [36] and $(0.05, 0, 1)$ [35], respectively. The measurements at $\mathbf{Q} = (0.95, -0.05, 0)$ probe mostly the T2-diffuse (butterfly diffuse), while those at $(0.05, 0, 1)$ have a larger contribution from the T1-diffuse. Clearly, these results suggest that both the T1 and T2 diffuse scattering intensities have a significant dynamic portion at high temperatures. The lifetime of the dynamic portion can also be extracted from the data. For example, in PZN-4.5%PT, at high temperature (550 K), the lifetime of the dynamic portion of the PNR is about 0.004 ns, corresponding to a energy width (half width at half maximu HWHM) of 0.16 meV. With cooling, not only the dynamic/static ratio decreases, but also the lifetime of the dynamic portion increases. At 200 K in PZN-4.5%PT, the lifetime of the dynamic portion increases to about 0.01 ns with an energy width HWHM $\sim$ 0.007 meV [35]. The situation is very similar in PMN [36]. These results indicate that the PNR gradually freezes with cooling, and become mostly static at low enough temperatures.

3.2. PNR interacting with phonons

There have long been suggestions that the diffuse scattering from PNR can couple to the transverse optic (TO) [47, 48] and transverse acoustic (TA) [49, 21, 50] phonon modes in relaxor ferroelectrics. Our neutron scattering measurements show that this coupling is strongest for PNR and phonons with the same polarizations. For instance, the "butterfly" diffuse (T2-diffuse) couples strongly to the TA phonon mode with the same $\langle 110 \rangle$ polarizations (T2 mode) [44], while the T1-diffuse couples to the T1 phonon mode [51].

The implications of the interaction between the PNR and phonons in the bulk goes beyond the coupling itself. Here we find that the PNR are related to the softening of a particular phonon mode [44]. A soft phonon mode is usually associated with a phase instability, which is necessary for a system to have a high piezoelectric response [52, 53, 54, 55, 56]. Our results suggest that there might be some connections between the local polar structures (PNR) and the extraordinaty bulk response in relaxor compounds.

4. Summary

Our neutron and x-ray scattering studies on PZN-xPT and PMN-xPT relaxor compounds have shown that these materials have complex local structures due to lattice-charge frustration. The local structures have both polar and strain characteristics. They are partially dynamic at high temperatures, gradually freeze with cooling, and can coexist with the long-range polar phase at low temperature. These local structures can also interact with the bulk through their coupling to phonons in the bulk material, which may be a key in understanding the extraordinary piezoelectric and electromechanical properties of these lead based perovskite relaxor systems.

5. acknowledgements

The work discussed in this article has been carried out with many collaborators. I would first like to give my special acknowledgment to Dr. Gen Shirane, who started our work on relaxor systems in the late 1990's. I would also like to thank all other collaborators including: F. Bai, Y. Bing, H. Cao, W. Chen, K. H. Conlon, J. R. D. Copley, J. S. Gardner, P. M. Gehring, M. J. Gutmann, H. Hiraka, K. Hirota, S.-H. Lee, J.-F. Li, H. Luo, M. Matsuura, K. Ohwada, C. Stock, I. Swainson, D. Viehland, S. Wakimoto, T. R. Welberry, J. Wen, H. Woo, Z.-G. Ye, Z. Xu, X. Zhao, and Z. Zhong. Financial support from the U.S. Department of Energy under contract No. DE-AC02-98CH10886 is also gratefully acknowledged.

References

[1] Park S E and Shrout T R 1997 *J. Appl. Phys.* **82** 1804
[2] Uchino K 1996 *Piezoelectric actuators and ultrasonic motors* (Kluwer Academic, Boston)
[3] Service R F 1997 *Science* **275** 1878
[4] Cross L E 1987 *Ferroelectrics* **76** 241
[5] Burton B P, Cockayne E, Tinte S and Waghmare U V 2006 *Phase Transitions* **79** 91
[6] Guangyong Xu 2010 *J. Phys. Soc. Japan* **79** 011011
[7] Imry Y and Ma S 1975 *Phys. Rev. Lett.* **35** 1399
[8] Halperin B I and Varma C M 1976 *Phys. Rev. B* **14** 4030
[9] Westphal V, Kleemann W and Glinchuk M D 1992 *Phys. Rev. Lett.* **68** 847
[10] Fisch R 2003 *Phys. Rev. B* **67** 094110
[11] Pirc R and Blinc R 1999 *Phys. Rev. B* **60** 13470
[12] Burns G and Dacol F H 1983 *Phys. Rev. B* **28** 2527
[13] You H and Zhang Q M 1997 *Phys. Rev. Lett.* **79** 3950
[14] Takesue N, Fujii Y and You H 2001 *Phys. Rev. B* **64** 184112
[15] Xu G, Zhong Z, Hiraka H and Shirane G 2004 *Phys. Rev. B* **70** 174109
[16] La-Orauttapong D, Toulouse J, Robertson J L and Ye Z G 2001 *Phys. Rev. B* **64** 212101
[17] La-Orauttapong D, Toulouse J, Ye Z G, Chen W, Erwin R and Robertson J L 2003 *Phys. Rev. B* **67** 134110
[18] Hirota K, Ye Z G, Wakimoto S, Gehring P M and Shirane G 2002 *Phys. Rev. B* **65** 104105
[19] Dkhil B, Kiat J M, Calvarin G, Baldinozzi G, Vakhrushev S B and Suard E 2001 *Phys. Rev. B* **65** 024104
[20] Vakhrushev S B, Naberezhnov A A, Okuneva N M and Savenko B N 1995 *Phys. Solid State* **37** 1993
[21] Hiraka H, Lee S H, Gehring P M, Xu G and Shirane G 2004 *Phys. Rev. B* **70** 184105
[22] Hlinka J, Kamba S, Petzelt J, Kulda J, Randall C A and Zhang S J 2003 *J. Phys.: Condens. Matter* **15** 4249
[23] Vakhrushev S B, Ivanov A and Kulda J 2005 *Phys. Chem. and Chem. Phys.* **7** 2340
[24] Xu G, Shirane G, Copley J R D and Gehring P M 2004 *Phys. Rev. B* **69** 064112
[25] Svitelskiy O, Toulouse J and Ye Z G 2003 *Phys. Rev. B* **68** 104107
[26] Bovtun V, Kamba S, Pashkin A, Savinov M, Samoukhina P, Petzelt J, Bykov I P and Glinchuk M D 2004 *Ferroelectrics* **298** 23

[27] Lehnen P, Kleemann W, Woike T and Pankrath R 2001 *Phys. Rev. B* **64** 224109
[28] Shvartsman V V and Kholkin A L 2004 *Phys. Rev. B* **69** 014102
[29] Xu G, Gehring P M and Shirane G 2005 *Phys. Rev. B* **72** 214106
[30] Xu G, Gehring P M and Shirane G 2006 *Phys. Rev. B* **74** 104110
[31] Vakhrushev S B, Naberezhnov A A, Okuneva N M and Savenko B N 1998 *Phys. Solid State* **40** 1728
[32] Gehring P M, Ohwada K and Shirane G 2004 *Phys. Rev. B* **70** 014110
[33] Xu G, Zhong Z, Bing Y, Ye Z G and Shirane G 2006 *Nature Materials* **5** 134
[34] Xu G, Zhong Z, Bing Y, Ye Z G, Stock C and Shirane G 2004 *Phys. Rev. B* **70** 064107–1–064107–6
[35] Zhijun Xu, Jinsheng Wen, Guangyong Xu, Stock C, Gardner J S and Gehring P M 2010 *Phys. Rev. B* **82** 134124
[36] Stock C, Eijck L V, Fouquet P, Maccarini M, Gehring P M, Guangyong Xu, Luo H, Zhao X, Li J F and Viehland D 2010 *Phys. Rev. B* **81** 144127
[37] Noheda B, Cox D E, Shirane G, Gao J and Ye Z G 2002 *Phys. Rev. B* **66** 054104
[38] La-Orauttapong D, Noheda B, Ye Z G, Gehring P M, Toulouse J, Cox D E and Shirane G 2002 *Phys. Rev. B* **65** 144101
[39] Matsuura M, Hirota K, Gehring P M, Ye Z G, Chen W and Shirane G 2006 *Phys. Rev. B* **74** 144107
[40] Welberry T R, Goossens D J and Gutmann M J 2006 *Phys. Rev. B* **74** 224108
[41] Stock C, Ellis D, Swainson I P, Xu G, Hiraka H, Zhong Z, Luo H, Zhao X, Viehland D, Birgeneau R J and Shirane G 2006 *Phys. Rev. B* **73** 064107
[42] Jinsheng Wen, Guangyong Xu, Stock C and Gehring P M 2008 *Appl. Phys. Lett.* **93** 082901
[43] Stock C, Xu G, Gehring P M, Luo H, Zhao X, Cao H, Li J F, Viehland D and Shirane G 2007 *Phys. Rev. B* **76** 064122
[44] Xu G, Wen J, Stock C and Gehring P M 2008 *Nature Materials* **7** 562
[45] Jeong I K, Darling T W, Lee J K, Proffen T, Heffner R H, Park J S, Hong K S, Dmowski W and Egami T 2005 *Phys. Rev. Lett.* **94** 147602
[46] Mezei F (ed) 1980 *Neutron Spin Echo* vol 128 (Springer, Berlin) lecture Notes in Physics
[47] Gehring P M, Wakimoto S, Ye Z G and Shirane G 2001 *Phys. Rev. Lett.* **87** 277601
[48] Gehring P M, Park S E and Shirane G 2000 *Phys. Rev. Lett.* **84** 5216
[49] Koo T Y, Gehring P M, Shirane G, Kiryukhin V, Lee S G and Cheong S W 2002 *Phys. Rev. B* **65** 144113
[50] Stock C, , Luo H, Viehland D, Li J F, Swainson I, Birgeneau R J and Shirane G 2005 *J. Phys. Soc. Japan* **74** 3002
[51] Xu Z, Wen J, Xu G, Stock C and Gehring P M Unpublished
[52] Fu H and Cohen R E 2000 *Nature* **403** 281
[53] Wu Z and Cohen R E 2005 *Phys. Rev. Lett.* **95** 037601
[54] Kutnjak Z, Petzelt J and Blinc R 2006 *Nature* **441** 956
[55] Frantti J, Fujioka Y and Nieminen R M 2007 *J. Phys. Chem. B* **111** 4287
[56] Budimir M, Damjanovic D and Setter N 2006 *Phys. Rev. B* **73** 174106

Numerical Simulation of Spin-Chirality Switching in Multiferroics via Intense Electromagnon Excitations

Masahito Mochizuki[1,2] **and Naoto Nagaosa**[1,3]

[1]Department of Applied Physics, University of Tokyo, Tokyo 113-8656, Japan
[2]Multiferroics Project, ERATO (JST), Tokyo 113-8656, Japan
[3]Cross-Correlated Materials Research Group (CMRG) and Correlated Electron Research Group (CERG), RIKEN-ASI, Wako 351-0198, Japan

E-mail: `mochizuki@erato-mf.t.u-tokyo.ac.jp`

Abstract. Chirality, i.e., the right- and left-handedness of structure, is one of the key concepts in many elds of science including biology, chemistry and physics, and its manipulation is an issue of vital importance. The electron spins in solids can form chiral con gurations. In perovskite manganites $R\text{MnO}_3$ (R=Tb, Dy,...etc), the Mn-spins form a cycloidal structure, which induces ferroelectric polarization ($\boldsymbol{P}$) through the relativistic spin-orbit interaction. This magnetism-induced ferroelectricity (multiferroics) and associated infrared-active spin waves (electromagnons) open a promising route to control the spins by purely electric means in a very short time. In this paper, we show theoretically with an accurate spin Hamiltonian for TbMnO_3 that a picosecond optical pulse can switch the spin chirality by intensely exciting the electromagnons with a terahertz frequency.

1. Introduction

Some structures are not invariant upon the mirror operation just like the right and left hands, and are called *chiral.* This concept is of fundamental importance in chemistry and biology. The selective chemical synthesis of chirality (called asymmetric synthesis) is crucial for reliable and safe productions of medicines and foods. Note that the chirality of molecules and crystals is hard to change once synthesized because it is determined by the chemical bonds. In physics, on the other hand, spins in magnets sometimes form a chiral order. The spin chirality is much more exible since the spin is a quantum-mechanical and relativistic object, which o ers an opportunity to control the chirality by external parameters.

The representative laboratory for this spin chirality is the multiferroic transition-metal oxides [1]. A model taking into account the spin-orbit interaction predicts the electric polarization $\boldsymbol{p}_{ij}$ produced by the chiral spin con guration [2, 3] as

$$\boldsymbol{p}_{ij} \propto \boldsymbol{e}_{ij} \quad (\boldsymbol{S}_i \quad \boldsymbol{S}_j), \tag{1}$$

with $\boldsymbol{e}_{ij}$ being the vector connecting ith and jth sites. Here the vector product $\boldsymbol{S}_i \quad \boldsymbol{S}_j$ is the spin chirality, which characterizes a direction of the spin rotation as one goes from i to j. This model explains the multiferroic behavior in $R\text{MnO}_3$ where the cycloidal Mn-spin order induces the ferroelectric polarization $\boldsymbol{P}$ [4]. For example, in TbMnO_3 [see Figs. 1(a) and (b)], the clockwise bc-plane spin spiral propagating along the $+\boldsymbol{b}$ direction induces $\boldsymbol{P}\|+\boldsymbol{c}$, while the counterclockwise one induces $\boldsymbol{P}\| \quad \boldsymbol{c}$ [5, 6]. For the a, b and c axes, we adopt the $Pbnm$ setting.

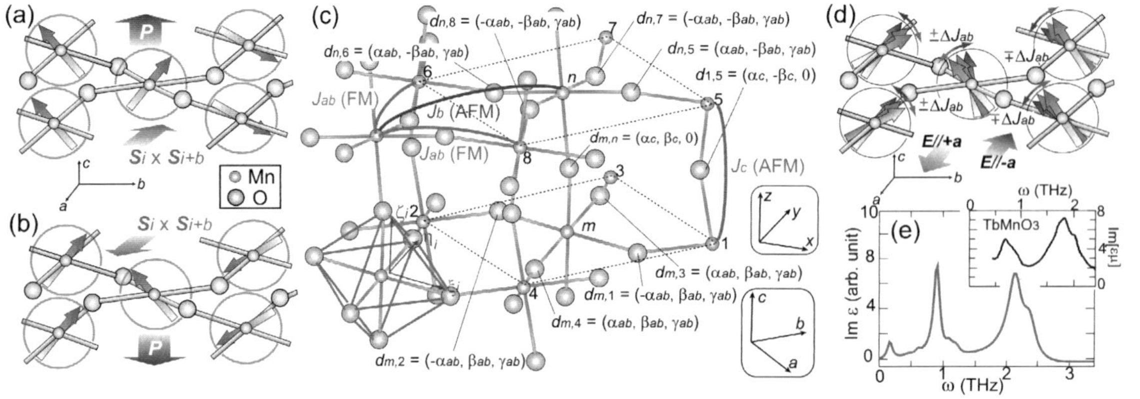

Figure 1. (a) Spin con guration and ferroelectric polarization $\boldsymbol{P}$ in the clockwise bc-plane spin spiral and (b) those in the counterclockwise one. (c) Crystal structure, spin exchanges, Dzyaloshinskii-Moriya vectors $\boldsymbol{d}_{i,j}$, and local axes attached to the ith MnO_6 octahedron in $R$$MnO_3$. Here FM (AFM) denotes (anti)ferromagnetic exchange. (d) Modulation of the in-plane ferromagnetic exchanges under $\boldsymbol{E}\|\pm\boldsymbol{a}$, which induces slight spin rotations indicated by red (blue) arrows under $\boldsymbol{E}\|+\boldsymbol{a}$ ($\boldsymbol{E}\|$ $\boldsymbol{a}$). Upper (lower) signs in front of J_{ab} (>0) correspond to the modulations under $\boldsymbol{E}\|+\boldsymbol{a}$ ($\boldsymbol{E}\|$ $\boldsymbol{a}$). (e) Calculated electromagnon optical spectrum. Inset shows the experimental spectrum for $TbMnO_3$ from Ref. [13].

When the electric polarization is driven by the spin order, it is naturally expected that collective excitations of spins (magnons) have an optical activity [7, 8, 9]. Indeed, strong optical absorptions were experimentally observed in $R$$MnO_3$ at terahertz frequencies, and they were ascribed to magnons activated by the electric- eld component of light i.e., electromagnons [8, 9, 10, 11, 12, 13, 14, 15]. In the early stage, the corresponding magnon modes were interpreted as rotation of the spin-spiral plane accompanied by oscillation of $\boldsymbol{p}_{ij}$ in Eq. (1) [9]. However, this interpretation contradicts to the experimental observation about the selection rule in terms of the light polarization [11, 12]. Afterwards, it turned out that the conventional magnetostriction mechanism where $\boldsymbol{p}_{ij}$ is given by

$$\boldsymbol{p}_{ij} = \boldsymbol{\pi}_{ij}(\boldsymbol{S}_i \quad \boldsymbol{S}_j) \tag{2}$$

is relevant to the electromagnons in $R$$MnO_3$ [16, 17]. Here the vector $\boldsymbol{\pi}_{ij}$ is nonzero because of the orthorhombic lattice distortion without inversion symmetry at the center of the Mn-O-Mn bond. The puzzling electromagnon optical spectrum with two speci c peaks was successfully explained by this mechanism [17].

Under this circumstance, the photo-induced phenomena become a challenging issue for the next step. Since $\boldsymbol{p}_{ij}$ in Eq. (2) does not require the spin-orbit interaction, its magnitude is much larger than that of $\boldsymbol{p}_{ij}$ in Eq. (1), which enables us to realize the intense optical excitations of magnons. In fact, the real-time dynamics of the photo-excitation is usually a notoriously complicated problem since the electronic excitations are the genuine many-body problem of various degrees of freedom. Therefore, a reliable theoretical analysis has been almost impossible thus far. However, we are now ready to attack such phenomena in $R$$MnO_3$ theoretically. This is because we know that the optical pulse activates mostly the spin systems only via the magnetostriction given in Eq. (2), which allows us to neglect electronic excitations at higher energies (>1.5 eV). In addition, we have an accurate spin Hamiltonian, which describes competitions among various phases in $R$$MnO_3$ [18], so that the optical switching among them can be simulated in a reliable way. Recent rapid experimental progress in intense THz-pulse generation also gives a growing importance to the theoretical study on this subject [19, 20, 21].

2. Model and Method

To describe the Mn-spin system in $TbMnO_3$, we employ a classical Heisenberg model on a cubic lattice, which is given by

$$\mathcal{H} = \sum_{<i,j>} J_{ij}\boldsymbol{S}_i \quad \boldsymbol{S}_j + D\sum_i S_{\zeta i}^2 + E\sum_i (\quad 1)^{i_x+i_y}(S_{\xi i}^2 \quad S_{\eta i}^2)$$

$$+ \sum_{<i,j>} \boldsymbol{d}_{i,j} \quad (\boldsymbol{S}_i \quad \boldsymbol{S}_j) \quad B_{\rm biq}\sum_{<i,j>}^{ab} (\boldsymbol{S}_i \quad \boldsymbol{S}_j)^2. \tag{3}$$

Here i_x, i_y, i_z represent the integer coordinates of the ith Mn ion with respect to the x, y and z axes. The rst term describes the spin-exchange interactions as shown in Fig. 1(c). The second and the third terms stand for the single-ion anisotropy de ned with local axes, ξ_i, $_i$ and ζ_i, attached to each MnO_6 octahedron tilted in the orthorhombic lattice [18]. The fourth term denotes the DM interaction where the DM vectors $\boldsymbol{d}_{i,j}$ are expressed using ve DM parameters, $_{ab}$, β_{ab}, $_{ab}$, $_c$ and β_c, as given in Ref. [22]. The last term represents the biquadratic interaction originating from the spin-phonon coupling [23]. The Monte-Carlo analysis of this model has successfully reproduced the rich phase diagram of RMnO$_3$ [18]. We perform calculations using following parameters: $J_{ab}=$ 0.74, J_b=0.64, J_c=1.0, ($_{ab}$,β_{ab}, $_{ab}$)=(0.1, 0.1, 0.14), ($_c$,β_c)=(0.48, 0.1), D=0.2, E=0.25, and $B_{\rm biq}$=0.025, where the energy unit is meV. This parameter set gives the bc-plane transverse spin spiral propagating along the b axis with a wave number q_b=0.3π at low temperatures, which resembles the bc-plane spin spiral in $TbMnO_3$ (q_b=0.29π) [5, 6].

We trace dynamics of the Mn spins by numerically solving the Landau-Lifshitz-Gilbert equation using the fourth-order Runge-Kutta method [27]. The equation is given by

$$\frac{\partial \boldsymbol{S}_i}{\partial t} = \quad \boldsymbol{S}_i \quad \boldsymbol{H}_i^{\rm eff} + \frac{_{\rm G}}{S}\boldsymbol{S}_i \quad \frac{\partial \boldsymbol{S}_i}{\partial t}, \tag{4}$$

where $_{\rm G}$(=0.05) is the Gilbert-damping coe cient. We derive an e ective magnetic eld $\boldsymbol{H}_i^{\rm eff}$ acting on the spin $\boldsymbol{S}_i$ from the Hamiltonian $\mathcal{H}$ as $\boldsymbol{H}_i^{\rm eff} = \quad \partial\mathcal{H}/\partial\boldsymbol{S}_i$. Considering the strongly reduced Mn moment revealed in the neutron-scattering experiment [24], we set the norm of the spin vector $|\boldsymbol{S}_i|$=1.4. The system used for calculations is 20 20 6 in size with the periodic boundary condition. For the magnetoelectric (ME) coupling, we consider $\boldsymbol{E} \quad \boldsymbol{p}_{ij} = \quad (\boldsymbol{E} \quad \boldsymbol{\pi}_{ij})\boldsymbol{S}_i \quad \boldsymbol{S}_j$ where $\boldsymbol{p}_{ij}$ is given in Eq. (2) [16, 17]. This coupling e ectively modulates the in-plane ferromagnetic exchanges from $J_{ab}\boldsymbol{S}_i \quad \boldsymbol{S}_j$ to $(J_{ab} \quad \boldsymbol{E} \quad \boldsymbol{\pi}_{ij})\boldsymbol{S}_i \quad \boldsymbol{S}_j$. More concretely, the applied $\boldsymbol{E}\|\pm\boldsymbol{a}$ modulates the spin exchanges as shown in Fig. 1(d). Here $|\pi_{ij}^a|$ is calculated to be 3.5 10 ^{26} μCm from the lattice parameters[25] and the observed P($5000\mu C/m^2$) for $HoMnO_3$ with an up-up-down-down spin order [26]. This means that E_a=1 MV/cm induces the modulation $|\quad J_{ab}|=|E_a\pi_{ij}^a|$=0.022 meV. Note that $\boldsymbol{p}_{ij}$ are perfectly canceled out in the equilibrium state, and never contribute to the static ferroelectricity.

The electromagnon spectrum is calculated as a response to the weak δ-function pulse. We trace the time evolution of the net polarization $\boldsymbol{P}(t) = \sum_{<i,j>}\boldsymbol{p}_{ij}(t)$ after application of the pulse where $\boldsymbol{p}_{ij}(t)$ is given by Eq. (2). Then we perform the Fourier transformation of $\boldsymbol{P}(t)$ to obtain Im $\varepsilon(\omega)$. The calculated spectrum is displayed in Fig. 1(e), which has two peaks at ω=0.94 and 2.1 THz, and reproduces well the experimental spectrum of $TbMnO_3$ [13].

Using this model, we theoretically demonstrate switching of the spin chirality by application of an intense optical pulse. Note rst that there are two kinds of bc-plane spirals with di erent spin chiralities, i.e. clockwise and counterclockwise turns as one propagates along the $+\boldsymbol{b}$ direction. Their spin chiralities $\boldsymbol{\chi}$ point in the $\boldsymbol{a}$ and $+\boldsymbol{a}$ directions ($\boldsymbol{\chi}\| \quad \boldsymbol{a}$ and $\boldsymbol{\chi}\|+\boldsymbol{a}$) so that they are

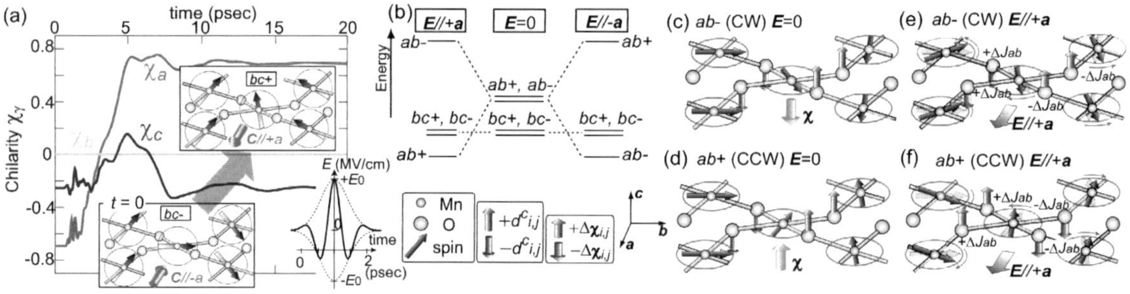

Figure 2. (a) Time evolution of the spin chirality χ=(χ_a, χ_b, χ_c) after application of the pulse with E_0=11 MV/cm, which shows the chirality reversal. Inset shows spin states before/after applying the pulse, and time pro le of the applied pulse $E_a(t)$. (b) Energy diagrams of the $ab_\pm$ and $bc_\pm$ states in the presence and absence of $\boldsymbol{E}\|\pm\boldsymbol{a}$. (c)[(d)] Direction of the net spin chirality χ and c-axis components of the DM vectors $\pm d^c_{i,j}$ on the in-plane bonds for the clockwise [counterclockwise] ab-plane spin spiral, i.e. the ab [ab_+] state when $\boldsymbol{E}$=0. (e) [(f)] Staggered modulation of the local spin chiralities $\chi_{i,j} \pm \ \chi_{i,j}$ under $\boldsymbol{E}\|+\boldsymbol{a}$ in the ab [ab_+] state.

referred to as bc and bc_+, respectively. The chirality χ is calculated as a sum of the local contributions $\chi_{i,i+\hat{x}} = \boldsymbol{S}_i \ \ \boldsymbol{S}_{i+\hat{x}}$ and $\chi_{i,i+\hat{y}} = \boldsymbol{S}_i \ \ \boldsymbol{S}_{i+\hat{y}}$ as $\chi = \frac{1}{2N}\sum_i(\chi_{i,i+\hat{x}} + \chi_{i,i+\hat{y}})/S^2$. Note that $\chi_{i,i+\hat{z}}$ is zero because of antiferromagnetic stacking in z-direction. The ab-plane spiral spin con guration is also possible in $TbMnO_3$ although it is slightly higher in energy than the bc-plane one without external elds. Again there are two kinds of ab-plane spirals, ab and ab_+, which have $\chi\|$ $\boldsymbol{c}$ and $\chi\|+\boldsymbol{c}$, respectively.

3. Results and Discussion

Starting with bc with $\chi\|$ $\boldsymbol{a}$, we apply an intense pulse of electric eld $\boldsymbol{E}(t)$=$[E_a(t), 0, 0]$ along the a axis. Chirality switching by the application of an intense optical pulse has been theoretically proposed in Ref. [27]. It has turned out that the switching processes show highly nonlinear behaviors with respect to the strength and shape of the pulse. Here we examine the pulse of

$$E_a(t) = E_0 \cos\omega t \exp[\ \ \frac{t^2}{2\sigma^2}], \tag{5}$$

which is di erent from the pulse with $\sin\omega t$ oscillation in the previous study [27]. Here ω and E_0 are xed at 2.1 THz and 11 MV/cm, while the full width of the half maximum of the Gaussian envelope is taken to be 0.5 psec [see inset of Fig. 2(a)]. In Fig. 2(a), we display calculated time evolution of the a-, b- and c-axis components of χ. We nd a reversal of the spin chirality χ from bc (χ_a<0) to bc_+ (χ_a>0).

To understand this phenomenon, let us rst consider energies of the four chirality states, i.e. $ab_\pm$ and $bc_\pm$. In Fig. 2(b), we summarize energy diagrams in the presence and absence of $\boldsymbol{E}$. When $\boldsymbol{E}$=0, the ab_+ and ab states are degenerate, and are higher in energy than the ground-state bc-plane spirals. Application of $\boldsymbol{E}\|\pm\boldsymbol{a}$ lifts this degeneracy, and if it is strong enough, either ab_+ or ab becomes the lowest-lying state depending on the sign of $\boldsymbol{E}$. Under the strong $\boldsymbol{E}\|+\boldsymbol{a}$ ($\boldsymbol{E}\|$ $\boldsymbol{a}$), the ab_+ (ab) state becomes the lowest in energy. In contrast, the energies of $bc_\pm$ are not a ected by $\boldsymbol{E}$.

More concretely, when $\boldsymbol{E}$=0, the spins in both ab_+ and ab states are rotating with nearly uniform turn angles, and all of the local spin chiralities $\chi_{i,j}$=$\boldsymbol{S}_i$ $\boldsymbol{S}_j$ have nearly the same amplitudes along $+\boldsymbol{c}$ and $\boldsymbol{c}$ directions, respectively as indicated by green arrows in Figs. 2(c) and (d). The local chiralities $\chi_{i,j}$ couple with the c-axis components of the

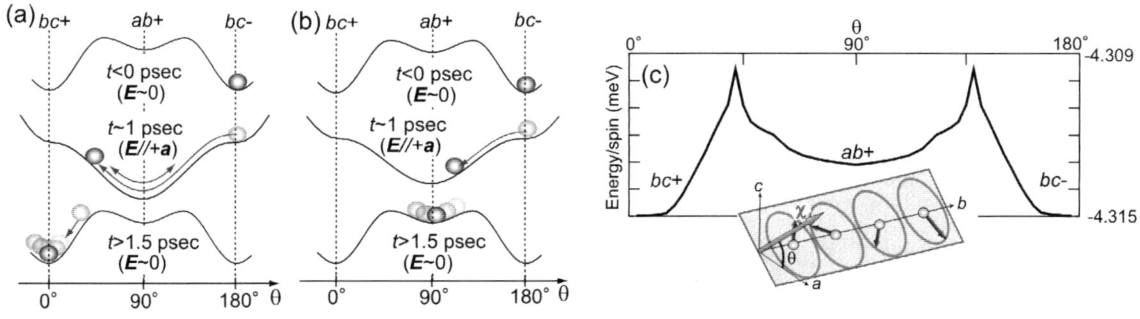

Figure 3. (a) Schematic gure for time evolution of the energy structure in the space for the chirality-reversal process. (b) That for the chirality- op process. (c) Calculated energy structure in the space at $\boldsymbol{E}$=0. Inset: Spin chirality $\boldsymbol{\chi}$ and the angle in the cycloidal spin state.

DM vectors, $\pm d^c_{i,j}$, on the in-plane Mn(i)-O-Mn(j) bonds, whose magnitudes are all equal to $_{ab}$. Since the signs of $\pm d^c_{i,j}$ are staggered as shown by blue arrows, the total DM energy, $E_{\rm DM} = \sum_i (\boldsymbol{d}_{i,i+\hat{x}} \;\; \boldsymbol{\chi}_{i,i+\hat{x}} + \boldsymbol{d}_{i,i+\hat{y}} \;\; \boldsymbol{\chi}_{i,i+\hat{y}})$, cancels out for both ab_+ and ab states, resulting in their degeneracy. When $\boldsymbol{E}$ is applied, the modulated ferromagnetic exchanges J_{ab} J_{ab} and $J_{ab}+$ J_{ab} cause slight spin rotations as indicated by thin red arrows in Figs. 2(e) and (f) to decrease and increase of the spin turn angles. Consequently the local spin chiralities become subject to a staggered modulation as $\boldsymbol{\chi}_{i,j} \pm$ $\boldsymbol{\chi}_{i,j}$. The way of the modulation is di erent between ab_+ and ab , namely depending on the sign of $\boldsymbol{\chi}$. When $\boldsymbol{E}\|+\boldsymbol{a}$, the + $\boldsymbol{\chi}_{i,j}$ and $\boldsymbol{\chi}_{i,j}$ are arranged antiparallel (parallel) to $\pm d^c_{i,j}$ in the ab_+ (ab) state, resulting in the decrease (increase) of DM energy. When this DM energy is large enough under the su ciently strong $\boldsymbol{E}\|+\boldsymbol{a}$, the ab_+ state becomes the lowest in energy as shown in Fig. 2(b). In contrast, under the strong $\boldsymbol{E}\|$ $\boldsymbol{a}$, the ab state becomes the lowest in energy. We note that energies of the $bc_\pm$ states never change even under $\boldsymbol{E}$ because the a-axis components of the DM vectors are alternately stacked, by which the DM energy cancels out despite the staggered modulation of the local chiralities, $\boldsymbol{\chi}_{i,j} \pm$ $\boldsymbol{\chi}_{i,j}$ ($\|a$).

Two key ingredients of this chirality switching are dynamical deformation of potential structure under the oscillating $\boldsymbol{E}$ via the above-discussed ME coupling and an inertial motion of the spin chirality. In Fig. 3(a), we depict a cartoon for time evolution of the potential structure in the space for the chirality-reversal process from =180 (bc) to =0 (bc_+). Here is the angle between the chirality $\boldsymbol{\chi}$ and the a axis as de ned in the inset of Fig. 3(c). When $E_a(t)$>0 as at t 1 psec, =90 (ab_+) becomes a new energy minimum, so that the chirality $\boldsymbol{\chi}$ starts rotating towards this minimum. Importantly the chirality does not stop its rotation at =90 immediately, but passes through that minimum or oscillates around it because of the inertial force. The DM interaction and the single-ion anisotropy originating from the spin-orbit coupling make the rotation of the spin-spiral plane massive, resulting in its inertial motion. When $E_a(t)$ becomes 0 again at t >1.5 psec, the system falls into the minimum at =0 (bc_+) if the system is within the domain of metastability of =0 . The system can also be trapped in the local minimum at =90 (ab_+), resulting in the 90 op of the spin chirality [see Fig. 3(b)].

Here the equation of motion for the chirality oscillation around an energy minimum at $=_0$ can be written by introducing the e ective mass $\tilde{m}$ as $\tilde{m}$ $+\tilde{m}$ $=$ $\frac{\partial V(\theta\ \theta_0)}{\partial\theta}$ where is the damping rate determined by $_G$. Note that the dimension of $\tilde{m}$ is not M but ML^2 since is a dimensionless variable in contrast to a variable appeared in the usual equation of motion. We calculate the dependence of energy by rotating the spin spiral plane around the b axis, which corresponds to $V($ $)$. We display calculated $V($ $)$ for $\boldsymbol{E}$=0 in Fig. 3(c). Since the typical frequency of the oscillation is a few terahertz, the order of $\tilde{m}$ is estimated as $10\ ^{18}$ sec^{2}eV.

This optical chirality switching requires a strong $\boldsymbol{E}$ ($|E_0|>11$ MV/cm) because it is necessary to lower the *ab*-spiral state energetically relative to the *bc*-spiral one. This threshold value can be reduced if we properly choose the materials: For example, the solid solutions $Tb_{1\ x}Gd_xMnO_3$ locating near the boundary between the *ab* and *bc* spiral phases [28] are promising candidates.

4. Summary

In summary, we have theoretically demonstrated the optical switching of spin chirality by intensely exciting the electromagnons in multiferroic $TbMnO_3$. We have revealed that the oscillating electric- eld component of light activates collective rotations of the spin-spiral plane with terahertz frequencies via the ME coupling, and their inertial motions result in the chirality reversal. The collective spin motion is crucial to this phenomenon, which supplies energy to cross the potential barriers. The mechanism here proposed is generic, and can be expected also in other spin-spiral-based multiferroics.

The authors are grateful to N. Kida, Y. Tokura, N. Furukawa, R. Shimano and I. Kezsmarki for fruitful discussions. This work was supported by Grant-in-Aid (Grants No. 22740214, No. 21244053, No. 17105002, No. 19048015, and No. 19048008) and G-COE Program "Physical Sciences Frontier" from MEXT Japan, and Funding Program for World-Leading Innovative R&D on Science and Technology (FIRST Program) from JSPS.

[1] Tokura Y 2007 *J. Magn. Magn. Mater.* **310** 1145
[2] Katsura H, Nagaosa N and Balatsky A V 2005 *Phys. Rev. Lett.* **95** 057205
[3] Mostovoy M 2006 *Phys. Rev. Lett.* **96** 067601
[4] Kimura T, Goto T, Shintani H, Ishizaka K, Arima T and Tokura Y 2003 *Nature* **426** 55
[5] Kenzelmann M, Harris A B, Jonas S, Broholm C, Schefer J, Kim S B, Zhang C L, Cheong S W, Vajk O P and Lynn J W, 2005 *Phys. Rev. Lett.* **95** 087206
[6] Yamasaki Y, Sagayama H, Goto T, Matsuura M, Hirota K, Arima T and Tokura Y 2007 *Phys. Rev. Lett.* **98** 147204
[7] Smolenski G A and Chupis I E 1982 *Sov. Phys. Usp.* **25** 475
[8] Pimenov A, Mukhin A A, Ivanov V Y, Travkin V D, Balbashov A M and Loidl A 2006 *Nat. Phys.* **2** 97
[9] Katsura H, Balatsky A V and Nagaosa N 2007 *Phys. Rev. Lett.* **98** 027203
[10] Pimenov A, Loidl A, Mukhin A A, Travkin V D, Ivanov V Y and Balbashov A M 2008 *Phys. Rev.* B **77** 014438
[11] Kida N, Ikebe Y, Takahashi Y, He J P, Kaneko Y, Yamasaki Y, Shimano R, Arima T, Nagaosa N and Tokura Y 2008 *Phys. Rev.* B **78** 104414
[12] Kida N, Yamasaki Y, Shimano R, Arima T and Tokura Y 2008 *J. Phys. Soc. Jpn.* **77** 123704
[13] Takahashi Y, Kida N, Yamasaki Y, Fujioka J, Arima T, Shimano R, Miyahara S, Mochizuki M, Furukawa N and Tokura Y 2008 *Phys. Rev. Lett.* **101** 187201
[14] Takahashi Y, Yamasaki Y, Kida N, Kaneko Y, Arima T, Shimano R and Tokura Y 2009 *Phys. Rev.* B **79** 214431
[15] Kida N, Takahashi Y, Lee J S, Shimano R, Yamasaki Y, Kaneko Y, Miyahara S, Furukawa N, Arima T and Tokura Y 2009 it J. Opt. Soc. Am. B **26** A35
[16] Aguilar R V, Mostovoy M, Sushkov A B, Zhang C L, Choi Y J, Cheong S W and Drew H D 2009 *Phys. Rev. Lett.* **102** 047203
[17] Mochizuki M, Furukawa N and N. Nagaosa 2010 *Phys. Rev. Lett.* **104** 177206
[18] Mochizuki M and Furukawa N 2009 *J. Phys. Soc. Jpn.* **78** 053704; 2009 *Phys. Rev.* B **80** 134416
[19] Xie X, Dai J and Zhang X C 2006 *Phys. Rev. Lett.* **96** 075005
[20] Gaal P, Kuehn W, Riemann K, Woerner M, Elsaesser T and Hey R 2007 *Nature* **450** 1210
[21] Sell A, Leitenstorfer A and Huber R 2008 *Optics Lett.* **33** 2767
[22] Solovyev I, Hamada N and Terakura K 1996 *Phys. Rev. Lett.* **76** 4825
[23] Kaplan T A 2009 *Phys. Rev. B* **80** 012407
[24] Arima T, Tokunaga A, Goto T, Kimura H, Noda Y and Tokura Y 2006 *Phys. Rev. Lett.* **96** 097202
[25] Alonso J A, Martínez-Lope M J, Casais M T and Fernández-Díaz M T 2000 *Inorg. Chem.* **39** 917
[26] Ishiwata S, Kaneko Y, Tokunaga Y, Taguchi Y, Arima T and Tokura Y 2010 *Phys. Rev.* B **81** 100411(R)
[27] Mochizuki M and Nagaosa N 2010 *Phys. Rev. Lett.* **105** 147202
[28] Goto T, Yamasaki Y, Watanabe H, Kimura T and Tokura Y 2005 *Phys. Rev.* B **72** 220403(R)

Charge dynamics of electronic ferroelectricity in geometrically frustrated lattice

Makoto Naka and Sumio Ishihara

Department of Physics, Tohoku University, Sendai, 980-8578, Japan

E-mail: naka@cmpt.phys.tohoku.ac.jp

Abstract. Charge dynamics in electronic ferroelectric materials, where electronic charge order is responsible for the electric polarization, is studied. Motivated from the layered iron oxide $LuFe_2O_4$ with a frustrated geometrical lattice, we introduce the spin-less fermion V-t model in a paired triangular lattice. This model is analyzed by utilizing the exact diagonalization method with the Lanczos algorithm in finite-size clusters. In particular, we focus on frustration effects on the charge dynamics. In the calculation of the ground state phase diagram, the non-polar two-fold charge order (CO) and the polar three-fold CO appear. In the optical conductivity spectra in the polar three-fold CO phase, multiple-peak structures appear in a wide energy range, in contrast to the spectra in the non-polar two-fold CO phase. This is originated from the non-equivalent two sublattices in the three-fold CO phase. The dynamical charge correlation function in the non-polar two-fold CO phase shows low-energy excitation at the wave vectors which correspond to the three-fold CO. This feature becomes remarkable near the phase boundary between the two-fold and three-fold CO phases. These precursor effects are attributed to the frustrated charge ordered system.

1. Introduction

Ferroelectricity and related phenomena have widely attracted much attention since discovery of the gigantic multiferroic phenomena in rare-earth manganese oxides. It is recently revealed that in the multiferroic materials, the electric polarization is driven by the inverse Dzyaloshinsky-Moriya interaction and/or the exchange striction mechanism under the non-collinear spin structures in the frustrated spin system. This is recognized to be a spin-driven ferroelectricity where the non-collinear spin order is the primary order parameter and the ferroelectric polarization is the secondary one, that is, the improper ferroelectricity. Recently, another type of the ferroelectricity is suggested theoretically and experimentally; the electronic charge order (CO) without the inversion symmetry is responsible for the electric polarization. This is termed the electronic ferroelectricty or the charge-driven ferroelectricity [1, 2, 3]. This is seen in some transition-metal oxides, e.g. $LuFe_2O_4$, and the charge-transfer organic salts such as κ-(BEDT-TTF)$_2$Cu$_2$(CN)$_3$. These materials are the so-called the 1/4 filling system where the electron per site is 0.5, and are the charge ordered insulator. This is in contrast to the spin-driven type ferroelectric materials, which are the Mott insulators of the 1/2 filling, and most of the displacive-type ferroelectric materials which are the band insulators.

As an example, we introduce the electronic structure and the dielectric properties in $LuFe_2O_4$. The crystal structure of the system consists of the Fe-O double layers and the Lu-O ones stacked

Published under licence by IOP Publishing Ltd

Journal of Physics: Conference Series **320** (2011) 012083 doi:10.1088/1742-6596/320/1/012083

along the c axis. The electric and magnetic properties are responsible for the Fe-O paired-triangular lattices which are termed the W-layer. The double triangular lattice is schematically shown in Fig. 1. The nominal valence of the Fe ions are 2.5+ which implies a coexistence of the Fe^{2+} and Fe^{3+} with equal amount. The three-dimensional CO of Fe ions occurs around 320K where large dielectric anomalies are observed in the dielectric constant. The 3-fold charge ordered state is confirmed by the resonant x-ray scattering experiments at the Fe K-edge. It is thought that the charge order pattern in the W-layer are mainly determined by the interactions inside of the W-layer, and the stacking pattern of the charge order along the c axis is described by the weak inter-W-layer interactions. Therefore, it is widely expected that the electric polarization is attributed to the CO in the W-layer without the inversion symmetry.

As well as the CO, the charge fluctuation in the electronic ferroelectric materials has much attention. Several characteristic charge dynamics is suggested experimentally. The relaxor-like dielectric fluctuations are observed in $LuFe_2O_4$ and κ-(BEDT-TTF)$_2$Cu$_2$(CN)$_3$ [4]. The optical absorption experiments in $LuFe_2O_4$ indicate that large charge fluctuation remains far below the charge ordering temperature [5]. A drastic change in the charge ordered patterns by introducing small amount of the oxygen deficiency is observed in rare-earth iron oxides. From the theoretical view points, the present authors and co-workers suggest that the polar three-fold charge ordered phase is stabilized by the thermal and quantum charge fluctuation [6, 7, 8, 9]. It is required to reveal origin of such characteristic charge dynamics in electronic ferroelectric materials.

Motivated from these recent experimental and theoretical results in the electronic ferroelectric materials, we study dynamical properties in the geometrically frustrated charge ordered system. We analyzed the extended V-t model in the paired triangular lattice by utilizing the exact diagonalization method. The optical conductivity spectra and the dynamical charge correlation function are calculated in a paired triangular lattice beyond the previous studied in the single-triangular lattice [10, 11]. It is shown that the multiple peak structure in the polar charge ordered phase is due to the two-sublattice structure in this charge ordered structure. This is directly related to origin of the electric polarization in this system. Near the phase boundary between different charge ordered phases, the low-energy charge excitations are observed in several momentum. Relation between the obtained results and the recent resonant x-ray scattering experiments is discussed.

2. Model and Method

Motivated from the crystal and electronic structure of the layered iron oxide $LuFe_2O_4$, we consider the double triangular lattices stacked along the c axis. As a minimal model to study the CO and dynamics in this crystal lattice, we adopt the half-filled extended V-t model for spinless fermions. The electron transfer and the long-range Coulomb interactions are introduced in the double triangular lattice. The model Hamiltonian is defined by

$$\mathcal{H}_{\rm Vt} = \mathcal{H}_{\rm t} + \mathcal{H}_{\rm V}, \quad (1)$$

where the first term is for the electron transfers given by

$$\mathcal{H}_{\rm t} = -t_{\rm ab}\sum_{\langle ij\rangle} c_i^\dagger c_j - t_{\rm c}\sum_{(ij)} c_i^\dagger c_j + H.c., \quad (2)$$

and the second term is for the Coulomb interactions given by

$$\mathcal{H}_{\rm V} = V_{\rm abNN}\sum_{\langle ij\rangle} n_i n_j + V_{\rm cNN}\sum_{(ij)} n_i n_j + V_{\rm cNNN}\sum_{[ij]} n_i n_j. \quad (3)$$

Here, c_i and $n_i(\equiv c_i^\dagger c_i)$ represent the annihilation operator and the number operator of a fermion,

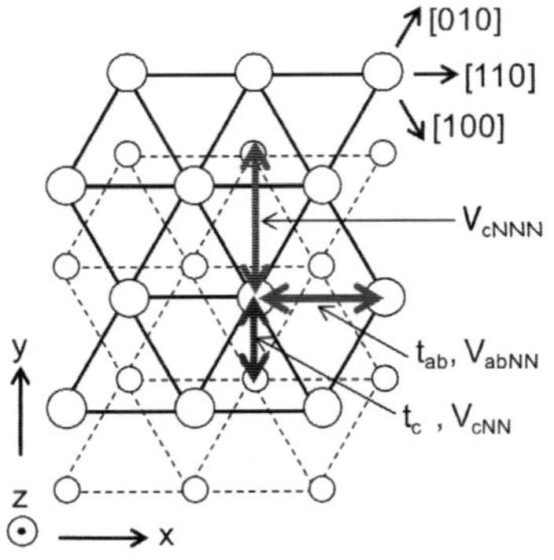

Figure 1. (Color online) A schematic view of a pair of the triangular lattices and the interactions in $\mathcal{H}_{\mathrm{Vt}}$.

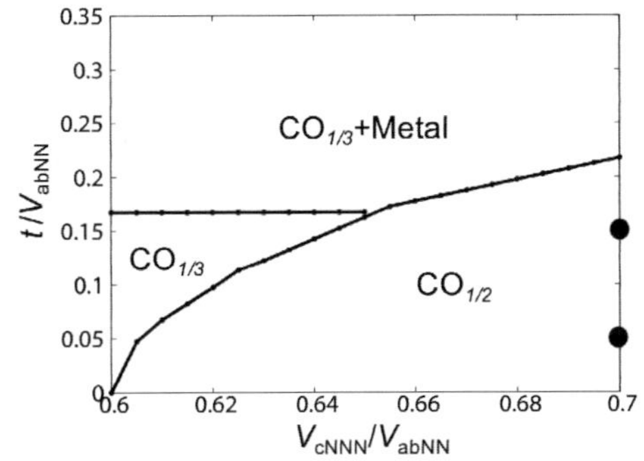

Figure 2. Ground state phase diagram of the extended V-t model with the half filling. The parameter sets are chosen to be $V_{\mathrm{cNN}} = 1.2V_{\mathrm{abNN}}$ and $t \equiv t_{\mathrm{ab}} = t_{\mathrm{c}}$. The dynamical charge correlation functions are calculated in parameter values indicated by black circles (see the text).

respectively. A schematic view of the crystal lattice and the interactions are shown in Fig. 1 where the large and small circles represent the sites in the upper and lower planes, respectively. We consider the inter-site Coulomb interactions between the in-plane NN sites (V_{abNN}), those between the inter-plane NN sites (V_{cNN}) and NNN sites (V_{cNNN}). The symbols $\sum_{\langle ij\rangle}$, $\sum_{(ij)}$ and $\sum_{[ij]}$ represent summation for the pairs corresponding to V_{abNN}, V_{cNN} and V_{cNNN}, respectively. As for the electron transfer term $\mathcal{H}_{\mathrm{t}}$, the transfer integral between the in-plane NN sites (t_{ab}) and that between the inter-plane NN site (t_{c}) are taken into account. The optical conductivity and the dynamical charge correlation functions are calculated by using the exact diagonalization method based on the Lanczos procedure. We adopt the finite size cluster of $4 \times 3 \times 2 = 24$ sites with the periodic boundary condition along the x and y directions (see Fig. 1), and the fermion number $N_e = 12$.

3. Results of Calculation

First we show the phase diagram in the plane of $V_{\mathrm{cNNN}}/V_{\mathrm{abNN}}$ and t/V_{abNN} in Fig 2. The parameter sets are chosen to be $V_{\mathrm{cNN}} = 1.2V_{\mathrm{abNN}}$ and $t_{\mathrm{ab}} = t_{\mathrm{c}} \equiv t$. The each phase is identified by the charge correlation function defined by $N(\mathbf{q}) = \langle n(\mathbf{q})n(-\mathbf{q})\rangle$ with $n(\mathbf{q}) = \frac{1}{N}\sum_i (n_i - \frac{1}{2})e^{-i\mathbf{q}\cdot\mathbf{r}_i}$ and the Drude weight in the optical conductivity introduced later. The two kinds of the charge ordered phases are realized in this parameter region of the phase diagram,

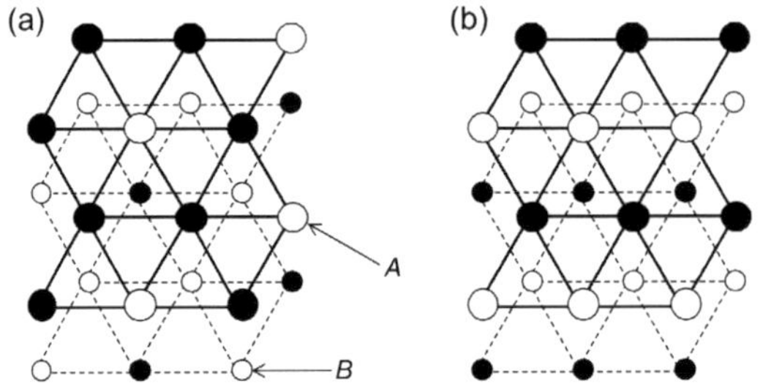

Figure 3. Schematic charge ordered structures of $CO_{1/3}$ in (a) and $CO_{1/2}$ in (b). The black and white circles are for the charge rich and poor sites, respectively. Symbols A and B represent the two kinds of the sublattices (see the text).

i.e. the three-fold CO ($CO_{1/3}$) and the two-fold one ($CO_{1/2}$) as schematically shown in Fig. 3(a) and (b), respectively, where black and white circles represent charge-rich and poor sites. The maximum values of $N(\mathbf{q})$ for $\mathbf{q} = (1/3, 1/3)$ in $CO_{1/3}$ and that for $\mathbf{q} = (1/2, 1/2)$ in $CO_{1/2}$ are 0.11 and 0.25, respectively. In $t/V_{\rm abNN} = 0.005$, we obtain $N(\mathbf{q}) = 0.10$ for $\mathbf{q} = (1/3, 1/3)$ and $N(\mathbf{q}) = 0.25$ for $\mathbf{q} = (1/2, 1/2)$ at $V_{\rm cNNN}/V_{\rm abNN} = 0.6$ and 0.7, respectively. Since the total charge in the upper and lower planes are different, a finite electric polarization exists in $CO_{1/3}$. At $t = 0$, where the present model is the classical charge model, it is exactly shown that the polar $CO_{1/3}$ phase is degenerated with the $CO_{1/2}$ phase at $V_{\rm cNNN}/V_{\rm abNN} = 0.6$. This point is recognized as a frustration point. When the finite electron transfer is introduced at the frustration point, the $CO_{1/3}$ phase is stabilized. In a region of large t, both the charge correlation $N(\mathbf{q})$ for the $CO_{1/3}$ phase and the Drude component in the optical conductivity spectra are observed. Thus, this phase is identified as a mixed phase of $CO_{1/3}$ and metal, although further studies are required to understand detailed character of this phase. The topological feature of the obtained phase diagram is consistent with the previous result where the present Hamiltonian of Eqs. (1), (2) and (3) are analyzed by utilizing variational Monte Carlo method in Ref. [8, 9].

Stability of the polar $CO_{1/3}$ phase is attributed to the large quantum charge fluctuation in the paired triangular lattice. A key issue is the two-sublattice structure in this charge ordered structure, as shown in Fig. 3(a). Each charge-poor site in the upper plane is surrounded by the six NN charge-rich sites in the same plane. Thus, the Coulomb potential, acting on these sites are strong, and the charge fluctuation hardly occurs. On the other hand, each charge-poor site in the lower plane is surrounded by three charge-rich sites and three charge-poor sites in the same plane. The Coulomb potentials at these sites from the in-plane NN sites are canceled out, and the charges easily fluctuate at these site. These two sublattices are termed A and B from now on. Therefore, in the case of finite electron transfer, there is large kinetic energy gain due to charge fluctuation. This is a unique character in the $CO_{1/3}$ phase. In the $CO_{1/2}$ phase, all sites are equivalent.

Now we introduce the dynamical properties in the V-t model in a paired triangular lattice. We calculate the optical conductivity spectra and the dynamical charge correlation function defined by

$$\sigma_\mu(\omega) = -\frac{e^2}{N\omega}{\rm Im}\langle 0|j_\mu^\dagger \frac{1}{\omega - \mathcal{H} + E_0 + i\eta} j_\mu|0\rangle, \tag{4}$$

and

$$N(\mathbf{q}, \omega) = -{\rm Im}\langle 0|n(\mathbf{q}) \frac{1}{\omega - \mathcal{H} + E_0 + i\eta} n(-\mathbf{q})|0\rangle, \tag{5}$$

respectively. Here, $|0\rangle$ and E_0 are the ground state and the ground state energy, respectively.

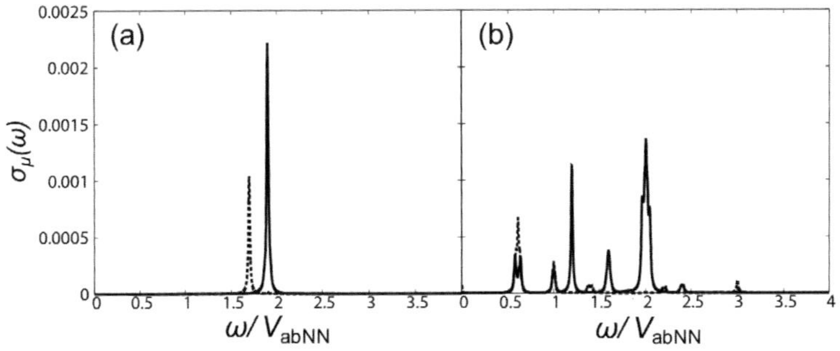

Figure 4. The optical conductivity spectra $\sigma_\mu(\omega)$ for $\mu = x$ (solid line) and $\mu = z$ (broken line). The parameters are chosen to be $V_{\rm cNNN}/V_{\rm abNN} = 0.7$ (the $\rm CO_{1/2}$ phase) in (a) and $V_{\rm cNNN}/V_{\rm abNN} = 0.6$ (the $\rm CO_{1/3}$ phase) (b). The other parameter value is chosen to be $t/V_{\rm abNN} = 0.01$.

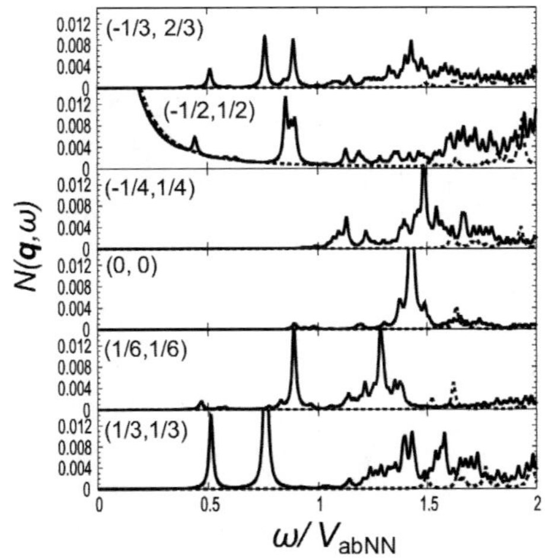

Figure 5. The dynamical charge correlation function $N(\mathbf{q}, \omega)$ in the $\rm CO_{1/2}$ phase. Broken lines and solid lines are for $t/V_{\rm abNN} = 0.05$ and $t/V_{\rm abNN} = 0.15$, respectively (see the symbols in Fig. 2).

We define the current operator as $j_\mu = \sum_{\langle ij\rangle} t_{ij}(\mathbf{r}_{ij}c_i^\dagger c_j + \mathbf{r}_{ji}c_j^\dagger c_i)$ for the $\mu = (x, z)$ direction where $\mathbf{r}_{ij}$ is a unit vector connecting site i and site j. We introduce an infinite decimal constant η for the recursion procedure in the Lanczos calculation.

In Fig. 4, we show the numerical results of the optical conductivity $\sigma_\mu(\omega)$ in the $\rm CO_{1/2}$ [Fig. 4(a)] and $\rm CO_{1/3}$ [Fig 4(b)] phases. Parameter values are chosen to be $V_{\rm cNNN}/V_{\rm abNN} = 0.6$ in Fig. 4(a) and $V_{\rm cNNN}/V_{\rm abNN} = 0.7$ in Fig. 4(b). In the $\rm CO_{1/2}$ phase, the spectra in both the x and z show simple peak structure. These peaks correspond to the charge excitations from the charge rich sites to the poor sites in the $\rm CO_{1/2}$ structure. On the other hand, in the $\rm CO_{1/3}$ phase, the multiple-peak structures are shown in a wide energy range. These structures are attributed to various excitation energies due to the sublattice structure of the $\rm CO_{1/3}$ phase explained above. For example, the lowest (highest) energy peak is due to the charge fluctuation between the sublattice B(A).

The calculated dynamical charge correlation function spectra $N(\mathbf{q}, \omega)$ in the $\rm CO_{1/2}$ phase are shown in Fig. 5. The parameter values of the electron transfer are chosen to be $t/V_{\rm abNN} = 0.05$ (broken lines) and 0.15 (solid lines) which correspond to the symbols in Fig. 2. The wave vectors are defined in the Brillouine zone of the triangular lattice and the phase between the upper and lower planes is taken to be π. In the case of $t/V_{\rm abNN} = 0.05$, a large peak structure appears around $\omega = 0$ at $\mathbf{q} = (-1/2, 1/2)$. This reflects the CO of the $CO_{1/2}$ structure. Other peak structures observed in the region of $\omega/V_{\rm abNN} = 1.7 \sim 1.9$ are not remarkable. On the other hand, in the case of $t/V_{\rm abNN} = 0.15$, which is close to the phase boundary between the $CO_{1/2}$ and the mixed phase of $CO_{1/3}$ and metal, the peaks in $N(\mathbf{q}, \omega)$ at the wave vectors $\mathbf{q} = (1/3, 1/3)$ and $\mathbf{q} = (-1/3, 2/3)$ are prominent in a low energy range. This is interpreted as a precursor of the phase transition from the $CO_{1/2}$ phase to the $CO_{1/3}$ one, and is resulted from the charge frustration.

4. Summary and discussion

Recently, the charge dynamics in $LuFe_2O_4$ is studied by the resonant inelastic x-ray scattering (RIXS) experiments [12]. The scattering intensity of RIXS is approximately given by the loss function as $I(\mathbf{q}, \omega) \sim -{\rm Im}[1/\varepsilon(\mathbf{q}, \omega)]$. It is shown that above the charge ordering temperature, the peak structures are observed in the several wave vectors correspond to the two-fold, three-fold and four-fold COs. This is in contrast to the results in other charge ordered systems such as $Sr_{14}Cu_{24}O_{41}$ where the spectra appear at the wave vector corresponding to the CO pattern. This may be related to the present calculated results in the dynamical charge correlation function presented in Fig. 5 where the peak structure appears at $CO_{1/3}$ even in the two-fold CO structure near the phase boundary. Further comparisons between theoretical and experimental results are required to reveal characteristic charge dynamics in electronic ferroelectricity.

In summary, we study the charge dynamics in the electronic ferroelectricty motivated from the layered iron oxides. It is shown that the multiple peak structure in the polar charge ordered phase is due to the two-sublattice structure in this charge ordered structure. This is directly related to origin of the electric polarization in this system. Near the phase boundary between different charge ordered phases, the low-energy charge excitations are observed in several momentum. This is a precursor of the phase transition due to the charge frustration. The obtained results and the recent resonant x-ray scattering experiments are discussed.

5. Acknowledgments

The authors would like to thank T. Watanabe, N. Ikeda and S. Mori for valuable discussions. This work was supported by JSPS KAKENHI, TOKUTEI from MEXT, CREST, a Grant-in-Aid for Scientific Research on Priority Areas from the Ministry of Education, Culture, Sports, Science and Technology, and Grand Challenges in Next-Generation Integrated Nanoscience. MN is supported by the global COE program of MEXT Japan.

6. References

[1] Ikeda N *et al.* 2005 *Nature* **436** 1136
[2] Brink J and Khomskii D 2008 *J. Phys.:Condens Matter* **20** 434217
[3] Ishihara S 2010 *J. Phys. Soc. Jpn.* **79** 011010
[4] Abdel-Jawad M, Terasaki I, Sasaki T, Yoneyama N, Kobayashi N, Useu Y and Hotta C 2010 *Phys. Rev.* B **82** 125119
[5] Xu X S *et al.* 2008 *Phys. Rev. Lett.* **99** 227602
[6] Nagano A, Naka M, Nasu J and Ishihara S 2007 *Phys. Rev. Lett.* **99** 217202
[7] Naka M, Nagano A and Ishihara S 2008 *Phys. Rev.* B **77** 224441
[8] Watanabe T and Ishihara S 2009 *J. Phys. Soc. Jpn.* **78** 113702
[9] Watanabe T and Ishihara S 2010 *J. Phys. Soc. Jpn.* **79** 114714
[10] Seo H, Tsutsui K, Ogata M and Merino J 2006 *J. Phys. Soc. Jpn.* **75** 114707
[11] Nishimoto S, Shingai M and Ohta Y 2008 *Phys. Rev.* B **78** 035113
[12] Ishii K, Murakami Y and Mizuki J (*private communication*)

Magnetism in Triangular Lattice System $LuFeCoO_4$ having Relaxor Property

Minoru Soda[1], Takuya Aoyama[2], Yusuke Wakabayashi[1], Kazuma Hirota[2]

[1]Division of Material Physics, Graduate School of Engineering Science, Osaka University, Toyonaka, Osaka 560-8531, Japan

[2]Department of Earth and Space Science, Graduate School of Science, Osaka University, Toyonaka, Osaka 560-0043, Japan

E-mail: soda@crystal.mp.es.osaka-u.ac.jp

Abstract. Electric and magnetic measurements were carried out on a single crystalline sample of triangular lattice system $LuFeCoO_4$. $LuFeCoO_4$ has a relaxor-like behavior in the dielectric constant, which corresponds with the existence of the polar nanoregion (PNR). A magnetic correlation length along *c*-plane was estimated to be nano-size by our neutron diffraction measurements, and an anomalous magnetization was found below 200 K. Similar electric and magnetic behaviors were also observed in relaxor ferroelectrics $2/3BiFeO_3$-$1/3BaTiO_3$, suggesting that the relaxor ferroelectrics having magnetic ions show commonly a nanomagnetism induced by PNRs.

In relaxor systems, temperature dependence of the dielectric permittivity shows a broad maximum and a characteristic frequency-dependence. To explain physical behaviors of relaxors, randomly oriented, very local polar regions, i.e.," Polar Nanoregions" (PNRs) are proposed.[1, 2] Physical properties of many relaxor ferroelectrics have been understood based on the behavior of PNRs. As a new variation of such relaxor systems, we have focused our attention on relaxor ferroelectrics having magnetic ions.

Previously we studied one of the relaxor ferroelectrics having the magnetic ions, $2/3BiFeO_3$-$1/3BaTiO_3$ (BFO-1/3BTO). We found that BFO-1/3BTO has PNR-sized nano magnetic domains, and their moments cause superparamagnetic behavior.[3] In the present study, we examined the triangular lattice system $LuFe\mathit{TM}O_4$ (TM=Cu, Mg, and Co). $LuFe_2O_4$ shows a ferroelectricity induced by a three-dimensional charge order of Fe^{2+} and Fe^{3+} below $\sim$340 K.[4] On the other hand, $LuFe\mathit{TM}O_4$, in which TM^{2+} substitutes the Fe site randomly, has a relaxor-like behavior in their dielectric properties.[5, 6] Relaxor-like dielectric property has hardly ever been reported in a triangular lattice system. Here, we report the interaction between the PNR and magnetism through both macroscopic property measurements and microscopic neutron-scattering measurements on single crystalline sample of $LuFeCoO_4$. Our results indicate that $LuFeCoO_4$ has nano-length magnetic correlation and an anomalous magnetism.

Single crystal of $LuFeCoO_4$ was grown by the floating zone method. Powder X-ray diffraction profiles (not shown) ensure that they did not have significant impurity phases. The dielectric permittivity and the magnetization M were measured using an LCR meter and a Quantum Design SQUID magnetometer, respectively. Neutron measurements were carried out using the

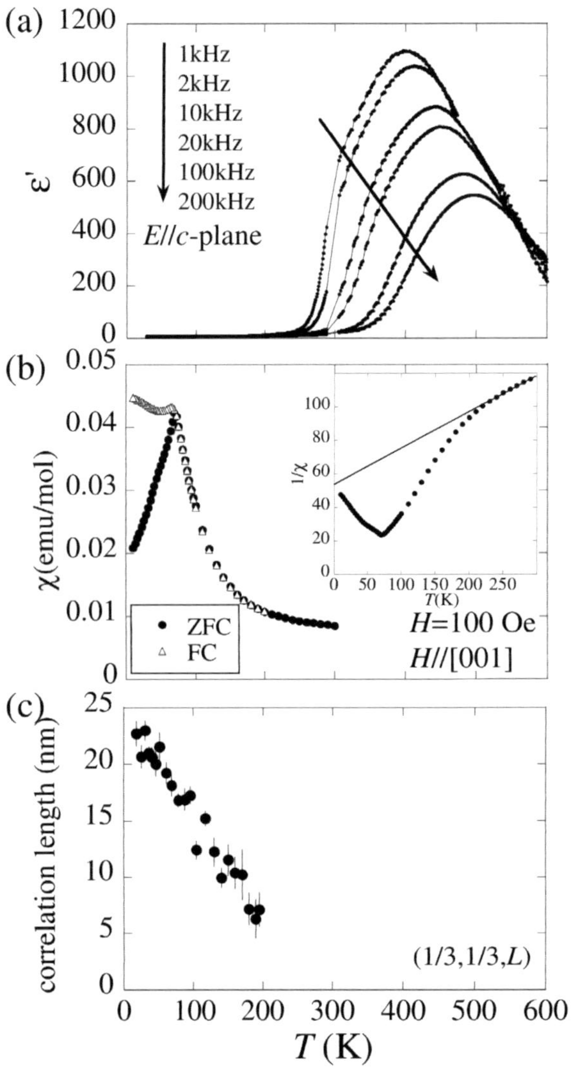

Figure 1. (a) T-dependence of ϵ' taken at several frequencies in $LuFeCoO_4$ with $\boldsymbol{E}//c$-plane. (b) T-dependence of the magnetic susceptibility of $LuFeCoO_4$ taken at H=100 Oe along [001] with conditions of zero-field cooling (ZFC) and field cooling (FC). (c) Magnetic correlation length within c-plane estimated by neutron measurements. (Inset) Inverse magnetic susceptibility of $LuFeCoO_4$ taken at H=100 Oe along [001].

ISSP triple axis spectrometer HQR installed at JRR-3 of JAEA, Japan. The crystal was oriented so that hhl plane is on the scattering plane.

The temperature (T) dependences of the real part of the dielectric constant ϵ' for the single crystal taken at various frequencies with electric field $\boldsymbol{E}//c$-plane are shown in Fig. 1(a). The dielectric constants of $LuFeCoO_4$ measured at different frequencies exhibit typical relaxor behavior; the dielectric peak shifts progressively toward higher temperature as the frequency increases. This suggests that $LuFeCoO_4$ has the polar domains, PNRs.

The T-dependences of the magnetic susceptibility χ taken at 100 Oe with $\boldsymbol{H}//[001]$ for the single crystal of $LuFeCoO_4$ is shown in Fig. 1(b). The hysteresis behavior was observed

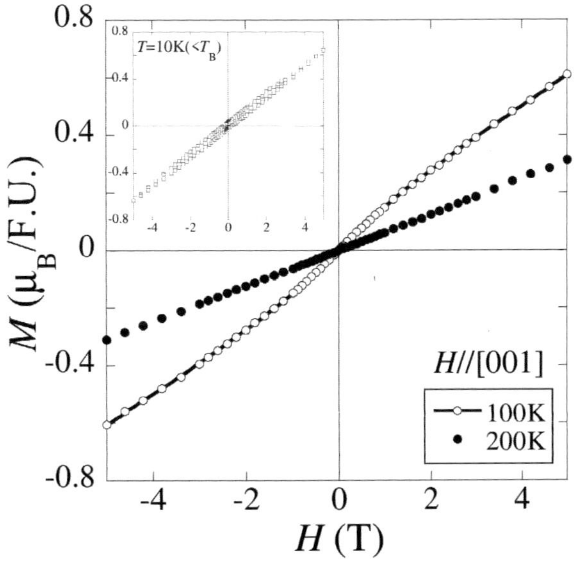

Figure 2. H-dependence of M of the single crystal of $LuFeCoO_4$ measured at 100 K and 200 K. At 100 K, the curves obtained by fitting with the combination of Langevin and linear functions is shown by the solid lines. (Inset) H-dependence of M of the single crystal of $LuFeCoO_4$ measured at 10 K ($<T_B$).

below 70 K. The inverse magnetic susceptibility shown in the inset of this panel indicates that the magnetic susceptibility deviates from the Curie-Weiss low below ~200 K. This deviation implies that the antiferromagnetic correlation is dominant above 200 K, and the ferromagnetic component appears at ~200 K as T decreases.

For neutron studies in $LuFeCoO_4$, rod-shaped magnetic reflections at (1/3,1/3,L) and (2/3,2/3,L) were observed below 200 K. The T-dependence of the magnetic correlation length estimated from the 1/3 1/3 L magnetic profile along (1,1,0) is shown in Fig. 1(c). This indicates that the magnetic correlation length along c-plane is nanometer size. On the other hand, the magnetic correlation along c-axis is almost absent, which corresponds with the two dimensional magnetic order. This quasi two dimensional magnetic order was also observed in $LuFeMgO_4$.[7] The appearance of these magnetic reflection coincides with the magnetic susceptibility which deviates from Curie-Weiss low below 200 K. However, magnetic reflections at (1/3,1/3,L) and (2/3,2/3,L) correspond with antiferromagnetic component although the magnetic susceptibility indicates the existence of the ferromagnetic component below 200 K.

We examined the magnetization curve in order to confirm the effect of the nano-sized antiferromagnetic correlation. The H-dependences of M with $\boldsymbol{H}$//[001] for the single crystal of $LuFeCoO_4$ are shown in Fig. 2. The H-dependence of M at 100 K has a non-linear behavior although that at 200 K shows the linearity. Although the H-dependences of M at 100 K suggests the existence of the ferromagnetic component, the residual magnetization was zero. This result might indicate that superparamagnetism-like behavior is realized in $LuFeCoO_4$ as well as BFO-1/3BTO. In the relaxor ferroelectrics having magnetic ions BFO-1/3BTO, the PNRs affect the magnetic ordering significantly, and the resulting magnetic nano-domains, in which the cant antiferromagnetic structure is realized, are the origin of superparamagnetism through the Dzyaloshinskii-Moriya (DM) interaction.[3] Considering a similar mechanism, the $LuFeCoO_4$ system is also expected to induce nano magnetic domains having a cant antiferromagnetic structure. Since the H-dependence of M at 100 K was not explained only by the Langevin function, Langevin function added by a linear function, which correspond with the cant antiferromagnetic structure, were employed. The experimental result is well reproduced by this function, as shown in Fig. 2. The average volume of the magnetic particle (V_{av}=total volume/domain number) estimated from the fitting with the Langevin function is (6 nm)^3 or (length between nearest layers 8 Å)$\times$(15 nm)^2 in considering three or two dimensional systems, respectively. On the other hand, the H-dependences of M shows discrepancy from the Langevin function below 70 K, where the hysteresis behavior was observed, as shown in Fig. 1(b) and the inset of Fig. 2. This characteristic temperature might be the blocking temperature (T_{B}) in the superparamagnetism.[8, 9] These results are consistent with the idea that $LuFeCoO_4$ has the cant antiferromagnetic structure within the nano magnetic domains.

In conclusion, electric and magnetic measurements were carried out on a single crystalline sample of $LuFeCoO_4$, which shows the relaxor dielectric property. In the magnetism, we found that $LuFeCoO_4$ has nano magnetic domains and an anomalous magnetization. The similar electric and magnetic behaviors were also observed in $2/3BiFeO_3$-$1/3BaTiO_3$, and the relaxor ferroelectrics having magnetic ions are expected to show the nanomagnetism induced by PNRs commonly.

Work at JRR-3 was supported by ISSP of the University of Tokyo. This work was supported by KAKENHI (19052002 and 20740171) and the Global COE program (G10).

References

[1] A. A. Bokov, and Z.-G. Ye, J. Mater. Sci. **41** 31 (2006) .
[2] G. Burns, and F. H. Dacol, Phys. Rev. B **28** 2527 (1983).
[3] M. Soda, M. Matsuura, Y. Wakabayashi, and K. Hirota, submitted.
[4] N. Ikeda, H. Ohsumi, K. Ohwada, K. Ishii, T. Inami, K. Kakurai, Y. Murakami, K. Yoshii, S. Mori, Y. Horibe, and H. Kito, Nature **436** 1136 (2005).
[5] K. Yoshii, N. Ikeda, Y. Matsuo, Y. Horibe, S. Mori, Phys. Rev. B **76** 024423 (2007).
[6] Y. Matsuo, M. Suzuki, Y. Noguchi, T. Yoshimura, N. Fujimura, K. Yoshii, N. Ikeda and S. Mori, Jpn. J. Appl. Phys. **47** 8464 (2008).
[7] A. Wiedenmann, W. Gunsser, J. Rossat-Mignod, M.O. Evrard, J. Magn. Magn. Mater. **31-34**, 1442 (1983).
[8] M. E. McHenry, S. A. Majetich, J. O. Artman, M. DeGraef, S. W. Staley, Phys. Rev. B **49** 11358 (1994).
[9] F. C. Fonseca, G. F. Goya, R. F. Jardim, R. Muccillo, N. L. V. Carreno, E. Longo, and E. R. Leite, Phys. Rev. B **66** 104406 (2002).

Collapsing processes of charge ordered structure in charge- and spin- frustrated ferrite $YbFe_2O_4$

K. Matsumoto, T. Koyama, S. Mori*, K. Yoshii[A], and N. Ikeda[B]

Department of Materials Science, Osaka Prefecture University, Japan,

[A]*Japan Atomic Energy Agency, Japan.*

[B]*Department of physics, Okayama University, Japan.*

*mori@mtr.osakafu-u.ac.jp

Abstract. In order to clarify the stability of charge ordered structure (CO) in charge- and spin-frustrated ferrite $YbFe_2O_4$, we have investigated changes of the CO structure by partial substitution of Mn^{2+} for Fe^{2+} in $YbFe_2O_4$ by a transmission electron microscope (TEM), in combination with conventional dielectric measurement. It is revealed that subtle substitution of Mn^{2+} for Fe^{2+} in $YbFe_2O_4$ destroyed drastically the CO structure with the wave vector of $\boldsymbol{q}$=<1/3 1/3 1/2> and, instead, induced polar clustering structure characterized by honeycomb-shaped diffuse scatterings in the reciprocal space. The formation of polar clustering structure should be responsible for the characteristic dielectric dispersion.

1. Introduction

Mixed-valence ferrite RFe_2O_4(R=Y, Yb, and Lu) is one of the multiferroic materials which is well-known for the presence of charge ordering (CO) of Fe^{2+} and Fe^{3+} ions on geometrically frustrated triangular lattices [1-4]. RFe_2O_4 has the rhombohedral crystal structure (space group: $R\bar{3}m$) characterized by the alternative stacking of triangular-lattice Fe-O bilayers and R-O layers along the [001] direction, and exhibits strong two-dimensionality (Fig. 1) [5]. The spin- and charge-ordering on Fe bilayers with the geometrically frustrated triangular lattice is expected to result in intriguing physical phenomena and structural changes in this system [6-9]. For example, charge imbalance within bilayers in the CO state of $LuFe_2O_4$ is suggested to be responsible for ferroelectricity below the two dimensional (2D) three-fold CO transition temperature at about 350 K [4,9]. This CO-driven ferroelectricity can be a new mechanism for ferroelectricity from the degree of freedom of spins and charges. The evolution of three-fold CO from 2D to three dimensional (3D) in the geometrically frustrated lattice leads to a structural phase transition with a development of double periodicity along the [001]* direction in $LuFe_2O_4$ and YFe_2O_4 [10].

On the other hand, stability of the 3D CO by partial substitution of other transition-metal elements such as Co, Mn and Cu for Fe in RFe_2O_4 have been examined so far. These compounds show the glass-like magnetic transition and the transition temperatures are much lower than the ferrimagnetic transition temperatures in RFe_2O_4 [11-13]. $LuFeMgO_4$ with the nonmagnetic ion (Mg^{2+}) shows a magnetic transition around 30K and shows a characteristic magnetic fluctuation [12]. Neutron diffraction experiments and Monte Carlo simulation revealed the presence of the short-ranged spin correlation [12]. It was suggested that a magnetic fluctuation in the $LuFeMgO_4$ has close relation with thermal fluctuation and freezing of the short-ranged magnetically ordered regions (magnetic clusters).

In this work, we have investigated changes of 3D CO structure by partial substitution of Mn^{2+} for Fe^{2+} in $YbFe_2O_4$ by a transmission electron microscopy (TEM). The 3D CO structure in $YbFe_2O_4$ is drastically destroyed by partial substitution of Mn^{2+} for Fe^{2+}. It was found that there exist characteristic honeycomb-shaped diffuse scatterings in the reciprocal space, which were caused by the presence of the polar clustering structure.

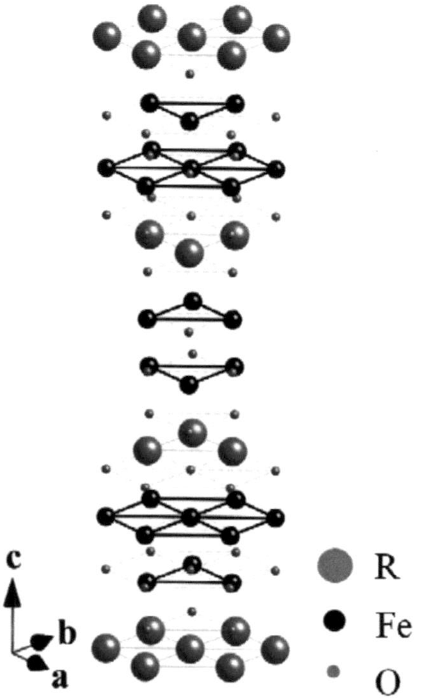

Figure 1: Crystal structure in the $R\bar{3}m$ rhombohedral structure of RFe_2O_4.

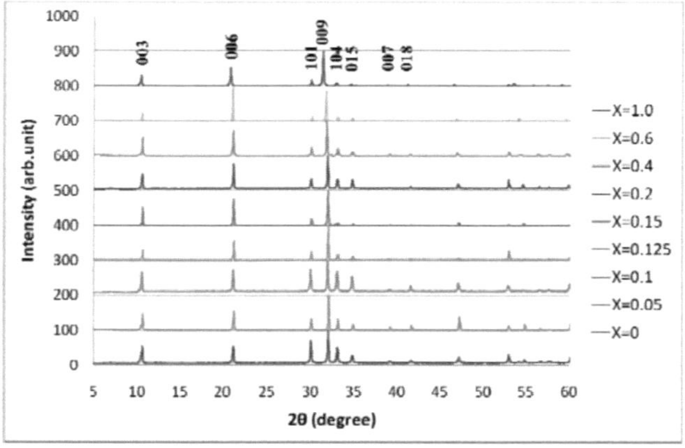

Figure 2: Powder x-ray diffraction profiles obtained in $YbFe_{2-x}Mn_xO_4$ with $0<x<1$ at 298 K.

2. Experimental procedure

We prepared single crystals of $YbFe_{2-x}Mn_xO_4$ with $0\leq x\leq 1$ by a conventional solid-state reaction from the mixture of Yb_2O_3, Fe_2O_3 and MnO and followed by the floating zone technique in reduction atmosphere under careful control of CO_2/CO gas ratio. Note that oxygen partial pressure in the reduction atmosphere, which is significant to synthesize the $YbFe_2O_4$ phase, can be controlled by changing the CO_2/CO mixture ratio in the chemical equilibrium state. Crystallographic information was obtained from powder x-ray diffraction at RT. The thin specimens for TEM observation were prepared by crushing methods, in order to avoid induction of oxygen deficiencies. The TEM observation was carried out with conventional JEM-2010 TEM equipped with a double-tilted holder. Note that the diffraction spots are indexed in a consistent manner among the electron diffraction patterns using the hexagonal notation with the space group of $R\bar{3}m$.

3. Results and discussion

We successfully synthesized single crystals of $YbFe_{2-x}Mn_xO_4$ with $0 \leq x \leq 1$. Crystal structures at 298 K were examined by powder x-ray diffraction experiment. Figure 2 shows powder x-ray diffraction profiles of $YbFe_{2-x}Mn_xO_4$ with $0<x<1$ and all the diffraction peaks can be indexed based on the hexagonal structure with the space group of $R\bar{3}m$ without any impurity phases. The lattice parameters are determined to be a_0=0.344 nm and c_0=2.52 nm in $YbFe_2O_4$. Figure 3 displays changes of lattice parameters as a function of Mn concentration (x) in $YbFe_{2-x}Mn_xO_4$ with $0<x<1$. The lattice parameter (c_0) along the c axis increases monotonically as the Mn concentration (x) increases and, on the other hand, the lattice parameter (a_0) in the in-plane remains to be almost constant. From this result, it is expected that the interaction between the Fe bilayers should become weak and the 3D CO structure should change into the 2D CO character as the Mn concentration (x) increases in $YbFe_{2-x}Mn_xO_4$.

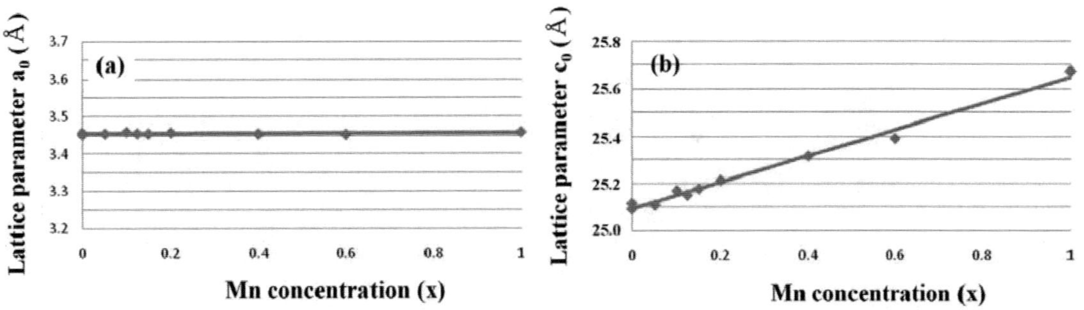

Figure 3: Variations of lattice parameters (a) a_0 and (b) c_0 as a function of Mn concentration (x) in $YbFe_{2-x}Mn_xO_4$ with $0<x<1$.

Thus, we have investigated changes of the 3D CO structure by partial substitution of Mn for Fe in $YbFe_2O_4$ by obtaining various ED patterns. Figure 4 shows a series of selected-area ED patterns at 298 K in $YbFe_{2-x}Mn_xO_4$. Superlattice diffraction spots at the 1/31/3 3/2 positions can be seen, as indicated by an arrow in Fig. 4(a). This implies that, as has been reported previously in $LuFe_2O_4$, the 3D CO structure characterized by the modulated structure with the modulation vector of $\boldsymbol{q}$=<1/3 1/3 1/2> is formed in $YbFe_2O_4$. When Mn ion is partially substituted for Fe ion in $YbFe_2O_4$, diffuse streaks elongating along the [001] direction appear, in addition to the superlattice spots at the 1/3 1/3 3/2 positions in the x=0.05 sample. The appearance of the diffuse streaks indicated the change from the 3D CO structure to the two dimensional (2D) CO structure. That is, in the x=0.05 sample, the coexisting state between the 3D and 2D CO structures is formed spatially. As the Mn concentration (x) increases in $YbFe_{2-x}Mn_xO_4$, the superlattice reflection spots at the 1/31/3 3/2 positions disappear and zig-zag shaped diffuse streak along the [001] direction appears, as indicated by arrows in Figs. 4(c) and 4(d). These experimental results imply that the interaction between the Fe bilayers becomes weak and short-ranged local structure with the 2D character is formed in Mn-doped $YbFe_2O_4$. It is suggested that partial substitution of Mn for Fe in $YbFe_2O_4$ destroyed the 3D CO structure drastically.

To clarify the spatial distribution of the diffuse scatterings in the reciprocal space, ED patterns with various zone-axis directions were examined in $YbFeMnO_4$ at 298 K and found that the shape of the diffuse scattering is honeycomb-shaped. Fig. 5(a) shows the ED pattern exhibiting the honeycomb-shaped diffuse scattering in $YbFeMnO_4$. Schematic representation of honeycomb-shaped diffuse scattering in $YbFeMnO_4$ was displayed in Fig. 5(b). These characteristic faint diffuse scatterings often have been observed in some binary alloys with the face-centered cubic structure such as Ni-Mo alloys [14,15]. In case of binary alloys, it was explained that characteristic diffuse scattering originates from the short-ranged ordered structure, which comprise the polyhedral clusters of atoms. It is speculated that the honeycomb-shaped diffuse scatterings found in Mn-doped $YbFe_2O_4$ should stem from the short-range ordered structure consisting of Fe^{3+} and Mn^{2+} ions in the triangular lattice. The detailed analysis of the honeycomb-shaped diffuse scatterings will be reported elsewhere [16].

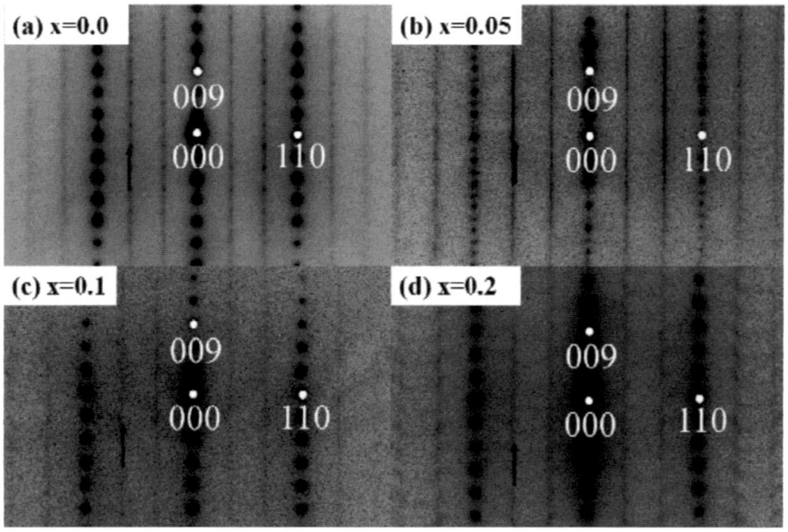

Figure 4; Changes of ED patterns taken along the [1-10] zone-axis as a function of the Mn concentration (x) in $YbFe_{2-x}Mn_xO_4$. (a) x=0.0 (b) 0.05 (c) 0.1 (d) 0.2.

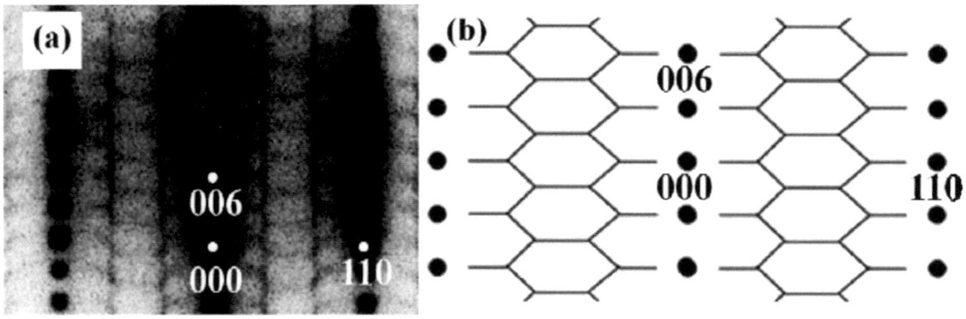

Figure 5; (a) ED pattern tilted to the [-111] direction slightly from the [1-10] zone-axis ED pattern (Fig. 4(d)) in $YbFe_{2-x}Mn_xO_4$ with x=0.2. (b) Schematic representation of honeycomb-shaped diffuse scattering.

In addition, we have examined dielectric properties in $YbFe_{2-x}Mn_xO_4$ with x~0.1. Figure 6 shows (a) the dielectric constant (ε') and (b) dielectric loss (tanδ) plotted in the temperature range between 25 K and 300 K. As shown in Fig. 6(a), ε' has a broad peaks in the temperature range between 150 K and 300 K and monotonically decreases with increasing the frequency of the ac electric field. This behaviour is known as a dielectric dispersion and can be observed in the RFe_2O_4 compounds [11]. According to our previous work in $LuFeCoO_4$, it is suggested that the dielectric dispersion originates from the presence of the polar clustering structure, which should give rise to the characteristic honeycomb-shaped diffuse scatterings in Fig. 5(b) [17]. The characteristic dielectric dispersion found in Mn-doped $YbFe_2O_4$ should originate of the polar clustering structure, which gives rise to the honeycomb-shaped diffuse scatterings in the reciprocal space.

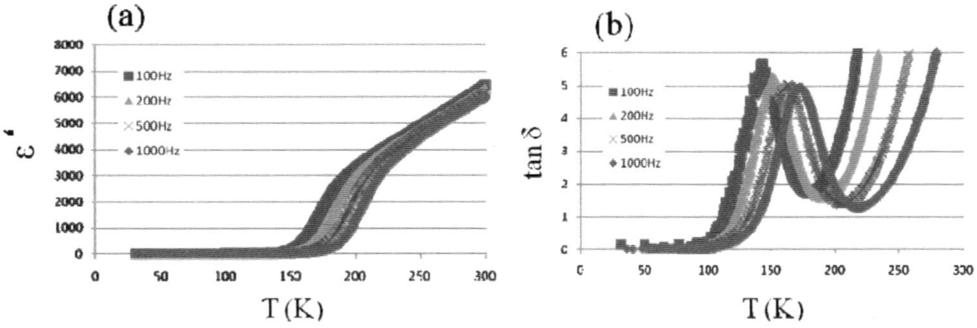

Figure 6; Temperature dependence of dielectric susceptibility (ε') in (a) $YbFe_2O_4$ and (b) $YbFe_{2-x}Mn_xO_4$ with x~0.1.

4. Summary

We have investigated collapsing processes of the 3D CO structure by partial substitution of Mn^{2+} for Fe^{2+} in $YbFe_2O_4$ by TEM experiment and conventional dielectric measurement. It is revealed that the 3D CO structure was collapsed by partial substitution of Mn^{2+} for Fe^{2+} in $YbFe_2O_4$. Electron diffraction experiments revealed that there appear honeycomb-type diffuse streaks in Mn-doped $YbFe_2O_4$. It is considered that short-range ordered structure consisting of Fe^{3+} and Mn^{2+} ions in the triangular lattice is formed, which gives rise to the honeycomb-type diffuse scatterings. The present results suggested that the polar clustering structure should play important roles in dielectric properties in Mn-doped $YbFe_2O_4$.

Acknowledgements

This work was partially supported by a Grant-in-Aid for Scientific Research on Priority Areas "Novel States of Matter Induced by Frustration" (No.19052002) and KIBAN (B) (No.20340078) from the Ministry of Education, Culture, Sports, Science, and Technology in Japan.

References

[1] N. Ikeda, H. Kito, J. Akimitsu, K. Kohn, and K. Siratori, J. Phys. Soc. Jpn. 63, 4556 (1994).
[2] Y. Yamada, S. Nohdo, and N. Ikeda, J. Phys. Soc. Jpn. 66, 3733 (1997).
[3] Y. Yamada, K. Kitsuda, S. Nohdo, and N. Ikeda, Phys. Rev. B 62, 12167 (2000).
[4] N. Ikeda *et al.*, Nature (London) 436, 1136 (2005).
[5] K. Kato, I Kawada, N. Kimizuka, and T. Katsura, Z. Kristallogr. 143, 314 (1975).
[6] N. Ikeda, K. Kohn, H. Kito, J. Akimitsu, and K. Shiratori, J. Phys. Soc. Jpn. 64, 1371 (1995).
[7] N. Ikeda, K. Odaka, E. Takahashi, K. Kohn, and K. Shiratori, Ferroelectrics 190, 191 (1997).
[8] N. Ikeda *et al.*, J. Phys. Soc. Jpn. 69, 1526 (2000).
[9] A. Nagano, M. Naka, N. Nasu, and S. Ishihara, Phys. Rev. Lett. 99, 217202 (2007).
[10] Y. Horibe, K. Yoshii, N. Ikeda, and S. Mori, Phys. Rev. B80, 092104 (2009).
[11] K. Yoshii, N. Ikeda, Y. Matsuo, Y. Horibe, and S. Mori, Phys. Rev. B 76 (2) 024423 (2007).
[12] J. Iida, M. Tanaka, and Y. Nakagawa, J. Phys. Soc. Jpn. 59, 4443-4448 (1997).
[13] K. Yoshii, N. Ikeda, T. Michiuchi, Y. Yokota, Y. Okajima, Y. Yoneda, Y. Matsuo, Y. Horibe, and S. Mori, J. Solid State Chem. 182, 1611-1618 (2009).
[14] R. De Ridder, G. Van Tendeloo and S. Amelinckx, Phys. State. Sol. (a) 38, 663-674 (1976).
[15] R. De Ridder, G. Van Tendeloo, and S. Amelinckx, Acta Cryst. A32 216-224 (1976).
[16] K. Matsumoto, T. Koyama, K. Yoshii, N. Ikeda and S. Mori (in preparation).
[17] Y. Matsuo *et al.*, Ferroelectrics 380, 56-62, (2009).

Journal of Physics: Conference Series **320** (2011) 012086 doi:10.1088/1742-6596/320/1/012086

X-ray photon correlation spectroscopy of structural fluctuations in relaxor ferroelectrics PZN-9%PT

Kenji Ohwada, Susumu Shimomura[1], Hironori Nakao[2], Mitsuyoshi Matsushita[3], Kazumichi Namikawa[4] and Jun'ichiro Mizuki

Quantum Beam Science Directorate, Japan Atomic Energy Agency, 1-1-1 Kouto, Sayo, Sayo, Hyogo 679-5148, Japan

[1]Department of Physics, Faculty of Science, Kyoto Sangyo University, Kyoto 603-8555, Japan

[2]Photon Factory, Institute of Materials Structure Science, High Energy Accelerator Research Organization, 1-1 Oho, Tsukuba, Ibaraki 305-0801, Japan

[3]Functional Materials Development Center, Research Laboratories, JFE Mineral Co., Ltd., 1, Nishihama, Chuou-ku, Chiba, Chiba 260-0826, Japan

[4]Tokyo University of Science, Research Institute of Science & Technology, 2641 Yamazaki, Noda, Chiba, 278-8510, Japan

E-mail: ohwada@spring8.or.jp

Abstract. We have performed an x-ray photon correlation spectroscopy measurement of 91%$Pb(Zn_{1/3}Nb_{2/3})O_3$-9%$PbTiO_3$ and found a slow structural fluctuation in time near T_C. The slow fluctuation consists of long-term and short-term fluctuations. We have derived the short-term fluctuation from the intensity fluctuations and found out the relaxation time to be an order of 10 s. It is suggested that the slow fluctuation is owing to the competition between macroscopic tetragonal symmetry and the microscopic orthorhombic symmetry. The competition should be a main reason for the low-frequency dielectric permittivity.

1. Introduction

Recent progress in imaging technology using coherent X-rays has enabled us to obtain spatial or time correlation function of domains. The dynamics — particularly slow dynamics — of mesoscale domains is rarely detectable by the conventional inelastic light-scattering techniques mostly used for measuring phonon-type dynamics with a characteristic time scale of $\sim 10^{-12}$ s, because the slow dynamics is integrated behind a quasi-elastic central peak. In this situation, the detection of the fluctuation is easier in momentum-and-time $(q-t)$ space than in momentum-and-energy $(q-\omega)$ space; the $q-\omega$ space is usually used in conventional inelastic techniques. The slow dynamics of materials is directly related to the low-frequency susceptibilities of materials and is important for understanding the fundamental physics of mesoscopic scale inside materials.

In this paper, we report a result of x-ray photon correlation spectroscopy (XPCS) measurement of the relaxor ferroelectrics 91%$Pb(Zn_{1/3}Nb_{2/3})O_3$-9%$PbTiO_3$ (PZN-9%PT) [1, 2, 3, 4]. The present target material PZN-9%PT is located near a morphotropic phase boundary (MPB) [5] in the phase diagram and exhibits extremely strong dielectric and piezoelectric responses that are one order of magnitude larger than those of conventional ferroelectric ceramics such as $Pb(Zr_xTi_{1-x})O_3$ (PZT). PZN-9%PT has a cubic perovskite-type structure at high

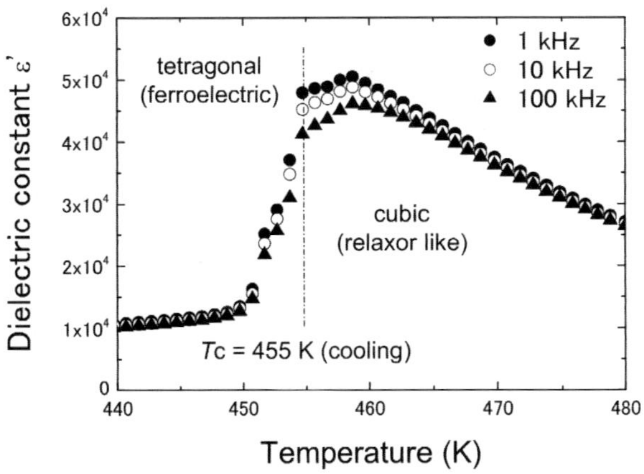

Figure 1. Frequency dependence of dielectric constant of PZN-9%PT measured upon cooling. T_c was determined as 455 K.

temperatures ($T_c = 455$ K on cooling) and shows relaxor-like phenomena. It is generally accepted that such relaxor properties are associated with nanosize (1–100 nm) ferroelectric domains.

2. X-ray Photon Correlation Spectroscopy

We prepared a $10 \times 10 \times 0.25$ mm^3 wafer of PZN-9%PT, which has a PT-concentration gradient as low as < 0.02 %/mm, from a large single crystal grown by JFE Mineral Co., Ltd. [8]. The wafer was cut to provide the (100) plane. Figure 1 shows dielectric constant (ε') of the sample. It shows relaxor-like phenomena with low frequency relaxation near T_c.

XPCS is a novel technique for studying the slow dynamics of various equilibrium and non-equilibrium processes occurring in materials. It is based on the generation of a speckle pattern by the scattered coherent x-ray originating from a material having static or dynamic nano-to-micro meter inhomogeneities.

With coherent x-rays, the coherent sum of scattered x-rays from inhomogeneous materials, i.e., having ordered regions (ORs), results in a speckle pattern. The changes in structure and the arrangement of ORs result in the fine changes to the speckle pattern. A Fourier transformation of the speckle pattern gives the spatial-autocorrelation function (S-ACF), which visualizes the structure and arrangement of ORs through a correlation function. The time evolution of the speckle pattern also gives the time-autocorrelation function (T-ACF) of materials.

The experiment was performed at BL22XU [6] of SPring-8. Incident x-ray energy was tuned to 9.0 keV (wavelength $\lambda = 1.377$ Å) with a liquid-nitrogen-cooled Si(111) double-crystal monochromator. As shown in Fig. 2 (a), the most important point is the relationship among beam size and two distances (aperture-to-sample (R_1) and sample-to-detector (R_2)) [7]. We put the $10(a_2) \times 10$ μm^2 aperture 0.15 m (= R_1) upstream from the sample for selecting coherent area of incident x-ray. The detector was put 1.3 m (= R_2) down stream from the sample. R_1 and R_2 almost satisfy the near-field ($R_1 < a_2/\lambda$) and far-field ($R_2 \gg a_2/\lambda$) ranges respectively. In the former case, the incident x-ray is pseudo-plane wave at the sample position. In the last case, the diffracted x-ray is far-field diffraction, where the speckle patterns show similar shape when the R_2 changes.

3. Result and Discussion

Two-dimensional *time averaged* speckle patterns (static information) were measured by a high-resolution x-ray CCD camera (4.7×4.7 μm^2 pixel size) with an exposure time of 60 s. Fig. 2

Journal of Physics: Conference Series **320** (2011) 012086 doi:10.1088/1742-6596/320/1/012086

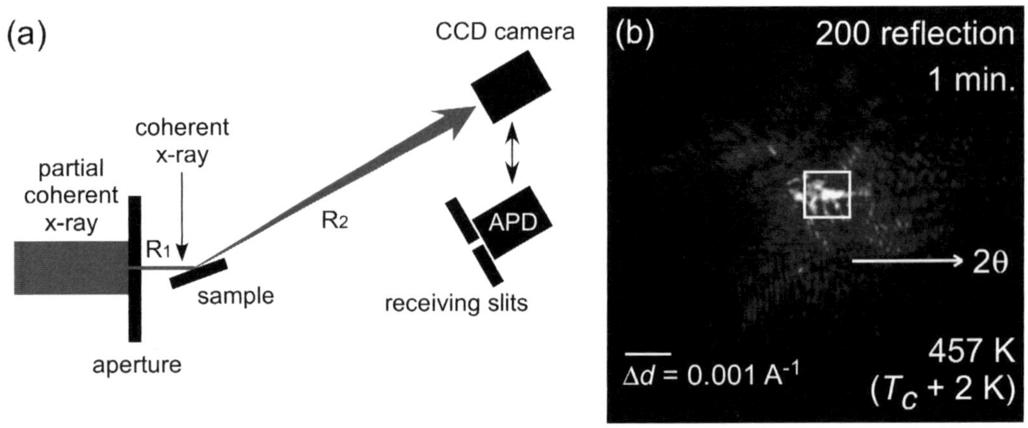

Figure 2. (a) Optical alignment for XPCS at BL22XU of SPring-8. (b) Speckle pattern from PZN-9%PT measured at 457 K.

(b) shows a speckle pattern of 200 reflection measured at 457 K on cooling, 2 K above T_c. In this case, the speckle pattern means the fine structure of the 200 Bragg reflection.

In principle, the time evolution of the speckle pattern can be obtained by the continuous measurement, like a "movie". It is, however, difficult to perform such a measurement due to lack of both coherent photon number and efficiency of the CCD camera, i.e., the minimum exposure time is limited to 60 s for this experiment.

We therefore measured a time evolution of the speckle pattern by a combination of receiving slits and point detector (avalanche photodiode (APD), the maximum count rate is 10 MHz), where some speckles are selected for the measurement. The size of 200 × 200 μm^2 receiving aperture before the point detector was schematically shown in Fig. 2 (b) by the open square [9]. The positive direction of 2θ angle is also shown.

Time-resolved scattered intensity was measured around (2, 0, 0). The temperature was set at a certain temperature and controlled within 0.04 K. Since the present measurement is sensitive to an extrinsic effect as the stage drift just after the temperature set, the measurement was started after the stabilization of the whole measurement system. Figure 3(a) shows a measurement result of 455 K. The intensities (I) are normalized by the monitor counts of the incident x-ray (I_0) as I/I_0. The count rate was ~200 kHz for I and ~20 kHz for I_0. The estimated fluctuation of the present normalized count rate is $\Delta I/I_0 \sim 0.07$. We also show the data of 454 K in Fig. 3 (a), where the fluctuation is no longer observable. The higher noise level compared to that of 455 K is due to the lower count rate compared to that of 455 K. Note that the intensity of 454 K is multiplied by a factor of 8.5. It is clear that the intensity exhibits large thermal fluctuations in time and the fluctuation consists of multiple time scale, i.e., long term fluctuation with about 1000 s and short-term fluctuation with tens of s. Such a multiple-time-scale fluctuation should strongly be correlated with the low-frequency dielectric permittivity as shown in Fig. 1. The existence of the large thermal fluctuation indicates an easy responsibility to the AC-electric field, which will give large dielectric permittivity.

As shown in Fig. 3 (b), we calculated T-ACF defined as $C(\tau) = < I(t)I(t+\tau) > / < I >^2$, where τ is the lag time. Short-term relaxation with 10-s-order can be seen in the total T-ACF. It also contains non-trivial long-and-middle-term relaxation over 100 s and hinders the estimation of the short-term component. In this paper, we have tried to obtain a typical relaxation time of the short-term fluctuation. As seen in the data of 455 K in Fig. 3 (c), we firstly eliminate the components of the long-and-middle-term fluctuations, which are estimated by the data

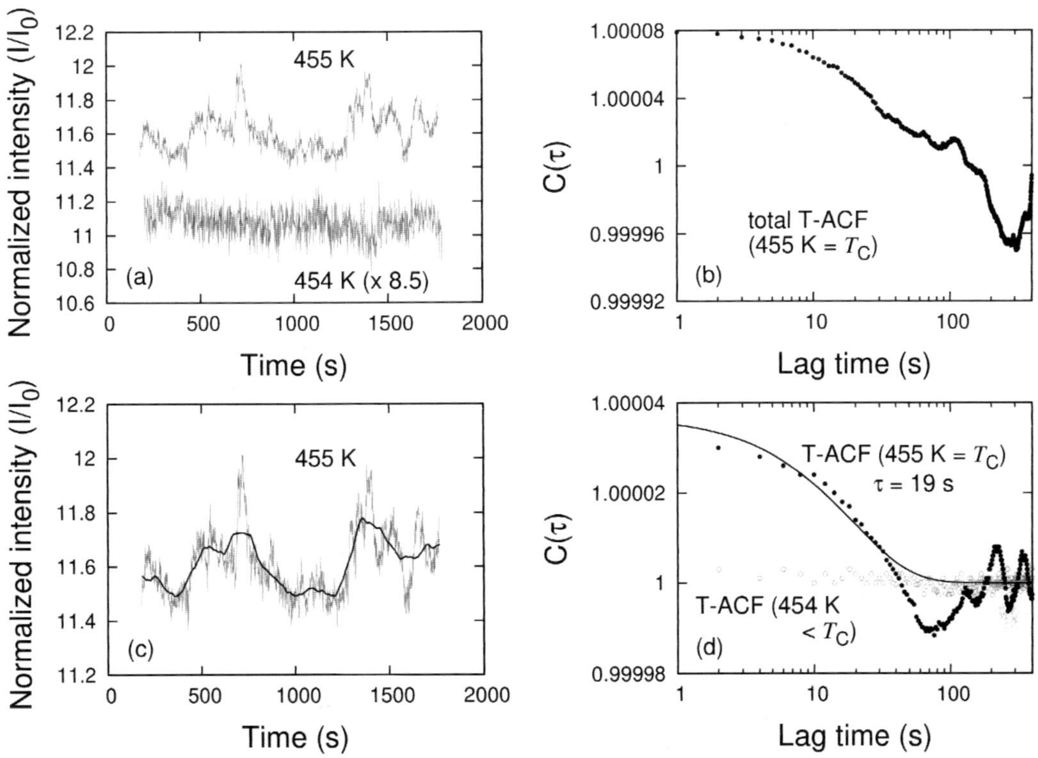

Figure 3. (a) Time dependence of the intensity measured around (2, 0, 0) at 455 K (T_c) and 454 K ($< T_c$). (b) Calculated total T-ACF of 455 K from (a). (c) Time dependence of the intensity and estimated long-and-middle-term fluctuations at 455 K. (d) Calculated T-ACF from (c).

smoothing over 100 s range (see solid line in Fig. 3 (c)). The difference between the data and the solid line approximately gives a component of the short-term fluctuation. By using the difference, we calculated the T-ACF as shown in Fig. 3 (d). $C(\tau)$ was roughly analyzed by the function $\propto e^{-t/\tau}$ (see solid line in Fig. 3 (d)) and the typical relaxation time of the short-term fluctuation was obtained as $\tau = 19$ s. The order is consistent with the result previously obtained by intensity fluctuation spectroscopy using nano-focused x-ray [10]. T-ACF of 454 K is also shown in Fig. 3 (d), which represents that the error level of the present T-ACF measurement is low and the data taken at 455 K is reliable.

As mentioned above, the measurement was done around (2, 0, 0) with the receiving aperture of 200×200 μm^2. The corresponding length scale inside the area is from hundreds of nm to tens of μm [11]. Thus the observed intensity fluctuation in that region mainly reflects the structural fluctuation with the length scale. The time scale is related to the size scale, i.e., the long-term fluctuation comes from the tens-of-μm region (whole region of the x-ray irradiated area), while the short-term fluctuation comes from the hundreds-nm regions.

We speculate that the slow fluctuation comes from the competition between macroscopic tetragonal symmetry and the microscopic orthorhombic symmetry. As is well known, the cubic phase macroscopically transforms into the tetragonal phase. On the other hand, there are local polarized regions with [110]-polarization (so-called polar nano region (PNR)) during the macroscopic cubic-to-tetragonal transition. This is suggested by diffuse scattering [12, 13] and thermal [14] measurement. We also mention the precise properties of the dielectric permittivity as shown in Fig. 1. The dielectric permittivity reaches maximum at 459 K, 4 K above T_c. The downturn before the phase transition should be caused by the competition between PNRs and macroscopic tetragonal symmetry that starts appearing. Note that the summation of the $\langle 110 \rangle$ and $\langle 001 \rangle$ polarization can give rhombohedral and monoclinic symmetries as observed in the MPB region [15].

4. Summary

We have performed an x-ray photon correlation spectroscopy measurement of PZN-9%PT and found a slow structural fluctuation in time near T_C. The slow fluctuation consists of long-term and short-term fluctuations. We have derived the short-term fluctuation from the intensity fluctuations and found out the relaxation time to be an order of 10 s. It is suggested that the slow fluctuation is owing to the competition between macroscopic tetragonal symmetry and the microscopic orthorhombic symmetry. The competition should be a main reason for the low-frequency dielectric permittivity.

We finally mention here that the data acquisition and analysis methods shown in this paper are not sophisticated yet. After the improvement of the methods, we will take the temperature and momentum dependence of the intensity fluctuation and derive the *real* multiple scales of time and space in relaxor ferroelectrics.

Acknowledgments

This work was partly supported by the following projects. Grant-in-Aid for Scientific Research on Priority Areas "Novel States of Matter Induced by Frustration" (19052002) and Grants-in-Aid for Scientific Research for Young Scientists B (16740177, 21710099) by MEXT and CREST "Novel Measuring and Analytical Technology Contributions to the Elucidation and Application of Material" by JST.

References

[1] Kuwata J, Uchino K and Nomura S 1981 *Ferroelectrics* **37** 579.
[2] Kuwata J, Uchino K and Nomura S 1982 *Jpn. J. Appl. Phys* **21** 1298.
[3] Park S-E and Shrout T R 1997 *J. Appl. Phys.* **82** 1804.
[4] Liu S F, Park S-E, Shrout T R and Cross L E 1999 *J. Appl. Phys.* **85** 2810.
[5] The ferroelectric region of (1-x)PZN-xPT is separated into two forms in the x ($PTconcentration$ %) $-T$ phase diagram by a morphotropic phase boundary (MPB), a nearly vertical phase boundary between the two forms, i.e., the boundary is composition-independent. The smaller-x side denotes rhombohedral symmetry (R), while the larger-x side denotes tetragonal symmetry (T). The MPB of (1-x)PZN-xPT is at $x = 9 \sim 10\%$. The MPB is commonly seen in some other solid solutions of lead-based perovskite ferroelectrics, such as $Pb(Zr_{1-x}Ti_x)O_3$ and (1-x)$Pb(Mg_{1/3}Nb_{2/3})O_3$-$x$$PbTiO_3$.
[6] Shobu T, Tozawa K, Shiwaku H, Konishi H, Inami T, Harami T and Mizuki J 2007 *AIP Conf. Proc.* **879** 902.
[7] Ohwada K, Namikawa K, Mizuki J, Shimomura S, Nakao H, Ito K, Matsushita M, Yoneda Y, Murakami Y and Hirota K 2007 *Trans. Mater. Res. Soc. Jpn.* **32** 7.
[8] Matsushita M, Tachi Y and Echizenya K. 2002 *J. Cryst. Growth.* **237-239** 853.
[9] It contains few dozens of speckles inside the receiving aperture. It will ensure the enough counting rate for getting the T-ACF, but on the contrary it will lower the contrast of the T-ACF. It is a trade-off between intensity and coherence.
[10] Ohwada K, Mizuki J, Namikawa K, Shimomura S, Nakao H, Mimura H, Yamauchi K, Matsushita M and Hirota K 2010 *Jpn. J. Appl. Phys.* **49** 020216.
[11] The size of the incident beam was 10 μm, while the present θ angle for the 200 Bragg reflection is 20 deg. Since the scattering plane lies on the horizontal plane at BL22XU, the *effective* horizontal-beam size of the x-ray on the sample at $\theta \sim 20$ deg. will be 30 μm.
[12] Xu G, Gehring P M and Shirane G 2006 *Phys. Rev.* B **74** 104110.
[13] Xu G 2010 *J. Phys. Soc. Jpn.* **79** 011011.
[14] Tachibana M, Sasame K, Kawaji H, Atake T and Takayama-Muromachi E 2009 *Phys. Rev.* B **80** 094115.
[15] Noheda B, Cox D E, Shirane G, Park S-E, Cross L E and Zhong Z 2001 *Phys. Rev. Lett.* **86** 3891.

Relationship between ferroelectricity and magnetic structure of $PbCuSO_4(OH)_2$ with CuO_2 ribbon chains

Yukio Yasui[1], Yudai Yanagisawa[1], Masatoshi Sato[1,2], and Ichiro Terasaki[1]

[1]Department of Physics, Division of Material Science, Nagoya University, Furo-cho, Chikusa-ku, Nagoya 464-8602, Japan
[2]Toyota Physical and Chemical Institute, Nagakute, Aichi 480-1192. Japan

E-mail: yasui.yukio@cc.nagoya-u.ac.jp

Abstract. Neutron diffraction, magnetization and dielectric constant have been measured for single-crystal samples of $PbCuSO_4(OH)_2$ with quasi one-dimensional spin-1/2 chains formed of edge-sharing CuO_4 square planes called CuO_2 ribbon chains. An antiferromagnetic transition with an incommensurate helical magnetic ordering and a ferroelectric transition have been found to appear simultaneously at T_N~2.8 K. From the neutron diffraction studies, 0 $k\pm\delta$ $l/2$ (k=even, l=integer) superlattice magnetic reflections with incommensurate modulation δ~0.189 have been found below T_N. Based on the neutron diffraction and magnetization measurements, the magnetic structure and the relationship to the ferroelectricity are discussed.

1. Introduction

Systems having both ferroelectric and magnetic orderings are called multiferroics and various such examples have been found [1]. Although multiferroics with high transition temperature attract much attention from the viewpoint of technological applications [2], we have been searching and studying systems with CuO_2 ribbon chains to investigate effects inherent to their quantum nature.

The ribbon chains are formed of edge-sharing CuO_4 square planes have one-dimensional Cu^{2+} spin (S=1/2) chains. Within the next-nearest-neighbor exchange interaction J_2 through the Cu-O-O-Cu exchange paths is antiferromagnetic ($J_2 > 0$), and rather strong as compared with the nearest-neighbor exchange interaction J_1 through the Cu-O-Cu paths, because the Cu-O-Cu angle is close to 90 °. In such system, the competition between J_1 and J_2 causes magnetic frustration, which induces a nontrivial magnetic structure in the magnetically ordered phase. Theoretically, a helical magnetic structure is expected for $|J_2/J_1| > 1/4$ within a classical spin model [3]. The J_1-J_2 quantum spin chains such as the CuO_2 ribbon chains can provide a good playground for exotic quantum phases driven by frustration. A spin nematic-(quadrupolar)-ordered phase is a prime example, where various phases are predicted as a function of external magnetic field H and J_1/J_2 [4,5].

$LiVCuO_4$ [6-8] and $LiCu_2O_2$ [9] are known as model systems of the CuO_2 ribbon chains, and have multiferroic nature. The magnetic structures of both systems are helical with modulation vector $\boldsymbol{Q}$ along the chain direction [7-8,10-12], and they exhibit a ferroelectric transition with the spontaneous polarization $\boldsymbol{P}$ described by the relation $\boldsymbol{P} \propto \boldsymbol{Q} \times \boldsymbol{e}_3$ as theoretically predicted [13-15], where $\boldsymbol{e}_3$ are the helical axis.

$PbCuSO_4(OH)_2$ is another example having CuO_2 ribbon chains. The structure is schematically shown in Fig. 1 (space group $P2_1/m$; monoclinic; a=9.701 Å; b=5.650 Å; c=4.690 Å; β=102.65 °) [16].

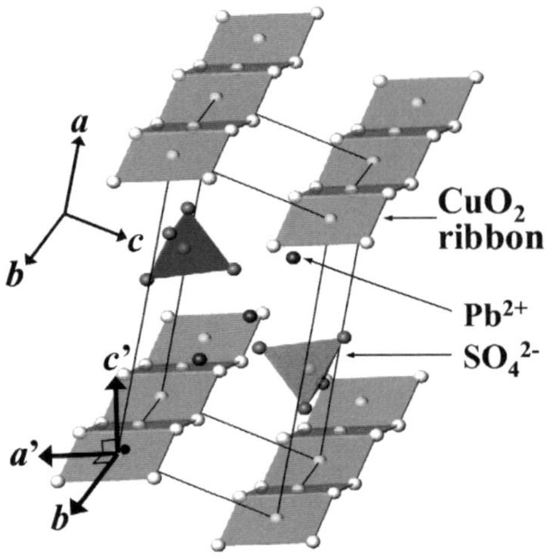

Figure 1. Crystal structure of $PbCuSO_4(OH)_2$. The edge-sharing chains of CuO_4 square planes (called CuO_2 ribbon chains) are separated by the Pb^{2+} ions and the SO_4^{2-} tetrahedra. The *a*'- and *c*'- axes are defined to be suitable for the local structure of the CuO_2 ribbon chains.

The unit cell contains only one equivalent chain which is separated by the nonmagnetic Pb^{2+} ions and SO_4^{2-} tetrahedra. The *a*'- and *c*'-axes are chosen to be convenient to describe the local structure of CuO_2 ribbon chains (see Fig. 1). We have found that the system exhibits a ferroelectric transition induced by the antiferromagnetic ordering at the Néel temperature T_N~2.8 K [17]. Here, in order to determine the magnetic structure of the system, magnetization measurements and neutron diffraction studies have been carried out on a single-crystal sample of $PbCuSO_4(OH)_2$, and on the basis of these results, a possible magnetic structure is proposed, and the relationship between the obtained magnetic structure and the ferroelectricity is discussed.

2. Experiments

Single crystals of a natural mineral $PbCuSO_4(OH)_2$ called linarite have been used in the present study. The crystal axes were determined by X-ray diffraction. The magnetic susceptibility χ and magnetization *M* were measured using a SQUID magnetometer (Quantum Design) in the temperature range from 2 to 350 K. The dielectric constant ε was measured using an ac capacitance bridge (Andeen Hagerling 2500A) at 1 kHz (typical sizes of $\sim 2\times1\times2$ mm^3) to which the electrodes were attached with silver paint.

Neutron scattering measurements were carried out using the triple axis spectrometer HQR (T1-1) of the thermal guide installed at JRR-3 of JAEA in Tokai. The horizontal collimations were 12′(effective)-40′-60′-60′ and the neutron wavelength was 2.4595 Å. The 002 reflection of Pyrolytic graphite (PG) was used as the monochromator. A PG filter was placed after the second collimator to suppress the higher-order contamination. The crystal was oriented with the [010] and [001] axes in the scattering plane. The crystal was set in an Al-can filled with exchange gas, and the can was mounted in a liquid ^4He cryostat.

3. Results and discussion

Figure 2 shows the temperature dependence of the magnetic susceptibility χ of $PbCuSO_4(OH)_2$ taken with the external field *H*=1 T along the three directions represented in the figure. All the χ-*T* curves take maxima at 4~5 K, which is attributed to the growth of the short-range spin correlation with decreasing *T*. They show abrupt drop at an antiferromagnetic transition temperature $T_N \sim$ 2.8 K, which is consistent with the previously reported data of refs. [18] and [19]. We can see the anisotropy in the χ-*T* curves; χ's for ***H*** // ***a***' and for ***H*** // ***b*** decrease below T_N more remarkably than χ for ***H*** // ***c***'. Since the *a*'- and *b*-axes are (nearly) parallel to the CuO_2-planes, which have buckling, in the strict sense, as shown in Fig. 1, the ordered spin components in the antiferromagnetic phase are (nearly) parallel to

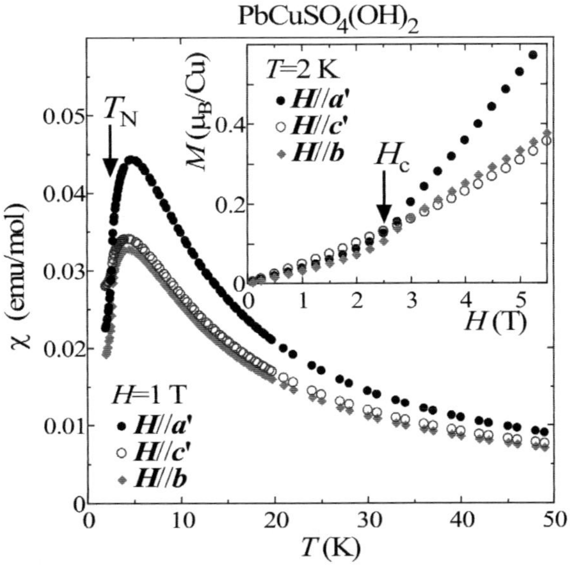

Figure 2. Temperature dependence of the magnetic susceptibility χ of $PbCuSO_4(OH)_2$. The magnetic field of H=1 T is applied along the three directions shown in the figure. The inset shows magnetization curves taken at T=2 K under various magnetic field directions.

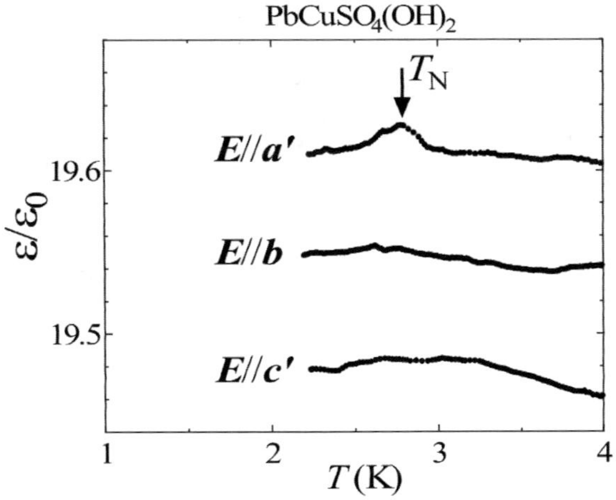

Figure 3. Temperature dependence of the dielectric constant ε of $PbCuSO_4(OH)_2$. The electric field ***E*** is applied along the three directions shown in the figure.

the CuO_2-plane ($a'b$-plane). From analyzing the χ-T curves using the high temperature expansion up to the fourth orders in the temperature region of 50 K<T<350 K, the exchange interactions are evaluated to be J_1=-13±3 K (ferromagnetic) and J_2=21±5 K (antiferromagnetic), which are consistent with those reported in ref. [20]. The values of $|J_1|$ and $|J_2|$ are smaller than those of $LiVCuO_4$ and $LiCu_2O_2$ (J_1=-19 K and J_2=49 K for $LiVCuO_4$ [21] and J_1=-81 K and J_2=44 K for $LiCu_2O_2$ [22]). Thus, the present system is worth exploring the exotic quantum phases in experimentally achievable static fields. The inset of Fig. 2 shows M-H curves of $PbCuSO_4(OH)_2$ at T=2 K (<T_N) along various magnetic field directions. The curves exhibit significant nonlinearity at around H_c~2.5 T for ***H*** // ***a***' and ***H*** // ***b***, whereas the magnetization for ***H*** // ***c***' is almost linear in *H*. Since the magnetic anisotropy of *S*=1/2 spins is expected to be small, the anomaly of *M-H* curves at H_c can be ascribed to the spin-flop transition. A similar spin-flop transition has also been confirmed by neutron diffraction measurements in $LiVCuO_4$ [7].

Figure 3 shows the temperature dependence of the dielectric constant ε for single-crystal samples of $PbCuSO_4(OH)_2$ taken in the electric fields ***E*** along the three directions represented in the figure. We have not found any anomaly of ε for ***E*** // ***b*** and ***E*** // ***c***'. Only for ***E*** // ***a***', an anomaly is seen at T_N~2.8 K. We have also observed that the electric polarization ***P*** appears at around T_N, and that its sign is reversed when the cooling electric field is reversed (not shown) [17]. These results indicate that the ferroelectric transition takes place simultaneously with the magnetic transition at T_N and ***P*** appears along the *a*' axis.

In order to clarify the magnetic structure, we have carried out neutron diffraction studies on a single crystal of $PbCuSO_4(OH)_2$. The neutron scattering intensities were measured at ***q***=(0, *k*, *l*) in the reciprocal space at *T*=1.7 K (<T_N) and 7.0 K (>T_N). At 1.7 K, we observed magnetic superlattice reflections at ***q***=(0, *k*±δ, *l*) (*k*=even, *l*=integer) with δ~0.189. Profiles of ω-scan (sample-angle scan) for 0 δ 1/2 and 0 δ 3/2 reflections are shown in Figs. 4(a) and 4(b), respectively. Figure 5 shows the temperature dependence of the peak intensities for 0 δ 1/2 reflection obtained by subtracting the background intensity from the raw data. From the figure, we find that the magnetic ordering grows with decreasing *T* below T_N. The determination of the magnetic structure of the present system is difficult by only the present neutron diffraction data, because the number of the observed magnetic reflections is insufficient to determine many parameters to describe the modulated magnetic structure. Therefore, the magnetic structure of the present system is discussed from the combined studies of neutron diffraction and detailed magnetization measurements in the following paragraph.

The behavior of the χ-*T* and *M-H* curves has following features. (i) The evaluated $|J_2/J_1|$ is ~ 0.62, which is theoretically expected to favor a helical type magnetic structure. (ii) The magnetic susceptibility χ below T_N indicates that the ordered spin components are (nearly) parallel to the CuO_2-plane (*a*'*b*-plane). (iii) The spin-flop transition at H_c~2.5 T for ***H*** // ***a***' and ***H*** // ***b*** also indicates that the ordered spin components in the antiferromagnetic phase lie in the CuO_2-plane (*a*'*b*-plane). The observed magnetic reflections at ***q***=(0, *k*±δ, *l*/2) (*k*=even, *l*=integer) have the following features. (iv) The δ value of ~0.189 indicates that the magnetic ordering is characterized by the incommensurate

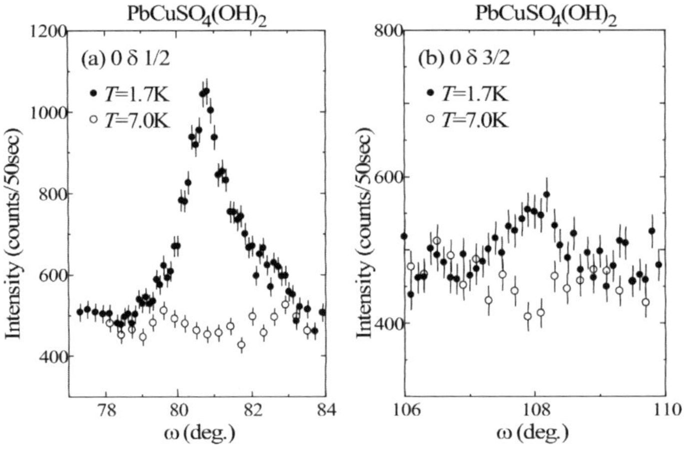

Figure 4. Profiles of the ω-scan (sample-angle scans) for (a) 0 δ 1/2 and (b) 0 δ 3/2 reflections (δ~0.189) taken at *T*=1.7 K and *T*=7.0 K.

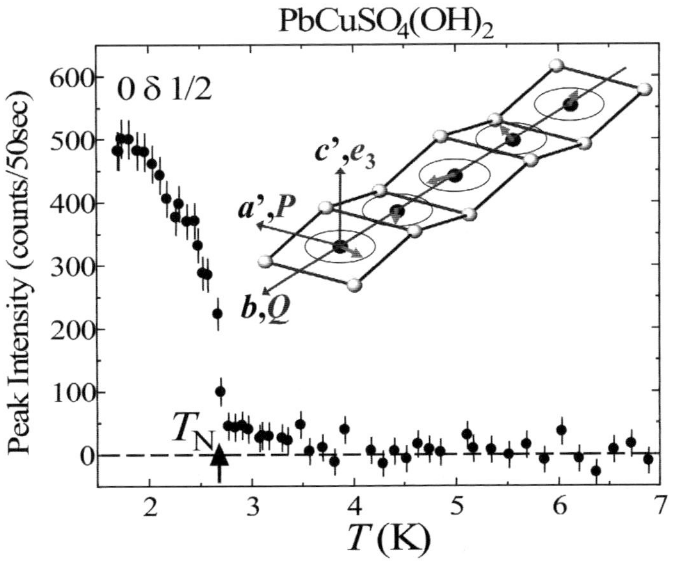

Figure 5. Temperature dependence of the peak intensities for 0 δ 1/2 reflection (δ~0.189) obtained by subtracting the background intensity from the raw data for $PbCuSO_4(OH)_2$. The inset show the magnetic structure of the Cu^{2+} spins.

modulation vector $\boldsymbol{Q}$, corresponding to the periodicity of ~5.29*b* (pitch angle Δϕ~34 °) along the Cu-chain direction. (v) The period of spin modulation along the *c*-axis is 2*c*, and spins at Cu^{2+} sites shifted by (0,0,1) are anti-parallel to those at the original sites. Considering the above results, the magnetic structure of $PbCuSO_4(OH)_2$ is found to be the helical type with the ordered spin components parallel (or nearly parallel) to the CuO_2-plane and the helical axis $\boldsymbol{e}_3$ parallel (or nearly parallel) to the *c*' axis (see the inset of Fig. 5). From the analyses of the neutron diffraction data, this structure is consistent with the observed intensities of the magnetic reflections, and the ordered moment at T=1.7 K is evaluated to be ~0.5 μ_B by comparing the magnetic scattering intensities with nuclear ones.

Here, we discuss the relationship between observed magnetic structure and ferroelectricity of $PbCuSO_4(OH)_2$. The magnetic structure is the helical with *a*'*b*-plane, where the incommensurate modulation vector $\boldsymbol{Q}$ is parallel to the *b** axis and the helical axis $\boldsymbol{e}_3$ is (nearly) parallel to the *c*' axis as shown in the inset of Fig. 5. The ferroelectric polarization *P* is observed along *a*'. Then, these properties indicate that the equation $\boldsymbol{P} \propto \boldsymbol{Q} \times \boldsymbol{e}_3$ holds, as was proposed by microscopic theoretically [13 15], where the ferroelectricity is originated from the helical type magnetic structure through the spin-orbit interaction. From the obtained data, the multiferroic nature of the present system is found to be basically similar to that of $LiVCuO_4$. We think that the characteristics of the $PbCuSO_4(OH)_2$ are small values of $|J_1|$ and $|J_2|$ and simple crystal structure containing one equivalent chain in a unit cell [17].

4. Conclusion

We have shown the experimental results of neutron magnetic reflection, magnetic susceptibility, magnetization, and dielectric constant for single-crystal samples of $PbCuSO_4(OH)_2$. The ferroelectric transition takes place simultaneously with the magnetic transition at T_N, and the ferroelectric polarization *P* appears along the *a*' axis. The 0 *k*±δ *l*/2 (*k*=even, *l*=integer) magnetic reflections (δ~0.189) have been observed below T_N. We have found that Cu^{2+} spins are ordered to the helical structure parallel to the *a*'*b*-plane, and the incommensurate modulation vector $\boldsymbol{Q}$ is along the Cu-chain, where the relation $\boldsymbol{P} \propto \boldsymbol{Q} \times \boldsymbol{e}_3$ is valid.

Acknowledgments
Neutron scattering was performed within the national user's program of the Neutron Science Laboratory of the Institute for Solid State Physics (NSL-ISSP). This study was supported by Grant-in-Aid for Scientific Research on Priority Areas ``Novel States of Matter Induced by Frustration" (22014005).

References

[1] Cheong S W and Mostovoy M 2007 Nat. Mater. **6** 13
[2] Kitagawa Y, Hiraoka Y, Honda T, Ishikura T, Nakamura H and Kimura T 2010 Nat. Mater. **9** 797
[3] Bursill R, Gehring G A, Farnell D J J, Parkinson J B, Xiang T and Zeng C 1995 J. Phys.; Condens. Matter **7** 8605
[4] Hikihara T, Kecke L, Momoi T and Furusaki A 2008 Phys. Rev. B **78** 144404
[5] Sato M, Momoi T and Furusaki A 2009 Phys. Rev. B **79** 060406
[6] Naito Y, Sato K, Yasui Y, Kobayashi Y, Kobayashi Y and Sato M 2007 J. Phys. Soc. Jpn. **76** 023708
[7] Yasui Y, Naito Y, Sato K, Moyoshi T, Sato M, and Kakurai K 2008 J. Phys. Soc. Jpn. **77** 023712
[8] Sato M, Yasui Y, Kobayashi Y, Sato K, Naito Y, Tarui Y and Kawamura Y 2010 Solid State Science **12** 670
[9] Park S, Choi Y J, Zhang C L and Cheong S W 2007 Phys. Rev. Lett. **98** 057601
[10] Gibson B J, Kremer R K, Prokofiev A V, Assmus W and McIntyre G J 2004 Physica B **350** e253
[11] Yasui Y, Sato K, Kobayashi Y and Sato M 2009 J. Phys. Soc. Jpn. **78** 084720
[12] Kobayashi Y, Sato K, Yasui Y, Moyoshi T, Sato M and Kakurai K 2009 J. Phys. Soc. Jpn. **78** 084721
[13] Katsura H, Nagaosa N and Balatsky A 2005 Phys. Rev. Lett. **95** 057205
[14] Mostovoy M 2006 Phys. Rev. Lett. **96** 067601
[15] Jia C, Onoda S, Nagaosa N and Han J H 2007 Phys. Rev. B **76** 144424
[16] Effenberger H 1987 Mineral. Petrol. **36** 3
[17] Yasui Y, Sato M and Terasaki I 2011 J. Phys. Soc. Jpn. **80** No.3
[18] Baran M, Jedrzejczak A, Szymczak H, Maltsev V, Kamieniarz G, Szukowski G, Loison C, Ormeci A, Drechsler S L and Rosner H 2006 Phys. Stat. Sol. **3** 220
[19] Szymczak R, Szymczak H, Kamieniarz G, Szukowski G, Jaśniewicz-Pacer K, Florek W, Maltsev V and Babonas G J 2009 Acta. Phys. Pol. A **115** 925
[20] Kamieniarz G, Bieliński M, Szukowski G, Szymczak R, Dyeyev S and Renard J P 2002 Comp. Phys. Commun. **147** 716
[21] Enderle M, Mukherjee C, Fak B, Kremer R K, Broto J M, Rosner H, Drechsler S L, Richter J, Malek J, Prokofiev A, Assmus W, Pujol S, Raggazzoni J L, Rakoto H, Rheinstadter M and Ronnow H M 2005 Europhys. Lett. **70** 237
[22] Masuda T, Zheludev A, Roessli B, Bush A, Markina M and Vasiliev A 2005 Phys. Rev. B **72** 014405.

International Conference on Frustration in Condensed Matter (ICFCM) IOP Publishing
Journal of Physics: Conference Series **320** (2011) 012088 doi:10.1088/1742-6596/320/1/012088

Correlation between magnetocapacitance effect and polarization flop direction in a slanted magnetic field in multiferroic helimagnets

Nobuyuki Abe[1], Kouji Taniguchi[2], Hajime Sagayama[1], Taka-hisa Arima[1]

[1] Institute of Multidisciplinary Research for Advanced Materials, Tohoku University, Sendai, Japan

[2] Department of Advanced Materials Science, University of Tokyo, Kashiwa, Japan

E-mail: `abe@tagen.tohoku.ac.jp`

Abstract. The relationship between the magnetocapacitance e ect and rotation direction of electric polarization (**P**) in a canted magnetic eld has been investigated for multiferroic RMnO$_3$ (R =Tb$_{1-x}$Dy$_x$ and Eu$_{0.6}$Y$_{0.4}$). We observed a clear correlation between the enhancement of the magnetocapacitance e ect and the rotation direction of **P** in a **P**- op transition. These results indicate that the mobility and the stability of the 90° domain wall in a **P**- op transition are dominated by its thickness.

1. Introduction

Magnetoelectric multiferroics, in which magnetic and ferroelectric orders coexist, have been intensively investigated because the cross correlation between ferroelectrics and magnetics is considered to provide a stage for new spintronic devices. For example, some multiferroic helimagnets show magnetic- eld-induced ferroelectric polarization (**P**) reversal and op[1, 2, 3, 4, 5, 6, 7, 8]. From the microscopic point of view, ferroelectric helimagnets have one-to-one correspondence between **P** and the magnetic structure[9, 10, 11]. The dynamics of the magnetic moment can therefore be probed by a measurement of electric permittivity ε. Ferroic materials generally show a large physical response based on the modi cation of the domain structure. Because a helimagnetic ferroelectric possesses ferroic ordering of the electric dipole moments and vector spin chirality ($\mathbf{S}_i$ $\mathbf{S}_j$), a large magnetoelectric response may arise from the motion of ferroelectric and helimagnetic domain walls (DWs). In the case of the prototypical ferroelectric helimagnet RMnO$_3$ (R = Tb, Dy), **P** rotates from along the c-axis to along the a-axis when a magnetic eld is applied along the b-axis[1, 12]. Moreover, if the magnetic eld is slightly canted from the b-axis, then the rotation direction of **P** can be controlled by the tilting direction of the magnetic eld[13]. Because the phase transition is of rst-order type, **P**$||c$ and **P**$||a$ phases can coexist near the phase transition, and hence four types of 90 DWs appear as their boundaries: $+P_c$ to $+P_a$, $+P_c$ to P_a, P_c to $+P_a$, and P_c to P_a. The selective rotation of **P** is ascribed to the selective nucleation of two types of DWs in a canted magnetic eld. Similar behavior has also been observed in another helimagnetic ferroelectric MnWO$_4$[14]. Another striking feature of RMnO$_3$ is the gigantic magnetocapacitance e ect at the **P**- op transition[12]. The enhancement of the dielectric constant is caused by the microscopic motion of multiferroic DWs at the **P**- op transition[15]. In this study, we show clear correlation between the magnetocapacitance e ect and the rotation direction of **P** in a canted magnetic eld in helimagetic ferroelectric $Tb_{1-x}Dy_xMnO_3$ and $Eu_{0.6}Y_{0.4}MnO_3$.

2. Experimental

Single crystals of $Tb_{1-x}Dy_xMnO_3$ and $Eu_{0.6}Y_{0.4}MnO_3$ were grown by a oating zone method. The crystal axes were determined by Laue photograph. To measure the rotation direction of **P**, the crystals were cut into plate-shaped samples with the normal vector forming an angle of 45 from both the *a*- and the *c*-axes[13]. Gold lms were deposited onto the widest surfaces to form the counter electrodes. Electric permittivity ε was measured at a frequency $f = 10$ KHz by using an LCR meter (Agilent 4263B). The electric polarization **P** was obtained by integrating the pyroelectric current measured on an electrometer (Keithley 6517A). The **P** and ε measurements were performed in a cryostat equipped with a 15 T superconducting magnet at the High Field Laboratory for Superconducting Materials, Institute for Material Research, Tohoku University, Japan.

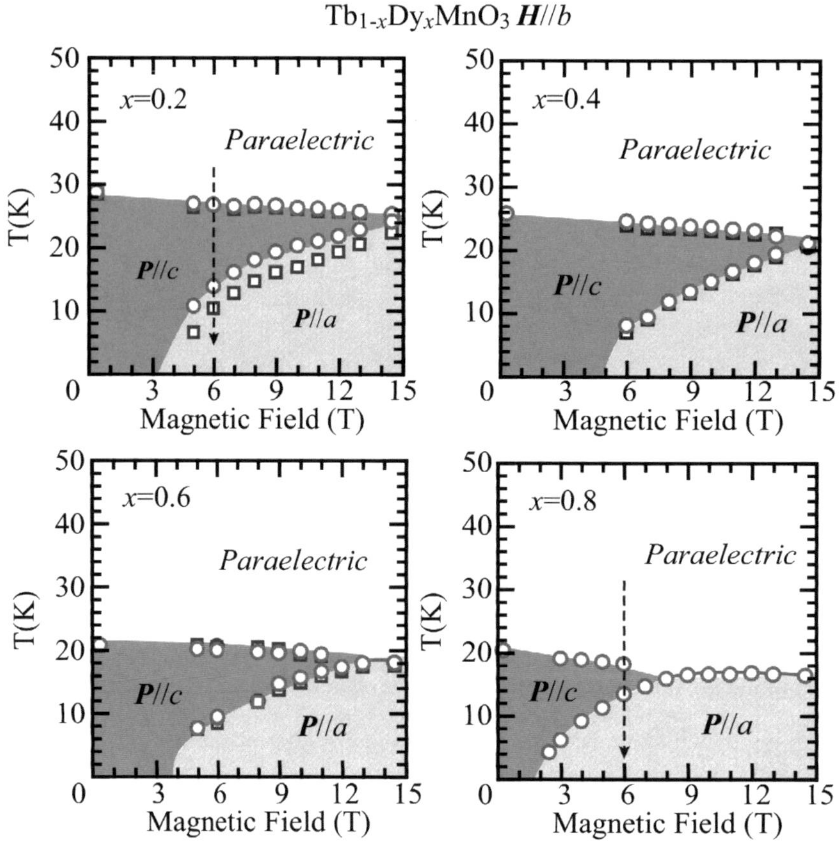

Figure 1. Magnetoelectric phase diagrams of $Tb_{1-x}Dy_xMnO_3$ in a magnetic eld applied along the *b*-axis. Circles and squares represent the data points in the warming and cooling runs of dielectric measurements, respectively. Arrows indicates the temperature sweep processes in Fig.2.

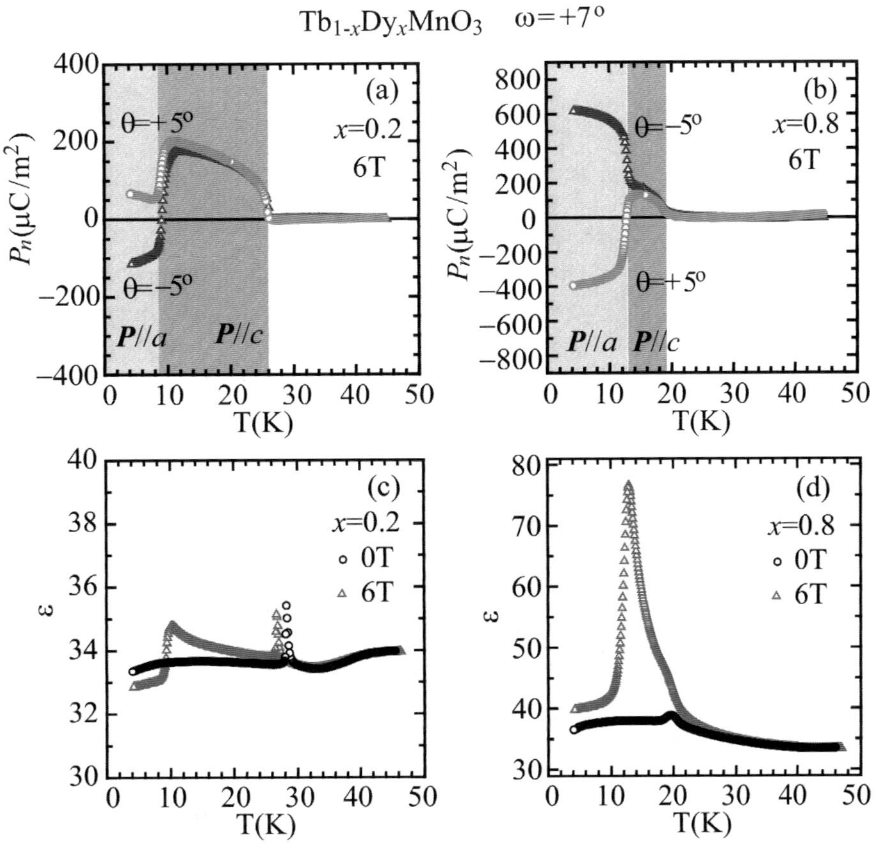

Figure 2. (a),(b) Selective rotation of **P** in a canted magnetic eld of 6 T. (c),(d) Temperature dependence of dielectric permittivity of 0 T and 6 T.

3. Results and Discussion

Figure 1 shows magnetoelectric phase diagrams of $Tb_{1\ x}Dy_xMnO_3$ in a magnetic eld applied along the *b*-axis. The paraelectric-to-ferroelectric transition temperature monotonically decreased as x was increased. The arrows in Fig. 1 indicate the temperature sweep process of the **P** and ε measurements. In both measurements shown in Fig. 2 (a) and (b), the sample was cooled from 45 K to below the $\mathbf{P}||c$ phase transition temperature in a positive electric eld to form a single-domain state with $P_c > 0$. The direction of the canted magnetic eld is represented by two angles, ω and [13]. Then, the *a*- and *c*-components of the magnetic eld are given by $H_a = H \sin$ and $H_c = H \cos$ $\sin\omega$, respectively. When the direction of the 6 T magnetic eld is canted from the *b*-axis with $\omega > 0$ and > 0, **P** rotates from the $+c$ direction to the $+a$ direction for x=0.2. This is the expected result for a thinner DW. On the other hand, the x=0.8 sample shows the opposite rotation of **P** from the $+c$ direction to the a direction. This response is expected for a thicker DW. A similar trend appears in the magneto-capacitance e ect. Figure 2 (c) and (d) shows the temperature dependence of ε at 0 T and 6 T. When $x = 0.8$, the enhancement of ε at the **P**- op transition is much larger than that when $x = 0.2$. Because the motion of the 90 DW contributes to ε in a lower frequency region[15], this result indicates that the mobility of DWs when $x = 0.8$ is much higher than that when $x = 0.2$.

To compare the thickness of 90 DWs between the systems with and without 4f moments, we measured the magnetic- eld-induced selective rotation of **P** in $Eu_{0.6}Y_{0.4}MnO_3$. Yamasaki

Journal of Physics: Conference Series **320** (2011) 012088 doi:10.1088/1742-6596/320/1/012088

et al. reported that the magnetic modulation wavenumber of $Eu_{0.6}Y_{0.4}MnO_3$ is the same as that of $TbMnO_3$ [16]. This result indicates that the $GdFeO_3$-type distortion of $Eu_{0.6}Y_{0.4}MnO_3$ is similar to the distortion of $TbMnO_3$. Figure 3 (d) shows the selective rotation of electric polarization in the **P**- op transition of $Eu_{0.6}Y_{0.4}MnO_3$. The sample is cooled, with a poling electric eld and a magnetic eld of 6 T canted from the a-axis to the c-axis by +5 or 5 . A **P**- op transition is induced by sweeping the magnetic eld from 6 T to 0 T. The rotation direction of **P** is completely determined by the direction of the canted magnetic eld. We nd that the relation between the rotation direction of **P** and the direction of the canted magnetic eld in $Eu_{0.6}Y_{0.4}MnO_3$ is opposite in the case of $TbMnO_3$, suggesting the importance of the interaction between Mn 3d and rare earth 4f magnetic moments. The enhancement of ε_a at the **P**- op transition is as large as that for $Tb_{0.2}Dy_{0.8}MnO_3$. These results indicate that the 90 DW in $Eu_{0.6}Y_{0.4}MnO_3$ is much thicker than the modulation of Mn^{3+} magnetic moments.

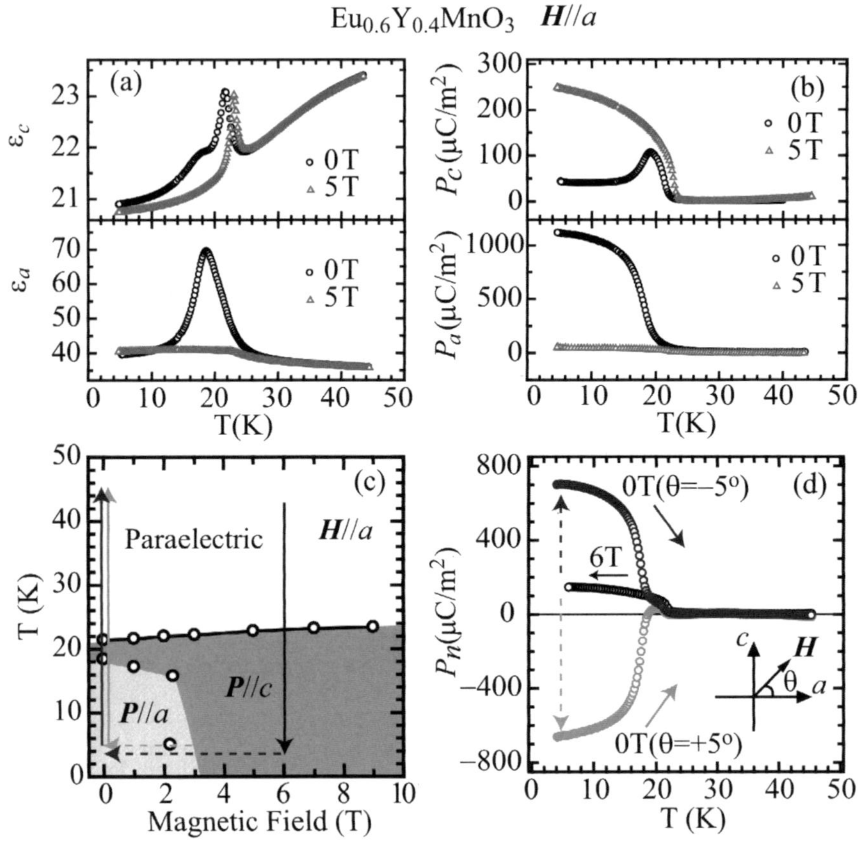

Figure 3. Selective rotation of **P** in canted magnetic elds in $Eu_{0.6}Y_{0.4}MnO_3$.(a) Temperature dependence of ε_c and ε_a. (b) Temperature dependence of $\mathbf{P}_c$ and $\mathbf{P}_a$. (c) Magnetoelectric phase diagram. Arrows show the sweeping path for the measurements.(d) Selective rotation of **P** in canted magnetic elds in $Eu_{0.6}Y_{0.4}MnO_3$. The magnetic eld is canted from the a-axis towards the c-axis with = +5 or 5 .

4. Conclusion

We show a clear correlation between selectivity of rotation direction of **P** in a canted magnetic eld and the enhancement of ε at the **P**- op transition in multiferroic $Tb_{1\ x}Dy_xMnO_3$ and

Journal of Physics: Conference Series **320** (2011) 012088 doi:10.1088/1742-6596/320/1/012088

$Eu_{0.6}Y_{0.4}MnO_3$. These results can be explained by assuming that the mobility of DWs would be dominated by the thickness of DWs relative to the magnetic modulation in a helical magnet.

References

[1] Kimura T, Goto T, Simotani H, Ishizaka K, Arima T and Tokura Y 2003 *Nature* **426** 55
[2] Hur N, Park S, Sharma P A, Ahn J S, Guha S and Cheong S-W 2004 *Nature* **429** 392
[3] Yamasaki Y, Miyasaka S, Kaneko Y, He J P, Arima T and Tokura Y 2006 *Phys. Rev. Lett.* **96** 207204
[4] Taniguchi K, Abe N, Takenobu T, Iwasa Y and Arima T 2006 *Phys. Rev. Lett.* **97** 097203
[5] Park S, Choi Y J, Zhang C L and Cheong S-W 2007 *Phys. Rev. Lett.* **98** 057601
[6] Taniguchi K, Abe N, Ohtani S, Umetsu H and Arima T 2008 *Appl. Phys. Express* **1** 031301
[7] Ishiwata S, Taguchi Y, Murakawa H, Onose S and Tokura Y 2008 *Science* **319** 1643
[8] Fukunaga M, Sakamoto Y, Kimura H, Noda Y, Abe N, Taniguchi K, Arima T, Wakimoto S, Takeda M, Kakurai K and Kohn K 2009 *Phys. Rev. Lett.* **103** 077204
[9] Katsura H, Nagaosa N and Balatsky A V 2005 *Phys. Rev. Lett.* **95** 057205
[10] Yamasaki Y, Sagayama H, Goto T, Matsuura M, Hirota K, Arima T and Tokura Y 2007 *Phys. Rev. Lett.* **98** 147204
[11] Yamasaki Y, Sagayama H, Abe N, Arima T, Sasai K, Matsuura M, Hirota K, Okuyama D, Noda Y and Tokura Y 2008 *Phys. Rev. Lett.* **101** 097204
[12] Goto T, Kimura T, Lawes G, Ramirez A P and Tokura Y 2004 *Phys. Rev. Lett.* **92** 257201
[13] Abe N, Taniguchi K, Ohtani S, Umetsu H and Arima T 2009 *Phys. Rev. B* **80** 020402(R)
[14] Taniguchi K, Abe N, Umetsu H, Aruga Katori H and Arima T 2008 *Phys. Rev. Lett.* **101** 207205
[15] Kagawa F, Mochizuki M, Onose Y, Murakawa H, Kaneko Y, Furukawa N and Tokura Y 2009 *Phys. Rev. Lett.* **102** 057604
[16] Yamasaki Y, Miyasaka S, Goto T, Sagayama H, Arima T and Tokura Y 2007 *Phys. Rev. B* **76** 184418

Multi-frequency ESR in the S = 5/2 triangular-lattice antiferromagnet $CuFe_{1-x}Ga_xO_2$

Takahito Fujita[1], Shojiro Kimura[1,2], Hironori Yamaguchi[1,3], Takanori Kida[1], Masayuki Hagiwara[1]
[1]KYOKUGEN, Osaka Univ., 1-3 Machikaneyama, Toyonaka, Osaka 560-8531, Japan

E-mail: fujita@mag.cqst.osaka-u.ac.jp

Abstract. We have performed multi-frequency electron spin resonance (ESR) and magnetization measurements on single crystals of the triangular-lattice antiferromagnet $CuFe_{1-x}Ga_xO_2$ (x=0.000-0.028) to clarify the origin of the shift of the magnetic transition field from an Ising-like four-sublattice phase into an incommensurate phase by doping of Ga^{3+}. As a result, we have succeeded in explaining the shift of transition field by broadening of the lowest excitation branch.

1. Introduction

Frustrated spin systems provide a rich variety of magnetic states, such as spin liquid [1], spin nematic [2], and spin ice [3]. In particular, magnetic substances with ferroelectricity induced by the spiral magnetic structure due to frustration have drawn attention to condensed matter scientists and engineers as multiferroic materials in recent years [4].

A delafossite compound $CuFeO_2$ is an extensively-studied triangular-lattice magnet [5]. Triangular-lattice antiferromagnets have attracted considerable interest as one of the most typical frustrated spin systems. $CuFeO_2$ consists of triangular layers of Fe^{3+} ($3d^5$) ions which are separated by nonmagnetic layers of Cu^+ and O^{2-} ions and are stacked perpendicularly to the c-axis. Since Fe^{3+} ions with $3d^5$ high-spin configuration have no orbital angular momentum, this is usually regarded as a Heisenberg-type spin system. However, $CuFeO_2$ exhibits an Ising-like four-sublattice (4SL) (↑ ↑ ↓ ↓) collinear order below 11 K at zero magnetic field. Moreover, the several successive phase transitions take place well below 11 K, when the external field H along the c-axis is increased. Above H_c = 7 T, ferroelectric and incommensurate (FEIC) order (7 T < H < 13 T) is realized, and then a five-sublattice (5SL) (↑ ↑ ↑ ↓ ↓) (13 T < H < 20 T) and a three-sublattice (3SL) (↑ ↑ ↓) (13 T < H < 20 T) orders are observed [6].

The changes of magnetic properties caused by substitution of nonmagnetic ions such as Al^{3+} and Ga^{3+} for Fe^{3+} ions have been studied. Such substitution reduces H_c with increasing the concentration of nonmagnetic ions. As a result, the FEIC phase can be induced at zero magnetic field by a certain amount of nonmagnetic ions [7]. This result indicates that the ferroelectric phase becomes stable and the realization of this phase can be controlled by substitution of nonmagnetic ions.

[2] Present address: IMR, Tohoku Univ., 2-1-1 Katahira, Aoba-ku, Sendai, Miyagi 980-8577, Japan
[3] Present address: Department of Physical Science, Osaka Prefecture Univ., 1-1 Gakuencho, Naka-ku, Sakai, Osaka 599-8531, Japan

ESR is one of the powerful tools to investigate magnetic excitations with high resolution. We have performed multi-frequency ESR measurements to clarify why the transition field H_c goes down with increasing x in $CuFe_{1-x}Ga_xO_2$ from the view point of magnetic excitations.

2. Experimental

In this study, we synthesized some single crystal samples of $CuFe_{1-x}Ga_xO_2$ with different Ga contents. The single crystals with different x ($x = 0.000$, 0.001, 0.008, 0.012, 0.013, and 0.028) were grown by a floating zone (FZ) technique [8]. We verified that the samples are in a single phase by powder X-ray diffraction. We also determined the crystal axes and Ga contents by Laue photography and inductively-coupled plasma mass spectrometer (ICP-MS), respectively. We performed ESR measurements at frequencies below 500 GHz in static magnetic fields of up to 8 T at 1.6 K by utilizing a 16 T superconducting magnet (Oxford Instruments) and a vector network analyzer (AB millimetre). High-field magnetization measurements were performed in pulsed magnetic fields of up to 20 T. The magnetization was measured with an induction method using a pick-up coil. The magnetization at 4.2 K in magnetic fields of up to 7 T was also measured with a superconducting quantum interference device (SQUID) magnetometer (Quantum Design MPMS-XL7). All the experiments were carried out at KYOKUGEN in Osaka University.

3. Experimental Results

Figures 1 (a) and (b) show magnetization curves at 4.2 K for $H//c$ of $CuFe_{1-x}Ga_xO_2$ with $x = 0.000$ and 0.028, respectively. In the magnetization curve for $x = 0.000$, steep increases, which correspond to the magnetic transitions reported previously, appear at about 7 T and 13 T. Large hysteresis observed in the curve indicates that the first order transitions occur at these fields. The field region between 7 T and 13 T is reported to be a FEIC phase. In Fig. 1 (b), we observed only one transition corresponding to higher transition in Fig. 1 (a), and thus the systematic shift of the FEIC state occurs by doping. It is depicted in Fig. 2 magnetic field (H) versus Ga^{3+} content (x) phase diagram determined by the magnetization measurements at 4.2 K and 1.3 K.

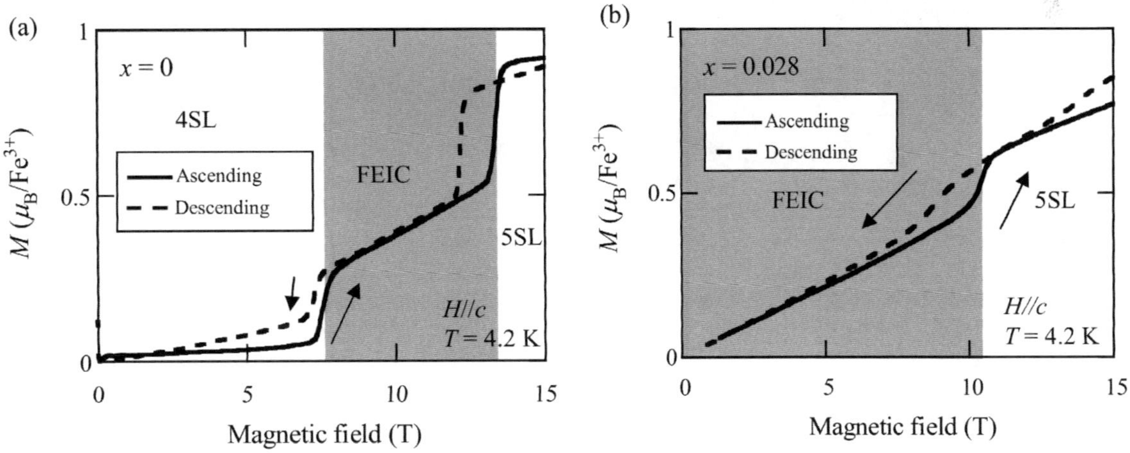

Figure 1. (a) Magnetization curves of $CuFe_{1-x}Ga_xO_2$ with $x = 0.000$ at 4.2 K for $H//c$. The steps at about 7 T and 13 T correspond to the transitions from a four-sublattice (4SL) to a ferroelectric incommensurate (FEIC) phase and from the FEIC to a five-sublattice (5SL) phase, respectively. (b) Magnetization curves of $CuFe_{1-x}Ga_xO_2$ with $x = 0.028$ at 4.2 K for $H//c$. The step at 10 T corresponds to the transition from the FEIC to the 5SL phase.

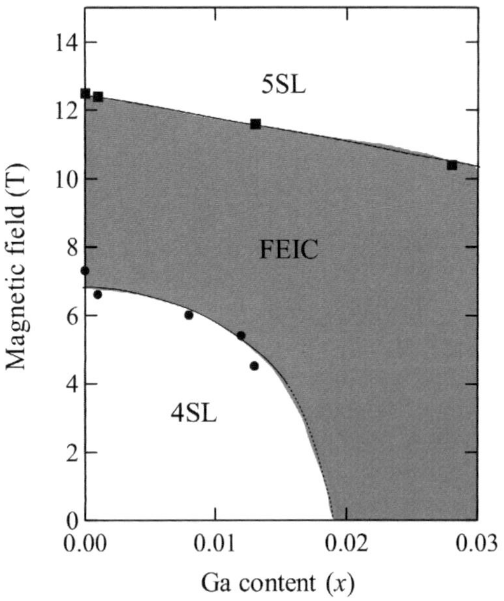

Figure 2. Phase diagram of magnetic field H versus Ga content x. Black circles show the transition field from the 4SL to the FEIC phase determined by the ascending magnetization process. Black squares show the transition field from the FEIC to the 5SL phase determined by the ascending magnetization process. As compared with the transition field from the FEIC to the 5SL phase, the transition field from the 4SL to the FEIC phase shifts rapidly with the increase of Ga^{3+} content.

This result suggests that ferroelectric phase becomes stable with the substitution of nonmagnetic ions for Fe^{3+} ions.

We observed some ESR signals in $CuFe_{1-x}Ga_xO_2$ (x less than 0.012) and the resonance fields for x = 0.000 and 0.012 are plotted on the frequency-field plane as shown in Fig. 3. The resonance branches for x = 0.012 slightly shift to the low frequency side compared with those for x = 0.000. The resonance mode at zero field changes only about 15 GHz. On the contrary, the linewidth changes dramatically with the increase of x in $CuFe_{1-x}Ga_xO_2$ as shown in Fig. 4. The linewidth broadens with increasing x. The full widths at half maximum for x = 0.000 and x = 0.012 are about 0.2 T and 1.3 T, respectively.

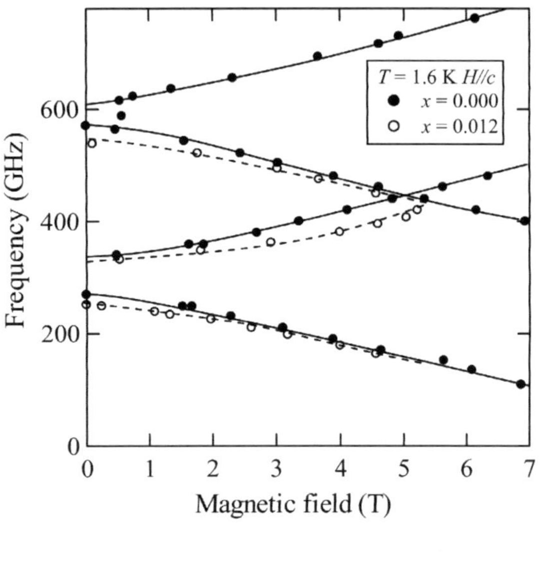

Figure 3. Frequency-field plot of the resonance fields of $CuFe_{1-x}Ga_xO_2$ with x = 0.000 and 0.012 at 1.6 K for $H//c$. Closed and open circles denote the ESR resonance points observed in $CuFe_{1-x}Ga_xO_2$ for x = 0.000 and 0.012, respectively. We found a small shift of the excitation mode by Ga doping.

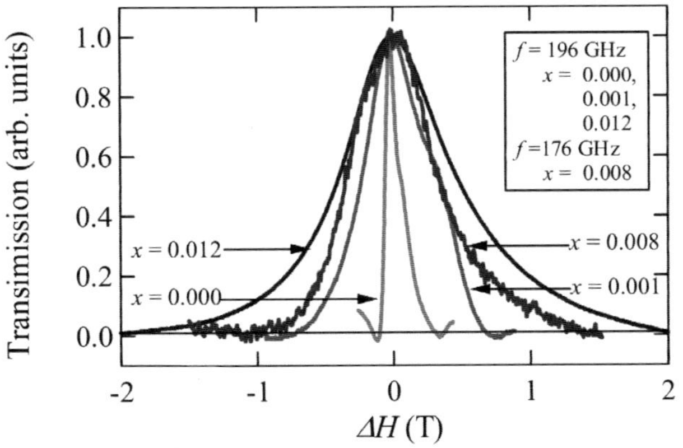

Figure. 4. Ga-content x dependence of the ESR spectra. These spectra, indicating the lowest excitation mode, were observed at close frequencies. We observed a rapid increase of the ESR linewidth with increasing x.

4. Discussion

The transition from the 4SL state to the FEIC one is understood as the softening of magnetic excitation mode [9]. As we can see in the previous section, Ga^{3+} doping gives a large effect on the excitation mode. First, we consider the shift of the lowest excitation mode. It seems too small to explain the large shift of H_c, which is defined by the change of the transition field from the $x = 0.000$ value in the field-ascending process of magnetization curve, and thus the shift of the excitation mode of only 15 GHz (the order of 0.1 T) is thought to be no origin of the large shift of H_c. Next, we consider the change of linewidth with the increase of x. The ESR linewidths are highly correlated to the shifts of H_c as shown in Fig. 5. Accordingly, we conclude that the dominant origin of the large shift of H_c is the ESR line broadening caused by doping of nonmagnetic ions.

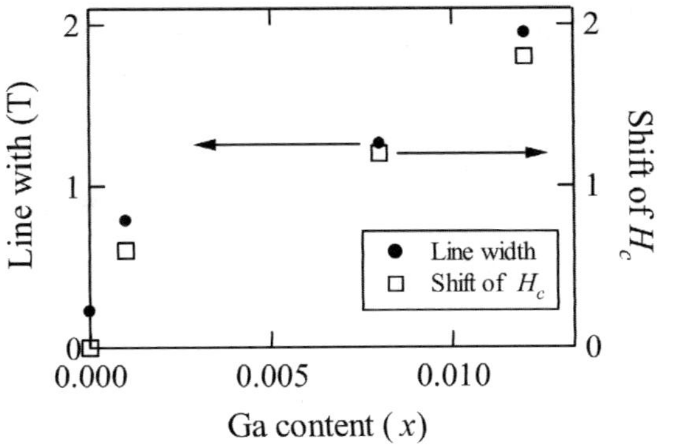

Figure 5. Ga-content x dependence of the ESR linewidth and the shift of H_c. The ESR linewidth is highly correlated with shift of H_c.

Unusual broadening of the linewidth can not be explained simply by the randomness due to the substitution of nonmagnetic ions for Fe^{3+}. This result, however, suggests that the substitution of nonmagnetic ions makes the excitation branch broad, and the softening of the lowest excitation branch by magnetic fields becomes easier. It was reported in a previous study [10] that $CuFeO_2$ has a spin system coupled strongly with the lattice. We, thus, speculate that the randomness of the local lattice distortions caused by the nonmagnetic-ion doping makes the excitation mode broad. Consequently, we conclude that the stabilization of the FEIC phase with the increase of x is caused by broadening of the excitation branch which probably induces the spin spiral correlation, resulting in ferroelectricity in the FEIC phase. Quite recently, inelastic neutron scattering (INS) measurements on a single crystal sample with x=0.035 which is larger than the x in our samples [11] were carried out and it was reported that the FEIC state is realized by softening of the excitation branch. This statement is not contradictory to our experimental results because of the experimental resolution. In our ESR measurements, the energy resolution is less than the order of 10^{-3} meV, while the resolution of INS measurements using cold neutron beam is at least 10^{-1} meV as you can see in Fig. 2(a) or Fig. 3(a) in Ref. [11]. Therefore, an INS measurement is not really suitable for investigating extremely low energy properties of condensed matters. Besides, the INS measurements were done on the 3.5 % Ga content sample which is in the FEIC phase at zero magnetic field and no phase transition in a magnetic field was observed in such a high Ga content sample.

5. Conclusion

We synthesized single crystals of the triangular-lattice antiferromagnet $CuFe_{1-x}Ga_xO_2$ with different x, and have performed multi-frequency ESR and magnetization measurements. Small changes in the ESR resonance modes and large broadening of the ESR spectra were observed from the ESR measurements. It is suggested that magnetic phase transition from the 4SL to the FEIC phases is not caused by the shift of the excitation mode but by its broadening. As a result, we conclude that the large shift of the transition field from the 4SL phase into the FEIC one is explained by the broadening of the excitation branch.

Acknowledgments

We thank Prof. T. Kimura at Osaka University for the use of Laue photography. This work was supported by a Grant-in-Aid for Scientific Research (No.20340089), and the Global COE Program (Core Research and Engineering of Advanced Materials- Interdisciplinary Education Center for Materials Science) (No. G10) from the MEXT, Japan.

References

[1] Anderson P W 1973 *Mat. Res. Bull.* **8** 153
[2] Chen H H and Levy P M 1971 *Phys. Rev. Lett.* **27** 1383
[3] Harris M J, Bramwell S T, McMorrow D F, Zeiske T, and Godfrey K W 1997 *Phys. Rev. Lett.* **79** 2554
[4] Kimura T, Lashley J C, and Ramirez A P 2003 Nature **426** 55
[5] Mekata M, Yaguchi N, Takagi T, Sugino T, Mitsuda S, Yoshizawa H, Hosoito N, and Shinjo T 1993 *J. Phys. Soc. Jpn.* **62** 4474
[6] Terada N, Narumi Y, Sawai Y, Katsumata K, Staub U, Tanaka Y, Kikkawa A, Fukui T, Kindo K, Yamamoto T, Kanmuri R, Hagiwara M, Toyokawa H, Ishikawa T, Kitamura H 2007 *Phys. Rev.* B **75** 224411
[7] Terada N, Nakajima T, Mitsuda S, Kitazawa H 2009 *J. Condens. Matter* **145** 012071
[8] Zhao T, Hasegawa M, Takei H 1995 *J. Cryst. Growth* **154** 322
[9] Ye F, Fernandez-Baca J A, Fishman R S, Ren Y, Kang, H J, Qiu Y, Kimura T 2007 *Phys. Rev. Lett.* **99** 157201
[10] Terada N, Mitsuda S, Ohsumi H, Tajima K 2006 *J. Phys. Soc. Jpn.* **75** 023602
[11] Haraldsen J T, Ye F, Fishman R S, Fernandez-Baca J A, Yamaguchi Y, Kimura K, and Kimura T 2010 *Phys. Rev.* B **82** 020404R

Measurement of complicated temperature-dependent polarization of multiferroic $R\mathrm{Mn_2O_5}$

Mamoru Fukunaga and Yukio Noda

Institute of Multidisciplinary Research for Advanced Materials, Tohoku University, Sendai 980-8577, Japan

E-mail: `fukunaga@tagen.tohoku.ac.jp`

Abstract. We have measured the temperature-dependent electric polarization $P(T)$ of multiferroic rare-earth (R) manganese oxides $R\mathrm{Mn_2O_5}$ using both typical pyroelectric measurements and hysteresis loops with the double-wave method (DWM), and revealed the complicated behavior of the $P(T)$. $R\mathrm{Mn_2O_5}$ single crystal samples often exhibit a tendency to macroscopically polarize without applying an external electric field. We have found that the tendency appeared in $P(T)$ by the pyroelectric measurement can be measured by the DWM loops. The tendency to polarize is equivalent to asymmetric non-hysteresis loops obtained by the DWM. We clarify the relationship between $P(T)$ by pyroelectric measurement and that by the DWM loops, which can warrant the measured $P(T)$.

1. Introduction

Orthorhombic rare-earth (R) manganese oxides of the form $R\mathrm{Mn_2O_5}$ are known as multiferroic materials with ferroelectricity driven by magnetic ordering [1]. Many magnetic interactions exist among Mn^{4+}, Mn^{3+}, and R^{3+}, which makes magnetic ordering accompanied by frustration. The magnetic ordering phases readily change with temperature T or applied magnetic field H due to the magnetic frustration, and the electric polarization P (and the permittivity) mainly along the b-axis change in accordance with the magnetic phase. Figure 1 shows the magnetic phases and $P(T)$ of several $R\mathrm{Mn_2O_5}$ for example [2, 3, 4, 5, 6]. The magnetic phases are found to be low-temperature incommensurate magnetic (LT-ICM), commensurate magnetic (CM), high-temperature incommensurate magnetic (HT-ICM), and paramagnetic (PM) phases. P in the CM phase is relatively large and almost common value regardless of R, while P in the LT-ICM phase is smaller and varies for each R [6].

To investigate the magnetic-order-induced polarization of $R\mathrm{Mn_2O_5}$, we have collected $P(T)$ data of $R\mathrm{Mn_2O_5}$ in the literature and by our own measurements [6], because the magnetic phases and P of $R\mathrm{Mn_2O_5}$ change by a temperature change as well as applying a magnetic field. $P(T)$ obtained by typical pyroelectric measurement depends on the measurement procedure, and reported $P(T)$ of $R\mathrm{Mn_2O_5}$ in the literature are confused because the change in $P(T)$ across the magnetic phase transition is not obvious. Moreover, $R\mathrm{Mn_2O_5}$ single crystal samples often exhibit a tendency to macroscopically polarize without applying an external electric field. Empirically, the tendency becomes stronger as the crystal quality becomes better with regard to diffraction. The tendency to polarize may arise from the nature of magnetic domains, which is an open question. This paper shows the complicated $P(T)$ of some $R\mathrm{Mn_2O_5}$ samples measured using

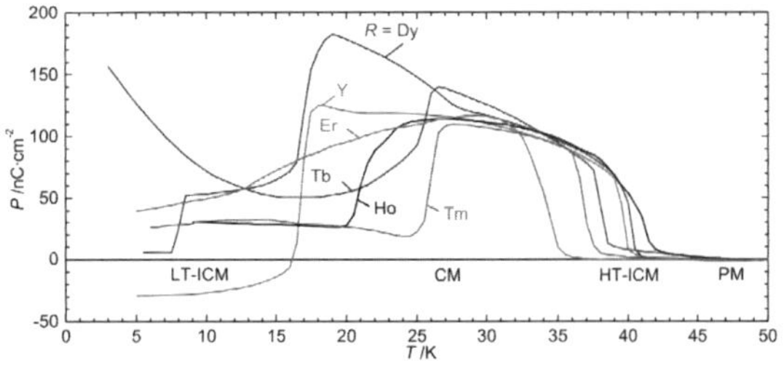

Figure 1. Magnetic phases and temperature dependence of the polarization $P(T)$ for several $R\mathrm{Mn_2O_5}$, where R = Dy [2], Y [3], Er [4], Tb [5], Ho [6], and Tm [6].

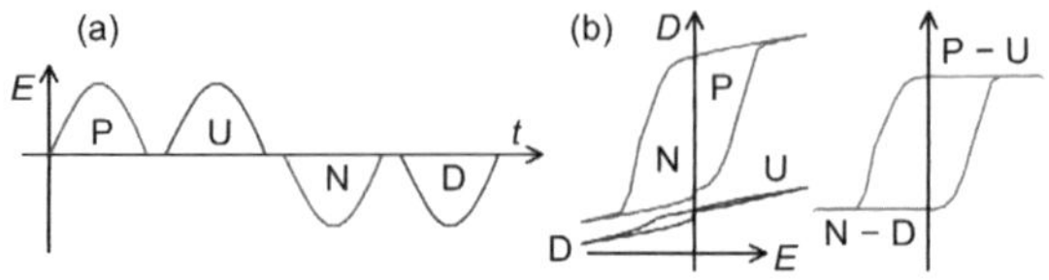

Figure 2. (a) Waveform of $E(t)$ for the DWM. (b) D–E loops obtained by the DWM.

both pyroelectric measurement and D–E hysteresis loops by the double-wave method (DWM) [7, 8], and explains the tendency and the relationship between the measurement results which we have recently found.

2. Experiment

$R\mathrm{Mn_2O_5}$ single crystal samples grown by the PbO-$\mathrm{PbF_2}$ flux method [9] were provided by coworkers. Conductive silver paste was painted on the *b*-faces as electrodes to apply the electric field along the *b*-axis of the samples in orthorhombic *Pbam* space group at room temperature. The sample temperature was controlled using a closed-cycle He-gas refrigerator. Pyroelectric measurements were performed using a picoammeter (Keithley 6485). The polarization was assumed to be zero above 50 K, and it was measured during zero-electric-field heating after electric-field cooling (poling) from 50 K. D–E hysteresis loops were obtained using a Sawyer–Tower circuit with a data acquisition board (NI6111) and a high-voltage bipolar amplifier (BOP1000M).

D–E loops were measured by the recently developed double-wave method (DWM) [7, 8]. The technique used in the DWM is known as positive-up-negative-down (PUND) measurement [10, 11]. Figure 2 explains the DWM in brief. Four half-waves corresponding to the P, U, N, and D pulses for the PUND measurement are applied as shown in Fig. 2(a). P and U are only $E > 0$, while N and D are only $E < 0$. The half-waves draws four half-loops as shown in Fig. 2(b). The half-loops by P and N, and those by U and D are connected. We call the pairs of half-loops by P and N as primary loops, and the pairs by U and D as secondary loops. The primary loops correspond to ordinarily-measured loops including all components, while the secondary loops correspond to loops due to non-hysteresis components such as linear dielectric, conductive, and antiferroelectric components, because the hysteresis components appear only in the P and N after D and U, respectively. We call the subtracted pairs of half-loops P − U and N − D as differential loops which include only the ferroelectric hysteresis components. The differential loops can simply be called as hysteresis loops, while the secondary loops as non-hysteresis loops.

3. Results and Discussion

3.1. Relationship between pyroelectric measurement and DWM loops

Figure 3(a) shows definitions of the polarization components from DWM loops. These loops were measured for an $\mathrm{EuMn_2O_5}$ sample at 30 K. Although $\mathrm{EuMn_2O_5}$ has peculiar magnetic phases unlike other $R\mathrm{Mn_2O_5}$ [12], we have obtained suitable data to see the relationship shown later with the sample which exhibits clear D–E loops [8]. The remanent polarization P_r is determined by the half of the height of the differential loop, and $2P_\mathrm{r}$ is drawn in Fig. 3(a). The DWM can

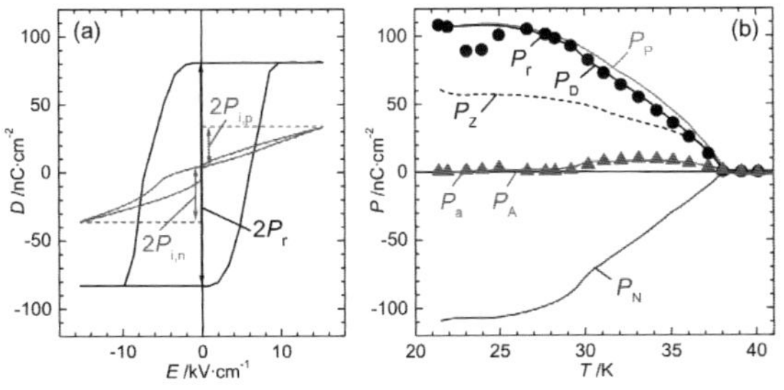

Figure 3. (a) DWM loops of an $EuMn_2O_5$ sample at 30 K, and definitions of the polarization components from DWM loops. (b) $P(T)$ of the $EuMn_2O_5$ sample; comparison between P_D and P_A (lines) from pyroelectric measurements and P_r and P_a (symbols) from DWM loops.

determine P_r more clearly than ordinary hysteresis-loop measurements. Similarly to P_r, the induced polarization $P_{i,p}$ and $P_{i,n}$ are defined by the secondary loops for positive and negative electric fields, respectively. $2P_{i,p} = D(E_m) - D(E=0)$ and $2P_{i,n} = D(E=0) - D(-E_m)$ are drawn in Fig. 3(a). where E_m is the amplitude (maximum) of the electric field for DWM loops, 15.6 kV/cm here. We use small-letter suffixes for P and E relating to DWM loops. Although the secondary loops may include the conductive component or leakage current which rounds the loops and fabricates a fake P_r, it is negligible for measuring the loops of RMn_2O_5 below 50 K, hence $P_{i,p}$ and $P_{i,n}$ here are regarded as the sum of the dielectric and electric-field-induced polarization components. When we define $P_a = P_{i,n} - P_{i,p}$, namely asymmetry in the secondary loop, the dielectric components in $P_{i,p}$ and $P_{i,n}$ corresponding to the slopes of the secondary loops are cancelled out, and P_a means asymmetry in the induced polarization.

Figure 3(b) shows the relationship between pyroelectric measurement results and the polarization components by DWM loops in the case of the $EuMn_2O_5$ sample. P_P and P_N are ordinary pyroelectric results measured after positive and negative poling, respectively. Here, usually $P_P > 0$ and $P_N < 0$. We use capital-letter suffixes for P and E relating to pyroelectric measurements. The poling electric field, E_P was 7.8 kV/cm here. Then we define $P_D = (P_P - P_N)/2$ and $P_A = (P_P + P_N)/2$, namely the half of the difference and the average of P_P and P_N, respectively. P_D means a reversible remanent polarization by an applied electric field, which ideally equals to P_r, and P_D may be rewritten as P_R. P_r exactly agrees with P_D above 26 K as shown in Fig. 3(b). The depression in P_r around 23 K arises from increasing coercive field and insufficient E_m.

On the other hand, P_A means an asymmetric component between P_P and P_N, which probably relates to the tendency to polarize. Although P_P and P_N in Fig. 3(b) seem almost symmetrical, non-zero P_A between 27 K and 38 K indicates that P_P and P_N are asymmetric. we found that P_A almost equals to P_a as shown in Fig. 3(b). In our previous papers, we regarded the pyroelectrically-measured P without poling at all, P_Z shown in Fig. 3(b), as the tendency to polarize. However, P_Z becomes quite large even if P_A is almost zero as shown in Fig. 3(b), and not P_Z but P_A corresponds to P_a. Hence we now regard P_A as the tendency. The finding $P_A = P_a$ means that the tendency to polarize determined by pyroelectric measurement is equivalent to asymmetric non-hysteresis loops obtained by the DWM.

The definitions and the relationship indicate that $P_P = P_A + P_D = +P_{i,n} + P_r - P_{i,p}$ and $P_N = P_A - P_D = +P_{i,n} - P_r - P_{i,p}$. Figure 4 illustrates the relationship with simple models. Note again that the variables with capital suffixes and small suffixes relate to pyroelctric measurements and DWM loops, respectively. The arrows indicate the sign and the amount of macroscopic polarization components, not microscpic polarization domains. The three polarization components, P_r, $P_{i,n}$, and $P_{i,p}$ can be drawn as reversible, always up, and always down arrows, respectively. Note that $P_{i,n}$ obtained by a negative E corresponds to a positive polarization component. It also should be noted that P_r, $P_{i,n}$, and $P_{i,p}$ are defined to be always positive, while $P_P < 0$ or $P_N > 0$ is possible, and it actually occurs due to the tendency. The polarization components shown in Fig. 4 are assumed to be measured completely. In principle,

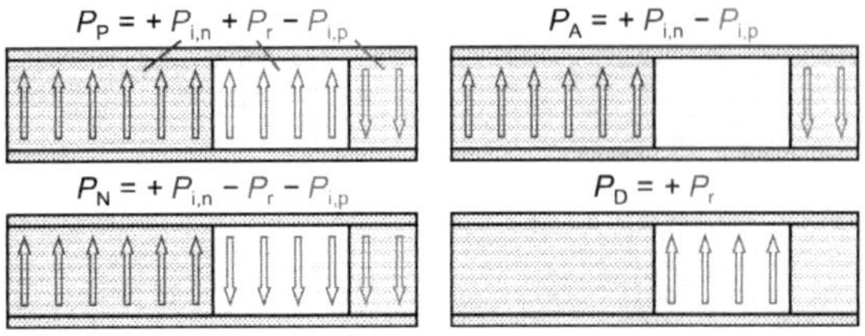

Figure 4. Models of the macroscopic polarization at $E = 0$ to explain the relationship.

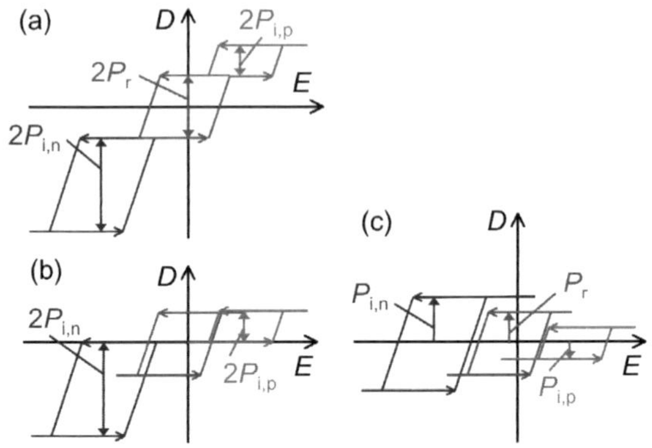

Figure 5. (a) Schematic image of the ferrielectric triple loop. (b) Redrawn image equivalent to (a). (c) Revised image of the triple loop corresponding to Fig. 4.

all of them can be determined from DWM loops. But in practice, it is difficult to obtain $P_{\mathrm{i,n}}$ and $P_{\mathrm{i,p}}$ completely because it often requires large E_{m}, which may cause the dielectric breakdown.

Although poling before pyroelectric measurement reverses only P_{r}, when P_{P} and P_{N} were measured, $P_{\mathrm{A}} = P_{\mathrm{i,n}} - P_{\mathrm{i,p}}$ can be obtained by pyroelectric measurement. If one of $P_{\mathrm{i,n}}$ and $P_{\mathrm{i,p}}$ was measured, the other can be obtained from P_{A} and the one. When P_{Z} is measured without poling, ideally the component corresponding to P_{r} becomes zero, and P_{Z} equals to P_{A}. However, P_{Z} does not equal to P_{A} in fact, and measured P_{Z} becomes between P_{P} and P_{N}. The tendency is regarded to be strong when P_{P} and P_{N} are close and P_{D} is small. In such a case, P_{Z} becomes close to both P_{P} and P_{N}, namely poling is ineffective.

Figure 5 shows a schematic image of the triple loop. Each loop corresponds to each component in the models in Fig. 4. We previously imaged the triple loop obtained by the DWM as shown in Fig. 5(a) [7], which means $D = \pm P_{\mathrm{r}}$ at $E = 0$ and a positive E induces $2P_{\mathrm{i,p}}$. Figure 5(b) shows an equivalent image to Fig. 5(a), though the three loops are separated by shifting the origin of D for each loop. In other words, $P_{\mathrm{i,p}}$ and $P_{\mathrm{i,n}}$ do not contribute to the measured polarization at $E = 0$ in Figs. 5(a) and 5(b). But according to Fig. 4, $P_{\mathrm{i,p}}$ and $P_{\mathrm{i,n}}$ exist even at $E = 0$, namely $D = P_{\mathrm{i,n}} \pm P_{\mathrm{r}} - P_{\mathrm{i,p}}$, and the triple loop should be considered as Fig. 5(c). Figure 5(c) suggests that the loops corresponding to $P_{\mathrm{i,p}}$ and $P_{\mathrm{i,n}}$ are considered to be ferroelectric loops with an internal bias electric field, and the bias causes non-zero P at $E = 0$. An internal electric field may cause a large polarization in the LT-ICM and CM phases and non-zero polarization in the HT-ICM phase [13].

3.2. P(T) measured after poling to the CM and LT-ICM phases

To verify the relationships $P_{\mathrm{A}} = P_{\mathrm{a}}$ and $P_{\mathrm{D}} = P_{\mathrm{r}}$, we have measured $P(T)$ of other $R\mathrm{Mn_2O_5}$ samples. Note that the tendency to polarize is sample-dependent. Figure 6 shows $P(T)$ of an $\mathrm{ErMn_2O_5}$ sample during heating. E_{P} and E_{m} for this sample were 12.8 kV/cm and 25.6 kV/cm, respectively. Pyroelectric measurements here were performed after poling to 30 K and 5 K, namely to the CM and LT-ICM phases, and the results are labeled [CM] and [ICM], respectively. The [CM] results are considered as usual pyroelectric measurement of ferroelectrics because the poling is performed from a paraelectric phase to a ferroelectric phase. On the other hand, the [ICM] results are complicated because the poling process across the CM to LT-ICM phase transition, and they are measured across the LT-ICM to CM transition during heating without applying E. But $P(T)$ of $R\mathrm{Mn_2O_5}$ in many reports were measured as the [ICM] only.

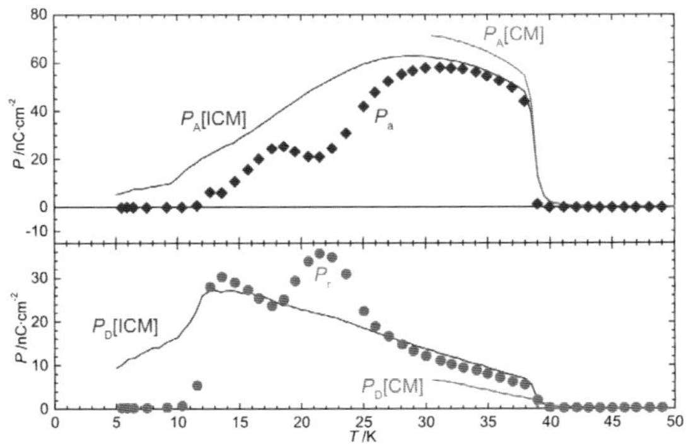

Figure 6. P_{A} and P_{D} (lines), and P_{a} and P_{r} (symbols) of an $\mathrm{ErMn_2O_5}$ sample.

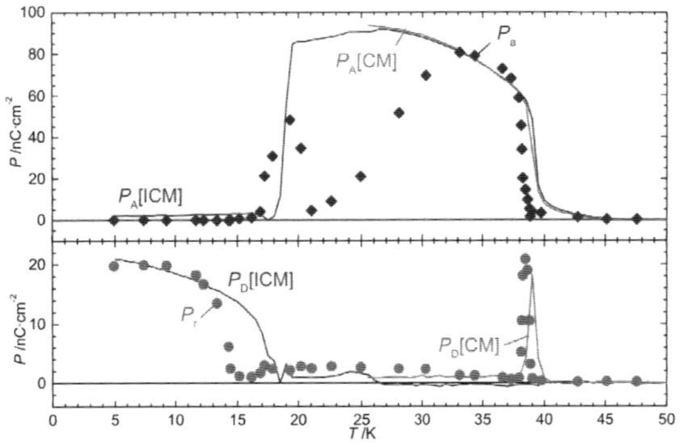

Figure 7. P_{A} and P_{D} (lines), and P_{a} and P_{r} (symbols) of a $\mathrm{HoMn_2O_5}$ sample.

As shown in Fig. 6, the results [ICM] and [CM] are different.

P_{A}[CM] > P_{A}[ICM] and P_{D}[CM] < P_{D}[ICM] above 30 K suggest that P_{D} increases and P_{A} decreases in the [ICM] process, though the reason is not clear. The differences between P_{D} and P_{r}, or between P_{A} and P_{a} may indicate that the measurement was incomplete. P_{A}[ICM] > P_{a} below 30 K means that E_{m} is insufficient to obtain P_{a}. But P_{D}[ICM] $\simeq$ P_{r} means that E_{m} is sufficient to obtain P_{r}. Although $P_{\mathrm{D}} > 10\ \mathrm{nC/cm^2}$ below 10 K, in the LT-ICM phase of $\mathrm{ErMn_2O_5}$, P_{r} is almost zero there probably because the coercive field E_{c} is very high, but a part of P_{D} in the CM phase was kept in the LT-ICM phase. $R\mathrm{Mn_2O_5}$ samples usually exhibit higher E_{c} in the LT-ICM phase than in the CM phase [6]. On the other hand, P_{D}[ICM] < P_{r} between 19 K and 26 K means that the poling was insufficient, or a part of P_{D} was lost through the magnetic phase transitions. Although P_{D}[ICM] is considered to be preserved at the transition to the CM phase [6], P_{D}[ICM] of the $\mathrm{ErMn_2O_5}$ sample increases there.

Figure 7 shows $P(T)$ of a $\mathrm{HoMn_2O_5}$ sample measured similarly to Fig. 6. Pyroelectric measurement were performed after poling to 25 K in the CM phase and 5 K in the LT-ICM phase. E_{P} and E_{m} for this sample were 21.0 kV/cm and 52.6 kV/cm, respectively. This sample has strong tendency to polarize in the CM phase, namely P_{D} is almost zero and P_{Z} (not shown) almost equals to P_{A}. However, P_{A}[ICM] below 18 K is much smaller than that above 18 K and P_{D}[ICM] below 18 K is much larger. The tendency for this sample is strong in the CM phase but weak in the LT-ICM phase. Similar results were obtained for some $R\mathrm{Mn_2O_5}$ samples, and perhaps the strong tendency in the CM phase rather than the LT-ICM phase may relate to the origin of the polarization arising from magnetic ordering.

P_{r} roughly agrees with P_{D}[ICM] below 25 K and P_{D}[CM] above 25 K. P_{a} roughly agrees with P_{D}[ICM] below 19 K and above 33 K. The peak in P_{r} at 38.5 K is accompanied by the decrease in P_{a}, which is considered to mean that the P_{a} component changes into P_{r} around the phase transition temperature. P_{D}[CM] also exhibits a peak at 39 K, while P_{D}[ICM] does not because P_{D}[ICM] disappears above 26 K.

4. Summary

We found that the relationship between pyroelectric measurement results and the polarization components determined by DWM D–E loops. Equality between $P(T)$ by pyroelectric measurement and that by D–E loops can warrant the measured $P(T)$. $P(T)$ of $R\mathrm{Mn_2O_5}$ depends on the measurement process, and it should be measured carefully. The tendency to polarize observed in $R\mathrm{Mn_2O_5}$ is equivalent to asymmetric non-hysteresis loops. It is considered that polarization domains with an internal bias electric field cause the tendency, and such domains may easily form in the CM phase of $R\mathrm{Mn_2O_5}$. The origin of such domains is a subject for further study.

Acknowledgments
The authors thank Y. Sakamoto, H. Kimura, K. Kohn, and Y. Tokura for providing the $R\mathrm{Mn_2O_5}$ samples. This work was supported by the Yamada Science Foundation, Grants-in-Aid for Scientific Research (Nos. 16340096 and 21244051), and a Grant-in-Aid for Scientific Research on Priority Areas "Novel State of Matter Induced by Frustration" (No. 19052001) from the Ministry of Education, Culture, Sports, Science and Technology, Japan.

References
[1] Noda Y, Kimura H, Fukunaga M, Kobayashi S, Kagomiya I and K. Kohn 2008 *J. Phys.: Condens. Matter* **20** 434206
[2] dela Cruz C R, Lorenz B and Chu C W 2008 *Physica* B **403** 1331
[3] Chaudhury R P, dela Cruz C R, Lorentz B, Sun Y, Chu C W, Park S and Cheong S W 2008 *Phys. Rev.* B **77** 220104
[4] Higashiyama D, Miyasaka S and Tokura Y 2005 *Phys. Rev.* B **72** 064421
[5] Haam S Y, Kim J W, Kim T H, Kim K H, Park S, Hur N, Cheong S W, Harrison N and Migliori A 2006 *Ferroelectrics* **336** 153
[6] Fukunaga M and Noda Y 2010 *J. Phys. Soc. Jpn.* **79** 054705
[7] Fukunaga M and Noda Y 2008 *J. Phys. Soc. Jpn.* **77** 064706
[8] Fukunaga M and Noda Y 2009 *J. Korean Phys. Soc.* **55** 888
[9] Wanklyn B 1972 *J. Mater. Sci.* **7** 813
[10] Scott J F, Araujo C A, Meadows H B, McMillan L D and Shawabkeh A 1989 *J. Appl. Phys.* **66** 1444
[11] Horiuchi S, Tokunaga Y, Giovannetti G, Picozzi S, Itoh H, Shimano R, Kumai R and Tokura Y 2010 *Nature* **463** 789
[12] Polyakov V, Plakhty V, Bonnet M, Burlet P, Regnault L P, Gavrilov S, Zobkalo I and Smirnov O 2001 *Physica* B **297** 208
[13] Fukunaga M, Sakamoto Y, Kimura H and Noda Y 2011 *J. Phys. Soc. Jpn.* **80** 014705

International Conference on Frustration in Condensed Matter (ICFCM) IOP Publishing
Journal of Physics: Conference Series **320** (2011) 012091 doi:10.1088/1742-6596/320/1/012091

Domain switching in nanometer scale in Bismuth-based relaxor solid solution

Naohiko Yasuda[1], Natsumi Takahashi[1], Nur Hidayah[1], Hidehiro Ohwa[1], Osamu Sakurada[2] Yoshihito Tachi[3], Kazuhiko Fujita[4], Yohachi Yamashita[5], Makoto Iwata[6] ,Hikaru Terauchi[7] and Yoshihiro Ishibashi[8]

[1]Electrical and Electronic Engineering Department, Gifu University, Gifu 501-1193, Japan
[2]Material Science and Technology Department, Gifu University, Gifu 501-1193, Japan
[3]Research Laboratories, JFE MINERAL Company, LTD. Chiba 260-0826, Japan
[4]Electronic and Control Engineering Department, Gifu National College of Technology, Gifu 501-0495, Japan
[5]Corporate R & D Center, Toshiba Corporation, Saiwai-ku, Kawasaki 212-8582, Japan
[6]Department of Engineering Physics, Electronics and Mechanics, Graduate School of Engineering,Nagoya Institute of Technology, Nagoya 466-8555, Japan
[7]Advanced Research Center of Science, School of Science, Kwansei-Gakuin University, Sanda 669-1337, Japan
[8]Department of Physics, Kyushu University, Hakozaki, Higashi-ku, Fukuoka, 812-8581, Japan

Corresponding author: yasuda-n@gifu-u.ac.jp

Abstract. Microscopic piezoresponse force microscopy observation in $0.9967(Na_{1/2}Bi_{1/2})TiO_3$(NBT)-$0.0033BaTiO_3$(BT) rhombohedral (001) plate reveals two kinds of piezoresponse images, one with domain switching characterized with superior piezoresponsibility and the other without one, whereas macroscopic observation reveals non-switching of domain. Such microscopic remarkable difference of domain switching depends on coercive field caused by localized nucleation of domains with reversed polarization. Such coercive fields depend on relaxor states induced by random fields due to built-in charge disorder resulting from lattice defects at A-site and oxygen vacancies in ABO_3 perovskite. The random fields based on lattice defects on {110} plane lower activation barrier for domain switching, leading to nucleation and growth of domains, while for non-switching of domains, oxygen vacancies on {110} and/or {001} plane play a role as clamping center for restriction of domain switching.

1.Introduction

Experimentally measured value of the ferroelectric coercive field, which indicates the magnitude of an external electric field required to reverse the macroscopic polarization, has been known [1] to be smaller than the intrinsic value predicted by the phenomenological theory. The low extrinsic coercive field observed in ferroelectrics has also been reported [2,3] to be caused by localized nucleation of domains with reversed polarization, which then grow in both the forward and side wide directions by domain wall motion. Relatively little energy has been reported [2.3] to be required to reverse polarization in a small nucleation volume around a defect compared to the energy needed for collective polarization reversal of crystal. Lattice defects has been known [3-6] to play an important role in the mechanism for the switching of polarization in ferroelectric properties. The piezoresponse force microscopy (PFM) has been applied to the study of ferroelectric materials on the nanometer scale [7-9]. The advantages of the PFM are a high spatial resolution and a high sensitivity to local polarization [7,10]. In this work, bismuth-based relaxor $(Na_{1/2}Bi_{1/2})TiO_3$ (NBT)- $BaTiO_3$(BT) solid solution single crystals were grown by the Bridgman method [5]. The obtained single crystals had good both optical quality and electrical properties [5]. The domain-switching on the rhombohedral NBT-BT (001) plate was microscopically investigated by using PFM. In this paper, the field-induced strain and the piezoelectric images on the rhombohedral NBT-BT (001) plate are presented, and discussed from the microscopic domain structure induced by random fields based on the lattice defects.

2. Experimental

The concentration of NBT-BT single crystals obtained was determined using an inductive charge plasma (ICP) analysis to be 0.487, 0.435 and 0.0033 for Bi, Na and Ba ions [5]. Single crystals obtained were cut into plates with a $(100)_{cub}$ orientation, as confirmed by X-ray diffraction method. 0.9967NBT-0.0033BT rhombohedral (001) plates were used. PFM studies were performed using a commercial PFM (SPA300/SPI3800N, Seiko Inc, Japan [7,9]. An external voltage was applied to one crystal surface through the Al coated cantilever (SI; DF3-R, f=25 kHz, C=1.4 N) with a tip apex radius of 30 nm as a top electrode. The other surface was electrode by silver paste and glued to the sample holder as a bottom electrode. The PFM images were taken over 3 μ m and 300 nm length scales on $(001)_{cub}$ faces at room temperature. When an ac voltage $V_{ac}=V_o \sin \omega t$ is applied between the tip top electrode and the bottom one of the sample, the alternating external electric field gives rise to the piezoelectric vibration of the sample or a change in the sample thickness due to the converse piezoelectric effect. In this experiment, an imaging ac voltage with amplitude V_o between 2 and 50 V and a frequency $\omega/2\pi$ of 5 kHz was applied [9]. The resolution limit of detected domain nucleus length and radius at applied voltage of 5 V for the tip radius R=30 nm were reported [10] to be about below 1 nm. ITO electrodes were deposited by conventional sputtering. The electrical capacitance and dielectric loss tangent tan δ were measured using an LCR meter at 1, 10 and 100 kHz with the probe field weaker than 10 V/cm at a rate of 1 K/min. The relationship between polarizations P and electric fields E was examined with an improved Sawyer-Tower circuit at 30 Hz.

3 . Results and Discussions

The electric field-induced strain behavior was microscopically observed using the PFM, based on contact-mode AFM. The two types of examples of strains taken by the surface-displacement detection are shown as a function of tip voltage on the rhombohedral (001) plate in the 0.9967NBT-0.0033BT single crystal as follows. One with domain switching was observed at the point marked with ● in figure 1(a) as shown in figure 1(b). In figure 1(a) the contrast was reported [8,9,11] to be determined by the out-of-plane component of polarization (P_z), where dark and bright areas correspond to the opposite directions of P_z. (see also figure 2 and figure 3). The other without one was observed at the point marked with ○ in figure 1(a), as shown in figure 1(c). The former image reveals the piezoresponse high Z versus a tip voltage V with domain switching, as seen in figure 1(b), where the curve shows the butterfly type hysteresis, corresponding to ferroelectric state. The coercive voltages were estimated to be about 11 V and -9 V as marked with arrows for forward and reverse direction of applied voltage, respectively. The latter one reveals the Z-V curve without domain switching, as seen in figure 1(c), where the curve shows the small hysteresis one characterized with electrostrictive effect, corresponding to relaxor state. The Z-V characteristics with domain switching in the rhombohedral (001) plate with thickness 100 μ m to estimate the magnitude in piezoresponse high Z was compared with the poled 0.68PMN-0.32PT rhombohedral (001) plate [12] with thickness of 100 μ m with $d_{33} \sim$ 2120 pC/N, as a reference, as shown in figure 1(d).

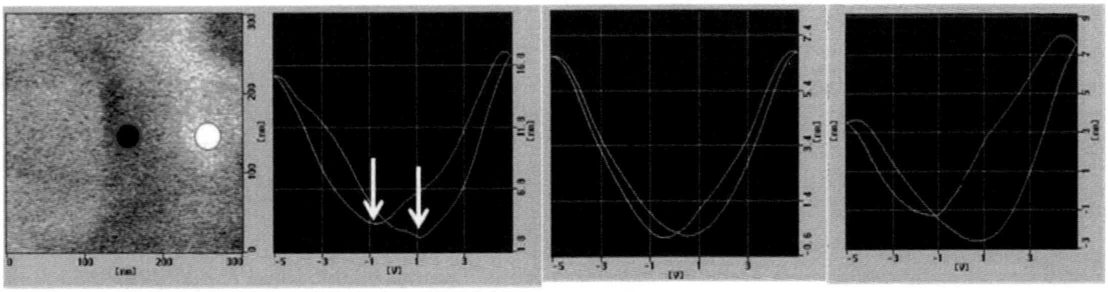

(a) (b) (c) (d)

Figure 1. Planar view of piezoresponse image (a), piezoresponse high Z vs. tip (apex radius of 30 nm) voltage V at point marked with ● (b), that at point marked with ○ (c), and that for poled 0.68PMN-0.32PT single crystal (d). V multiplied by 10 for horizontal axis.

The value of Z=15.0 nm estimated at V=50 V was compared with the Z=10.6 nm at 50 V for poled 0.68PMN-0.32PT (001) plate. The remarkable difference in Z-V curves between switching and non-switching in the 0.9967NBT-0.0033BT rhombohedral (001) plate depends on the ferroic state induced by the random fields based on the lattice defects of Bi, Na and Ba ions and oxygen vacancies [3,5,8]. The substitution of Ba ion for the A-site ion seems to give effect of larger off-center to enhance piezoelectric property [13]. For lattice defects in the case of NBT, the planar defect of Guinier-Preston zones (GPZ's) was pointed out [14] as follows from the x-ray diffuse scattering study. The asymmetry in the x-ray diffuse scattering intensity was reported [14] to result from the relaxor states induced by random fields based on the planar defects of the GPZ's characterized with the deformation of the structure around segregation planes in form of monoclinic platelets in random matrix of rhombohedral symmetry. In the case of domain-switching, the planar defects on the {110} plane may contribute to switching sites of domain nucleation, resulting in polarization rotation on the {110} plane, leading to larger piezoresponse high Z [14]. Polar clusters with a reversed polarization within ferroelectric domains seems to be nucleated under applied field in the vicinity of random fields around quenched defects [3]. The random fields was known [3] to lower the activation barrier required for nucleation. Such local random fields due to built-in charge disorder in relaxors were considered as one of the main reasons responsible for the formation of the polar nanoregions [8,11,15].

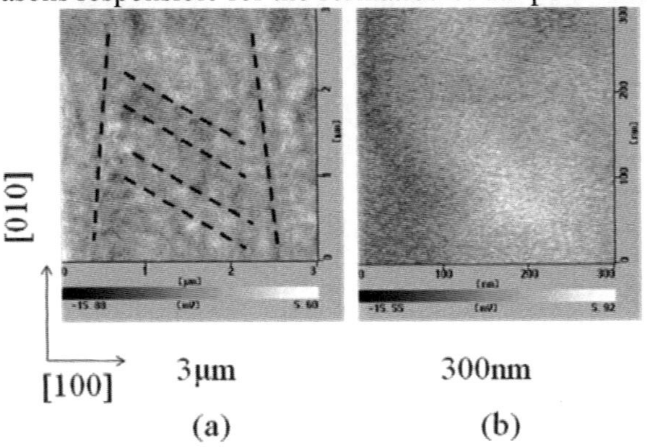

Figure 2. Piezoresponse images taken over (a) 3 μ m and (b) 300 nm length scales on $(001)_{cub}$ faces in imaging ac voltage at 30 V in 0.9967NBT-0.0033BT single crystal at 30 ℃.

Figures 2(a) and 2(b) show piezoresponse images taken over 3 μ m and 300 nm length scales on $(001)_{cub}$ faces in 0.9967NBT-0.0033BT single crystal in the imaging ac voltage at 30 V by PFM at 30 ℃, respectively. The domains are not continuous, some of them disappear or merge, forming labyrinth structures [16], indicating the preferred orientation of domain walls close to the [110] and [100] directions as marked with black dotted line in the figure 2(a). The domain structures characterized with spherical-like piezoresponse images in some range from about 10 to 100 nm were found as revealed as dark and bright domains in the rhombohedral phase. Micron-sized domains consist of much finer irregular domain structure in nanometer scale. Seemingly dark areas include numerous bright inclusions and vice versa. Appearance of such complex domain structures was known [15] to be a distinctive feature of relaxor crystals, which is associated with the distribution of random electric fields. The random electric fields was reported [15,17] to be Gaussian distributed. The example in the case of 300 nm for the statistical distribution of the domain area inside micron-sized domain at the tip voltage of 5 V is shown in figure 3(a).

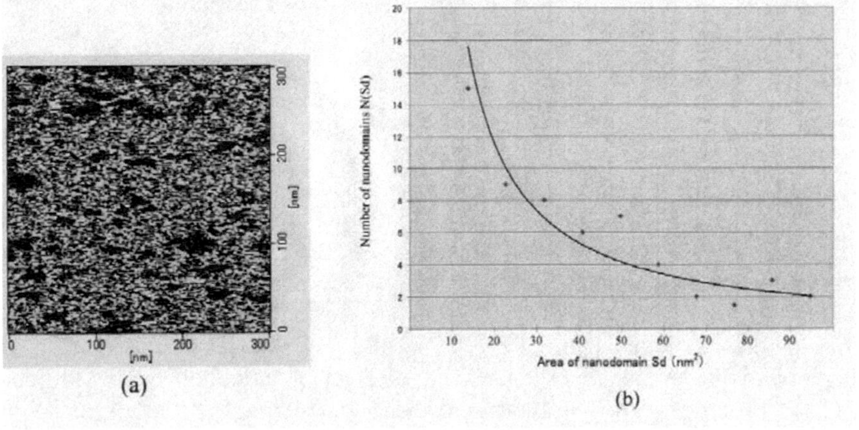

Figure 3. (a) Piezoelectric images in 0.9967NBT-0.0033BT rhombohedral (001) plate and (b) its nanodomain size distribution. The solid line represents the fit to the equation (1).

Domain distribution with polarizations $\pm$Pz as indicated by dark and bright areas in gray tones was shown in figure 3 (a). The average size of fine domains is about 6.5 nanometer in the present scale of 300 nm. The measured piezoresponse signal distribution was reported [11] to be treated as a (001) cut of the three dimensional domain structure existing near the surface because of the rhombohedral symmetry of the 0.9967NBT-0.0033BT (001) plate. The measured piezoresponse signal was reported [11,15] to have only two values of +Pz and –Pz component of polarization, and then the piezoresponse signal in the 0.9967NBT-0.0033BT rhombohedral (001) plate may be considered as an uniaxial relaxor, similar to $Sr_{0.61-x}Ce_xBa_{0.39}Nb_2O_6$. (SBN61;Ce) [15]. It has been reported [11,15] that the relaxor properties are well described by the random-field Ising Model (RFIM). In the case of a two-dimensional (2D) RFIM model with which the (001) projection of structure is represented, the domain size distribution follows a power law with exponential cutoff [11,15,17];

$$N_d(S_d) \sim S_d^{-\delta} \exp\left(-\frac{S_d}{S_0}\right) \qquad (1),$$

where N_d is the number of nanodomains having size S_d. The observed nanodomain size distribution could be fitted with the equation (1) with the parameter, the upper cutoff of nanodomain size, S_0= 94.8 nm^2 as shown as a solid line in figure 3(b). The fitted value of δ is estimated to be 0.7, compared with δ = 0.5 for 0.80PMN-0.2PT [11] and δ =1.5 for uniaxial relaxor SBN61;Ce [15]. These values of δ in relaxor ferroelectrics correspond to the variation of the number of nanodomains with size. The smaller values was pointed out to correspond to slower variation of number of nanodomain with size [11]. On the other hand, X-ray diffuse scattering patterns for NBT was reported [14] to be asymmetric and characterized with the planar (2D) defect platelet (GPZ's), whereas those for PMN to be symmetric and isotropic [18]. Such different parameter, δ = 0.7, 0.5 and 1.5 for NBT (2D), PMN (isotropic), and SBN61;Ce (1D), respectively, which is closely related to the distribution of random fields, may reflect a different distinctive feature of relaxor single crystals [11,15,17]. On the other hand, in oxide perovskite, it has been reported [19] that increasing numbers of oxygen vacancies in perovskites cause a self-ordering into planar structures, which with increasing vacancy concentrations, grow from an isolated point defect to a linear chain of oxygen vacancies along [101] in the cubic perovskite, resulting in one-dimensional chains whose lengths increase and above a certain vacancy density, in planes equivalent to the {110} plane. On the other hand, the preferential electromigration of oxygen vacancies into sites in planes parallel to the electrodes has also been reported [19]. Such planes seem to be of {001} ones, because of the sample with the rhombohedral (001) plate as a electrode. The different relaxor states for domain-switching and non-switching induced by such cations as Bi^{3+}, Na^{1+} and Ba^{2+} which have charge and size differences in the A site and oxygen vacancies significantly influence the domain switching, leading to nucleation or clamping effects on domain switching. On the other hand, macroscopic observation of P-E hysteresis loops on the 0.9967NBT-0.0033BT rhombohedral (001) plate exhibits non-switching polarization as shown in figure 4.

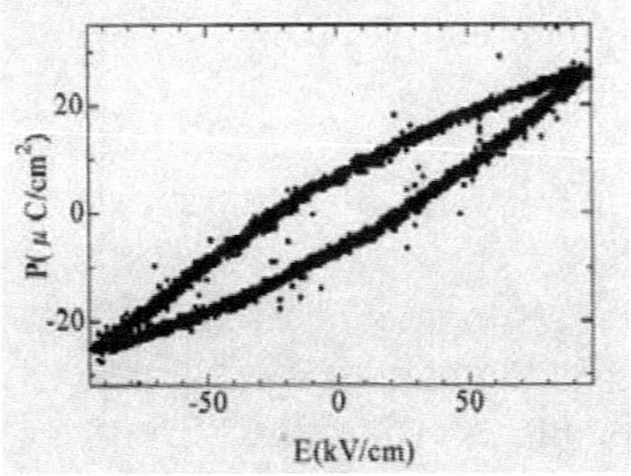

Figure 4. Relationships between P vs. E along [001] direction for rhombohedral 0.9967NBT-0.0033BT single crystal (001) plate at 30 ℃.

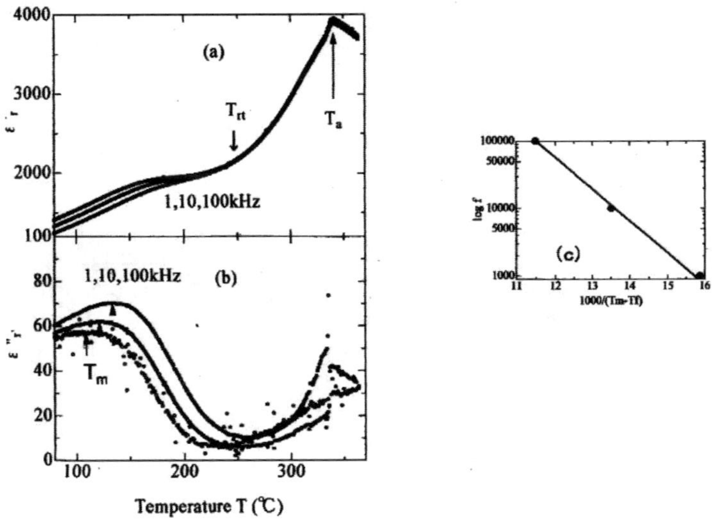

Figure 5. Temperature dependence of (a) real ε_r' and (b) imaginary parts ε_r" of complex relative permittivity and (c) reciprocal temperature dependence of relaxational frequency f at 1,10 and 100 kHz along [001] direction in 0.9967NBT-0.0033BT single crystal.

As for the domain walls observed along the [110] and [001] direction at 30 ℃, the [110] type domain walls disappeared with heating on going to T_{rt}, while the [100] type ones remained in the rhombohedral phase [5]. The [110] type domain walls was reported [6] to vary sensitively with the variation of oxygen pressure at a high oxygen pressure crystal growth, and also the switched spontaneous polarization to increase with increasing oxygen pressure. The temperature dependence of the real ε_r' and imaginary parts ε_r" of complex relative permittivity in the 0.9967NBT-0.0033BT (001) plate for different frequencies are shown in figure 5 (a) and 5 (b), respectively.

Dielectric dispersion in audio frequency range appeared clearly at temperatures below the $T_{rt.}$ and then its dielectric dispersion disappeared at temperatures above T_{rt}. Such dielectric dispersion below T_{rt} in the 0.9967NBT-0.0033BT single crystal was described with Vogel-Fulcher [13] type one with $f=f_o$
$\exp\{-E_a/k(T_m-T_f)\}$, where the Boltzmann constant k, the Debye frequency $f_o=2.2x10^{10}$ Hz, activation energy $E_a=92$ meV, freezing temperature $T_f=320$ K, as shown in figure 5 (c). Such dielectric dispersion is considered to origin from the domain wall movement based on hierarchical domain structures from μ m to nanometer scale. Judging from such phenomena, the isotropization appeared at temperatures above T_{rt} [5] is considered to be due to the domain breaking from macroscopic domain to polar region in nanometer scale due to random fields. Thus, there are little defects in this NBT-BT single crystal [5] relative to other single crystals grown by Bridgman method reported previously [4].

5. Conclusions

Superior piezoresponce image on the 0.9967NBT-0.0033BT rhombohedral (001) plate was observed as well as the poled 0.68PMN-0.32PT rhombohedral (001) plate. This origin is considered to be the random field due to Bi, Na and Ba lattice defects at A-site and oxygen vacancies in ABO_3 perovskite. Preferential planes of lattice defects such as {001} and {110} ones play an important role for domain switching mechanisms. The {110} planar defects may play a role as a nucleation sites to lower the activation barrier for domain switching, leading to the polarization rotation from the [111] direction in the rhombohedral phase to the [001] one in the tetragonal phase through the [110] one in the intermediate monoclinic plane under electric fields applied along the [001] direction, whereas oxygen vacancies of planar defects on the {001} and/or {110} plane have the clamping effect of pinning domain wall movement, leading to non-switching polarization.

This work was supported in part by the Nippon Sheet Glass Foundation for Materials Science and Technology and Grant-in-Aid for Scientific Research (B) (No.16360150) and (C) (No.20560633) from the Japan Society for the Promotion of Science.

References

[1] Gerra G, Tagantsev A K, and Setter N 2005 *Phys. Rev. Lett.* **94** 107602
[2] Bratkovsky A M and Levanyuk A P 2000 *Phys. Rev. Lett.* **85** 4614
[3] Viehland D and Chen Y H 2000 *J. Appl. Phys.* **88** 6696
[4] Ge W, Liu H, Xiangyong Z, Bijun F, Xiaobing L, Feifei W, Dan Z, Ping Y, Xiaoming P, Di L, and Haosu L 2008 *J. Phys. D : Appl. Phys.* **41** 115403.
[5] Yasuda N, Hashimoto S, Ohwa H, Sakurada O, Fujita K, Yamashita Y, Iwata M, and Ishibashi Y 2009 *Jpn. J. Appl. Phys.* **48** 09KC06
[6] Yamamoto K, Suzuki M, Noguchi Y, and Miyayama M 2008 *Jpn. J. Appl. Phys.* **47**, 7623.
[7] Tanaka K, Kushida K, Torii K, and Miki H 1994 *Jpn. J. Appl. Phys.* **33** 3193
[8] Bdikin I K, Shvartsman V V, and Kholkin A L 2003 *Appl. Phys. Lett.* **83** 4232
[9] Rahman Md M and Yasuda N 2009 *Solid State Commun.* **149** 630
[10] Molotskii M I and Shvebelman M M 2005 *Phil. Mag.* **85** 1637
[11] Shvartsman V V and Kholkin A L 2004 *Phys. Rev.* **B69** 014102
[12] Matsushita M, Tachi Y and Ebchizenya K 2001 *Abstract American Ceramic Society,* 103rd Annual Meeting p.247
[13] Samara G A 2003 *J. Phys. Condens. Matt.* **15** R367
[14] Kreisel J, Bouvier P, Dkhil B, Thomas P A, Glazer A M, Welberry T R, Chaabane B, and Mezouar M 2003 *Phys. Rev.* **B 68** 014113
[15] Kleemann W 2006 *J. Mater. Sci.,* **41** 129
[16] Bai F, Li J, and D. Viehland 2005 *J. Appl. Phys.* **97** 054103
[17] Esser J, Nowak U, and Usadel K D 1997 *Phys. Rev.* **B55** 5866
[18] Kreisel K, Bouvier P, Dkhil B, Chaabane B, Glazer A M, Thomas P A, and Welberry T R 2004 *Ferroelectrics* **302** 293

[19] Scott J F and Dawber M 2000 *Appl. Phys. Lett.* **76** 3801

AUTHOR INDEX

AUTHOR INDEX

AUTHOR INDEX